MATHEMATICS OF PUBLIC KEY CRYPTOGRAPHY

Public key cryptography is a major interdisciplinary subject with many real-world applications, such as digital signatures. A strong background in the mathematics underlying public key cryptography is essential for a deep understanding of the subject, and this book provides exactly that for students and researchers in mathematics, computer science and electrical engineering.

Carefully written to communicate the major ideas and techniques of public key cryptography to a wide readership, this text is enlivened throughout with historical remarks and insightful perspectives on the development of the subject. Numerous examples, proofs and exercises make it suitable as a textbook for an advanced course, as well as for self-study. For more experienced researchers, it serves as a convenient reference for many important topics: the Pollard algorithms, Maurer reduction, isogenies, algebraic tori, hyperelliptic curves, lattices and many more.

STEVEN D. GALBRAITH is a leading international authority on the mathematics of public key cryptography. He is an Associate Professor in the Department of Mathematics at the University of Auckland.

MATHEMATICS OF PUBLIC KEY CRYPTOGRAPHY

STEVEN D. GALBRAITH
University of Auckland

CAMBRIDGE
UNIVERSITY PRESS

University Printing House, Cambridge CB2 8BS, United Kingdom

One Liberty Plaza, 20th Floor, New York, NY 10006, USA

477 Williamstown Road, Port Melbourne, VIC 3207, Australia

314-321, 3rd Floor, Plot 3, Splendor Forum, Jasola District Centre, New Delhi - 110025, India

79 Anson Road, #06-04/06, Singapore 079906

Cambridge University Press is part of the University of Cambridge.

It furthers the University's mission by disseminating knowledge in the pursuit of education, learning and research at the highest international levels of excellence.

www.cambridge.org
Information on this title: www.cambridge.org/9781107013926

First published 2012

A catalogue record for this publication is available from the British Library

Library of Congress Cataloging in Publication data
Galbraith, Steven D.
Mathematics of public key cryptography / Steven D. Galbraith.
p. cm.
Includes bibliographical references and index.
ISBN 978-1-107-01392-6 (hardback)
1. Coding theory. 2. Cryptography – Mathematics. I. Title.
QA268.G35 2012
003 .5′4 – dc23 2011042606

ISBN 978-1-107-01392-6 Hardback

Additional resources for this publication at
www.math.auckland.ac.nz/~sgal018/crypto-book/crypto-book.html

Contents

Preface

The book has grown from lecture notes of a Master's level course in mathematics, for students who have already attended a cryptography course along the lines of Stinson's or Smart's books. The book is therefore suitable as a teaching tool or for self-study. However, it is not expected that the book will be read linearly. Indeed, we discourage anyone to start reading with either Part I, Part II or Part III. The best place to start, for an understanding of mathematical cryptography, is probably Part V (replacing all references to "algebraic group G" by $\mathbb{F}_p^*$). For an introduction to RSA and Rabin one could start reading at Part VI and ignore most references to the earlier parts.

Exercises are distributed throughout the book so that the reader performing self-study can do them at precisely the right point in their learning. Readers may find exercises denoted by ★ somewhat more difficult than the others, but it would be dangerous to assume that everyone's experience of the exercises will be the same.

Despite our best efforts, it is inevitable that the book will contain errors and misleading statements. Errata will be listed on the author's webpage for the book at www.math.auckland.ac.nz/~sgal018/crypto-book/crypto-book.html. Readers are encouraged to bring any errors to the attention of the author.

I would like to thank Royal Holloway, University of London and the University of Auckland, each of which in turn was my employer for a substantial time while I was writing the book. I also thank the EPSRC, who supported my research with an advanced fellowship for the first few years of writing the book.

The book is dedicated to Siouxsie and Eve, both of whom tolerated my obsession with writing for the last four years.

Steven Galbraith
Auckland

Acknowledgements

The book grew out of my lecture notes from the Master's course "Public key cryptography" at Royal Holloway. I thank the students who took that course for asking questions and doing their homework in unexpected ways.

The staff at Cambridge University Press have been very helpful during the preparation of this book.

I also thank the following people for answering my questions, pointing out errors in drafts of the book, helping with LaTeX, examples, proofs, exercises, etc: José de Jesús Angel Angel, Olivier Bernard, Nicolas Bonifas, Nils Bruin, Ilya Chevyrev, Bart Coppens, Alex Dent, Claus Diem, Marion Duporté, Andreas Enge, Victor Flynn, David Freeman, Pierrick Gaudry, Takuya Hayashi, Nadia Heninger, Florian Hess, Mark Holmes, Everett Howe, David Jao, Jonathan Katz, Eike Kiltz, Kitae Kim, David Kohel, Cong Ling, Alexander May, Esmaeil Mehrabi, Ciaran Mullan, Mats Näslund, Francisco Monteiro, James McKee, James Nelson, Samuel Neves, Phong Nguyen, TaeHun Oh, Chris Peikert, Michael Phillips, John Pollard, Francesco Pretto, Oded Regev, Christophe Ritzenthaler, Karl Rubin, Raminder Ruprai, Takakazu Satoh, Leanne Scheepers, Davide Schipani, Michael Schneider, Peter Schwabe, Reza Sepahi, Victor Shoup, Igor Shparlinski, Andrew Shallue, Francesco Sica, Alice Silverberg, Benjamin Smith, Martijn Stam, Damien Stehlé, Anton Stolbunov, Drew Sutherland, Garry Tee, Emmanuel Thomé, Frederik Vercauteren, Timothy Vogel, Anastasia Zaytseva, Chang-An Zhao, Paul Zimmermann.

Any remaining errors and omissions are the author's responsibility.

1
Introduction

Cryptography is an interdisciplinary field of great practical importance. The subfield of public key cryptography has notable applications, such as digital signatures. The security of a public key cryptosystem depends on the difficulty of certain computational problems in mathematics. A deep understanding of the security and efficient implementation of public key cryptography requires significant background in algebra, number theory and geometry.

This book gives a rigorous presentation of most of the mathematics underlying public key cryptography. Our main focus is mathematics. We put mathematical precision and rigour ahead of generality, practical issues in real-world cryptography or algorithmic optimality. It is infeasible to cover all the mathematics of public key cryptography in one book. Hence, we primarily discuss the mathematics most relevant to cryptosystems that are currently in use, or that are expected to be used in the near future. More precisely, we focus on discrete logarithms (especially on elliptic curves), factoring based cryptography (e.g., RSA and Rabin), lattices and pairings. We cover many topics that have never had a detailed presentation in any textbook.

Due to lack of space some topics are not covered in as much detail as others. For example, we do not give a complete presentation of algorithms for integer factorisation, primality testing and discrete logarithms in finite fields, as there are several good references for these subjects. Some other topics that are not covered in the book include hardware implementation, side-channel attacks, lattice-based cryptography, cryptosystems based on coding theory, multivariate cryptosystems and cryptography in non-Abelian groups. In the future quantum cryptography or post-quantum cryptography (see the book [48] by Bernstein, Buchmann and Dahmen) may be used in practice, but these topics are also not discussed in this book.

The reader is assumed to have at least a standard undergraduate background in groups, rings, fields and cryptography. Some experience with algorithms and complexity is also assumed. For a basic introduction to public key cryptography and the relevant mathematics the reader is recommended to consult Smart [513], Stinson [532] or Vaudenay [553].

An aim of the present book is to collect in one place all the necessary background and results for a deep understanding of public key cryptography. Ultimately, the text presents what I believe is the “core” mathematics required for current research in public key cryptography and it is what I would want my PhD students to know.

The remainder of this chapter states some fundamental definitions in public key cryptography and illustrates them using the RSA cryptosystem.

1.1 Public key cryptography

Two fundamental goals of cryptography are to provide privacy of communication between two entities and to provide authentication of one entity to another. Both goals can be achieved with symmetric cryptography. However, symmetric cryptography is not convenient in some applications for the following reasons. First, each pair of communicating entities needs to have a shared key. Second, these keys must be transmitted securely. Third, it is difficult to obtain signatures with non-repudiation (e.g., suitable for signing contracts).

In the mid 1970s Merkle, Diffie and Hellman proposed the idea of **public key cryptography** (also sometimes called **asymmetric cryptography**). This idea was also proposed by Ellis at GCHQ, under the name "non-secret encryption". One of the earliest and most important public key cryptosystems is RSA, invented by Adleman, Rivest and Shamir in 1977 (essentially the same scheme was also invented by Cocks at GCHQ in 1973).

As noted above, a major application of public key cryptography is to provide authentication. An extremely important example of this in the real world is digital signatures for authenticating automatic software updates. The public key of the software developer is stored in the application or operating system, and the software update is only performed if the digital signature on the update is verified for that public key (see Section 11.1 of Katz and Lindell [300] for more details). Signature schemes also provide message integrity, message authentication and non-repudiation (see Section 9.2 of Smart [513]). Other important applications of public key cryptography are key exchange and key transport for secure communication (e.g., in SSL or TLS).

1.2 The textbook RSA cryptosystem

We briefly describe the "textbook" RSA cryptosystem. The word "textbook" indicates that, although the RSA cryptosystem as presented below appears in many papers and books, this is definitely not how it should be used in the real world. In particular, public key encryption is most commonly used to transmit keys (the functionality is often called key transport or key encapsulation) rather than to encrypt data. Chapter 24 gives many more details about RSA including, in Section 24.7, a very brief discussion of padding schemes for use in real applications.

Alice chooses two large primes p and q of similar size and computes $N = pq$. Alice also chooses $e \in \mathbb{N}$ coprime to $\varphi(N) = (p-1)(q-1)$ and computes $d \in \mathbb{N}$ such that

$$ed \equiv 1 \pmod{\varphi(N)}.$$

Alice's **public key** is the pair of integers (N, e) and her **private key** is the integer d. To **encrypt** a message to Alice, Bob does the following:

1. Obtain an authentic copy of Alice's public key (N, e). This step may require trusted third parties and public key infrastructures, which are outside the scope of this book; see Chapter 11 of Smart [513] or Chapter 11 of Stinson [532]. We suppress this issue in this book.
2. Encode the message as an integer $1 \leq \mathsf{m} < N$.
   Note that m does not necessarily lie in $(\mathbb{Z}/N\mathbb{Z})^*$. However, if $p, q \approx \sqrt{N}$ then the probability that $\gcd(\mathsf{m}, N) > 1$ is $(p + q - 1)/(N - 1) \approx 2/\sqrt{N}$. Hence, in practice one may assume that $\mathsf{m} \in (\mathbb{Z}/N\mathbb{Z})^*$.[1]
3. Compute and transmit the **ciphertext**

$$\mathsf{c} = \mathsf{m}^e \pmod{N}.$$

To **decrypt** the ciphertext, Alice computes $\mathsf{m} = \mathsf{c}^d \pmod{N}$ and decodes this to obtain the original message.

Exercise 1.2.1 Show that if $\gcd(\mathsf{m}, N) = 1$ then $(\mathsf{m}^e)^d \equiv \mathsf{m} \pmod{N}$. Show that if $\gcd(\mathsf{m}, N) \neq 1$ then $(\mathsf{m}^e)^d \equiv \mathsf{m} \pmod{N}$.

The RSA system can also be used as a digital signature algorithm. When sending a message m to Bob, Alice computes the signature $\mathsf{s} = \mathsf{m}^d \pmod{N}$. When Bob receives (m, s) he obtains an authentic copy of Alice's public key and then verifies that $\mathsf{m} \equiv \mathsf{s}^e \pmod{N}$. If the verification equation holds then Bob believes that the message m does come from Alice. The value m is not usually an actual message or document (which might be huge) but a short integer that is the output of some (non-injective) compression function (such as a hash function). We sometimes call m a **message digest**.

The idea is that exponentiation to the power e modulo N is a **one-way function**: a function that is easy to compute but such that it is hard to compute preimages. Indeed, exponentiation modulo N is a **one-way permutation** on $(\mathbb{Z}/N\mathbb{Z})^*$ when e is co-prime to $\varphi(N)$. The private key d allows the permutation to be efficiently inverted and is known as a **trapdoor**. Therefore, RSA is often described as a **trapdoor one-way permutation**.

A number of practical issues must be considered:

1. Can public keys be efficiently generated?
2. Is the cryptosystem efficient in the sense of computation time and ciphertext size?
3. How does Bob know that Alice's public key is authentic?
4. Is the scheme secure?
5. What does "security" mean anyway?

One aim of this book is to explore the above issues in depth. We will study RSA (and some other cryptosystems based on integer factorisation) as well as cryptosystems based on the discrete logarithm problem.

[1] If N is a product of two 150 digit primes (which is the minimum size for an RSA modulus) then the expected number of trials to find $1 \leq \mathsf{m} < N$ with $\gcd(\mathsf{m}, N) > 1$ is therefore $\approx 10^{150}$. Note that the age of the universe is believed to be less than 10^{18} seconds.

To indicate some of the potential problems with the "textbook" RSA cryptosystem as described above we present three simple attacks.

1. Suppose the RSA cryptosystem is being used for an online election to provide privacy of an individual's vote to everyone outside the electoral office.[2] Each voter encrypts their vote under the public key of the electoral office and then sends their vote by email. Voters do not want any other member of the public to know who they voted for.

   Suppose the eavesdropper Eve is monitoring internet traffic from Alice's computer and makes a copy of the ciphertext corresponding to her vote. Since encryption is deterministic and there is only a short list of possible candidates, it is possible for Eve to compute each possible vote by encrypting each candidate's name under the public key. Hence, Eve can deduce who Alice voted for.
2. To speed up encryption it is tempting to use small encryption exponents, such as $e = 3$ (assuming that $N = pq$ where $p \equiv q \equiv 2 \pmod 3$). Now suppose Bob is only sending a very small message $0 < \mathsf{m} < N^{1/3}$ to Alice; this is quite likely, since public key cryptography is most often used to securely transmit symmetric keys. Then $\mathsf{c} = \mathsf{m}^3$ in $\mathbb{N}$, i.e. no modular reduction has taken place. An adversary can therefore compute the message m from the ciphertext c by taking cube roots in $\mathbb{N}$ (using numerical analysis techniques).
3. A good encryption scheme should allow an adversary to learn absolutely nothing about a message from the ciphertext. But, with the RSA cryptosystem, one can compute the Jacobi symbol $(\frac{\mathsf{m}}{N})$ of the message by computing $(\frac{\mathsf{c}}{N})$ (this can be computed efficiently without knowing the factorisation of N; see Section 2.4). The details are given in Exercise 24.1.8.

The above three attacks may be serious for some applications, but not for others. However, a cryptosystem designer often has little control over the applications in which their system is to be used. Hence, it is preferable to have systems that are not vulnerable to attacks of the above form. In Section 24.7 we will explain how to secure RSA against these sorts of attacks, by making the encryption process randomised and by using padding schemes that encode short messages as sufficiently large integers and that destroy algebraic relationships between messages.

1.3 Formal definition of public key cryptography

To study public key cryptography using mathematical techniques it is necessary to give a precise definition of an encryption scheme. The following definition uses terminology about algorithms that is recalled in Section 2.1. Note that the problem of obtaining an

[2] Much more interesting electronic voting schemes have been invented. This unnatural example is chosen purely for pedagogical purposes.

authentic copy of the public key is not covered by this definition; the public key is an input to the encryption function.

Definition 1.3.1 Let $\kappa \in \mathbb{N}$ be a **security parameter** (note that κ is not necessarily the same as the "key length"; see Example 1.3.2). An **encryption scheme** is defined by the following spaces (all depending on the security parameter κ) and algorithms:

M_κ	the space of all possible messages;
PK_κ	the space of all possible public keys;
SK_κ	the space of all possible private keys;
C_κ	the space of all possible ciphertexts;
KeyGen	a randomised algorithm that takes the security parameter κ, runs in expected polynomial-time (i.e., $O(\kappa^c)$ bit operations for some constant $c \in \mathbb{N}$) and outputs a public key $\mathsf{pk} \in \mathsf{PK}_\kappa$ and a private key $\mathsf{sk} \in \mathsf{SK}_\kappa$;
Encrypt	a randomised algorithm that takes as input $\mathsf{m} \in \mathsf{M}_\kappa$ and pk, runs in expected polynomial-time (i.e., $O(\kappa^c)$ bit operations for some constant $c \in \mathbb{N}$) and outputs a ciphertext $\mathsf{c} \in \mathsf{C}_\kappa$;
Decrypt	an algorithm (not usually randomised) that takes $\mathsf{c} \in \mathsf{C}_\kappa$ and sk, runs in polynomial-time and outputs either $\mathsf{m} \in \mathsf{M}_\kappa$ or the invalid ciphertext symbol $\perp$.

It is required that

$$\text{Decrypt}(\text{Encrypt}(\mathsf{m}, \mathsf{pk}), \mathsf{sk}) = \mathsf{m}$$

if (pk, sk) is a matching key pair. Typically, we require that the fastest known attack on the system requires at least 2^κ bit operations.

Example 1.3.2 We sketch how to write "textbook" RSA encryption in the format of Definition 1.3.1. The KeyGen algorithm takes input κ and outputs a modulus N that is a product of two randomly chosen primes of a certain length, as well as an encryption exponent e. Giving a precise recipe for the bit-length of the primes as a function of the security parameter is non-trivial for RSA; we merely note here that if $\kappa = 128$ (i.e., so that there is no known attack on the system performing fewer than 2^{128} bit operations) then it is standard to use 1536-bit primes. As we will discuss in Chapter 12, one can generate primes in expected polynomial-time and hence KeyGen is a randomised algorithm with expected polynomial-time complexity.

The message space M_κ depends on the randomised padding scheme being used. The ciphertext space C_κ in this case is $(\mathbb{Z}/N\mathbb{Z})^*$, which does not agree with Definition 1.3.1 as it does not depend only on κ. Instead, one usually takes C_κ to be the set of $\lceil \log_2(N) \rceil$-bit strings.

The Encrypt and Decrypt algorithms are straightforward (though the details depend on the padding scheme). The correctness condition is easily checked.

1.3.1 Security of encryption

We now give precise definitions for the security of public key encryption. An **adversary** is a randomised polynomial-time algorithm that interacts with the cryptosystem in some way. It is necessary to define the **attack model**, which specifies the way the adversary can interact with the cryptosystem. It is also necessary to define the **attack goal** of the adversary. For further details of these issues see Sections 10.2 and 10.6 of Katz and Lindell [300], Section 1.13 of Menezes, van Oorschot and Vanstone [376] or Section 15.1 of Smart [513].

We first list the **attack goals for public key encryption**. The most severe one is the **total break**, where the adversary computes a private key. There are three other commonly studied attacks, and they are usually formulated as **security properties** (the security property is the failure of an adversary to achieve its attack goal).

The word **oracle** is used below. This is just a fancy name for a magic box that takes some input and then outputs a correct answer in constant time. Precise definitions are given in Section 2.1.3.

- **One-way encryption (OWE):** Given a challenge ciphertext c the adversary cannot compute the corresponding message m.
- **Semantic security:** An adversary learns no information at all about a message from its ciphertext, apart from possibly the length of the message.

  This concept is made precise as follows: assume all messages in M_κ have the same length. A **semantic security adversary** is a randomised polynomial-time algorithm A that first chooses a function $f : \mathsf{M}_\kappa \to \{0, 1\}$ such that the probability, over uniformly chosen $\mathsf{m} \in \mathsf{M}_\kappa$, that $f(\mathsf{m}) = 1$ is $1/2$. The adversary A then takes as input a challenge $(\mathsf{c}, \mathsf{pk})$ where c is the encryption of a randomly chosen message $\mathsf{m} \in \mathsf{M}_\kappa$, and outputs a bit b. The adversary is **successful** if $b = f(\mathsf{m})$.

  Note that the standard definition of semantic security allows messages $\mathsf{m} \in \mathsf{M}_\kappa$ to be drawn according to any probability distribution. We have simplified to the case of the uniform distribution on M_κ.
- **Indistinguishability (IND):** An adversary cannot distinguish the encryption of any two messages m_0 and m_1, chosen by the adversary, of the same length.

  This concept is made precise by defining an **indistinguishability adversary** to be a randomised polynomial-time algorithm A that plays the following game with a challenger: first, the challenger generates a public key and gives it to A. Then (this is the "first phase" of the attack) A performs some computations (and possibly queries to oracles) and outputs two equal length messages m_0 and m_1. The challenger computes the **challenge ciphertext** c (which is an encryption of m_b where $b \in \{0, 1\}$ is randomly chosen) and gives it to A. In the "second phase", the adversary A performs more calculations (and possibly oracle queries) and outputs a bit b'. The adversary is **successful** if $b = b'$.

For a fixed value κ one can consider the probability that an adversary is successful over all public keys pk output by KeyGen, and (except when studying a total break adversary) all challenge ciphertexts c output by Encrypt, and over all random choices made by the adversary. The adversary breaks the security property if its success probability is noticeable as a function of κ (see Definition 2.1.10 for the terms noticeable and negligible). The cryptosystem achieves the security property if every polynomial-time adversary has negligible success probability as a function of κ. An adversary that works with probability 1 is called a **perfect adversary**.

We now list the three main attack models for public key cryptography.

- **Passive attack/chosen plaintext attack (CPA):** The adversary is given the public key.
- **Lunchtime attack (CCA1):**[3] The adversary has the public key and can also ask for decryptions of ciphertexts of its choosing during the first stage of the attack (i.e., before the challenge ciphertext is received).
- **Adaptive chosen-ciphertext attack (CCA)** (also denoted CCA2): The adversary has the public key and is given access to a decryption oracle O that will provide decryptions of any ciphertext of its choosing, with the restriction that O outputs $\perp$ in the second phase of the attack if the challenge ciphertext is submitted to O.

One can consider an adversary against any of the above security properties in any of the above attack models. For example, the strongest security notion is indistinguishability under an adaptive chosen-ciphertext attack. A cryptosystem that achieves this security level is said to have **IND-CCA security**. It has become standard in theoretical cryptography to insist that all cryptosystems have IND-CCA security. This is not because CCA attacks occur frequently in the real world, but because a scheme that has IND-CCA security should also be secure against any real-world attacker.[4]

Exercise 1.3.3 Show that the "textbook" RSA cryptosystem does not have IND-CPA security.

Exercise 1.3.4 Show that the "textbook" RSA cryptosystem does not have OWE-CCA security.

Exercise 1.3.5 Prove that if a cryptosystem has IND security under some attack model then it has semantic security under the same attack model.

1.3.2 Security of signatures

Definition 1.3.6 A **signature scheme** is defined, analogously to encryption, by message, signature and key spaces depending on a security parameter κ. There is a KeyGen algorithm and algorithms:

[3] The name comes from an adversary who breaks into someone's office during their lunch break, interacts with their private key in some way and then later in the day tries to decrypt a ciphertext.

[4] Of course, there are attacks that lie outside the attack model we are considering, such as side-channel attacks or attacks by dishonest system administrators.

Sign: A randomised algorithm that runs in polynomial-time (i.e., $O(\kappa^c)$ for some constant $c \in \mathbb{N}$), takes as input a message m and a private key sk, and outputs a signature s.

Verify: An algorithm (usually deterministic) that runs in polynomial-time, takes as input a message m, a signature s and a public key pk, and outputs "valid" or "invalid".

We require that Verify(m, Sign(m, sk), pk) = "valid". Typically, we require that all known algorithms to break the signature scheme require at least 2^κ bit operations.

The main **attack goals for signatures** are the following (for more discussion see Goldwasser, Micali and Rivest [235], Section 12.2 of Katz and Lindell [300], Section 15.4 of Smart [513], or Section 7.2 of Stinson [532]):

- **Total break:** An adversary can obtain the private key for the given public key.
- **Selective forgery** (also called **target message forgery**): An adversary can generate a valid signature for the given public key on any message.
- **Existential forgery:** An adversary can generate a pair (m, s) where m is a message and s is a signature for the given public key on that message.

  The acronym **UF** stands for the security property "unforgeable". In other words, a signature scheme has UF security if every polynomial-time existential forgery algorithm succeeds with only negligible probability. Be warned that some authors use UF to denote "universal forgery", which is another name for selective forgery.

As with encryption, there are various attack models.

- **Passive attack:** The adversary is given the public key only. This is also called a "public key only" attack.
- **Known message attack:** The adversary is given various sample message–signature pairs for the public key.
- **Adaptive chosen-message attack (CMA):** The adversary is given a signing oracle that generates signatures for the public key on messages of their choosing.

  In this case, **signature forgery** usually means producing a valid signature s for the public key pk on a message m such that m was not already queried to the signing oracle for key pk. Another notion, which we do not consider further in this book, is **strong forgery**; namely, to output a valid signature s on m for public key pk such that s is not equal to any of the outputs of the signing oracle on m.

As with encryption, one says the signature scheme has the stated security property under the stated attack model if there is no polynomial-time algorithm A that solves the problem with noticeable success probability under the appropriate game. The standard notion of security for digital signatures is **UF-CMA** security.

Exercise 1.3.7 Give a precise definition for UF-CMA security.

Exercise 1.3.8 Do "textbook" RSA signatures have selective forgery security under a passive attack?

Exercise 1.3.9 Show that there is a passive existential forgery attack on "textbook" RSA signatures.

Exercise 1.3.10 Show that, under a chosen-message attack, one can selectively forge "textbook" RSA signatures.

PART I

BACKGROUND

2
Basic algorithmic number theory

The aim of this chapter is to give a brief summary of some fundamental algorithms for arithmetic in finite fields. The intention is not to provide an implementation guide; instead, we sketch some important concepts and state some complexity results that will be used later in the book. We do not give a consistent level of detail for all algorithms; instead, we only give full details for algorithms that will play a significant role in later chapters of the book.

More details of these subjects can be found in Crandall and Pomerance [150], Shoup [497], Buhler and Stevenhagen [106], Brent and Zimmermann [95], Knuth [308], von zur Gathen and Gerhard [220], Bach and Shallit [21] and the handbooks [16, 376].

The chapter begins with some remarks about computational problems, algorithms and complexity theory. We then present methods for fast integer and modular arithmetic. Next we present some fundamental algorithms in computational number theory such as Euclid's algorithm, computing Legendre symbols and taking square roots modulo p. Finally, we discuss polynomial arithmetic, constructing finite fields and some computational problems in finite fields.

2.1 Algorithms and complexity

We assume the reader is already familiar with computers, computation and algorithms. General references for this section are Chapter 1 of Cormen *et al.* [136], Davis and Weyuker [154], Hopcroft and Ullman [265], Section 3.1 of Shoup [497], Sipser [509] and Talbot and Welsh [539].

Rather than using a fully abstract model of computation, such as Turing machines, we consider all algorithms as running on a digital computer with a typical instruction set, an infinite number of bits of memory and constant-time memory access. This is similar to the random access machine (or register machine) model; see Section 3.6 of [21], [129], Section 2.2 of [136], Section 7.6 of [265] or Section 3.2 of [497]. We think of an algorithm as a sequence of bit operations, though it is more realistic to consider word operations.

A **computational problem** is specified by an input (of a certain form) and an output (satisfying certain properties relative to the input). An **instance** of a computational problem is a specific input. The **input size** of an instance of a computational problem is the number of bits required to represent the instance. The **output size** of an instance of a computational problem is the number of bits necessary to represent the output. A **decision problem** is a computational problem where the output is either "yes" or "no". As an example, we give one of the most important definitions in the book.

Definition 2.1.1 Let G be a group written in multiplicative notation. The **discrete logarithm problem (DLP)** is: given $g, h \in G$ find a, if it exists, such that $h = g^a$.

In Definition 2.1.1 the input is a description of the group G together with the group elements g and h and the output is a or the failure symbol $\perp$ (to indicate that $h \notin \langle g \rangle$). Typically, G is an algebraic group over a finite field and the order of g is assumed to be known. We stress that an instance of the DLP, according to Definition 2.1.1, includes the specification of G, g and h, so one must understand that they are all allowed to vary (note that in many cryptographic applications one considers the group G and element g as being fixed; we discuss this in Exercise 21.1.2). As explained in Section 2.1.2, a computational problem should be defined with respect to an instance generator; in the absence of any further information it is usual to assume that the instances are chosen uniformly from the space of all possible inputs of a given size. In particular, for the DLP it is usual to denote the order of g by r and to assume that $h = g^a$ where a is chosen uniformly in $\mathbb{Z}/r\mathbb{Z}$. The output is the integer a (e.g., written in binary). The input size depends on the specific group G and the method used to represent it. If h can take all values in $\langle g \rangle$ then one needs at least $\log_2(r)$ bits to specify h from among the r possibilities. Hence, the input size is at least $\log_2(r)$ bits. Similarly, if the output a is uniformly distributed in $\mathbb{Z}/r\mathbb{Z}$ then the output size is at least $\log_2(r)$ bits.

An algorithm to solve a computational problem is called **deterministic** if it does not make use of any randomness. We will study the asymptotic complexity of deterministic algorithms by counting the number of bit operations performed by the algorithm expressed as a function of the input size. Upper bounds on the complexity are presented using "big O" notation. When giving complexity estimates using big O notation we implicitly assume that there is a countably infinite number of possible inputs to the algorithm.

Definition 2.1.2 Let $f, g : \mathbb{N} \to \mathbb{R}_{>0}$. Write $f = O(g)$ if there are $c \in \mathbb{R}_{>0}$ and $N \in \mathbb{N}$ such that

$$f(n) \le cg(n)$$

for all $n \ge N$.

Similarly, if $f(n_1, \dots, n_m)$ and $g(n_1, \dots, n_m)$ are functions from $\mathbb{N}^m$ to $\mathbb{R}_{>0}$ then we write $f = O(g)$ if there are $c \in \mathbb{R}_{>0}$ and $N_1, \dots, N_m \in \mathbb{N}$ such that $f(n_1, \dots, n_m) \le cg(n_1, \dots, n_m)$ for all $(n_1, \dots, n_m) \in \mathbb{N}^m$ with $n_i \ge N_i$ for all $1 \le i \le m$.

Example 2.1.3 $3n^2 + 2n + 1 = O(n^2)$, $n + \sin(n) = O(n)$, $n^{100} + 2^n = O(2^n)$, $\log_{10}(n) = O(\log(n))$.

Exercise 2.1.4 Show that if $f(n) = O(\log(n)^a)$ and $g(n) = O(\log(n)^b)$ then $(f + g)(n) = f(n) + g(n) = O(\log(n)^{\max\{a,b\}})$ and $(fg)(n) = f(n)g(n) = O(\log(n)^{a+b})$. Show that $O(n^c) = O(2^{c\log(n)})$.

We also present the "little o", "soft O", "big Omega" and "big Theta" notation. These will only ever be used in this book for functions of a single argument.

Definition 2.1.5 Let $f, g : \mathbb{N} \to \mathbb{R}_{>0}$. Write $f(n) = o(g(n))$ if

$$\lim_{n\to\infty} f(n)/g(n) = 0.$$

Write $f(n) = \tilde{O}(g(n))$ if there is some $m \in \mathbb{N}$ such that $f(n) = O(g(n)\log(g(n))^m)$. Write $f(n) = \Omega(g(n))$ if $g(n) = O(f(n))$. Write $f(n) = \Theta(g(n))$ if $f(n) = O(g(n))$ and $g(n) = O(f(n))$.

Exercise 2.1.6 Show that if $g(n) = O(n)$ and $f(n) = \tilde{O}(g(n))$ then there is some $m \in \mathbb{N}$ such that $f(n) = O(n\log(n)^m)$.

Definition 2.1.7 (Worst-case asymptotic complexity.) Let A be a deterministic algorithm and let $t(n)$ be a bound on the running time of A on every problem of input size n bits.

- A is **polynomial-time** if there is an integer $k \in \mathbb{N}$ such that $t(n) = O(n^k)$.
- A is **superpolynomial-time** if $t(n) = \Omega(n^c)$ for all $c \in \mathbb{R}_{>1}$.
- A is **exponential-time** if there is a constant $c_2 > 1$ such that $t(n) = O(c_2^n)$.
- A is **subexponential-time** if $t(n) = O(c^n)$ for all $c \in \mathbb{R}_{>1}$.

Exercise 2.1.8 Show that $n^{a\log(\log(n))}$ and $n^{a\log(n)}$, for some $a \in \mathbb{R}_{>0}$, are functions that are $\Omega(n^c)$ and $O(c^n)$ for all $c \in \mathbb{R}_{>1}$.

For more information about computational complexity, including the definitions of complexity classes such as P and NP, see Chapters 2 to 4 of [539], Chapter 13 of [265], Chapter 15 of [154], Chapter 7 of [509] or Chapter 34 of [136]. Definition 2.1.7 is for **uniform** complexity, as a single algorithm A solves all problem instances. One can also consider **non-uniform complexity**, where one has an algorithm A and, for each $n \in \mathbb{N}$, polynomially sized auxiliary input $h(n)$ (the hint) such that if x is an n-bit instance of the computational problem then $A(x, h(n))$ solves the instance. An alternative definition is a sequence A_n of algorithms, one for each input size $n \in \mathbb{N}$, and such that the description of the algorithm is polynomially bounded. We stress that the hint is not required to be efficiently computable. We refer to Section 4.6 of Talbot and Welsh [539] for details.

Complexity theory is an excellent tool for comparing algorithms, but one should always be aware that the results can be misleading. For example, it can happen that there are several algorithms to solve a computational problem and that the one with the best complexity is slower than the others for the specific problem instance one is interested in (e.g., see Remark 2.2.5).

2.1.1 Randomised algorithms

All our algorithms may be **randomised**, in the sense that they have access to a random number generator. A deterministic algorithm should terminate after a finite number of steps, but a randomised algorithm can run forever if an infinite sequence of "unlucky" random choices is made.[1] Also, a randomised algorithm may output an incorrect answer for some choices of randomness. A **Las Vegas algorithm** is a randomised algorithm which, if it terminates,[2] outputs a correct solution to the problem. A randomised algorithm for a decision problem is a **Monte Carlo algorithm** if it always terminates, and if the output is "yes" then it is correct and if the output is "no" then it is correct with "noticeable" probability (see the next section for a formal definition of noticeable success probability).

An example of a Las Vegas algorithm is choosing a random quadratic non-residue modulo p by choosing random integers modulo p and computing the Legendre symbol (see Exercise 2.4.6 in Section 2.4); the algorithm could be extremely unlucky forever. Of course, there is a deterministic algorithm for this problem, but its complexity is worse than the randomised algorithm. An example of a Monte Carlo algorithm is testing primality of an integer N using the Miller–Rabin test (see Section 12.1.2). Many of the algorithms in the book are randomised Monte Carlo or Las Vegas algorithms. We will often omit the words "Las Vegas" or "Monte Carlo".

Deterministic algorithms have a well-defined running time on any given problem instance. For a randomised algorithm the running time for a fixed instance of the problem is not necessarily well-defined. Instead, one considers the expected value of the running time over all choices of the randomness. We usually consider **worst-case complexity**. The worst-case complexity for input size n is the maximum, over all problem instances of size n, of the expected running time of the algorithm. As always, when considering asymptotic complexity it is necessary that the computational problem has a countably infinite number of problem instances.

A randomised algorithm is **expected polynomial-time** if the worst case over all problem instances of size n bits of the expected value of the running time is $O(n^c)$ for some $c \in \mathbb{R}_{>0}$. (An expected polynomial-time algorithm can run longer than polynomial-time if it makes many "unlucky" choices.) A randomised algorithm is **expected exponential-time** (respectively, **expected subexponential-time**) if there exists $c \in \mathbb{R}_{>1}$ (respectively, for all $c \in \mathbb{R}_{>1}$) such that the expected value of the running time on problem instances of size n bits is $O(c^n)$.

One can also consider **average-case complexity**, which is the average, over all problem instances of size n, of the expected running time of the algorithm. Equivalently, the average-case complexity is the expected number of bit operations of the algorithm where the

[1] In algorithmic number theory it is traditional to allow algorithms that do not necessarily terminate, whereas in cryptography it is traditional to consider algorithms whose running time is bounded (typically by a polynomial in the input size). Indeed, in security reductions it is crucial that an adversary (i.e. randomised algorithm) always terminates. Hence, some of the definitions in this section (e.g., Las Vegas algorithms) mainly arise in the algorithmic number theory literature.

[2] An alternative definition is that a Las Vegas algorithm has finite expected running time, and outputs either a correct result or the failure symbol $\perp$.

expectation is taken over all problem instances of size n as well as all choices of the randomness. For more details see Section 4.2 of Talbot and Welsh [539].

2.1.2 Success probability of a randomised algorithm

Throughout the book we give very simple definitions (like Definition 2.1.1) for computational problems. However, it is more subtle to define what it means for a randomised algorithm A to solve a computational problem. A **perfect algorithm** is one where the output is always correct (i.e., it always succeeds). We also consider algorithms that give the correct answer only for some subset of the problem instances, or for all instances but only with a certain probability.

The issue of whether an algorithm is successful is handled somewhat differently by the two communities whose work is surveyed in this book. In the computational number theory community, algorithms are expected to solve all problem instances with probability of success close to 1. In the cryptography community, it is usual to consider algorithms that only solve some noticeable (see Definition 2.1.10) proportion of problem instances, and even then only with some noticeable probability. The motivation for the latter community is that an algorithm to break a cryptosystem is considered devastating even if only a relatively small proportion of ciphertexts are vulnerable to the attack. Two examples of attacks that only apply to a small proportion of ciphertexts are the attack by Boneh, Joux and Nguyen on textbook Elgamal (see Exercise 20.4.9) and the Desmedt–Odlyzko signature forgery method (see Section 24.4.3).

We give general definitions for the success probability of an algorithm in this section, but rarely use the formalism in our later discussion. Instead, for most of the book, we focus on the case of algorithms that always succeed (or, at least, that succeed with probability extremely close to 1). This choice allows shorter and simpler proofs of many facts. In any case, for most computational problems the success probability can be increased by running the algorithm repeatedly, see Section 2.1.4.

The success of an algorithm to solve a computational problem is defined with respect to an **instance generator**, which is a randomised algorithm that takes as input $\kappa \in \mathbb{N}$ (often κ is called the **security parameter**), runs in polynomial-time in the output size and outputs an instance of the computational problem (or fails to generate an instance with some negligible probability). The output is usually assumed to be $\Theta(\kappa)$ bits,[3] so "polynomial-time" in the previous sentence means $O(\kappa^m)$ bit operations for some $m \in \mathbb{N}$. We give an example of an instance generator for the DLP in Example 2.1.9.

Let A be a randomised algorithm that takes an input $\kappa \in \mathbb{N}$. Write $\mathcal{S}_\kappa$ for the set of possible outputs of $A(\kappa)$. The **output distribution** of A on input κ is the distribution on $\mathcal{S}_\kappa$ such that $\Pr(x)$ for $x \in \mathcal{S}_\kappa$ is the probability, over the random choices made by A, that the output of $A(\kappa)$ is x.

[3] Hence, for problems related to RSA or factoring, κ is now assumed to be the bit-length of the modulus.

Example 2.1.9 Let a security parameter $\kappa \in \mathbb{N}$ be given. First, generate a random prime number r such that $2^{2\kappa} < r < 2^{2\kappa+1}$ (by choosing uniformly at random $(2\kappa+1)$-bit integers and testing each for primality). Next, try consecutive small integers k until $p = kr + 1$ is prime. Then, choose a random integer $1 < u < p$ and set $g = u^{(p-1)/r}$ and repeat if $g = 1$. It follows that g is a generator of a cyclic subgroup of $G = \mathbb{F}_p^*$ of order r. Finally, choose uniformly at random an integer $0 < a < r$ and set $h = g^a$. Output (p, r, g, h), which can be achieved using $3\lceil \log_2(p) \rceil + \lceil \log_2(r) \rceil$ bits.

One sees that there are finitely many problem instances for a given value of the security parameter κ, but infinitely many instances in total. The output distribution has (r, g, h) uniform in the appropriate sets, but p is not uniform.

When considering an algorithm A to solve a computational problem we assume that A has been customised for a particular instance generator. Hence, a problem might be easy with respect to some instance generators and hard for others. Thus it makes no sense to claim that "DLP is a hard problem"; instead, one should conjecture that DLP is hard for certain instance generators.

We now define what is meant by the word negligible.

Definition 2.1.10 A function $\epsilon : \mathbb{N} \to \mathbb{R}_{>0}$ is **negligible** if for every polynomial $p(x) \in \mathbb{R}[x]$ there is some $K \in \mathbb{N}$ such that for all $\kappa > K$ with $p(\kappa) \neq 0$ we have $\epsilon(\kappa) < 1/|p(\kappa)|$.

A function $\epsilon : \mathbb{N} \to \mathbb{R}_{>0}$ is **noticeable** if there exists a polynomial $p(x) \in \mathbb{R}[x]$ and an integer K such that $\epsilon(\kappa) > 1/|p(\kappa)|$ for all $\kappa > K$ with $p(\kappa) \neq 0$.

Let $[0, 1] = \{x \in \mathbb{R} : 0 \le x \le 1\}$. A function $p : \mathbb{N} \to [0, 1]$ is **overwhelming** if $1 - p(\kappa)$ is negligible.

Note that noticeable is not the logical negation of negligible. There are functions that are neither negligible nor noticeable.

Example 2.1.11 The function $\epsilon(\kappa) = 1/2^\kappa$ is negligible.

Exercise 2.1.12 Let $f_1(\kappa)$ and $f_2(\kappa)$ be negligible functions. Prove that $f_1 + f_2$ is a negligible function and that $p(\kappa) f_1(\kappa)$ is a negligible function for any polynomial $p(x) \in \mathbb{R}[x]$ such that $p(x) > 0$ for all sufficiently large x.

Definition 2.1.13 Let A be a randomised algorithm to solve instances of a computational problem generated by a specific instance generator. The **success probability** of the algorithm A is the function $f : \mathbb{N} \to [0, 1]$ such that, for $\kappa \in \mathbb{N}$, $f(\kappa)$ is the probability that A outputs the correct answer, where the probability is taken over the randomness used by A and according to the output distribution of the instance generator on input κ. An algorithm A with respect to an instance generator **succeeds** if its success probability is a noticeable function.

Note that the success probability is taken over both the random choices made by A and the distribution of problem instances. In particular, an algorithm that succeeds does not necessarily solve a specific problem instance even when run repeatedly with different random choices.

Example 2.1.14 Consider an algorithm A for the DLP with respect to the instance generator of Example 2.1.9. Suppose A simply outputs an integer a chosen uniformly at random in the range $0 < a < r$. Since $r > 2^{2\kappa}$ the probability that A is correct is $1/(r-1) \leq 1/2^{2\kappa}$. For any polynomial $p(x)$ there are $X, c \in \mathbb{R}_{>0}$ and $n \in \mathbb{N}$ such that $|p(x)| \leq cx^n$ for $x \geq X$. Similarly, there is some $K \geq X$ such that $cK^n \leq 2^{2K}$. Hence, the success probability of A is negligible.

Certain decision problems (e.g., decision Diffie–Hellman) require an algorithm to behave differently when given inputs drawn from different distributions on the same underlying set. In this case, the success probability is not the right concept and one instead uses the **advantage**. We refer to Definition 20.2.4 for an example.

Chapter 7 of Shoup [497] gives further discussion of randomised algorithms and success probabilities.

2.1.3 Reductions

An **oracle** for a computational problem takes one unit of running time, independent of the size of the instance, and returns an output. An oracle that always outputs a correct answer is called a **perfect oracle**. One can consider oracles that only output a correct answer with a certain noticeable probability (or advantage). For simplicity we usually assume that oracles are perfect and leave the details in the general case as an exercise for the reader. We sometimes use the word **reliable** for an oracle whose success probability is overwhelming (i.e., success probability $1-\epsilon$ where ϵ is negligible) and **unreliable** for an oracle whose success probability is small (but still noticeable).

Note that the behaviour of an oracle is only defined if its input is a valid instance of the computational problem it solves. Similarly, the oracle performs with the stated success probability only if it is given problem instances drawn with the correct distribution from the set of all problem instances.

Definition 2.1.15 A **reduction** from problem A to problem B is a randomised algorithm to solve problem A (running in expected polynomial-time and having noticeable success probability) by making queries to an oracle (which succeeds with noticeable probability) to solve problem B.

If there is a reduction from problem A to problem B then we write[4]

$$\text{A} \leq_R \text{B}.$$

Theorem 2.1.16 *Let A and B be computational problems such that $A \leq_R B$. If there is a polynomial-time randomised algorithm to solve B then there is a polynomial-time randomised algorithm to solve A.*

[4] The subscript R denotes the word "reduction" and should also remind the reader that our reductions are randomised algorithms.

A reduction between problems A and B therefore explains that "if you can solve B then you can solve A". This means that solving A has been "reduced" to solving problem B and we can infer that problem B is "at least as hard as" problem A or that problem A is "no harder than" problem B.

Since oracle queries take one unit of running time and reductions are polynomial-time algorithms, a reduction makes only polynomially many oracle queries.

Definition 2.1.17 If there is a reduction from A to B and a reduction from B to A then we say that problems A and B are **equivalent** and write $A \equiv_R B$.

Some authors use the phrases **polynomial-time reduction** and **polynomial-time equivalent** in place of reduction and equivalence. However, these terms have a technical meaning in complexity theory that is different from reduction (see Section 34.3 of [136]). Definition 2.1.15 is closer to the notion of Turing reduction, except that we allow randomised algorithms, whereas a Turing reduction is a deterministic algorithm. We abuse terminology and define the terms **subexponential-time reduction** and **exponential-time reduction** by relaxing the condition in Definition 2.1.15 that the algorithm be polynomial-time (these terms are used in Section 21.4.3).

2.1.4 Random self-reducibility

There are two different ways that an algorithm or oracle can be unreliable. First, it may be randomised and only output the correct answer with some probability; such a situation is relatively easy to deal with by repeatedly running the algorithm/oracle on the same input. The second situation, which is more difficult to handle, is when there is a subset of problem instances for which the algorithm or oracle extremely rarely or never outputs the correct solution; for this situation random self-reducibility is essential. We give a definition only for the special case of computational problems in groups.

Definition 2.1.18 Let P be a computational problem for which every instance of the problem is an n_1-tuple of elements of some cyclic group G of order r and such that the solution is an n_2-tuple of elements of G together with an n_3-tuple of elements of $\mathbb{Z}/r\mathbb{Z}$ (where n_2 or n_3 may be zero).

The computational problem P is **random self-reducible** if there is a polynomial-time algorithm that transforms an instance of the problem (with elements in a group G) into a uniformly random instance of the problem (with elements in the *same* group G) such that the solution to the original problem can be obtained in polynomial-time from the solution to the new instance.

We stress that a random self-reduction of a computational problem in a group G gives instances of the same computational problem in the same group. In general, there is no way to use information about instances of a computational problem in a group G' to solve computational problems in G if $G' \neq G$ (unless perhaps G' is a subgroup of G or vice versa).

Lemma 2.1.19 *Let G be a group and let $g \in G$ have prime order r. Then the DLP in $\langle g \rangle$ is random self-reducible.*

Proof First, note that the DLP fits the framework of computational problems in Definition 2.1.18. Denote by $\mathcal{X}$ the set $(\langle g \rangle - \{1\}) \times \langle g \rangle$. Let $(g, h) \in \mathcal{X}$ be an instance of the DLP.

Choose $1 \le x < r$ and $0 \le y < r$ uniformly at random and consider the pair $(g^x, h^x g^{xy}) \in \mathcal{X}$. Every pair $(g_1, g_2) \in \mathcal{X}$ arises in this way for exactly one pair (x, y). Hence, we have produced a DLP instance uniformly at random.

If a is the solution to the new DLP instance, i.e. $h^x g^{xy} = (g^x)^a$, then the solution to the original instance is

$$a - y \pmod{r}.$$

This completes the proof. □

A useful feature of random self-reducible problems is that if A is an algorithm that solves an instance of the problem in a group G with probability (or advantage) ϵ then one can obtain an algorithm A' that repeatedly calls A and solves any instance in G of the problem with overwhelming probability. This is called **amplifying** the success probability (or advantage). An algorithm to transform an unreliable oracle into a reliable one is sometimes called a **self-corrector**.

Lemma 2.1.20 *Let g have prime order r and let $G = \langle g \rangle$. Let A be an algorithm that solves the DLP in G with probability at least $\epsilon > 0$. Let $\epsilon' > 0$ and define $n = \lceil \log(1/\epsilon')/\epsilon \rceil$ (where* log *denotes the natural logarithm). Then there is an algorithm A' that solves the DLP in G with probability at least $1 - \epsilon'$. The running time of A' is $O(n \log(r))$ group operations plus n times the running time of A.*

Proof Run A on n random self-reduced versions of the original DLP. One convenient feature of the DLP is that one can check whether a solution is correct (this takes $O(\log(r))$ group operations for each guess for the DLP).

The probability that all n trials are incorrect is at most $(1 - \epsilon)^n < (e^{-\epsilon})^{\log(1/\epsilon')/\epsilon} = e^{\log(\epsilon')} = \epsilon'$. Hence, A' outputs the correct answer with probability at least $1 - \epsilon'$. □

2.2 Integer operations

We now begin our survey of efficient computer arithmetic. General references for this topic are Section 9.1 of Crandall and Pomerance [150], Section 3.3 of Shoup [497], Section 4.3.1 of Knuth [308], Chapter 1 of Brent and Zimmermann [95] and von zur Gathen and Gerhard [220].

Integers are represented as a sequence of binary words. Operations like add or multiply may correspond to many bit or word operations. The **length** of an unsigned integer a represented in binary is

$$\operatorname{len}(a) = \begin{cases} \lfloor \log_2(a) \rfloor + 1 & \text{if } a \neq 0 \\ 1 & \text{if } a = 0. \end{cases}$$

For a signed integer we define $\operatorname{len}(a) = \operatorname{len}(|a|) + 1$.

The complexity of algorithms manipulating integers depends on the length of the integers; hence, one should express the complexity in terms of the function len. However, it is traditional to just use $\log_2$ or the natural logarithm log.

Exercise 2.2.1 Show that, for $a \in \mathbb{N}$, $\operatorname{len}(a) = O(\log(a))$ and $\log(a) = O(\operatorname{len}(a))$.

Lemma 2.2.2 *Let $a, b \in \mathbb{Z}$ be represented as a sequence of binary words.*

1. *It requires $O(\log(a))$ bit operations to write a out in binary.*
2. *One can compute $a \pm b$ in $O(\max\{\log(a), \log(b)\})$ bit operations.*
3. *One can compute ab in $O(\log(a)\log(b))$ bit operations.*
4. *Suppose $|a| > |b|$. One can compute q and r such that $a = bq + r$ and $0 \leq r < |b|$ in $O(\log(b)\log(q)) = O(\log(b)(\log(a) - \log(b) + 1))$ bit operations.*

Proof Only the final statement is non-trivial. The school method of long division computes q and r simultaneously and requires $O(\log(q)\log(a))$ bit operations. It is more efficient to compute q first by considering only the most significant $\log_2(q)$ bits of a, and then to compute r as $a - bq$. For more details see Section 4.3.1 of [308], Section 2.4 of [220] or Section 3.3.4 of [497]. □

2.2.1 Faster integer multiplication

An important discovery is that it is possible to multiply integers more quickly than the "school method". General references for this subject include Section 9.5 of [150], Section 4.3.3 of [308], Section 3.5 of [497] and Section 1.3 of [95].

Karatsuba multiplication is based on the observation that one can compute $(a_0 + 2^n a_1)(b_0 + 2^n b_1)$, where a_0, a_1, b_0 and b_1 are n-bit integers, in three multiplications of n-bit integers rather than four.

Exercise 2.2.3 Prove that the complexity of Karatsuba multiplication of n bit integers is $O(n^{\log_2(3)}) = O(n^{1.585})$ bit operations.

[Hint: Assume n is a power of 2.]

Toom–Cook multiplication is a generalisation of Karatsuba. Fix a value k and suppose $a = a_0 + a_1 2^n + a_2 2^{2n} + \cdots a_k 2^{kn}$ and similarly for b. One can think of a and b as being polynomials in x of degree k evaluated at 2^n and we want to compute the product $c = ab$, which is a polynomial of degree $2k$ in x evaluated at $x = 2^n$. The idea is to compute the coefficients of the polynomial c using polynomial interpolation and therefore to recover c. The arithmetic is fast if the polynomials are evaluated at small integer values. Hence, we compute $c(1) = a(1)b(1)$, $c(-1) = a(-1)b(-1)$, $c(2) = a(2)b(2)$ etc. The complexity of Toom–Cook multiplication for n-bit integers is $O(n^{\log_{k+1}(2k+1)})$ (e.g., when $k = 3$ the complexity is $O(n^{1.465})$). For more details see Section 9.5.1 of [150].

Exercise 2.2.4★ Give an algorithm for Toom–Cook multiplication with $k = 3$.

Schönhage–Strassen multiplication multiplies n-bit integers in nearly linear time, namely $O(n \log(n) \log(\log(n)))$ bit operations, using the fast Fourier transform (FFT). The Fürer algorithm is slightly better. These algorithms are not currently used in the implementation of RSA or discrete logarithm cryptosystems so we do not describe them in this book. We refer to Sections 9.5.2 to 9.5.7 of [150], Chapter 8 of [220], Chapter 2 of [95], [549] and Chapter 4 of [83] for details.

Remark 2.2.5 In practice, the "school" method is fastest for small numbers. The crossover point (i.e., when Karatsuba becomes faster than the school method) depends on the word size of the processor and many other issues, but seems to be for numbers of around 300–1000 bits (i.e., 90–300 digits) for most computing platforms. For a popular 32 bit processor, Zimmermann [575] reports reports that Karatsuba beats the school method for integers of 20 words (640 bits) and Toom–Cook with $k = 3$ beats Karatsuba at 77 words (2464 bits). Bentahar [40] reports crossovers of 23 words (i.e., about 700 bits) and 133 words (approximately 4200 bits) respectively. The crossover point for the FFT methods is much larger. Hence, for elliptic curve cryptography at current security levels the "school" method is usually used, while for RSA cryptography the Karatsuba method is usually used.

Definition 2.2.6 Denote by $M(n)$ the number of bit operations to perform a multiplication of n bit integers.

For the remainder of the book we assume that $M(n) = c_1 n^2$ for some constant c_1 when talking about elliptic curve arithmetic, and that $M(n) = c_2 n^{1.585}$ for some constant c_2 when talking about RSA .

Applications of Newton's method

Recall that if $F : \mathbb{R} \to \mathbb{R}$ is differentiable and if x_0 is an approximation to a zero of $F(x)$ then one can efficiently get a very close approximation to the zero by running Newton's iteration

$$x_{n+1} = x_n - F(x_n)/F'(x_n).$$

Newton's method has quadratic convergence, in general, so the precision of the approximation roughly doubles at each iteration.

It is beyond the scope of this book to give all applications of Newton's method for fast arithmetic. Instead, we briefly mention some complexity results.

First, one can use Newton's method to compute rational approximations to $1/a \in \mathbb{Q}$ of the form $b/2^e$. The complexity to get an approximation such that $|1/a - b/2^e| < 1/2^m$ is $O(M(m))$ bit operations. For details see Section 3.5 of [497] (especially Exercise 3.35), Chapter 9 of [220] or, for a slightly different formulation, Section 9.2.2 of [150]. For applications of this idea to efficient modular arithmetic see Section 2.5 or Section 10.5 of [16]; an important result is that one can compute $a \pmod{b}$, where a and b are m-bit integers, in $O(M(m))$ bit operations.

Another application is to compute integer roots of polynomials $F(x) \in \mathbb{Z}[x]$. One uses Newton's method to find an approximation to a root $x \in \mathbb{R}$, and then rounds to the nearest integer. As a special case, integer square roots of m-bit numbers can be computed in $O(M(m))$ bit operations. Similarly, other roots (such as cube roots) can be computed in polynomial-time. It follows that one can test, in polynomial-time, whether an integer is a **perfect power** (i.e., whether the integer is a^m for some $a, m \in \mathbb{N}, m > 1$).

Exercise 2.2.7 Show that if $N = p^e$ where p is prime and $e \geq 1$ then one can factor N in polynomial-time.

2.3 Euclid's algorithm

For $a, b \in \mathbb{N}$, Euclid's algorithm computes $d = \gcd(a, b)$. A simple way to express Euclid's algorithm is by the recursive formula

$$\gcd(a, b) = \begin{cases} \gcd(a, 0) = a \\ \gcd(b, a \pmod{b}) & \text{if } b \neq 0. \end{cases}$$

The traditional approach is to work with positive integers a and b throughout the algorithm and to choose $a \pmod{b}$ to be in the set $\{0, 1, \ldots, b-1\}$. In practice, the algorithm can be used with $a, b \in \mathbb{Z}$ and it runs faster if we choose remainders in the range $\{-\lceil |b|/2 \rceil + 1, \ldots, -1, 0, 1, \ldots, \lceil |b|/2 \rceil\}$. However, for some applications (especially, those related to diophantine approximation) the version with positive remainders is the desired choice.

In practice, we often want to compute integers (s, t) such that $d = as + bt$, in which case we use the extended Euclidean algorithm (due to Lagrange). This is presented in Algorithm 1, where the integers r_i, s_i, t_i always satisfy $r_i = s_i a + t_i b$.

Theorem 2.3.1 *The complexity of Euclid's algorithm is $O(\log(a)\log(b))$ bit operations.*

Proof Each iteration of Euclid's algorithm involves computing the quotient and remainder of division of r_{i-2} by r_{i-1} where we may assume $|r_{i-2}| > |r_{i-1}|$ (except maybe for $i = 1$).

Algorithm 1 Extended Euclidean algorithm

INPUT: $a, b \in \mathbb{Z}$

OUTPUT: $d = \gcd(a, b)$ and $s, t \in \mathbb{Z}$ such that $d = sa + tb$

1: $r_{-1} = a,\ s_{-1} = 1,\ t_{-1} = 0$
2: $r_0 = b,\ s_0 = 0,\ t_0 = 1$
3: $i = 0$
4: **while** $(r_i \neq 0)$ **do**
5: $\quad i = i + 1$
6: $\quad$ find $q_i, r_i \in \mathbb{Z}$ such that $-|r_{i-1}|/2 < r_i \leq |r_{i-1}|/2$ and $r_{i-2} = q_i r_{i-1} + r_i$
7: $\quad s_i = s_{i-2} - q_i s_{i-1}$
8: $\quad t_i = t_{i-2} - q_i t_{i-1}$
9: **end while**
10: **return** $r_{i-1},\ s_{i-1},\ t_{i-1}$

By Lemma 2.2.2 this requires $\leq c \log(r_{i-1})(\log(r_{i-2}) - \log(r_{i-1}) + 1)$ bit operations for some constant $c \in \mathbb{R}_{>0}$. Hence, the total running time is at most

$$c \sum_{i \geq 1} \log(r_{i-1})(\log(r_{i-2}) - \log(r_{i-1}) + 1).$$

Rearranging terms gives

$$c \log(r_{-1}) \log(r_0) + c \sum_{i \geq 1} \log(r_{i-1})(1 + \log(r_i) - \log(r_{i-1})).$$

Now, $2|r_i| \leq |r_{i-1}|$ so $1 + \log(r_i) \leq \log(r_{i-1})$; hence, all the terms in the above sum are ≤ 0. It follows that the algorithm performs $O(\log(a)\log(b))$ bit operations. □

Exercise 2.3.2 Show that the complexity of Algorithm 1 is still $O(\log(a)\log(b))$ bit operations even when the remainders in line 6 are chosen in the range $0 \leq r_i < r_{i-1}$.

A more convenient method for fast computer implementation is the binary Euclidean algorithm (originally due to Stein). This uses bit operations such as division by 2 rather than taking general quotients; see Section 4.5.2 of [308], Section 4.7 of [21], Chapter 3 of [220], Section 9.4.1 of [150] or Section 14.4.3 of [376].

There are subquadratic versions of Euclid's algorithm. One can compute the extended gcd of two n-bit integers in $O(M(n)\log(n))$ bit operations. We refer to Section 9.4 of [150], [524] or Section 11.1 of [220].

The rest of the section gives some results about diophantine approximation that are used later (e.g., in the Wiener attack on RSA, see Section 24.5.1). We assume that $a, b > 0$ and that the extended Euclidean algorithm with positive remainders is used to generate the sequence of values (r_i, s_i, t_i).

The integers s_i and t_i arising from the extended Euclidean algorithm are equal, up to sign, to the convergents of the continued fraction expansion of a/b. To be precise, if the convergents of a/b are denoted h_i/k_i for $i = 0, 1, \dots$ then, for $i \geq 1$, $s_i = (-1)^{i-1}k_{i-1}$ and $t_i = (-1)^i h_{i-1}$. Therefore, the values (s_i, t_i) satisfy various equations, summarised below, that will be used later in the book. We refer to Chapter 10 of [250] or Chapter 7 of [420] for details on continued fractions.

Lemma 2.3.3 *Let $a, b \in \mathbb{N}$ and let $r_i, s_i, t_i \in \mathbb{Z}$ be the triples generated by running Algorithm 1 in the case of positive remainders $0 \leq r_i < r_{i-1}$.*

1. *For $i \geq 1$, $|s_i| < |s_{i+1}|$ and $|t_i| < |t_{i+1}|$.*
2. *If $a, b > 0$ then $t_i > 0$ when $i \geq 1$ is even and $t_i < 0$ when i is odd (and vice versa for s_i).*
3. *$t_{i+1}s_i - t_i s_{i+1} = (-1)^{i+1}$.*
4. *$r_i s_{i-1} - r_{i-1}s_i = (-1)^i b$ and $r_i t_{i-1} - r_{i-1}t_i = (-1)^{i-1}a$. In other words, $r_i|s_{i-1}| + r_{i-1}|s_i| = b$ and $r_i|t_{i-1}| + r_{i-1}|t_i| = a$.*
5. *$|a/b + t_i/s_i| \leq 1/|s_i s_{i+1}|$.*
6. *$|r_i s_i| < |r_i s_{i+1}| \leq |b|$ and $|r_i t_i| < |r_i t_{i+1}| \leq |a|$.*
7. *If $s, t \in \mathbb{Z}$ are such that $|a/b + t/s| < 1/(2s^2)$ then (s, t) is (up to sign) one of the pairs (s_i, t_i) computed by Euclid's algorithm.*
8. *If $r, s, t \in \mathbb{Z}$ satisfy $r = as + bt$ and $|rs| < |b|/2$ then (r, s, t) is (up to sign) one of the triples (r_i, s_i, t_i) computed by Euclid's algorithm.*

Proof Statements 1, 2 and 3 are proved using the relation $s_i = (-1)^{i-1}k_{i-1}$ and $t_i = (-1)^i h_{i-1}$ where h_i/k_i are the continued fraction convergents to a/b. From Chapter 10 of [250] and Chapter 7 of [420] one knows that $h_m = q_{m+1}h_{m-1} + h_{m-2}$ and $k_m = q_{m+1}k_{m-1} + k_{m-2}$ where q_{m+1} is the quotient in iteration $m + 1$ of Euclid's algorithm. The first statement follows immediately and the third statement follows from the fact that $h_m k_{m-1} - h_{m-1}k_m = (-1)^{m-1}$. The second statement follows since $a, b > 0$ implies $h_i, k_i > 0$.

Statement 4 can be proved by induction, using the fact that $r_{i+1}s_i - r_i s_{i+1} = (r_{i-1} - q_i r_i)s_i - r_i(s_{i-1} - q_i s_i) = -(r_i s_{i-1} - r_{i-1}s_i)$. Statement 5 is the standard result (equation (10.7.7) of [250], Theorem 7.11 of [420]) that the convergents of a/b satisfy $|a/b - h_m/k_m| < 1/|k_m k_{m+1}|$. Statement 6 follows directly from statements 2 and 4. For example, $a = r_i(-1)^{i-1}t_i + r_{i-1}(-1)^i t_i$ and both terms on the right-hand side are positive.

Statement 7 is also a standard result in diophantine approximation; see Theorem 184 of [250] or Theorem 7.14 of [420].

Finally, to prove statement 8 suppose $r, s, t \in \mathbb{Z}$ are such that $r = as + bt$ and $|rs| < |b|/2$. Then

$$|a/b + t/s| = |(as + bt)/bs| = |r|/|bs| = |rs|/|bs^2| < 1/(2s^2).$$

The result follows from statement 7. □

Example 2.3.4 The first few terms of Euclid's algorithm on $a = 513$ and $b = 311$ give

i	r_i	q_i	s_i	t_i	$\lvert r_i s_i \rvert$	$\lvert r_i t_i \rvert$
−1	513	–	1	0	513	0
0	311	–	0	1	0	311
1	202	1	1	−1	202	202
2	109	1	−1	2	109	218
3	93	1	2	−3	186	279
4	16	1	−3	5	48	80
5	13	5	17	−28	221	364

One can verify that $|r_i s_i| \le |b|$ and $|r_i t_i| \le |a|$. Indeed, $|r_i s_{i+1}| \le |b|$ and $|r_i t_{i+1}| \le |a|$ as stated in part 6 of Lemma 2.3.3.

Diophantine approximation is the study of approximating real numbers by rationals. Statement 7 in Lemma 2.3.3 is a special case of one of the famous results; namely, that the "best" rational approximations to real numbers are given by the convergents in their continued fraction expansion. Lemma 2.3.5 shows how the result can be relaxed slightly, giving "less good" rational approximations in terms of convergents to continued fractions.

Lemma 2.3.5 *Let $\alpha \in \mathbb{R}$, $c \in \mathbb{R}_{>0}$ and let $s, t \in \mathbb{N}$ be such that $|\alpha - t/s| < c/s^2$. Then $(t, s) = (uh_{n+1} \pm vh_n, uk_{n+1} \pm vk_n)$ for some $n, u, v \in \mathbb{Z}_{\ge 0}$ such that $uv < 2c$.*

Proof See Theorem 1 and Remark 2 of Dujella [170]. □

2.4 Computing Legendre and Jacobi symbols

The Legendre symbol tells us when an integer is a square modulo p. It is a non-trivial group homomorphism from $(\mathbb{Z}/p\mathbb{Z})^*$ to the multiplicative group $\{-1, 1\}$.

Definition 2.4.1 Let p be an odd prime and $a \in \mathbb{Z}$. The **Legendre symbol** $(\frac{a}{p})$ is

$$\left(\frac{a}{p}\right) = \begin{cases} 1 & \text{if } x^2 \equiv a \pmod{p} \text{ has a solution} \\ 0 & \text{if } p \mid a \\ -1 & \text{otherwise.} \end{cases}$$

If p is prime and $a \in \mathbb{Z}$ satisfies $(\frac{a}{p}) = 1$ then a is a **quadratic residue**, while if $(\frac{a}{p}) = -1$ then a is a **quadratic non-residue**.

Let $n = \prod_i p_i^{e_i}$ be odd. The **Jacobi symbol** is

$$\left(\frac{a}{n}\right) = \prod_i \left(\frac{a}{p_i}\right)^{e_i}.$$

A further generalisation is the **Kronecker symbol** $(\frac{a}{n})$, which allows n to be even. This is defined in equation (25.4), which is the only place in the book that it is used.

Exercise 2.4.2 Show that if p is an odd prime then $(\frac{a}{p}) = 1$ for exactly half the integers $1 \le a \le p-1$.

Theorem 2.4.3 *Let $n \in \mathbb{N}$ be odd and $a \in \mathbb{Z}$. The Legendre and Jacobi symbols satisfy the following properties:*

- $(\frac{a}{n}) = (\frac{a \pmod{n}}{n})$ *and* $(\frac{1}{n}) = 1$.
- *(Euler's criterion) If n is prime then* $(\frac{a}{n}) = a^{(n-1)/2} \pmod{n}$.
- *(Multiplicative)* $(\frac{ab}{n}) = (\frac{a}{n})(\frac{b}{n})$ *for all $a, b \in \mathbb{Z}$.*
- $(\frac{-1}{n}) = (-1)^{(n-1)/2}$. *In other words*

$$\left(\frac{-1}{n}\right) = \begin{cases} 1 & \text{if } n \equiv 1 \pmod{4} \\ -1 & \text{otherwise.} \end{cases}$$

- $(\frac{2}{n}) = (-1)^{(n^2-1)/8}$. *In other words*

$$\left(\frac{2}{n}\right) = \begin{cases} 1 & \text{if } n \equiv 1, 7 \pmod{8} \\ -1 & \text{otherwise.} \end{cases}$$

- *(Quadratic reciprocity) Let n and m be odd integers with $\gcd(m, n) = 1$. Then*

$$\left(\frac{n}{m}\right) = (-1)^{(m-1)(n-1)/4}\left(\frac{m}{n}\right).$$

In other words, $(\frac{n}{m}) = (\frac{m}{n})$ *unless* $m \equiv n \equiv 3 \pmod{4}$.

Proof See Section II.2 of [313], Sections 3.1, 3.2 and 3.3 of [420] or Chapter 6 of [250]. □

An important fact is that it is not necessary to factor integers to compute the Jacobi symbol.

Exercise 2.4.4 Write down an algorithm to compute Legendre and Jacobi symbols using quadratic reciprocity.

Exercise 2.4.5 Prove that the complexity of computing $(\frac{m}{n})$ is $O(\log(m)\log(n))$ bit operations.

Exercise 2.4.6 Give a randomised algorithm to compute a quadratic non-residue modulo p. What is the expected complexity of this algorithm?

Exercise 2.4.7 Several applications require knowing a quadratic non-residue modulo a prime p. Prove that the values a in the following table satisfy $(\frac{a}{p}) = -1$.

p	a
$p \equiv 3 \pmod 4$	-1
$p \equiv 1 \pmod 4$, $p \equiv 2 \pmod 3$	3
$p \equiv 1 \pmod 4$, $p \not\equiv 1 \pmod 8$	$\sqrt{-1}$
$p \equiv 1 \pmod 8$, $p \not\equiv 1 \pmod{16}$	$(1 + \sqrt{-1})/\sqrt{2}$

Remark 2.4.8 The problem of computing quadratic non-residues has several algorithmic implications. One conjectures that the least quadratic non-residue modulo p is $O(\log(p)\log(\log(p)))$. Burgess proved that the least quadratic non-residue modulo p is at most $p^{1/(4\sqrt{e})+o(1)} \approx p^{0.151633+o(1)}$, while Ankeny showed, assuming the extended Riemann hypothesis, that it is $O(\log(p)^2)$. We refer to Section 8.5 of Bach and Shallit [21] for details and references. It follows that one can compute a quadratic non-residue in $O(\log(p)^4)$ bit operations, assuming the extended Riemann hypothesis.

Exercise 2.4.9 Give a Las Vegas algorithm to test whether $a \in \mathbb{N}$ is a square by computing $(\frac{a}{p})$ for some random small primes p. What is the complexity of this algorithm?

Exercise 2.4.10 Let p be prime. In Section 2.8 we give algorithms to compute modular exponentiation quickly. Compare the cost of computing $(\frac{a}{p})$ using quadratic reciprocity versus using Euler's criterion.

Remark 2.4.11 An interesting computational problem (considered, for example, by Damgård [152]) is: given a prime p, an integer k and the sequence $(\frac{a}{p}), (\frac{a+1}{p}), \dots, (\frac{a+k-1}{p})$ to output $(\frac{a+k}{p})$. A potentially harder problem is to determine a given the sequence of values. It is known that if k is a little larger than $\log_2(p)$ then a is usually uniquely determined modulo p and so both problems make sense. No efficient algorithms are known to solve either of these problems. One can also consider the natural analogue for Jacobi symbols. We refer to [152] for further details. This is also discussed as Conjecture 2.1 of Boneh and Lipton [78].

Finally, we remark that one can compute the Legendre or Jacobi symbol of n-bit integers in $O(M(n)\log(n))$ operations using an analogue of fast algorithms for computing gcds. We refer to Exercise 5.52 (also see pages 343–344) of Bach and Shallit [21] or Brent and Zimmermann [96] for the details.

2.5 Modular arithmetic

In cryptography, modular arithmetic (i.e., arithmetic modulo $n \in \mathbb{N}$) is a fundamental building block. We represent elements of $\mathbb{Z}/n\mathbb{Z}$ as integers from the set $\{0, 1, \dots, n-1\}$. We first summarise the complexity of standard "school" methods for modular arithmetic.

Lemma 2.5.1 *Let $a, b \in \mathbb{Z}/n\mathbb{Z}$.*

1. *Computing $a \pm b \pmod{n}$ can be done in $O(\log(n))$ bit operations.*
2. *Computing $ab \pmod{n}$ can be done in $O(\log(n)^2)$ bit operations.*
3. *Computing $a^{-1} \pmod{n}$ can be done in $O(\log(n)^2)$ bit operations.*
4. *For $a \in \mathbb{Z}$, computing $a \pmod{n}$ can be done in $O(\log(n)(\log(a) - \log(n) + 1))$ bit operations.*

Montgomery multiplication

This method[5] is useful when one needs to perform an operation such as $a^m \pmod{n}$ when n is odd. It is based on the fact that arithmetic modulo 2^s is easier than arithmetic modulo n. Let $R = 2^s > n$ (where s is typically a multiple of the word size).

Definition 2.5.2 Let $n \in \mathbb{N}$ be odd and $R = 2^s > n$. The **Montgomery representation** of $a \in (\mathbb{Z}/n\mathbb{Z})$ is $\overline{a} = aR \pmod{n}$ such that $0 \le \overline{a} < n$.

To transform a into the Montgomery representation requires a standard modular multiplication. However, Lemma 2.5.3 shows that transforming back from the Montgomery representation to standard representation may be performed more efficiently.

Lemma 2.5.3 *(Montgomery reduction) Let $n \in \mathbb{N}$ be odd and $R = 2^s > n$. Let $n' = -n^{-1} \pmod{R}$ be such that $1 \le n' < R$. Let $\overline{a}$ be an element of $(\mathbb{Z}/n\mathbb{Z})$ in Montgomery representation. Let $u = \overline{a}n' \pmod{R}$. Then $w = (\overline{a} + un)/R$ lies in $\mathbb{Z}$ and satisfies $w \equiv \overline{a}R^{-1} \pmod{n}$.*

Proof Write $w' = \overline{a} + un$. Clearly, $w' \equiv 0 \pmod{R}$ so $w \in \mathbb{Z}$. Further, $0 \le w' \le (n-1) + (R-1)n = Rn - 1$ and hence $w < n$. Finally, it is clear that $w \equiv \overline{a}R^{-1} \pmod{n}$. □

The reason why this is efficient is that division by R is easy. The computation of n' is also easier than a general modular inversion (see Algorithm II.5 of [60]) and, in many applications, it can be precomputed.

We now sketch the **Montgomery multiplication** algorithm. If $\overline{a}$ and $\overline{b}$ are in Montgomery representation then we want to compute the Montgomery representation of ab, which is $\overline{a}\overline{b}R^{-1} \pmod{n}$. Compute $x = \overline{a}\overline{b} \in \mathbb{Z}$ so that $0 \le x < n^2 < nR$, then compute $u = xn' \pmod{R}$ and $w' = x + nu \in \mathbb{Z}$. As in Lemma 2.5.3, we have $w' \equiv 0 \pmod{R}$ and can compute $w = w'/R$. It follows that $w \equiv \overline{a}\overline{b}R^{-1} \pmod{n}$ and $0 \le w < 2n$ so $\overline{ab}$ is either w or $w - n$.

Lemma 2.5.4 *The complexity of the Montgomery multiplication modulo n is $O(M(\log(n)))$ bit operations.*

For further details see Section 9.2.1 of [150], Section II.1.4 of [60], Section 11.1.2.b of [16] or Section 2.2.4 of [248].

[5] Credited to Montgomery [391], but apparently a similar idea was used by Hensel.

Faster modular reduction

Using Newton's method to compute $\lfloor a/n \rfloor$ one can compute $a \pmod n$ using only multiplication of integers. If $a = O(n^2)$ then the complexity is $O(M(\log(n)))$. See Exercises 3.35, 3.36 of [497] and Section 9.1 of [220] for details. For large a, the cost of computing $a \pmod n$ remains $O(\log(a)\log(n))$ as before. This idea gives rise to Barret reduction; see Section 9.2.2 of [150], Section 2.3.1 of [95], Section 14.3.3 of [376], Section II.1.3 of [60] or Section 10.4.1 of [16].

Special moduli

For cryptography based on discrete logarithms, especially elliptic curve cryptography, it is recommended to use primes of a special form to speed up arithmetic modulo p. Commonly used primes are of the form $p = 2^k - c$ for some small $c \in \mathbb{N}$ or the **NIST primes** $p = 2^{n_k w} \pm 2^{n_{k-1} w} \pm \cdots \pm 2^{n_1 w} \pm 1$ where $w = 16, 32$ or 64. In these cases, it is possible to compute reduction modulo p much more quickly than for general p. See Section 2.2.6 of [248], Section 14.3.4 of [376] or Section 10.4.3 of [16] for examples and details.

Modular inversion

Suppose $a, n \in \mathbb{N}$ are such that $\gcd(a, n) = 1$. One can compute $a^{-1} \pmod n$ using the extended Euclidean algorithm: computing integers $s, t \in \mathbb{Z}$ such that $as + nt = 1$ gives $a^{-1} \equiv s \pmod n$. Hence, if $0 < a < n$ then one can compute $a^{-1} \pmod n$ in $O(\log(n)^2)$ bit operations, or faster using subquadratic versions of the extended Euclidean algorithm.

In practice, modular inversion is significantly slower than modular multiplication. For example, when implementing elliptic curve cryptography it is usual to assume that the cost of an inversion in $\mathbb{F}_p$ is between 8 and 50 times slower than the cost of a multiplication in $\mathbb{F}_p$ (the actual figure depends on the platform and algorithms used).

Simultaneous modular inversion

One can compute $a_1^{-1} \pmod n, \ldots, a_m^{-1} \pmod n$ with a single inversion modulo n and a number of multiplications modulo n using a trick due to Montgomery. Namely, one computes $b = a_1 \cdots a_m \pmod n$, computes $b^{-1} \pmod n$ and then recovers the individual a_i^{-1}.

Exercise 2.5.5 Give pseudocode for simultaneous modular inversion and show that it requires one inversion and $3(m - 1)$ modular multiplications.

2.6 Chinese remainder theorem

The Chinese remainder theorem (CRT) states that if $\gcd(m_1, m_2) = 1$ then there is a unique solution $0 \le x < m_1 m_2$ to $x \equiv c_i \pmod{m_i}$ for $i = 1, 2$. Computing x can be done in polynomial-time in various ways. One method is to use the formula

$$x = c_1 + (c_2 - c_1)(m_1^{-1} \pmod{m_2})m_1.$$

This is a special case of Garner's algorithm (see Section 14.5.2 of [376] or Section 10.6.4 of [16]).

Exercise 2.6.1 Suppose $m_1 < m_2$ and $0 \le c_i < m_i$. What is the input size of the instance of the CRT? What is the complexity of computing the solution?

Exercise 2.6.2 Let $n > 2$ and suppose coprime integers $2 \le m_1 < \cdots < m_n$ and integers $c_1, \ldots, c_n$ such that $0 \le c_i < m_i$ for $1 \le i \le n$ are given. Let $N = \prod_{i=1}^n m_i$. For $1 \le i \le n$ define $N_i = N/m_i$ and $u_i = N_i^{-1} \pmod{m_i}$. Show that

$$x = \sum_{i=1}^{n} c_i u_i N_i \tag{2.1}$$

satisfies $x \equiv c_i \pmod{m_i}$ for all $1 \le i \le n$.

Show that one can compute the integer x in equation (2.1) in $O(n^2 \log(m_n)^2)$ bit operations.

Exercise 2.6.3 Show that a special case of Exercise 2.6.2 (which is recommended when many computations are required for the same pair of moduli) is to precompute the integers $A = u_1 N_1$ and $B = u_2 N_2$ so that $x = c_1 A + c_2 B \pmod{N}$.

Algorithm 10.22 of [220] gives an asymptotically fast CRT algorithm.

2.7 Linear algebra

Let A be an $n \times n$ matrix over a field $\Bbbk$. One can perform Gaussian elimination to solve the linear system $A\underline{x} = \underline{b}$ (or determine there are no solutions), to compute $\det(A)$ or to compute A^{-1} in $O(n^3)$ field operations. When working over $\mathbb{R}$ a number of issues arise due to rounding errors, but no such problems arise when working over finite fields. We refer to Section 3.3 of Joux [283] for details.

A matrix is called **sparse** if almost all entries of each row are zero. To make this precise one usually considers the asymptotic complexity of an algorithm on $m \times n$ matrices, as m and/or n tends to infinity, and where the number of non-zero entries in each row is bounded by $O(\log(n))$ or $O(\log(m))$.

One can compute the kernel (i.e., a vector $\underline{x}$ such that $A\underline{x} = \underline{0}$) of an $n \times n$ sparse matrix A over a field in $O(n^2)$ field operations using the algorithms of Wiedemann [562] or Lanczos [325]. We refer to Section 3.4 of [283] or Section 12.4 of [220] for details.

Hermite normal form

When working over a ring the Hermite normal form (HNF) is an important tool for solving or simplifying systems of equations. Some properties of the Hermite normal form are mentioned in Section A.11.

Algorithms to compute the HNF of a matrix are given in Section 2.4.2 of Cohen [127], Hafner and McCurley [247], Section 3.3.3 of Joux [283], Algorithm 16.26 of von zur

Gathen and Gerhard [220], Section 5.3 of Schrijver [478], Kannan and Bachem [297], Storjohann and Labahn [533] and Micciancio and Warinschi [381]. Naive algorithms to compute the HNF suffer from coefficient explosion, so computing the HNF efficiently in practice, and determining the complexity of the algorithm, is non-trivial. One solution is to work modulo the determinant (or a subdeterminant) of the matrix A (see Section 2.4.2 of [127], [247] or [533] for further details). Let $A = (A_{i,j})$ be an $n \times m$ matrix over $\mathbb{Z}$ and define $\|A\|_\infty = \max_{i,j}\{|A_{i,j}|\}$. The complexity of the HNF algorithm of Storjohann and Labahn on A (using naive integer and matrix multiplication) is $O(nm^4 \log(\|A\|_\infty)^2)$ bit operations.

One can also use lattice reduction to compute the HNF of a matrix. For details see page 74 of [478], Havas, Majewski and Matthews [254], or van der Kallen [294].

2.8 Modular exponentiation

Exponentiation modulo n can be performed in polynomial-time by the "square-and-multiply" method. This method is presented in Algorithm 2; it is called a "left-to-right" algorithm as it processes the bits of the exponent m starting with the most significant bits. Algorithm 2 can be applied in any group, in which case the complexity is $O(\log(m))$ times the complexity of the group operation. In this section we give some basic techniques to speed-up the algorithm; further tricks are described in Chapter 11.

Algorithm 2 Square-and-multiply algorithm for modular exponentiation

INPUT: $g, n, m \in \mathbb{N}$
OUTPUT: $b \equiv g^m \pmod{n}$

1: $i = \lfloor \log_2(m) \rfloor - 1$
2: Write m in binary as $(1m_i \ldots m_1 m_0)_2$
3: $b = g$
4: **while** $(i \geq 0)$ **do**
5: $\quad b = b^2 \pmod{n}$
6: $\quad$ **if** $m_i = 1$ **then**
7: $\quad\quad b = bg \pmod{n}$
8: $\quad$ **end if**
9: $\quad i = i - 1$
10: **end while**
11: **return** b

Lemma 2.8.1 *The complexity of Algorithm 2 using naive modular arithmetic is $O(\log(m)\log(n)^2)$ bit operations.*

Exercise 2.8.2 Prove Lemma 2.8.1.

Lemma 2.8.3 *If Montgomery multiplication (see Section 2.5) is used then the complexity of Algorithm 2.5 is $O(\log(n)^2 + \log(m)M(\log(n)))$.*

Proof Convert g to Montgomery representation $\overline{g}$ in $O(\log(n)^2)$ bit operations. Algorithm 2 then proceeds using Montgomery multiplication in lines 5 and 7, which requires $O(\log(m)M(\log(n)))$ bit operations. Finally, Montgomery reduction is used to convert the output to standard form. □

The algorithm using Montgomery multiplication is usually better than the naive version, especially when fast multiplication is available. An application of the above algorithm, where Karatsuba multiplication would be appropriate, is RSA decryption (either the standard method, or using the CRT). Since $\log(m) = \Omega(\log(n))$ in this case, decryption requires $O(\log(n)^{2.585})$ bit operations.

Corollary 2.8.4 *One can compute the Legendre symbol* $(\frac{a}{p})$ *using Euler's criterion in* $O(\log(p)M(\log(p)))$ *bit operations.*

When storage for precomputed group elements is available there are many ways to speed up exponentiation. These methods are particularly appropriate when many exponentiations of a fixed element g are required. The methods fall naturally into two types: those that reduce the number of squarings in Algorithm 2 and those that reduce the number of multiplications. An extreme example of the first type is to precompute and store $u_i = g^{2^i} \pmod{n}$ for $2 \le i \le \log(n)$. Given $2^l \le m < 2^{l+1}$ with binary expansion $(1m_{l-1} \dots m_1 m_0)_2$ one computes $\prod_{i=0:m_i=1}^{l} u_i \pmod{n}$. Obviously, this method is not more efficient than Algorithm 2 if g varies. An example of the second type is **sliding window methods** that we now briefly describe. Note that there is a simpler but less efficient "non-sliding" window method, also called the 2^k-ary method, which can be found in many books. Sliding window methods can be useful even in the case where g varies (e.g., Algorithm 3 below).

Given a **window length** w one precomputes $u_i = g^i \pmod{n}$ for all odd integers $1 \le i < 2^w$. Then one runs a variant of Algorithm 2 where w (or more) squarings are performed followed by one multiplication corresponding to a w-bit substring of the binary expansion of m that corresponds to an odd integer. One subtlety is that algorithms based on the "square-and-multiply" idea and which use precomputation must parse the exponent starting with the most significant bits (i.e., from left to right), whereas to work out sliding windows one needs to parse the exponent from the least significant bits (i.e., right to left).

Example 2.8.5 Let $w = 2$ so that one precomputes $u_1 = g$ and $u_3 = g^3$. Suppose m has binary expansion $(10011011)_2$. By parsing the binary expansion starting with the least significant bits, one obtains the representation 10003003 (we stress that this is still a representation in base 2). One then performs the usual square-and-multiply algorithm by parsing the exponent from left to right; the steps of the sliding window algorithm are (omitting the $\pmod{n}$ notation)

$$b = u_1,\ b = b^2;\ b = b^2,\ b = b^2,\ b = b^2,\ b = bu_3,\ b = b^2,\ b = b^2,\ b = b^2,\ b = bu_3.$$

Exercise 2.8.6 Write pseudocode for the sliding window method. Show that the precomputation stage requires one squaring and $2^{w-1} - 1$ multiplications.

Exercise 2.8.7 Show that the expected number of squarings between each multiply in the sliding window algorithm is $w+1$. Hence, show that (ignoring the precomputation) exponentiation using sliding windows requires $\log(m)$ squarings and, on average, $\log(m)/(w+1)$ multiplications.

Exercise 2.8.8 Consider running the sliding window method in a group, with varying g and m (so the powers of g must be computed for every exponentiation) but with unlimited storage. For a given bound on len(m) one can compute the value for w that minimises the total cost. Verify that the choices in the following table are optimal.

len(m)	80	160	300	800	2000
w	3	4	5	6	7

Exercise 2.8.9 Algorithm 2 processes the bits of the exponent m from left to right. Give pseudocode for a modular exponentiation algorithm that processes the bits of the exponent m from right to left.

[Hint: Have two variables in the main loop; one that stores g^{2^i} in the ith iteration, and the other that stores the value $g^{\sum_{j=0}^{i} a_j 2^j}$.]

Exercise 2.8.10 Write pseudocode for a right to left sliding window algorithm for computing $g^m \pmod{n}$, extending Exercise 2.8.9. Explain why this variant is not so effective when computing $g^m \pmod{n}$ for many random m but when g is fixed.

One can also consider the opposite scenario where one is computing $g^m \pmod{n}$ for a fixed value m and varying g. Again, with some precomputation, and if there is sufficient storage available, one can get an improvement over the naive algorithm. The idea is to determine an efficient **addition chain** for m. This is a sequence of squarings and multiplications, depending on m, that minimises the number of group operations. More precisely, an addition chain of length l for m is a sequence $m_1, m_2, \ldots, m_l$ of integers such that $m_1 = 1$, $m_l = m$ and for each $2 \le i \le l$ we have $m_i = m_j + m_k$ for some $1 \le j \le k < i$. One computes each of the intermediate values g^{m_i} for $2 \le i \le l$ with one group operation. Note that *all* these intermediate values are stored. The algorithm requires l group operations and l group elements of storage.

It is conjectured by Stolarsky that every integer m has an addition chain of length $\log_2(m) + \log_2(\mathrm{wt}(m))$ where wt(m) is the **Hamming weight** of m (i.e., the number of ones in the binary expansion of m). There is a vast literature on addition chains, we refer to Section C6 of [246], Section 4.6.3 of [308] and Section 9.2 of [16] for discussion and references.

Exercise 2.8.11 Prove that an addition chain has length at least $\log_2(m)$.

2.9 Square roots modulo p

There are a number of situations in this book that require computing square roots modulo a prime. Let p be an odd prime and let $a \in \mathbb{N}$. We have already shown that Legendre symbols can be computed in polynomial-time. Hence, the decision problem "Is a a square modulo p?" is soluble in polynomial-time. But this fact does not imply that the computational problem "Find a square root of a modulo p" is easy.

We present two methods in this section. The Tonelli–Shanks algorithm is the best method in practice. The Cipolla algorithm actually has better asymptotic complexity, but is usually slower than Tonelli–Shanks.

Recall that half the integers $1 \le a < p$ are squares modulo p and, if so, there are two solutions $\pm x$ to the equation $x^2 \equiv a \pmod{p}$.

Lemma 2.9.1 *Let $p \equiv 3 \pmod{4}$ be prime and $a \in \mathbb{N}$. If $(\frac{a}{p}) = 1$ then $x = a^{(p+1)/4} \pmod{p}$ satisfies $x^2 \equiv a \pmod{p}$.*

This result can be verified directly by computing x^2, but we give a more group-theoretic proof that helps to motivate the general algorithm.

Proof Since $p \equiv 3 \pmod{4}$ it follows that $q = (p-1)/2$ is odd. The assumption $(\frac{a}{p}) = 1$ implies that $a^q = a^{(p-1)/2} \equiv 1 \pmod{p}$ and so the order of a is odd. Therefore, a square root of a is given by

$$x = a^{2^{-1} \pmod{q}} \pmod{p}.$$

Now, $2^{-1} \pmod{q}$ is just $(q+1)/2 = (p+1)/4$. □

Lemma 2.9.2 *Let p be a prime and suppose a is a square modulo p. Write $p - 1 = 2^e q$ where q is odd. Let $w = a^{(q+1)/2} \pmod{p}$. Then $w^2 \equiv ab \pmod{p}$ where b has order dividing 2^{e-1}.*

Proof We have

$$w^2 \equiv a^{q+1} \equiv aa^q \pmod{p}$$

so $b \equiv a^q \pmod{p}$. Now a has order dividing $(p-1)/2 = 2^{e-1}q$ so b has order dividing 2^{e-1}. □

The value w is like a "first approximation" to the square root of a modulo p. To complete the computation it is therefore sufficient to compute a square root of b.

Lemma 2.9.3 *Suppose $1 < n < p$ is such that $(\frac{n}{p}) = -1$. Then $y \equiv n^q \pmod{p}$ has order 2^e.*

Proof The order of y is a divisor of 2^e. The fact $n^{(p-1)/2} \equiv -1 \pmod{p}$ implies that y satisfies $y^{2^{e-1}} \equiv -1 \pmod{p}$. Hence, the order of y is equal to 2^e. □

Since $\mathbb{Z}_p^*$ is a cyclic group, it follows that y generates the full subgroup of elements of order dividing 2^e. Hence, $b = y^i \pmod{p}$ for some $1 \le i \le 2^e$. Furthermore, since the order of b divides 2^{e-1}, it follows that i is even.

Writing $i = 2j$ and $x = w/y^j \pmod{p}$ then

$$x^2 \equiv w^2/y^{2j} \equiv ab/b \equiv a \pmod{p}.$$

Hence, if one can compute i then one can compute the square root of a.

If e is small then the value i can be found by a brute-force search. A more advanced method is to use the Pohlig–Hellman method to solve the discrete logarithm of b to the base y (see Section 13.2 for an explanation of these terms). This idea leads to the Tonelli–Shanks algorithm for computing square roots modulo p (see Section 1.3.3 of [60] or Section 1.5 of [127]).

Algorithm 3 Tonelli–Shanks algorithm

INPUT: a, p such that $(\frac{a}{p}) = 1$
OUTPUT: x such that $x^2 \equiv a \pmod{p}$
1: Write $p - 1 = 2^e q$ where q is odd
2: Choose random integers $1 < n < p$ until $(\frac{n}{p}) = -1$
3: Set $y = n^q \pmod{p}$
4: Set $w = a^{(q+1)/2} \pmod{p}$ and $b = a^q \pmod{p}$
5: Compute an integer j such that $b \equiv y^{2j} \pmod{p}$
6: **return** $w/y^j \pmod{p}$

Exercise 2.9.4 Compute $\sqrt{3}$ modulo 61 using the Tonelli–Shanks algorithm.

Line 2 of Algorithm 3 is randomised, but the rest of the algorithm is deterministic. The complexity is dominated by the cost of computing exponentiations such as $n^q \pmod{p}$ in lines 3 and 4 (which take $O(\log(q)M(\log(p)))$ bit operations), and the cost of computing j in line 5. When e is large the latter cost dominates. A complexity of $O(\log(p)^2 M(\log(p)))$ bit operations for the Pohlig–Hellman method is determined in Exercise 13.2.6 (and an improved complexity is given in equation (13.1)). In practice, when e is large, one uses the Cipolla method (see Exercise 2.9.7 below, Section 7.2 of [21] or Section 3.5 of [376]). The Cipolla method is expected to perform $O(\log(p)M(\log(p)))$ bit operations. This complexity can also be obtained using general-purpose polynomial factorisation (see Exercise 2.12.6).

Exercise 2.9.5 Let $r \in \mathbb{N}$. Generalise the Tonelli–Shanks algorithm so that it computes rth roots in $\mathbb{F}_p$ (the only non-trivial case being when $p \equiv 1 \pmod{r}$).

Exercise 2.9.6 (Atkin) Let $p \equiv 5 \pmod{8}$ be prime and $a \in \mathbb{Z}$ such that $(\frac{a}{p}) = 1$. Let $z = (2a)^{(p-5)/8} \pmod{p}$ and $i = 2az^2 \pmod{p}$. Show that $i^2 = -1 \pmod{p}$ and that $w = az(i-1)$ satisfies $w^2 \equiv a \pmod{p}$.

Exercise 2.9.7 (Cipolla) Let p be prime and $a \in \mathbb{Z}$. Show that if $t \in \mathbb{Z}$ is such that $(\frac{t^2-4a}{p}) = -1$ then $x^{(p+1)/2}$ in $\mathbb{F}_p[x]/(x^2 - tx + a)$ is a square root of a modulo p. Hence, write down an algorithm to compute square roots modulo p and show that it has expected running time $O(\log(p)M(\log(p)))$ bit operations.

We remark that, in some applications, one wants to compute a Legendre symbol to test whether an element is a square and, if so, compute the square root. If one computes the Legendre symbol using Euler's criterion as $a^{(p-1)/2} \pmod{p}$ then one will have already computed $a^q \pmod{p}$ and so this value can be recycled. This is not usually faster than using quadratic reciprocity for large p, but it is relevant for applications such as Lemma 21.4.9.

A related topic is, given a prime p and an integer $d > 0$, to find integer solutions (x, y), if they exist, to the equation $x^2 + dy^2 = p$. The **Cornacchia algorithm** achieves this. The algorithm is given in Section 2.3.4 of Crandall and Pomerance [150], and a proof of correctness is given in Section 4 of Schoof [477] or Morain and Nicolas [393]. In brief, the algorithm computes $p/2 < x_0 < p$ such that $x_0^2 \equiv -d \pmod{p}$, then runs the Euclidean algorithm on $2p$ and x_0 stopping at the first remainder $r < \sqrt{p}$, then computes $s = \sqrt{(p - r^2)/d}$ if this is an integer. The output is $(x, y) = (r, s)$. The complexity is dominated by computing the square root modulo p, and so is an expected $O(\log(p)^2 M(\log(p)))$ bit operations. A closely related algorithm finds solutions to $x^2 + dy^2 = 4p$.

2.10 Polynomial arithmetic

Let R be a commutative ring. A polynomial in $R[x]$ of degree d is represented[6] as a $(d + 1)$-tuple over R. A polynomial of degree d over $\mathbb{F}_q$ therefore requires $(d + 1)\lceil\log_2(q)\rceil$ bits for its representation. An algorithm on polynomials will be polynomial-time if the number of bit operations is bounded above by a polynomial in $d\log(q)$.

Arithmetic on polynomials is analogous to integer arithmetic (indeed, it is simpler as there are no carries to deal with). We refer to Chapter 2 of [220], Chapter 18 of [497], Section 4.6 of [308] or Section 9.6 of [150] for details.

Lemma 2.10.1 *Let R be a commutative ring and $F_1(x), F_2(x) \in R[x]$.*

1. *One can compute $F_1(x) + F_2(x)$ in $O(\max\{\deg(F_1), \deg(F_2)\})$ additions in R.*
2. *One can compute $F_1(x)F_2(x)$ in $O(\deg(F_1)\deg(F_2))$ additions and multiplications in R.*
3. *If R is a field then one can compute the quotient and remainder of division of $F_1(x)$ by $F_2(x)$ in $O(\deg(F_2)(\deg(F_1) - \deg(F_2) + 1))$ operations (i.e., additions, multiplications and inversions) in R.*
4. *If R is a field then one can compute $F(x) = \gcd(F_1(x), F_2(x))$ and polynomials $s(x), t(x) \in R[x]$ such that $F(x) = s(x)F_1(x) + t(x)F_2(x)$, using the extended Euclidean algorithm in $O(\deg(F_1)\deg(F_2))$ operations in R.*

[6] We restrict attention in this and the following section to univariate polynomials. There are alternative representations for sparse and/or multivariate polynomials, but we do not consider this issue further.

Exercise 2.10.2 Prove Lemma 2.10.1.

Exercise 2.10.3★ Describe the Karatsuba and 3-Toom–Cook algorithms for multiplication of polynomials of degree d in $\mathbb{F}_q[x]$. Show that these algorithms have complexity $O(d^{1.585})$ and $O(d^{1.404})$ multiplications in $\mathbb{F}_q$.

Asymptotically fast multiplication of polynomials, analogous to the algorithms mentioned in Section 2.2, are given in Chapter 8 of [220] or Section 18.6 of [497]. Multiplication of polynomials in $\Bbbk[x]$ of degree bounded by d can be done in $O(M(d))$ multiplications in $\Bbbk$. The methods mentioned in Section 2.5 for efficiently computing remainders $F(x) \pmod{G(x)}$ in $\Bbbk[x]$ can also be used with polynomials; see Section 9.6.2 of [150] or Section 11.1 of [220] for details. Fast variants of the extended Euclidean algorithm for polynomials in $\Bbbk[x]$ of degree bounded by d require $O(M(d)\log(d))$ multiplications in $\Bbbk$ and $O(d)$ inversions in $\Bbbk$ (Corollary 11.6 of [220]).

Kronecker substitution is a general technique that transforms polynomial multiplication into integer multiplication. It allows multiplication of two-degree d polynomials in $\mathbb{F}_q[x]$ (where q is prime) in $O(M(d(\log(q) + \log(d)))) = O(M(d\log(dq)))$ bit operations; see Section 1.3 of [95], Section 8.4 of [220] or Section 18.6 of [497]. Kronecker substitution can be generalised to bivariate polynomials and hence to polynomials over $\mathbb{F}_q$ where q is a prime power. We write $M(d, q) = M(d\log(dq))$ for the number of bit operations to multiply two-degree d polynomials over $\mathbb{F}_q$.

Exercise 2.10.4 Show that Montgomery reduction and multiplication can be implemented for arithmetic modulo a polynomial $F(x) \in \mathbb{F}_q[x]$ of degree d.

Exercise 2.10.5 One can evaluate a polynomial $F(x) \in R[x]$ at a value $a \in R$ efficiently using **Horner's rule**. More precisely, if $F(x) = \sum_{i=0}^{d} F_i x^i$ then one computes $F(a)$ as $(\cdots((F_d a)a + F_{d-1})a + \cdots + F_1)a + F_0$. Write pseudocode for Horner's rule and show that the method requires d additions and d multiplications if $\deg(F(x)) = d$.

2.11 Arithmetic in finite fields

Efficient algorithms for arithmetic modulo p have been presented, but we now consider arithmetic in finite fields $\mathbb{F}_{p^m}$ when $m > 1$. We assume $\mathbb{F}_{p^m}$ is represented using either a polynomial basis (i.e., as $\mathbb{F}_p[x]/(F(x))$) or a normal basis. Our main focus is when either p is large and m is small, or vice versa. Optimal asymptotic complexities for the case when both p and m grow large require some care.

Exercise 2.11.1 Show that addition and subtraction of elements in $\mathbb{F}_{p^m}$ requires $O(m)$ additions in $\mathbb{F}_p$. Show that multiplication in $\mathbb{F}_{p^m}$, represented by a polynomial basis and using naive methods, requires $O(m^2)$ multiplications modulo p and $O(m)$ reductions modulo p.

If p is constant and m grows then multiplication in $\mathbb{F}_{p^m}$ requires $O(m^2)$ bit operations or, using fast polynomial arithmetic, $O(M(m))$ bit operations. If m is fixed and p goes to infinity then the complexity is $O(M(\log(p)))$ bit operations.

Inversion of elements in $\mathbb{F}_{p^m} = \mathbb{F}_p[x]/(F(x))$ can be done using the extended Euclidean algorithm in $O(m^2)$ operations in $\mathbb{F}_p$. If p is fixed and m grows then one can invert elements in $\mathbb{F}_{p^m}$ in $O(M(m)\log(m))$ bit operations.

Alternatively, for any vector space basis $\{\theta_1, \dots, \theta_m\}$ for $\mathbb{F}_{q^m}$ over $\mathbb{F}_q$ there is an $m \times m$ matrix M over $\mathbb{F}_q$ such that the product ab for $a, b \in \mathbb{F}_{q^m}$ is given by

$$(a_1, \dots, a_m)M(b_1, \dots, b_m)^t = \sum_{i=1}^{m}\sum_{j=1}^{m} M_{i,j}a_i b_j$$

where $(a_1, \dots, a_m)$ and $(b_1, \dots, b_m)$ are the coefficient vectors for the representation of a and b with respect to the basis.

In particular, if $\mathbb{F}_{q^m}$ is represented by a normal basis $\{\theta, \theta^q, \dots, \theta^{q^{m-1}}\}$ then multiplication of elements in normal basis representation is given by

$$\left(\sum_{i=0}^{m-1} a_i \theta^{q^i}\right)\left(\sum_{j=0}^{m-1} b_j \theta^{q^j}\right) = \sum_{i=0}^{m-1}\sum_{j=0}^{m-1} a_i b_j \theta^{q^i + q^j}$$

so it is necessary to precompute the representation of each term $M_{i,j} = \theta^{q^i+q^j}$ over the normal basis. Multiplication in $\mathbb{F}_{p^m}$ using a normal basis representation is typically slower than multiplication with a polynomial basis; indeed, the complexity can be as bad as $O(m^3)$ operations in $\mathbb{F}_p$. An **optimal normal basis** is a normal basis for which the number of non-zero coefficients in the product is minimal (see Section II.2.2 of [60] for the case of $\mathbb{F}_{2^m}$). Much work has been done on speeding up multiplication with optimal normal bases; for example see Bernstein and Lange [52] for discussion and references.

Raising an element of $\mathbb{F}_{q^m}$ to the qth power is always fast since it is a linear operation. Taking qth powers (respectively, qth roots) is especially fast for normal bases as it is a rotation; this is the main motivation for considering normal bases. This fact has a number of important applications, see for example, Exercise 14.4.7.

2.12 Factoring polynomials over finite fields

There is a large literature on polynomial factorisation and we only give a very brief sketch of the main concepts. The basic ideas go back to Berlekamp and others. For full discussion, historical background and extensive references see Chapter 7 of Bach and Shallit [21] or Chapter 14 of von zur Gathen and Gerhard [220]. One should be aware that for polynomials over fields of small characteristic the algorithm by Niederreiter [418] can be useful.

Let $F(x) \in \mathbb{F}_q[x]$ have degree d. If there exists $G(x) \in \mathbb{F}_q[x]$ such that $G(x)^2 \mid F(x)$ then $G(x) \mid F'(x)$ where $F'(x)$ is the derivative of $F(x)$. A polynomial is **square-free** if it has no repeated factor. It follows that $F(x)$ is square-free if $F'(x) \neq 0$ and $\gcd(F(x), F'(x)) = 1$. If $F(x) \in \mathbb{F}_q[x]$ and $S(x) = \gcd(F(x), F'(x))$ then $F(x)/S(x)$ is square-free.

Exercise 2.12.1 Determine the complexity of testing whether a polynomial $F(x) \in \mathbb{F}_q[x]$ is square-free.

Exercise 2.12.2 Show that one can reduce polynomial factorisation over finite fields to the case of factoring square-free polynomials.

Finding roots of polynomials in finite fields

Let $F(x) \in \mathbb{F}_q[x]$ have degree d. The roots of $F(x)$ in $\mathbb{F}_q$ are precisely the roots of

$$R_1(x) = \gcd(F(x), x^q - x). \tag{2.2}$$

If q is much larger than d then the efficient way to compute $R_1(x)$ is to compute $x^q \pmod{F(x)}$ using a square-and-multiply algorithm and then run Euclid's algorithm.

Exercise 2.12.3 Determine the complexity of computing $R_1(x)$ in equation (2.2). Hence, explain why the decision problem "Does $F(x)$ have a root in $\mathbb{F}_q$?" has a polynomial-time solution.

The basic idea of root-finding algorithms is to note that, if q is odd, $x^q - x = x(x^{(q-1)/2} + 1)(x^{(q-1)/2} - 1)$. Hence, one can try to split[7] $R_1(x)$ by computing

$$\gcd(R_1(x), x^{(q-1)/2} - 1). \tag{2.3}$$

Similar ideas can be used when q is even (see Section 2.14.2).

Exercise 2.12.4 Show that the roots of the polynomial in equation (2.3) are precisely the $\alpha \in \mathbb{F}_q$ such that $F(\alpha) = 0$ and α is a square in $\mathbb{F}_q^*$.

To obtain a randomised (Las Vegas) algorithm to factor $R_1(x)$ completely when q is odd one simply chooses a random polynomial $u(x) \in \mathbb{F}_q[x]$ of degree $< d$ and computes

$$\gcd(R_1(x), u(x)^{(q-1)/2} - 1).$$

This computation selects those roots α of $R_1(x)$ such that $u(\alpha)$ is a square in $\mathbb{F}_q$. In practice, it suffices to choose $u(x)$ to be linear. Performing this computation sufficiently many times on the resulting factors of $R_1(x)$ and taking gcds (greatest common divisors) eventually leads to the complete factorisation of $R_1(x)$.

Exercise 2.12.5 Write down pseudocode for the above root-finding algorithm and show that its expected complexity (without using a fast Euclidean algorithm) is bounded by $O(\log(d)(\log(q)M(d) + d^2)) = O(\log(q)\log(d)d^2)$ field operations.

[7] We reserve the word "factor" for giving the full decomposition into irreducibles, whereas we use the word "split" to mean breaking into two pieces.

Exercise 2.12.6 Let q be an odd prime power and $R(x) = x^2 + ax + b \in \mathbb{F}_q[x]$. Show that the expected complexity of finding roots of $R(x)$ using polynomial factorisation is $O(\log(q)M(\log(q)))$ bit operations.

Exercise 2.12.7★ Show, using Kronecker substitution, fast versions of Euclid's algorithm and other tricks, that one can compute one root in $\mathbb{F}_q$ (if any exist) of a degree d polynomial in $\mathbb{F}_q[x]$ in an expected $O(\log(qd)M(d, q))$ bit operations.

When q is even (i.e., $q = 2^m$) then, instead of $x^{(q-1)/2}$, one considers the **trace polynomial** $T(x) = \sum_{i=0}^{m-1} x^{2^i}$. (A similar idea can be used over any field of small characteristic.)

Exercise 2.12.8 Show that the roots of the polynomial $\gcd(R_1(x), T(x))$ are precisely the $\alpha \in \mathbb{F}_q$ such that $R_1(\alpha) = 0$ and $\mathrm{Tr}_{\mathbb{F}_{2^m}/\mathbb{F}_2}(\alpha) = 0$.

Taking random $u(x) \in \mathbb{F}_{2^m}[x]$ of degree $< d$ and then computing $\gcd(R_1(x), T(u(x)))$ gives a Las Vegas root-finding algorithm as before. See Section 21.3.2 of [497] for details.

Higher degree factors

Having found the roots in $\mathbb{F}_q$ one can try to find factors of larger degree. The same ideas can be used. Let

$$R_2(x) = \gcd(F(x)/R_1(x), x^{q^2} - x),\ \ R_3(x) = \gcd(F(x)/(R_1(x)R_2(x)), x^{q^3} - x), \ldots.$$

Exercise 2.12.9 Show that all irreducible factors of $R_i(x)$ over $\mathbb{F}_q[x]$ have degree i.

Exercise 2.12.10 Give an algorithm to test whether a polynomial $F(x) \in \mathbb{F}_q[x]$ of degree d is irreducible. What is the complexity?

When q is odd, one can factor $R_i(x)$ using similar ideas to the above, i.e. by computing

$$\gcd(R_i(x), u(x)^{(q^i-1)/2} - 1).$$

These techniques lead to the **Cantor–Zassenhaus algorithm**. It factors polynomials of degree d over $\mathbb{F}_q$ in an expected $O(d\log(d)\log(q)M(d))$ field operations. For many more details about polynomial factorisation see Section 7.4 of [21], Sections 21.3 and 21.4 of [497], Chapter 14 of [220], [332], Chapter 4 of [350, 351] or Section 4.6.2 of [308].

Exercise 2.12.11 Let $d \in \mathbb{N}$ and $F(x) \in \mathbb{F}_q[x]$ of degree d. Given $1 < b < n$ suppose we wish to output all irreducible factors of $F(x)$ of degree at most b. Show that the expected complexity is $O(b\log(d)\log(q)M(d))$ operations in $\mathbb{F}_q$. Hence, one can factor $F(x)$ completely in $O(d\log(d)\log(q)M(d))$ operations in $\mathbb{F}_q$.

Exercise 2.12.12★ Using the same methods as Exercise 2.12.7, show that one can find an irreducible factor of degree $1 < b < d$ of a degree d polynomial in $\mathbb{F}_q[x]$ in an expected $O(b\log(dq)M(d, q))$ bit operations.

2.13 Hensel lifting

Hensel lifting is a tool for solving polynomial equations of the form $F(x) \equiv 0 \pmod{p^e}$, where p is prime and $e \in \mathbb{N}_{>1}$. One application of Hensel lifting is the Takagi variant of RSA, see Example 24.1.5. The key idea is given in the following Lemma.

Lemma 2.13.1 *Let $F(x) \in \mathbb{Z}[x]$ be a polynomial and p a prime. Let $x_k \in \mathbb{Z}$ satisfy $F(x_k) \equiv 0 \pmod{p^k}$ where $k \in \mathbb{N}$. Suppose $F'(x_k) \not\equiv 0 \pmod{p}$. Then one can compute $x_{k+1} \in \mathbb{Z}$ in polynomial-time such that $F(x_{k+1}) \equiv 0 \pmod{p^{k+1}}$.*

Proof Write $x_{k+1} = x_k + p^k z$ where z is a variable. Note that $F(x_{k+1}) \equiv 0 \pmod{p^k}$. One has

$$F(x_{k+1}) \equiv F(x_k) + p^k F'(x_k) z \pmod{p^{k+1}}.$$

Setting $z = -(F(x_k)/p^k)F'(x_k)^{-1} \pmod{p}$ gives $F(x_{k+1}) \equiv 0 \pmod{p^{k+1}}$. □

Example 2.13.2 We solve the equation

$$x^2 \equiv 7 \pmod{3^3}.$$

Let $f(x) = x^2 - 7$. First, the equation $f(x) \equiv x^2 - 1 \pmod{3}$ has solutions $x \equiv 1, 2 \pmod{3}$. We take $x_1 = 1$. Since $f'(1) = 2 \not\equiv 0 \pmod{3}$, we can "lift" this to a solution modulo 3^2. Write $x_2 = 1 + 3z$. Then

$$f(x_2) = x_2^2 - 7 \equiv -6 + 6z \pmod{3^2}$$

or, in other words, $1 - z \equiv 0 \pmod{3}$. This has the solution $z = 1$ giving $x_2 = 4$.

Now lift to a solution modulo 3^3. Write $x_3 = 4 + 9z$. Then $f(x_3) \equiv 9 + 72z \pmod{3^3}$ and dividing by 9 yields $1 - z \equiv 0 \pmod{3}$. This has solution $z = 1$ giving $x_3 = 13$ as one solution to the original equation.

Exercise 2.13.3 The equation $x^3 \equiv 3 \pmod{5}$ has the solution $x \equiv 2 \pmod{5}$. Use Hensel lifting to find a solution to the equation $x^3 \equiv 3 \pmod{5^3}$.

2.14 Algorithms in finite fields

We present some algorithms for constructing finite fields $\mathbb{F}_{p^m}$ when $m > 1$, solving equations in them and transforming between different representations of them.

2.14.1 Constructing finite fields

Lemma A.5.1 implies a randomly chosen monic polynomial in $\mathbb{F}_q[x]$ of degree m is irreducible with probability $\geq 1/(2m)$. Hence, using the algorithm of Exercise 2.12.10, one can generate a random irreducible polynomial $F(x) \in \mathbb{F}_q[x]$ of degree m, using naive arithmetic, in $O(m^4 \log(q))$ operations in $\mathbb{F}_q$. In other words, one can construct a polynomial basis for $\mathbb{F}_{q^m}$ in $O(m^4 \log(q))$ operations in $\mathbb{F}_q$. This complexity is not the best known.

Constructing a normal basis

We briefly survey the literature on constructing normal bases for finite fields. We assume that a polynomial basis for $\mathbb{F}_{q^m}$ over $\mathbb{F}_q$ has already been computed.

The simplest randomised algorithm is to choose $\theta \in \mathbb{F}_{q^m}$ at random and test whether the set $\{\theta, \theta^q, \dots, \theta^{q^{m-1}}\}$ is linearly independent over $\mathbb{F}_q$. Corollary 3.6 of von zur Gathen and Giesbrecht [221] (also see Theorem 3.73 and Exercise 3.76 of [350, 351]) shows that a randomly chosen θ is normal with probability at least $1/34$ if $m < q^4$ and probability at least $1/(16\log_q(m))$ if $m \geq q^4$.

Exercise 2.14.1 Determine the complexity of constructing a normal basis by randomly choosing θ.

When $q > m(m-1)$ there is a better randomised algorithm based on the following result.

Theorem 2.14.2 *Let $F(x) \in \mathbb{F}_q[x]$ be irreducible of degree m and let $\alpha \in \mathbb{F}_{q^m}$ be any root of $F(x)$. Define $G(x) = F(x)/((x-\alpha)F'(\alpha)) \in \mathbb{F}_{q^m}[x]$. Then there are $q - m(m-1)$ elements $u \in \mathbb{F}_q$ such that $\theta = G(u)$ generates a normal basis.*

Proof See Theorem 28 of Section II.N of Artin [14] or Section 3.1 of Gao [218]. □

Deterministic algorithms for constructing a normal basis have been given by Lüneburg [360] and Lenstra [341] (also see Gao [218]).

2.14.2 Solving quadratic equations in finite fields

This section is about solving quadratic equations $x^2 + ax + b = 0$ over $\mathbb{F}_q$. One can apply any of the algorithms for polynomial factorisation mentioned earlier. As we saw in Exercise 2.12.6, when q is odd, the basic method computes roots in $O(\log(q)M(\log(q)))$ bit operations. When q is odd it is also natural to use the quadratic formula and a square-roots algorithm (see Section 2.9).

Exercise 2.14.3 Generalise the Tonelli–Shanks algorithm from Section 2.9 to work for any finite field $\mathbb{F}_q$ where q is odd. Show that the complexity remains an expected $O(\log(q)^2 M(\log(q)))$ bit operations.

Exercise 2.14.4 Suppose $\mathbb{F}_{q^2}$ is represented as $\mathbb{F}_q(\theta)$ where $\theta^2 \in \mathbb{F}_q$. Show that one can compute square roots in $\mathbb{F}_{q^2}$ using two square roots in $\mathbb{F}_q$ and a small number of multiplications in $\mathbb{F}_q$.

Since squaring in $\mathbb{F}_{2^m}$ is a linear operation, one can take square roots in $\mathbb{F}_{2^m}$ using linear algebra in $O(m^3)$ bit operations. The following exercise gives a method that is more efficient.

Exercise 2.14.5 Suppose one represents $\mathbb{F}_{2^m}$ using a polynomial basis $\mathbb{F}_2[x]/(F(x))$. Precompute $\sqrt{x}$ as a polynomial in x. Let $g = \sum_{i=0}^{m-1} a_i x^i$. To compute $\sqrt{g}$ write (assuming

m is odd; the case of m even is similar)

$$g = \left(a_0 + a_2x^2 + \cdots + a_{m-1}x^{m-1}\right) + x\left(a_1 + a_3x^2 + \cdots + a_{m-2}x^{m-3}\right).$$

Show that

$$\sqrt{g} = \left(a_0 + a_2x + \cdots x_{m-1}x^{(m-1)/2}\right) + \sqrt{x}\left(a_1 + a_3x + \cdots + a_{m-2}x^{(m-3)/2}\right).$$

Show that this computation takes roughly half the cost of one field multiplication, and hence $O(m^2)$ bit operations.

Exercise 2.14.6 Generalise Exercise 2.14.5 to computing pth roots in $\mathbb{F}_{p^m}$. Show that the method requires $(p-1)$ multiplications in $\mathbb{F}_{p^m}$.

We now consider how to solve quadratic equations of the form

$$x^2 + x = \alpha \tag{2.4}$$

where $\alpha \in \mathbb{F}_{2^m}$.

Exercise 2.14.7★ Prove that the equation $x^2 + x = \alpha$ has a solution $x \in \mathbb{F}_{2^m}$ if and only if $\mathrm{Tr}_{\mathbb{F}_{2^m}/\mathbb{F}_2}(\alpha) = 0$.

Lemma 2.14.8 *If m is odd (we refer to Section II.2.4 of [60] for the case where m is even) then a solution to equation (2.4) is given by the* ***half trace***

$$x = \sum_{i=0}^{(m-1)/2} \alpha^{2^{2i}}. \tag{2.5}$$

Exercise 2.14.9 Prove Lemma 2.14.8. Show that the complexity of solving quadratic equations in $\mathbb{F}_q$ when $q = 2^m$ and m is odd is an expected $O(m^3)$ bit operations (or $O(m^2)$ bit operations when a normal basis is being used).

The expected complexity of solving quadratic equations in $\mathbb{F}_{2^m}$ when m is even is $O(m^4)$ bit operations, or $O(m^3)$ when a normal basis is being used. Hence, we can make the statement that the complexity of solving a quadratic equation over any field $\mathbb{F}_q$ is an expected $O(\log(q)^4)$ bit operations.

2.14.3 Isomorphisms between finite fields

Computing the minimal polynomial of an element

Given $g \in \mathbb{F}_{q^m}$ one can compute the minimal polynomial $F(x)$ of g over $\mathbb{F}_q$ using linear algebra. To do this, one considers the set $S_n = \{1, g, g^2, \ldots, g^n\}$ for $n = 1, \ldots, m$. Let n be the smallest integer such that S_n is linearly dependent over $\mathbb{F}_q$. Then there are $a_0, \ldots, a_n \in \mathbb{F}_q$ such that $\sum_{i=0}^n a_i g^i = 0$. Since S_{n-1} is linearly independent, it follows that $F(x) = \sum_{i=0}^n a_i x^i$ is the minimal polynomial for g.

Exercise 2.14.10 Show that the above algorithm requires $O(m^3)$ operations in $\mathbb{F}_q$.

Computing a polynomial basis for a finite field

Suppose $\mathbb{F}_{q^m}$ is given by some basis that is not a polynomial basis. We now give a method to compute a polynomial basis for $\mathbb{F}_{q^m}$.

If $g \in \mathbb{F}_{q^m}$ is chosen uniformly at random, then, by Lemma A.8.4, with probability at least $1/q$ the element g does not lie in a subfield of $\mathbb{F}_{q^m}$ that contains $\mathbb{F}_q$. Hence, the minimal polynomal $F(x)$ of g over $\mathbb{F}_q$ has degree m and the algorithm of the previous subsection computes $F(x)$. One therefore has a polynomial basis $\{1, x, \ldots, x^{m-1}\}$ for $\mathbb{F}_{q^m}$ over $\mathbb{F}_q$.

Exercise 2.14.11 Determine the complexity of this algorithm.

Computing isomorphisms between finite fields

Suppose one has two representations for $\mathbb{F}_{q^m}$ as a vector space over $\mathbb{F}_q$ and wants to compute an isomorphism between them. We do this in two stages: first, we compute an isomorphism from any representation to a polynomial basis, and second we compute isomorphisms between any two polynomial bases. We assume that one already has an isomorphism between the corresponding representations of the subfield $\mathbb{F}_q$.

Let $\{\theta_1, \ldots, \theta_m\}$ be the vector space basis over $\mathbb{F}_q$ for one of the representations of $\mathbb{F}_{q^m}$. The first task is to compute an isomorphism from this representation to a polynomial representation. To do this, one computes a polynomial basis for $\mathbb{F}_{q^m}$ over $\mathbb{F}_q$ using the method above. One now has a monic irreducible polynomial $F(x) \in \mathbb{F}_q[x]$ of degree m and a representation $x = \sum_{i=1}^{m} a_i \theta_i$ for a root of $F(x)$ in $\mathbb{F}_{q^m}$. Determine the representations of $x^2, x^3, \ldots, x^m$ over the basis $\{\theta_1, \ldots, \theta_m\}$. This gives an isomorphism from $\mathbb{F}_q[x]/(F(x))$ to the original representation of $\mathbb{F}_{q^m}$. By solving a system of linear equations, one can express each of $\theta_1, \ldots, \theta_m$ with respect to the polynomial basis; this gives the isomorphism from the original representation to $\mathbb{F}_q[x]/(F(x))$. The above ideas appear in a special case in the work of Zierler [574].

Exercise 2.14.12 Determine the complexity of the above algorithm to give an isomorphism between an arbitrary vector space representation of $\mathbb{F}_{q^m}$ and a polynomial basis for $\mathbb{F}_{q^m}$.

Finally, it remains to compute an isomorphism between any two polynomial representations $\mathbb{F}_q[x]/(F_1(x))$ and $\mathbb{F}_q[y]/(F_2(y))$ for $\mathbb{F}_{q^m}$. This is done by finding a root $a(y) \in \mathbb{F}_q[y]/(F_2(y))$ of the polynomial $F_1(x)$. The function $x \mapsto a(y)$ extends to a field isomorphism from $\mathbb{F}_q[x]/(F_1(x))$ to $\mathbb{F}_q[y]/(F_2(y))$. The inverse to this isomorphism is computed by linear algebra.

Exercise 2.14.13 Determine the complexity of the above algorithm to give an isomorphism between an arbitrary vector space representation of $\mathbb{F}_{q^m}$ and a polynomial basis for $\mathbb{F}_{q^m}$.

See Lenstra [341] for deterministic algorithms to solve this problem.

Random sampling of finite fields

Let $\mathbb{F}_{p^m}$ be represented as a vector space over $\mathbb{F}_p$ with basis $\{\theta_1, \ldots, \theta_m\}$. Generating an element $g \in \mathbb{F}_{p^m}$ uniformly at random can be done by selecting m integers $a_1, \ldots, a_m$

uniformly at random in the range $0 \le a_i < p$ and taking $g = \sum_{i=1}^m a_i \theta_i$. Section 11.4 mentions some methods to get random integers modulo p from random bits.

To sample uniformly from $\mathbb{F}_{p^m}^*$ one can use the above method, repeating the process if $a_i = 0$ for all $1 \le i \le m$. This is much more efficient than choosing $0 \le a < p^m - 1$ uniformly at random and computing $g = \gamma^a$ where γ is a primitive root.

2.15 Computing orders of elements and primitive roots

We first consider how to determine the order of an element $g \in \mathbb{F}_q^*$. Assume the factorisation $q - 1 = \prod_{i=1}^m l_i^{e_i}$ is known.[8] Then it is sufficient to determine, for each i, the smallest $0 \le f \le e_i$ such that

$$g^{(q-1)/l_i^f} = 1.$$

This leads to a simple algorithm for computing the order of g that requires $O(\log(q)^4)$ bit operations.

Exercise 2.15.1 Write pseudocode for the basic algorithm for determining the order of g and determine the complexity.

The next subsection gives an algorithm (also used in other parts of the book) that leads to an improvement of the basic algorithm.

2.15.1 *Sets of exponentials of products*

We now explain how to compute sets of the form $\{g^{(q-1)/l} : l \mid (q-1)\}$ efficiently. We generalise the problem as follows. Let $N_1, \ldots, N_m \in \mathbb{N}$ and $N = \prod_{i=1}^m N_i$ (typically the integers N_i will be coprime, but it is not necessary to assume this). Let $k = \lceil \log_2(m) \rceil$ and, for $m < i \le 2^k$ set $N_i = 1$. Let G be a group and $g \in G$ (where g typically has order $\ge N$). The goal is to efficiently compute

$$\{g^{N/N_i} : 1 \le i \le m\}.$$

The naive approach (computing each term separately and not using any window methods etc.) requires at least

$$\sum_{i=1}^m \log(N/N_i) = m\log(N) - \sum_{i=1}^m \log(N_i) = (m-1)\log(N)$$

operations in G and at most $2m\log(N)$ operations in G.

For the improved solution one re-uses intermediate values. The basic idea can be seen in the following example. Computing products in such a way is often called using a **product tree**.

[8] As far as I am aware, it has not been proved that computing the order of an element in $\mathbb{F}_q^*$ is equivalent to factoring $q-1$; or even that computing the order of an element in $\mathbb{Z}_N^*$ is equivalent to factoring $\varphi(N)$. Yet it seems to be impossible to correctly determine the order of every $g \in \mathbb{F}_q^*$ without knowing the factorisation of $q-1$.

Example 2.15.2 Let $N = N_1N_2N_3N_4$ and suppose one needs to compute

$$g^{N_1N_2N_3}, g^{N_1N_2N_4}, g^{N_1N_3N_4}, g^{N_2N_3N_4}.$$

We first compute

$$h_{1,1} = g^{N_3N_4} \qquad \text{and} \qquad h_{1,2} = g^{N_1N_2}$$

in $\le 2(\log_2(N_1N_2) + \log_2(N_3N_4)) = 2\log_2(N)$ operations. One can then compute the result

$$g^{N_1N_2N_3} = h_{1,2}^{N_3},\ g^{N_1N_2N_4} = h_{1,2}^{N_4},\ g^{N_1N_3N_4} = h_{1,1}^{N_1},\ g^{N_2N_3N_4} = h_{1,1}^{N_2}.$$

This final step requires at most $2(\log_2(N_3) + \log_2(N_4) + \log_2(N_1) + \log_2(N_2)) = 2\log_2(N)$ operations. The total complexity is at most $4\log_2(N)$ operations in the group.

The algorithm has a compact recursive description. Let F be the function that on input $(g, m, N_1, \dots, N_m)$ outputs the list of m values g^{N/N_i} for $1 \le i \le m$ where $N = N_1 \cdots N_m$. Then $F(g, 1, N_1) = g$. For $m > 1$ one computes $F(g, m, N_1, \dots, N_m)$ as follows: Let $l = \lfloor m/2 \rfloor$ and let $h_1 = g^{N_1 \cdots N_l}$ and $h_2 = g^{N_{l+1} \cdots N_m}$. Then $F(g, m, N_1, \dots, N_m)$ is equal to the concatenation of $F(h_1, (m-l), N_{l+1}, \dots, N_m)$ and $F(h_2, l, N_1, \dots, N_l)$.

We introduce some notation to express the algorithm in a non-recursive format.

Definition 2.15.3 Define $S = \{1, 2, \dots, 2^k\}$. For $1 \le l \le k$ and $1 \le j \le 2^l$ define

$$S_{l,j} = \{i \in S : (j-1)2^{k-l} + 1 \le i \le j2^{k-l}\}.$$

Lemma 2.15.4 *Let $1 \le l \le k$ and $1 \le j \le 2^l$. The sets $S_{l,j}$ satisfy:*

1. $\#S_{l,j} = 2^{k-l}$;
2. $S_{l,j} \cap S_{l,j'} = \varnothing$ *if* $j \ne j'$;
3. $\cup_{j=1}^{2^l} S_{l,j} = S$ *for all* $1 \le l \le k$;
4. *If* $l \ge 2$ *and* $1 \le j \le 2^{l-1}$ *then* $S_{l-1,j} = S_{l,2j-1} \cup S_{l,2j}$;
5. $S_{k,j} = \{j\}$ *for* $1 \le j \le 2^k$.

Exercise 2.15.5 Prove Lemma 2.15.4.

Definition 2.15.6 For $1 \le l \le k$ and $1 \le j \le 2^l$ define

$$h_{l,j} = g^{\prod_{j \in S - S_{l,j}} N_i}.$$

To compute $\{h_{k,j} : 1 \le j \le m\}$ efficiently, one notes that if $l \ge 2$ and $1 \le j \le 2^l$ then writing $j_1 = \lceil j/2 \rceil$

$$h_{l,j} = h_{l-1,j_1}^{\prod_{i \in S_{l-1,j_1} - S_{l,j}} N_i}.$$

This leads to Algorithm 4.

Algorithm 4 Computing set of exponentials of products

INPUT: $N_1, \ldots, N_m$
OUTPUT: $\{g^{N/N_i} : 1 \le i \le m\}$

1: $k = \lceil \log_2(m) \rceil$
2: $h_{1,1} = g^{N_{2^{k-1}+1} \cdots N_{2^k}}, h_{1,2} = g^{N_1 \cdots N_{2^{k-1}}}$
3: **for** $l = 2$ to k **do**
4:     **for** $j = 1$ to 2^l **do**
5:         $j_1 = \lceil j/2 \rceil$
6:         $h_{l,j} = h_{l-1,j_1}^{\prod_{i \in S_{l-1,j_1} - S_{l,j}} N_i}$
7:     **end for**
8: **end for**
9: **return** $\{h_{k,1}, \ldots, h_{k,m}\}$

Lemma 2.15.7 *Algorithm 4 is correct and requires* $\le 2\lceil \log_2(m) \rceil \log(N)$ *group operations.*

Proof Almost everything is left as an exercise. The important observation is that lines 4 to 7 involve raising to the power N_i for all $i \in S$. Hence, the cost for each iteration of the loop in line 3 is at most $2\sum_{i=1}^{2^k} \log_2(N_i) = 2\log_2(N)$. □

This method works efficiently in all cases (i.e., it does not require m to be large). However, Exercise 2.15.8 shows that for small values of m there may be more efficient solutions.

Exercise 2.15.8 Let $N = N_1 N_2 N_3$ where $N_i \approx N^{1/3}$ for $1 \le i \le 3$. One can compute g^{N/N_i} for $1 \le i \le 3$ using Algorithm 4 or in the "naive" way. Suppose one uses the standard square-and-multiply method for exponentiation and assume that each of N_1, N_2 and N_3 has Hamming weight about half their bit-length.

Note that the exponentiations in the naive solution are all with respect to the fixed base g. A simple optimisation is therefore to precompute all g^{2^j} for $1 \le j \le \log_2(N^{2/3})$. Determine the number of group operations for each algorithm if this optimisation is performed. Which is better?

Remark 2.15.9 Sutherland gives an improved algorithm (which he calls the **snowball algorithm**) as Algorithm 7.4 of [536]. Proposition 7.3 of [536] states that the complexity is

$$O(\log(N)\log(m)/\log(\log(m))) \tag{2.6}$$

group operations.

2.15.2 Computing the order of a group element

We can now return to the original problem of computing the order of an element in a finite field.

Theorem 2.15.10 *Let $g \in \mathbb{F}_q^*$ and assume that the factorisation $q - 1 = \prod_{i=1}^m l_i^{e_i}$ is known. Then one can determine the order of g in $O(\log(q) \log\log(q))$ multiplications in $\mathbb{F}_q$.*

Proof The idea is to use Algorithm 4 to compute all $h_i = g^{(q-1)/l_i^{e_i}}$. Since $m = O(\log(q))$ this requires $O(\log(q) \log\log(q))$ multiplications in $\mathbb{F}_q$. One can then compute all $h_i^{l_i^f}$ for $1 \le f < e_i$ and, since $\prod_{i=1}^m l_i^{e_i} = q - 1$ this requires a further $O(\log(q))$ multiplications. □

The complexity in Theorem 2.15.10 cannot be improved to $O(\log(q) \log\log(q)/\log(\log(\log(q))))$ using the result of equation (2.6) because the value m is not always $\Theta(\log(q))$.

2.15.3 Computing primitive roots

Recall that $\mathbb{F}_q^*$ is a cyclic group and that a primitive root in $\mathbb{F}_q^*$ is an element of order $q - 1$. We assume in this section that the factorisation of $q - 1$ is known.

One algorithm to generate primitive roots is to choose $g \in \mathbb{F}_q^*$ uniformly at random and to compute the order of g using the method of Theorem 2.15.10 until an element of order $q - 1$ is found. The probability that a random $g \in \mathbb{F}_q^*$ is a primitive root is $\varphi(q-1)/(q-1)$. Using Theorem A.3.1, this probability is at least $1/(6\log(\log(q)))$. Hence, this gives an algorithm that requires $O(\log(q)(\log(\log(q)))^2)$ field multiplications in $\mathbb{F}_q$.

We now present a better algorithm for this problem, which works by considering the prime powers dividing $q - 1$ individually. See Exercise 11.2 of Section 11.1 of [497] for further details.

Theorem 2.15.11 *Algorithm 5 outputs a primitive root. The complexity of the algorithm is $O(\log(q) \log\log(q))$ multiplications in $\mathbb{F}_q$.*

Proof The values g_i are elements of order dividing $l_i^{e_i}$. If $g_i^{l_i^{e_i-1}} \neq 1$ then g_i has order exactly $l_i^{e_i}$. On completion of the while loop the value t is the product of m elements of maximal coprime orders $l_i^{e_i}$. Hence, t is a primitive root.

Each iteration of the while loop requires $O(\log(q) \log\log(q))$ multiplications in $\mathbb{F}_q$. It remains to bound the number of iterations of the loop. First, note that, by the Chinese remainder theorem, the g_i are independent and uniformly at random in subgroups of order $l_i^{e_i}$. Hence, the probability that $g_i^{l_i^{e_i-1}} = 1$ is $1/l_i \le 1/2$ and the expected number of trials for any given value g_i less than or equal to 2. Hence, the expected number of iterations of the while loop is less than or equal to 2. This completes the proof. □

Algorithm 5 Computing a primitive root in $\mathbb{F}_q^*$

INPUT: $q, m, \{(l_i, e_i)\}$ such that $q - 1 = \prod_{i=1}^{m} l_i^{e_i}$ and the l_i are distinct primes
OUTPUT: primitive root g

```
1:  Let S = {1, ..., m}
2:  t = 1
3:  while S ≠ ∅ do
4:      Choose g ∈ F_q^* uniformly at random
5:      Compute g_i = g^((q−1)/l_i^(e_i)) for 1 ≤ i ≤ m using Algorithm 4
6:      for i ∈ S do
7:          if g_i^(l_i^(e_i − 1)) ≠ 1 then
8:              t = t g_i
9:              Remove i from S
10:         end if
11:     end for
12: end while
13: return t
```

2.16 Fast evaluation of polynomials at multiple points

We have seen that one can evaluate a univariate polynomial at a field element efficiently using Horner's rule. For some applications, for example the attack on small CRT exponents for RSA in Section 24.5.2, one must evaluate a fixed polynomial repeatedly at lots of field elements. Naively repeating Horner's rule n times would give a total cost of n^2 multiplications. This section shows that one can solve this problem more efficiently than the naive method.

Theorem 2.16.1 *Let $F(x) \in \Bbbk[x]$ have degree n and let $x_1, \ldots, x_n \in \Bbbk$. Then one can compute $\{F(x_1), \ldots, F(x_n)\}$ in $O(M(n)\log(n))$ field operations. The storage requirement is $O(n \log(n))$ elements of $\Bbbk$.*

Proof (Sketch) Let $t = \lceil \log_2(n) \rceil$ and set $x_i = 0$ for $n < i \le 2^t$. For $0 \le i \le t$ and $1 \le j \le 2^{t-i}$ define

$$G_{i,j}(x) = \prod_{k=(j-1)2^i+1}^{j2^i} (x - x_k).$$

One computes the $G_{i,j}(x)$ for $i = 0, 1, \ldots, t$ using the formula $G_{i,j}(x) = G_{i-1,2j-1}(x)G_{i-1,2j}(x)$. (This is essentially the same trick as Section 2.15.1.)

Once all the $G_{i,j}(x)$ have been computed one defines, for $0 \le i \le t$, $1 \le j \le 2^{t-i}$ the polynomials $F_{i,j}(x) = F(x) \pmod{G_{i,j}(x)}$. One computes $F_{t,0}(x) = F(x) \pmod{G_{t,0}(x)}$ and then computes $F_{i,j}(x)$ efficiently as $F_{i+1,\lfloor (j+1)/2 \rfloor}(x) \pmod{G_{i,j}(x)}$ for $i = t - 1$ downto 0. Note that $F_{0,j}(x) = F(x) \pmod{(x - x_j)} = F(x_j)$ as required.

Table 2.1 *Expected complexity of basic algorithms for numbers of size relevant for cryptography and related applications. The symbol $*$ indicates that better asymptotic complexities are known.*

Computational problem	Expected complexity for cryptography
Multiplication of m-bit integers, $M(m)$	$O(m^2)$ or $O(m^{1.585})$ bit operations
Compute $\lfloor a/n \rfloor$, $a \pmod{n}$	$O((\log(\|a\|) - \log(n))\log(n))$ or $O(M(\log(\|a\|))$ bit operations
Compute $\lfloor \sqrt{\|a\|} \rfloor$	$O(M(\log(\|a\|)))$ bit operations
Extended gcd(a, b) where a and b are m-bit integers	$O(m^2)$ or $O(M(m)\log(m))$ bit operations
Legendre/Jacobi symbol $(\frac{a}{n})$, $\|a\| < n$	$O(\log(n)^2)$ or $O(M(\log(n))\log(\log(n)))$ bit operations
Multiplication modulo n	$O(M(\log(n)))$ bit operations
Inversion modulo n	$O(\log(n)^2)$ or $O(M(\log(n))\log(n))$ bit operations
Compute $g^m \pmod{n}$	$O(\log(m)M(\log(n)))$ bit operations
Compute square roots in $\mathbb{F}_q^*$ (q odd)	$O(\log(q)M(\log(q)))$ bit operations
Multiplication of two-degree d polys in $\Bbbk[x]$	$O(M(d))$ $\Bbbk$-multiplications
Multiplication of two-degree d polys in $\mathbb{F}_q[x]$, $M(d, q)$	$O(M(d\log(dq)))$ bit operations
Inversion in $\Bbbk[x]/(F(x))$ where $\deg(F(x)) = d$	$O(d^2)$ or $O(M(d)\log(d))$ $\Bbbk$-operations
Multiplication in $\mathbb{F}_{q^m}$	$O(M(m))$ operations in $\mathbb{F}_q$ $*$
Evaluate a degree d polynomial at $\alpha \in \Bbbk$	$O(d)$ $\Bbbk$-operations
Find all roots in $\mathbb{F}_q$ of a degree d polynomial in $\mathbb{F}_q[x]$	$O(\log(d)\log(q)d^2)$ $\mathbb{F}_q$-operations $*$
Find one root in $\mathbb{F}_q$ of a degree d polynomial in $\mathbb{F}_q[x]$	$O(\log(dq)M(d, q))$ bit operations
Determine if degree d poly over $\mathbb{F}_q$ is irreducible	$O(d^3\log(q))$ $\mathbb{F}_q$-operations $*$
Factor degree d polynomial over $\mathbb{F}_q$	$O(d^3\log(q))$ $\mathbb{F}_q$-operations $*$
Construct polynomial basis for $\mathbb{F}_{q^m}$	$O(m^4\log(q))$ $\mathbb{F}_q$-operations $*$
Construct normal basis for $\mathbb{F}_{q^m}$ given a poly basis	$O(m^3\log_q(m))$ $\mathbb{F}_q$-operations $*$
Solve quadratic equations in $\mathbb{F}_q$	$O(\log(q)^4)$ bit operations $*$
Compute the minimal poly over $\mathbb{F}_q$ of $\alpha \in \mathbb{F}_{q^m}$	$O(m^3)$ $\mathbb{F}_q$-operations
Compute an isomorphism between repns of $\mathbb{F}_{q^m}$	$O(m^3)$ $\mathbb{F}_q$-operations
Compute order of $\alpha \in \mathbb{F}_q^*$ given factorisation of $q - 1$	$O(\log(q)\log(\log(q)))$ $\mathbb{F}_q$-multiplications
Compute primitive root in $\mathbb{F}_q^*$ given factorisation of $q - 1$	$O(\log(q)\log(\log(q)))$ $\mathbb{F}_q$-multiplications
Compute $f(\alpha_j) \in \Bbbk$ for $f \in \Bbbk[x]$ of degree n and $\alpha_1, \dots, \alpha_n \in \Bbbk$	$O(M(n)\log(n))$ $\Bbbk$-multiplications

One can show that the complexity is $O(M(n)\log(n))$ operations in $\Bbbk$. For details see Theorem 4 of [549], Section 10.1 of [220] or Corollary 4.5.4 of [83]. □

Exercise 2.16.2 Show that Theorem 2.16.1 also holds when the field $\Bbbk$ is replaced by a ring.

2.17 Pseudorandom generation

Many of the above algorithms, and also many cryptographic systems, require generation of random or pseudorandom numbers. The precise definitions for random and pseudorandom are out of the scope of this book, as is a full discussion of methods to extract almost perfect randomness from the environment and methods to generate pseudorandom sequences from a short random seed.

There are pseudorandom number generators related to RSA (the Blum-Blum-Shub generator) and discrete logarithms. Readers interested to learn more about this topic should consult Chapter 5 of [376], Chapter 3 of [308], Chapter 30 of [16], or [359].

2.18 Summary

Table 2.16 gives a brief summary of the complexities for the algorithms discussed in this chapter. The notation used in the table is $n \in \mathbb{N}$, $a, b \in \mathbb{Z}$, p is a prime, q is a prime power and $\Bbbk$ is a field. Recall that $M(m)$ is the number of bit operations to multiply two m-bit integers (which is also the number of operations in $\Bbbk$ to multiply two degree m polynomials over a field $\Bbbk$). Similarly, $M(d, q)$ is the number of bit operations to multiply two degree d polynomials in $\mathbb{F}_q[x]$.

Table 2.16 gives the asymptotic complexity for the algorithms that are used in cryptographic applications (i.e., for integers of, say, at most 10 000 bits). Many of the algorithms are randomised and so the complexity in those cases is the expected complexity. The reader is warned that the best possible asymptotic complexity may be different: sometimes it is sufficient to replace $M(m)$ by $m\log(m)\log(\log(m))$ to get the best complexity, but in other cases (such as constructing a polynomial basis for $\mathbb{F}_{q^m}$) there are totally different methods that have better asymptotic complexity. In cryptographic applications, $M(m)$ typically behaves as $M(m) = O(m^2)$ or $M(m) = O(m^{1.585})$.

The words "$\Bbbk$-operations" includes additions, multiplications and inversions in $\Bbbk$. If inversions in $\Bbbk$ are not required in the algorithm then we say "$\Bbbk$ multiplications".

3

Hash functions and MACs

Hash functions are an important tool in cryptography. In public key cryptography, they are used in key derivation functions, digital signatures and message authentication codes. We are unable to give a thorough presentation of hash functions. Instead, we refer to Chapter 4 of Katz and Lindell [300], Chapter 9 of Menezes, van Oorschot and Vanstone [376], Chapter 4 of Stinson [532] or Chapter 3 of Vaudenay [553].

3.1 Security properties of hash functions

Definition 3.1.1 A **cryptographic hash function** is a deterministic algorithm H that maps bitstrings of arbitrary finite length (we denote the set of arbitrary finite length bitstrings by $\{0, 1\}^*$) to bitstrings of a fixed length l (e.g., $l = 160$ or $l = 256$). A **cryptographic hash family** is a set of functions $\{H_k : k \in \mathbb{K}\}$, for some finite set $\mathbb{K}$, such that each function in the family is of the form $H_k : \{0, 1\}^* \to \{0, 1\}^l$.

The value k that specifies a hash function H_k from a hash family is called a key, but in many applications the key is not kept secret (an exception is message authentication codes). We now give an informal description of the typical security properties for hash functions.

1. **Preimage resistance**: Given an l-bit string y it should be computationally infeasible to compute a bitstring x such that $H(x) = y$.
2. **Second-preimage resistance**: Given a bitstring x and a bitstring $y = H(x)$ it should be computationally infeasible to compute a bitstring $x' \neq x$ such that $H(x') = y$.
3. **Collision resistance**: It should be computationally infeasible to compute bitstrings $x \neq x'$ such that $H(x) = H(x')$.

In general, one expects that for any $y \in \{0, 1\}^l$ there are infinitely many bitstrings x such that $H(x) = y$. Hence, all the above problems will have many solutions.

To obtain a meaningful definition for collision resistance it is necessary to consider hash families rather than hash functions. The problem is that an efficient adversary for collision resistance against a fixed hash function H is only required to output a pair $\{x, x'\}$ of messages. As long as such a collision exists then there exists an efficient algorithm that outputs one (namely, an algorithm that has the values x and x' hard-coded into it). Note that there is an important distinction here between the running time of the algorithm and the running time of the programmer (who is obliged to compute the collision as part of their task). A full discussion of this issue is given by Rogaway [450]; also see Section 4.6.1 of Katz and Lindell [300].

Intuitively, if one can compute preimages then one can compute second-preimages (though some care is needed here to be certain that the value x' output by a preimage oracle is not just x again; Note 9.20 of Menezes, van Oorschot and Vanstone [376] gives an artificial hash function that is second-preimage resistant but not preimage resistant). Similarly, if one can compute second-preimages then one can find collisions. Hence, in practice we prefer to study hash families that offer collision resistance. For more details about these relations see Section 4.6.2 of [300], Section 4.2 of [532] or Section 10.3 of [513].

Another requirement of hash families is that they be **entropy smoothing**: if G is a "sufficiently large" finite set (i.e., $\#G \gg 2^l$) with a "sufficiently nice" distribution on it (but not necessarily uniform) then the distribution on $\{0,1\}^l$ given by $\Pr(y) = \sum_{x \in G : H(x) = y} \Pr(x)$ is "close" to uniform. We do not make this notion precise, but refer to Section 6.9 of Shoup [497].

In Section 23.2 the following notion (which is just a restatement of second-preimage resistance) will be used.

Definition 3.1.2 Let X and Y be finite sets. A hash family $\{H_k : X \to Y : k \in \mathbb{K}\}$ is called **target collision resistant** if there is no polynomial-time adversary A with non-negligible advantage in the following game: A receives $x \in X$ and a key $k \in \mathbb{K}$, both chosen uniformly at random, then outputs an $x' \in X$ such that $x' \neq x$ and $H_k(x') = H_k(x)$.

For more details about target collision resistant hash families we refer to Section 5 of Cramer and Shoup [149].

3.2 Birthday attack

Computing preimages for a general hash function with l-bit output is expected to take approximately 2^l computations of the hash algorithm, but one can find collisions much more efficiently. Indeed, one can find collisions in roughly $\sqrt{\pi 2^{l-1}}$ applications of the hash function using a randomised algorithm as follows: choose a subset $\mathcal{D} \subset \{0,1\}^l$ of distinguished points (e.g., those whose κ least significant bits are all zero, where $0 < \kappa < l/4$). Choose random starting values $x_0 \in \{0,1\}^l$ (Joux [283] suggests that these should be distinguished points) and compute the sequence $x_n = H(x_{n-1})$ for $n = 1, 2, \ldots$ until $x_n \in \mathcal{D}$. Store (x_0, x_n) (i.e., the starting point and the ending distinguished point) and repeat. When the same distinguished point x is found twice then, assuming the starting points x_0 and x_0' are distinct, one can find a collision in the hash function by computing the full sequences x_i and x_j' and determining the smallest integers i and j such that $x_i = x_j'$ and hence the collision is $H(x_{i-1}) = H(x_{j-1}')$.

If we assume that values x_i are close to uniformly distributed in $\{0,1\}^l$ then, by the birthday paradox, one expects to have a collision after $\sqrt{\pi 2^l / 2}$ strings have been encountered (i.e., that many evaluations of the hash function). The storage required is expected to be

$$\sqrt{\pi 2^{l-1}}\frac{\#\mathcal{D}}{2^l}$$

pairs (x_0, x_n). For the choice of $\mathcal{D}$ as above this would be about $2^{l/2-\kappa}$ bitstrings of storage. For many more details on this topic see Section 7.5 of Joux [283], Section 9.7.1 of Menezes, van Oorschot and Vanstone [376] or Section 3.2 of Vaudenay [553].

This approach also works if one wants to find collisions under some constraint on the messages (e.g., all messages have a fixed prefix or suffix).

3.3 Message authentication codes

Message authentication codes are a form of symmetric cryptography. The main purpose is for a sender and receiver who share a secret key k to determine whether a communication between them has been tampered with.

A **message authentication code** (**MAC**) is a set of functions $\{\mathsf{MAC}_k(x) : k \in \mathbb{K}\}$ such that $\mathsf{MAC}_k : \{0, 1\}^* \to \{0, 1\}^l$. Note that this is exactly the same definition as a hash family. The difference between MACs and hash families lies in the security requirement; in particular the security model for MACs assumes the adversary does not know the key k. Informally, a MAC is secure against forgery if there is no efficient adversary that, given pairs $(x_i, y_i) \in \{0, 1\}^* \times \{0, 1\}^l$ such that $y_i = \mathsf{MAC}_k(x_i)$ (for some fixed, but secret, key k) for $1 \le i \le n$, can output a pair $(x, y) \in \{0, 1\}^* \times \{0, 1\}^l$ such that $y = \mathsf{MAC}_k(x)$ but $(x, y) \neq (x_i, y_i)$ for all $1 \le i \le n$. For precise definitions and further details of MACs see Section 4.3 of Katz and Lindell [300], Section 9.5 of Menezes, van Oorschot and Vanstone [376], Section 6.7.2 of Shoup [497], Section 4.4 of Stinson [532] or Section 3.4 of Vaudenay [553].

There are well-known constructions of MACs from hash functions (such as HMAC, see Section 4.7 of [300], Section 4.4.1 of [532] or Section 3.4.6 of [553]) and from block ciphers (such as CBC-MAC, see Section 4.5 of [300], Section 4.4.2 of [532] or Section 3.4.4 of [553]).

3.4 Constructions of hash functions

There is a large literature on constructions of hash functions and it is beyond the scope of this book to give the details. The basic process is to first define a **compression function** (namely, a function that takes bitstrings of a fixed length to bitstrings of some shorter fixed length) and then to build a hash function on arbitrary length bitstrings by iterating the compression function (e.g., using the Merkle–Damgård construction). We refer to Chapter 4 of Katz and Lindell [300], Sections 9.3 and 9.4 of Menezes, van Oorschot and Vanstone [376], Chapter 10 of Smart [513], Chapter 4 of Stinson [532] or Chapter 3 of Vaudenay [553] for the details.

3.5 Number-theoretic hash functions

We briefly mention some compression functions and hash functions that are based on algebraic groups and number theory. These schemes are not usually used in practice as the computational overhead is usually much too high.

An early proposal for hashing based on number theory, due to Davies and Price, was to use the function $H(x) = x^2 \pmod{N}$ where N is an RSA modulus whose factorisation is not known. Inverting such a function or finding collisions (apart from the trivial collisions $H(x) = H(\pm x + yN)$ for $y \in \mathbb{Z}$) is as hard as factoring N. There are a number of papers that build on this idea.

Another approach to hash functions based on factoring is to let N be an RSA modulus whose factorisation is unknown and let $g \in (\mathbb{Z}/N\mathbb{Z})^*$ be fixed. One can define the compression function $H : \mathbb{N} \to (\mathbb{Z}/N\mathbb{Z})^*$ by

$$H(x) = g^x \pmod{N}.$$

Finding a collision $H(x) = H(x')$ is equivalent to finding a multiple of the order of g. This can be shown to be hard if factoring is hard. Finding preimages is the discrete logarithm problem modulo N, which is also as hard as factoring. Hence, we have a collision resistant compression function. More generally, fix $g, h \in (\mathbb{Z}/N\mathbb{Z})^*$ and consider the compression function $H : \mathbb{N} \times \mathbb{N} \to (\mathbb{Z}/N\mathbb{Z})^*$ defined by $H(x, y) = g^x h^y \pmod{N}$. A collision leads to either finding the order of g or h, or essentially finding the discrete logarithm of h with respect to g, and all these problems are as hard as factoring.

One can also base hash functions on the discrete logarithm problem in any group G. Let $g, h \in G$ having order r. One can now consider the compression function $H : \{0, \dots, r-1\}^2 \to G$ by $H(x, y) = g^x h^y \pmod{p}$. It is necessary to fix the domain of the function since $H(x, y) = H(x + r, y) = H(x, y + r)$. If one can find collisions in this function then one can compute the discrete logarithm of h to the base g. A reference for this scheme is Chaum, van Heijst and Pfitzmann [121].

3.6 Full domain hash

Hash functions usually output binary strings of some fixed length l. Some cryptosystems, such as full domain hash RSA signatures, require hashing uniformly (or, at least, very close to uniformly) to $\mathbb{Z}/N\mathbb{Z}$, where N is large.

Bellare and Rogaway gave two methods to do this (one in Section 6 of [35] and another in Appendix A of [38]). We briefly recall the latter. The idea is to take some hash function H with fixed length output and define a new function $h(x)$ using a constant bitstring c and a counter i as

$$h(x) = H(c\|0\|x) \parallel H(c\|1\|x) \parallel \cdots \parallel H(c\|i\|x).$$

For the RSA application, one can construct a bitstring that is a small amount larger than N and then reduce the resulting integer modulo N (as in Example 11.4.2).

These approaches have been critically analysed by Leurent and Nguyen [347]. They give a number of results that demonstrate that care is needed in assessing the security level of a hash function with "full domain" output.

3.7 Random oracle model

The random oracle model is a tool for the security analysis of cryptographic systems. It is a computational model that includes the **standard model** (i.e., the computational model mentioned in Section 2.1) together with an oracle that computes a "random" function from the set $\{0, 1\}^*$ (i.e., binary strings of arbitrary finite length) to $\{0, 1\}^\infty$ (i.e., bitstrings of countably infinite length). Since the number of such functions is uncountable, care must be taken when defining the word "random". In any given application, one has a fixed bit-length l in mind for the output, and one also can bound the length of the inputs. Hence, one is considering functions $H : \{0, 1\}^n \to \{0, 1\}^l$ and, since there are $l2^n$ such functions, we can define "random" to mean uniformly chosen from the set of all possible functions. We stress that a random oracle is a function: if it is queried twice on the same input then the output is the same.

A **cryptosystem in the random oracle model** is a cryptosystem where one or more hash functions are replaced by oracle queries to the random function. A cryptosystem is **secure in the random oracle model** if the cryptosystem in the random oracle model is secure. This does not imply that the cryptosystem in the standard model is secure, since there may be an attack that exploits some feature of the hash function. Indeed, there are "artificial" cryptosystems that are proven secure in the random oracle model, but that are insecure for any instantiation of the hash function (see Canetti, Goldreich and Halevi [108]).

The random oracle model enables security proofs in several ways. We list three of these ways in increasing order of power.

1. It ensures that the output of H is truly random (rather than merely pseudorandom).
2. It allows the security proof to "look inside" the working of the adversary by learning the values that are inputs to the hash function.
3. It allows the security proof to "programme" the hash function so that it outputs a specific value at a crucial stage in the security game.

A classic example of a proof in the random oracle model is Theorem 20.4.11. An extensive discussion of the random oracle model is given in Section 13.1 of Katz and Lindell [300].

Since a general function from $\{0, 1\}^n$ to $\{0, 1\}^l$ cannot be represented more compactly than by a table of values, a random oracle requires $l2^n$ bits to describe. It follows that a random oracle cannot be implemented in polynomial space. However, the crucial observation that is used in security proofs is that a random oracle can be simulated in polynomial-time and space (assuming only polynomially many queries to the oracle are made) by creating, on-the-fly, a table giving the pairs (x, y) such that $H(x) = y$.

PART II
ALGEBRAIC GROUPS

4
Preliminary remarks on algebraic groups

For efficient public key cryptography based on discrete logarithms one would like to have groups for which computing g^n is as fast as possible, the representation of group elements is as small as possible and for which the DLP (see Definition 2.1.1 or 13.0.1) is (at least conjecturally) as hard as possible.

If g is a group element of order r then one needs at least $\log_2(r)$ bits to represent an arbitrary element of $\langle g \rangle$. This optimal size can be achieved by using the **exponent representation**, i.e. represent g^a as $a \in \mathbb{Z}/r\mathbb{Z}$. However, the DLP is not hard when this representation is used.

Ideally, for any cyclic group G of order r, one would like to be able to represent arbitrary group elements (in a manner which does not then render the DLP trivial) using roughly $\log_2(r)$ bits. This can be done in some cases (e.g., elliptic curves over finite fields with a prime number of points), but it is unlikely that it can always be done. Using algebraic groups over finite fields is a good way to achieve these conflicting objectives.

4.1 Informal definition of an algebraic group

The subject of algebraic groups is large and has an extensive literature. Instead of presenting the full theory, in this book we present only the algebraic groups that are currently believed to be useful in public key cryptography. Informally,[1] an **algebraic group** over a field $\Bbbk$ is a group such that:

- Group elements are specified as n-tuples of elements in a field $\Bbbk$.
- The group operations (multiplication and inversion) can be performed using only polynomial equations (or ratios of polynomials) defined over $\Bbbk$. In other words, we have polynomial or rational maps mult : $\Bbbk^{2n} \to \Bbbk^n$ and inverse : $\Bbbk^n \to \Bbbk^n$. There is not necessarily a single n-tuple of polynomial equations that defines mult for all possible pairs of group elements.

An algebraic group quotient is the set of equivalence classes of an algebraic group under some equivalence relation (see Section 4.3 for an example). Note that, in general, an algebraic group quotient is not a group.

[1] We refrain from giving a formal definition of algebraic groups; mainly as it requires defining products of projective varieties.

We stress that being an algebraic group is not a group-theoretic property, it is a property of a particular description of the group. Perhaps it helps to give an example of a group whose usual representation is not an algebraic group.

Example 4.1.1 Let $n \in \mathbb{N}$ and let S_n be the group of permutations on n symbols. Permutations can be represented as an n-tuple of distinct integers from the set $\{1, 2, \ldots, n\}$. The composition $(x_1, \ldots, x_n) \circ (y_1, \ldots, y_n)$ of two permutations is $(x_{y_1}, x_{y_2}, \ldots, x_{y_n})$. Since x_{y_1} is not a polynomial, the usual representation of S_n is not an algebraic group. However, S_n can be represented as a subgroup of the matrix group $GL_n(\Bbbk)$ (for any field $\Bbbk$), which is an algebraic group.

Our main interest is algebraic groups over finite fields $\mathbb{F}_q$. For each example of an algebraic group (or quotient) G we will explain how to achieve the following basic functionalities:

- efficient group operations in G (typically requiring $O(\log(q)^2)$ bit operations);
- compact representation of elements of G (typically $O(\log(q))$ bits);
- generating cryptographically suitable G in polynomial-time (i.e., $O(\log(q)^c)$ for some (small) $c \in \mathbb{N}$);
- generating random elements in G in polynomial-time;
- hashing from $\{0, 1\}^l$ to G or from G to $\{0, 1\}^l$ in polynomial-time.

In order to be able to use an algebraic group (or quotient) G for cryptographic applications, we need some or all of these functionalities, as well as requiring the discrete logarithm problem (and possibly other computational problems) to be hard.

We sometimes use the notation AG to mean "algebraic group in the context of this book"; similarly, AGQ means "algebraic group quotient in the context of this book". The aim of this part of the book is to describe the algebraic groups of relevance for public key cryptography (namely, multiplicative groups, algebraic tori, elliptic curves and divisor class groups). As is traditional, we will use multiplicative notation for the group operation in multiplicative groups and tori, and additive notation for the group operation on elliptic curves and divisor class groups of hyperelliptic curves. In Parts III and V, when we discuss cryptographic applications, we will always use multiplicative notation for algebraic groups.

The purpose of this chapter is to give the simplest examples of algebraic groups and quotients. The later chapters introduce enough algebraic geometry to be able to define the algebraic groups of interest in this book and prove some important facts about them.

4.2 Examples of algebraic groups

The simplest examples of algebraic groups are the **additive group** G_a and **multiplicative group** G_m of a field $\Bbbk$. For $G_a(\Bbbk)$ the set of points is $\Bbbk$ and the group operation is given by the polynomial $\text{mult}(x, y) = x + y$ (for computing the group operation) and $\text{inverse}(x) = -x$ (for computing inverses). For $G_m(\Bbbk)$ the set of points is $\Bbbk^* = \Bbbk - \{0\}$ and the group operation is given by the polynomial $\text{mult}(x, y) = xy$ and the rational function $\text{inverse}(x) = 1/x$ (Example 5.1.4 shows how to express $G_m(\Bbbk)$ as an algebraic set).

The additive group is useless for cryptography since the discrete logarithm problem is easy. The discrete logarithm problem is also easy for the multiplicative group over certain fields (e.g., if $g \in \mathbb{R}^*$ then the discrete logarithm problem in $\langle g \rangle \subseteq \mathbb{R}^*$ is easy due to algorithms that compute approximations to the natural logarithm function). However, $G_m(\mathbb{F}_q)$ is useful for cryptography and will be one of the main examples used in this book.

The other main examples of algebraic groups in public key cryptography are algebraic tori (see Chapter 6), elliptic curves and divisor class groups of hyperelliptic curves.

4.3 Algebraic group quotients

Quotients of algebraic groups are used to reduce the storage and communication requirements of public key cryptosystems. Let G be a group with an automorphism ψ such that $\psi^n = 1$ (where $1 : G \to G$ is the identity map and ψ^n is the n-fold composition $\psi \circ \cdots \circ \psi$). We define $\psi^0 = 1$. Define the **orbit** or **equivalence class** of $g \in G$ under ψ to be $[g] = \{\psi^i(g) : 0 \leq i < n\}$. Define the **quotient** as the set of orbits under ψ. In other words

$$G/\psi = \{[g] : g \in G\}.$$

We call G the **covering group** of a quotient G/ψ. In general, the group structure of G does not induce a group structure on the quotient G/ψ. Nevertheless, we can define exponentiation on the quotient by $[g]^n = [g^n]$ for $n \in \mathbb{Z}$. Since exponentiation is the fundamental operation for many cryptographic applications it follows that quotients of algebraic groups are sufficient for many cryptographic applications.

Lemma 4.3.1 *Let $n \in \mathbb{Z}$ and $[g] \in G/\psi$, then $[g]^n$ is well-defined.*

Proof Since ψ is a group homomorphism we have $\psi^i(g)^n = \psi^i(g^n)$ and so for each $g_1 \in [g]$ we have $g_1^n \in [g^n]$. □

The advantage of algebraic group quotients G/ψ is that they can require less storage than the original algebraic group G. We now give an example of this.

Example 4.3.2 Let p be an odd prime. Consider the subgroup $G \subset \mathbb{F}_{p^2}^*$ of order $p + 1$. Note that $\gcd(p - 1, p + 1) = 2$ so $G \cap \mathbb{F}_p^* = \{1, -1\}$. If $g \in G$ then we have $g^{p+1} = 1$, which is equivalent to $g^p = g^{-1}$. Let ψ be the automorphism $\psi(g) = g^p$. Then $\psi^2 = 1$ in $\mathbb{F}_{p^2}$ and the orbits $[g]$ in G/ψ all have size 2 except for $[1]$ and $[-1]$.

The natural representation for elements of $G \subseteq \mathbb{F}_{p^2}$ is a pair of elements of $\mathbb{F}_p$. However, since $\#(G/\psi) = 2 + (p - 1)/2$ one might expect to be able to represent elements of G/ψ using just one element of $\mathbb{F}_p$.

Let $g \in G$. Then the elements of $[g] = \{g, g^p\}$ are the roots of the equation $x^2 - tx + 1$ in $\mathbb{F}_{p^2}$ where $t = g + g^p \in \mathbb{F}_p$. Conversely, each $t \in \mathbb{F}_p$ such that the roots of $x^2 - tx + 1$ are Galois conjugates corresponds to a class $[g]$ (the values $t = \pm 2$ correspond to $[1]$ and $[-1]$). Hence, one can represent an element of G/ψ by the trace t. We therefore require half the storage compared with the standard representation of $G \subset \mathbb{F}_{p^2}$.

In Section 6.3.2 we show that, given the trace t of g, one can compute the trace t_n of g^n efficiently using Lucas sequences (though there is a slight catch, namely that we have to work with a pair (t_n, t_{n-1}) of traces).

Another important example of an algebraic group quotient is elliptic curve arithmetic using x-coordinates only. This is the quotient of an elliptic curve by the equivalence relation $P \equiv -P$.

4.4 Algebraic groups over rings

Algebraic geometry is traditionally studied over fields. However, several applications (both algorithmic and cryptographic) will exploit algebraic groups or algebraic group quotients over $\mathbb{Z}/N\mathbb{Z}$ (we do not consider general rings).

Let $N = \prod_{i=1}^k p_i$ be square-free (the non-square-free case is often more subtle). By the Chinese remainder theorem, $\mathbb{Z}/N\mathbb{Z}$ is isomorphic as a ring to $\oplus_{i=1}^k \mathbb{F}_{p_i}$ (where $\oplus$ denotes the direct sum of rings). Hence, if G is an algebraic group then it is natural to define

$$G(\mathbb{Z}/N\mathbb{Z}) = \bigoplus_{i=1}^{k} G(\mathbb{F}_{p_i}) \tag{4.1}$$

(where $\oplus$ now denotes the direct sum of groups). A problem is that this representation for $G(\mathbb{Z}/N\mathbb{Z})$ does not satisfy the natural generalisation to rings of our informal definition of an algebraic group. For example, group elements are not n-tuples over the ring, but over a collection of different fields. Also the value n is no longer bounded.

The challenge is to find a representation for $G(\mathbb{Z}/N\mathbb{Z})$ that uses n-tuples over $\mathbb{Z}/N\mathbb{Z}$ and satisfies the other properties of the informal definition. Example 4.4.1 shows that this holds for the additive and multiplicative groups.

Example 4.4.1 Let $N = \prod_i p_i$ where the p_i are distinct primes. Then, using the definition in equation (4.1),

$$G_a(\mathbb{Z}/N\mathbb{Z}) \cong \bigoplus_i G_a(\mathbb{F}_{p_i}) \cong \bigoplus_i \mathbb{F}_{p_i} \cong \mathbb{Z}/N\mathbb{Z}.$$

Similarly

$$G_m(\mathbb{Z}/N\mathbb{Z}) \cong \bigoplus_i G_m(\mathbb{F}_{p_i}) \cong \bigoplus_i \mathbb{F}_{p_i}^* \cong (\mathbb{Z}/N\mathbb{Z})^*.$$

Hence, both groups can naturally be considered as algebraic groups over $\mathbb{Z}/N\mathbb{Z}$.

Note that $G_m(\mathbb{Z}/N\mathbb{Z})$ is not cyclic when N is square-free but not prime.

To deal with non-square-free N it is necessary to define $G(\mathbb{Z}/p^n\mathbb{Z})$. The details of this depend on the algebraic group. For G_a and G_m it is straightforward and we still have $G_a(\mathbb{Z}/N\mathbb{Z}) = \mathbb{Z}/N\mathbb{Z}$ and $G_m(\mathbb{Z}/N\mathbb{Z}) = (\mathbb{Z}/N\mathbb{Z})^*$. For other groups it can be more complicated.

5

Varieties

The purpose of this chapter is to state some basic definitions and results from algebraic geometry that are required for the main part of the book. In particular, we define algebraic sets, irreducibility, function fields, rational maps and dimension. The chapter is not intended as a self-contained introduction to algebraic geometry. Many proofs are omitted. We make the following recommendations to the reader:

1. Readers who want a very elementary introduction to elliptic curves are advised to consult one or more of Koblitz [313], Silverman and Tate [508], Washington [560], Smart [513] or Stinson [532].
2. Readers who wish to learn algebraic geometry properly should first read a basic text such as Reid [447] or Fulton [199]. They can then skim this chapter and consult Stichtenoth [529], Moreno [395], Hartshorne [252], Lorenzini [355] or Shafarevich [489] for detailed proofs and discussion.

5.1 Affine algebraic sets

Let $\mathbb{k}$ be a perfect field contained in a fixed algebraic closure $\overline{\mathbb{k}}$. All algebraic extensions $\mathbb{k}'/\mathbb{k}$ are implicitly assumed to be subfields of $\overline{\mathbb{k}}$. We use the notation $\mathbb{k}[\underline{x}] = \mathbb{k}[x_1, \ldots, x_n]$ (in later sections we also use $\mathbb{k}[\underline{x}] = \mathbb{k}[x_0, \ldots, x_n]$). When $n = 2$ or 3 we often write $\mathbb{k}[x, y]$ or $\mathbb{k}[x, y, z]$.

Define **affine n-space over** $\mathbb{k}$ as $\mathbb{A}^n(\mathbb{k}) = \mathbb{k}^n$. We call $\mathbb{A}^1(\mathbb{k})$ the **affine line** and $\mathbb{A}^2(\mathbb{k})$ the **affine plane** over $\mathbb{k}$. If $\mathbb{k} \subseteq \mathbb{k}'$ then we have the natural inclusion $\mathbb{A}^n(\mathbb{k}) \subseteq \mathbb{A}^n(\mathbb{k}')$. We write $\mathbb{A}^n$ for $\mathbb{A}^n(\overline{\mathbb{k}})$ and so $\mathbb{A}^n(\mathbb{k}) \subseteq \mathbb{A}^n$.

Definition 5.1.1 Let $S \subseteq \mathbb{k}[\underline{x}]$. Define

$$V(S) = \{P \in \mathbb{A}^n(\overline{\mathbb{k}}) : f(P) = 0 \text{ for all } f \in S\}.$$

If $S = \{f_1, \ldots, f_m\}$ then we write $V(f_1, \ldots, f_m)$ for $V(S)$. An **affine algebraic set** is a set $X = V(S) \subset \mathbb{A}^n$ where $S \subset \mathbb{k}[\underline{x}]$.

Let $\Bbbk'/\Bbbk$ be an algebraic extension. The $\Bbbk'$-**rational points** of $X = V(S)$ are

$$X(\Bbbk') = X \cap \mathbb{A}^n(\Bbbk') = \{P \in \mathbb{A}^n(\Bbbk') : f(P) = 0 \text{ for all } f \in S\}.$$

An algebraic set $V(f)$, where $f \in \Bbbk[\underline{x}]$, is a **hypersurface**. If $f(\underline{x})$ is a polynomial of total degree 1 then $V(f)$ is a **hyperplane**.

Informally we often write "the algebraic set $f = 0$" instead of $V(f)$. For example, $y^2 = x^3$ instead of $V(y^2 - x^3)$. We stress that, as is standard, $V(S)$ is the set of solutions over an algebraically closed field.

When an algebraic set is defined as the vanishing of a set of polynomials with coefficients in $\Bbbk$ then it is called a $\Bbbk$-algebraic set. The phrase "defined over $\Bbbk$" has a different meaning and the relation between them will be explained in Remark 5.3.7.

Example 5.1.2 If $X = V(x_1^2 + x_2^2 + 1) \subseteq \mathbb{A}^2$ over $\mathbb{Q}$ then $X(\mathbb{Q}) = \varnothing$. Let $\Bbbk = \mathbb{F}_2$ and let $X = V(y^8 + x^6 y + x^3 + 1) \subseteq \mathbb{A}^2$. Then $X(\mathbb{F}_2) = \{(0, 1), (1, 0), (1, 1)\}$.

Exercise 5.1.3 Let $\Bbbk$ be a field. Show that $\{(t, t^2) : t \in \overline{\Bbbk}\} \subseteq \mathbb{A}^2$, $\{(t, \pm\sqrt{t}) : t \in \overline{\Bbbk}\} \subseteq \mathbb{A}^2$ and $\{(t^2 + 1, t^3) : t \in \overline{\Bbbk}\} \subseteq \mathbb{A}^2$ are affine algebraic sets.

Example 5.1.4 Let $\Bbbk$ be a field. There is a one-to-one correspondence between the set $\Bbbk^*$ and the $\Bbbk$-rational points $X(\Bbbk)$ of the affine algebraic set $X = V(xy - 1) \subset \mathbb{A}^2$. Multiplication in the field $\Bbbk$ corresponds to the function $\text{mult} : X \times X \to X$ given by $\text{mult}((x_1, y_1), (x_2, y_2)) = (x_1x_2, y_1y_2)$. Similarly, inversion in $\Bbbk^*$ corresponds to the function $\text{inverse}(x, y) = (y, x)$. Hence, we have represented $\Bbbk^*$ as an algebraic group, which we call $G_m(\Bbbk)$.

Example 5.1.5 Another elementary example of an algebraic group is the affine algebraic set $X = V(x^2 + y^2 - 1) \subset \mathbb{A}^2$ with the group operation $\text{mult}((x_1, y_1), (x_2, y_2)) = (x_1x_2 - y_1y_2, x_1y_2 + x_2y_1)$. (These formulae are analogous to the angle addition rules for sine and cosine as, over $\mathbb{R}$, one can identify (x, y) with $(\cos(\theta), \sin(\theta))$.) The reader should verify that the image of mult is contained in X. The identity element is $(1, 0)$ and the inverse of (x, y) is $(x, -y)$. One can verify that the axioms of a group are satisfied. This group is sometimes called the **circle group**.

Exercise 5.1.6 Let $p \equiv 3 \pmod 4$ be prime and define $\mathbb{F}_{p^2} = \mathbb{F}_p(i)$ where $i^2 = -1$. Show that the group $X(\mathbb{F}_p)$, where X is the circle group from Example 5.1.5, is isomorphic as a group to the subgroup $G \subseteq \mathbb{F}_{p^2}^*$ of order $p + 1$.

Proposition 5.1.7 *Let $S \subseteq \Bbbk[x_1, \ldots, x_n]$.*

1. *$V(S) = V((S))$ where (S) is the $\Bbbk[\underline{x}]$-ideal generated by S.*
2. *$V(\Bbbk[\underline{x}]) = \varnothing$ and $V(\{0\}) = \mathbb{A}^n$ where $\varnothing$ denotes the empty set.*
3. *If $S_1 \subseteq S_2$ then $V(S_2) \subseteq V(S_1)$.*

4. $V(fg) = V(f) \cup V(g)$.
5. $V(f) \cap V(g) = V(f, g)$.

Exercise 5.1.8 Prove Proposition 5.1.7.

The following result assumes a knowledge of Galois theory. See Section A.7 for background.

Lemma 5.1.9 *Let $X = V(S)$ be an algebraic set with $S \subseteq \Bbbk[\underline{x}]$ (i.e., X is a $\Bbbk$-algebraic set). Let $\Bbbk'$ be an algebraic extension of $\Bbbk$. Let $\sigma \in \mathrm{Gal}(\overline{\Bbbk}/\Bbbk')$. For $P = (P_1, \ldots, P_n)$ define $\sigma(P) = (\sigma(P_1), \ldots, \sigma(P_n))$.*

1. *If $P \in X(\overline{\Bbbk})$ then $\sigma(P) \in X(\overline{\Bbbk})$.*
2. *$X(\Bbbk') = \{P \in X(\overline{\Bbbk}) : \sigma(P) = P$ for all $\sigma \in \mathrm{Gal}(\overline{\Bbbk}/\Bbbk')\}$.*

Exercise 5.1.10 Prove Lemma 5.1.9.

Definition 5.1.11 The **ideal** over $\Bbbk$ of a set $X \subseteq \mathbb{A}^n(\overline{\Bbbk})$ is

$$I_{\Bbbk}(X) = \{f \in \Bbbk[\underline{x}] : f(P) = 0 \text{ for all } P \in X(\overline{\Bbbk})\}.$$

We define $I(X) = I_{\overline{\Bbbk}}(X)$.[1]

An algebraic set X is **defined over** $\Bbbk$ (sometimes abbreviated to "X over $\Bbbk$") if $I(X)$ can be generated by elements of $\Bbbk[\underline{x}]$.

Perhaps surprisingly, it is not necessarily true that an algebraic set described by polynomials defined over $\Bbbk$ is an algebraic set defined over $\Bbbk$. In Remark 5.3.7 we will explain that these concepts are equivalent for the objects of interest in this book.

Exercise 5.1.12 Show that $I_{\Bbbk}(X) = I(X) \cap \Bbbk[\underline{x}]$.

Proposition 5.1.13 *Let $X, Y \subseteq \mathbb{A}^n$ be sets and J a $\Bbbk[\underline{x}]$-ideal. Then:*

1. *$I_{\Bbbk}(X)$ is a $\Bbbk[\underline{x}]$-ideal.*
2. *$X \subseteq V(I_{\Bbbk}(X))$.*
3. *If $X \subseteq Y$ then $I_{\Bbbk}(Y) \subseteq I_{\Bbbk}(X)$.*
4. *$I_{\Bbbk}(X \cup Y) = I_{\Bbbk}(X) \cap I_{\Bbbk}(Y)$.*
5. *If X is an algebraic set defined over $\Bbbk$ then $V(I_{\Bbbk}(X)) = X$.*
6. *If X and Y are algebraic sets defined over $\Bbbk$ and $I_{\Bbbk}(X) = I_{\Bbbk}(Y)$ then $X = Y$.*
7. *$J \subseteq I_{\Bbbk}(V(J))$.*
8. *$I_{\Bbbk}(\emptyset) = \Bbbk[\underline{x}]$.*

Exercise 5.1.14 Prove Proposition 5.1.13.

[1] The notation $I_{\Bbbk}(X)$ is not standard (Silverman [505] calls it $I(X/\Bbbk)$), but the notation $I(X)$ agrees with many elementary books on algebraic geometry, since they work over an algebraically closed field.

Definition 5.1.15 The **affine coordinate ring** over $\Bbbk$ of an affine algebraic set $X \subseteq \mathbb{A}^n$ defined over $\Bbbk$ is

$$\Bbbk[X] = \Bbbk[x_1, \dots, x_n]/I_{\Bbbk}(X).$$

Warning: Here $\Bbbk[X]$ does not denote polynomials in the variable X. Hartshorne and Fulton write $A(X)$ and $\Gamma(X)$ respectively for the affine coordinate ring.

Exercise 5.1.16 Prove that $\Bbbk[X]$ is a commutative ring with an identity.

Note that $\Bbbk[X]$ is isomorphic to the ring of all functions $f : X \to \Bbbk$ given by polynomials defined over $\Bbbk$.

Hilbert's Nullstellensatz is a powerful tool for understanding $I_{\overline{\Bbbk}}(X)$ and it has several other applications (e.g., we use it in Section 7.5). We follow the presentation of Fulton [199]. Note that it is necessary to work over $\overline{\Bbbk}$.

Theorem 5.1.17 *(Weak Nullstellensatz) Let $X \subseteq \mathbb{A}^n$ be an affine algebraic set defined over $\overline{\Bbbk}$ and let $\mathfrak{m}$ be a maximal ideal of the affine coordinate ring $\overline{\Bbbk}[X]$. Then $V(\mathfrak{m}) = \{P\}$ for some $P = (P_1, \dots, P_n) \in X(\overline{\Bbbk})$ and $\mathfrak{m} = (x_1 - P_1, \dots, x_n - P_n)$.*

Proof See Section 1.7 of Fulton [199]. □

Corollary 5.1.18 *If I is a proper ideal in $\overline{\Bbbk}[x_1, \dots, x_n]$ then $V(I) \neq \emptyset$.*

We can now state the **Hilbert Nullstellensatz**. This form of the theorem (which applies to $I_{\Bbbk}(V(I))$ where $\Bbbk$ is not necessarily algebraically closed, appears as Proposition VIII.7.4 of [271].

Theorem 5.1.19 *Let I be an ideal in $R = \Bbbk[x_1, \dots, x_n]$. Then $I_{\Bbbk}(V(I)) = \mathrm{rad}_R(I)$ (see Section A.9 for the definition of the radical ideal).*

Proof See Section 1.7 of Fulton [199]. □

Corollary 5.1.20 *Let $f(x, y) \in \Bbbk[x, y]$ be irreducible over $\overline{\Bbbk}$ and let $X = V(f(x, y)) \subset \mathbb{A}^2(\overline{\Bbbk})$. Then $I(X) = (f(x, y))$, i.e. the ideal over $\overline{\Bbbk}[x, y]$ generated by $f(x, y)$.*

Proof By Theorem 5.1.19 we have $I(X) = \mathrm{rad}_{\overline{\Bbbk}}((f(x, y)))$. Since $\overline{\Bbbk}[x, y]$ is a unique factorisation domain and $f(x, y)$ is irreducible, then $f(x, y)$ is prime. So $g(x, y) \in \mathrm{rad}_{\overline{\Bbbk}}((f(x, y)))$ implies $g(x, y)^n = f(x, y)h(x, y)$ for some $h(x, y) \in \overline{\Bbbk}[x, y]$, which implies $f(x, y) \mid g(x, y)$ and $g(x, y) \in (f(x, y))$. □

5.2 Projective algebraic sets

Studying affine algebraic sets is not sufficient for our applications. In particular, the set of affine points of the Weierstrass equation of an elliptic curve (see Section 7.2) does not form a group. Projective geometry is a way to "complete" the picture by adding certain "points at infinity".

For example, consider the hyperbola $xy = 1$ in $\mathbb{A}^2(\mathbb{R})$. Projective geometry allows an interpretation of the behaviour of the curve at $x = 0$ or $y = 0$; see Example 5.2.7.

Definition 5.2.1 **Projective space** over $\Bbbk$ of dimension n is

$$\mathbb{P}^n(\Bbbk) = \{\text{lines through } (0, \dots, 0) \text{ in } \mathbb{A}^{n+1}(\Bbbk)\}.$$

A convenient way to represent points of $\mathbb{P}^n(\Bbbk)$ is using **homogeneous coordinates**. Let $a_0, a_1, \dots, a_n \in \Bbbk$ with not all $a_j = 0$ and define $(a_0 : a_1 : \cdots : a_n)$ to be the equivalence class of $(n+1)$-tuples under the equivalence relation

$$(a_0, a_1, \cdots, a_n) \equiv (\lambda a_0, \lambda a_1, \cdots, \lambda a_n)$$

for any $\lambda \in \Bbbk^*$. Thus $\mathbb{P}^n(\Bbbk) = \{(a_0 : \cdots : a_n) : a_i \in \Bbbk \text{ for } 0 \le i \le n \text{ and } a_i \ne 0 \text{ for some } 0 \le i \le n\}$. Write $\mathbb{P}^n = \mathbb{P}^n(\overline{\Bbbk})$.

In other words, the equivalence class $(a_0 : \cdots : a_n)$ is the set of points on the line between $(0, \dots, 0)$ and $(a_0, \dots, a_n)$ with the point $(0, \dots, 0)$ removed.

There is a map $\varphi : \mathbb{A}^n \to \mathbb{P}^n$ given by $\varphi(x_1, \dots, x_n) = (x_1 : \cdots : x_n : 1)$. Hence, $\mathbb{A}^n$ is identified with a subset of $\mathbb{P}^n$.

Example 5.2.2 The **projective line** $\mathbb{P}^1(\Bbbk)$ is in one-to-one correspondence with $\mathbb{A}^1(\Bbbk) \cup \{\infty\}$ since $\mathbb{P}^1(\Bbbk) = \{(a_0 : 1) : a_0 \in \Bbbk\} \cup \{(1 : 0)\}$. The **projective plane** $\mathbb{P}^2(\Bbbk)$ is in one-to-one correspondence with $\mathbb{A}^2(\Bbbk) \cup \mathbb{P}^1(\Bbbk)$.

Definition 5.2.3 A point $P = (P_0 : P_1 : \cdots : P_n) \in \mathbb{P}^n(\overline{\Bbbk})$ is **defined over** $\Bbbk$ if there is some $\lambda \in \overline{\Bbbk}^*$ such that $\lambda P_j \in \Bbbk$ for all $0 \le j \le n$. If $P \in \mathbb{P}^n$ and $\sigma \in \mathrm{Gal}(\overline{\Bbbk}/\Bbbk)$ then $\sigma(P) = (\sigma(P_0) : \cdots : \sigma(P_n))$.

Exercise 5.2.4 Show that P is defined over $\Bbbk$ if and only if there is some $0 \le i \le n$ such that $P_i \ne 0$ and $P_j / P_i \in \Bbbk$ for all $0 \le j \le n$. Show that $\mathbb{P}^n(\Bbbk)$ is equal to the set of points $P \in \mathbb{P}^n(\overline{\Bbbk})$ that are defined over $\Bbbk$. Show that $\sigma(P)$ in Definition 5.2.3 is well-defined (i.e., if $P = (P_0, \dots, P_n) \equiv P' = (P'_0, \dots, P'_n)$ then $\sigma(P) \equiv \sigma(P')$).

Lemma 5.2.5 *A point $P \in \mathbb{P}^n(\overline{\Bbbk})$ is defined over $\Bbbk$ if and only if $\sigma(P) = P$ for all $\sigma \in \mathrm{Gal}(\overline{\Bbbk}/\Bbbk)$.*

Proof Let $P = (P_0 : \cdots : P_n) \in \mathbb{P}^n(\overline{\Bbbk})$ and suppose $\sigma(P) \equiv P$ for all $\sigma \in \mathrm{Gal}(\overline{\Bbbk}/\Bbbk)$. Then there is some $\xi : \mathrm{Gal}(\overline{\Bbbk}/\Bbbk) \to \overline{\Bbbk}^*$ such that $\sigma(P_i) = \xi(\sigma) P_i$ for all $0 \le i \le n$. One can verify[2] that ξ is a 1-cocycle in $\overline{\Bbbk}^*$. It follows by Theorem A.7.2 (Hilbert 90) that $\xi(\sigma) = \sigma(\gamma)/\gamma$ for some $\gamma \in \overline{\Bbbk}^*$. Hence, $\sigma(P_i/\gamma) = P_i/\gamma$ for all $0 \le i \le n$ and all $\sigma \in \mathrm{Gal}(\overline{\Bbbk}/\Bbbk)$. Hence, $P_i/\gamma \in \Bbbk$ for all $0 \le i \le n$ and the proof is complete. □

Recall that if f is a homogeneous polynomial of degree d then $f(\lambda x_0, \dots, \lambda x_n) = \lambda^d f(x_0, \dots, x_n)$ for all $\lambda \in \Bbbk$ and all $(x_0, \dots, x_n) \in \mathbb{A}^{n+1}(\Bbbk)$.

[2] At least, one can verify the formula $\xi(\sigma\tau) = \sigma(\xi(\tau))\xi(\sigma)$. The topological condition also holds, but we do not discuss this.

Definition 5.2.6 Let $f \in \Bbbk[x_0, \ldots, x_n]$ be a homogeneous polynomial. A point $P = (x_0 : \cdots : x_n) \in \mathbb{P}^n(\Bbbk)$ is a **zero** of f if $f(x_0, \ldots, x_n) = 0$ for some (hence, every) point $(x_0, \ldots, x_n)$ in the equivalence class $(x_0 : \cdots : x_n)$. We therefore write $f(P) = 0$. Let S be a set of polynomials and define

$$V(S) = \{P \in \mathbb{P}^n(\overline{\Bbbk}) : P \text{ is a zero of } f(\underline{x}) \text{ for all homogeneous } f(\underline{x}) \in S\}.$$

A **projective algebraic set** is a set $X = V(S) \subseteq \mathbb{P}^n(\overline{\Bbbk})$ for some $S \subseteq \Bbbk[\underline{x}]$. Such a set is also called a $\Bbbk$-algebraic set. For $X = V(S)$ and $\Bbbk'$ an algebraic extension of $\Bbbk$ define

$$X(\Bbbk') = \{P \in \mathbb{P}^n(\Bbbk') : f(P) = 0 \text{ for all homogeneous } f(\underline{x}) \in S\}.$$

Example 5.2.7 The hyperbola $y = 1/x$ can be described as the affine algebraic set $X = V(xy - 1) \subset \mathbb{A}^2$ over $\mathbb{R}$. One can consider the corresponding projective algebraic set $V(xy - z^2) \subseteq \mathbb{P}^2$ over $\mathbb{R}$ whose points consist of the points of X together with the points $(1 : 0 : 0)$ and $(0 : 1 : 0)$. These two points correspond to the asymptotes $x = 0$ and $y = 0$ of the hyperbola and they essentially "tie together" the disconnected components of the affine curve to make a single closed curve in projective space.

Exercise 5.2.8 Describe the sets $V(x^2 + y^2 - z^2)(\mathbb{R}) \subset \mathbb{P}^2(\mathbb{R})$ and $V(yz - x^2)(\mathbb{R}) \subseteq \mathbb{P}^2(\mathbb{R})$.

A set of homogeneous polynomials does not in general form an ideal as one cannot simultaneously have closure under multiplication and addition. Hence, it is necessary to introduce the following definition.

Definition 5.2.9 A $\Bbbk[x_0, \ldots, x_n]$-ideal $I \subseteq \Bbbk[x_0, \ldots, x_n]$ is a **homogeneous ideal** if for every $f \in I$ with homogeneous decomposition $f = \sum_i f_i$ we have $f_i \in I$.

Exercise 5.2.10 Let $S \subset \Bbbk[\underline{x}]$ be a set of homogeneous polynomials. Define (S) to be the $\Bbbk[\underline{x}]$-ideal generated by S in the usual way, i.e. $(S) = \{\sum_{j=1}^{n} f_j(\underline{x}) s_j(\underline{x}) : n \in \mathbb{N}, f_j(\underline{x}) \in \Bbbk[x_0, \ldots, x_n], s_j(\underline{x}) \in S\}$. Prove that (S) is a homogeneous ideal. Prove that if I is a homogeneous ideal then $I = (S)$ for some set of homogeneous polynomials S.

Definition 5.2.11 For any set $X \subseteq \mathbb{P}^n(\overline{\Bbbk})$ define

$$I_{\Bbbk}(X) = \big(\{f \in \Bbbk[x_0, \ldots, x_n] : f \text{ is homogeneous and } f(P) = 0 \text{ for all } P \in X\}\big).$$

We stress that $I_{\Bbbk}(X)$ is not the stated set of homogeneous polynomials but the ideal generated by them. We write $I(X) = I_{\overline{\Bbbk}}(X)$.

An algebraic set $X \subseteq \mathbb{P}^n$ is **defined over** $\Bbbk$ if $I(X)$ can be generated by homogeneous polynomials in $\Bbbk[\underline{x}]$.

Proposition 5.2.12 *Let $\Bbbk$ be a field:*

1. *If $S_1 \subseteq S_2 \subseteq \Bbbk[x_0, \ldots, x_n]$ then $V(S_2) \subseteq V(S_1) \subseteq \mathbb{P}^n(\overline{\Bbbk})$.*
2. *If fg is a homogeneous polynomial then $V(fg) = V(f) \cup V(g)$ (recall from Lemma A.5.4 that f and g are both homogeneous).*

3. *$V(f) \cap V(g) = V(f, g)$.*
4. *If $X_1 \subseteq X_2 \subseteq \mathbb{P}^n(\Bbbk)$ then $I_{\Bbbk}(X_2) \subseteq I_{\Bbbk}(X_1) \subseteq \Bbbk[x_0, \dots, x_n]$.*
5. *$I_{\Bbbk}(X_1 \cup X_2) = I_{\Bbbk}(X_1) \cap I_{\Bbbk}(X_2)$.*
6. *If J is a homogeneous ideal then $J \subseteq I_{\Bbbk}(V(J))$.*
7. *If X is a projective algebraic set defined over $\Bbbk$ then $V(I_{\Bbbk}(X)) = X$. If Y is another projective algebraic set defined over $\Bbbk$ and $I_{\Bbbk}(Y) = I_{\Bbbk}(X)$ then $Y = X$.*

Exercise 5.2.13 Prove Proposition 5.2.12.

Definition 5.2.14 If X is a projective algebraic set defined over $\Bbbk$ then the **homogeneous coordinate ring** of X over $\Bbbk$ is $\Bbbk[X] = \Bbbk[x_0, \dots, x_n]/I_{\Bbbk}(X)$.

Note that elements of $\Bbbk[X]$ are not necessarily homogeneous polynomials.

Definition 5.2.15 Let X be an algebraic set in $\mathbb{A}^n$ (respectively, $\mathbb{P}^n$) The **Zariski topology** is the topology on X defined as follows: the closed sets are $X \cap Y$ for every algebraic set $Y \subseteq \mathbb{A}^n$ (respectively, $Y \subseteq \mathbb{P}^n$).

Exercise 5.2.16 Show that the Zariski topology satisfies the axioms of a topology.

Definition 5.2.17 For $0 \le i \le n$ define $U_i = \{(x_0 : \cdots : x_n) \in \mathbb{P}^n : x_i \neq 0\} = \mathbb{P}^n - V(x_i)$. (These are open sets in the Zariski topology.)

Exercise 5.2.18 Show that $\mathbb{P}^n = \cup_{i=0}^n U_i$ (not a disjoint union).

We already mentioned the map $\varphi : \mathbb{A}^n \to \mathbb{P}^n$ given by $\varphi(x_1, \dots, x_n) = (x_1 : \cdots : x_n : 1)$, which has image equal to U_n. A useful way to study a projective algebraic set X is to consider $X \cap U_i$ for $0 \le i \le n$ and interpret $X \cap U_i$ as an affine algebraic set. We now introduce the notation for this.

Definition 5.2.19 Let $\varphi_i : \mathbb{A}^n(\Bbbk) \to U_i$ be the one-to-one correspondence

$$\varphi_i(y_1, \dots, y_n) = (y_1 : \cdots : y_i : 1 : y_{i+1} : \cdots : y_n).$$

We write φ for φ_n. Let

$$\varphi_i^{-1}(x_0 : \cdots : x_n) = (x_0/x_i, \dots, x_{i-1}/x_i, x_{i+1}/x_i, \dots, x_n/x_i)$$

be the map $\varphi_i^{-1} : \mathbb{P}^n(\Bbbk) \to \mathbb{A}^n(\Bbbk)$, which is defined only on U_i (i.e., $\varphi_i^{-1}(X) = \varphi_i^{-1}(X \cap U_i)$).[3]

We write $X \cap \mathbb{A}^n$ as an abbreviation for $\varphi_n^{-1}(X \cap U_n)$.

Let $\varphi_i^* : \Bbbk[x_0, \dots, x_n] \to \Bbbk[y_1, \dots, y_n]$ be the **de-homogenisation** map[4]

$$\varphi_i^*(f)(y_1, \dots, y_n) = f \circ \varphi_i(y_1, \dots, y_n) = f(y_1, \dots, y_i, 1, y_{i+1}, \dots, y_n).$$

We write φ^* for φ_n^*.

[3] This notation does not seem to be standard. Our notation agrees with Silverman [505], but Hartshorne [252] has φ_i and φ_i^{-1} the other way around.

[4] The upper star notation is extended in Definition 5.5.20.

Let $\varphi_i^{-1*} : \mathbb{k}[y_1, \dots, y_n] \to \mathbb{k}[x_0, \dots, x_n]$ be the **homogenisation**

$$\varphi_i^{-1*}(f)(x_0, \dots, x_n) = x_i^{\deg(f)} f(x_0/x_i, \dots, x_{i-1}/x_i, x_{i+1}/x_i, \dots, x_n/x_i)$$

where $\deg(f)$ is the total degree.

We write $\overline{f}$ as an abbreviation for $\varphi_n^{-1*}(f)$. For notational simplicity we often consider polynomials $f(x, y)$; in this case, we define $\overline{f} = z^{\deg(f)} f(x/z, y/z)$.

We now state some elementary relations between projective algebraic sets X and their affine parts $X \cap U_i$.

Lemma 5.2.20 *Let the notation be as above.*

1. *$\varphi_i^* : \mathbb{k}[x_0, \dots, x_n] \to \mathbb{k}[y_1, \dots, y_n]$ and $\varphi_i^{-1*} : \mathbb{k}[y_1, \dots, y_n] \to \mathbb{k}[x_0, \dots, x_n]$ are $\mathbb{k}$-algebra homomorphisms.*
2. *Let $P = (P_0 : \cdots : P_n) \in \mathbb{P}^n(\mathbb{k})$ with $P_i \neq 0$ and let $f \in \mathbb{k}[x_0, \dots, x_n]$ be homogeneous. Then $f(P) = 0$ implies $\varphi_i^*(f)(\varphi_i^{-1}(P)) = 0$.*
3. *Let $f \in \mathbb{k}[x_0, \dots, x_n]$ be homogeneous. Then $\varphi_i^{-1}(V(f)) = V(\varphi_i^*(f))$. In particular, $V(f) \cap \mathbb{A}^n = V(f \circ \varphi)$.*
4. *Let $X \subseteq \mathbb{P}^n(\mathbb{k})$. Then $f \in I_{\mathbb{k}}(X)$ implies $\varphi_i^*(f) \in I_{\mathbb{k}}(\varphi_i^{-1}(X))$. In particular, $f \in I_{\mathbb{k}}(X)$ implies $f \circ \varphi \in I_{\mathbb{k}}(X \cap \mathbb{A}^n)$.*
5. *If $P \in \mathbb{A}^n(\mathbb{k})$ and $f \in \mathbb{k}[y_1, \dots, y_n]$ then $f(P) = 0$ implies $\varphi_i^{-1*}(f)(\varphi_i(P)) = 0$. In particular, $f(P) = 0$ implies $\overline{f}(\varphi(P)) = 0$.*
6. *For $f \in \mathbb{k}[x_0, \dots, x_n]$ then $\varphi_i^{-1*}(\varphi_i^*(f)) \mid f$. Furthermore, if f has a monomial that does not include x_i then $\varphi_i^{-1*}(\varphi_i^*(f)) = f$ (in particular, $\overline{f \circ \varphi} = f$).*

Exercise 5.2.21 Prove Lemma 5.2.20.

Definition 5.2.22 Let $I \subseteq \mathbb{k}[y_1, \dots, y_n]$. Define the **homogenisation** $\overline{I}$ to be the $\mathbb{k}[x_0, \dots, x_n]$-ideal generated by the set $\{\overline{f}(x_0, \dots, x_n) : f \in I\}$.

Exercise 5.2.23 Let $I \subseteq \mathbb{k}[x_1, \dots, x_n]$. Show that $\overline{I}$ is a homogeneous ideal.

Definition 5.2.24 Let $X \subseteq \mathbb{A}^n(\mathbb{k})$. Define the **projective closure** of X to be $\overline{X} = V\left(\overline{I(X)}\right) \subseteq \mathbb{P}^n$.

Lemma 5.2.25 *Let the notation be as above.*

1. *Let $X \subseteq \mathbb{A}^n$. Then $\varphi(X) \subseteq \overline{X}$ and $\overline{X} \cap \mathbb{A}^n = X$.*
2. *Let $X \subseteq \mathbb{A}^n(\overline{\mathbb{k}})$ be non-empty. Then $I_{\mathbb{k}}\left(\overline{X}\right) = \overline{I_{\mathbb{k}}(X)}$.*

Proof Part 1 follows directly from the definitions.

Part 2 is essentially that the homogenisation of a radical ideal is a radical ideal, we give a direct proof. Let $f \in \mathbb{k}[x_0, \dots, x_n]$ be such that $f(\overline{X}) = 0$. Write $f = x_0^d g$ where $g \in \mathbb{k}[x_0, \dots, x_n]$ has a monomial that does not include x_0. By part 1, g is not constant.

Then $g \circ \varphi \in I_{\Bbbk}(X)$ and so $g = \overline{g \circ \varphi} \in \overline{I_{\Bbbk}(X)}$. It follows from part 6 of Lemma 5.2.20 that $f \in \overline{I_{\Bbbk}(X)}$. □

Theorem 5.2.26 *Let $f(x_0, x_1, x_2) \in \Bbbk[x_0, x_1, x_2]$ be a $\overline{\Bbbk}$-irreducible homogeneous polynomial. Let*

$$X = V(f(x_0, x_1, x_2)) \subseteq \mathbb{P}^2.$$

Then $I_{\overline{\Bbbk}}(X) = (f(x_0, x_1, x_2))$.

Proof Let $0 \leq i \leq 2$ be such that $f(x_0, x_1, x_2)$ has a monomial that does not feature x_i (such an i must exist since f is irreducible). Without loss of generality, suppose $i = 2$. Write $g(y_1, y_2) = \varphi^*(f) = f(y_1, y_2, 1)$. By part 6 of Lemma 5.2.20 the homogenisation of g is f.

Let $Y = X \cap \mathbb{A}^2 = V(g)$. Note that g is $\overline{\Bbbk}$-irreducible (since $g = g_1 g_2$ implies, by taking homogenisation, $f = \overline{g_1}\,\overline{g_2}$). Let $h \in I_{\overline{\Bbbk}}(X)$. Then $h \circ \varphi \in I_{\overline{\Bbbk}}(Y)$, and so, by Corollary 5.1.20, $h \circ \varphi \in (g)$. In other words, there is some $h_1(y_1, y_2)$ such that $h \circ \varphi = g h_1$. Taking homogenisations gives $f \overline{h_1} \mid h$ and so $h \in (f)$. □

Corollary 5.2.27 *Let $f(x, y) \in \Bbbk[x, y]$ be a $\overline{\Bbbk}$-irreducible polynomial and let $X = V(f) \subseteq \mathbb{A}^2$. Then $\overline{X} = V(\overline{f}) \subseteq \mathbb{P}^2$.*

Exercise 5.2.28 Prove Corollary 5.2.27.

Example 5.2.29 The projective closure of $V(y^2 = x^3 + Ax + B) \subseteq \mathbb{A}^n$ is $V(y^2 z = x^3 + Axz^2 + Bz^3)$.

Exercise 5.2.30 Let $X = V(f(x_0, x_1)) \subseteq \mathbb{A}^2$ and let $\overline{X} \subseteq \mathbb{P}^2$ be the projective closure of X. Show that $\overline{X} - X$ is finite (in other words, there are only finitely many points at infinity).

A generalisation of projective space, called **weighted projective space**, is defined as follows: for $i_0, \ldots, i_n \in \mathbb{N}$ denote by $(a_0 : a_1 : \cdots : a_n)$ the equivalence class of elements in $\Bbbk^{n+1}$ under the equivalence relation

$$(a_0, a_1, \cdots, a_n) \equiv (\lambda^{i_0} a_0, \lambda^{i_1} a_1, \cdots, \lambda^{i_n} a_n)$$

for any $\lambda \in \Bbbk^*$. The set of equivalence classes is denoted $\mathbb{P}(i_0, \ldots, i_n)(\Bbbk)$. For example, it makes sense to consider the curve $y^2 = x^4 + ax^2z^2 + z^4$ as lying in $\mathbb{P}(1, 2, 1)$. We will not discuss this topic further in the book (we refer to Reid [448] for details), but it should be noted that certain coordinate systems used for efficient elliptic curve cryptography naturally live in weighted projective space.

5.3 Irreducibility

We have seen that $V(fg)$ decomposes as $V(f) \cup V(g)$ and it is natural to consider $V(f)$ and $V(g)$ as being 'components' of $V(fg)$. It is easier to deal with algebraic sets that cannot be decomposed in this way. This concept is most useful when working over an algebraically closed field, but we give some of the theory in greater generality.

Definition 5.3.1 An affine algebraic set $X \subseteq \mathbb{A}^n$ defined over $\Bbbk$ is **$\Bbbk$-reducible** if $X = X_1 \cup X_2$ with X_1 and X_2 algebraic sets defined over $\Bbbk$ and $X_i \neq X$ for $i = 1, 2$. An affine algebraic set is **$\Bbbk$-irreducible** if there is no such decomposition. An affine algebraic set is **geometrically irreducible** if X is $\overline{\Bbbk}$-irreducible. An **affine variety** over $\Bbbk$ is a geometrically irreducible algebraic set defined over $\Bbbk$.

A projective algebraic set $X \subseteq \mathbb{P}^n$ defined over $\Bbbk$ is **$\Bbbk$-irreducible** (respectively, **geometrically irreducible**) if X is not the union $X_1 \cup X_2$ of projective algebraic sets $X_1, X_2 \subseteq \mathbb{P}^n$ defined over $\Bbbk$ (respectively, $\overline{\Bbbk}$) such that $X_i \neq X$ for $i = 1, 2$. A **projective variety** over $\Bbbk$ is a geometrically irreducible projective algebraic set defined over $\Bbbk$.

Let X be a variety (affine or projective). A **subvariety** of X over $\Bbbk$ is a subset $Y \subseteq X$ that is a variety (affine or projective) defined over $\Bbbk$.

This definition matches the usual topological definition of a set being irreducible if it is not a union of proper closed subsets.

Example 5.3.2 The algebraic set $X = V(x^2 + y^2) \subseteq \mathbb{A}^2$ over $\mathbb{R}$ is $\mathbb{R}$-irreducible. However, over $\mathbb{C}$ we have $X = V(x + iy) \cup V(x - iy)$ and so X is $\mathbb{C}$-reducible.

Exercise 5.3.3 Show that $X = V(wx - yz, x^2 - yz) \subseteq \mathbb{P}^3$ is not irreducible.

It is often easy to determine that a reducible algebraic set is reducible, just by exhibiting the non-trivial union. However, it is not necessarily easy to show that an irreducible algebraic set is irreducible. We now give an algebraic criterion for irreducibility and some applications of this result.

Theorem 5.3.4 *Let X be an algebraic set (affine or projective). Then X is $\Bbbk$-irreducible if and only if $I_{\Bbbk}(X)$ is a prime ideal.*

Proof ($\Rightarrow$): Suppose X is irreducible and that there are elements $f, g \in \Bbbk[\underline{x}]$ such that $fg \in I_{\Bbbk}(X)$. Then $X \subseteq V(fg) = V(f) \cup V(g)$, so $X = (X \cap V(f)) \cup (X \cap V(g))$. Since $X \cap V(f)$ and $X \cap V(g)$ are algebraic sets it follows that either $X = X \cap V(f)$ or $X = X \cap V(g)$, and so $f \in I_{\Bbbk}(X)$ or $g \in I_{\Bbbk}(X)$.

($\Leftarrow$): Suppose $I = I_{\Bbbk}(X)$ is a prime ideal and that $X = X_1 \cup X_2$ where X_1 and X_2 are $\Bbbk$-algebraic sets. Let $I_1 = I_{\Bbbk}(X_1)$ and $I_2 = I_{\Bbbk}(X_2)$. By parts 3 and 4 of Proposition 5.1.13 or parts 4 and 5 of Proposition 5.2.12 we have $I \subseteq I_1$, $I \subseteq I_2$ and $I = I_1 \cap I_2$. Since $I_1 I_2 \subseteq I_1 \cap I_2 = I$ and I is a prime ideal, then either $I_1 \subseteq I$ or $I_2 \subseteq I$. Hence, either $I = I_1$ or $I = I_2$ and so, by part 6 of Proposition 5.1.13 or part 7 of Proposition 5.2.12, $X = X_1$ or $X = X_2$. □

Exercise 5.3.5 Show that $V(y - x^2)$ is irreducible in $\mathbb{A}^2(\Bbbk)$.

Exercise 5.3.6 Let $X \subset \mathbb{A}^n$ be an algebraic set over $\Bbbk$. Suppose there exist polynomials $f_1, \dots, f_n \in \Bbbk[t]$ such that $X = \{(f_1(t), f_2(t), \dots, f_n(t)) : t \in \overline{\Bbbk}\}$. Prove that X is geometrically irreducible.

Remark 5.3.7 A $\Bbbk$-algebraic set X is the vanishing of polynomials in $\Bbbk[x_1, \dots, x_n]$. However, we say X is defined over $\Bbbk$ if $I_{\overline{\Bbbk}}(X)$ is generated by polynomials in $\Bbbk[x_1, \dots, x_n]$. Hence, it is clear that an algebraic set defined over $\Bbbk$ is a $\Bbbk$-algebraic set. The converse does not hold in general. However, if X is absolutely irreducible and $\Bbbk$ is a perfect field then these notions are equivalent (see Corollary 10.2.2 of Fried and Jarden [197] and use the fact that when X is absolutely irreducible then the algebraic closure of $\Bbbk$ in $\Bbbk(X)$ is $\Bbbk$). Note that Corollary 5.1.20 proves a special case of this result.

Theorem 5.3.8 *Let $X = V(f(x, y)) \subset \mathbb{A}^2$ or $X = V(f(x, y, z)) \subseteq \mathbb{P}^2$. Then X is geometrically irreducible if and only if f is irreducible over $\overline{\Bbbk}$.*

Proof If $f = gh$ is a non-trivial factorisation of f then $X = V(f) = V(g) \cup V(h)$ is reducible. Hence, if X is geometrically irreducible then f is $\overline{\Bbbk}$-irreducible.

Conversely, by Corollary 5.1.20 (respectively, Theorem 5.2.26) we have $I_{\overline{\Bbbk}}(V(f)) = (f)$. Since f is irreducible it follows that (f) is a prime ideal and so X is irreducible. □

Example 5.3.9 It is necessary to work over $\overline{\Bbbk}$ for Theorem 5.3.8. For example, let $f(x, y) = y^2 + x^2(x - 1)^2$. Then $V(f(x, y)) \subseteq \mathbb{A}^2(\mathbb{R})$ consists of two points and so is reducible, even though $f(x, y)$ is $\mathbb{R}$-irreducible.

Lemma 5.3.10 *Let X be a variety and U a non-empty open subset of X. Then $I_{\Bbbk}(U) = I_{\Bbbk}(X)$.*

Proof Since $U \subseteq X$ we have $I_{\Bbbk}(X) \subseteq I_{\Bbbk}(U)$. Now let $f \in I_{\Bbbk}(U)$. Then $U \subseteq V(f) \cap X$. Write $X_1 = V(f) \cap X$, which is an algebraic set, and $X_2 = X - U$, which is also an algebraic set. Then $X = X_1 \cup X_2$ and, since X is irreducible and $X_2 \neq X$, $X = X_1$. In other words, $f \in I_{\Bbbk}(X)$. □

Exercise 5.3.11 Let X be an irreducible variety. Prove that if $U_1, U_2 \subseteq X$ are non-empty open sets then $U_1 \cap U_2 \neq \varnothing$.

5.4 Function fields

If X is a variety defined over $\Bbbk$ then $I_{\Bbbk}(X)$ is a prime ideal and so the affine or homogeneous coordinate ring is an integral domain. One can therefore consider its field of fractions. If X is affine then the field of fractions has a natural interpretation as a set of maps $X \to \Bbbk$. When X is projective then a ratio f/g of polynomials does not give a well-defined function on X unless f and g are homogeneous of the same degree.

Definition 5.4.1 Let X be an affine variety defined over $\Bbbk$. The **function field** $\Bbbk(X)$ is the set

$$\Bbbk(X) = \{f_1/f_2 : f_1, f_2 \in \Bbbk[X], f_2 \notin I_{\Bbbk}(X)\}$$

of classes under the equivalence relation $f_1/f_2 \equiv f_3/f_4$ if and only if $f_1 f_4 - f_2 f_3 \in I_{\Bbbk}(X)$. In other words, $\Bbbk(X)$ is the field of fractions of the affine coordinate ring $\Bbbk[X]$ over $\Bbbk$.

Let X be a projective variety. The **function field** is

$$\Bbbk(X) = \{f_1/f_2 : f_1, f_2 \in \Bbbk[X] \text{ homogeneous of the same degree, } f_2 \notin I_{\Bbbk}(X)\}$$

with the equivalence relation $f_1/f_2 \equiv f_3/f_4$ if and only if $f_1 f_4 - f_2 f_3 \in I_{\Bbbk}(X)$.

Elements of $\Bbbk(X)$ are called **rational functions.** For $a \in \Bbbk$ the rational function $f : X \to \Bbbk$ given by $f(P) = a$ is called a **constant function.**

Exercise 5.4.2 Prove that the field of fractions of an integral domain is a field. Hence, deduce that if X is an affine variety then $\Bbbk(X)$ is a field. Prove also that if X is a projective variety then $\Bbbk(X)$ is a field.

We stress that, when X is projective, $\Bbbk(X)$ is not the field of fractions of $\Bbbk[X]$ and that $\Bbbk[X] \not\subseteq \Bbbk(X)$. Also note that elements of the function field are not functions $X \to \Bbbk$ but maps $X \to \Bbbk$ (i.e., they are not necessarily defined everywhere).

Example 5.4.3 One has $\Bbbk(\mathbb{A}^2) \cong \Bbbk(x, y)$ and $\Bbbk(\mathbb{P}^2) \cong \Bbbk(x, y)$.

Definition 5.4.4 Let X be a variety and let $f_1, f_2 \in \Bbbk[X]$. Then f_1/f_2 is **defined** or **regular** at P if $f_2(P) \neq 0$. An equivalence class $f \in \Bbbk(X)$ is **regular** at P if it contains some f_1/f_2 with $f_1, f_2 \in \Bbbk[X]$ (if X is projective then necessarily $\deg(f_1) = \deg(f_2)$) such that f_1/f_2 is regular at P.

Note that there may be many choices of representative for the equivalence class of f, and only some of them may be defined at P.

Example 5.4.5 Let $\Bbbk$ be a field of characteristic not equal to 2. Let X be the algebraic set $V(y^2 - x(x-1)(x+1)) \subset \mathbb{A}^2(\Bbbk)$. Consider the functions

$$f_1 = \frac{x(x-1)}{y} \quad \text{and} \quad f_2 = \frac{y}{x+1}.$$

One can check that f_1 is equivalent to f_2. Note that f_1 is not defined at $(0, 0)$, $(1, 0)$ or $(-1, 0)$, while f_2 is defined at $(0, 0)$ and $(1, 0)$ but not at $(-1, 0)$. The equivalence class of f_1 is therefore regular at $(0, 0)$ and $(1, 0)$. Section 7.3 gives techniques to deal with these issues for curves, from which one can deduce that no function in the equivalence class of f_1 is defined at $(-1, 0)$.

Exercise 5.4.6 Let X be a variety over $\Bbbk$. Suppose f_1/f_2 and f_3/f_4 are equivalent functions on X that are both defined at $P \in X(\Bbbk)$. Show that $(f_1/f_2)(P) = (f_3/f_4)(P)$.

Hence, if f is a function that is defined at a point P then it makes sense to speak of the **value** of the function at P. If the value of f at P is zero then P is called a **zero** of f.[5]

[5] For curves we will later define the notion of a function f having a pole at a point P. This notion does not make sense for general varieties, as shown by the function x/y on $\mathbb{A}^2$ at $(0, 0)$ for example.

Exercise 5.4.7 Let $X = V(w^2x^2 - w^2z^2 - y^2z^2 + x^2z^2) \subseteq \mathbb{P}^3(\Bbbk)$. Show that $(x^2 + yz)/(x^2 - w^2) \equiv (x^2 - z^2)/(x^2 - yz)$ in $\Bbbk(X)$. Hence, find the value of $(x^2 - z^2)/(x^2 - yz)$ at the point $(w : x : y : z) = (0 : 1 : 1 : 1)$. Show that both representations of the function have the same value on the point $(w : x : y : z) = (2 : 1 : -1 : 1)$.

Theorem 5.4.8 *Let X be a variety and let f be a rational function. Then there is a non-empty open set $U \subset X$ such that f is regular on U. Conversely, if $U \subset X$ is non-empty and open and $f : U \to \Bbbk$ is a function given by a ratio f_1/f_2 of polynomials (homogeneous polynomials of the same degree if X is projective) that is defined for all $P \in U$ then f extends uniquely to a rational function $f : X \to \Bbbk$.*

Proof Let $f = f_1/f_2$ where $f_1, f_2 \in \Bbbk[X]$. Define $U = X - V(f_2)$. Since $f_2 \neq 0$ in $\Bbbk[X]$ we have U a non-empty open set, and f is regular on U.

For the converse let $f = f_1/f_2$ be a function on U given as a ratio of polynomials. Then one can consider f_1 and f_2 as elements of $\Bbbk[X]$ and f_2 non-zero on U implies $f_2 \neq 0$ in $\Bbbk[X]$. Hence, f_1/f_2 corresponds to an element of $\Bbbk(X)$. Finally, suppose f_1/f_2 and f_3/f_4 are functions on X (where f_1, f_2, f_3, f_4 are polynomials) such that the restrictions $(f_1/f_2)|_U$ and $(f_3/f_4)|_U$ are equal. Then $f_1 f_4 - f_2 f_3$ is zero on U and, by Lemma 5.3.10, $(f_1 f_4 - f_2 f_3) \in I_{\Bbbk}(X)$ and $f_1/f_2 \equiv f_3/f_4$ on X. □

Corollary 5.4.9 *If X is a projective variety and $X \cap \mathbb{A}^n \neq \varnothing$ then $\Bbbk(X) \cong \Bbbk(X \cap \mathbb{A}^n)$. If X is non-empty affine variety then $\Bbbk(X) \cong \Bbbk(\overline{X})$.*

Proof The result follows since $X \cap \mathbb{A}^n = X - V(x_n)$ is open in X and X is open in $\overline{X}$. □

Definition 5.4.10 Let X be a variety and $U \subseteq X$. Define $\mathcal{O}(U)$ to be the elements of $\overline{\Bbbk}(X)$ that are regular on all $P \in U(\overline{\Bbbk})$.

Lemma 5.4.11 *If X is an affine variety over $\Bbbk$ then $\mathcal{O}(X) = \overline{\Bbbk}[X]$.*

Proof Theorem I.3.2 of Hartshorne [252]. □

Definition 5.4.12 Let X be a variety over $\Bbbk$ and $f \in \overline{\Bbbk}(X)$. Let $\sigma \in \mathrm{Gal}(\overline{\Bbbk}/\Bbbk)$. If $f = f_1/f_2$ where $f_1, f_2 \in \overline{\Bbbk}[\underline{x}]$ define $\sigma(f) = \sigma(f_1)/\sigma(f_2)$ where $\sigma(f_1)$ and $\sigma(f_2)$ denote the natural Galois action on polynomials (i.e., $\sigma(\sum_i a_i x^i) = \sum_i \sigma(a_i) x^i$). Some authors write this as f^σ.

Exercise 5.4.13 Prove that $\sigma(f)$ is well-defined (i.e., if $f \equiv f'$ then $\sigma(f) \equiv \sigma(f')$). Let $P \in X(\overline{\Bbbk})$. Prove that $f(P) = 0$ if and only if $\sigma(f)(\sigma(P)) = 0$.

Remark 5.4.14 Having defined an action of $G = \mathrm{Gal}(\overline{\Bbbk}/\Bbbk)$ on $\overline{\Bbbk}(X)$ it is natural to ask whether $\overline{\Bbbk}(X)^G = \{f \in \overline{\Bbbk}(X) : \sigma(f) = f \ \forall \sigma \in \mathrm{Gal}(\overline{\Bbbk}/\Bbbk)\}$ is the same as $\Bbbk(X)$. The issue is whether a function being "defined over $\Bbbk$" is the same as "can be written with coefficients in $\Bbbk$". Indeed, this is true, but not completely trivial.

A sketch of the argument is given in Exercise 1.12 of Silverman [505] and we give a few extra hints here. Let X be a projective variety. One first shows that if X is defined over $\Bbbk$ and if $\Bbbk'$ is a finite Galois extension of $\Bbbk$ then $I_{\Bbbk'}(X)$ is an induced Galois module (see page 110 of Serre [488]) for $\mathrm{Gal}(\Bbbk'/\Bbbk)$. It follows from Section VII.1 of [488] that the Galois cohomology group $H^1(\mathrm{Gal}(\Bbbk'/\Bbbk), I_{\Bbbk'}(X))$ is trivial and hence, by Section X.3 of [488], that $H^1(G, I_{\Bbbk'}(X)) = 0$. One can therefore deduce, as in Exercise 1.12(a) of [505], that $\overline{\Bbbk}[X]^G = \Bbbk[X]$.

To show that $\overline{\Bbbk}(X)^G = \Bbbk(X)$ let $(f_0 : f_1) : X \to \mathbb{P}^1$ and let $\sigma \in G$. Then $\sigma(f_0) = \lambda_\sigma f_0 + G_{0,\sigma}$ and $\sigma(f_1) = \lambda_\sigma f_1 + G_{1,\sigma}$ where $\lambda_\sigma \in \overline{\Bbbk}^*$ and $G_{0,\sigma}, G_{1,\sigma} \in I_{\overline{\Bbbk}}(X)$. One shows first that $\lambda_\sigma \in H^1(G, \overline{\Bbbk}^*)$, which is trivial by Hilbert 90, and so $\lambda_\sigma = \sigma(\alpha)/\alpha$ for some $\alpha \in \overline{\Bbbk}$. Replacing f_0 by αf_0 and f_1 by αf_1 gives $\lambda_\sigma = 1$ and one can proceed to showing that $G_{0,\sigma}, G_{1,\sigma} \in H^1(G, I_{\overline{\Bbbk}}(X)) = 0$ as above. The result follows.

For a different approach see Theorem 7.8.3 and Remark 8.4.8 below, or Corollary 2, Section VI.5 (page 178) of Lang [326].

5.5 Rational maps and morphisms

Definition 5.5.1 Let X be an affine or projective variety over a field $\Bbbk$ and Y an affine variety in $\mathbb{A}^n$ over $\Bbbk$. Let $\phi_1, \ldots, \phi_n \in \Bbbk(X)$. A map $\phi : X \to \mathbb{A}^n$ of the form

$$\phi(P) = (\phi_1(P), \ldots, \phi_n(P)) \tag{5.1}$$

is **regular** at a point $P \in X(\overline{\Bbbk})$ if all ϕ_i, for $1 \le i \le n$, are regular at P. A **rational map** $\phi : X \to Y$ defined over $\Bbbk$ is a map of the form (5.1) such that, for all $P \in X(\overline{\Bbbk})$ for which ϕ is regular at P, $\phi(P) \in Y(\overline{\Bbbk})$.

Let X be an affine or projective variety over a field $\Bbbk$ and Y a projective variety in $\mathbb{P}^n$ over $\Bbbk$. Let $\phi_0, \ldots, \phi_n \in \Bbbk(X)$. A map $\phi : X \to \mathbb{P}^n$ of the form

$$\phi(P) = (\phi_0(P) : \cdots : \phi_n(P)) \tag{5.2}$$

is **regular** at a point $P \in X(\overline{\Bbbk})$ if there is some function $g \in \Bbbk(X)$ such that all $g\phi_i$, for $0 \le i \le n$, are regular at P and, for some $0 \le i \le n$, one has $(g\phi_i)(P) \ne 0$.[6] A **rational map** $\phi : X \to Y$ defined over $\Bbbk$ is a map of the form (5.2) such that, for all $P \in X(\overline{\Bbbk})$ for which ϕ is regular at P, $\phi(P) \in Y(\overline{\Bbbk})$.

We stress that a rational map is not necessarily defined at every point of the domain. In other words, it is not necessarily a function.

Exercise 5.5.2 Let X and Y be projective varieties. Show that one can write a rational map in the form $\phi(P) = (\phi_0(P) : \cdots : \phi_n(P))$ where the $\phi_i(\underline{x}) \in \Bbbk[\underline{x}]$ are all homogeneous polynomials of the same degree, not all $\phi_i(\underline{x}) \in I_\Bbbk(X)$, and for every $f \in I_\Bbbk(Y)$ we have $f(\phi_0(\underline{x}), \ldots, \phi_n(\underline{x})) \in I_\Bbbk(X)$.

[6] This last condition is to prevent ϕ mapping to $(0 : \cdots : 0)$, which is not a point in $\mathbb{P}^n$.

Example 5.5.3 Let $X = V(x - y) \subseteq \mathbb{A}^2$ and $Y = V(x - z) \subseteq \mathbb{P}^2$. Then

$$\phi(x, y) = (x : xy : y)$$

is a rational map from X to Y. Note that this formula for ϕ is not defined at $(0, 0)$. However, ϕ is regular at $(0, 0)$ since taking $g = x^{-1}$ gives the equivalent form $\phi(x, y) = (x^{-1}x : x^{-1}xy : x^{-1}y) = (1 : y : y/x)$ and $y/x \equiv 1$ in $\Bbbk(X)$. Also note that the image of ϕ is not equal to $Y(\Bbbk)$ as it misses the point $(0 : 1 : 0)$.

Similarly, $\psi(x : y : z) = (x/y, z/y)$ is a rational map from Y to X. This map is not regular at $(1 : 0 : 1)$, but it is surjective to X. The composition $\psi \circ \phi$ maps (x, y) to $(1/y, 1/x)$.

Example 5.5.4 Let $X = V(y^2 z - (x^3 + Axz^2)) \subseteq \mathbb{P}^2$ and $Y = \mathbb{P}^1$. Consider the rational map

$$\phi(x : y : z) = (x/z : 1).$$

Note that this formula for ϕ is defined at all points of X except $P_0 = (0 : 1 : 0)$. Let $g(x : y : z) = (x^2 + Az^2)/y^2 \in \Bbbk(X)$. Then the map $(x : y : z) \mapsto (gx/z : g)$ can be written as $(x : y : z) \mapsto (1 : g)$ and this is defined at $(0 : 1 : 0)$. It follows that ϕ is regular at P_0 and that $\phi(P_0) = (1 : 0)$.

Lemma 5.5.5 *Let X and Y be varieties over $\Bbbk$ and let $\phi : X \to Y$ be a rational map. Then there is an open set $U \subseteq X$ such that ϕ is regular on U.*

It immediately follows that Theorem 5.4.8 generalises to rational maps.

Theorem 5.5.6 *Let X and Y be varieties. Suppose $\phi_1, \phi_2 : X \to Y$ are rational maps that are regular on non-empty open sets $U_1, U_2 \subseteq X$. Suppose further that $\phi_1|_{U_1 \cap U_2} = \phi_2|_{U_1 \cap U_2}$. Then $\phi_1 = \phi_2$.*

Exercise 5.5.7 Prove Theorem 5.5.6.

Definition 5.5.8 Let X and Y be algebraic varieties over $\Bbbk$. A rational map $\phi : X \to Y$ defined over $\Bbbk$ is a **birational equivalence** over $\Bbbk$ if there exists a rational map $\psi : Y \to X$ over $\Bbbk$ such that:

1. $\psi \circ \phi(P) = P$ for all points $P \in X(\overline{\Bbbk})$ such that $\psi \circ \phi(P)$ is defined;
2. $\phi \circ \psi(Q) = Q$ for all points $Q \in Y(\overline{\Bbbk})$ such that $\phi \circ \psi(Q) = Q$ is defined.

Varieties X and Y are **birationally equivalent** if there is a birational equivalence $\phi : X \to Y$ between them.

Exercise 5.5.9 Show that $X = V(xy - 1) \subseteq \mathbb{A}^2$ and $Y = V(x_1 - x_2) \subseteq \mathbb{P}^2$ are birationally equivalent.

Exercise 5.5.10 Verify that birational equivalence is an equivalence relation.

Example 5.5.11 The maps $\varphi_i : \mathbb{A}^n \to \mathbb{P}^n$ and $\varphi_i^{-1} : \mathbb{P}^n \to \mathbb{A}^n$ from Definition 5.2.19 are rational maps. Hence, $\mathbb{A}^n$ and $\mathbb{P}^n$ are birationally equivalent.

Definition 5.5.12 Let X and Y be varieties over $\Bbbk$ and let $U \subseteq X$ be open. A rational map $\phi : U \to Y$ over $\Bbbk$, which is regular at every point $P \in U(\overline{\Bbbk})$, is called a **morphism** over $\Bbbk$.

Let $U \subseteq X$ and $V \subseteq Y$ be open. If $\phi : U \to Y$ is a morphism over $\Bbbk$ and $\psi : V \to X$ is a morphism over $\Bbbk$ such that $\phi \circ \psi$ and $\psi \circ \phi$ are the identity on V and U respectively then we say that U and V are **isomorphic** over $\Bbbk$. If U and V are isomorphic we write $U \cong V$.

Example 5.5.13 Let X be $V(xy - z^2) \subseteq \mathbb{P}^2$ and let $\phi : X \to \mathbb{P}^1$ be given by $\phi(x : y : z) = (x/z : 1)$. Then ϕ is a morphism (for $(1 : 0 : 0)$ replace ϕ by the equivalent form $\phi(x : y : z) = (1 : z/x)$ and for $(0 : 1 : 0)$ use $\phi(x : y : z) = (z/y : 1)$). Indeed, ϕ is an isomorphism with inverse $\psi(x : z) = (x/z : z/x : 1)$.

Lemma 5.5.5 shows that every rational map $\phi : X \to Y$ restricts to a morphism $\phi : U \to Y$ on some open set $U \subseteq X$.

Exercise 5.5.14 Let $X = V(x^2 + y^2 - 1) \subset \mathbb{A}^2$ over $\Bbbk$. By taking a line of slope $t \in \Bbbk$ through $(-1, 0)$, give a formula for a rational map $\phi : \mathbb{A}^1 \to X$. Explain how to extend this to a morphism from $\mathbb{P}^1$ to $V(x^2 + y^2 - 1)$. Show that this is an isomorphism.

We now give the notion of a dominant rational map (or morphism). This is the appropriate analogue of surjectivity for maps between varieties. Essentially, a rational map from X to Y is dominant if its image is not contained in a proper subvariety of Y. For example, the map $\varphi_i : \mathbb{A}^n \to \mathbb{P}^n$ is dominant.

Definition 5.5.15 Let X and Y be varieties over $\Bbbk$. A set $U \subseteq Y(\overline{\Bbbk})$ is **dense** if its closure in the Zariski topology in $Y(\overline{\Bbbk})$ is equal to $Y(\overline{\Bbbk})$. A rational map $\phi : X \to Y$ is **dominant** if $\phi(X(\overline{\Bbbk}))$ is **dense** in $Y(\overline{\Bbbk})$.

Example 5.5.16 Let $\phi : \mathbb{A}^2 \to \mathbb{A}^2$ be given by $\phi(x, y) = (x, x)$. Then ϕ is not dominant (though it is dominant to $V(x - y) \subseteq \mathbb{A}^2$). Let $\phi : \mathbb{A}^2 \to \mathbb{A}^2$ be given by $\phi(x, y) = (x, xy)$. Then ϕ is dominant, even though it is not surjective.

Lemma 5.5.17 *Let X and Y be affine varieties over $\Bbbk$, let $U \subseteq X$ be open and let $\phi : U \to Y$ be a morphism. Then the composition $f \circ \phi$ induces a well-defined ring homomorphism from $\overline{\Bbbk}[Y]$ to $\mathcal{O}(U)$.*

Proof Proposition 6.2 of Fulton [199] or Proposition I.3.5 of Hartshorne [252]. □

Definition 5.5.18 Let X and Y be affine varieties over $\Bbbk$, let $U \subseteq X$ be open, and let $\phi : U \to Y$ be a morphism. The **pullback**[7] is the ring homomorphism $\phi^* : \overline{\Bbbk}[Y] \to \mathcal{O}(U)$ defined by $\phi^*(f) = f \circ \phi$.

[7] Pullback is just a fancy name for "composition", but we think of it as "pulling" a structure from the image of ϕ back to the domain.

Lemma 5.5.19 *Let X and Y be affine varieties over $\Bbbk$, let $U \subseteq X$ be open and let $\phi : U \to Y$ be a morphism. Then ϕ is dominant if and only if ϕ^* is injective.*

Note that if X and ϕ are defined over $\Bbbk$ then $\phi^* : \overline{\Bbbk}[Y] \to \mathcal{O}(U)$ restricts to $\phi^* : \Bbbk[Y] \to \Bbbk(X)$. If ϕ^* is injective then one can extend it to get a homomorphism of the field of fractions of $\Bbbk[Y]$ to $\Bbbk(X)$.

Definition 5.5.20 Let X and Y be varieties over $\Bbbk$ and let $\phi : X \to Y$ be a dominant rational map defined over $\Bbbk$. Define the **pullback** $\phi^* : \Bbbk(Y) \to \Bbbk(X)$ by $\phi^*(f) = f \circ \phi$.

We will now state that ϕ^* is a $\Bbbk$-algebra homomorphism. Recall that a $\Bbbk$-algebra homomorphism of fields is a field homomorphism that is the identity map on $\Bbbk$.

Theorem 5.5.21 *Let X and Y be varieties over $\Bbbk$ and let $\phi : X \to Y$ be a dominant rational map defined over $\Bbbk$. Then the pullback $\phi^* : \Bbbk(Y) \to \Bbbk(X)$ is an injective $\Bbbk$-algebra homomorphism.*

Proof Proposition 6.11 of Fulton [199] or Theorem I.4.4 of Hartshorne [252]. □

Example 5.5.22 Consider the rational maps from Example 5.5.16. The map $\phi(x, y) = (x, x)$ is not dominant and does not induce a well-defined function from $\Bbbk(x, y)$ to $\Bbbk(x, y)$ since, for example, $\phi^*(1/(x - y)) = 1/(x - x) = 1/0$.

The map $\phi(x, y) = (x, xy)$ is dominant and $\phi^*(f(x, y)) = f(x, xy)$ is a field isomorphism.

Exercise 5.5.23 Let K_1, K_2 be fields containing a field $\Bbbk$. Let $\theta : K_1 \to K_2$ be a $\Bbbk$-algebra homomorphism. Show that θ is injective.

Theorem 5.5.24 *Let X and Y be varieties over $\Bbbk$ and let $\theta : \Bbbk(Y) \to \Bbbk(X)$ be a $\Bbbk$-algebra homomorphism. Then θ induces a dominant rational map $\phi : X \to Y$ defined over $\Bbbk$.*

Proof Proposition 6.11 of Fulton [199] or Theorem I.4.4 of Hartshorne [252]. □

Theorem 5.5.25 *Let X and Y be varieties over $\Bbbk$. Then X and Y are birationally equivalent over $\Bbbk$ if and only if $\Bbbk(X) \cong \Bbbk(Y)$ (isomorphic as fields).*

Proof Proposition 6.12 of Fulton [199] or Corollary I.4.5 of Hartshorne [252]. □

Some authors prefer to study function fields rather than varieties, especially in the case of dimension 1 (there are notable classical texts that take this point of view by Chevalley and Deuring; see Stichtenoth [529] for a more recent version). By Theorem 5.5.25 (and other results), the study of function fields up to isomorphism is the study of varieties up to birational equivalence. A specific set of equations to describe a variety is called a **model**.

5.6 Dimension

The natural notion of dimension (a point has dimension 0, a line has dimension 1, a plane has dimension 2, etc.) generalises to algebraic varieties. There are algebraic and topological ways to define dimension. We use an algebraic approach.[8]

We stress that the notion of dimension only applies to irreducible algebraic sets. For example, $X = V(x, y) \cup V(x - 1) = V(x(x - 1), y(x - 1)) \subseteq \mathbb{A}^2$ is the union of a point and a line so has components of different dimension.

Recall the notion of transcendence degree of an extension $\Bbbk(X)$ over $\Bbbk$ from Definition A.6.3.

Definition 5.6.1 Let X be a variety over $\Bbbk$. The **dimension** of X, denoted $\dim(X)$, is the transcendence degree of $\Bbbk(X)$ over $\Bbbk$.

Example 5.6.2 The dimension of $\mathbb{A}^n$ is n. The dimension of $\mathbb{P}^n$ is n.

Theorem 5.6.3 *Let X and Y be varieties. If X and Y are birationally equivalent then* $\dim(X) = \dim(Y)$.

Proof Immediate from Theorem 5.5.25. □

Corollary 5.6.4 *Let X be a projective variety such that $X \cap \mathbb{A}^n$ is non-empty. Then* $\dim(X) = \dim(X \cap \mathbb{A}^n)$. *Let X be an affine variety. Then* $\dim(X) = \dim(\overline{X})$.

Exercise 5.6.5 Let f be a non-constant polynomial and let $X = V(f)$ be a variety in $\mathbb{A}^n$. Show that $\dim(X) = n - 1$.

Exercise 5.6.6 Show that if X is a non-empty variety of dimension 0 then $X = \{P\}$ is a single point.

A useful alternative formulation of dimension is as follows.

Definition 5.6.7 Let R be a ring. The **Krull dimension** of R is the supremum of $n \in \mathbb{Z}_{\geq 0}$ such that there exists a chain $I_0 \subset I_1 \subset \cdots \subset I_n$ of prime R-ideals such that $I_{j-1} \neq I_j$ for $1 \leq j \leq n$.

Theorem 5.6.8 *Let X be an affine variety over $\Bbbk$. Then* $\dim(X)$ *is equal to the Krull dimension of the affine coordinate ring* $\Bbbk[X]$.

Proof See Proposition I.1.7 and Theorem I.1.8A of [252]. □

Corollary 5.6.9 *Let X and Y be affine varieties over $\Bbbk$ such that Y is a proper subset of X. Then* $\dim(Y) < \dim(X)$.

Proof Since $Y \neq X$ we have $I_\Bbbk(X) \subsetneq I_\Bbbk(Y)$ and both ideals are prime since X and Y are irreducible. It follows that the Krull dimension of $\Bbbk[X]$ is at least one more than the Krull dimension of $\Bbbk[Y]$. □

Exercise 5.6.10 Show that a proper closed subset of a variety of dimension 1 is finite.

[8] See Chapter 8 of Eisenbud [176] for a clear criticism of this approach.

5.7 Weil restriction of scalars

Weil restriction of scalars is simply the process of re-writing a system of polynomial equations over a finite algebraic extension $\Bbbk'/\Bbbk$ as a system of equations in more variables over $\Bbbk$. The canonical example is identifying the complex numbers $\mathbb{A}^1(\mathbb{C})$ with $\mathbb{A}^2(\mathbb{R})$ via $z = x + iy \in \mathbb{A}^1(\mathbb{C}) \mapsto (x, y) \in \mathbb{A}^2(\mathbb{R})$. We only need to introduce this concept in the special case of affine algebraic sets over finite fields.

Lemma 5.7.1 *Let q be a prime power, $m \in \mathbb{N}$ and fix a vector space basis $\{\theta_1, \dots, \theta_m\}$ for $\mathbb{F}_{q^m}$ over $\mathbb{F}_q$. Let $x_1, \dots, x_n$ be coordinates for $\mathbb{A}^n$ and let $y_{1,1}, \dots, y_{1,m}, \dots, y_{n,1}, \dots, y_{n,m}$ be coordinates for $\mathbb{A}^{nm}$. The map $\phi : \mathbb{A}^{nm} \to \mathbb{A}^n$ given by*

$$\phi(y_{1,1}, \dots, y_{n,m}) = (y_{1,1}\theta_1 + \cdots + y_{1,m}\theta_m, y_{2,1}\theta_1 + \cdots + y_{2,m}\theta_m, \dots, y_{n,1}\theta_1 + \cdots + y_{n,m}\theta_m)$$

gives a bijection between $\mathbb{A}^{nm}(\mathbb{F}_q)$ and $\mathbb{A}^n(\mathbb{F}_{q^m})$.

Exercise 5.7.2 Prove Lemma 5.7.1.

Definition 5.7.3 Let $X = V(S) \subseteq \mathbb{A}^n$ be an affine algebraic set over $\mathbb{F}_{q^m}$. Let ϕ be as in Lemma 5.7.1. For each polynomial $f(x_1, \dots, x_n) \in S \subseteq \mathbb{F}_{q^m}[x_1, \dots, x_n]$ write

$$\phi^*(f) = f \circ \phi = f(y_{1,1}\theta_1 + \cdots + y_{1,m}\theta_m, y_{2,1}\theta_1 + \cdots + y_{2,m}\theta_m, \dots, y_{n,1}\theta_1 + \cdots + y_{n,m}\theta_m) \quad (5.3)$$

as

$$f_1(y_{1,1}, \dots, y_{n,m})\theta_1 + f_2(y_{1,1}, \dots, y_{n,m})\theta_2 + \cdots + f_m(y_{1,1}, \dots, y_{n,m})\theta_m \quad (5.4)$$

where each $f_j \in \mathbb{F}_q[y_{1,1}, \dots, y_{n,m}]$. Define $S' \subseteq \mathbb{F}_q[y_{1,1}, \dots, y_{n,m}]$ to be the set of all such polynomials f_j over all $f \in S$. The **Weil restriction of scalars** of X with respect to $\mathbb{F}_{q^m}/\mathbb{F}_q$ is the affine algebraic set $Y \subseteq \mathbb{A}^{mn}$ defined by

$$Y = V(S').$$

Example 5.7.4 Let $p \equiv 3 \pmod 4$ and define $\mathbb{F}_{p^2} = \mathbb{F}_p(i)$ where $i^2 = -1$. Consider the algebraic set $X = V(x_1x_2 - 1) \subseteq \mathbb{A}^2$. The Weil restriction of scalars of X with respect to $\mathbb{F}_{p^2}/\mathbb{F}_p$ with basis $\{1, i\}$ is

$$Y = V(y_{1,1}y_{2,1} - y_{1,2}y_{2,2} - 1, y_{1,1}y_{2,2} + y_{1,2}y_{2,1}) \subseteq \mathbb{A}^4.$$

Recall from Example 5.1.4 that X is an algebraic group. The multiplication operation $\mathrm{mult}((x_1, x_2), (x_1', x_2')) = (x_1x_1', x_2x_2')$ on X corresponds to the operation

$$\mathrm{mult}((y_{1,1}, y_{1,2}, y_{2,1}, y_{2,2}), (y_{1,1}', y_{1,2}', y_{2,1}', y_{2,2}'))$$
$$= (y_{1,1}y_{1,1}' - y_{1,2}y_{1,2}', y_{1,1}y_{1,2}' + y_{1,2}y_{1,1}', y_{2,1}y_{2,1}' - y_{2,2}y_{2,2}', y_{2,1}y_{2,2}' + y_{2,2}y_{2,1}')$$

on Y.

Exercise 5.7.5 Let $p \equiv 3 \pmod 4$. Write down the Weil restriction of scalars of $X = V(x^2 - 2i) \subset \mathbb{A}^1$ with respect to $\mathbb{F}_{p^2}/\mathbb{F}_p$.

Exercise 5.7.6 Let $p \equiv 3 \pmod 4$. Write down the Weil restriction of scalars of $V(x_1^2 + x_2^2 - (1+2i)) \subset \mathbb{A}^2$ with respect to $\mathbb{F}_{p^2}/\mathbb{F}_p$.

Theorem 5.7.7 *Let $X \subseteq \mathbb{A}^n$ be an affine algebraic set over $\mathbb{F}_{q^m}$. Let $Y \subseteq \mathbb{A}^{mn}$ be the Weil restriction of X. Let $k \in \mathbb{N}$ be coprime to m. Then there is a bijection between $X(\mathbb{F}_{q^{mk}})$ and $Y(\mathbb{F}_{q^k})$.*

Proof When $\gcd(k, m) = 1$ it is easily checked that the map ϕ of Lemma 5.7.1 gives a a one-to-one correspondence between $\mathbb{A}^{nm}(\mathbb{F}_{q^k})$ and $\mathbb{A}^n(\mathbb{F}_{q^{mk}})$.

Now, let $P = (x_1, \ldots, x_n) \in X$ and write $Q = (y_{1,1}, \ldots, y_{n,m})$ for the corresponding point in $\mathbb{A}^{mn}$. For any $f \in S$ we have $f(P) = 0$. Writing $f_1, \ldots, f_m$ for the polynomials in equation (5.4) we have

$$f_1(Q)\theta_1 + f_2(Q)\theta_2 + \cdots + f_m(Q)\theta_m = 0.$$

Since $\{\theta_1, \ldots, \theta_m\}$ is also a vector space basis for $\mathbb{F}_{q^{mk}}$ over $\mathbb{F}_{q^k}$ we have

$$f_1(Q) = f_2(Q) = \cdots = f_m(Q) = 0.$$

Hence, $f(Q) = 0$ for all $f \in S'$ and so $Q \in Y$. Similarly, if $Q \in Y$ then $f_j(Q) = 0$ for all such f_j and so $f(P) = 0$ for all $f \in S$. □

Note that, as the following example indicates, when k is not coprime to m then $X(\mathbb{F}_{q^{mk}})$ is not usually in one-to-one correspondence with $Y(\mathbb{F}_{q^k})$.

Exercise 5.7.8 Consider the algebraic set X from Exercise 5.7.5. Show that $X(\mathbb{F}_{p^4}) = \{1 + i, -1 - i\}$. Let Y be the Weil restriction of X with respect to $\mathbb{F}_{p^2}/\mathbb{F}_p$. Show that $Y(\mathbb{F}_{p^2}) = \{(1, 1), (-1, -1), (i, -i), (-i, i)\}$.

Note that the Weil restriction of $\mathbb{P}^n$ with respect to $\mathbb{F}_{q^m}/\mathbb{F}_q$ is not the projective closure of $\mathbb{A}^{mn}$. For example, considering the case $n = 1$, $\mathbb{P}^1$ has one point not contained in $\mathbb{A}^1$, whereas the projective closure of $\mathbb{A}^m$ has an $(m-1)$-dimensional algebraic set of points at infinity.

Exercise 5.7.9 Recall from Exercise 5.5.14 that there is a morphism from $\mathbb{P}^1$ to $Y = V(x^2 + y^2 - 1) \subseteq \mathbb{A}^2$. Determine the Weil restriction of scalars of Y with respect to $\mathbb{F}_{p^2}/\mathbb{F}_p$. It makes sense to call this algebraic set the Weil restriction of $\mathbb{P}^1$ with respect to $\mathbb{F}_{p^2}/\mathbb{F}_p$.

6

Tori, LUC and XTR

Recall from Example 5.1.4 that $\mathbb{F}_q^*$ satisfies our informal notion of an algebraic group. This chapter concerns certain subgroups of the multiplicative group of finite fields of the form $\mathbb{F}_{q^n}$ with $n > 1$. The main goal is to find short representations for elements. Algebraic tori give short representations of elements of certain subgroups of $\mathbb{F}_{q^n}^*$. Traces can be used to give short representations of certain algebraic group quotients in $\mathbb{F}_{q^n}^*$, and the most successful implementations of this are called LUC and XTR. These ideas are sometimes called **torus-based cryptography** or **trace-based-cryptography**, though this is misleading: the issue is only about representation of elements and is independent of any specific cryptosystem.

6.1 Cyclotomic subgroups of finite fields

Definition 6.1.1 Let $n \in \mathbb{N}$. A complex number z is an nth **root of unity** if $z^n = 1$, and is a **primitive** nth root of unity if $z^n = 1$ and $z^d \neq 1$ for any divisor $d \mid n$ with $1 \leq d < n$.

The nth **cyclotomic polynomial** $\Phi_n(x)$ is the product $(x - z)$ over all primitive nth roots of unity z.

Lemma 6.1.2 *Let $n \in \mathbb{N}$. Then:*

1. $\deg(\Phi_n(x)) = \varphi(n)$.
2. $\Phi_n(x) \in \mathbb{Z}[x]$.
3. $$x^n - 1 = \prod_{d \mid n, 1 \leq d \leq n} \Phi_d(x).$$
4. *If $m \in \mathbb{N}$ is such that $m \neq n$ then* $\gcd(\Phi_n(x), \Phi_m(x)) = 1$.
5. $$\Phi_n(x) = \prod_{d \mid n} (x^{n/d} - 1)^{\mu(d)}$$

where $\mu(d)$ is the Möbius function (Definition 4.3 of [420]).

Proof Let z be a primitive nth root of unity. Then every nth root of unity is a power of z and, for $0 \le i < n$, z^i is a primitive nth root of unity if and only if $\gcd(n, i) = 1$. Therefore

$$\Phi_n(x) = \prod_{0 \le i < n, \gcd(n,i)=1} (x - z^i)$$

and so $\deg(\Phi_n(x)) = \varphi(n)$.

Galois theory implies $\Phi_n(x) \in \mathbb{Q}[x]$ and, since z is an algebraic integer, it follows that $\Phi_n(x) \in \mathbb{Z}[x]$.[1]

The third fact follows since $x^n - 1 = \prod_{i=0}^{n-1}(x - z^i)$ and each z^i has some order $d \mid n$.

Let z be a root of $\gcd(\Phi_n(x), \Phi_m(x))$. Then z has order equal to both n and m, which is impossible if $n \neq m$.

Finally, writing z_d for some primitive dth root of unity, note that

$$\begin{aligned}\prod_{d \mid n}(x^{n/d} - 1)^{\mu(d)} &= \prod_{d \mid n}\prod_{j=1}^{n/d}\left(x - z_{n/d}^{j}\right)^{\mu(d)} \\ &= \prod_{d \mid n}\prod_{j=1}^{n/d}(x - z_n^{dj})^{\mu(d)} \\ &= \prod_{i=1}^{n}(x - z_n^i)^{\sum_{d \mid \gcd(n,i)} \mu(d)}.\end{aligned}$$

Since $\sum_{d \mid n} \mu(d)$ is 0 when $n > 1$ and is 1 when $n = 1$ (Theorem 4.7 of [420]) the result follows. □

Exercise 6.1.3 Show that $\Phi_1(x) = x - 1$, $\Phi_2(x) = x + 1$, $\Phi_6(x) = x^2 - x + 1$ and $\Phi_l(x) = x^{l-1} + x^{l-2} + \cdots + x + 1$ if l is prime.

Exercise 6.1.4 Prove that if $p \mid n$ then $\Phi_{pn}(x) = \Phi_n(x^p)$ and that if $p \nmid n$ then $\Phi_{pn}(x) = \Phi_n(x^p)/\Phi_n(x)$. Prove that if n is odd then $\Phi_{2n}(x) = \Phi_n(-x)$.

[Hint: Use part 5 of Lemma 6.1.2.]

It is well-known that $\Phi_n(x)$ is irreducible over $\mathbb{Q}$; we do not need this result, so we omit the proof.

Lemma 6.1.5 *Let $n \in \mathbb{N}$. The greatest common divisor of the polynomials $(x^n - 1)/(x^d - 1)$ over all $1 \le d < n$ such that $d \mid n$ is $\Phi_n(x)$.*

Proof Define $I = \{d \in \mathbb{N} : 1 \le d < n, d \mid n\}$. By part 3 of Lemma 6.1.2, we have $\Phi_n(x) = (x^n - 1)/f(x)$ where $f(x) = \prod_{d \in I} \Phi_d(x) = \operatorname{lcm}(x^d - 1 : d \in I)$. Hence

$$\Phi_n(x) = \frac{x^n - 1}{\operatorname{lcm}(x^d - 1 : d \in I)} = \gcd\left(\frac{x^n - 1}{x^d - 1} : d \in I\right).$$

□

[1] One can find more elementary proofs of this fact in any book on polynomials.

Definition 6.1.6 Let $n \in \mathbb{N}$ and q a prime power. Define the **cyclotomic subgroup** $G_{q,n}$ to be the subgroup of $\mathbb{F}_{q^n}^*$ of order $\Phi_n(q)$.

The subgroups $G_{q,n}$ are of interest as most elements of $G_{q,n}$ do not lie in any subfield of $\mathbb{F}_{q^n}$ (see Corollary 6.2.3 below). In other words, $G_{q,n}$ is the "hardest part" of $\mathbb{F}_{q^n}^*$ from the point of view of the DLP. Note that $G_{q,n}$ is trivially an algebraic group, by virtue of being a subgroup of the algebraic group $\mathbb{F}_{q^n}^* = G_m(\mathbb{F}_{q^n})$ (see Example 5.1.4). The goal of this subject area is to develop compact representations for the groups $G_{q,n}$ and efficient methods to compute with them.

The two most important cases are $G_{q,2}$, which is the subgroup of $\mathbb{F}_{q^2}^*$ of order $q+1$, and $G_{q,6}$, which is the subgroup of $\mathbb{F}_{q^6}^*$ of order $q^2 - q + 1$. We give compact representations of these groups in Sections 6.3 and 6.4.

6.2 Algebraic tori

Algebraic tori are a classical object in algebraic geometry and their relevance to cryptography was first explained by Rubin and Silverberg [452]. An excellent survey of this area is [453].

Recall from Theorem 5.7.7 that the Weil restriction of scalars of $\mathbb{A}^1$ with respect to $\mathbb{F}_{q^n}/\mathbb{F}_q$ is $\mathbb{A}^n$. Let $n > 1$ and let $f : \mathbb{A}^n(\mathbb{F}_q) \to \mathbb{F}_{q^n}$ be a bijective $\mathbb{F}_q$-linear function (e.g., corresponding to the fact that $\mathbb{F}_{q^n}$ is a vector space of dimension n over $\mathbb{F}_q$). For any $d \mid n$ define the norm $\mathrm{N}_{\mathbb{F}_{q^n}/\mathbb{F}_{q^d}}(g) = \prod_{i=0}^{n/d-1} g^{q^{di}}$. The equation $\mathrm{N}_{\mathbb{F}_{q^n}/\mathbb{F}_{q^d}}(f(x_1, \dots, x_n)) = 1$ defines an algebraic set in $\mathbb{A}^n$.

Definition 6.2.1 The **algebraic torus**[2] $\mathbb{T}_n$ is the algebraic set

$$V(\{\mathrm{N}_{\mathbb{F}_{q^n}/\mathbb{F}_{q^d}}(f(x_1, \dots, x_n)) - 1 : 1 \le d < n, d \mid n\}) \subset \mathbb{A}^n.$$

Note that there is a group operation on $\mathbb{T}_n(\mathbb{F}_q)$, given by polynomials, inherited from multiplication in $\mathbb{F}_{q^n}^*$. Hence, ignoring the inverse map, $\mathbb{T}_n(\mathbb{F}_q)$ satisfies our informal definition of an algebraic group.

Lemma 6.2.2 *Let the notation be as above.*

1. $G_{q,n} = \{g \in \mathbb{F}_{q^n}^* : \mathrm{N}_{\mathbb{F}_{q^n}/\mathbb{F}_{q^d}}(g) = 1$ *for all* $1 \le d < n$ *such that* $d \mid n\}$.
2. $\mathbb{T}_n(\mathbb{F}_q)$ *is isomorphic as a group to* $G_{q,n}$.
3. $\#\mathbb{T}_n(\mathbb{F}_q) = \Phi_n(q)$.

Proof For the first statement note that

$$\mathrm{N}_{\mathbb{F}_{q^n}/\mathbb{F}_{q^d}}(g) = \prod_{i=0}^{n/d-1} g^{q^{di}} = g^{(q^n-1)/(q^d-1)}.$$

[2] The plural of "torus" is "tori".

Recall that $\Phi_n(q) \mid (q^n - 1)/(q^d - 1)$ and, by Lemma 6.1.5, $\gcd((q^n - 1)/(q^d - 1) : 1 \le d < n, d \mid n) = \Phi_n(q)$. Hence, all the norms are 1 if and only if $g^{\Phi_n(q)} = 1$, which proves the first claim. The second and third statements follow immediately. □

Corollary 6.2.3 *Let $n \in \mathbb{N}$ and q a prime power. Suppose $g \in G_{q,n}$ has order $r > n$. Then g does not lie in any proper subfield of $\mathbb{F}_{q^n}$.*

Proof Suppose $g \in \mathbb{F}_{q^d}$ for some $1 \le d < n$ such that $d \mid n$. Then $1 = \mathrm{N}_{\mathbb{F}_{q^n}/\mathbb{F}_{q^d}}(g) = g^{n/d}$, but this contradicts the order of g being $> n$. □

It follows from the general theory that $\mathbb{T}_n$ is irreducible and of dimension $\varphi(n)$. Hence, $\mathbb{T}_n$ is a variety and one can speak of birational maps from $\mathbb{T}_n$ to another algebraic set. We refer to Section 5 of [453] for details and references.

Definition 6.2.4 The torus $\mathbb{T}_n$ is **rational** if there is a birational map from $\mathbb{T}_n$ to $\mathbb{A}^{\varphi(n)}$.

If $\mathbb{T}_n$ is rational then $\mathbb{A}^{\varphi(n)}(\mathbb{F}_q)$ is a compact representation for $G_{q,n}$. Performing discrete logarithm cryptography by transmitting elements of $\mathbb{A}^{\varphi(n)}(\mathbb{F}_q)$ is called **torus-based cryptography** and was developed by Rubin and Silverberg [452].

If $\mathbb{T}_n$ is rational then there is an induced "partial" group operation on $\mathbb{A}^{\varphi(n)}$, given by rational functions. This is not an algebraic group in general since there is not usually a one-to-one correspondence between $\mathbb{A}^{\varphi(n)}(\mathbb{F}_q)$ and $G_{q,n}$. Nevertheless, "most" of the elements of the group $G_{q,n}$ appear in $\mathbb{A}^{\varphi(n)}(\mathbb{F}_q)$ and, for many cryptographic purposes, the partial group law is sufficient. In practice, however, working with the partial group operation on $\mathbb{A}^{\varphi(n)}$ is not usually as efficient as using other representations for the group. The main application of tori is therefore the compact representation for elements of certain subgroups of $\mathbb{F}_{q^n}^*$.

It is not known if $\mathbb{T}_n$ is rational for all $n \in \mathbb{N}$ (we refer to [453] for more details and references about when $\mathbb{T}_n$ is known to be rational). The cryptographic applications of $\mathbb{T}_2$ and $\mathbb{T}_6$ rely on the well-known fact that these tori are both rational. The details are given in the following sections.

As mentioned in Section 4.3, sometimes it is convenient to consider quotients of algebraic groups by an equivalence relation. In the following sections we describe algebraic group quotients (more commonly known by the names LUC and XTR) for $G_{q,2}$ and $G_{q,6}$, but we construct them directly without using the theory of tori.

6.3 The group $G_{q,2}$

Define $\mathbb{F}_{q^2} = \mathbb{F}_q(\theta)$ where

$$\theta^2 + A\theta + B = 0 \tag{6.1}$$

for some $A, B \in \mathbb{F}_q$ such that $x^2 + Ax + B$ is irreducible over $\mathbb{F}_q$ (e.g., if q is odd then $A^2 - 4B$ is not a square in $\mathbb{F}_q$). In practice, there are performance advantages from using a simpler equation, such as $\theta^2 = B$ or $\theta^2 + \theta = B$ where B is "small". Every element of $\mathbb{F}_{q^2}$ is of the form $u + v\theta$ where $u, v \in \mathbb{F}_q$.

The **conjugate** of θ is $\overline{\theta} = \theta^q = -A - \theta$. We have $\theta + \overline{\theta} = -A$ and $\theta\overline{\theta} = B$. The conjugate of an element $g = u + v\theta \in \mathbb{F}_{q^2}$ is $u + v\overline{\theta}$ and g has **norm**

$$\mathrm{N}_{\mathbb{F}_{q^2}/\mathbb{F}_q}(g) = (u + v\theta)(u + v\overline{\theta}) = u^2 - Auv + Bv^2. \tag{6.2}$$

The group $G_{q,2}$ is defined to be the set of elements $g \in \mathbb{F}_{q^2}$ such that $g^{q+1} = 1$. Equivalently, this is the set of $u + v\theta$ such that $u^2 - Auv + Bv^2 = 1$.

Exercise 6.3.1 Show that if $g = u + v\theta \in G_{q,2}$ then $g^{-1} = g^q = u + v\overline{\theta} = (u - Av) + (-v)\theta$. Hence, inversion in $G_{q,2}$ is cheaper than a general group operation (especially if $A = 0$ or A is "small").

Exercise 6.3.2 Suppose q is not a power of 2. Suppose $\mathbb{F}_{q^2} = \mathbb{F}_q(\theta)$ where $\theta^2 + A\theta + B = 0$ and multiplying an element of $\mathbb{F}_q$ by A or B has negligible cost (e.g., $A = 0$ and $B = 1$). Show that one can compute a product (respectively: squaring; inversion) in $\mathbb{F}_{q^2}^*$ using 3 multiplications (respectively: 3 squarings; one inversion, 3 multiplications and 2 squarings) in $\mathbb{F}_q$. Ignore the cost of additions and multiplication by small constants such as 2 (since they are significantly faster to perform than multiplications etc).

Exercise 6.3.3★ Suppose $q \equiv 3 \pmod 4$ is prime. Show that one can represent $\mathbb{F}_{q^2}$ as $\mathbb{F}_q(\theta)$ where $\theta^2 + 1 = 0$. Show that, using this representation, one can compute a product (respectively: squaring; inversion; square root) in $\mathbb{F}_{q^2}^*$ using 3 multiplications (respectively: 2 multiplications; one inversion, 2 squarings and 2 multiplications; 2 square roots, one inversion, one Legendre symbol, one multiplication and 2 squarings) in $\mathbb{F}_q$. Ignore the cost of additions.

6.3.1 The torus $\mathbb{T}_2$

Recall that $G_{q,2}$ can be represented as the $\mathbb{F}_q$-points of the algebraic torus $\mathbb{T}_2 = V(\mathrm{N}_{\mathbb{F}_{q^2}/\mathbb{F}_q}(f(x, y)) - 1) \subset \mathbb{A}^2$, where $f : \mathbb{A}^2(\mathbb{F}_q) \to \mathbb{F}_{q^2}$. By equation (6.2), an affine equation for $\mathbb{T}_2$ is $V(x^2 - Axy + By^2 - 1)$. Being a conic with a rational point, it is immediate from general results in geometry (see Exercise 5.5.14 for a special case) that $\mathbb{T}_2$ is birational with $\mathbb{A}^1$.

The next two results give a more algebraic way to show that $\mathbb{T}_2$ is rational. Rather than directly constructing a birational map from $\mathbb{T}_2$ to $\mathbb{A}^1$, we go via $G_{q,2}$. Lemma 6.3.4 provides a map from $\mathbb{A}^1(\mathbb{F}_q)$ to $G_{q,2}$, while Lemma 6.3.6 provides a map from $G_{q,2}$ to $\mathbb{A}^1(\mathbb{F}_q)$.

Lemma 6.3.4 *The set $G_{q,2} \subseteq \mathbb{F}_{q^2}^*$ is equal to the set*

$$\{(a + \theta)/(a + \overline{\theta}) : a \in \mathbb{F}_q\} \cup \{1\}.$$

Proof Clearly, every element $g = (a+\theta)/(a+\overline{\theta})$ satisfies $g\overline{g} = 1$. It is also easy to check that $(a+\theta)/(a+\overline{\theta}) = (a'+\theta)/(a'+\overline{\theta})$ implies $a = a'$. Hence, we have obtained q distinct elements of $G_{q,2}$. The missing element is evidently 1 and the result follows. □

Exercise 6.3.5 Determine the value for a such that $(a+\theta)/(a+\overline{\theta}) = -1$.

Lemma 6.3.6 *Let $g = u + v\theta \in G_{q,2}$, $g \neq \pm 1$. Then $u + v\theta = (a+\theta)/(a+\overline{\theta})$ for the unique value $a = (u+1)/v$.*

Proof The value a must satisfy

$$a + \theta = (u + v\theta)(a + \overline{\theta}) = ua + u\overline{\theta} + av\theta + v\theta\overline{\theta} = (ua - Au + Bv) + \theta(av - u).$$

Equating coefficients of θ gives $av = u + 1$ and the result follows as long as $v \neq 0$ (i.e., $g \neq \pm 1$). □

The above results motivate the following definition.

Definition 6.3.7 The $\mathbb{T}_2$ **decompression map** is the function $\mathrm{decomp}_2 : \mathbb{A}^1 \to G_{q,2}$ given by $\mathrm{decomp}_2(a) = (a+\theta)/(a+\overline{\theta})$.

The $\mathbb{T}_2$ **compression map** is the function $\mathrm{comp}_2 : G_{q,2} - \{1, -1\} \to \mathbb{A}^1$ given by $\mathrm{comp}_2(u + v\theta) = (u+1)/v$.

Lemma 6.3.8 *The maps* comp_2 *and* decomp_2 *are injective. The compression map is not defined at* ± 1. *If $g \in G_{q,2} - \{1, -1\}$ then* $\mathrm{decomp}_2(\mathrm{comp}_2(g)) = g$.

Exercise 6.3.9 Prove Lemma 6.3.8.

Alert readers will notice that the maps comp_2 and decomp_2 are between $G_{q,2}$ and $\mathbb{A}^1$, rather than between $\mathbb{T}_2$ and $\mathbb{A}^1$. For completeness we now give a map from $G_{q,2}$ to $\mathbb{T}_2 \subset \mathbb{A}^2$. From this, one can deduce birational maps between $\mathbb{T}_2$ and $\mathbb{A}^1$, which prove that $\mathbb{T}_2$ is indeed rational.

Lemma 6.3.10 *An element of the form $(a+\theta)/(a+\overline{\theta}) \in G_{q,2}$ corresponds to the element*

$$\left(\frac{a^2 - B}{a^2 - aA + B}, \frac{2a - A}{a^2 - aA + B} \right)$$

of $\mathbb{T}_2$.

Proof Let (x, y) be the image point in $\mathbb{T}_2$. In other words

$$(a+\theta)/(a+\overline{\theta}) = x + y\theta$$

and so $a + \theta = (x + y\theta)(a + \overline{\theta}) = (ax + By - Ax) + \theta(ay - x)$. Equating coefficients gives the result. □

Exercise 6.3.11 Prove that $\mathbb{T}_2$ is rational.

We now present the partial group operations on $\mathbb{A}^1$ induced by the map from $\mathbb{A}^1$ to $G_{q,2}$. We stress that $\mathbb{A}^1$ is not a group with respect to these operations, since the identity element of $G_{q,2}$ is not represented as an element of $\mathbb{A}^1$.

Lemma 6.3.12 *Let the notation be as above. For $a, b \in \mathbb{A}^1$ define $a \star b = (ab - B)/(a + b - A)$ and $a' = A - a$. Then $a \star b$ is the product and a' is the inverse for the partial group law.*

Proof The partial group law on $\mathbb{A}^1$ is defined by $\mathrm{comp}_2(\mathrm{decomp}_2(a)\mathrm{decomp}_2(b))$. Now

$$\mathrm{decomp}_2(a)\mathrm{decomp}_2(b) = \left(\frac{a+\theta}{a+\overline{\theta}}\right)\left(\frac{b+\theta}{b+\overline{\theta}}\right) = \frac{ab - B + (a+b-A)\theta}{ab - B + (a+b-A)\overline{\theta}}.$$

The formula for $a \star b$ follows.

Similarly,

$$\mathrm{decomp}_2(a)^{-1} = \frac{a+\overline{\theta}}{a+\theta} = \frac{a + (-A-\theta)}{a + (-A-\overline{\theta})},$$

which gives the formula for a'. □

It follows that one can compute directly with the compressed representation of elements of $\mathbb{T}_2(\mathbb{F}_q)$. Note that computing the partial group law on $\mathbb{A}^1$ requires an inversion, so is not very efficient. For cryptographic applications one is usually computing $\mathrm{comp}_2(g^n)$ from $\mathrm{comp}_2(g)$; to do this one decompresses to obtain $g \in G_{q,2}$, then computes g^n using any one of a number of techniques and finally applies comp_2 to obtain a compact representation.[3]

6.3.2 Lucas sequences

Lucas sequences[4] can be used for efficient computation in quadratic fields. We give the details for $G_{q,2} \subset \mathbb{F}_{q^2}^*$. The name LUC cryptosystem is applied to any cryptosystem using Lucas sequences to represent elements in an algebraic group quotient of $G_{2,q}$. Recall the **trace** $\mathrm{Tr}_{\mathbb{F}_{q^2}/\mathbb{F}_q}(g) = g + g^q$ for $g \in \mathbb{F}_{q^2}$.

Definition 6.3.13 Let $g \in \mathbb{F}_{q^2}^*$ satisfy $g^{q+1} = 1$. For $i \in \mathbb{Z}$ define $V_i = \mathrm{Tr}_{\mathbb{F}_{q^2}/\mathbb{F}_q}(g^i)$.

Lemma 6.3.14 *Let $g = v_1 + w_1\theta$ with $v_1, w_1 \in \mathbb{F}_q$ and θ as in equation (6.1). Suppose $g^{q+1} = 1$ and let V_i be as in Definition 6.3.13. Then, for $i, j \in \mathbb{Z}$:*

1. $V_0 = 2$ *and* $V_1 = \mathrm{Tr}_{\mathbb{F}_{q^2}/\mathbb{F}_q}(g) = 2v_1 - Aw_1$.
2. $V_{-i} = V_i$.
3. $V_{i+1} = V_1 V_i - V_{i-1}$.
4. $V_{2i} = V_i^2 - 2$.
5. $V_{2i-1} = V_i V_{i-1} - V_1$.

[3] This is analgous to using projective coordinates for efficient elliptic curve arithmetic; see Exercise 9.1.5.

[4] They are named after Edouard Lucas (1842–1891); who apparently died due to a freak accident involving broken crockery. Lucas sequences were used for primality testing and factorisation before their cryptographic application was recognised.

6. $V_{2i+1} = V_i V_{i+1} - V_1$.
7. $V_{2i+1} = V_1 V_i^2 - V_i V_{i-1} - V_1$.
8. $V_{i+j} = V_i V_j - V_{i-j}$.

Proof Let $\overline{g} = g^q = v_1 + w_1\overline{\theta}$. Then $\mathrm{Tr}_{\mathbb{F}_{q^2}/\mathbb{F}_q}(g) = g + \overline{g} = (v_1 + w_1\theta) + (v_1 + w_1(-\theta - A)) = 2v_1 - Aw_1$. Similarly, $g^0 = 1$ and the first statement is proven. The second statement follows from $g^{-1} = \overline{g}$. Statements 3 to 6 are all special cases of statement 8, which follows from the equation

$$V_{i+j} = g^{i+j} + \overline{g}^{i+j} = (g^i + \overline{g}^i)(g^j + \overline{g}^j) - g^j\overline{g}^j(g^{i-j} + \overline{g}^{i-j}).$$

(An alternative proof of Statement 3 is to use the fact that g satisfies $g^2 = V_1 g - 1$.) Statement 7 then follows from 3 and 6. □

Exercise 6.3.15 Define $U_i = (g^i - \overline{g}^i)/(g - \overline{g})$. Prove that $U_{i+1} = \mathrm{Tr}_{\mathbb{F}_{q^2}/\mathbb{F}_q}(g)U_i - U_{i-1}$, $U_{2i} = V_i U_i$, $U_{i+j} = U_i U_{j+1} - U_{i-1}U_j$.

Definition 6.3.16 Denote by $G_{q,2}/\langle\sigma\rangle$ the set of equivalence classes of $G_{q,2}$ under the equivalence relation $g \equiv \sigma(g) = g^q = g^{-1}$. Denote the class of $g \in G_{q,2}$ by $[g] = \{g, g^q\}$.

The main observation is that $\mathrm{Tr}_{\mathbb{F}_{q^2}/\mathbb{F}_q}(g) = \mathrm{Tr}_{\mathbb{F}_{q^2}/\mathbb{F}_q}(g^q)$ and so a class $[g]$ can be identified with the value $V = \mathrm{Tr}_{\mathbb{F}_{q^2}/\mathbb{F}_q}(g)$. This motivates Definition 6.3.18. When q is odd, the classes $[1]$ and $[-1]$ correspond to $V = 2$ and $V = -2$ respectively; apart from these cases, the other possible values for V are those for which the polynomial $x^2 - Vx + 1$ is irreducible over $\mathbb{F}_q$.

Exercise 6.3.17 Prove that if $\mathrm{Tr}_{\mathbb{F}_{q^2}/\mathbb{F}_q}(g) = \mathrm{Tr}_{\mathbb{F}_{q^2}/\mathbb{F}_q}(g')$ for $g, g' \in G_{q,2}$ then $g' \in \{g, g^q\}$. Hence, show that when q is odd there are $2 + (q-1)/2$ values for $\mathrm{Tr}_{\mathbb{F}_{q^2}/\mathbb{F}_q}(g)$ over $g \in G_{q,2}$.

The set $G_{q,2}/\langle\sigma\rangle$ is not a group; however, for a class $[g] \in G_{q,2}/\langle\sigma\rangle$ and $n \in \mathbb{N}$ one can define $[g]^n$ to be $[g^n]$.

Definition 6.3.18 Let $G'_{q,2} = \{\mathrm{Tr}_{\mathbb{F}_{q^2}/\mathbb{F}_q}(g) : g \in G_{q,2}\}$. For $V \in G'_{q,2}$ and $n \in \mathbb{N}$ define $[n]V = \mathrm{Tr}_{\mathbb{F}_{q^2}/\mathbb{F}_q}(g^n)$ for any $g \in G_{q,2}$ such that $V = \mathrm{Tr}_{\mathbb{F}_{q^2}/\mathbb{F}_q}(g)$.

It follows that we may treat the set $G'_{q,2}$ as an algebraic group quotient. One method to efficiently compute $[n]V$ for $n \in \mathbb{N}$ is to take a root $g \in \mathbb{F}_{q^2}$ of $x^2 - Vx + 1 = 0$, compute g^n in $\mathbb{F}_{q^2}$ using the square-and-multiply method and then compute $\mathrm{Tr}_{\mathbb{F}_{q^2}/\mathbb{F}_q}(g^n)$. However, we want to be able to compute $[n]V$ directly using an analogue of the square-and-multiply method.[5] Lemma 6.3.14 shows that, although V_{2n} is determined by V_n and n, V_{n+1} is not determined by V_n alone. Hence, it is necessary to develop an algorithm that works on a pair (V_n, V_{n-1}) of consecutive values. Such algorithms are known as **ladder methods**. One starts the ladder computation with $(V_1, V_0) = (V, 2)$.

[5] In practice, it is often more efficient to use other processes instead of the traditional square-and-multiply method. We refer to Chapter 3 of [520] for more details.

Lemma 6.3.19 *Given (V_i, V_{i-1}) and V one can compute (V_{2i}, V_{2i-1}) (i.e., "squaring") or (V_{2i+1}, V_{2i}) (i.e., "square-and-multiply") in one multiplication, one squaring and two or three additions in $\mathbb{F}_q$.*

Proof One must compute V_i^2 and $V_i V_{i-1}$ and then apply part 4 and either part 5 or 7 of Lemma 6.3.14. □

Exercise 6.3.20 Write the ladder algorithm for computing $[n]V$ using Lucas sequences in detail.

The storage requirement of the ladder algorithm is the same as when working in $\mathbb{F}_{q^2}$, although the output value is compressed to a single element of $\mathbb{F}_q$. Note however that computing a squaring alone in $\mathbb{F}_{q^2}$ already requires more computation (at least when q is not a power of 2) than Lemma 6.3.19.

We have shown that for $V \in G'_{q,2}$ one can compute $[n]V$ using polynomial operations starting with the pair $(V, 2)$. Since $G'_{q,2}$ is in one-to-one correspondence with $G_{q,2}/\langle\sigma\rangle$, it is natural to consider $G'_{q,2}$ as being an algebraic group quotient.

Performing discrete logarithm based cryptography in $G'_{q,2}$ is sometimes called the LUC cryptosystem.[6] To solve the discrete logarithm problem in $G'_{q,2}$, one usually lifts the problem to the **covering group** $G_{q,2} \subset \mathbb{F}_{q^2}^*$ by taking one of the roots in $\mathbb{F}_{q^2}$ of the polynomial $x^2 - Vx + 1$.

Example 6.3.21 Define $\mathbb{F}_{37^2} = \mathbb{F}_{37}(\theta)$ where $\theta^2 - 3\theta + 1 = 0$. The element $g = -1 + 3\theta$ has order 19 and lies in $G_{37,2}$. Write $V = \mathrm{Tr}_{\mathbb{F}_{37^2}/\mathbb{F}_{37}}(g) = 7$. To compute $[6]V$, one uses the addition chain $(V_1, V_0) = (7, 2) \to (V_3, V_2) = (26, 10) \to (V_6, V_5) = (8, 31)$; this is because $6 = (110)_2$ in binary so the intermediate values for i are $(1)_2 = 1$ and $(11)_2 = 3$.

Exercise 6.3.22 Using the same values as Example 6.3.21 compute $[10]V$.

Exercise 6.3.23★ Compare the number of $\mathbb{F}_q$ multiplications and squarings to compute a squaring or a squaring-and-multiplication in the quotient $G'_{q,2}$ using Lucas sequences with the cost for general arithmetic in $G_{q,2} \subset \mathbb{F}_{q^2}$.

6.4 The group $G_{q,6}$

The group $G_{q,6}$ is the subgroup of $\mathbb{F}_{q^6}^*$ of order $\Phi_6(q) = q^2 - q + 1$. The natural representation of elements of $G_{q,6}$ requires 6 elements of $\mathbb{F}_q$.

Assume (without loss of generality) that $\mathbb{F}_{q^6} = \mathbb{F}_{q^3}(\theta)$ where $\theta \in \mathbb{F}_{q^2}$ and $\theta^2 + A\theta + B = 0$ for some $A, B \in \mathbb{F}_q$.

[6] The original LUC cryptosystem due to Smith and Lennon [515] was using Lucas sequences modulo a composite integer N; we refer to Section 6.6 for further discussion. The finite field version is only very briefly mentioned in [515], but is further developed in [516].

6.4.1 The torus $\mathbb{T}_6$

Recall that $\mathbb{T}_6$ is a two-dimensional algebraic set in $\mathbb{A}^6$ defined by the intersection of the kernels of the norm maps $\mathrm{N}_{\mathbb{F}_{q^6}/\mathbb{F}_{q^3}}$ and $\mathrm{N}_{\mathbb{F}_{q^6}/\mathbb{F}_{q^2}}$. It is known that $\mathbb{T}_6$ is rational, so the goal is to represent elements of $G_{q,6}$ using only two elements of $\mathbb{F}_q$.

The kernel of the norm map $\mathrm{N}_{\mathbb{F}_{q^6}/\mathbb{F}_{q^3}}$ is identified with $\mathbb{T}_2(\mathbb{F}_{q^3}) \subset \mathbb{A}^2(\mathbb{F}_{q^3})$. As in Section 6.3.1, $\mathbb{T}_2$ is birational to $\mathbb{A}^1(\mathbb{F}_{q^3})$ (which can then be identified with $\mathbb{A}^3(\mathbb{F}_q)$) via the map $\mathrm{decomp}_2(a) = (a+\theta)/(a+\overline{\theta})$ where $\mathbb{F}_{q^6} = \mathbb{F}_{q^3}(\theta)$. The next step is to compute the kernel of the norm map with respect to $\mathbb{F}_{q^6}/\mathbb{F}_{q^2}$.

Lemma 6.4.1 *The Weil restriction of the kernel of* $\mathrm{N}_{\mathbb{F}_{q^6}/\mathbb{F}_{q^2}}$ *on* $\mathbb{T}_2(\mathbb{F}_{q^3})$ *is birational with a quadratic hypersurface* U *in* $\mathbb{A}^3(\mathbb{F}_q)$*.*

Proof First, we represent an element of $\mathbb{T}_2(\mathbb{F}_{q^3}) - \{1\}$ as a single value $a \in \mathbb{F}_{q^3}$. Now impose the norm equation on the image of $\mathrm{decomp}_2(a)$

$$\mathrm{N}_{\mathbb{F}_{q^6}/\mathbb{F}_{q^2}}(\mathrm{decomp}_2(a)) = \left(\frac{a+\theta}{a+\overline{\theta}}\right)\left(\frac{a+\theta}{a+\overline{\theta}}\right)^{q^2}\left(\frac{a+\theta}{a+\overline{\theta}}\right)^{q^4}$$

$$= \left(\frac{a+\theta}{a+\overline{\theta}}\right)\left(\frac{a^{q^2}+\theta}{a^{q^2}+\overline{\theta}}\right)\left(\frac{a^{q^4}+\theta}{a^{q^4}+\overline{\theta}}\right).$$

To solve $\mathrm{N}_{\mathbb{F}_{q^6}/\mathbb{F}_{q^2}}(\mathrm{decomp}_2(a)) = 1$ one clears the denominator and equates coefficients of θ, giving

$$a^{1+q^2+q^4} + \theta(a^{1+q^2} + a^{1+q^4} + a^{q^2+q^4}) + \theta^2(a + a^{q^2} + a^{q^4}) + \theta^3$$
$$= a^{1+q^2+q^4} + \overline{\theta}(a^{1+q^2} + a^{1+q^4} + a^{q^2+q^4}) + \overline{\theta}^2(a + a^{q^2} + a^{q^4}) + \overline{\theta}^3.$$

The crucial observations are that the cubic terms in a cancel and that $\theta^2 - \overline{\theta}^2 = -A(\theta - \overline{\theta})$ and $\theta^3 - \overline{\theta}^3 = (A^2 - B)(\theta - \overline{\theta})$. Hence, we obtain a single equation in a.

Now, we identify $a \in \mathbb{A}^1(\mathbb{F}_{q^3})$ with a 3-tuple $(a_0, a_1, a_2) \in \mathbb{A}^3(\mathbb{F}_q)$. Using the fact that $a \mapsto a^q$ corresponds to an $\mathbb{F}_q$-linear map on $\mathbb{A}^3(\mathbb{F}_q)$, it follows that the single equation given above is actually a quadratic polynomial in (a_0, a_1, a_2). In other words, the values (a_0, a_1, a_2) corresponding to solutions of the norm equation are points on a quadratic hypersurface in $\mathbb{A}^3(\mathbb{F}_q)$, which we call U. □

The general theory (see Rubin and Silverberg [453]) implies that U is irreducible, but we do not prove this. It remains to give a rational parameterisation $p_U : U \to \mathbb{A}^2$ of the hypersurface. This is done using essentially the same method as Example 5.5.14.

Lemma 6.4.2 *An irreducible quadratic hypersurface* $U \subset \mathbb{A}^3$ *over a field* $\mathbb{k}$ *is birational over* $\mathbb{k}$ *to* $\mathbb{A}^2$*.*

Proof (Sketch) Let $P = (x_P, y_P, z_P)$ be a point on U and change variables so that the tangent plane T to U at P is $x = x_P$. We have not discussed T in this book; the only property we need is that T contains every line through P that is not contained in U and that intersects U at P with multiplicity 2.

Let $Q \in U(\Bbbk)$ be such that $Q \neq P$ and such that the line between P and Q is not contained in U (this is generically the case for an irreducible quadratic hypersurface). Then the line between P and Q does not lie in T and so is given by an equation of the form[7]

$$(x, y, z) = P + t(1, a, b) \tag{6.3}$$

for some $a, b \in \Bbbk$ (in other words, the equations $x = x_P + t$, $y = y_P + at$, etc). Such a line hits U at precisely one point $Q \in U(\Bbbk)$ with $Q \neq P$. Writing $U = V(F(x, y, z))$ it follows that $F(x_P + t, y_P + at, z_P + bt) = 0$ has the form $t(h(a, b)t - g(a, b)) = 0$ for some quadratic polynomial $h(a, b) \in \Bbbk[a, b]$ and some linear polynomial $g(a, b) \in \Bbbk[a, b]$. Hence, we have a rational map $\mathbb{A}^2 \to U$ given by

$$(a, b) \mapsto P + \tfrac{g(a,b)}{h(a,b)}(1, a, b).$$

The inverse is the rational map

$$p_U(x_Q, y_Q, z_Q) = ((y_Q - y_P)/(x_Q - x_P), (z_Q - z_P)/(x_Q - x_P))$$

such that $p_U : U \to \mathbb{A}^2$. □

Recall the map $\mathrm{comp}_2 : G_{q^3,2} \to \mathbb{A}^1(\mathbb{F}_{q^3})$ from the study of $\mathbb{T}_2$. We identify $\mathbb{A}^1(\mathbb{F}_{q^3})$ with $\mathbb{A}^3(\mathbb{F}_q)$. The image of comp_2 is U, which is birational via p_U to $\mathbb{A}^2$. This motivates the following definition.

Definition 6.4.3 The $\mathbb{T}_6$ **compression map** is $\mathrm{comp}_6 : G_{q,6} \to \mathbb{A}^2$ is given by $\mathrm{comp}_6 = p_U \mathrm{comp}_2$. The inverse of comp_6 is the $\mathbb{T}_6$ **decompression map** $\mathrm{decomp}_6 = \mathrm{decomp}_2\, p_U^{-1}$.

Example 6.4.4 Let $q \equiv 2, 5 \pmod 9$ be an odd prime power so that $\mathbb{F}_{q^6} = \mathbb{F}_q(\zeta_9)$ where ζ_9 is a primitive 9th root of unity (see Exercise 6.4.5). Let $\theta = \zeta_9^3$ and $\alpha = \zeta_9 + \zeta_9^{-1}$. Then $\mathbb{F}_{q^2} = \mathbb{F}_q(\theta)$ and $\mathbb{F}_{q^3} = \mathbb{F}_q(\alpha)$. Note that $\alpha^3 - 3\alpha + 1 = 0$. Identify $\mathbb{A}^3(\mathbb{F}_q)$ with $\mathbb{A}^1(\mathbb{F}_{q^3})$ by $f : (x, y, z) \mapsto x + y\alpha + z(\alpha^2 - 2)$. As in the proof of Lemma 6.4.1 one can verify that the equation

$$\mathrm{N}_{\mathbb{F}_{q^6}/\mathbb{F}_{q^2}}((f(x, y, z) + \theta)/(f(x, y, z) + \overline{\theta})) = 1$$

is equivalent to

$$F(x, y, z) = x^2 - x - y^2 + yz - z^2 = 0.$$

Denote by U the hyperplane $V(F(x, y, z))$ in $\mathbb{A}^3$. Let $P = (0, 0, 0)$. The tangent plane to U at P is given by the equation $x = 0$. Note that, since -3 is not a square in $\mathbb{F}_q$, the only solution to $F(0, y, z) = 0$ over $\mathbb{F}_q$ is $(y, z) = (0, 0)$ (but this statement is not true over $\overline{\mathbb{F}}_q$; U contains, for example, the line $(0, -\zeta_3 t, t)$). Given $a, b \in \mathbb{F}_q$ the line (t, at, bt) hits U at $t = 0$ and

$$t = 1/(1 - a^2 + ab - b^2).$$

[7] Here, and below, $P + Q$ denotes the usual coordinate-wise addition of 3-tuples over a field.

One therefore defines a birational map $g : \mathbb{A}^2 \to \mathbb{A}^3$ by

$$g : (a,b) \mapsto \left(\frac{1}{1-a^2+ab-b^2}, \frac{a}{1-a^2+ab-b^2}, \frac{b}{1-a^2+ab-b^2} \right).$$

Finally, the map decomp_6 from $\mathbb{A}^2$ to $G_{q,6}$ is $(f(g(a,b))+\theta)/((f(g(a,b))+\overline{\theta})$. It is then straightforward to compute comp_6.

Exercise 6.4.5 Let q be a prime power and ζ_9 a primitive 9th root of unity in $\overline{\mathbb{F}}_q$. Show that $\mathbb{F}_q(\zeta_9) = \mathbb{F}_{q^6}$ if and only if $q \equiv 2, 5 \pmod 9$.

In principle, one can write down the partial group operations on $\mathbb{A}^2$ induced from $G_{q,6}$, but this is not an efficient way to compute. Instead, to compute $\mathrm{comp}_6(g^n)$ from $\mathrm{comp}_6(g)$ one decompresses to obtain an element $g \in G_{q,6}$ (or $G_{q^3,2}$), computes g^n, and then compresses again.

6.4.2 XTR

An excellent survey of work in this area is the thesis of Stam [520].

The Galois group of $\mathbb{F}_{q^6}/\mathbb{F}_{q^2}$ is cyclic of order 3 and generated by the q^2-power Frobenius map σ. One can consider the set $G_{q,6}/\langle\sigma\rangle = G_{q,6}/\mathrm{Gal}(\mathbb{F}_{q^6}/\mathbb{F}_{q^2})$ of equivalence classes under the relation $g \equiv \sigma^i(g)$ for $0 \le i \le 2$. This gives an algebraic group quotient, which was named **XTR**[8] by Lenstra and Verheul. The goal is to give a compressed representation for this quotient; this is achieved by using the trace with respect to $\mathbb{F}_{q^6}/\mathbb{F}_{q^2}$.

Lemma 6.4.6 *Let $g \in G_{q,6}$. Let $t = \mathrm{Tr}_{\mathbb{F}_{q^6}/\mathbb{F}_{q^2}}(g) \in \mathbb{F}_{q^2}$. Then $\mathrm{N}_{\mathbb{F}_{q^6}/\mathbb{F}_{q^2}}(g) = g^{1+q^2+q^4} = 1$ and the characteristic polynomial of g over $\mathbb{F}_{q^2}$ is $\chi_g(x) = x^3 - tx^2 + t^q x - 1$.*

Proof The first claim follows since $g^{q^2-q+1} = 1$ and $(q^2-q+1)(q^2+q+1) = q^4 + q^2 + 1$. Now, write $(x-g)(x-g^{q^2})(x-g^{q^4}) = x^3 - tx^2 + sx - 1$. Since this polynomial is fixed by $\mathrm{Gal}(\mathbb{F}_{q^6}/\mathbb{F}_{q^2})$ it follows that $s, t \in \mathbb{F}_{q^2}$. Indeed, $t = \mathrm{Tr}_{\mathbb{F}_{q^6}/\mathbb{F}_{q^2}}(g) = g + g^{q^2} + g^{q^4} = g + g^{q-1} + g^{-q}$. Also

$$s = g^{1+q^2} + g^{1+q^4} + g^{q^2+q^4} = g^q + g^{1-q} + g^{-1}.$$

Finally, $s^q = g^{q^2} + g^{q^2-q} + g^{-q} = t$, from which we have $s = t^q$. □

This result shows that one can represent an equivalence class of $g \in G_{q,6}/\mathrm{Gal}(\mathbb{F}_{q^6}/\mathbb{F}_{q^2})$ using a single element $t \in \mathbb{F}_{q^2}$, as desired. It remains to explain how to perform exponentiation in the quotient (as usual, the quotient structure is not a group and so it makes no sense to try to compute a general group operation on it).

Exercise 6.4.7 Write $f(x) = x^3 - tx^2 + t^q x - 1$ for $t \in \mathbb{F}_{q^2}$. Prove that if $f(a) = 0$ for $a \in \overline{\mathbb{F}}_q$ then $f(a^{-q}) = 0$. Hence, prove that either $f(x)$ is irreducible over $\mathbb{F}_{q^2}$ or splits completely over $\mathbb{F}_{q^2}$.

[8] XTR is an abbreviation for ECSTR, which stands for "Efficient and Compact Subgroup Trace Representation".

Definition 6.4.8 Fix $g \in G_{q,6}$. For $n \in \mathbb{Z}$ write $t_n = \mathrm{Tr}_{\mathbb{F}_{q^6}/\mathbb{F}_{q^2}}(g^n)$.

Lemma 6.4.9 *Let the notation be as above. Then, for $n, m \in \mathbb{Z}$:*

1. $t_{-n} = t_{nq} = t_n^q$.
2. $t_{n+m} = t_n t_m - t_m^q t_{n-m} + t_{n-2m}$.

Proof We have $t_n = g^n + g^{n(q-1)} + g^{n(-q)}$ The first statement follows from the proof of Lemma 6.4.6, where it is proved that $t^q = g^q + g^{1-q} + g^{-1} = \mathrm{Tr}_{\mathbb{F}_{q^6}/\mathbb{F}_{q^2}}(g^{-1})$.

For the second statement an elementary calculation verifies that

$$\begin{aligned} t_n t_m - t_{n+m} &= (g^n + g^{n(q-1)} + g^{-nq})(g^m + g^{m(q-1)} + g^{-mq}) \\ &\quad - (g^{n+m} + g^{(n+m)(q-1)} + g^{-(n+m)q}) \\ &= g^{n+m(q-1)} + g^{n-mq} + g^{n(q-1)+m} + g^{n(q-1)-mq} + g^{-nq+m} + g^{-nq+m(q-1)}. \end{aligned}$$

This is equal to $t_m^q t_n - t_{n-2m}$. □

It remains to give a ladder algorithm to compute t_n. In this case, one can work with triples (t_{n+1}, t_n, t_{n-1}) of 'adjacent' values centered at t_n. This is the **XTR representation** of Lenstra and Verheul [324]. Note that, given $t_1 = \mathrm{Tr}_{\mathbb{F}_{q^6}/\mathbb{F}_{q^2}}(g)$ one can compute the triple $(t_1, t_0, t_{-1}) = (t_1, 3, t_1^q)$. Given a triple (t_{n+1}, t_n, t_{n-1}) and t_1 one can compute the triple centered at t_{2n} or t_{2n+1} using the following exercise.

Exercise 6.4.10 Prove that

1. $t_{2n-1} = t_{n-1}t_n - t_1^q t_n^q + t_{n+1}^q$;
2. $t_{2n} = t_n^2 - 2t_n^q$;
3. $t_{2n+1} = t_{n+1}t_n - t_1 t_n^q + t_{n-1}^q$.

Exercise 6.4.11 If one uses triples (t_{n+1}, t_n, t_{n-1}) as above then what is the cost of a square or square-and-multiply in $G_{q,6}$?

Exercise 6.4.12★ Give a more efficient ladder for XTR, for which the cost of squaring and square-and-multiply are the same.

In other words, one can compute $\mathrm{Tr}_{\mathbb{F}_{q^6}/\mathbb{F}_{q^2}}(g^n)$ from $t = \mathrm{Tr}_{\mathbb{F}_{q^6}/\mathbb{F}_{q^2}}(g)$ using polynomial arithmetic and so $G_{q,6}/\mathrm{Gal}(\mathbb{F}_{q^6}/\mathbb{F}_{q^2})$ is an algebraic group quotient. Performing discrete logarithm based cryptography in this setting is called the **XTR cryptosystem**. To solve the discrete logarithm problem in $G_{q,6}/\mathrm{Gal}(\mathbb{F}_{q^6}/\mathbb{F}_{q^2})$ one usually lifts the problem to the **covering group** $G_{q,6} \subset \mathbb{F}_{q^6}^*$ by taking any root of the polynomial $x^3 - tx^2 + t^q x - 1$. For further details about efficient arithmetic using XTR we refer to [520].

Exercise 6.4.13 Represent $\mathbb{F}_{67^2}$ as $\mathbb{F}_{67}(i)$ where $i^2 = -1$. Given that $t_1 = \mathrm{Tr}_{\mathbb{F}_{67^6}/\mathbb{F}_{67^2}}(g) = 48 + 63i$ for some $g \in G_{67,6}$ compute $t_7 = \mathrm{Tr}_{\mathbb{F}_{67^6}/\mathbb{F}_{67^2}}(g^7)$.

Exercise 6.4.14 (The Gong-Harn cryptosystem [235]) Consider the quotient $G'_{q,3} = G_{q,3}/\langle\sigma\rangle$ where σ is the q-power Frobenius in $\mathbb{F}_{q^3}$. Fix $g \in G_{q,3}$ and define $t_n = g^n + g^{nq} + g^{nq^2} \in \mathbb{F}_q$. Show that the characteristic polynomial for g is $x^3 - t_1 x^2 + t_{-1} x - 1$. Hence, show that an element of $G'_{q,3}$ can be represented using two elements of $\mathbb{F}_q$. Show that

$$t_{n+m} = t_n t_m - t_{n-m} t_{-m} + t_{n-2m}.$$

Hence, develop a ladder algorithm for exponentiation in $G'_{q,3}$.

Exercise 6.4.15 (Shirase, Han, Hibino, Kim and Takagi [476]) Let $q = 3^m$ with m odd. Show that $(q - \sqrt{3q} + 1)(q + \sqrt{3q} + 1) = q^2 - q + 1$. Let $g \in \mathbb{F}_{3^{6m}}^*$ have order dividing $q - \sqrt{3q} + 1$. Show that $g^{q+1} = g^{\sqrt{3q}}$ and $g^{q^3+1} = 1$. Let $t = \mathrm{Tr}_{\mathbb{F}_{q^6}/\mathbb{F}_{q^2}}(g)$ and $s = \mathrm{Tr}_{\mathbb{F}_{q^6}/\mathbb{F}_q}(g)$. Show that the roots of $x^2 - sx + s^{\sqrt{3q}}$ are t and t^q.

Hence, one can use s as a compressed representative for g; requiring only half the storage of XTR. To compute $\mathrm{Tr}_{\mathbb{F}_{q^6}/\mathbb{F}_q}(g^n)$ one solves the quadratic to obtain t, computes $\mathrm{Tr}_{\mathbb{F}_{q^6}/\mathbb{F}_{q^2}}(g^n)$ using the XTR formulae, and then performs the further compression.

6.5 Further remarks

Granger and Vercauteren [242] have proposed an index calculus algorithm for $\mathbb{T}_n(\mathbb{F}_{p^m})$ where $m > 1$. Kohel [316] has shown that one might map the discrete logarithm problem in an algebraic torus $\mathbb{T}_n(\mathbb{F}_q)$ to the discrete logarithm problem in the generalised Jacobian (which is a certain type of divisor class group) of a singular hyperelliptic curve over $\mathbb{F}_q$. This latter problem might be attacked using an index calculus method such as Gaudry's algorithm (see Section 15.6.3). It seems this approach will not be faster than performing index calculus methods in $\mathbb{F}_{p^n}^*$, but further investigation would be of interest.

6.6 Algebraic tori over rings

Applications in factoring and primality testing motivate the study of tori over $\mathbb{Z}/N\mathbb{Z}$. As mentioned in Section 4.4, the simplest approach is to restrict to N being square-free and to use the Chinese remainder theorem to define the groups. First, we explain how to construct rings isomorphic to the direct product of finite fields.

Example 6.6.1 Let $N = \prod_{i=1}^k p_i$ be square-free. Let $F(x) = x^2 + Ax + B \in \mathbb{Z}[x]$ be a quadratic polynomial such that $F(x)$ is irreducible modulo p_i for all $1 \le i \le k$. Define $R = (\mathbb{Z}/N\mathbb{Z})[x]/(F(x))$. By the Chinese remainder theorem, $R \cong \oplus \mathbb{F}_{p_i^2}$. We will usually write θ for the image of x in R and $\overline{\theta} = -A - x = Bx^{-1}$.

Define $G_{N,2}$ to be the subgroup of R^* of order $\prod_{i=1}^k (p_i + 1)$ isomorphic to the direct sum of the groups $G_{p_i,2}$. Note that $G_{N,2}$ is not usually cyclic.

We would like to represent a "general" element of $G_{N,2}$ using a single element of $\mathbb{Z}/N\mathbb{Z}$. In other words, we would like to have a map from $\mathbb{Z}/N\mathbb{Z}$ to $G_{N,2}$. One can immediately apply Definition 6.3.7 to obtain the map $a \mapsto (a+\theta)/(a+\overline{\theta})$. Since the reduction modulo p_i of this map correctly maps to $G_{p_i,2}$, for each prime p_i, it follows that it correctly maps to $G_{N,2}$. Hence, we can identify $\mathbb{T}_2(\mathbb{Z}/N\mathbb{Z})$ with $\mathbb{Z}/N\mathbb{Z}$. The group operation $\star$ from Lemma 6.3.12 can also be applied in $\mathbb{Z}/N\mathbb{Z}$ and its correctness follows from the Chinese remainder theorem.

Note that the image of $\mathbb{Z}/N\mathbb{Z}$ in $G_{N,2}$ under this map has size $N = \prod p_i$, whereas $G_{N,2}$ has order $\prod_i (p_i + 1)$. Hence, there are many elements of $G_{N,2}$ that are missed by the decompression map. Note that these "missed" elements are those which correspond to the identity element of $G_{p_i,2}$ for at least one prime p_i. In other words, they are of the form $g = u + v\theta$ where $\gcd(v, N) > 1$.

Similarly, Lucas sequences can be used modulo N when N is square-free, and their properties follow from the properties modulo p_i for all prime factors p_i of N. However, one should be careful when interpreting the Galois theory. In Section 6.3.2 the non-trivial element of $\mathrm{Gal}(\mathbb{F}_{q^2}/\mathbb{F}_q)$ is written as $\sigma(g) = g^q$, but this formulation does not naturally generalise to the ring R of Example 6.6.1. Instead, define $\sigma(u + v\theta) = u + v\overline{\theta}$ so that $\sigma : R \to R$ is a ring homomorphism and $\sigma(g) \pmod{p_i} = \sigma(g \pmod{p_i})$. One can then define the trace map $\mathrm{Tr}_{R/(\mathbb{Z}/N\mathbb{Z})}(g) = g + \sigma(g)$. The theory of Section 6.3.2 can then immediately be adapted to give Lucas sequences modulo N.

Exercise 6.6.2 Let $N = \prod_{i=1}^{k} p_i$ be a square-free integer and let R be as in Example 6.6.1. Let $g \in G_{N,2}$. Determine how many elements $h \in G_{N,2}$, in general, satisfy $\mathrm{Tr}_{R/(\mathbb{Z}/N\mathbb{Z})}(h) = \mathrm{Tr}_{R/(\mathbb{Z}/N\mathbb{Z})}(g)$. Show that roughly $N/2^k$ of the values $V \in \mathbb{Z}/N\mathbb{Z}$ correspond to the trace of an element in $G_{N,2}$.

Using similar methods to the above it is straightforward to adapt the torus $\mathbb{T}_6$ and XTR and to the ring $\mathbb{Z}/N\mathbb{Z}$ when N is square-free. We leave the details to the reader.

7

Curves and divisor class groups

The purpose of this chapter is to develop some basic theory of divisors and functions on curves. We use this theory to prove that the set of points on an elliptic curve over a field is a group. There exist more elementary proofs of this fact, but I feel the approach via divisor class groups gives a deeper understanding of the subject.

We start by introducing the theory of singular points on varieties. Then we define uniformisers and the valuation of a function at a point on a curve. When working over a field $\Bbbk$ that is not algebraically closed it turns out to be necessary to consider not just points on C defined over $\Bbbk$ but also those defined over $\overline{\Bbbk}$ (alternatively, one can generalise the notion of point to places of degree greater than one; see [529] for details). We then discuss divisors, principal divisors and the divisor class group. The hardest result is that the divisor of a function has degree zero; the proof for general curves is given in Chapter 8. Finally, we discuss the "chord and tangent" group law on elliptic curves.

7.1 Non-singular varieties

The word "local" is used throughout analysis and topology to describe any property that holds in a neighbourhood of a point. We now develop some tools to study "local" properties of points of varieties. The algebraic concept of "localisation" is the main technique used.

Definition 7.1.1 Let X be a variety over $\Bbbk$. The **local ring** over $\Bbbk$ of X at a point $P \in X(\Bbbk)$ is

$$\mathcal{O}_{P,\Bbbk}(X) = \{f \in \Bbbk(X) : f \text{ is regular at } P\}.$$

Define

$$\mathfrak{m}_{P,\Bbbk}(X) = \{f \in \mathcal{O}_{P,\Bbbk}(X) : f(P) = 0\} \subseteq \mathcal{O}_{P,\Bbbk}(X).$$

When the variety X and field $\Bbbk$ are clear from the context we simply write $\mathcal{O}_P$ and $\mathfrak{m}_P$.

Lemma 7.1.2 *Let the notation be as above. Then:*

1. *$\mathcal{O}_{P,\Bbbk}(X)$ is a ring;*
2. *$\mathfrak{m}_{P,\Bbbk}(X)$ is an $\mathcal{O}_{P,\Bbbk}(X)$-ideal;*

3. *$\mathfrak{m}_{P,\Bbbk}(X)$ is a maximal ideal;*
4. *$\mathcal{O}_{P,\Bbbk}(X)$ is a Noetherian local ring.*

Proof The first three parts are straightforward. The fourth part follows from the fact that, if X is affine, $\mathcal{O}_{P,\Bbbk}(X)$ is the localisation of $\Bbbk[X]$ (which is Noetherian) at the maximal ideal $\mathfrak{m} = \{f \in \Bbbk[X] : f(P) = 0\}$. Lemma A.9.5 shows that the localisation of a Noetherian ring at a maximal ideal is Noetherian. Similarly, if X is projective then $\mathcal{O}_{P,\Bbbk}(X)$ is isomorphic to a localisation of $R = \Bbbk[\varphi_i^{-1}(X)]$ (again, Noetherian) where i is such that $P \in U_i$. □

Note that, for an affine variety X

$$\Bbbk \subseteq \Bbbk[X] \subseteq \mathcal{O}_P(X) \subseteq \Bbbk(X).$$

Remark 7.1.3 We remark that $\mathcal{O}_{P,\Bbbk}(X)$ and $\mathfrak{m}_{P,\Bbbk}(X)$ are defined in terms of $\Bbbk(X)$ rather than any particular model for X. Hence, if $\phi : X \to Y$ is a birational map over $\Bbbk$ of varieties over $\Bbbk$ and ϕ is defined at $P \in X(\Bbbk)$ then $\mathcal{O}_{P,\Bbbk}(X)$ is isomorphic as a ring to $\mathcal{O}_{\phi(P),\Bbbk}(Y)$ (precisely, if $f \in \mathcal{O}_{\phi(P),\Bbbk}(Y)$ then $\phi^*(f) = f \circ \phi \in \mathcal{O}_{P,\Bbbk}(X)$). Similarly, $\mathfrak{m}_{P,\Bbbk}(X)$ and $\mathfrak{m}_{\phi(P),\Bbbk}(Y)$ are isomorphic.

Let X be a projective variety, let $P \in X(\Bbbk)$, and let i such that $P \in U_i$. By Corollary 5.4.9, $\Bbbk(X) \cong \Bbbk(\varphi_i^{-1}(X))$ and so $\mathcal{O}_{P,\Bbbk}(X) \cong \mathcal{O}_{\varphi_i^{-1}(P),\Bbbk}(\varphi_i^{-1}(X \cap U_i))$. It is therefore sufficient to consider affine varieties when determining local properties of a variety.

Example 7.1.4 Let $X \subseteq \mathbb{A}^n$ be an affine variety and suppose $P = (0, \ldots, 0) \in X(\Bbbk)$. Then $\mathcal{O}_P = \mathcal{O}_{P,\Bbbk}(X)$ is the set of equivalence classes

$$\{f_1(x_1, \ldots, x_n)/f_2(x_1, \ldots, x_n) : f_1, f_2 \in \Bbbk[x_1, \ldots, x_n], f_2(0, \ldots, 0) \neq 0\}.$$

In other words, the ratios of polynomials such that the denominators always have non-zero constant coefficient. Similarly, $\mathfrak{m}_P$ is the $\mathcal{O}_P$-ideal generated by $x_1, \ldots, x_n$. Since $f_1(x_1, \ldots, x_n)$ can be written in the form $f_1 = c + h(x_1, \ldots, x_n)$ where $c \in \Bbbk$ is the constant coefficient and $h(x_1, \ldots, x_n) \in \mathfrak{m}_P$, it follows that $\mathcal{O}_P/(x_1, \ldots, x_n) \cong \Bbbk$. Hence, $\mathfrak{m}_P$ is a maximal ideal.

Exercise 7.1.5 Let $X \subseteq \mathbb{A}^n$ be a variety over $\Bbbk$ and let $P = (P_1, \ldots, P_n) \in X(\Bbbk)$. Consider the **translation** morphism $\phi : X \to \mathbb{A}^n$ given by $\phi(x_1, \ldots, x_n) = (x_1 - P_1, \ldots, x_n - P_n)$. Show that $\phi(P) = (0, \ldots, 0)$ and that ϕ maps X to a variety Y that is isomorphic to X. Show further that $\mathcal{O}_{\phi(P),\Bbbk}(\phi(X))$ is isomorphic to $\mathcal{O}_{P,\Bbbk}(X)$ as a $\Bbbk$-algebra.

We now introduce the notion of singular points and non-singular varieties. These concepts are crucial in our discussion of curves: on a non-singular curve, one can define the order of a pole or zero of a function in a well-behaved way. Since singularity is a local property of a point (i.e., it can be defined in terms of $\mathcal{O}_P$) it is sufficient to restrict attention to affine varieties. Before stating the definition we need a lemma.

Lemma 7.1.6 *Let $X \subseteq \mathbb{A}^n$ be an affine variety over $\Bbbk$ and let $P \in X(\Bbbk)$. Then the quotient ring $\mathcal{O}_{P,\Bbbk}(X)/\mathfrak{m}_{P,\Bbbk}(X)$ is isomorphic to $\Bbbk$ as a $\Bbbk$-algebra. Furthermore, the quotient $\mathfrak{m}_{P,\Bbbk}(X)/\mathfrak{m}_{P,\Bbbk}(X)^2$ of $\mathcal{O}_{P,\Bbbk}(X)$-ideals is a $\Bbbk$-vector space of dimension at most n.*

Exercise 7.1.7 Prove Lemma 7.1.6.

As the following example shows, the dimension of the vector space $\mathfrak{m}_{P,\Bbbk}(X)/\mathfrak{m}_{P,\Bbbk}(X)^2$ carries information about the local geometry of X at the point P.

Example 7.1.8 Let $X = \mathbb{A}^2$ and $P = (0, 0) \in X(\Bbbk)$. We have $\mathfrak{m}_P = (x, y)$, $\mathfrak{m}_P^2 = (x^2, xy, y^2)$ and so the $\Bbbk$-vector space $\mathfrak{m}_P/\mathfrak{m}_P^2$ has dimension 2. Note that X has dimension 2.

Let $X = V(y^2 - x) \subseteq \mathbb{A}^2$, which has dimension 1. Let $P = (0, 0) \in X(\Bbbk)$. Then $\mathfrak{m}_P = (x, y)$ and $\{x, y\}$ span the $\Bbbk$-vector space $\mathfrak{m}_P/\mathfrak{m}_P^2$. Since $x = y^2$ in $\Bbbk(X)$ it follows that $x \in \mathfrak{m}_P^2$ and so $x = 0$ in $\mathfrak{m}_P/\mathfrak{m}_P^2$. Hence, $\mathfrak{m}_P/\mathfrak{m}_P^2$ is a one-dimensional vector space over $\Bbbk$ with basis vector y.

Consider now $X = V(y^2 - x^3) \subseteq \mathbb{A}^2$, which has dimension 1. Let $P = (0, 0)$. Again, $\{x, y\}$ spans $\mathfrak{m}_P/\mathfrak{m}_P^2$ over $\Bbbk$. Unlike the previous example, there is no linear dependence among the elements $\{x, y\}$ (as there is no polynomial relation between x and y having a non-zero linear component). Hence, $\mathfrak{m}_P/\mathfrak{m}_P^2$ has basis $\{x, y\}$ and has dimension 2.

Exercise 7.1.9 Let $X = V(x^4 + x + yx - y^2) \subseteq \mathbb{A}^2$ over $\Bbbk$ and let $P = (0, 0)$. Find a basis for the $\Bbbk$-vector space $\mathfrak{m}_{P,\Bbbk}(X)/\mathfrak{m}_{P,\Bbbk}(X)^2$. Repeat the exercise for $X = V(x^4 + x^3 + yx - y^2)$.

Example 7.1.8 motivates the following definition. One important feature of this definition is that it is in terms of the local ring at a point P and so applies equally to affine and projective varieties.

Definition 7.1.10 Let X be a variety (affine or projective) over $\Bbbk$ and let $P \in X(\overline{\Bbbk})$ be point. Then P is **non-singular** if $\dim_{\overline{\Bbbk}} \mathfrak{m}_{P,\overline{\Bbbk}}(X)/\mathfrak{m}_{P,\overline{\Bbbk}}(X)^2 = \dim(X)$ and is **singular** otherwise.[1] The variety X is **non-singular** or **smooth** if every point $P \in X(\overline{\Bbbk})$ is non-singular.

Indeed, it follows from the arguments in this section that if $P \in X(\Bbbk)$ then P is non-singular if and only if $\dim_{\Bbbk} \mathfrak{m}_{P,\Bbbk}(X)/\mathfrak{m}_{P,\Bbbk}(X)^2 = \dim(X)$. The condition of Definition 7.1.10 is inconvenient for practical computation. Hence, we now give an equivalent condition (Corollary 7.1.12) for a point to be singular.

Theorem 7.1.11 *Let $X = V(f_1, \ldots, f_m) \subseteq \mathbb{A}^n$ be a variety defined over $\Bbbk$ and let $P \in X(\Bbbk)$. Let d_1 be the dimension of the $\Bbbk$-vector space $\mathfrak{m}_{P,\Bbbk}/\mathfrak{m}_{P,\Bbbk}^2$. Let d_2 be the rank of the* ***Jacobian matrix***

$$J_{X,P} = \left(\frac{\partial f_i}{\partial x_j}(P) \right)_{\substack{1 \le i \le m \\ 1 \le j \le n}}.$$

Then $d_1 + d_2 = n$.

[1] The dimension of the vector space $\mathfrak{m}_{P,\Bbbk}(X)/\mathfrak{m}_{P,\Bbbk}(X)^2$ is always greater than or equal to $\dim(X)$, but we do not need this.

Proof Theorem I.5.1 of Hartshorne [252]. □

Corollary 7.1.12 *Let $X = V(f_1(\underline{x}), \dots, f_m(\underline{x})) \subseteq \mathbb{A}^n$ be an affine variety over $\Bbbk$ of dimension d. Let $P \in X(\Bbbk)$. Then $P \in X(\Bbbk)$ is a* **singular point** *of X if and only if the Jacobian matrix $J_{X,P}$ has rank not equal to $n - d$. The point is* **non-singular** *if the rank of $J_{X,P}$ is equal to $n - d$.*

Corollary 7.1.13 *Let $X = V(f(x_1, \dots, x_n)) \subseteq \mathbb{A}^n$ be irreducible and let $P \in X(\Bbbk)$. Then P is singular if and only if*

$$\frac{\partial f}{\partial x_j}(P) = 0$$

for all $1 \le j \le n$

Exercise 7.1.14 Let $\Bbbk$ be a field such that $\mathrm{char}(\Bbbk) \neq 2$ and let $F(x) \in \Bbbk[x]$ be such that $\gcd(F(x), F'(x)) = 1$. Show that

$$X : y^2 = F(x)$$

is non-singular as an affine algebraic set. Now consider the projective closure $\overline{X} \subseteq \mathbb{P}^2$. Show that if $\deg(F(x)) \ge 4$ then there is a unique point in $\overline{X} - X$ and that it is a singular point.

Finally we can define what we mean by a curve.

Definition 7.1.15 A **curve** is a projective non-singular variety of dimension 1. A **plane curve** is a curve that is given by an equation $V(F(x, y, z)) \subseteq \mathbb{P}^2$.

Remark 7.1.16 We stress that in this book a curve is always projective and non-singular. Note that many authors (including Hartshorne [252] and Silverman [505]) allow affine and/or singular dimension 1 varieties X to be curves. A fact that we will not prove is that every finitely generated, transcendence degree 1 extension K of an algebraic closed field $\overline{\Bbbk}$ is the function field $\overline{\Bbbk}(C)$ of a curve (see Theorem I.6.9 of Hartshorne [252]; note that working over $\overline{\Bbbk}$ is essential as there are finitely generated, transcendence degree 1 extensions of $\Bbbk$ that are not $\Bbbk(C)$ for a curve C defined over $\Bbbk$). It follows that every irreducible algebraic set of dimension 1 over $\Bbbk$ is birational over $\overline{\Bbbk}$ to a non-singular curve (see Theorem 1.1 of Moreno [395] for the details). Hence, in practice one often writes down an affine and/or singular equation X that is birational to the projective, non-singular curve C one has in mind. In our notation, the commonly used phrase "singular curve" is an oxymoron. Instead, one can say "singular equation for a curve" or "singular model for a curve".

The following result is needed in a later proof.

Lemma 7.1.17 *Let C be a curve over $\Bbbk$. Let $P, Q \in C(\overline{\Bbbk})$. Then $\mathcal{O}_{P,\overline{\Bbbk}} \subseteq \mathcal{O}_{Q,\overline{\Bbbk}}$ implies $P = Q$.*

Proof Lemma I.6.4 of Hartshorne [252]. □

7.2 Weierstrass equations

Definition 7.2.1 Let $a_1, a_2, a_3, a_4, a_6 \in \Bbbk$. A **Weierstrass equation** is a projective algebraic set E over $\Bbbk$ given by

$$y^2z + a_1xyz + a_3yz^2 = x^3 + a_2x^2z + a_4xz^2 + a_6z^3. \tag{7.1}$$

The **affine Weierstrass equation** is

$$y^2 + a_1xy + a_3y = x^3 + a_2x^2 + a_4x + a_6. \tag{7.2}$$

Exercise 7.2.2 Let E be a Weierstrass equation as in Definition 7.2.1. Let $\iota(x : y : z) = (x : -y - a_1x - a_3z : z)$. Show that if $P \in E(\Bbbk)$ then $\iota(P) \in E(\Bbbk)$ and that ι is an isomorphism over $\Bbbk$ from E to itself.

Lemma 7.2.3 *Let* $H(x), F(x) \in \Bbbk[x]$, $\deg(F) = 3$, $\deg(H) \leq 1$. *Then* $E(x, y) = y^2 + H(x)y - F(x)$ *is irreducible over* $\Bbbk$.

Theorem 5.3.8 therefore implies that a Weierstrass equation describes a projective variety. By Exercise 5.6.5 the variety has dimension 1. Not every Weierstrass equation gives a curve, since some of them are singular. We now give conditions for when a Weierstrass equation is non-singular.

Exercise 7.2.4 Show that a Weierstrass equation has a unique point with $z = 0$. Show that this point is not a singular point.

Definition 7.2.5 Let E be a Weierstrass equation over $\Bbbk$. The point $(0 : 1 : 0) \in E(\Bbbk)$ is denoted by $\mathcal{O}_E$ and is called the **point at infinity**.

Exercise 7.2.6 Show that if $\mathrm{char}(\Bbbk) \neq 2, 3$ then every Weierstrass equation over $\Bbbk$ is isomorphic over $\Bbbk$ to a Weierstrass equation

$$y^2z = x^3 + a_4xz^2 + a_6z^3 \tag{7.3}$$

for some $a_4, a_6 \in \Bbbk$. This is called the **short Weierstrass form**. Show that this equation is non-singular if and only if the **discriminant** $4a_4^3 + 27a_6^2 \neq 0$ in $\Bbbk$.

Exercise 7.2.7 Show that if $\mathrm{char}(\Bbbk) = 2$ then every Weierstrass equation over $\Bbbk$ is isomorphic over $\Bbbk$ to a Weierstrass equation

$$y^2z + xyz = x^3 + a_2x^2z + a_6z^3 \quad \text{or} \quad y^2z + yz^2 = x^3 + a_4xz^2 + a_6z^3. \tag{7.4}$$

The former is non-singular if $a_6 \neq 0$ and the latter is non-singular for all $a_4, a_6 \in \Bbbk$.

Formulae to determine whether a general Weierstrass equation is singular are given in Section III.1 of [505].

Definition 7.2.8 An **elliptic curve** is a curve given by a non-singular Weierstrass equation.

The following easy result is useful for explicit calculations.

Lemma 7.2.9 *Let E be an elliptic curve over $\Bbbk$. Then every function $f \in \Bbbk(E)$ restricts to a function on the affine Weierstrass equation of E that is equivalent to a function of the form*

$$\frac{a(x) + b(x)y}{c(x)} \tag{7.5}$$

where $a(x), b(x), c(x) \in \Bbbk[x]$. Conversely, every such function on the affine curve corresponds to a unique[2] *function on the projective curve.*

7.3 Uniformisers on curves

Let C be a curve over $\Bbbk$ with function field $\Bbbk(C)$. It is necessary to formalise the notion of multiplicity of a zero or pole of a function at a point. The basic definition will be that $f \in \mathcal{O}_{P,\overline{\Bbbk}}(C)$ has multiplicity m at P if $f \in \mathfrak{m}_{P,\overline{\Bbbk}}^m$ and $f \notin \mathfrak{m}_{P,\overline{\Bbbk}}^{m+1}$. However, there are a number of technicalities to be dealt with before we can be sure this definition makes sense. We introduce uniformisers in this section as a step towards the rigorous treatment of multiplicity of functions.

First we recall the definition of non-singular from Definition 7.1.10: let C be a non-singular curve over $\Bbbk$ and $P \in C(\Bbbk)$, then the quotient $\mathfrak{m}_{P,\Bbbk}(C)/\mathfrak{m}_{P,\Bbbk}(C)^2$ (which is a $\Bbbk$-vector space by Lemma 7.1.6) has dimension one as a $\Bbbk$-vector space.

Lemma 7.3.1 *Let C be a non-singular curve over a field $\Bbbk$ and let $P \in C(\Bbbk)$. Then the ideal $\mathfrak{m}_{P,\Bbbk}(C)$ is principal as an $\mathcal{O}_{P,\Bbbk}(C)$-ideal.*

Proof This result needs Nakayama's Lemma. We refer to Section I.6 of Hartshorne [252] for the details; also see Proposition II.1.1 of Silverman [505] for references. □

Definition 7.3.2 Let C be a non-singular curve over $\Bbbk$ and $P \in C(\overline{\Bbbk})$. A **uniformiser** (or **uniformising parameter**) at P is an element $t_P \in \mathcal{O}_{P,\overline{\Bbbk}}(C)$ such that $\mathfrak{m}_{P,\overline{\Bbbk}}(C) = (t_P)$ as an $\mathcal{O}_{P,\overline{\Bbbk}}(C)$-ideal.

One can choose t_P to be any element of $\mathfrak{m}_{P,\overline{\Bbbk}}(C) - \mathfrak{m}_{P,\overline{\Bbbk}}(C)^2$; in other words, the uniformiser is not unique. If P is defined over $\Bbbk$ then one can take $t_P \in \mathfrak{m}_{P,\Bbbk}(C) - \mathfrak{m}_{P,\Bbbk}(C)^2$, i.e. take the uniformiser to be defined over $\Bbbk$; this is typically what one does in practice.

For our presentation it is necessary to know uniformisers on $\mathbb{P}^1$ and on a Weierstrass equation. The next two examples determine such uniformisers.

Example 7.3.3 Let $C = \mathbb{P}^1$. For a point $(a : 1) \in U_1 \subseteq \mathbb{P}^1$ one can work instead with the point a on the affine curve $\mathbb{A}^1 = \varphi_1^{-1}(U_1)$. One has $\mathfrak{m}_a = (x - a)$ and so $t_a = (x - a)$ is

[2] By unique we mean that there is only one function on the projective curve corresponding to a given function on the affine curve. The actual polynomials $a(x)$, $b(x)$ and $c(x)$ are, of course, not unique.

a uniformiser at a. In terms of the projective equation, one has $t_a = (x - az)/z$ being a uniformiser. For the point $\infty = (1:0) \in U_0 \subseteq \mathbb{P}^1$ one again works with the corresponding point $0 \in \varphi_0^{-1}(U_0)$. The uniformiser is $t_a = z$ which, projectively, is $t_a = z/x$. A common abuse of notation is to say that $1/x$ is a uniformiser at ∞ on $\mathbb{A}^1 = \varphi_1^{-1}(U_1)$.

Example 7.3.4 We determine uniformisers for the points on an elliptic curve. First, consider points (x_P, y_P) on the affine equation

$$E(x, y) = y^2 + a_1xy + a_3y - \left(x^3 + a_2x^2 + a_4x + a_6\right).$$

Without loss of generality, we can translate the point to $P_0 = (0, 0)$, in which case write $a_1', \dots, a_6'$ for the coefficients of the translated equation $E'(x, y) = 0$ (i.e., $E'(x, y) = E(x + x_P, y + y_P)$). One can verify that $a_6' = 0$, $a_3' = (\partial E/\partial y)(P)$ and $a_4' = (\partial E/\partial x)(P)$. Then $\mathfrak{m}_{P_0} = (x, y)$ and, since the curve is not singular, at least one of a_3' or a_4' is non-zero.

If $a_3' = 0$ then[3]

$$x(x^2 + a_2'x + a_4' - a_1'y) = y^2.$$

Since $(x^2 + a_2'x + a_4' - a_1'y)(P_0) = a_4' \neq 0$ we have $(x^2 + a_2'x + a_4' - a_1'y)^{-1} \in \mathcal{O}_{P_0}$ and so

$$x = y^2(a_4' + a_2'x + x^2 - a_1'y)^{-1}.$$

In other words, $x \in (y^2) \subseteq \mathfrak{m}_{P_0}^2$ and y is a uniformiser at P_0.

Similarly, if $a_4' = 0$ then $y(a_3' + a_1'x + y) = x^2(x + a_2')$ and so $y \in (x^2) \subseteq \mathfrak{m}_{P_0}^2$ and x is a uniformiser at P_0. If $a_3', a_4' \neq 0$ then either x or y can be used as a uniformiser. (Indeed, any linear combination $ax + by$ except $a_3'y - a_4'x$ can be used as a uniformiser; geometrically, any line through P, except the line which is tangent to the curve at P, is a uniformiser.)

Now consider the point at infinity $\mathcal{O}_E = (x : y : z) = (0 : 1 : 0)$ on E. Taking $y = 1$ transforms the point to $(0, 0)$ on the affine curve

$$z + a_1xz + a_3z^2 = x^3 + a_2x^2z + a_4xz^2 + a_6z^3. \tag{7.6}$$

It follows that

$$z(1 + a_1x + a_3z - a_2x^2 - a_4xz - a_6z^2) = x^3$$

and so $z \in (x^3) \subseteq \mathfrak{m}_P^3$ and so x is a uniformiser. This is written as x/y in homogeneous coordinates.

In practice, it is not necessary to move P to $(0, 0)$ and compute the a_i'. We have shown that if $P = (x_P, y_P)$ then $t_P = x - x_P$ is a uniformiser unless $P = \mathcal{O}_E$, in which case $t_P = x/y$, or $P = \iota(P)$,[4] in which case $t_P = y - y_P$.

Lemma 7.3.5 *Let C be a curve over $\Bbbk$, let $P \in C(\overline{\Bbbk})$ and let t_P be a uniformiser at P. Let $\sigma \in \mathrm{Gal}(\overline{\Bbbk}/\Bbbk)$. Then $\sigma(t_P)$ is a uniformiser at $\sigma(P)$.*

[3] We will see later that $a_3' = 0$ implies $(0, 0)$ has order 2 (since $-(x, y) = (x, -y - a_1'x - a_3')$).
[4] i.e., has order 2.

Proof Since $\sigma(f)(\sigma(P)) = \sigma(f(P))$ the map $f \mapsto \sigma(f)$ is an isomorphism of local rings $\sigma : \mathcal{O}_{P,\overline{\Bbbk}}(C) \to \mathcal{O}_{\sigma(P),\overline{\Bbbk}}(C)$. It also follows that $\sigma(\mathfrak{m}_P) = \mathfrak{m}_{\sigma(P)}$. Since $\mathfrak{m}_P = (t_P)$ one has $\mathfrak{m}_{\sigma(P)} = (\sigma(t_P))$, which completes the proof. □

We now give an application of uniformisers. It will be used in several later results.

Lemma 7.3.6 *Let C be a non-singular curve over $\Bbbk$ and let $\phi : C \to Y \subseteq \mathbb{P}^n$ be a rational map for any projective variety Y. Then ϕ is a morphism.*

Exercise 7.3.7 Prove Lemma 7.3.6.

7.4 Valuation at a point on a curve

The aim of this section is to define the multiplicity of a zero or pole of a function on a curve. For background on discrete valuation rings see Chapter 1 of Serre [488], Section I.7 of Lang [327] or Sections XII.4 and XII.6 of Lang [329].

Definition 7.4.1 Let K be a field. A **discrete valuation** on K is a function $v : K^* \to \mathbb{Z}$ such that:

1. for all $f, g \in K^*$, $v(fg) = v(f) + v(g)$;
2. for all $f, g \in K^*$ such that $f + g \neq 0$, $v(f + g) \geq \min\{v(f), v(g)\}$;
3. there is some $f \in K^*$ such that $v(f) = 1$ (equivalently, v is surjective to $\mathbb{Z}$).

Lemma 7.4.2 *Let K be a field and v a discrete valuation.*

1. *$v(1) = 0$.*
2. *If $f \in K^*$ then $v(1/f) = -v(f)$.*
3. *$R_v = \{f \in K^* : v(f) \geq 0\} \cup \{0\}$ is a ring, called the* ***valuation ring****.*
4. *$\mathfrak{m}_v = \{f \in K^* : v(f) > 0\}$ is a maximal ideal in R_v, called the* ***maximal ideal*** *of the valuation.*
5. *If $f \in K$ is such that $f \notin R_v$ then $1/f \in \mathfrak{m}_v$.*
6. *R_v is a local ring.*

Exercise 7.4.3 Prove Lemma 7.4.2.

Lemma 7.4.4 *Let C be a curve over $\Bbbk$ and $P \in C(\Bbbk)$. For every non-zero function $f \in \mathcal{O}_{P,\Bbbk}(C)$ there is some $m \in \mathbb{N}$ such that $f \notin \mathfrak{m}_{P,\Bbbk}^m$.*

Definition 7.4.5 Let C be a curve over $\Bbbk$ and $P \in C(\Bbbk)$. Let $\mathfrak{m}_P = \mathfrak{m}_{P,\Bbbk}(C)$ be as in Definition 7.1.1 and define $\mathfrak{m}_P^0 = \mathcal{O}_{P,\Bbbk}(C)$. Let $f \in \mathcal{O}_{P,\Bbbk}(C)$ be such that $f \neq 0$ and define the **order** of f at P to be $v_P(f) = \max\{m \in \mathbb{Z}_{\geq 0} : f \in \mathfrak{m}_P^m\}$. If $v_P(f) = 1$ then f has a **simple zero** at P. (We exclude the constant function $f = 0$, though one could define $v_P(0) = \infty$.)

We stress that $v_P(f)$ is well-defined. If $f, h \in \mathcal{O}_{P,\Bbbk}(C)$ and $f \equiv h$ then $f - h = 0$ in $\mathcal{O}_{P,\Bbbk}(C)$. Hence, if $f \in \mathfrak{m}_P^m$ then $h \in \mathfrak{m}_P^m$ (and vice versa).

Exercise 7.4.6 Show that $v_P(f)$ does not depend on the underlying field. In other words, if $\Bbbk'$ is an algebraic extension of $\Bbbk$ in $\overline{\Bbbk}$ then $v_P(f) = \max\{m \in \mathbb{Z}_{\geq 0} : f \in \mathfrak{m}_{P,\Bbbk'}(C)^m\}$.

Lemma 7.4.7 *Let C be a curve over $\Bbbk$ and $P \in C(\overline{\Bbbk})$. Let $t_P \in \mathcal{O}_{P,\overline{\Bbbk}}(C)$ be any uniformiser at P. Let $f \in \mathcal{O}_{P,\overline{\Bbbk}}(C)$ be such that $f \neq 0$. Then $v_P(f) = \max\{m \in \mathbb{Z}_{\geq 0} : f/t_P^m \in \mathcal{O}_{P,\overline{\Bbbk}}(C)\}$ and $f = t_P^{v_P(f)} u$ for some $u \in \mathcal{O}_{P,\overline{\Bbbk}}(C)^*$.*

Exercise 7.4.8 Prove Lemma 7.4.7.

Writing a function f as $t_P^{v_P(f)} u$ for some $u \in \mathcal{O}_{P,\overline{\Bbbk}}(C)^*$ is analogous to writing a polynomial $F(x) \in \Bbbk[x]$ in the form $F(x) = (x-a)^m G(x)$ where $G(x) \in \Bbbk[x]$ satisfies $G(a) \neq 0$. Hopefully, the reader is convinced that this is a powerful tool. For example, it enables a simple proof of Exercise 7.4.9. Further, one can represent a function f as a formal power series $\sum_{n=v_P(f)}^{\infty} a_n t_P^n$ where $a_n \in \Bbbk$; see Exercises 2-30 to 2-32 of Fulton [199]. Such expansions will used in Chapters 25 and 26, but we do not develop the theory rigorously.

Exercise 7.4.9 Let C be a curve over $\Bbbk$ and $P \in C(\overline{\Bbbk})$. Let $f, h \in \mathcal{O}_{P,\overline{\Bbbk}}(C)$ be such that $f, h \neq 0$. Show that $v_P(fh) = v_P(f) + v_P(h)$.

Lemma 7.4.10 *Let C be a curve over $\Bbbk$, let $P \in C(\overline{\Bbbk})$ and let $f \in \Bbbk(C)$. Then f can be written as f_1/f_2 where $f_1, f_2 \in \mathcal{O}_{P,\overline{\Bbbk}}(C)$.*

Proof Without loss of generality, C is affine. By definition, $f = f_1/f_2$ where $f_1, f_2 \in \Bbbk[C]$. Since $\Bbbk[C] \subset \overline{\Bbbk}[C] \subset \mathcal{O}_{P,\overline{\Bbbk}}(C)$ the result follows. □

Definition 7.4.11 Let C be a curve over $\Bbbk$ and let $f \in \Bbbk(C)$. A point $P \in C(\overline{\Bbbk})$ is called a **pole** of f if $f \notin \mathcal{O}_{P,\overline{\Bbbk}}(C)$. If $f = f_1/f_2 \in \Bbbk(C)$ where $f_1, f_2 \in \mathcal{O}_{P,\overline{\Bbbk}}(C)$ then define $v_P(f) = v_P(f_1) - v_P(f_2)$.

Exercise 7.4.12 Show that if $P \in C(\overline{\Bbbk})$ is a pole of $f \in \Bbbk(C)$ then $v_P(f) < 0$ and P is a zero of $1/f$.

Lemma 7.4.13 *For every function $f \in \Bbbk(C)$ the order $v_P(f)$ of f at P is independent of the choice of representative of f.*

We now give some properties of $v_P(f)$.

Lemma 7.4.14 *Let $P \in C(\overline{\Bbbk})$. Then v_P is a discrete valuation on $\Bbbk(C)$. Furthermore, the following properties hold.*

1. *If $f \in \overline{\Bbbk}^*$ then $v_P(f) = 0$.*
2. *If $c \in \overline{\Bbbk}$ and if $v_P(f) < 0$ then $v_P(f + c) = v_P(f)$.*
3. *If $f_1, f_2 \in \Bbbk(C)^*$ are such that $v_P(f_1) \neq v_P(f_2)$ then $v_P(f_1 + f_2) = \min\{v_P(f_1), v_P(f_2)\}$.*
4. *Suppose C is defined over $\Bbbk$ and let $P \in C(\overline{\Bbbk})$. Let $\sigma \in \mathrm{Gal}(\overline{\Bbbk}/\Bbbk)$. Then $v_P(f) = v_{\sigma(P)}(\sigma(f))$.*

Having shown that every v_P is a discrete valuation on $\Bbbk(C)$, it is natural to ask whether every discrete valuation on $\Bbbk(C)$ is v_P for some point $P \in C(\Bbbk)$. To make this true over fields that are not algebraically closed requires a more general notion of a point of C defined over $\Bbbk$. Instead of doing this, we continue to work with points over $\overline{\Bbbk}$ and show in Theorem 7.5.2 that every discrete valuation on $\overline{\Bbbk}(C)$ is v_P for some $P \in C(\overline{\Bbbk})$. But first we give some examples.

Example 7.4.15 Let $E : y^2 = x(x-1)(x+1)$ over $\Bbbk$ and let $P = (1, 0) \in E(\Bbbk)$. We determine $v_P(x)$, $v_P(x-1)$, $v_P(y)$ and $v_P(x+y-1)$.

First, $x(P) = 1$ so $v_P(x) = 0$. For the rest, since $P = \iota(P)$ we take the uniformiser to be $t_P = y$. Hence, $v_P(y) = 1$. Since

$$x - 1 = y^2/(x(x+1))$$

and $1/(x(x+1)) \in \mathcal{O}_P$ we have $v_P(x-1) = 2$.

Finally, $f(x, y) = x + y - 1 = y + (x-1)$ so $v_P(f(x, y)) = \min\{v_P(y), v_P(x-1)\} = \min\{1, 2\} = 1$. One can see this directly by writing $f(x, y) = y(1 + y/x(x+1))$.

Lemma 7.4.16 *Let E be an elliptic curve. Then* $v_{\mathcal{O}_E}(x) = -2$ *and* $v_{\mathcal{O}_E}(y) = -3$.

Proof We consider the projective equation, so that the functions become x/z and y/z then set $y = 1$ so that we are considering x/z and $1/z$ on

$$z + a_1xz + a_3z^2 = x^3 + a_2x^2z + a_4xz^2 + a_6z^3.$$

As in Example 7.3.4, we have $z \in (x^3)$ and so $v_{\mathcal{O}_E}(x) = 1$, $v_{\mathcal{O}_E}(z) = 3$. This implies $v_{\mathcal{O}_E}(1/z) = -3$ and $v_{\mathcal{O}_E}(x/z) = -2$ as claimed. □

7.5 Valuations and points on curves

Let C be a curve over $\Bbbk$ and $P \in C(\overline{\Bbbk})$. We have shown that $v_P(f)$ is a discrete valuation on $\overline{\Bbbk}(C)$. The aim of this section is to show (using the weak Nullstellensatz) that every discrete valuation v on $\overline{\Bbbk}(C)$ arises as v_P for some point $P \in C(\overline{\Bbbk})$.

Lemma 7.5.1 *Let C be a curve over* $\Bbbk$ *and let v be a discrete valuation on* $\Bbbk(C)$*. Write* R_v, $\mathfrak{m}_v$ *for the corresponding valuation ring and maximal ideal (over* $\overline{\Bbbk}$*). Suppose* $C \subset \mathbb{P}^n$ *with coordinates* $(x_0 : \cdots : x_n)$*. Then there exists some* $0 \le i \le n$ *such that* $\overline{\Bbbk}[\varphi_i^{-1}(C)]$ *is a subring of* R_v *(where* φ_i^{-1} *is as in Definition 5.2.19).*

Theorem 7.5.2 *Let C be a curve over* $\Bbbk$ *and let v be a discrete valuation on* $\overline{\Bbbk}(C)$*. Then* $v = v_P$ *for some* $P \in C(\overline{\Bbbk})$.

Proof See Corollary I.6.6 of Hartshorne [252] or Theorem VI.9.1 of Lorenzini [355]. □

7.6 Divisors

A divisor is just a notation for a finite multi-set of points. As always, we work with points over an algebraically closed field $\overline{\mathbb{k}}$.

Definition 7.6.1 Let C be a curve over $\mathbb{k}$ (necessarily non-singular and projective). A **divisor** on C is a formal sum

$$D = \sum_{P \in C(\overline{\mathbb{k}})} n_P(P) \tag{7.7}$$

where $n_P \in \mathbb{Z}$ and only finitely many $n_P \neq 0$. The divisor with all $n_P = 0$ is written 0. The **support** of the divisor D in equation (7.7) is $\mathrm{Supp}(D) = \{P \in C(\overline{\mathbb{k}}) : n_P \neq 0\}$. Note that many authors use the notation $|D|$ for the support of D. Denote by $\mathrm{Div}_{\overline{\mathbb{k}}}(C)$ the set of all divisors on C. If $D' = \sum_{P \in C(\overline{\mathbb{k}})} n'_P(P)$ then define

$$D - D' = \sum_{P \in C(\overline{\mathbb{k}})} (n_P - n'_P)(P).$$

Write $D \geq D'$ if $n_P \geq n'_P$ for all P. So $D \geq 0$ if $n_P \geq 0$ for all P, and such a divisor is called **effective**.

Example 7.6.2 Let $E : y^2 = x^3 + 2x - 3$ over $\mathbb{Q}$ and let $P = (2, 3)$, $Q = (1, 0) \in E(\mathbb{Q})$. Then

$$D = 5(P) - 7(Q)$$

is a divisor on E. The support of D is $\mathrm{Supp}(D) = \{P, Q\}$ and D is not effective.

Definition 7.6.3 The **degree** of a divisor $D = \sum_P n_P(P)$ is the integer

$$\deg(D) = \sum_{P \in C(\overline{\mathbb{k}})} n_P.$$

(We stress that this is a finite sum.) We write $\mathrm{Div}^0_{\overline{\mathbb{k}}}(C) = \{D \in \mathrm{Div}_{\overline{\mathbb{k}}}(C) : \deg(D) = 0\}$.

Lemma 7.6.4 $\mathrm{Div}_{\overline{\mathbb{k}}}(C)$ *is a group under addition, and* $\mathrm{Div}^0_{\overline{\mathbb{k}}}(C)$ *is a subgroup.*

Exercise 7.6.5 Prove Lemma 7.6.4.

Definition 7.6.6 Let C be a curve over $\mathbb{k}$ and let $D = \sum_{P \in C(\overline{\mathbb{k}})} n_P(P)$ be a divisor on C. For $\sigma \in \mathrm{Gal}(\overline{\mathbb{k}}/\mathbb{k})$ define $\sigma(D) = \sum_P n_P(\sigma(P))$. Then D is **defined over** $\mathbb{k}$ if $\sigma(D) = D$ for all $\sigma \in \mathrm{Gal}(\overline{\mathbb{k}}/\mathbb{k})$. Write $\mathrm{Div}_{\mathbb{k}}(C)$ for the set of divisors on C that are defined over $\mathbb{k}$.

Since $\mathrm{Gal}(\overline{\mathbb{k}}/\mathbb{k})$ is an enormous and complicated object it is important to realise that testing the field of definition of any specific divisor is a finite task. There is an extension $\mathbb{k}'/\mathbb{k}$ of finite degree containing the coordinates of all points in the support of D. Let $\mathbb{k}''$ be the Galois closure of $\mathbb{k}'$. Since $\mathbb{k}''$ is normal over $\mathbb{k}$, any $\sigma \in \mathrm{Gal}(\overline{\mathbb{k}}/\mathbb{k})$ is such that $\sigma(\mathbb{k}'') = \mathbb{k}''$. Hence, it is sufficient to study the behaviour of D under $\sigma \in \mathrm{Gal}(\mathbb{k}''/\mathbb{k})$.

Example 7.6.7 Let $C : x^2 + y^2 = 6$ over $\mathbb{Q}$ and let $P = (1 + \sqrt{2}, 1 - \sqrt{2})$, $Q = (1 - \sqrt{2}, 1 + \sqrt{2}) \in C(\mathbb{Q}(\sqrt{2})) \subseteq C(\overline{\mathbb{Q}})$. Define

$$D = (P) + (Q).$$

It is sufficient to consider $\sigma(D)$ for $\sigma \in \mathrm{Gal}(\mathbb{Q}(\sqrt{2})/\mathbb{Q})$. The only non-trivial element is $\sigma(\sqrt{2}) = -\sqrt{2}$ and one sees that $\sigma(P) = Q$ and $\sigma(Q) = P$. Hence, $\sigma(D) = D$ for all $\sigma \in \mathrm{Gal}(\mathbb{Q}(\sqrt{2})/\mathbb{Q})$ and D is defined over $\mathbb{Q}$. Note that $C(\mathbb{Q}) = \varnothing$, so this example shows it is possible to have $\mathrm{Div}_{\Bbbk}(C) \neq \{0\}$ even if $C(\Bbbk) = \varnothing$.

7.7 Principal divisors

This section contains an important and rather difficult result, namely that the number of poles of a function on a curve (counted according to multiplicity) is finite and equal to the number of zeroes (counted according to multiplicity). The finiteness condition is essential to be able to represent the poles and zeroes of a function as a divisor. The other condition is required to show that the set of all divisors of functions is a subgroup of $\mathrm{Div}^0_{\Bbbk}(C)$.

In this chapter, finite poles and finite zeroes is only proved for plane curves and $\deg(\mathrm{div}(f)) = 0$ is proved only for elliptic curves. The general results are given in Section 8.3 in the next chapter.

Theorem 7.7.1 *Let C be a curve over $\Bbbk$ and $f \in \Bbbk(C)^*$. Then f has finitely many poles and zeroes.*

Proof (Special case of plane curves.) Let $C = V(F(x, y, z)) \subseteq \mathbb{P}^2$ where F is irreducible. By Exercise 5.2.30 there are only finitely many points at infinity, so we can restrict to the affine case $C = V(F(x, y))$.

Let $f = f_1(x, y)/f_2(x, y)$ with $f_1, f_2 \in \Bbbk[x, y]$. Then f is regular whenever $f_2(P) \neq 0$ so the poles of f are contained in $C \cap V(f_2)$. Without loss of generality $f_2(x, y)$ contains monomials featuring x. The resultant $R_x(f_2(x, y), F(x, y))$ is a polynomial in y with a finite number of roots; hence, $C \cap V(f_2)$ is finite.

To show there are finitely many zeroes write $f = f_1/f_2$. The zeroes of f are contained in $C \cap (V(f_1) \cup V(f_2))$ and the argument above applies. □

Definition 7.7.2 Let $f \in \overline{\Bbbk}(C)^*$ and define the **divisor of a function** (this is a divisor by Theorem 7.7.1)

$$\mathrm{div}(f) = \sum_{P \in C(\overline{\Bbbk})} v_P(f)(P).$$

The divisor of a function is also called a **principal divisor**. Note that some authors write $\mathrm{div}(f)$ as (f). Let

$$\mathrm{Prin}_{\Bbbk}(C) = \{\mathrm{div}(f) : f \in \Bbbk(C)^*\}.$$

Exercise 7.7.3 Show that the zero element of $\mathrm{Div}_{\Bbbk}(C)$ lies in $\mathrm{Prin}_{\Bbbk}(C)$.

Lemma 7.7.4 *Let C be a curve over $\Bbbk$ and let $f, f' \in \Bbbk(C)^*$.*

1. $\operatorname{div}(ff') = \operatorname{div}(f) + \operatorname{div}(f')$.
2. $\operatorname{div}(1/f) = -\operatorname{div}(f)$.
3. $\operatorname{div}(f + f') \geq \sum_P \min\{v_P(f), v_P(f')\}(P)$.
4. $\operatorname{div}(f^n) = n\operatorname{div}(f)$ *for* $n \in \mathbb{Z}$.
5. *Let* $f \in \overline{\Bbbk}(C)$ *and let* $\sigma \in \operatorname{Gal}(\overline{\Bbbk}/\Bbbk)$. *Then* $\operatorname{div}(\sigma(f)) = \sigma(\operatorname{div}(f))$.

Exercise 7.7.5 Prove Lemma 7.7.4.

Lemma 7.7.6 *With notation as above,* $\operatorname{Prin}_{\Bbbk}(C)$ *is a subgroup of* $\operatorname{Div}_{\Bbbk}(C)$ *under addition.*

Exercise 7.7.7 Prove Lemma 7.7.6.

Lemma 7.7.8 *In $\mathbb{P}^1(\Bbbk)$, every degree zero divisor is principal.*

Proof Let $D = \sum_{i=1}^n e_i(x_i : z_i)$ where $\sum_{i=1}^n e_i = 0$. Define

$$f(x, z) = \prod_{i=1}^{n} (xz_i - zx_i)^{e_i}. \tag{7.8}$$

Since $\sum_{i=1}^n e_i = 0$ it follows that $f(x, z)$ is a ratio of homogeneous polynomials of the same degree and therefore a rational function on $\mathbb{P}^1$. Using the uniformisers on $\mathbb{P}^1$ from Example 7.3.3 one can verify that $v_{P_i}(f) = e_i$ when $P_i = (x_i : z_i)$ and hence that $D = \operatorname{div}(f)$. □

Note that if D is defined over $\Bbbk$ then one can show that the function $f(x, z)$ in equation (7.8) is defined over $\Bbbk$.

Exercise 7.7.9 Prove that if $f \in \Bbbk(\mathbb{P}^1)$ then $\deg(\operatorname{div}(f)) = 0$.

Lemma 7.7.10 *Let $E : y^2 + H(x)y = F(x)$ be a Weierstrass equation over $\Bbbk$ and let $P = (x_i, y_i) \in E(\overline{\Bbbk})$ be a non-singular point. Then* $\operatorname{div}(x - x_i) = (P) + (\iota(P)) - 2(\mathcal{O}_E)$.

Proof There are one or two points $P \in E(\overline{\Bbbk})$ with x-coordinate equal to x_i, namely $P = (x_i, y_i)$ and $\iota(P) = \iota(P) = (x_i, -y_i - H(x_i))$ (and these are equal if and only if $2y_i + H(x_i) = 0$). By Example 7.3.4 one can take the uniformiser $t_P = t_{\iota(P)} = (x - x_i)$ unless $(\partial E/\partial y)(P) = 2y_i + H(x_i) = 0$, in which case the uniformiser is $t_P = (y - y_i)$. In the former case, we have $v_P(x - x_i) = v_{\iota(P)}(x - x_i) = 1$. In the latter case, write $F(x) = (x - x_i)g(x) + F(x_i) = (x - x_i)g(x) + y_i^2 + H(x_i)y_i$ and $H(x) = (x - x_i)a_1 + H(x_i)$. Note that $a_1 y_i - g(x_i) = (\partial E/\partial x)(P) \neq 0$ and so $g_1(x) := 1/(a_1 y - g(x)) \in \mathcal{O}_P$. Then

$$\begin{aligned}
0 &= y^2 + H(x)y - F(x)\\
&= (y - y_i)^2 + 2yy_i - y_i^2 + (x - x_i)a_1 y + H(x_i)y - (x - x_i)g(x) - y_i^2 - H(x_i)y_i\\
&= (y - y_i)^2 + (x - x_i)(a_1 y - g(x)) + (y - y_i)(2y_i + H(x_i)).
\end{aligned}$$

Hence, $x - x_i = (y - y_i)^2 g_1(x)$ and $v_P(x - x_i) = 2$. Finally, the function $(x - x_i)$ corresponds to

$$\frac{x - x_i z}{z} = \frac{x}{z} - x_i$$

on the projective curve E. Since $v_{\mathcal{O}_E}(x/z) = -2$ it follows from part 2 of Lemma 7.4.14 that $v_{\mathcal{O}_E}(x - x_i) = -2$. Hence, if $P = (x_i, y_i)$ then, in all cases, $\mathrm{div}(x - x_i) = (P) + (\iota(P)) - 2(\mathcal{O}_E)$ and $\deg(\mathrm{div}(x - x_i)) = 0$. □

Exercise 7.7.9 and Lemma 7.7.10 determine the divisor of certain functions, and in both cases they turn out to have degree zero. This is not a coincidence. Indeed, we now state a fundamental[5] result which motivates the definition of the divisor class group.

Theorem 7.7.11 *Let C be a curve over $\Bbbk$. Let $f \in \Bbbk(C)^*$. Then* $\deg(\mathrm{div}(f)) = 0$.

Proof See Theorem 8.3.14. □

Corollary 7.7.12 *Let C be a curve over $\Bbbk$ and let $f \in \Bbbk(C)^*$. The following are equivalent:*

1. $\mathrm{div}(f) \geq 0$.
2. $f \in \Bbbk^*$.
3. $\mathrm{div}(f) = 0$.

Proof Certainly statement 2 implies statement 3 and 3 implies 1. So it suffices to prove 1 implies 2. Let $f \in \Bbbk(C)^*$ be such that $\mathrm{div}(f) \geq 0$. Then f is regular everywhere, so choose some $P_0 \in C(\overline{\Bbbk})$ and define $h = f - f(P_0) \in \overline{\Bbbk}(C)$. Then $h(P_0) = 0$. If $h = 0$ then f is the constant function $f(P_0)$ and, since f is defined over $\Bbbk$, it follows that $f \in \Bbbk^*$. To complete the proof suppose that $h \neq 0$ in $\overline{\Bbbk}(C)$. Since $\deg(\mathrm{div}(h)) = 0$ by Theorem 7.7.11 it follows that h must have at least one pole. But then f has a pole, which contradicts $\mathrm{div}(f) \geq 0$. □

Corollary 7.7.13 *Let C be a curve over $\Bbbk$. Let $f, h \in \Bbbk(C)^*$. Then* $\mathrm{div}(f) = \mathrm{div}(h)$ *if and only if $f = ch$ for some $c \in \Bbbk^*$.*

Exercise 7.7.14 Prove Corollary 7.7.13.

7.8 Divisor class group

We have seen that $\mathrm{Prin}_{\Bbbk}(C) = \{\mathrm{div}(f) : f \in \Bbbk(C)^*\}$ is a subgroup of $\mathrm{Div}^0_{\Bbbk}(C)$. Hence, since all the groups are Abelian, one can define the quotient group; we call this the divisor class group. It is common to use the notation Pic for the divisor class group since the divisor

[5] This innocent-looking fact is actually the hardest result in this chapter to prove. There are several accessible proofs of the general result: Stichtenoth (Theorem I.4.11 of [529]; also see Moreno [395] Lemma 2.2) gives a proof based on "weak approximation" of valuations and this is probably the simplest proof for a reader who has already got this far through the current book; Fulton [199] gives a proof for projective plane curves based on Bézout's theorem; Silverman [505], Shafarevich [489], Hartshorne [252] and Lorenzini [355] all give proofs that boil down to ramification theory of $f : C \to \mathbb{P}^1$, and this is the argument we will give in the next chapter.

class group of a curve is isomorphic to the Picard group of a curve (even though the Picard group is usually defined differently, in terms of line bundles).

Definition 7.8.1 The (degree zero) **divisor class group** of a curve C over $\Bbbk$ is $\mathrm{Pic}^0_{\Bbbk}(C) = \mathrm{Div}^0_{\Bbbk}(C)/\mathrm{Prin}_{\Bbbk}(C)$.

We call two divisors $D_1, D_2 \in \mathrm{Div}^0_{\Bbbk}(C)$ **linearly equivalent** and write $D_1 \equiv D_2$ if $D_1 - D_2 \in \mathrm{Prin}_{\Bbbk}(C)$. The equivalence class (called a **divisor class**) of a divisor $D \in \mathrm{Div}^0_{\Bbbk}(C)$ under linear equivalence is denoted $\overline{D}$.

Example 7.8.2 By Lemma 7.7.8, $\mathrm{Pic}^0_{\Bbbk}(\mathbb{P}^1) = \{0\}$.

Theorem 7.8.3 *Let C be a curve over $\Bbbk$ and let $f \in \overline{\Bbbk}(C)$. If $\sigma(f) = f$ for all $\sigma \in \mathrm{Gal}(\overline{\Bbbk}/\Bbbk)$ then $f \in \Bbbk(C)$. If $\mathrm{div}(f)$ is defined over $\Bbbk$ then $f = ch$ for some $c \in \overline{\Bbbk}$ and $h \in \Bbbk(C)$.*

Proof The first claim follows from Remark 5.4.14 (also see Remark 8.4.8 of Section 8.4).

For the second statement let $\mathrm{div}(f)$ be defined over $\Bbbk$. Then $\mathrm{div}(f) = \sigma(\mathrm{div}(f)) = \mathrm{div}(\sigma(f))$ where the second equality follows from part 4 of Lemma 7.4.14. Corollary 7.7.13 implies $\sigma(f) = c(\sigma)f$ for some $c(\sigma) \in \overline{\Bbbk}^*$. The function $c : \mathrm{Gal}(\overline{\Bbbk}/\Bbbk) \to \overline{\Bbbk}^*$ is a 1-cocycle (the fact that $c(\sigma\tau) = \sigma(c(\tau))c(\sigma)$ is immediate, the fact that $c : \mathrm{Gal}(\overline{\Bbbk}/\Bbbk) \to \overline{\Bbbk}^*$ is continuous also follows). Hence, Theorem A.7.2 (Hilbert 90) implies that $c(\sigma) = \sigma(\gamma)/\gamma$ for some $\gamma \in \overline{\Bbbk}^*$. In other words, taking $h = f/\gamma \in \overline{\Bbbk}(C)$, we have

$$\sigma(h) = \sigma(f)/\sigma(\gamma) = f/\gamma = h.$$

By the first part of the theorem $h \in \Bbbk(C)$. □

Theorem 7.8.3 has the following important corollary, namely that $\mathrm{Pic}^0_{\Bbbk}(C)$ is a subgroup of $\mathrm{Pic}^0_{\Bbbk'}(C)$ for every extension $\Bbbk'/\Bbbk$.

Corollary 7.8.4 *Let C be a curve over $\Bbbk$ and let $\Bbbk'/\Bbbk$ be an algebraic extension. Then $\mathrm{Pic}^0_{\Bbbk}(C)$ injects into $\mathrm{Pic}^0_{\Bbbk'}(C)$.*

Proof Suppose a divisor class $\overline{D} \in \mathrm{Pic}^0_{\Bbbk}(C)$ becomes trivial in $\mathrm{Pic}^0_{\Bbbk'}(C)$. Then there is some divisor D on C defined over $\Bbbk$ such that $D = \mathrm{div}(f)$ for some $f \in \Bbbk'(C)^*$. But Theorem 7.8.3 implies $D = \mathrm{div}(h)$ for some $h \in \Bbbk(C)$ and so the divisor class is trivial in $\mathrm{Pic}^0_{\Bbbk}(C)$. □

Corollary 7.8.5 *Let $\Bbbk$ be a finite field. Let C be a curve over $\Bbbk$. Define*

$$\mathrm{Pic}^0_{\overline{\Bbbk}}(C)^{\mathrm{Gal}(\overline{\Bbbk}/\Bbbk)} = \{\overline{D} \in \mathrm{Pic}^0_{\overline{\Bbbk}}(C) : \sigma(\overline{D}) = \overline{D} \text{ for all } \sigma \in \mathrm{Gal}(\overline{\Bbbk}/\Bbbk)\}.$$

Then $\mathrm{Pic}^0_{\overline{\Bbbk}}(C)^{\mathrm{Gal}(\overline{\Bbbk}/\Bbbk)} = \mathrm{Pic}^0_{\Bbbk}(C)$.

Proof (Sketch) Let $G = \mathrm{Gal}(\overline{\Bbbk}/\Bbbk)$. Take Galois cohomology of the exact sequence

$$1 \to \overline{\Bbbk}^* \to \overline{\Bbbk}(C)^* \to \mathrm{Prin}_{\overline{\Bbbk}}(C) \to 0$$

to get

$$1 \to \Bbbk^* \to (\overline{\Bbbk}(C)^*)^G \to \mathrm{Prin}_{\overline{\Bbbk}}(C)^G \to H^1(G, \overline{\Bbbk}^*) \to H^1(G, \overline{\Bbbk}(C)^*)$$
$$\to H^1(G, \mathrm{Prin}_{\overline{\Bbbk}}(C)) \to H^2(G, \overline{\Bbbk}^*).$$

Since $(\overline{\Bbbk}(C)^*)^G = \Bbbk(C)$ (Theorem 7.8.3) and $H^1(G, \overline{\Bbbk}^*) = 0$ (Hilbert 90) we have $\mathrm{Prin}_{\overline{\Bbbk}}(C)^G = \mathrm{Prin}_{\Bbbk}(C)$. Further, $H^2(G, \overline{\Bbbk}^*) = 0$ when $\Bbbk$ is finite (see Section X.7 of [488]) and $H^1(G, \overline{\Bbbk}(C)^*) = 0$ (see Silverman Exercise X.10). Hence, $H^1(G, \mathrm{Prin}_{\overline{\Bbbk}}(C)) = 0$.

Now, take Galois cohomology of the exact sequence

$$1 \to \mathrm{Prin}_{\overline{\Bbbk}}(C) \to \mathrm{Div}^0_{\overline{\Bbbk}}(C) \to \mathrm{Pic}^0_{\overline{\Bbbk}}(C) \to 0$$

to get

$$\mathrm{Prin}_{\Bbbk}(C) \to \mathrm{Div}^0_{\overline{\Bbbk}}(C)^G \to \mathrm{Pic}^0_{\overline{\Bbbk}}(C)^G \to H^1(G, \mathrm{Prin}_{\overline{\Bbbk}}(C)) = 0.$$

Now, $\mathrm{Div}^0_{\overline{\Bbbk}}(C)^G = \mathrm{Div}^0_{\Bbbk}(C)$ by definition and so the result follows. □

We minimise the use of the word Jacobian in this book; however, we make a few remarks here. We have associated to a curve C over a field $\Bbbk$ the divisor class group $\mathrm{Pic}^0_{\Bbbk}(C)$. This group can be considered as an algebraic group. To be precise there is a variety J_C (called the **Jacobian variety** of C) that is an algebraic group (i.e., there is a morphism[6] $+ : J_C \times J_C \to J_C$) and such that, for any extension $\mathbb{K}/\Bbbk$, there is a bijective map between $\mathrm{Pic}^0_{\mathbb{K}}(C)$ and $J_C(\mathbb{K})$ that is a group homomorphism.

One can think of Pic^0 as a functor that, given a curve C over $\Bbbk$, associates with every field extension $\Bbbk'/\Bbbk$ a group $\mathrm{Pic}^0_{\Bbbk'}(C)$. The Jacobian variety of the curve is a variety J_C over $\Bbbk$ whose $\Bbbk'$-rational points $J_C(\Bbbk')$ are in one-to-one correspondence with the elements of $\mathrm{Pic}^0_{\Bbbk'}(C)$ for all $\Bbbk'/\Bbbk$. For most applications it is sufficient to work in the language of divisor class groups rather than Jacobians (despite our remarks about algebraic groups in Chapter 4).

7.9 Elliptic curves

The goal of this section is to show that the 'traditional' **chord-and-tangent rule** for elliptic curves does give a group operation. Our approach is to show that this operation coincides with addition in the divisor class group of an elliptic curve. Hence, elliptic curves are an algebraic group.

First we state the chord-and-tangent rule without justifying any of the claims or assumptions made in the description. The results later in the section will justify these claims (see Remark 7.9.4). For more details about the chord-and-tangent rule see Washington [560], Cassels [114], Reid [447] or Silverman and Tate [508].

Let $P_1 = (x_1, y_1)$ and $P_2 = (x_2, y_2)$ be points on the affine part of an elliptic curve E. Draw the line $l(x, y) = 0$ between P_1 and P_2 (if $P_1 \neq P_2$ then this is called a chord; if

[6] To make this statement precise requires showing that $J_C \times J_C$ is a variety.

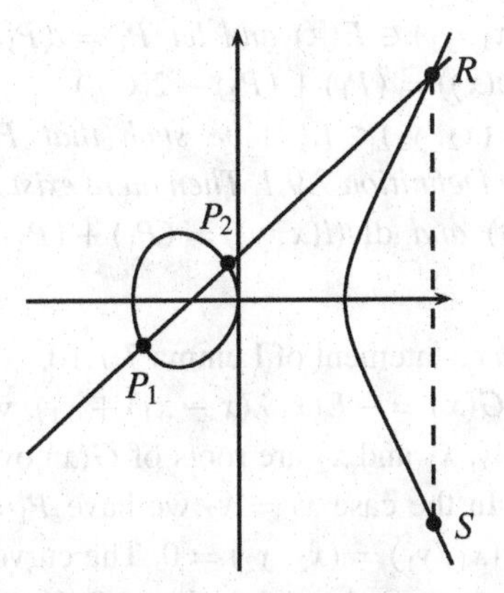

Figure 7.1 Chord and tangent rule for elliptic curve addition.

$P_1 = P_2$ then let the line be the tangent to the curve at P_1). Denote by R the third point[7] of intersection (counted according to multiplicities) of the line with the curve E. Now draw the line $v(x) = 0$ between $\mathcal{O}_E$ and R (if $R = \mathcal{O}_E$ then this is the "line at infinity" and if R is an affine point this is a vertical line so a function of x only). Denote by S the third point of intersection of this line with the curve E. Then one defines $P_1 + P_2$ to be S. Over the real numbers this operation is illustrated in Figure 7.1.

We now transform the above geometric description into algebra, and show that the points R and S do exist. The first step is to write down the equation of the line between $P_1 = (x_1, y_1)$ and $P_2 = (x_2, y_2)$. We state the equation of the line as a definition and then show that it corresponds to a function with the correct divisor.

Definition 7.9.1 Let $E(x, y)$ be a Weierstrass equation for an elliptic curve over $\Bbbk$. Let $P_1 = (x_1, y_1)$, $P_2 = (x_2, y_2) \in E(\Bbbk) \cap \mathbb{A}^2$. If $P_1 = \iota(P_2)$ then the line between P_1 and P_2 is[8] $v(x) = x - x_1$.

If $P_1 \neq \iota(P_2)$ then there are two subcases. If $P_1 = P_2$ then define $\lambda = (3x_1^2 + 2a_2x_1 + a_4)/(2y_1 + a_1x_1 + a_3)$ and if $P_1 \neq P_2$ then define $\lambda = (y_2 - y_1)/(x_2 - x_1)$. The line between P_1 and P_2 is then

$$l(x, y) = y - \lambda(x - x_1) - y_1.$$

We stress that whenever we write $l(x, y)$ then we are implicitly assuming that it is not a vertical line $v(x)$.

Warning: Do not confuse the line $v(x)$ with the valuation v_P. The notation $v(P)$ means the line evaluated at the point P. The notation $v_P(x)$ means the valuation of the function x at the point P.

Exercise 7.9.2 Let the notation be as in Definition 7.9.1. Show that if $P_1 = \iota(P_2)$ then $v(P_1) = v(P_2) = 0$ and if $P_1 \neq \iota(P_2)$ then $l(P_1) = l(P_2) = 0$.

[7] Possibly this point is at infinity.
[8] This includes the case $P_1 = P_2 = \iota(P_1)$.

Lemma 7.9.3 *Let* $P_1 = (x_1, y_1) \in E(\Bbbk)$ *and let* $P_2 = \iota(P_1)$*. Let* $v(x) = (x - x_1)$ *as in Definition 7.9.1. Then* $\mathrm{div}(v(x)) = (P_1) + (P_2) - 2(\mathcal{O}_E)$.

Let $P_1 = (x_1, y_1), P_2 = (x_2, y_2) \in E(\Bbbk)$ *be such that* $P_1 \neq \iota(P_2)$ *and let* $l(x, y) = y - \lambda(x - x_1) - y_1$ *be as in Definition 7.9.1. Then there exists* $x_3 \in \Bbbk$ *such that* $E(x, \lambda(x - x_1) + y_1) = -\prod_{i=1}^{3}(x - x_i)$ *and* $\mathrm{div}(l(x, y)) = (P_1) + (P_2) + (R) - 3(\mathcal{O}_E)$ *where* $R = (x_3, \lambda(x_3 - x_1) + y_1)$.

Proof The first part is just a restatement of Lemma 7.7.10.

For the second part set $G(x) = -E(x, \lambda(x - x_1) + y_1)$, which is a monic polynomial over $\Bbbk$ of degree 3. Certainly, x_1 and x_2 are roots of $G(x)$ over $\Bbbk$ so if $x_1 \neq x_2$ then $G(x)$ has a third root x_3 over $\Bbbk$. In the case $x_1 = x_2$ we have $P_1 = P_2 \neq \iota(P_2)$. Make a linear change of variables so that $(x_1, y_1) = (x_2, y_2) = 0$. The curve equation is $E(x, y) = y^2 + a_1xy + a_3y - (x^3 + a_2x^2 + a_4x)$ and $a_3 \neq 0$ since $(0, 0) \neq \iota(0, 0)$. Now, by definition, $l(x, y) = a_4x/a_3$ and one has

$$G(x) = E(x, a_4x/a_3) = (a_4x/a_3)^2 + a_1x(a_4x/a_3) + a_4x - (x^3 + a_2x^2 + a_4x)$$

which is divisible by x^2. Hence, $G(x)$ splits completely over $\Bbbk$.

For the final part we consider $l(x, y)$ as a function on the affine curve. By Lemma 7.4.14 and Lemma 7.4.16 we have $v_{\mathcal{O}_E}(l(x, y)) = \min\{v_{\mathcal{O}_E}(y), v_{\mathcal{O}_E}(x), v_{\mathcal{O}_E}(1)\} = -3$. Since $\deg(\mathrm{div}(l(x, y))) = 0$ there are three affine zeroes counted according to multiplicity.

Define $\bar{l}(x, y) = y + (a_1x + a_3) + \lambda(x - x_1) + y_1$. Note that $\bar{l} = -l \circ \iota$ so $v_P(l(x, y)) = v_{\iota(P)}(\bar{l}(x, y))$. One can check that

$$l(x, y)\bar{l}(x, y) = -E(x, \lambda(x - x_1) + y_1) = \prod_{i=1}^{3}(x - x_i) \tag{7.9}$$

where the first equality is equivalence modulo $E(x, y)$, not equality of polynomials. Hence, for any point $P \in E(\Bbbk)$,

$$v_P(l(x, y)) + v_P(\bar{l}(x, y)) = v_P\left(\prod_{i=1}^{3}(x - x_i)\right).$$

Write $P_i = (x_i, y_i)$, let e_i be the multiplicity of x_i in the right-hand side of equation (7.9) and recall that $v_{P_i}(x - x_i) = 1$ if $P_i \neq \iota(P_i)$ and 2 otherwise. Also note that $l(P_i) = 0$ implies $\bar{l}(P_i) \neq 0$ unless $P_i = \iota(P_i)$, in which case $v_{P_i}(l(x, y)) = v_{P_i}(\bar{l}(x, y))$. It follows that $v_{P_i}(l(x, y)) = e_i$, which proves the result. □

Remark 7.9.4 It follows from the above results that it does make sense to speak of the "third point of intersection" R of $l(x, y)$ with E and to call $l(x, y)$ a tangent line in the case when $P_1 = P_2$. Hence, we have justified the assumptions made in the informal description of the chord-and-tangent rule.

Exercise 7.9.5 Let $E(x, y, z)$ be a Weierstrass equation for an elliptic curve. The line $z = 0$ is called the line at infinity on E. Show that $z = 0$ only passes through $(0, 0)$ on the affine curve given by the equation $E(x, 1, z) = 0$.

Exercise 7.9.6 Prove that the following algebraic formulae for the chord-and-tangent rule are correct. Let $P_1, P_2 \in E(\Bbbk)$, we want to compute $S = P_1 + P_2$. If $P_1 = \mathcal{O}_E$ then $S = P_2$ and if $P_2 = \mathcal{O}_E$ then $S = P_1$. Hence, we may now assume that $P_1 = (x_1, y_1)$ and $P_2 = (x_2, y_2)$ are affine. If $y_2 = -y_1 - H(x_1)$ then $S = \mathcal{O}_E$. Otherwise, set λ to be as in Definition 7.9.1 and compute $x_3 = \lambda^2 + a_1\lambda - a_2 - x_1 - x_2$ and $y_3 = -\lambda(x_S - x_1) - y_1$. The sum is $S = (x_3, y_3)$.

Before proving the main theorem, we state the following technical result, whose proof is postponed to the next chapter (Corollary 8.6.5).

Theorem 7.9.7 *Let $P_1, P_2 \in E(\Bbbk)$ be a points on an elliptic curve such that $P_1 \neq P_2$. Then $(P_1) - (P_2)$ is not a principal divisor.*

We now consider the divisor class group $\mathrm{Pic}^0_{\Bbbk}(E)$. The following result is usually obtained as a corollary to the Riemann–Roch theorem, but we give an ad-hoc proof for elliptic curves. One can consider this result as the Abel–Jacobi map in the case of genus 1 curves.

Theorem 7.9.8 *There is a one-to-one correspondence between $E(\Bbbk)$ and $\mathrm{Pic}^0_{\Bbbk}(E)$, namely $P \mapsto (P) - (\mathcal{O}_E)$.*

Proof We first show that the map is injective. Suppose $(P_1) - (\mathcal{O}_E) \equiv (P_2) - (\mathcal{O}_E)$. Then $(P_1) - (P_2)$ is principal, and so Theorem 7.9.7 implies $P_1 = P_2$.

It remains to show that the map is surjective. Let $D = \sum_P n_P(P)$ be any effective divisor on E. We prove that D is equivalent to a divisor of the form

$$(P) + (\deg(D) - 1)(\mathcal{O}_E). \tag{7.10}$$

We will do this by replacing any term $(P_1) + (P_2)$ by a term of the form $(S) + (\mathcal{O}_E)$ for some point S.

The key equations are $(P) + (\iota(P)) = 2(\mathcal{O}_E) + \mathrm{div}(v(x))$ where $v(x)$ is as in Definition 7.9.1, or, if $P_1 \neq \iota(P_2)$, $(P_1) + (P_2) = (S) + (\mathcal{O}_E) + \mathrm{div}(l(x, y)/v(x))$. The first equation allows us to replace any pair $(P) + (\iota(P))$, including the case $P = \iota(P)$, by $2(\mathcal{O}_E)$. The second equation allows us to replace any pair $(P_1) + (P_2)$, where $P_1 \neq \iota(P_2)$ (but including the case $P_1 = P_2$) with $(S) + (\mathcal{O}_E)$. It is clear that any pair of affine points is included in one of these two cases, and so repeating these operations a finite number of times reduces any effective divisor to the form in equation (7.10).

Finally, let D be a degree zero divisor on E. Write $D = D_1 - D_2$ where D_1 and D_2 are effective divisors of the same degree. By the above argument, we can write $D_1 \equiv (S_1) + (\deg(D_1) - 1)(\mathcal{O}_E)$ and $D_2 \equiv (S_2) + (\deg(D_1) - 1)(\mathcal{O}_E)$. Hence, $D \equiv (S_1) - (S_2)$. Finally, adding the divisor of the vertical line function through S_2 and subtracting the divisor of the line between S_1 and $\iota(S_2)$ gives $D \equiv (S) - (\mathcal{O}_E)$ for some point S as required. □

Since $E(\Bbbk)$ is in bijection with the group $\mathrm{Pic}^0_{\Bbbk}(E)$ it follows that $E(\Bbbk)$ is a group, with the group law coming from the divisor class group structure of E. It remains to show that the group law is just the chord-and-tangent rule. In other words, this result shows that the chord-and-tangent rule is associative. Note that many texts prove that both $E(\Bbbk)$ and $\mathrm{Pic}^0_{\Bbbk}(E)$ are groups and then prove that the map $P \mapsto (P) - (\mathcal{O}_E)$ is a group homomorphism; whereas we use this map to prove that $E(\Bbbk)$ is a group.

Theorem 7.9.9 *Let E be an elliptic curve over a field $\Bbbk$. The group law induced on $E(\Bbbk)$ by pulling back the divisor class group operations via the bijection of Theorem 7.9.8 is the chord-and-tangent rule.*

Proof Let $P_1, P_2 \in E(\Bbbk)$. To add these points, we map them to divisor classes $(P_1) - (\mathcal{O}_E)$ and $(P_2) - (\mathcal{O}_E)$ in $\mathrm{Pic}^0_{\Bbbk}(E)$. Their sum is $(P_1) + (P_2) - 2(\mathcal{O}_E)$, which is reduced to the form $(S) - (\mathcal{O}_E)$ by applying the rules in the proof of Theorem 7.9.8. In other words, we get $(P_1) + (P_2) - 2(\mathcal{O}_E) = (S) - (\mathcal{O}_E) + \mathrm{div}(f(x, y))$ where $f(x, y) = v(x)$ if $P_1 = \iota(P_2)$ or $f(x, y) = l(x, y)/v(x)$ in the general case, where $l(x, y)$ and $v(x)$ are the lines from Definition 7.9.1. Since these are precisely the same lines as in the description of the chord-and-tangent rule it follows that the point S is the same point as produced by the chord-and-tangent rules. □

A succinct way to describe the elliptic curve addition law (since there is a single point at infinity) is that three points sum to zero if they lie on a line. This is simply a restatement of the fact that if P, Q and R line on the line $l(x, y, z) = 0$ then the divisor $(P) + (Q) + (R) - 3(\mathcal{O}_E)$ is a principal divisor.

Exercise 7.9.10 One can choose any $\Bbbk$-rational point $P_0 \in E(\Bbbk)$ and define a group law on $E(\Bbbk)$ such that P_0 is the identity element. The sum of points P and Q is defined as follows: let l be the line through P and Q (taking the tangent if $P = Q$, which uniquely exists since E is non-singular). Then l hits E at a third point (counting multiplicities) R. Draw a line v between P_0 and R. This hits E at a third point (again counting with multiplicities) S. Then $P + Q$ is defined to be the point S. Show that this operation satisfies the axioms of a group.

8
Rational maps on curves and divisors

The purpose of this chapter is to develop some tools in the theory of algebraic curves that are needed for the applications (especially, hyperelliptic curve cryptography). The technical machinery in this chapter is somewhat deeper than the previous one and readers can skip this chapter if they wish.

The reader should note that the word "curve" in this chapter always refers to a non-singular curve.

8.1 Rational maps of curves and the degree

Lemma 8.1.1 *Let C be a curve over $\Bbbk$ and $f \in \Bbbk(C)$. One can associate with f a rational map $\phi : C \to \mathbb{P}^1$ over $\Bbbk$ by $\phi = (f : 1)$. (Indeed, this is a morphism by Lemma 7.3.6.) Denote by ∞ the constant map $\infty(P) = (1 : 0)$. Then there is a one-to-one correspondence between $\Bbbk(C) \cup \{\infty\}$ and the set of morphisms $\phi : C \to \mathbb{P}^1$.*

Exercise 8.1.2 Prove Lemma 8.1.1.

Note that since $\Bbbk(C) \cup \{\infty\}$ is not a field, it does not make sense to interpret the set of rational maps $\phi : C \to \mathbb{P}^1$ as a field.

Lemma 8.1.3 *Let C_1 and C_2 be curves over $\Bbbk$ (in particular, non-singular and projective) and let $\phi : C_1 \to C_2$ be a non-constant rational map over $\Bbbk$. Then ϕ is a dominant morphism.*

Proof See Proposition II.6.8 of Hartshorne [252] or Proposition II.2.1 of Silverman [505]. □

The notion of degree of a mapping is fundamental in algebra and topology; a degree d map is "d-to-one on most points".

Example 8.1.4 Let $\Bbbk$ be a field of characteristic not equal to 2. The morphism $\phi : \mathbb{A}^1(\Bbbk) \to \mathbb{A}^1(\Bbbk)$ given by $\phi(x) = x^2$ is clearly two-to-one away from the point $x = 0$. We say that ϕ has degree 2.

Example 8.1.4 suggests several possible definitions for degree: the first in terms of the number of preimages of a general point in the image; the second in terms of the degrees of the polynomials defining the map. A third definition is to recall the injective field homomorphism $\phi^* : \Bbbk(\mathbb{A}^1) \to \Bbbk(\mathbb{A}^1)$. One sees that $\phi^*(\Bbbk(\mathbb{A}^1)) = \Bbbk(x^2) \subseteq \Bbbk(x)$ and that $[\Bbbk(x) : \Bbbk(x^2)] = 2$. This latter formulation turns out to be a suitable definition for degree.

Theorem 8.1.5 *Let C_1, C_2 be curves over $\Bbbk$. Let $\phi : C_1 \to C_2$ be a non-constant rational map over $\Bbbk$. Then $\Bbbk(C_1)$ is a finite algebraic extension of $\phi^*(\Bbbk(C_2))$.*

Proof Theorem II.2.4(a) of Silverman [505]. □

Definition 8.1.6 Let $\phi : C_1 \to C_2$ be a non-constant rational map of curves over $\Bbbk$. The **degree** of ϕ is $[\Bbbk(C_1) : \phi^*(\Bbbk(C_2))]$.

Let F be a field such that $\phi^*(\Bbbk(C_2)) \subset F \subset \Bbbk(C_1)$ and $\Bbbk(C_1)/F$ is separable and $F/\phi^*(\Bbbk(C_2))$ is purely inseparable (recall the notion of separability from Section A.6). The **separable degree** of ϕ is $\deg_s(\phi) = [\Bbbk(C_1) : F]$ and the **inseparable degree** of ϕ is $\deg_i(\phi) = [F : \phi^*(\Bbbk(C_2))]$.

A non-constant rational map of curves is called **separable** (respectively, **inseparable**) if its inseparable (respectively, separable) degree is 1.

Example 8.1.7 Let $\Bbbk = \mathbb{F}_p$. The **Frobenius map** $\pi_p : \mathbb{A}^1(\overline{\Bbbk}) \to \mathbb{A}^1(\overline{\Bbbk})$ is given by $\pi_p(x) = x^p$. Since $\Bbbk(\mathbb{A}^1) = \Bbbk(x)$ and $\pi_p^*(\Bbbk(\mathbb{A}^1)) = \Bbbk(x^p)$ it follows that $\Bbbk(x)/\pi_p^*(\Bbbk(\mathbb{A}^1)) = \Bbbk(x)/\Bbbk(x^p)$ is inseparable of degree p. Hence, $\deg_s(\pi_p) = 1$ and $\deg(\pi_p) = \deg_i(\pi_p) = p$. Note that π_p is one-to-one on $\mathbb{A}^1(\overline{\mathbb{F}}_p)$.

Lemma 8.1.8 *Let $\phi : \mathbb{A}^1 \to \mathbb{A}^1$ be a non-constant morphism over $\Bbbk$ given by $\phi(x) = a(x)$ for some polynomial $a(x) \in \Bbbk[x]$. Then $\deg(\phi) = \deg_x(a(x))$.*

Proof Let $\theta = a(x)$. We have $\phi^*(\Bbbk(\mathbb{A}^1)) = \Bbbk(\theta) \subseteq \Bbbk(x)$ and we are required to determine $[\Bbbk(x) : \Bbbk(\theta)]$. We claim the minimal polynomial of x over $\Bbbk(\theta)$ is given by

$$F(T) = a(T) - \theta.$$

First, it is clear that $F(x) = 0$. Second, it follows from Eisenstein's criteria (see Proposition III.1.14 of [529], Theorem IV.3.1 of [329] or Theorem III.6.15 of [271]), taking for example the place (i.e., valuation) at infinity in $\Bbbk(\theta)$, that $F(T)$ is irreducible. Since $\deg_T(F(T)) = \deg_x(a(x))$ the result follows. □

Lemma 8.1.9 *Let $\phi : \mathbb{A}^1 \to \mathbb{A}^1$ be a non-constant rational map over $\Bbbk$ given by $\phi(x) = a(x)/b(x)$ where $\gcd(a(x), b(x)) = 1$. Then $\deg(\phi) = \max\{\deg_x(a(x)), \deg_x(b(x))\}$.*

Proof Let $\theta = a(x)/b(x)$ so that $\phi^*(\Bbbk(\mathbb{A}^1)) = \Bbbk(\theta) \subseteq \Bbbk(x)$. Since $\Bbbk(\theta) = \Bbbk(1/\theta)$ we may assume $\deg_x(a(x)) \geq \deg_x(b(x))$. If these degrees are equal then one can reduce the degree of $a(x)$ by using $\Bbbk(a(x)/b(x)) = \Bbbk((a(x) - cb(x))/b(x))$ for a suitable $c \in \Bbbk$; replacing θ by $1/\theta$ again we may assume that $\deg_x(a(x)) > \deg_x(b(x))$. We may also assume that $a(x)$ and $b(x)$ are monic.

We claim the minimal polynomial of x over $\Bbbk(\theta)$ is given by

$$F(T) = a(T) - \theta b(T).$$

To see this, first note that $F(x) = 0$. Now, $a(T) - \theta b(T)$ is irreducible in $\Bbbk[\theta, T]$ since it is linear in θ. The irreducibility of $F(T)$ in $\Bbbk(\theta)[T]$ then follows from the Gauss Lemma (see, for example, Lemma III.6.13 of Hungerford [271]). □

Exercise 8.1.10 Let $C_1 : y^2 = x^3$ and $C_2 : Y^2 = X$ over a field $\Bbbk$ of characteristic not equal to 2 and consider the map $\phi : C_1 \to C_2$ such that $\phi(x, y) = (x, y/x)$. Show that $\deg(\phi) = 1$.

Exercise 8.1.11 Let $C_1 : y^2 = x^6 + 2x^2 + 1$ and $C_2 : Y^2 = X^3 + 2X + 1$ over a field $\Bbbk$ of characteristic not equal to 2 and consider the map $\phi : C_1 \to C_2$ such that $\phi(x, y) = (x^2, y)$. Show that $\deg(\phi) = 2$.

Exercise 8.1.12 Let C_1, C_2 and C_3 be curves over $\Bbbk$ and let $\psi : C_1 \to C_2$ and $\phi : C_2 \to C_3$ be morphisms over $\Bbbk$. Show that $\deg(\phi \circ \psi) = \deg(\phi)\deg(\psi)$.

Lemma 8.1.13 *Let C_1 and C_2 be curves over $\Bbbk$ (in particular, smooth and projective). Let $\phi : C_1 \to C_2$ be a birational map over $\Bbbk$. Then ϕ has degree 1.*

For Lemma 8.1.15 (and Lemma 8.2.3) we need the following technical result. This is a special case of weak approximation; see Stichtenoth [529] for a presentation that uses similar techniques to obtain most of the results in this chapter.

Lemma 8.1.14 *Let C be a curve over $\Bbbk$ and let $Q, Q' \in C(\overline{\Bbbk})$ be distinct points. Then there is a function $f \in \overline{\Bbbk}(C)$ such that $v_Q(f) = 0$ and $v_{Q'}(f) > 0$.*

Proof By Lemma 7.1.17 we have $\mathcal{O}_{Q',\overline{\Bbbk}} \not\subseteq \mathcal{O}_{Q,\overline{\Bbbk}}$ (and vice versa). Hence, there exists a function $u \in \mathcal{O}_{Q,\overline{\Bbbk}} - \mathcal{O}_{Q',\overline{\Bbbk}}$. Then $v_Q(u) \geq 0$ while $v_{Q'}(u) < 0$. If $u(Q) = -1$ then set $f = 1/(1 + u^2)$ else set $f = 1/(1 + u)$. Then $v_Q(f) = 0$ and $v_{Q'}(f) > 0$ as required. □

Lemma 8.1.15 *Let C_1 and C_2 be curves over $\Bbbk$ (in particular, smooth and projective). Let $\phi : C_1 \to C_2$ be a rational map over $\Bbbk$ of degree 1. Then ϕ is an isomorphism.*

Proof Corollary II.2.4 of Silverman [505]. □

8.2 Extensions of valuations

Let $\phi : C_1 \to C_2$ be a non-constant morphism of curves over $\Bbbk$. Then $F_1 = \Bbbk(C_1)$ is a finite extension of $F_2 = \phi^*(\Bbbk(C_2))$. We now study the preimages of points $Q \in C_2(\overline{\Bbbk})$ under ϕ and a notion of multiplicity of preimages of Q (namely, ramification indices). The main result is Theorem 8.2.9.

There are several approaches to these results in the literature. One method, which unifies algebraic number theory and the theory of curves, is to note that if U is an open subset of C then $\Bbbk[U]$ is a Dedekind domain. The splitting of the maximal ideal $\mathfrak{m}_Q$ of $\Bbbk[U]$

(for $Q \in U$) in the integral closure of $\phi^*(\Bbbk[U])$ in $\Bbbk(C_1)$ yields the results. Details of this approach are given in Section VII.5 of Lorenzini [355], Section I.4 of Serre [488] (especially Propositions I.10 and I.11), Chapter 1 of Lang [327] and Chapter XII of Lang [329]. An analogous ring-theoretic formulation is used in Proposition II.6.9 of Hartshorne [252]. A different method is to study extensions of valuations directly, for example see Section III.1 of Stichtenoth [529]. Note that, since we consider points over $\overline{\Bbbk}$, the notion of residue degree does not arise, which simplifies the presentation compared with many texts.

Definition 8.2.1 Let F_2 be a field of transcendence degree 1 over $\overline{\Bbbk}$. Let F_1/F_2 be a finite extension. Let v be a discrete valuation on F_2. A valuation v' on F_1 is an **extension** of v (or, v is the **restriction** of v') if $\{f \in F_2 : v(f) \geq 0\} = \{f \in F_2 : v'(f) \geq 0\}$. We write $v' \mid v$ if this is the case.

Note that if v' is an extension of v as above then one does not necessarily have $v'(f) = v(f)$ for all $f \in F_2$ (indeed, we will see later that $v'(f) = ev(f)$ for some $e \in \mathbb{N}$).

Let $\phi : C_1 \rightarrow C_2$ be a morphism of curves and let $F_2 = \phi^*(\Bbbk(C_2))$ and $F_1 = \Bbbk(C_1)$. We now explain the relation between extensions of valuations from F_2 to F_1 and preimages of points under ϕ.

Lemma 8.2.2 *Let $\phi : C_1 \rightarrow C_2$ be a non-constant morphism of curves over $\Bbbk$ (this is shorthand for C_1, C_2 and ϕ all being defined over $\Bbbk$). Let $P \in C_1(\overline{\Bbbk})$ and $Q \in C_2(\overline{\Bbbk})$. Denote by v the valuation on $\phi^*(\Bbbk(C_2)) \subseteq \Bbbk(C_1)$ defined by $v(\phi^*(f)) = v_Q(f)$ for $f \in \Bbbk(C_2)$. If $\phi(P) = Q$ then v_P is an extension of v.*

Proof Let $f \in \Bbbk(C_2)$. Since $\phi(P) = Q$ we have $\phi^*(f) = f \circ \phi$ regular at P if and only if f is regular at Q. Hence, $v_P(\phi^*(f)) \geq 0$ if and only if $v_Q(f) \geq 0$. It follows that $v_P \mid v$. □

Lemma 8.2.3 *Let the notation be as in Lemma 8.2.2. In particular, $P \in C_1(\overline{\Bbbk})$, $Q \in C_2(\overline{\Bbbk})$, v_P is the corresponding valuation on $F_1 = \overline{\Bbbk}(C_1)$ and v is the valuation on $\phi^*(\overline{\Bbbk}(C_2))$ corresponding to v_Q on $\overline{\Bbbk}(C_2)$. Then $v_P \mid v$ implies $\phi(P) = Q$.*

In other words, Lemmas 8.2.2 and 8.2.3 show that $\phi(P) = Q$ if and only if the maximal ideal $\mathfrak{m}_P$ in $\mathcal{O}_P \subseteq \Bbbk(C_1)$ contains $\phi^*(\mathfrak{m}_Q)$ where $\mathfrak{m}_Q$ is the maximal ideal in $\mathcal{O}_Q \subseteq \Bbbk(C_2)$. This is the connection between the behaviour of points under morphisms and the splitting of ideals in Dedekind domains.

We already know that a non-constant morphism of curves is dominant, but the next result makes the even stronger statement that a morphism is surjective.

Theorem 8.2.4 *Let C_1 and C_2 be curves over $\Bbbk$ (in particular, they are projective and non-singular). Let $\phi : C_1 \rightarrow C_2$ be a non-constant morphism of curves over $\Bbbk$. Then ϕ is surjective from $C_1(\overline{\Bbbk})$ to $C_2(\overline{\Bbbk})$.*

Proof Proposition VII.5.7 of Lorenzini [355]. □

Definition 8.2.5 Let C_1 and C_2 be curves over $\Bbbk$ and let $\phi : C_1 \to C_2$ be a non-constant rational map over $\Bbbk$. Let $P \in C_1(\overline{\Bbbk})$. The **ramification index** of ϕ at P is

$$e_\phi(P) = v_P(\phi^*(t_{\phi(P)}))$$

where $t_{\phi(P)}$ is a uniformiser on C_2 at $\phi(P)$. If $e_\phi(P) = 1$ for all $P \in C_1(\overline{\Bbbk})$ then ϕ is **unramified**.

We now show that this definition agrees with Definition III.1.5 of Stichtenoth [529].

Lemma 8.2.6 *Let $\phi : C_1 \to C_2$ be a non-constant morphism of curves over $\Bbbk$. Let $P \in C_1(\overline{\Bbbk})$, $Q = \phi(P) \in C_2(\overline{\Bbbk})$ and $f \in \overline{\Bbbk}(C_2)$. Then*

$$v_P(\phi^*(f)) = e_\phi(P) v_Q(f).$$

Proof Let $v_Q(f) = n$ and write $f = t_Q^n h$ for some $h \in \overline{\Bbbk}(C_2)$ such that $h(Q) \neq 0$. Then $\phi^*(f) = \phi^*(t_Q)^n \phi^*(h)$ and $v_P(\phi^*(h))) = 0$. The result follows since $v_P(\phi^*(t_Q)^n) = n v_P(\phi^*(t_Q))$. □

Exercise 8.2.7 Let $\phi : C_1 \to C_2$ be a non-constant rational map of curves over $\Bbbk$. Let $P \in C_1(\overline{\Bbbk})$, $Q = \phi(P)$, and suppose $e_\phi(P) = 1$. Show that $t \in \overline{\Bbbk}(C_2)$ is a uniformiser at Q if and only if $\phi^*(t)$ is a uniformiser at P.

Exercise 8.2.8 Let $\phi : C_1 \to C_2$ be an isomorphism of curves over $\Bbbk$. Show that ϕ is unramified.

The following result is of fundamental importance.

Theorem 8.2.9 *Let C_1 and C_2 be curves over $\Bbbk$ and let $\phi : C_1 \to C_2$ be a non-constant rational map over $\Bbbk$. Then for all $Q \in C_2(\overline{\Bbbk})$ we have*

$$\sum_{P \in C_1(\overline{\Bbbk}) : \phi(P) = Q} e_\phi(P) = \deg(\phi).$$

Proof As mentioned above, one can see this by noting that $\phi^*(\mathcal{O}_Q)$ and $\phi^*(\Bbbk[U])$ (for an open set $U \subseteq C_2$ with $Q \in U$) are Dedekind domains and studying the splitting of $\mathfrak{m}_Q$ in their integral closures in $\Bbbk(C_1)$. For details see any of Proposition 1.10 and 1.11 of Serre [488], Corollary XII.6.3 of Lang [329], Proposition I.21 of Lang [327], Theorem III.3.5 of Lorenzini [355], Proposition II.6.9 of Hartshorne [252], or Theorem III.1.11 of Stichtenoth [529]. □

Corollary 8.2.10 *If $\phi : C_1 \to C_2$ is a rational map of degree d and $Q \in C_2(\overline{\Bbbk})$ then there are at most d points $P \in C_1(\overline{\Bbbk})$ such that $\phi(P) = Q$.*

Furthermore, if ϕ is separable then there is an open subset $U \subseteq C_2$ such that for all $Q \in U$ one has $\#\phi^{-1}(Q) = d$.

Proof The first statement is immediate. The second follows by choosing U to be the complement of points corresponding to factors of the discriminant of $\Bbbk(C_1)/\phi^*(\Bbbk(C_2))$; see Proposition VII.5.7 of Lorenzini [355]. □

Example 8.2.11 Consider $\phi : \mathbb{A}^1 \to \mathbb{A}^1$ given by $\phi(x) = x^2$ as in Example 8.1.4. This extends to the morphism $\phi : \mathbb{P}^1 \to \mathbb{P}^1$ given by $\phi((x : z)) = (x^2/z^2 : 1)$, which is regular at $\infty = (1 : 0)$ via the equivalent formula $(1 : z^2/x^2)$. One has $\phi^{-1}((a : 1)) = \{(\sqrt{a} : 1), (-\sqrt{a} : 1)\}, \phi^{-1}((0 : 1)) = \{(0 : 1)\}$ and $\phi^{-1}((1 : 0)) = \{(1 : 0)\}$. At a point $Q = (a : 1)$ with $a \neq 0$ one has uniformiser $t_Q = x/z - a$ and

$$\phi^*(t_Q) = x^2/z^2 - a = (x/z - \sqrt{a})(x/z + \sqrt{a}).$$

Writing $P = (\sqrt{a} : 1)$ one has $\phi(P) = Q$ and $e_\phi(P) = 1$. However, one can verify that $e_\phi((0 : 1)) = e_\phi((1 : 0)) = 2$.

Lemma 8.2.12 *Let $\phi : C_1 \to C_2$ and $\psi : C_2 \to C_3$ be non-constant morphisms of curves over $\Bbbk$. Let $P \in C_1(\overline{\Bbbk})$. Then $e_{\psi\circ\phi}(P) = e_\phi(P)e_\psi(\phi(P))$.*

Exercise 8.2.13 Prove Lemma 8.2.12.

Exercise 8.2.14 Let $\phi : C_1 \to C_2$ be defined over $\Bbbk$. Let $P \in C_1(\overline{\Bbbk})$ and let $\sigma \in \mathrm{Gal}(\overline{\Bbbk}/\Bbbk)$. Show that $e_\phi(\sigma(P)) = e_\phi(P)$.

8.3 Maps on divisor classes

We can now define some important maps on divisors that will be used in several proofs later. In particular, this will enable an elegant proof of Theorem 7.7.11 for general curves.

Definition 8.3.1 Let $\phi : C_1 \to C_2$ be a non-constant morphism over $\Bbbk$. Define the **pullback**

$$\phi^* : \mathrm{Div}_{\overline{\Bbbk}}(C_2) \to \mathrm{Div}_{\overline{\Bbbk}}(C_1)$$

as follows. For $Q \in C_2(\overline{\Bbbk})$ define $\phi^*(Q) = \sum_{P\in\phi^{-1}(Q)} e_\phi(P)(P)$ and extend ϕ^* to $\mathrm{Div}_{\overline{\Bbbk}}(C_2)$ by linearity, i.e.

$$\phi^*\left(\sum_{Q\in C_2(\overline{\Bbbk})} n_Q(Q)\right) = \sum_{Q\in C_2(\overline{\Bbbk})} n_Q\phi^*(Q).$$

Note that, since $\mathrm{Div}_{\overline{\Bbbk}}(C_2)$ and $\mathrm{Div}_{\overline{\Bbbk}}(C_1)$ are not varieties, it does not make sense to ask whether ϕ^* is a rational map or morphism.

Example 8.3.2 Consider $\phi : \mathbb{A}^1 \to \mathbb{A}^1$ given by $\phi(x) = x^2$. Let $D = (0) + (1)$ be a divisor on $\mathbb{A}^1$. Then $\phi^*(D) = 2(0) + (1) + (-1)$.

Let $\phi : C_1 \to C_2$ be a non-constant morphism over $\Bbbk$ and let $P \in C_2(\overline{\Bbbk})$. Then the divisor $\phi^*(P)$ is also called the **conorm** of P with respect to $\Bbbk(C_1)/\phi^*(\Bbbk(C_2))$ (see Definition III.1.8 of Stichtenoth [529]).

Lemma 8.3.3 *Let C be a curve over $\Bbbk$ and let $f \in \Bbbk(C)^*$ be a non-constant rational function. Define the rational map $\phi : C \to \mathbb{P}^1$ by $\phi = (f : 1)$ (in future we will write f instead of ϕ). Then ϕ is a morphism and $\mathrm{div}(f) = \phi^*((0 : 1) - (1 : 0))$.*

There is a natural map ϕ_* on divisors that is called the **pushforward** (it is called the **divisor-norm map** in Section VII.7 of Lorenzini [355]).

Definition 8.3.4 Let $\phi : C_1 \to C_2$ be a non-constant morphism of curves. Define the **pushforward**

$$\phi_* : \mathrm{Div}_{\overline{\Bbbk}}(C_1) \to \mathrm{Div}_{\overline{\Bbbk}}(C_2)$$

by $\phi_*(P) = \phi(P)$ and extend to the whole of $\mathrm{Div}_{\overline{\Bbbk}}(C_1)$ by linearity.

It remains to find a map from $\Bbbk(C_1)$ to $\Bbbk(C_2)$ that corresponds (in the sense of property 4 of Theorem 8.3.8) to the pushforward. This is achieved using the norm map with respect to the extension $\Bbbk(C_1)/\phi^*(\Bbbk(C_2))$. As we will show in Lemma 8.3.13, this norm satisfies, for $f \in \Bbbk(C_1)$ and $Q \in C_2(\overline{\Bbbk})$, $\mathrm{N}_{\Bbbk(C_1)/\phi^*(\Bbbk(C_2))}(f)(Q) = \prod_{\phi(P)=Q} f(P)^{e_\phi(P)}$.

Definition 8.3.5 Let C_1, C_2 be curves over $\Bbbk$ and let $\phi : C_1 \to C_2$ be a non-constant rational map. Let $N_{\Bbbk(C_1)/\phi^*\Bbbk(C_2)}$ be the usual norm map in field theory (see Section A.6). Define

$$\phi_* : \Bbbk(C_1) \to \Bbbk(C_2)$$

by $\phi_*(f) = (\phi^*)^{-1}(\mathrm{N}_{\Bbbk(C_1)/\phi^*(\Bbbk(C_2))}(f))$.

Note that the definition of $\phi_*(f)$ makes sense since $\mathrm{N}_{\Bbbk(C_1)/\phi^*\Bbbk(C_2)}(f) \in \phi^*(\Bbbk(C_2))$ and so is of the form $h \circ \phi$ for some $h \in \Bbbk(C_2)$. So $\phi_*(f) = h$.

Example 8.3.6 Let $C_1 = C_2 = \mathbb{A}^1$ and $\phi : C_1 \to C_2$ be given by $\phi(x) = x^2$. Then $\phi^*(\Bbbk(C_2)) = \Bbbk(x^2)$ and $\Bbbk(C_1) = \phi^*(\Bbbk(C_2))(x)$. Let $f(x) = x^2/(x-1)$. Then

$$\mathrm{N}_{\Bbbk(C_1)/\phi^*\Bbbk(C_2)}(f) = f(x)f(-x) = \frac{x^2}{(x-1)}\frac{(-x)^2}{(-x-1)} = \frac{x^4}{-x^2+1},$$

which is $h \circ \phi$ for $h(X) = X^2/(1-X)$. Hence, $\phi_*(f(x)) = -f(x)$.

Exercise 8.3.7 Let $C_1 = V(y^2 = x^2 + 1) \subseteq \mathbb{A}^2$, $C_2 = \mathbb{A}^1$ and let $\phi : C_1 \to C_2$ be given by $\phi(x, y) = x$. Let $f(x, y) = x/y$. Show that

$$\mathrm{N}_{\Bbbk(C_1)/\phi^*\Bbbk(C_2)}(f) = \frac{-x^2}{x^2+1}$$

and so $\phi_*(f) = h(X)$ where $h(X) = -X^2/(X^2+1)$.

We now state the main properties of the pullback and pushforward.

Theorem 8.3.8 *Let $\phi : C_1 \to C_2$ be a non-constant morphism of curves over $\Bbbk$. Then:*

1. $\deg(\phi^*(D)) = \deg(\phi)\deg(D)$ *for all* $D \in \mathrm{Div}(C_2)$.
2. $\phi^*(\mathrm{div}(f)) = \mathrm{div}(\phi^* f)$ *for all* $f \in \Bbbk(C_2)^*$.
3. $\deg(\phi_*(D)) = \deg(D)$ *for all* $D \in \mathrm{Div}(C_1)$.
4. $\phi_*(\mathrm{div}(f)) = \mathrm{div}(\phi_*(f))$ *for all* $f \in \Bbbk(C_1)^*$.
5. $\phi_*(\phi^*(D)) = \deg(\phi)D$ *for* $D \in \mathrm{Div}(C_2)$.

6. *If* $\psi : C_2 \to C_3$ *is another non-constant rational map of curves over* $\Bbbk$ *then* $(\psi \circ \phi)^* = \phi^* \circ \psi^*$ *and* $(\psi \circ \phi)_* = \psi_* \circ \phi_*$.

Proof Most of the claims follow from earlier results such as Lemma 8.2.6, Theorem 8.2.9 and Lemma 8.2.12; also see Proposition II.3.6 of Silverman [505]. Claim 4 is harder; see Proposition VII.7.8 and Lemma VII.7.16 of Lorenzini [355]. □

Exercise 8.3.9 Give all the details in the proof of Theorem 8.3.8.

Corollary 8.3.10 *Let* $\phi : C_1 \to C_2$ *be a non-constant morphism of curves over* $\Bbbk$. *Then the induced maps* $\phi^* : \mathrm{Pic}^0_{\Bbbk}(C_2) \to \mathrm{Pic}^0_{\Bbbk}(C_1)$ *and* $\phi_* : \mathrm{Pic}^0_{\Bbbk}(C_1) \to \mathrm{Pic}^0_{\Bbbk}(C_2)$ *on divisor class groups are well-defined group homomorphisms.*

Proof The maps ϕ^* and ϕ_* are well-defined on divisor classes by parts 2 and 4 of Theorem 8.3.8. The homomorphic property follows from the linearity of the definitions. □

Exercise 8.3.11 Show that if $\phi : C_1 \to C_2$ is an isomorphism of curves over $\Bbbk$ then $\mathrm{Pic}^0_{\Bbbk}(C_1) \cong \mathrm{Pic}^0_{k}(C_2)$ (isomorphic as groups). Give an example to show that the converse is not true.

A further corollary of this result is that a rational map $\phi : E_1 \to E_2$ between elliptic curves such that $\phi(\mathcal{O}_{E_1}) = \mathcal{O}_{E_2}$ is automatically a group homomorphism (see Theorem 9.2.1).

Exercise 8.3.12 Let $\phi : \mathbb{P}^1 \to \mathbb{P}^1$ be defined by $\phi((x : z)) = (x^2/z^2 : 1)$. Let $D = (-1 : 1) + (1 : 0) - (0 : 1)$. Compute $\phi_*(D)$, $\phi^*(D)$, $\phi_*\phi^*(D)$ and $\phi^*\phi_*(D)$.

We now make an observation that was mentioned when we defined ϕ_* on $\Bbbk(C_1)$.

Lemma 8.3.13 *Let* $\phi : C_1 \to C_2$ *be a non-constant morphism of curves over* $\Bbbk$. *Let* $f \in \Bbbk(C_1)^*$ *and* $Q \in C_2(\overline{\Bbbk})$. *Suppose that* $v_P(f) = 0$ *for all points* $P \in C_1(\overline{\Bbbk})$ *such that* $\phi(P) = Q$. *Then*

$$\mathrm{N}_{\Bbbk(C_1)/\phi^*(\Bbbk(C_2))}(f)(Q) = \prod_{P \in C_1(\overline{\Bbbk}) : \phi(P) = Q} f(P)^{e_\phi(P)} = f(\phi^*(Q)).$$

Another formulation would be: f *of conorm of* Q *equals norm of* f *at* Q.

Proof (Sketch) This uses similar ideas to the proof of part 4 of Theorem 8.3.8. We work over $\overline{\Bbbk}$.

As always, $\overline{\Bbbk}(C_1)$ is a finite extension of $\phi^*(\overline{\Bbbk}(C_2))$. Let $A = \phi^*(\mathcal{O}_Q(C_2))$ and let B be the integral closure of A in $\Bbbk(C_1)$. Then B is a Dedekind domain and the ideal $\phi^*(\mathfrak{m}_Q)$ splits as a product $\prod_i \mathfrak{m}_{P_i}^{e_\phi(P_i)}$ where $P_i \in C_1(\overline{\Bbbk})$ are distinct points such that $\phi(P_i) = Q$.

By assumption, f has no poles at P_i and so $f \in B$. Note that $f(P_i) = c_i \in \overline{\Bbbk}$ if and only if $f \equiv c_i \pmod{\mathfrak{m}_{P_i}}$. Hence, the right-hand side is

$$\prod_i f(P_i)^{e_\phi(P_i)} = \prod_i c_i^{e_\phi(P_i)} = \prod_i (f \pmod{\mathfrak{m}_{P_i}})^{e_\phi(P_i)}.$$

It remains to prove that this is equal to the norm of f evaluated at Q and we sketch this when the extension is Galois and cyclic (the general case is simple linear algebra). The elements $\sigma \in \mathrm{Gal}(\overline{\Bbbk}(C_1)/\phi^*(\overline{\Bbbk}(C_2)))$ permute the $\mathfrak{m}_{P_i}$ and the ramification indices $e_\phi(P_i)$ are all equal. Since $c_i \in \overline{\Bbbk} \subset \phi^*(\Bbbk(C_2))$ we have $f \equiv c_i \pmod{\mathfrak{m}_{P_i}}$ if and only if $\sigma(f) \equiv c_i \pmod{\sigma(\mathfrak{m}_{P_i})}$. Hence

$$\mathrm{N}_{\Bbbk(C_1)/\phi^*(\Bbbk(C_2))}(f) = \prod_{\sigma \in \mathrm{Gal}(\overline{\Bbbk}(C_1)/\phi^*(\overline{\Bbbk}(C_2)))} \sigma(f) \equiv \prod_i c_i^{e_\phi(P_i)} \pmod{\mathfrak{m}_{P_1}}$$

and since $\mathrm{N}_{\Bbbk(C_1)/\phi^*(\Bbbk(C_2))}(f) \in \phi^*(\Bbbk(C_2))$ this congruence holds modulo $\phi^*(\mathfrak{m}_Q)$. The result follows. □

We now give an important application of Theorem 8.3.8, already stated as Theorem 7.7.11.

Theorem 8.3.14 *Let C be a curve over $\Bbbk$ and let $f \in \Bbbk(C)^*$. Then f has only finitely many zeroes and poles (i.e.,* $\mathrm{div}(f)$ *is a divisor) and* $\deg(\mathrm{div}(f)) = 0$.

Proof Let $D = (0:1) - (1:0)$ on $\mathbb{P}^1$. Interpreting f as a rational map $f : C \to \mathbb{P}^1$ as in Lemma 8.1.1 we have $\mathrm{div}(f) = f^*(D)$ and, by part 1 of Theorem 8.3.8, $\deg(f^*(D)) = \deg(f)\deg(D) = 0$. One also deduces that f has, counting with multiplicity, $\deg(f)$ poles and zeroes. □

Exercise 8.3.15 Let $\phi : C_1 \to C_2$ be a rational map over $\Bbbk$. Show that if $D \in \mathrm{Div}_\Bbbk(C_1)$ (respectively, $D \in \mathrm{Div}_\Bbbk(C_2)$) then $\phi_*(D)$ (respectively, $\phi^*(D)$) is defined over $\Bbbk$.

8.4 Riemann–Roch spaces

Definition 8.4.1 Let C be a curve over $\Bbbk$ and let $D = \sum_P n_P(P)$ be a divisor on C. The **Riemann–Roch space** of D is

$$\mathcal{L}_\Bbbk(D) = \{f \in \Bbbk(C)^* : v_P(f) \geq -n_P \text{ for all } P \in C(\overline{\Bbbk})\} \cup \{0\}.$$

We denote $\mathcal{L}_{\overline{\Bbbk}}(D)$ by $\mathcal{L}(D)$.

Lemma 8.4.2 *Let C be a curve over $\Bbbk$ and let D be a divisor on C. Then:*

1. *$\mathcal{L}_\Bbbk(D)$ is a $\Bbbk$-vector space.*
2. *$D \leq D'$ implies $\mathcal{L}_\Bbbk(D) \subseteq \mathcal{L}_\Bbbk(D')$.*
3. *$\mathcal{L}_\Bbbk(0) = \Bbbk$, $\mathcal{L}_\Bbbk(D) = \{0\}$ if* $\deg(D) < 0$.
4. *Let $P_0 \in C(\overline{\Bbbk})$. Then* $\dim_\Bbbk(\mathcal{L}_\Bbbk(D + P_0)/\mathcal{L}_\Bbbk(D)) \leq 1$ *and if $D' \geq D$ then* $\dim_\Bbbk(\mathcal{L}_\Bbbk(D')/\mathcal{L}_\Bbbk(D)) \leq \deg(D') - \deg(D)$.
5. *$\mathcal{L}_\Bbbk(D)$ is finite dimensional and if $D = D_+ - D_-$, where D_+, D_- are effective, then* $\dim_\Bbbk \mathcal{L}_\Bbbk(D) \leq \deg(D_+) + 1$.
6. *If $D' = D + \mathrm{div}(f)$ for some $f \in \Bbbk(C)^*$ then $\mathcal{L}_\Bbbk(D)$ and $\mathcal{L}_\Bbbk(D')$ are isomorphic as $\Bbbk$-vector spaces.*

Proof Proposition 8.3 of Fulton [199]. □

Exercise 8.4.3 Let $D = \sum_{P \in C(\overline{\Bbbk})} n_P(P)$ be a divisor on C. Explain why $\{f \in \Bbbk(C)^* : v_P(f) = n_P \text{ for all } P \in C(\overline{\Bbbk})\} \cup \{0\}$ is not usually a $\Bbbk$-vector space.

Definition 8.4.4 Let C be a curve over $\Bbbk$ and let D be a divisor on C. Define

$$\ell_{\Bbbk}(D) = \dim_{\Bbbk} \mathcal{L}_{\Bbbk}(D).$$

Write $\ell(D) = \ell_{\overline{\Bbbk}}(D)$.

Exercise 8.4.5 Show that $\ell_{\Bbbk}(0) = 1$ and, for $f \in \Bbbk(C)$, $\ell_{\Bbbk}(\operatorname{div}(f)) = 1$.

Theorem 8.4.6 *(Riemann's theorem) Let C be a curve over $\Bbbk$ (in particular, non-singular and projective). Then there exists a unique minimal integer g such that, for all divisors D on C over $\overline{\Bbbk}$*

$$\ell_{\overline{\Bbbk}}(D) \geq \deg(D) + 1 - g.$$

Proof See Proposition I.4.14 of Stichtenoth [529], Section 8.3 (page 196) of Fulton [199] or Theorem 2.3 of Moreno [395]. □

Definition 8.4.7 The number g in Theorem 8.4.6 is called the **genus** of C.

Note that the genus is independent of the model of the curve C and so one can associate the genus with the function field or birational equivalence class of the curve.

Remark 8.4.8 We give an alternative justification for Remark 5.4.14. Suppose $f \in \overline{\Bbbk}(C)$ is such that $\sigma(f) = f$ for all $\sigma \in \operatorname{Gal}(\overline{\Bbbk}/\Bbbk)$. Write $D = \operatorname{div}(f)$. Note that D is defined over $\Bbbk$. Then $f \in \mathcal{L}_{\overline{\Bbbk}}(D)$, which has dimension 1 by Exercise 8.4.5. Now, performing the Brill–Noether proof of Riemann's theorem (e.g., see Section 8.5 of Fulton [199]), one can show that $\mathcal{L}_{\Bbbk}(D)$ contains a function $h \in \Bbbk(C)$. It follows that $\operatorname{div}(h) = D$ and that $f = ch$ for some $c \in \Bbbk$. Hence, Theorem 7.8.3 is proved.

8.5 Derivations and differentials

Differentials arise in differential geometry: a manifold is described by open patches homeomorphic to $\mathbb{R}^n$ (or $\mathbb{C}^n$ for complex manifolds) with coordinate functions $x_1, \ldots, x_n$ and the differentials dx_i arise naturally. It turns out that differentials can be described in a purely formal way (i.e., without reference to limits).

When working over general fields (such as finite fields) it no longer makes sense to consider differentiation as a process defined by limits. But the formal description of differentials makes sense and the concept turns out to be useful.

We first explain how to generalise partial differentiation to functions on curves. We can then define differentials. Throughout this section, if $F(x, y)$ is a polynomial or rational function then $\partial F/\partial x$ denotes standard undergraduate partial differentiation.

Definition 8.5.1 Let C be a curve over $\Bbbk$. A **derivation** on $\Bbbk(C)$ is a $\Bbbk$-linear map (treating $\Bbbk(C)$ as a $\Bbbk$-vector space) $\delta : \Bbbk(C) \to \Bbbk(C)$ such that $\delta(f_1 f_2) = f_1\delta(f_2) + f_2\delta(f_1)$.

Lemma 8.5.2 *Let $\delta : \Bbbk(C) \to \Bbbk(C)$ be a derivation. Then:*

1. *If $c \in \Bbbk$ then $\delta(c) = 0$.*
2. *If $x \in \Bbbk(C)$ and $n \in \mathbb{Z}$ then $\delta(x^n) = nx^{n-1}\delta(x)$.*
3. *If $\mathrm{char}(\Bbbk) = p$ and $x \in \Bbbk(C)$ then $\delta(x^p) = 0$.*
4. *If $h \in \Bbbk(C)$ then $\delta'(f) = h\delta(f)$ is a derivation.*
5. *If $x, y \in \Bbbk(C)$ then $\delta(x/y) = (y\delta(x) - x\delta(y))/y^2$.*
6. *If $x, y \in \Bbbk(C)$ and $F(u, v) \in \Bbbk[u, v]$ is a polynomial then $\delta(F(x, y)) = (\partial F/\partial x)\delta(x) + (\partial F/\partial y)\delta(y)$.*

Exercise 8.5.3 Prove Lemma 8.5.2.

Definition 8.5.4 Let C be a curve over $\Bbbk$. A function $x \in \Bbbk(C)$ is a **separating element** (or **separating variable**) if $\Bbbk(C)$ is a finite separable extension of $\Bbbk(x)$.

Note that if $x \in \Bbbk(C)$ is such that $x \notin \overline{\Bbbk}$ then $\Bbbk(C)/\Bbbk(x)$ is finite; hence, the non-trivial condition is that $\Bbbk(C)/\Bbbk(x)$ is separable.

Example 8.5.5 For $\mathbb{P}^1(\mathbb{F}_p)$, x is a separating element (since $\Bbbk(\mathbb{P}^1) = \Bbbk(x)$) and x^p is not a separating element (since $\Bbbk(\mathbb{P}^1)/\Bbbk(x^p) = \Bbbk(x)/\Bbbk(x^p)$ is not separable). The mapping $\delta(f) = \partial f/\partial x$ is a derivation.

The following exercise shows that separating elements exist for elliptic and hyperelliptic curves. For general curves we need Lemma 8.5.7.

Exercise 8.5.6 Let $\Bbbk$ be any field and let C be a curve given by an equation of the form $y^2 + H(x)y = F(x)$ with $H(x), F(x) \in \Bbbk[x]$. Show that if either $H(x) \neq 0$ or if $\mathrm{char}(\Bbbk) \neq 2$ then x is a separating element of $\Bbbk(C)$.

Lemma 8.5.7 *Let C be a curve over $\Bbbk$, where $\Bbbk$ is a perfect field. Then there exists a separating element $x \in \Bbbk(C)$.*

Proof See the full version of the book or Proposition III.9.2 of Stichtenoth [529]. □

Lemma 8.5.8 *Let C be a curve over $\Bbbk$, let $P \in C(\Bbbk)$ and let t_P be a uniformiser at P. Then t_p is a separating element of $\Bbbk(C)$.*

Proof Proposition III.9.2 of Stichtenoth [529]. □

Suppose now that C is a curve over $\Bbbk$ and x is a separating element. We wish to extend $\delta(f) = \partial f/\partial x$ from $\Bbbk(x)$ to the whole of $\Bbbk(C)$. The natural approach is to use property 6 of Lemma 8.5.2: If $f \in \Bbbk(C)$ then $\Bbbk(x, f)/\Bbbk(x)$ is finite and separable; write $F(T)$ for the minimal polynomial of f over $\Bbbk(x)$ in $\Bbbk(C)$; since the extension is separable we have

$\partial F/\partial T \neq 0$; as a function on C we have $F(x, f) = 0$ and so

$$0 = \delta(F(x, f)) = \frac{\partial F}{\partial x}\delta(x) + \frac{\partial F}{\partial T}\delta(f). \tag{8.1}$$

This motivates the following definition.

Definition 8.5.9 Let C be a curve over $\Bbbk$ and let $x \in \Bbbk(C)$ be a separating element. Let $y \in \Bbbk(C)$. Let $F(x, T)$ be a rational function such that $F(x, y) = 0$. Define

$$\frac{\partial y}{\partial x} = -(\partial F/\partial x)/(\partial F/\partial T)$$

evaluated at y.

Lemma 8.5.10 *The value $\partial y/\partial x$ in Definition 8.5.9 is well-defined. More precisely, if F and F' are rational functions such that $F(x, y) = F'(x, y) = 0$ then $(\partial F/\partial x)/(\partial F/\partial y) = (\partial F'/\partial x)/(\partial F'/\partial y)$ and if $z \equiv y$ in $\Bbbk(C)$ then $\partial z/\partial x \equiv \partial y/\partial x$.*

Proof The first claim follows from equation (8.1). For the second claim, if $z = y$ in $\Bbbk(C)$ then they satisfy the same minimal polynomial. □

It remains to show that this construction does give a derivation.

Lemma 8.5.11 *Let C be a curve over $\Bbbk$ and $x \in \Bbbk(C)$ a separating element. The function $\delta : \Bbbk(C) \to \Bbbk(C)$ defined by $\delta(y) = \partial y/\partial x$ as in Definition 8.5.9 is $\Bbbk$-linear and satisfies the product rule.*

Furthermore, if $f = H(y) \in \Bbbk(C)$ is another function, where $H(T) \in \Bbbk(x)[T]$ is a polynomial, then

$$\delta(f) = \frac{\partial H}{\partial x} - (\partial F/\partial x)/(\partial F/\partial T)\frac{\partial H}{\partial T} \tag{8.2}$$

evaluated at y, where F is as in Definition 8.5.9.

Proof See Proposition IV.1.4 and Lemma IV.1.3 of Stichtenoth [529]. □

Example 8.5.12 Let $C : y^2 = x^3 + x + 1$ over $\mathbb{Q}$. Note that x is a separating element. To compute $\partial y/\partial x$ one uses the fact that $F(x, y) = y^2 - (x^3 + x + 1) = 0$ in $\Bbbk(C)$ and so $\partial y/\partial x = (3x^2 + 1)/(2y)$.

Consider the function $f(x, y) = xy$ and let $\delta(f) = \partial f/\partial x$. Then $\delta(f) = x\delta(y) + y = x(3x^2 + 1)/(2y) + y = (3x^3 + x + 2y^2)/(2y) = (5x^3 + 3x + 2)/(2y)$.

Exercise 8.5.13 Let $\Bbbk(C)$ be as in Example 8.5.12. Show that $\delta(y/x) = (x^3 - x - 2)/(2yx^2)$.

Lemma 8.5.14 *Let C be a curve over $\Bbbk$ and let $x, y \in \Bbbk(C)$ be separating elements. Then the corresponding derivations on $\Bbbk(C)$ satisfy the chain rule, namely*

$$\frac{\partial f}{\partial y} = \frac{\partial f}{\partial x}\frac{\partial x}{\partial y}.$$

In particular, if $x, y \in \Bbbk(C)$ are separating elements then $\partial x/\partial y = 1/(\partial y/\partial x) \neq 0$.

Let $t \in \Bbbk(C)$. Then $\partial t/\partial x = 0$ if and only if t is not a separating element.

Proof See Lemma IV.1.6 of Stichtenoth [529]. □

Exercise 8.5.15 Let $C = \mathbb{P}^1$ over $\mathbb{F}_p$ with variable x and let $\delta(f) = \partial f/\partial x$. Show that $\delta(x^p) = 0$.

Now we have defined $\partial f/\partial x$ for general $f \in \Bbbk(C)$ we can introduce the differentials on a curve over a field. Our definition is purely formal and the symbol dx is not assumed to have any intrinsic meaning. We essentially follow Section IV.1 of Stichtenoth [529]; for a slightly different approach see Section 8.4 of Fulton [199].

Definition 8.5.16 Let C be a curve over $\Bbbk$. The set of **differentials** $\Omega_{\Bbbk}(C)$ (some authors write $\Omega^1_{\Bbbk}(C)$) is the quotient of the free $\Bbbk(C)$-module on symbols dx for $x \in \Bbbk(C)$ under the relations:

1. $dx \neq 0$ if x is a separating element.
2. If x is a separating element and $h_1, h_2 \in \Bbbk(C)$ then $h_1 dx + h_2 dx = (h_1 + h_2)dx$.
3. If x is a separating element and $y \in \Bbbk(C)$ then $dy = (\partial y/\partial x)dx$.

In other words, differentials are equivalence classes of formal symbols

$$\left\{ \sum_{i=1}^{m} h_i dx_i : x_i, h_i \in \Bbbk(C) \right\}$$

where one may assume the x_i are all separating elements.

Lemma 8.5.17 *Let C be a curve over $\Bbbk$ and $x, y \in \Bbbk(C)$ separating elements.*

1. *$dx = 0$ if x is not a separating element.*
2. *$d(x + y) = dx + dy$.*
3. *$d(\lambda x) = \lambda dx$ and $d\lambda = 0$ for all $\lambda \in \Bbbk$.*
4. *$d(xy) = xdy + ydx$.*
5. *If x is a separating element and $y \in \Bbbk(C)$ then $dx + dy = (1 + (\partial y/\partial x))dx$.*
6. *For $n \in \mathbb{Z}$, $d(x^n) = nx^{n-1}dx$.*
7. *$d(x/y) = (ydx - xdy)/y^2$.*
8. *If $f \in \Bbbk(C)$ then $d(f(x)) = (\partial f/\partial x)dx$.*
9. *For $i \in \mathbb{Z}$, $d(f(x)y^i) = (\partial f/\partial x)y^i dx + f(x)iy^{i-1}dy$.*
10. *If $F(x, y)$ is a rational function in x and y then $dF(x, y) = (\partial F/\partial x)dx + (\partial F/\partial y)dy$.*

Exercise 8.5.18 Prove Lemma 8.5.17.

Exercise 8.5.19 Let C be a curve over $\Bbbk$. Let $x_1, x_2 \in \Bbbk(C)$ be separating elements and $h_1, h_2 \in \Bbbk(C)$. Show that $h_1 dx_1$ is equivalent to $h_2 dx_2$ if and only if

$$h_2 = h_1 \frac{\partial x_1}{\partial x_2}.$$

Example 8.5.20 We determine $\Omega_{\Bbbk}(\mathbb{P}^1)$. Since $\Bbbk(\mathbb{P}^1) = \Bbbk(x)$ the differentials are $d(f(x)) = (\partial f/\partial x)dx$ for $f(x) \in \Bbbk(x)$.

The following theorem, that all differentials on a curve are multiples of dx where x is a separating element, is a direct consequence of the definition.

Theorem 8.5.21 *Let C be a curve over $\Bbbk$ and let x be a separating element. Let $\omega \in \Omega_{\Bbbk}(C)$. Then $\omega = hdx$ for some $h \in \Bbbk(C)$.*

Exercise 8.5.22 Prove Theorem 8.5.21.

This result shows that $\Omega_{\Bbbk}(C)$ is a $\Bbbk(C)$-vector space of dimension 1 (we know that $\Omega_{\Bbbk}(C) \neq \{0\}$ since $dx \neq 0$ if x is a separating element). Therefore, for any $\omega_1, \omega_2 \in \Omega_{\Bbbk}(C)$ with $\omega_2 \neq 0$ there is a unique function $f \in \Bbbk(C)$ such that $\omega_1 = f\omega_2$. We define ω_1/ω_2 to be f (see Proposition II.4.3 of Silverman [505]).

We now define the divisor of a differential by using uniformisers. Recall from Lemma 8.5.8 that a uniformiser t_P is a separating element and so $dt_P \neq 0$.

Definition 8.5.23 Let C be a curve over $\Bbbk$. Let $\omega \in \Omega_{\Bbbk}(C)$, $\omega \neq 0$ and let $P \in C(\overline{\Bbbk})$ have uniformiser $t_P \in \overline{\Bbbk}(C)$. Then the **order** of ω at P is $v_P(\omega) := v_P(\omega/dt_P)$. The **divisor of a differential** is

$$\operatorname{div}(\omega) = \sum_{P \in C(\overline{\Bbbk})} v_P(\omega)(P).$$

Lemma 8.5.24 *Let C be a curve over $\Bbbk$ and let ω be a differential on C. Then $v_P(\omega) \neq 0$ for only finitely many $P \in C(\overline{\Bbbk})$ and so $\operatorname{div}(\omega)$ is a divisor.*

Proof See Proposition II.4.3(e) of Silverman [505]. □

Exercise 8.5.25 Show that $v_P(hdx) = v_P(h) + v_P(dx)$ and $v_P(df) = v_P(\partial f/\partial t_P)$.

Lemma 8.5.26 *The functions $v_P(\omega)$ and $\operatorname{div}(\omega)$ in Definition 8.5.23 are well-defined (both with respect to the choice of representative for ω and choice of t_P).*

Exercise 8.5.27 Prove Lemma 8.5.26.

Lemma 8.5.28 *Let C be a curve over $\Bbbk$ and $\omega, \omega' \in \Omega_{\Bbbk}(C)$. Then:*

1. $\deg(\operatorname{div}(\omega)) = \deg(\operatorname{div}(\omega'))$.
2. $\operatorname{div}(\omega)$ *is well-defined up to principal divisors.*

Exercise 8.5.29 Prove Lemma 8.5.28.

Definition 8.5.30 Any divisor $\operatorname{div}(\omega)$ is called a **canonical divisor**. The set $\{\operatorname{div}(\omega) : \omega \in \Omega_{\Bbbk}(C)\}$ is the **canonical divisor class**.

Example 8.5.31 We determine the canonical class of $C = \mathbb{P}^1$.

Let $\omega = dx$. Since x is a uniformiser at the point 0 we have $v_0(\omega) = v_0(dx/dx) = 0$. More generally, for $P \in \Bbbk$ we have $(x - P)$ a uniformiser and $v_P(\omega) = v_P(dx/d(x - P)) = v_P(1) = 0$. Finally, a uniformiser at ∞ is $t = 1/x$ and $dt = (-x^{-2})dx$ so $v_\infty(\omega) = v_\infty(-x^2) = -2$. Hence, $\operatorname{div}(\omega) = -2\infty$ and the degree of $\operatorname{div}(\omega)$ is -2.

Example 8.5.32 We determine the divisor of a differential on an elliptic curve E in Weierstrass form. Rather than computing $\operatorname{div}(dx)$ it is easier to compute $\operatorname{div}(\omega)$ for

$$\omega = \frac{dx}{2y + a_1x + a_3}.$$

Let $P \in E(\overline{\Bbbk})$. There are three cases, if $P = \mathcal{O}_E$ then one can take uniformiser $t = x/y$, if $P = (x_P, y_P) = \iota(P)$ then take uniformiser $(y - y_P)$ (and note that $v_P(2y + a_1x + a_3) = 1$ in this case) and otherwise take uniformiser $(x - x_P)$ and note that $v_P(2y + a_1x + a_3) = 0$.

We deal with the general case first. Since $dx/d(x - x_P) = \partial x/\partial(x - x_P) = 1$ it follows that $v_P(\omega) = 0$. For the case $P = \mathcal{O}_E$ write $x = t^{-2}f$ and $y = t^{-3}h$ for some functions $f, h \in \Bbbk(E)$ regular at $\mathcal{O}_E$ and with $f(\mathcal{O}_E), h(\mathcal{O}_E) \neq 0$. One can verify that

$$\frac{\omega}{dt} = \frac{-2t^{-3}f + t^{-2}f'}{2t^{-3}h + a_1t^{-2}f + a_3} = \frac{-2f + tf'}{2h + a_1tf + a_3t^3}$$

and so $v_{\mathcal{O}_E}(\omega) = 0$. Finally, when $P = \iota(P)$ we must consider

$$\frac{dx}{d(y - y_P)} = \frac{1}{\partial y/\partial x} = \frac{2y + a_1x + a_3}{3x^2 + 2a_2x + a_4}.$$

It follows that $\omega = (1/(3x^2 + 2a_2x + a_4))d(y - y_P)$ and, since P is not a singular point, $3x_P^2 + 2a_2x_P + a_4 \neq 0$ and so $v_P(\omega) = 0$.

In other words, we have shown that $\operatorname{div}(\omega) = 0$. One can verify that

$$\operatorname{div}(dx) = (P_1) + (P_2) + (P_3) - 3(\mathcal{O}_E)$$

where P_1, P_2, P_3 are the three non-trivial points of order 2 in $E(\overline{\Bbbk})$.

Exercise 8.5.33 Show that

$$\frac{dx}{2y + a_1x + a_3} = \frac{dy}{3x^2 + 2a_2x + a_4 - a_1y}$$

on an elliptic curve.

Definition 8.5.34 Let $\phi : C_1 \to C_2$ be a non-constant morphism of curves over $\Bbbk$. Define the function $\phi^* : \Omega_{\Bbbk}(C_2) \to \Omega_{\Bbbk}(C_1)$ by

$$\phi^*(f\,dx) = \phi^*(f)d(\phi^*(x)).$$

Lemma 8.5.35 *The function ϕ^* of Definition 8.5.34 is $\Bbbk$-linear and ϕ^* is injective (= non-zero) if and only if ϕ is separable.*

Proof The linearity follows since dx is $\Bbbk$-linear. The second part follows since if x is separating for $\Bbbk(C_2)$ and ϕ is separable then $\Bbbk(C_1)/\phi^*\Bbbk(C_2)$ and $\phi^*\Bbbk(C_2)/\Bbbk(\phi^*(x))$ are separable. Hence, $\phi^*(x)$ is a separating element for $\Bbbk(C_1)$ and $d\phi^*(x) \neq 0$. The reverse implication is also straightforward. □

Lemma 8.5.36 *Let $\phi : C_1 \to C_2$ be an unramified morphism of curves over $\Bbbk$ and let $\omega \in \Omega_\Bbbk(C_2)$. Then $\phi^*(\mathrm{div}(\omega)) = \mathrm{div}(\phi^*(\omega))$.*

Proof Let $P \in C_1(\overline{\Bbbk})$ and $Q = \phi(P)$. Let t_Q be a uniformiser at Q. Since ϕ is unramified it follows that $t_P = \phi^*(t_Q)$ is a uniformiser at P. Let $f \in \Bbbk(C_2)$. It suffices to show that $v_P(\phi^*(df)) = v_Q(df)$.

Recall from Exercise 8.5.25 that $v_Q(df) = v_Q(\partial f/\partial t_Q)$. If $F(x, y)$ is a rational function such that $F(t_Q, f) = 0$ then $0 = F(t_Q, f) \circ \phi = F(t_Q \circ \phi, f \circ \phi) = F(t_P, \phi^*(f)) = 0$. Hence, by definition,

$$\partial\phi^*(f)/\partial t_P = -(\partial F/\partial x)/(\partial F/\partial y) = \partial f/\partial t_Q$$

and so $v_P(d\phi^*(f)) = v_P(\partial\phi^*(f)/\partial t_P) = v_Q(f)$. □

Corollary 8.5.37 *Let $\phi : C_1 \to C_2$ be an isomorphism of curves over $\Bbbk$ and let $\omega \in \Omega_\Bbbk(C_2)$. Then $\deg(\mathrm{div}(\omega)) = \deg(\mathrm{div}(\phi^*(\omega)))$.*

8.6 Genus zero curves

Theorem 8.6.1 *Let C be a curve over $\Bbbk$ (i.e., projective non-singular). The following are equivalent.*

1. *C is birationally equivalent over $\Bbbk$ to $\mathbb{P}^1$.*
2. *The divisor class group of C over $\Bbbk$ is trivial and $\#C(\Bbbk) \geq 2$.*
3. *There is a point $P \in C(\Bbbk)$ with $\ell_\Bbbk(P) \geq 2$.*

Proof See Section 8.2 of Fulton [199]. □

Definition 8.6.2 A curve satisfying any of the above equivalent conditions is called a **genus 0 curve**.

Exercise 8.6.3 Write down a curve C over a field $\Bbbk$ such that the divisor class group $\mathrm{Pic}^0_\Bbbk(C)$ is trivial but C is not birationally equivalent over $\Bbbk$ to $\mathbb{P}^1$.

Theorem 8.6.4 *An elliptic curve does not have genus 0.*

Proof See Lemma 11.3 or Lemma 11.5 of Washington [560]. □

Corollary 8.6.5 *Let E be an elliptic curve and $P_1, P_2 \in E(\Bbbk)$. If $P_1 \neq P_2$ then $(P_1) - (P_2)$ is not a principal divisor.*

8.7 Riemann–Roch theorem and Hurwitz genus formula

In this section we state, without proof, two very important results in algebraic geometry. Neither will play a crucial role in this book.

Lemma 8.7.1 *Let C be a curve over $\Bbbk$ of genus g and let $\omega \in \Omega_{\Bbbk}(C)$. Then*

1. $\deg(\operatorname{div}(\omega)) = 2g - 2$.
2. $\ell_{\Bbbk}(\operatorname{div}(\omega)) = g$.

Proof See Corollary I.5.16 of Stichtenoth [529] or Corollary 11.16 of Washington [560]. For non-singular plane curves see Sections 8.5 and 8.6 of Fulton [199]. □

Theorem 8.7.2 *(Riemann–Roch) Let C be a non-singular projective curve over $\Bbbk$ of genus g, $\omega \in \Omega_{\Bbbk}(C)$ a differential and D a divisor. Then*

$$\ell_{\Bbbk}(D) = \deg(D) + 1 - g + \ell_{\Bbbk}(\operatorname{div}(\omega) - D).$$

Proof There are several proofs. Section 8.6 of Fulton [199] gives the Brill–Noether proof for non-singular plane curves. Theorem I.5.15 of Stichtenoth [529] and Theorem 2.5 of Moreno [395] give proofs using repartitions. □

Some standard applications of the Riemann–Roch theorem are to prove that every genus 1 curve with a rational point is birational to an elliptic curve in Weierstrass form, and to prove that every hyperelliptic curve of genus g is birational to an affine curve of the form $y^2 + H(x)y = F(x)$ with $\deg(H(x)) \leq g + 1$ and $\deg(F(x)) \leq 2g + 2$.

Theorem 8.7.3 *(Hurwitz genus formula) Let $\phi : C_1 \to C_2$ be a rational map of curves over $\Bbbk$. Let g_i be the genus of C_i. Suppose that $\Bbbk$ is a field of characteristic zero or characteristic coprime to all $e_\phi(P)$. Then*

$$2g_1 - 2 = \deg(\phi)(2g_2 - 2) + \sum_{P \in C_1(\overline{\Bbbk})} (e_\phi(P) - 1).$$

Proof See Theorem III.4.12 and Corollary III.5.6 of Stichtenoth [529], Theorem II.5.9 of Silverman [505] or Exercise 8.36 of Fulton [199]. □

A variant of the above formula is known in the case where some of the $e_\phi(P)$ are divisible by $\operatorname{char}(\Bbbk)$.

9

Elliptic curves

This chapter summarises the theory of elliptic curves. Since there are already many outstanding textbooks on elliptic curves (such as Silverman [505] and Washington [560]) we do not give all the details. Our focus is on facts relevant for the cryptographic applications, especially those for which there is not already a suitable reference.

9.1 Group law

Recall that an elliptic curve over a field $\Bbbk$ is given by a non-singular affine Weierstrass equation

$$E : y^2 + a_1 xy + a_3 y = x^3 + a_2 x^2 + a_4 x + a_6 \tag{9.1}$$

where $a_1, a_2, a_3, a_4, a_6 \in \Bbbk$. There is a unique point $\mathcal{O}_E$ on the projective closure that does not lie on the affine curve.

We recall the formulae for the elliptic curve group law with identity element $\mathcal{O}_E$: For all $P \in E(\Bbbk)$ we have $P + \mathcal{O}_E = \mathcal{O}_E + P = P$ so it remains to consider the case where $P_1, P_2 \in E(\Bbbk)$ are such that $P_1, P_2 \neq \mathcal{O}_E$. In other words, P_1 and P_2 are affine points and so write $P_1 = (x_1, y_1)$ and $P_2 = (x_2, y_2)$. Recall that Lemma 7.7.10 shows the inverse of $P_1 = (x_1, y_1)$ is $\iota(P_1) = (x_1, -y_1 - a_1 x_1 - a_3)$. Hence, if $x_1 = x_2$ and $y_2 = -y_1 - a_1 x_1 - a_3$ (i.e., $P_2 = -P_1$) then $P_1 + P_2 = \mathcal{O}_E$. In the remaining cases, let

$$\lambda = \begin{cases} \dfrac{3x_1^2 + 2a_2 x_1 + a_4 - a_1 y_1}{2y_1 + a_1 x_1 + a_3} & \text{if } P_1 = P_2 \\[2ex] \dfrac{y_2 - y_1}{x_2 - x_1} & \text{if } P_1 \neq \pm P_2. \end{cases} \tag{9.2}$$

and set $x_3 = \lambda^2 + a_1\lambda - x_1 - x_2 - a_2$ and $y_3 = -\lambda(x_3 - x_1) - y_1 - a_1 x_3 - a_3$. Then $P_1 + P_2 = (x_3, y_3)$.

Exercise 9.1.1 It is possible to "unify" the two cases in equation (9.2). Show that if $P_1 = (x_1, y_1)$ and $P_2 = (x_2, y_2)$ lie on $y^2 + (a_1 x + a_3)y = x^3 + a_2 x^2 + a_4 x + a_6$ and

$y_2 \neq -y_1 - a_1x_1 - a_3$ then $P_1 + P_2$ can be computed using the fomula

$$\lambda = \frac{x_1^2 + x_1x_2 + x_2^2 + a_2(x_1 + x_2) + a_4 - a_1y_1}{y_1 + y_2 + a_1x_2 + a_3} \tag{9.3}$$

instead of equation (9.2).

Definition 9.1.2 Let E be an elliptic curve over a field $\Bbbk$ and let $P \in E(\Bbbk)$. For $n \in \mathbb{N}$ define $[n]P$ to be $P + \cdots + P$ where P appears n times. In particular, $[1]$ is the identity map. Define $[0]P = \mathcal{O}_E$ and $[-n]P = [n](-P)$.

The n**-torsion subgroup** is

$$E[n] = \{P \in E(\overline{\Bbbk}) : [n]P = \mathcal{O}_E\}.$$

We write $E(\Bbbk)[n]$ for $E[n] \cap E(\Bbbk)$.

Exercise 9.1.3 Let $E : y^2 + y = x^3$ be an elliptic curve over $\mathbb{F}_2$. Let $m \in \mathbb{N}$ and $P = (x_P, y_P) \in E(\mathbb{F}_{2^m})$. Show that $[2]P = (x_P^4, y_P^4 + 1)$. (We will show in Example 9.11.6 that this curve is supersingular.)

Exercise 9.1.4 Let $E : y^2 + xy = x^3 + a_2x^2 + a_6$ be an elliptic curve over $\mathbb{F}_{2^m}$ for $m \in \mathbb{N}$. Show that there is a point $P = (x_P, y_P) \in E(\mathbb{F}_{2^m})$ if and only if $\mathrm{Tr}_{\mathbb{F}_{2^m}/\mathbb{F}_2}(x_P + a_2 + a_6/x_P^2) = 0$. Given $Q = (x_Q, y_Q) \in E(\mathbb{F}_{2^m})$ show that the slope of the tangent line to E at Q is $\lambda_Q = x_Q + y_Q/x_Q$. Show that $y_Q = x_Q(\lambda_Q + x_Q)$. Hence, show that if $P = [2]Q$ then $\mathrm{Tr}_{\mathbb{F}_{2^m}/\mathbb{F}_2}(x_p) = \mathrm{Tr}_{\mathbb{F}_{2^m}/\mathbb{F}_2}(a_2)$, $x_P = x_Q^2 + a_6/x_Q^2$ and $\mathrm{Tr}_{\mathbb{F}_{2^m}/\mathbb{F}_2}(a_6/x_P^2) = 0$. Conversely, show that if $P \in E(\mathbb{F}_{2^m})$ is such that $\mathrm{Tr}_{\mathbb{F}_{2^m}/\mathbb{F}_2}(x_P) = \mathrm{Tr}_{\mathbb{F}_{2^m}/\mathbb{F}_2}(a_2)$ and $\mathrm{Tr}_{\mathbb{F}_{2^m}/\mathbb{F}_2}(a_6/x_P^2) = 0$ then $P = [2]Q$ for some $Q \in E(\mathbb{F}_{2^m})$.

(Point halving) Given $P = (x_P, y_P) \in E(\mathbb{F}_{2^m})$ such that $\mathrm{Tr}_{\mathbb{F}_{2^m}/\mathbb{F}_2}(x_P) = \mathrm{Tr}_{\mathbb{F}_{2^m}/\mathbb{F}_2}(a_2)$ show that there are two solutions λ_Q to the equation $\lambda_Q^2 + \lambda_Q = x_P + a_2$. For either solution let $x_Q = \sqrt{y_P + x_P\lambda_Q + x_P}$, and $y_Q = x_Q(\lambda_Q + x_Q)$. Show that $[2](x_Q, y_Q) = P$.

One can consider Weierstrass equations over $\Bbbk$ that have a singular point in the affine plane (recall that there is a unique point at infinity $\mathcal{O}_E$ and it is non-singular). By a change of variable one may assume that the singular point is $(0, 0)$ and the equation is $C : y^2 + a_1xy = x^3 + a_2x^2$. Let $G = C(\Bbbk) \cup \{\mathcal{O}_E\} - \{(0, 0)\}$. It turns out that the elliptic curve group law formulae give rise to a group law on G. There is a morphism over $\overline{\Bbbk}$ from C to $\mathbb{P}^1$ and the group law on G corresponds to either the additive group G_a or the multiplicative group G_m; see Section 9 of [114], Section 2.10 of [560] or Proposition III.2.5 of [505] for details.

Since an elliptic curve is a projective variety it is natural to consider addition formulae on projective coordinates. In the applications there are good reasons to do this (e.g., to minimise the number of inversions in fast implementations of elliptic curve cryptography, or in the elliptic curve factoring method).

Exercise 9.1.5 Let $P_1 = (x_1 : y_1 : z_1)$ and $P_2 = (x_2 : y_2 : z_2)$ be points on the elliptic curve $E : y^2 z = x^3 + a_4 x z^2 + a_6 z^3$ over $\Bbbk$. Let

$$u = x_1 z_2 - x_2 z_1.$$

Show that $(x_3 : y_3 : z_3)$ is a projective representation for $P_1 + P_2$ where

$$x_3 = z_1 z_2 (y_1 z_2 - y_2 z_1)^2 u - (x_1 z_2 + x_2 z_1) u^3 \tag{9.4}$$
$$y_3 = -z_1 z_2 (y_1 z_2 - y_2 z_1)^3 + (2 x_1 z_2 + x_2 z_1)(y_1 z_2 - y_2 z_1) u^2 - y_1 z_2 u^3 \tag{9.5}$$
$$z_3 = z_1 z_2 u^3 \tag{9.6}$$

(as long as the resulting point is not $(0, 0, 0)$).

The elliptic curve addition formula of equations (9.3) and (9.4)-(9.6) are undefined on certain inputs (such as $P = \mathcal{O}_E$ or $P_2 = -P_1$) and so one currently needs to make decisions (i.e., use "if" statements) to compute on elliptic curves. This does not agree with the definition of an algebraic group (informally, that the group operation is given by polynomial equations; formally that there is a morphism $E \times E \to E$). However, it can be shown (see Theorem III.3.6 of Silverman [505]) that elliptic curves are algebraic groups.

To make this concrete let E be an elliptic curve over $\Bbbk$ written projectively. A **complete system of addition laws** for $E(\Bbbk)$ is a set of triples of polynomials $\{(f_{i,x}(x_1, y_1, z_1, x_2, y_2, z_2), f_{i,y}(x_1, y_1, z_1, x_2, y_2, z_2), f_{i,z}(x_1, y_1, z_1, x_2, y_2, z_2)) : 1 \le i \le k\}$ such that, for all points $P, Q \in E(\Bbbk)$, at least one of $(f_{i,x}(P, Q), f_{i,y}(P, Q), f_{i,z}(P, Q))$ is defined and all triples defined at (P, Q) give a projective representation of the point $P + Q$.

A rather surprising fact, due to Bosma and Lenstra [87], is that one can give a complete system of addition laws for $E(\overline{\Bbbk})$ using only two triples of polynomials. The resulting equations are unpleasant and not useful for practical computation.

9.2 Morphisms between elliptic curves

The goal of this section is to show that a morphism between elliptic curves is the composition of a group homomorphism and a translation. In other words, all geometric maps between elliptic curves have a group-theoretic interpretation.

Theorem 9.2.1 *Let E_1 and E_2 be elliptic curves over $\Bbbk$ and let $\phi : E_1 \to E_2$ be a morphism of varieties such that $\phi(\mathcal{O}_{E_1}) = \mathcal{O}_{E_2}$. Then ϕ is a group homomorphism.*

Proof (Sketch) The basic idea is to note that $\phi_* : \mathrm{Pic}^0_{\Bbbk}(E_1) \to \mathrm{Pic}^0_{\Bbbk}(E_2)$ (where $\mathrm{Pic}^0_{\Bbbk}(E_i)$ denotes the degree zero divisor class group of E_i over $\Bbbk$) is a group homomorphism and $\phi_*((P) - (\mathcal{O}_{E_1})) = (\phi(P)) - (\mathcal{O}_{E_2})$. We refer to Theorem III.4.8 of [505] for the details. □

Definition 9.2.2 Let E be an elliptic curve over $\Bbbk$ and let $Q \in E(\Bbbk)$. We define the translation map to be the function $\tau_Q : E \to E$ given by $\tau_Q(P) = P + Q$.

Clearly, τ_Q is a rational map that is defined everywhere on E and so is a morphism. Since τ_Q has inverse map τ_{-Q} it follows that τ_Q is an isomorphism of the curve E to itself (though be warned that in the next section we will define isomorphism for pointed curves and τ_Q will not be an isomorphism in this sense).

Corollary 9.2.3 *Let E_1 and E_2 be elliptic curves over $\Bbbk$ and let $\phi : E_1 \to E_2$ be a rational map. Then ϕ is the composition of a group homomorphism and a translation map.*

Proof First, by Lemma 7.3.6 a rational map to a projective curve is a morphism. Now let $\phi(\mathcal{O}_{E_1}) = Q \in E_2(\Bbbk)$. The composition $\psi = \tau_{-Q} \circ \phi$ is therefore a morphism. As in Theorem 9.2.1 it is a group homomorphism. □

Hence, every rational map between elliptic curves corresponds naturally to a map of groups. Theorem 9.6.19 gives a partial converse.

Example 9.2.4 Let $E : y^2 = x^3 + x$ and $Q = (0, 0)$. We determine the map τ_Q on E.

Let $P = (x, y) \in E(\Bbbk)$ be a point such that P is neither Q nor $\mathcal{O}_E$. To add P and Q to obtain (x_3, y_3) we compute $\lambda = (y - 0)/(x - 0) = y/x$. It follows that

$$x_3 = \lambda^2 - x - 0 = \frac{y^2}{x^2} - x = \frac{y^2 - x^3}{x^2} = \frac{1}{x}$$

and

$$y_3 = -\lambda(x_3 - 0) - 0 = \frac{-y}{x^2}.$$

Hence, $\tau_Q(x, y) = (1/x, -y/x^2)$ away from $\{\mathcal{O}_E, Q\}$. It is clear that τ_Q is a rational map of degree 1 and hence an isomorphism of curves by Lemma 8.1.15. Indeed, it is easy to see that the inverse of τ_Q is itself (this is because Q has order 2).

One might wish to write τ_Q projectively (we write the rational map in the form mentioned in Exercise 5.5.2). Replacing x by x/z and y by y/z gives $\tau_Q(x/z, y/z) = (z/x, -yz/x^2)$ from which we deduce

$$\tau_Q(x : y : z) = (xz : -yz : x^2). \tag{9.7}$$

Note that this map is not defined at either $\mathcal{O}_E = (0 : 1 : 0)$ or $Q = (0 : 0 : 1)$, in the sense that evaluating at either point gives $(0 : 0 : 0)$.

To get a map defined at Q one can multiply the right-hand side of equation (9.7) through by y to get

$$(xyz : -y^2z : x^2y) = (xyz : -x^3 - xz^2 : x^2y)$$

and dividing by x gives $\tau_Q(x : y : z) = (yz : -x^2 - z^2 : xy)$. One can check that $\tau_Q(0 : 0 : 1) = (0 : -1 : 0) = (0 : 1 : 0)$ as desired. Similarly, to get a map defined at $\mathcal{O}_E$ one can multiply (9.7) by x, rearrange, and divide by z to get

$$\tau_Q(x : y : z) = (x^2 : -xy : y^2 - xz),$$

which gives $\tau_Q(0 : 1 : 0) = (0 : 0 : 1)$ as desired.

9.3 Isomorphisms of elliptic curves

We have already defined isomorphisms of algebraic varieties. It is natural to ask when two Weierstrass equations are isomorphic. Since one can compose any isomorphism with a translation map it is sufficient to restrict attention to isomorphisms $\phi : E \to \widetilde{E}$ such that $\phi(\mathcal{O}_E) = \mathcal{O}_{\widetilde{E}}$.

Formally, one defines a **pointed curve** to be a curve C over a field $\Bbbk$ together with a fixed $\Bbbk$-rational point P_0. An **isomorphism of pointed curves** $\phi : (C, P_0) \to (\widetilde{C}, \widetilde{P}_0)$ is an isomorphism $\phi : C \to \widetilde{C}$ over $\overline{\Bbbk}$ of varieties such that $\phi(P_0) = \widetilde{P}_0$. When one refers to an elliptic curve one usually means the pointed curve $(E, \mathcal{O}_E)$.

Definition 9.3.1 Let $(E, \mathcal{O}_E)$ and $(\widetilde{E}, \mathcal{O}_{\widetilde{E}})$ be elliptic curves over $\Bbbk$. An **isomorphism of elliptic curves** $\phi : E \to \widetilde{E}$ is an isomorphism over $\overline{\Bbbk}$ of algebraic varieties such that $\phi(\mathcal{O}_E) = \mathcal{O}_{\widetilde{E}}$. If there is an isomorphism from E to $\widetilde{E}$ then we write $E \cong \widetilde{E}$.

By Theorem 9.2.1, an isomorphism of elliptic curves is a group homomorphism over $\overline{\Bbbk}$.

Exercise 9.3.2 Let E_1 and E_2 be elliptic curves over $\Bbbk$. Show that if E_1 is isomorphic over $\Bbbk$ to E_2 then $E_1(\Bbbk)$ is isomorphic as a group to $E_2(\Bbbk)$. In particular, if $\Bbbk = \mathbb{F}_q$ is a finite field then $\#E_1(\mathbb{F}_q) = \#E_2(\mathbb{F}_q)$.

Note that the translation map τ_Q is not considered to be an isomorphism of the pointed curve $(E, \mathcal{O}_E)$ to itself, unless $Q = \mathcal{O}_E$ in which case τ_Q is the identity map.

Exercise 9.3.3 Exercises 7.2.6 and 7.2.7 give simplified Weierstrass models for elliptic curves when $\mathrm{char}(\Bbbk) \neq 3$. Verify that there are isomorphisms, from a general Weierstrass equation to these models that fix $\mathcal{O}_E$.

Theorem 9.3.4 *Let $\Bbbk$ be a field and E_1, E_2 elliptic curves over $\Bbbk$. Every isomorphism from E_1 to E_2 defined over $\overline{\Bbbk}$ restricts to an affine isomorphism of the form*

$$\phi(x, y) = (u^2x + r, u^3y + su^2x + t) \tag{9.8}$$

where $u, r, s, t \in \overline{\Bbbk}$. The isomorphism is defined over $\Bbbk$ if and only if $u, r, s, t \in \Bbbk$.

Proof See Proposition III.3.1(b) of [505]. □

Definition 9.3.5 Suppose $\mathrm{char}(\Bbbk) \neq 2, 3$ and let $a_4, a_6 \in \Bbbk$ be such that $4a_4^3 + 27a_6^2 \neq 0$. For the short Weierstrass equation $y^2z = x^3 + a_4xz^2 + a_6z^3$, define the **$j$-invariant**

$$j(E) = 1728\frac{4a_4^3}{4a_4^3 + 27a_6^2}.$$

Suppose $\mathrm{char}(\Bbbk) = 2$ and $a_2, a_6 \in \Bbbk$ with $a_6 \neq 0$. For the short Weierstrass equation $y^2z + xyz = x^3 + a_2x^2z + a_6z^3$ define the j-invariant

$$j(E) = 1/a_6$$

and for $E : y^2z + yz^2 = x^3 + a_4xz^2 + a_6z^3$ (we now allow $a_6 = 0$) define $j(E) = 0$.

We refer to Section III.1 of [505] for the definition of the j-invariant for general Weierstrass equations.

Theorem 9.3.6 *Let $\Bbbk$ be a field and E_1, E_2 elliptic curves over $\Bbbk$. Then there is an isomorphism from E_1 to E_2 defined over $\overline{\Bbbk}$ if and only if $j(E_1) = j(E_2)$.*

Proof See Proposition III.1.4(b) of [505] or Theorem 2.19 of [560]. □

Exercise 9.3.7 Let $E : y^2 = x^3 + a_4x + a_6$ be an elliptic curve. Show that $j(E) = 0$ if and only if $a_4 = 0$ and $j(E) = 1728$ if and only if $a_6 = 0$. Suppose $\mathrm{char}(\Bbbk) \neq 2, 3$. Let $j \in \Bbbk$, $j \neq 0, 1728$. Show that the elliptic curve

$$E : y^2 = x^3 + \frac{3j}{1728 - j}x + \frac{2j}{1728 - j}$$

has $j(E) = j$.

Exercise 9.3.8 Let $E_1 : y^2 + y = x^3$, $E_2 : y^2 + y = x^3 + 1$ and $E_3 : y^2 + y = x^3 + x$ be elliptic curves over $\mathbb{F}_2$. Since $j(E_1) = j(E_2) = j(E_3) = 0$ it follows that there are isomorphisms over $\overline{\mathbb{F}}_2$ from E_1 to E_2 and from E_1 to E_3. Write down such isomorphisms.

Exercise 9.3.9 Let E_1, E_2 be elliptic curves over $\mathbb{F}_q$ that are isomorphic over $\mathbb{F}_q$. Show that the discrete logarithm problem in $E_1(\mathbb{F}_q)$ is equivalent to the discrete logarithm problem in $E_2(\mathbb{F}_q)$. In other words, the discrete logarithm problem on $\mathbb{F}_q$-isomorphic curves has exactly the same security.

9.4 Automorphisms

Definition 9.4.1 Let E be an elliptic curve over $\Bbbk$. An **automorphism** of E is an isomorphism from $(E, \mathcal{O}_E)$ to itself defined over $\overline{\Bbbk}$. The set of all automorphisms of E is denoted $\mathrm{Aut}(E)$.

We stress that an automorphism maps $\mathcal{O}_E$ to $\mathcal{O}_E$. Under composition, $\mathrm{Aut}(E)$ forms a group, The identity of the group is the identity map.

Example 9.4.2 The map $\phi(P) = -P$ is an automorphism that is not the identity map. On $y^2 = x^3 + 1$ over $\Bbbk$, the map $\rho(x, y) = (\zeta_3 x, y)$ is an automorphism where $\zeta_3 \in \overline{\Bbbk}$ satisfies $\zeta_3^3 = 1$.

Exercise 9.4.3 Show that if E_1 and E_2 are elliptic curves over $\Bbbk$ that are isomorphic over $\overline{\Bbbk}$ then $\mathrm{Aut}(E_1) \cong \mathrm{Aut}(E_2)$.

Theorem 9.4.4 *Let E be an elliptic curve over $\Bbbk$. Then $\#\mathrm{Aut}(E)$ is even and $\#\mathrm{Aut}(E) \mid 24$. More precisely:*

- $\#\mathrm{Aut}(E) = 2$ *if* $j(E) \neq 0, 1728$,
- $\#\mathrm{Aut}(E) = 4$ *if* $j(E) = 1728$ *and* $\mathrm{char}(\Bbbk) \neq 2, 3$,
- $\#\mathrm{Aut}(E) = 6$ *if* $j(E) = 0$ *and* $\mathrm{char}(\Bbbk) \neq 2, 3$,

- $\#\mathrm{Aut}(E) = 12$ *if* $j(E) = 0$ *and* $\mathrm{char}(\Bbbk) = 3$,
- $\#\mathrm{Aut}(E) = 24$ *if* $j(E) = 0$ *and* $\mathrm{char}(\Bbbk) = 2$.

(Note that when $\mathrm{char}(\Bbbk) = 2$ *or* 3 *then* $0 = 1728$ *in* $\Bbbk$.*)*

Proof See Theorem III.10.1 and Proposition A.1.2 of [505]. □

Exercise 9.4.5 Consider $E : y^2 + y = x^3$ over $\mathbb{F}_2$. Let $u \in \overline{\mathbb{F}}_2$ satisfy $u^3 = 1$, $s \in \overline{\mathbb{F}}_2$ satisfy $s^4 + s = 0$ and $t \in \overline{\mathbb{F}}_2$ satisfy $t^2 + t = s^6$. Show that $u, s \in \mathbb{F}_{2^2}$, $t \in \mathbb{F}_{2^4}$ and that

$$\phi(x, y) = (u^2 x + s^2, y + u^2 s x + t)$$

is an automorphism of E. Show that every automorphism arises this way and so $\#\mathrm{Aut}(E) = 24$. Show that if $\phi \in \mathrm{Aut}(E)$ then either $\phi^2 = \pm 1$ or $\phi^3 = \pm 1$. Show that $\mathrm{Aut}(E)$ is non-Abelian.

9.5 Twists

Twists of elliptic curves have several important applications such as point compression in pairing-based cryptography, and efficient endomorphisms on elliptic curves (see Exercise 11.3.24).

Definition 9.5.1 Let E be an elliptic curve over $\Bbbk$. A **twist** of E is an elliptic curve $\widetilde{E}$ over $\Bbbk$ such that there is an isomorphism $\phi : E \to \widetilde{E}$ over $\overline{\Bbbk}$ of pointed curves (i.e., such that $\phi(\mathcal{O}_E) = \mathcal{O}_{\widetilde{E}}$). Two twists $\widetilde{E}_1$ and $\widetilde{E}_2$ of E are **equivalent** if there is an isomorphism from $\widetilde{E}_1$ to $\widetilde{E}_2$ defined over $\Bbbk$. A twist $\widetilde{E}$ of E is called a **trivial twist** if $\widetilde{E}$ is equivalent to E. Denote by $\mathrm{Twist}(E)$ the set of equivalence classes of twists of E.

Example 9.5.2 Let $\Bbbk$ be a field such that $\mathrm{char}(\Bbbk) \neq 2$. Let $E : y^2 = x^3 + a_4 x + a_6$ over $\Bbbk$ and let $d \in \Bbbk^*$. Define the elliptic curve $E^{(d)} : y^2 = x^3 + d^2 a_4 x + d^3 a_6$. The map $\phi(x, y) = (dx, d^{3/2} y)$ is an isomorphism from E to $E^{(d)}$. Hence, $E^{(d)}$ is a twist of E. Note that $E^{(d)}$ is a trivial twist if $\sqrt{d} \in \Bbbk^*$.

If $\Bbbk = \mathbb{Q}$ then there are infinitely many non-equivalent twists $E^{(d)}$, since one can let d run over the square-free elements in $\mathbb{N}$.

Exercise 9.5.3 Let q be an odd prime power and let $E : y^2 = x^3 + a_4 x + a_6$ over $\mathbb{F}_q$. Let $d \in \mathbb{F}_q^*$. Show that the twist $E^{(d)}$ of E by d is not isomorphic over $\mathbb{F}_q$ to E if and only if d is a non-square (i.e., the equation $u^2 = d$ has no solution in $\mathbb{F}_q$). Show also that if d_1 and d_2 are non-squares in $\mathbb{F}_q^*$ then $E^{(d_1)}$ and $E^{(d_2)}$ are isomorphic over $\mathbb{F}_q$. Hence, there is a unique $\mathbb{F}_q$-isomorphism class of elliptic curves arising in this way. Any curve in this isomorphism class is called a **quadratic twist** of E.

Exercise 9.5.4 Show that if $E : y^2 = x^3 + a_4 x + a_6$ over $\mathbb{F}_q$ has $q + 1 - t$ points then a quadratic twist of E has $q + 1 + t$ points over $\mathbb{F}_q$.

Exercise 9.5.5 Let $F(x) = x^3 + a_2x^2 + a_4x + a_6$ and let $E : y^2 + (a_1x + a_3)y = F(x)$ be an elliptic curve over $\mathbb{F}_{2^n}$. Let $\alpha \in \mathbb{F}_{2^n}$ be such that $\mathrm{Tr}_{\mathbb{F}_{2^n}/\mathbb{F}_2}(\alpha) = 1$. Define $\widetilde{E} : y^2 + (a_1x + a_3)y = F(x) + \alpha(a_1x + a_3)^2$. For the special case (see Exercise 7.2.7) $E : y^2 + xy = x^3 + a_2x^2 + a_6$, this is $\widetilde{E} : y^2 + xy = x^3 + (a_2 + \alpha)x^2 + a_6$.

Show that $\widetilde{E}$ is isomorphic to E over $\mathbb{F}_{2^{2n}}$ but not over $\mathbb{F}_{2^n}$. Hence, it makes sense to call $\widetilde{E}$ a **quadratic twist** of E. Show, using Exercise 2.14.7, that $\#E(\mathbb{F}_{2^n}) + \#\widetilde{E}(\mathbb{F}_{2^n}) = 2(2^n + 1)$. Hence, if $\#E(\mathbb{F}_{2^n}) = 2^n + 1 - t$ then $\#\widetilde{E}(\mathbb{F}_{2^n}) = 2^n + 1 + t$.

Let E and $\widetilde{E}$ be elliptic curves over $\Bbbk$. Let $\phi : E \to \widetilde{E}$ be an isomorphism that is not defined over $\Bbbk$. Then $\phi^{-1} : \widetilde{E} \to E$ is also an isomorphism that is not defined over $\Bbbk$. One can therefore consider $\sigma(\phi^{-1}) : \widetilde{E} \to E$ for any $\sigma \in \mathrm{Gal}(\overline{\Bbbk}/\Bbbk)$. The composition $\psi(\sigma) = \sigma(\phi^{-1}) \circ \phi$ is therefore an automorphism of E.

Exercise 9.5.6 Let E and $\widetilde{E}$ be elliptic curves over $\Bbbk$. Show that if $\phi : E \to \widetilde{E}$ is an isomorphism that is not defined over $\Bbbk$ then there exists some $\sigma \in \mathrm{Gal}(\overline{\Bbbk}/\Bbbk)$ such that $\sigma(\phi^{-1}) \circ \phi$ is not the identity.

One can show that $\sigma \mapsto \sigma(\phi^{-1}) \circ \phi$ is a 1-cocycle with values in $\mathrm{Aut}(E)$. We refer to Section X.2 of Silverman [505] for further discussion of this aspect of the theory (note that Silverman considers twists for general curves C and his definition of $\mathrm{Twist}(C)$ is not for pointed curves).

Lemma 9.5.7 *Let E be an elliptic curve over a finite field $\Bbbk$ where* $\mathrm{char}(\Bbbk) \neq 2, 3$ *and* $j(E) \neq 0, 1728$. *Then* $\#\mathrm{Twist}(E) = 2$.

Proof Let $\widetilde{E}/\Bbbk$ be isomorphic to E. Without loss of generality, E and $\widetilde{E}$ are given in short Weierstrass form $y^2 = x^3 + a_4x + a_6$ and $Y^2 = X^3 + a'_4X + a'_6$ with $a_4, a'_4, a_6, a'_6 \neq 0$. Since $\widetilde{E} \cong E$ over $\overline{\Bbbk}$ it follows from Theorem 9.3.4 that $a'_4 = u^4a_4$ and $a'_6 = u^6a_6$ for some $u \in \overline{\Bbbk}^*$. Hence, $u^2 = a'_6a_4/(a_6a'_4) \in \Bbbk^*$. Since $\Bbbk$ is finite and $\mathrm{char}(\Bbbk) \neq 2$ the result follows from the fact that $[\Bbbk^* : (\Bbbk^*)^2] = 2$. □

An immediate consequence of Lemma 9.5.7 is that the number of $\mathbb{F}_q$-isomorphism classes of elliptic curves over $\mathbb{F}_q$ is approximately $2q$.

Exercise 9.5.8★ Let $\Bbbk$ be a finite field such that $\mathrm{char}(\Bbbk) \geq 5$ and let E be an elliptic curve over $\Bbbk$. Show that $\#\mathrm{Twist}(E) = d$ where $d = 2$ if $j(E) \neq 0, 1728$, $d = 4$ if $j(E) = 1728$, $d = 6$ if $j(E) = 0$.

Due to Theorem 9.4.4 one might be tempted to phrase Lemma 9.5.7 and Exercise 9.5.8 as $\#\mathrm{Twist}(E) = \#\mathrm{Aut}(E)$, but the following example shows that this statement is not true in general.

Exercise 9.5.9 Let $E : y^2 + y = x^3$ over $\mathbb{F}_2$. Show that the number of non-equivalent twists of E over $\mathbb{F}_2$ is 4, whereas $\#\mathrm{Aut}(E) = 24$.

9.6 Isogenies

We now return to more general maps between elliptic curves. Recall from Theorem 9.2.1 that a morphism $\phi : E_1 \to E_2$ of elliptic curves such that $\phi(\mathcal{O}_{E_1}) = \mathcal{O}_{E_2}$ is a group homomorphism. Hence, isogenies are group homomorphisms. Chapter 25 discusses isogenies in further detail. In particular, Chapter 25 describes algorithms to compute isogenies efficiently.

Definition 9.6.1 Let E_1 and E_2 be elliptic curves over $\Bbbk$. An **isogeny** over $\Bbbk$ is a morphism $\phi : E_1 \to E_2$ over $\Bbbk$ such that $\phi(\mathcal{O}_{E_1}) = \mathcal{O}_{E_2}$. The **zero isogeny** is the constant map $\phi : E_1 \to E_2$ given by $\phi(P) = \mathcal{O}_{E_2}$ for all $P \in E_1(\overline{\Bbbk})$. If $\phi(x, y) = (\phi_1(x, y), \phi_2(x, y))$ is an isogeny then define $-\phi$ by $(-\phi)(x, y) = -(\phi_1(x, y), \phi_2(x, y))$. where $-(X, Y)$ denotes, as always, the inverse for the group law. The **kernel** of an isogeny is $\ker(\phi) = \{P \in E_1(\overline{\Bbbk}) : \phi(P) = \mathcal{O}_{E_2}\}$. The **degree** of a non-zero isogeny is the degree of the morphism. The degree of the zero isogeny is zero. If there is an isogeny (respectively, isogeny of degree d) between two elliptic curves E_1 and E_2 then we say that E_1 and E_2 are **isogenous** (respectively, d**-isogenous**). A non-zero isogeny is **separable** if it is separable as a morphism (see Definition 8.1.6). Denote by $\mathrm{Hom}_{\Bbbk}(E_1, E_2)$ the set of isogenies from E_1 to E_2 defined over $\Bbbk$. Denote by $\mathrm{End}_{\Bbbk}(E_1)$ the set of isogenies from E_1 to E_1 defined over $\Bbbk$; this is called the **endomorphism ring** of the elliptic curve.

Exercise 9.6.2 Show that if $\phi : E_1 \to E_2$ is an isogeny then so is $-\phi$.

Theorem 9.6.3 *Let E_1 and E_2 be elliptic curves over $\Bbbk$. If $\phi : E_1 \to E_2$ is a non-zero isogeny over $\overline{\Bbbk}$ then $\phi : E_1(\overline{\Bbbk}) \to E_2(\overline{\Bbbk})$ is surjective.*

Proof This follows from Theorem 8.2.4. □

We now relate the degree to the number of points in the kernel. First, we remark the standard group theoretical fact that, for all $Q \in E_2(\overline{\Bbbk})$, $\#\phi^{-1}(Q) = \#\ker(\phi)$ (this is just the fact that all cosets have the same size).

Lemma 9.6.4 *A non-zero separable isogeny $\phi : E_1 \to E_2$ over $\Bbbk$ of degree d has $\#\ker(\phi) = d$.*

Proof It follows from Corollary 8.2.10 that a separable degree d map ϕ has $\#\phi^{-1}(Q) = d$ for a generic point $Q \in E_2(\overline{\Bbbk})$. Hence, by the above remark, $\#\phi^{-1}(Q) = d$ for all points Q and $\#\ker(\phi) = d$. (Also see Proposition 2.21 of [560] for an elementary proof.) □

A morphism of curves $\phi : C_1 \to C_2$ is called **unramified** if $e_\phi(P) = 1$ for all $P \in C_1(\overline{\Bbbk})$. Let $\phi : E_1 \to E_2$ be a separable isogeny over $\Bbbk$ and let $P \in E_1(\overline{\Bbbk})$. Since $\phi(P) = \phi(P + R)$ for all $R \in \ker(\phi)$ it follows that a separable morphism of elliptic curves is unramified (this also follows from the Hurwitz genus formula).

Exercise 9.6.5 Let E_1 and E_2 be elliptic curves over $\Bbbk$ and suppose $\phi : E_1 \to E_2$ is an isogeny over $\Bbbk$. Show that $\ker(\phi)$ is defined over $\Bbbk$ (in the sense that $P \in \ker(\phi)$ implies $\sigma(P) \in \ker(\phi)$ for all $\sigma \in \mathrm{Gal}(\overline{\Bbbk}/\Bbbk)$).

Lemma 9.6.6 *Let E_1 and E_2 be elliptic curves over $\Bbbk$. Then* $\mathrm{Hom}_{\Bbbk}(E_1, E_2)$ *is a group with addition defined by* $(\phi_1 + \phi_2)(P) = \phi_1(P) + \phi_2(P)$. *Furthermore,* $\mathrm{End}_{\Bbbk}(E_1) = \mathrm{Hom}_{\Bbbk}(E_1, E_1)$ *is a ring with addition defined in the same way and with multiplication defined by composition.*

Proof The main task is to show that if $\phi_1, \phi_2 : E_1 \to E_2$ are morphisms then so is $(\phi_1 + \phi_2)$. The case $\phi_2 = -\phi_1$ is trivial, so assume $\phi_2 \neq -\phi_1$. Let U be an open set such that: ϕ_1 and ϕ_2 are regular on U; $P \in U$ implies $\phi_1(P) \neq \mathcal{O}_{E_2}$ and $\phi_2(P) \neq \mathcal{O}_{E_2}$; $\phi_1(P) \neq -\phi_2(P)$. That such an open set exists is immediate for all but the final requirement, but one can also show that the points such that $\phi_1(x, y) = -\phi_2(x, y)$ form a closed subset of E_1 as long as $\phi_1 \neq -\phi_2$. Then using equation (9.3) one obtains a rational map $(\phi_1 + \phi_2) : E_1 \to E_2$. Finally, since composition of morphisms is a morphism it is easy to check that $\mathrm{End}_{\Bbbk}(E_1)$ is a ring. □

By Exercise 8.1.12, if $\phi_1 : E_1 \to E_2$ and $\phi_2 : E_2 \to E_3$ are non-constant isogenies then $\deg(\phi_2 \circ \phi_1) = \deg(\phi_2)\deg(\phi_1)$. This fact will often be used.

An important example of an isogeny is the multiplication by n map.

Exercise 9.6.7 Show that $[n]$ is an isogeny.

Example 9.6.8 Let $E : y^2 = x^3 + x$. Then the map $[2] : E \to E$ is given by the rational function

$$[2](x, y) = \left(\frac{(x^2 - 1)^2}{4(x^3 + x)}, \frac{y(x^6 + 5x^4 - 5x^2 - 1)}{8(x^3 + x)^2} \right).$$

The kernel of $[2]$ is $\mathcal{O}_E$ together with the three points $(x_P, 0)$ such that $x_P^3 + x_P = 0$. In other words, the kernel is the set of four points of order dividing 2.

We now give a simple example of an isogeny that is not $[n]$ for some $n \in \mathbb{N}$. The derivation of a special case of this example is given in Example 25.1.5.

Example 9.6.9 Let $A, B \in \Bbbk$ be such that $B \neq 0$ and $D = A^2 - 4B \neq 0$. Consider the elliptic curve $E : y^2 = x(x^2 + Ax + B)$ over $\Bbbk$, which has the point $(0, 0)$ of order 2. There is an elliptic curve $\widetilde{E}$ and an isogeny $\phi : E \to \widetilde{E}$ such that $\ker(\phi) = \{\mathcal{O}_E, (0, 0)\}$. One can verify that

$$\phi(x, y) = \left(\frac{y^2}{x^2}, \frac{y(B - x^2)}{x^2} \right) = \left(\frac{x^2 + Ax + B}{x}, y\frac{B - x^2}{x^2} \right)$$

has the desired kernel, and the image curve is $\widetilde{E} : Y^2 = X(X^2 - 2AX + D)$.

Before proving the next result we need one exercise (which will also be used later).

Exercise 9.6.10 Let $y^2 + a_1xy + a_3y = x^3 + a_2x^2 + a_4x + a_6$ be an elliptic curve over $\Bbbk$. Show that if $\mathrm{char}(\Bbbk) = 2$ and $a_1 = 0$ then there are no points (x, y) of order 2. Show that if $\mathrm{char}(\Bbbk) = 2$ and $a_1 \neq 0$ then (x, y) has order 2 if and only if $x = a_3/a_1$. Hence, if $\mathrm{char}(\Bbbk) = 2$ then $\#E[2] \in \{1, 2\}$.

Show that if $\mathrm{char}(\Bbbk) \neq 2$ then (x, y) has order 2 if and only if $2y + a_1x + a_3 = 0$. Show that this is also equivalent to

$$4x^3 + (a_1^2 + 4a_2)x^2 + (2a_1a_3 + 4a_4)x + (a_3^2 + 4a_6) = 0. \tag{9.9}$$

Note that when $a_1 = a_3 = 0$ this polynomial is simply 4 times the right-hand side of the elliptic curve equation. Show that this polynomial has distinct roots and so if $\mathrm{char}(\Bbbk) \neq 2$ then $\#E[2] = 4$.

Lemma 9.6.11 *Let E and $\widetilde{E}$ be elliptic curves over $\Bbbk$. If $n \in \mathbb{N}$ then $[n]$ is not the zero isogeny. Further,* $\mathrm{Hom}_{\Bbbk}(E, \widetilde{E})$ *is torsion-free as a $\mathbb{Z}$-module (i.e., if $\phi \in \mathrm{Hom}_{\Bbbk}(E, \widetilde{E})$ is non-zero then $[n] \circ \phi$ is non-zero for all $n \in \mathbb{Z}$, $n \neq 0$) and* $\mathrm{End}_{\Bbbk}(E)$ *has no zero divisors.*

Proof First, suppose $\phi_1, \phi_2 : E \to \widetilde{E}$ are non-zero isogenies such that $[0] = \phi_1 \circ \phi_2$. By Theorem 9.6.3, ϕ_1, ϕ_2 and hence $\phi_1 \circ \phi_2$ are surjective over $\overline{\Bbbk}$. Since the zero isogeny is not surjective it follows that there are no zero divisors in $\mathrm{End}_{\Bbbk}(E)$.

Now, consider any $n \in \mathbb{N}$ and note that $n = 2^k m$ for some $k \in \mathbb{Z}_{\geq 0}$ and some odd $m \in \mathbb{N}$. By Exercise 9.6.10 we know that $[2]$ is not zero over $\overline{\Bbbk}$ (when $\mathrm{char}(\Bbbk) \neq 2$ this is immediate since there are at most three points of order 2; when $\mathrm{char}(\Bbbk) = 2$ one must show that if equation (9.9) is identically zero then the Weierstrass equation is singular). It follows that $[2^k] = [2] \circ [2] \circ \cdots \circ [2]$ is not zero either (since if $[2]$ is non-zero then it is surjective on $E(\overline{\Bbbk})$). Finally, since there exists $P \in E(\overline{\Bbbk})$ such that $P \neq \mathcal{O}_E$ but $[2]P = \mathcal{O}_E$ we have $[m]P = P \neq \mathcal{O}_E$ and so $[m]$ is not the zero isogeny. It follows that $[n] = [m] \circ [2^k]$ is not the zero isogeny.

Similarly, if $[0] = [n]\phi$ for $\phi \in \mathrm{Hom}_{\Bbbk}(E, \widetilde{E})$ then either $[n]$ or ϕ is the zero isogeny. □

Lemma 9.6.12 *Let $E : y^2 + a_1xy + a_3y = x^3 + a_2x^2 + a_4x + a_6$ and $\widetilde{E} : Y^2 + \widetilde{a}_1XY + \widetilde{a}_3Y = X^3 + \widetilde{a}_2X^2 + \widetilde{a}_4X + \widetilde{a}_6$ be elliptic curves over $\Bbbk$. Let $\phi : E \to \widetilde{E}$ be an isogeny of elliptic curves over $\Bbbk$. Then ϕ may be expressed by a rational function in the form*

$$\phi(x, y) = (\phi_1(x), y\phi_2(x) + \phi_3(x))$$

where

$$2\phi_3(x) = -\widetilde{a}_1\phi_1(x) - \widetilde{a}_3 + (a_1x + a_3)\phi_2(x).$$

In particular, if $\mathrm{char}(\Bbbk) \neq 2$ *and $a_1 = a_3 = \widetilde{a}_1 = \widetilde{a}_3 = 0$ then $\phi_3(x) = 0$, while if* $\mathrm{char}(\Bbbk) = 2$ *then $\phi_2(x) = (\widetilde{a}_1\phi_1(x) + \widetilde{a}_3)/(a_1x + a_3)$.*

Proof Certainly, ϕ may be written as $\phi(x, y) = (\phi_1(x) + yf(x), y\phi_2(x) + \phi_3(x))$ where $\phi_1(x)$, $f(x)$, $\phi_2(x)$ and $\phi_3(x)$ are rational functions.

Since ϕ is a group homomorphism it satisfies $\phi(-P) = -\phi(P)$. Writing $P = (x, y)$ the left-hand side is

$$\begin{aligned}\phi(-(x, y)) &= \phi(x, -y - a_1x - a_3)\\ &= (\phi_1(x) + (-y - a_1x - a_3)f(x), (-y - a_1x - a_3)\phi_2(x) + \phi_3(x))\end{aligned}$$

while the right-hand side is

$$-\phi(P) = (\phi_1(x) + yf(x), -y\phi_2(x) - \phi_3(x) - \widetilde{a}_1(\phi_1(x) + yf(x)) - \widetilde{a}_3).$$

It follows that $(2y + a_1x + a_3)f(x)$ is a function that is zero for all points $(x, y) \in E(\overline{\Bbbk})$. Since $2y + a_1x + a_3$ is not the zero function (if it was zero then $\Bbbk(E) = \Bbbk(x, y) = \Bbbk(y)$, which contradicts Theorem 8.6.4) it follows that $f(x) = 0$.

It then follows that

$$2\phi_3(x) = -\widetilde{a}_1\phi_1(x) - \widetilde{a}_3 + (a_1x + a_3)\phi_2(x).$$ □

Lemma 9.6.12 will be refined in Theorem 9.7.5.

Lemma 9.6.13 *Let $\phi : E \to \widetilde{E}$ be as in Lemma 9.6.12 where $\phi_1(x) = a(x)/b(x)$. Then the degree of ϕ is* $\max\{\deg_x(a(x)), \deg_x(b(x))\}$.

Corollary 25.1.8 will give a more precise version of this result in a special case.

Proof We have $\Bbbk(E) = \Bbbk(x, y)$ being a quadratic extension of $\Bbbk(x)$, and $\Bbbk(\widetilde{E}) = \Bbbk(\widetilde{x}, \widetilde{y})$ being a quadratic extension of $\Bbbk(\widetilde{x})$. Now $\phi_1(x)$ gives a morphism $\phi_1 : \mathbb{P}^1 \to \mathbb{P}^1$ and this morphism has degree $d = \max\{\deg_x(a(x)), \deg_x(b(x))\}$ by Lemma 8.1.9. It follows that $\Bbbk(x)$ is a degree d extension of $\phi_1^*\Bbbk(\widetilde{x})$. We therefore have the following diagram of field extensions

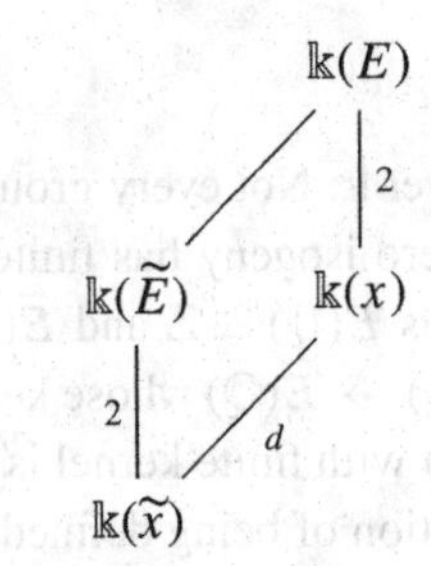

and it follows that $[\Bbbk(E) : \phi^*\Bbbk(\widetilde{E})] = d$. □

Example 9.6.14 Let p be a prime and let $q = p^m$ for some $m \in \mathbb{N}$. Let E be an elliptic curve over $\mathbb{F}_q$. The q-power **Frobenius map** is the rational map $\pi_q : E \to E$ such that $\pi_q(\mathcal{O}_E) = \mathcal{O}_E$ and $\pi_q(x, y) = (x^q, y^q)$. Since π_q is a morphism that fixes $\mathcal{O}_E$ it is an isogeny (this can also be easily seen by explicit computation). If E has equation $y^2 = F(x)$ (and so q is odd) then one can write π_q in the form of Lemma 9.6.12 as $\pi_q(x, y) = (x^q, yF(x)^{(q-1)/2})$. Note that π_q is the identity map on $E(\mathbb{F}_q)$ but is not the identity on $E(\overline{\mathbb{F}}_q)$.

Corollary 9.6.15 *Let the notation be as in Example 9.6.14. The q-power Frobenius map is inseparable of degree q.*

Exercise 9.6.16 Prove Corollary 9.6.15.

Theorem 9.6.17 *Let p be a prime and $E, \widetilde{E}$ elliptic curves over $\overline{\mathbb{F}}_p$. Let $\psi : E \to \widetilde{E}$ be a non-zero isogeny. Then there is an integer m and an elliptic curve $E^{(q)}$ (namely, the curve obtained by applying the $q = p^m$-power Frobenius map to the coefficients of E; the reader should not confuse the notation $E^{(q)}$ with the quadratic twist $E^{(d)}$) and a separable isogeny $\phi : E^{(q)} \to \widetilde{E}$ of degree $\deg(\psi)/q$ such that $\psi = \phi \circ \pi_q$ where $\pi_q : E \to E^{(q)}$ is the q-power Frobenius morphism.*

Proof See Corollary II.2.12 of [505]. □

The following result is needed to obtain many useful results in this chapter and in Chapter 25.

Theorem 9.6.18 *Let E_1, E_2, E_3 be elliptic curves over $\Bbbk$ and $\phi : E_1 \to E_2$, $\psi : E_1 \to E_3$ isogenies over $\Bbbk$. Suppose $\ker(\phi) \subseteq \ker(\psi)$ and that ψ is separable. Then there is a unique isogeny $\lambda : E_2 \to E_3$ defined over $\Bbbk$ such that $\psi = \lambda \circ \phi$.*

Proof (Sketch) See Corollary III.4.11 of [505] for the case where $\Bbbk$ is algebraically closed. The proof uses the fact that $\overline{\Bbbk}(E_1)$ is a Galois extension of $\phi^*(\overline{\Bbbk}(E_2))$ (with Galois group isomorphic to $\ker(\phi)$). Furthermore, one has $\psi^*(\overline{\Bbbk}(E_3)) \subseteq \phi^*(\overline{\Bbbk}(E_2)) \subseteq \overline{\Bbbk}(E_1)$. The existence and uniqueness of the morphism λ follows from the Galois extension $\phi^*(\overline{\Bbbk}(E_2))/\psi^*(\overline{\Bbbk}(E_3))$ and Theorem 5.5.24. The uniqueness of λ implies it is actually defined over $\Bbbk$, since

$$\psi = \sigma(\psi) = \sigma(\lambda) \circ \sigma(\phi) = \sigma(\lambda) \circ \phi.$$

for all $\sigma \in \mathrm{Gal}(\overline{\Bbbk}/\Bbbk)$. □

Let E and $\widetilde{E}$ be elliptic curves over $\Bbbk$. Not every group homomorphism $E(\Bbbk) \to \widetilde{E}(\Bbbk)$ is an isogeny. In particular, a non-zero isogeny has finite degree and hence finite kernel, whereas one can have groups such as $E(\mathbb{Q}) \cong \mathbb{Z}$ and $\widetilde{E}(\mathbb{Q}) \cong \mathbb{Z}/2\mathbb{Z}$ for which there is a non-zero group homomorphism $E(\mathbb{Q}) \to \widetilde{E}(\mathbb{Q})$ whose kernel is infinite. It is natural to ask whether every group homomorphism with finite kernel is an isogeny. The following result shows that this is the case (the condition of being defined over $\Bbbk$ can be ignored by taking a field extension).

Theorem 9.6.19 *Let E be an elliptic curve over $\Bbbk$. Let $G \subseteq E(\overline{\Bbbk})$ be a finite group that is defined over $\Bbbk$ (i.e., $\sigma(P) \in G$ for all $P \in G$ and $\sigma \in \mathrm{Gal}(\overline{\Bbbk}/\Bbbk)$). Then there is a unique (up to isomorphism over $\overline{\Bbbk}$) elliptic curve $\widetilde{E}$ over $\Bbbk$ and a (not necessarily unique) isogeny $\phi : E \to \widetilde{E}$ over $\Bbbk$ such that $\ker(\phi) = G$.*

Proof See Proposition III.4.12 and Exercise 3.13(e) of [505]. We will give a constructive proof (Vélu's formulae) in Section 25.1.1, which also proves that the isogeny is defined over $\Bbbk$. □

The elliptic curve $\widetilde{E}$ in Theorem 9.6.19 is sometimes written E/G. As noted, the isogeny in Theorem 9.6.19 is not necessarily unique, but Exercise 9.6.20 shows the only way that non-uniqueness can arise.

Exercise 9.6.20 Let the notation be as in Theorem 9.6.19. Let $\psi : E \to \widetilde{E}$ be another isogeny over $\Bbbk$ such that $\ker(\psi) = G$. Show that $\psi = \lambda \circ \phi$ where λ is an automorphism of $\widetilde{E}$ (or, if $\Bbbk$ is finite, the composition of an isogeny and a Frobenius map). Similarly, if $\psi : E \to E_2$ is an isogeny over $\Bbbk$ with $\ker(\psi) = G$ then show that $\psi = \lambda \circ \phi$ where $\lambda : \widetilde{E} \to E_2$ is an isomorphism over $\Bbbk$ of elliptic curves.

We now present the dual isogeny. Let $\phi : E \to \widetilde{E}$ be an isogeny over $\Bbbk$. Then there is a group homomorphism $\phi^* : \mathrm{Pic}^0_{\overline{\Bbbk}}(\widetilde{E}) \to \mathrm{Pic}^0_{\overline{\Bbbk}}(E)$. Since $\mathrm{Pic}^0_{\overline{\Bbbk}}(E)$ is identified with $E(\overline{\Bbbk})$ in a standard way (and similarly for $\widetilde{E}$) one gets a group homomorphism from $\widetilde{E}(\overline{\Bbbk})$ to $E(\overline{\Bbbk})$. Indeed, the next result shows that this is an isogeny of elliptic curves; this is not trivial as ϕ^* is defined set-theoretically and it is not possible to interpret it as a rational map in general.

Theorem 9.6.21 *Let E and $\widetilde{E}$ be elliptic curves over $\Bbbk$. Let $\phi : E \to \widetilde{E}$ be a non-zero isogeny over $\Bbbk$ of degree m. Then there is a non-zero isogeny $\widehat{\phi} : \widetilde{E} \to E$ over $\Bbbk$ such that*

$$\widehat{\phi} \circ \phi = [m] : E \to E.$$

Indeed, $\widehat{\phi}$ is unique (see Exercise 9.6.22).

Proof Let $\alpha_1 : E(\Bbbk) \to \mathrm{Pic}^0_{\Bbbk}(E)$ be the canonical map $P \mapsto (P) - (\mathcal{O}_E)$ and similarly for $\alpha_2 : \widetilde{E} \to \mathrm{Pic}^0_{\Bbbk}(\widetilde{E})$. We have $\widehat{\phi} = \alpha_1 \circ \phi^* \circ \alpha_2^{-1}$ as above. We refer to Theorem III.6.1 of [505] (or Section 21.1 of [560] for elliptic curves over $\mathbb{C}$) for the details. □

Exercise 9.6.22 Suppose as in Theorem 9.6.21 that $\phi : E \to \widetilde{E}$ is a non-zero isogeny over $\Bbbk$ of degree m. Show that if $\psi : \widetilde{E} \to E$ is any isogeny such that $\psi \circ \phi = [m]$ then $\psi = \widehat{\phi}$.

Definition 9.6.23 Let E and $\widetilde{E}$ be elliptic curves over $\Bbbk$ and let $\phi : E \to \widetilde{E}$ be a non-zero isogeny over $\Bbbk$. The isogeny $\widehat{\phi} : \widetilde{E} \to E$ of Theorem 9.6.21 is called the **dual isogeny**.

Example 9.6.24 Let E be an elliptic curve over $\mathbb{F}_q$ and $\pi_q : E \to E$ the q-power Frobenius map. The dual isogeny $\widehat{\pi_q}$ is called the **Verschiebung**. Since $\widehat{\pi_q} \circ \pi_q = [q]$ it follows that $\widehat{\pi_q}(x, y) = [q](x^{1/q}, y^{1/q})$. Example 9.10.2 gives another way to write the Verschiebung.

Exercise 9.6.25 Let $E : y^2 = x^3 + a_6$ over $\Bbbk$ with $\mathrm{char}(\Bbbk) \neq 2, 3$. Let $\zeta_3 \in \overline{\Bbbk}$ be such that $\zeta_3^2 + \zeta_3 + 1 = 0$ and define the isogeny $\rho(\mathcal{O}_E) = \mathcal{O}_E$ and $\rho(x, y) = (\zeta_3 x, y)$. Show that $\widehat{\rho} = \rho^2$ (where ρ^2 means $\rho \circ \rho$).

Exercise 9.6.26 Recall E, $\widetilde{E}$ and ϕ from Example 9.6.9. Show that $\widehat{\phi} : \widetilde{E} \to E$ is given by

$$\widehat{\phi}(X, Y) = \left(\frac{Y^2}{4X^2}, \frac{Y(D - X^2)}{8X^2} \right)$$

and that $\widehat{\phi} \circ \phi(x, y) = [2](x, y)$.

We list some properties of the dual isogeny.

Theorem 9.6.27 *Let $\phi : E \to \widetilde{E}$ be a non-zero isogeny of elliptic curves over $\Bbbk$.*

1. *Let $d = \deg(\phi)$. Then $\deg(\widehat{\phi}) = d$, $\widehat{\phi} \circ \phi = [d]$ on E and $\phi \circ \widehat{\phi} = [d]$ on $\widetilde{E}$.*
2. *Let $\psi : \widetilde{E} \to E_3$ be an isogeny. Then $\widehat{\psi \circ \phi} = \widehat{\phi} \circ \widehat{\psi}$.*
3. *Let $\psi : E \to \widetilde{E}$ be an isogeny. Then $\widehat{\phi + \psi} = \widehat{\phi} + \widehat{\psi}$.*
4. *$\widehat{\widehat{\phi}} = \phi$.*

Proof See Theorem III.6.2 of [505]. □

Corollary 9.6.28 *Let E be an elliptic curve over $\Bbbk$ and let $m \in \mathbb{Z}$. Then $\widehat{[m]} = [m]$ and $\deg([m]) = m^2$.*

Proof The first claim follows by induction from part 3 of Theorem 9.6.27. The second claim follows from part 1 of Theorem 9.6.27 and since $\widehat{[1]} = [1]$: write $d = \deg([m])$ and use $[d] = \widehat{[m]}[m] = [m^2]$; since $\mathrm{End}_{\Bbbk}(E)$ is torsion-free (Lemma 9.6.11) it follows that $d = m^2$. □

An important consequence of Corollary 9.6.28 is that it determines the possible group structures of elliptic curves over finite fields. We return to this topic in Theorem 9.8.1.

We end this section with another example of an isogeny.

Exercise 9.6.29 Let $\Bbbk$ be a field such that $\mathrm{char}(\Bbbk) \neq 2, 3$. Let E be an elliptic curve with a subgroup of order 3 defined over $\Bbbk$. Show that, after a suitable change of variable, one has a point $P = (0, v)$ such that $[2]P = (0, -v)$ and $v^2 \in \Bbbk$. Show that E is $\Bbbk$-isomorphic to a curve of the form

$$y^2 = x^3 + \tfrac{1}{a_6}\left(\tfrac{a_4}{2}x + a_6\right)^2.$$

Show that there is a $\overline{\Bbbk}$-isomorphism to a curve of the form

$$Y^2 = X^3 + A(X+1)^2$$

where $A \neq 0, \frac{27}{4}$.

Exercise 9.6.30 (Doche, Icart and Kohel [161]) Let $\Bbbk$ be a field such that $\mathrm{char}(\Bbbk) \neq 2, 3$. Let $u \in \Bbbk$ be such that $u \neq 0, \frac{9}{4}$. Consider the elliptic curve $E : y^2 = x^3 + 3u(x+1)^2$ as in Exercise 9.6.29. Then $(0, \sqrt{3u})$ has order 3 and $G = \{\mathcal{O}_E, (0, \pm\sqrt{3u})\}$ is a $\Bbbk$-rational subgroup of $E(\Bbbk)$. Show that

$$\phi(x, y) = \left(\frac{x^3 + 4ux^2 + 12u(x+1)}{x^2}, y\left(1 - 12u\frac{x+2}{x^3}\right)\right)$$

is an isogeny from E to $\widetilde{E} : Y^2 = X^3 - u(3X - 4u + 9)^2$ with $\ker(\phi) = G$. Determine the dual isogeny to ϕ and show that $\widehat{\phi} \circ \phi = [3]$.

9.7 The invariant differential

Let E over $\Bbbk$ be an elliptic curve. Recall the differential

$$\omega_E = \frac{dx}{2y + a_1x + a_3} \tag{9.10}$$

on the Weierstrass equation for E, which was studied in Example 8.5.32. We showed that the divisor of ω_E is 0. Let $Q \in E(\Bbbk)$ and τ_Q be the translation map. Then $\tau_Q^*(\omega_E) \in \Omega_{\Bbbk}(E)$ and so, by Theorem 8.5.21, $\tau_Q^*(\omega_E) = f\omega_E$ for some $f \in \Bbbk(E)$. Lemma 8.5.36 implies $\tau_Q^*(\mathrm{div}(\omega_E)) = 0$ and so $\mathrm{div}(f) = 0$. It follows that $\tau_Q^*(\omega_E) = c\omega_E$ for some $c \in \Bbbk^*$. The following result shows that $c = 1$ and so ω_E is fixed by translation maps. This explains why ω_E is called the **invariant differential**.

Theorem 9.7.1 *Let E be an elliptic curve in Weierstrass form and let ω_E be the differential in equation (9.10). Then $\tau_Q^*(\omega_E) = \omega_E$ for all $Q \in E(\Bbbk)$.*

Proof See Proposition III.5.1 of [505]. □

An important fact is that the action of isogenies on differentials is linear.

Theorem 9.7.2 *Let $E, \widetilde{E}$ be elliptic curves over $\Bbbk$ and $\omega_{\widetilde{E}}$ the invariant differential on $\widetilde{E}$. Suppose $\phi, \psi : E \to \widetilde{E}$ are isogenies. Then*

$$(\phi + \psi)^*(\omega_{\widetilde{E}}) = \phi^*(\omega_{\widetilde{E}}) + \psi^*(\omega_{\widetilde{E}}).$$

Proof See Theorem III.5.2 of [505]. □

A crucial application is to determine when certain isogenies are separable. In particular, if E is an elliptic curve over $\mathbb{F}_{p^n}$ then $[p]$ is inseparable on E, while $\pi_{p^n} - 1$ is separable (where π_{p^n} is the p^n-power Frobenius).

Corollary 9.7.3 *Let E be an elliptic curve over $\Bbbk$. Let $m \in \mathbb{Z}$. Then $[m]$ is separable if and only if m is coprime to the characteristic of $\Bbbk$. Let $\Bbbk = \mathbb{F}_q$ and π_q be the q-power Frobenius. Let $m, n \in \mathbb{Z}$. Then $m + n\pi_q$ is separable if and only if m is coprime to q.*

Proof (Sketch) Theorem 9.7.2 implies $[m]^*(\omega_E) = m\omega_E$. So $[m]^*$ maps $\Omega_{\Bbbk}(E)$ to $\{0\}$ if and only if the characteristic of $\Bbbk$ divides m. The first part then follows from Lemma 8.5.35. The second part follows by the same argument, using the fact that π_q is inseparable and so $\pi_q^*(\omega_E) = 0$. For full details see Corollaries III.5.3 to III.5.5 of [505]. □

This result has the following important consequence.

Theorem 9.7.4 *Let E and $\widetilde{E}$ be elliptic curves over a finite field $\mathbb{F}_q$. If $\phi : E \to \widetilde{E}$ is an isogeny over $\mathbb{F}_q$ then $\#E(\mathbb{F}_q) = \#\widetilde{E}(\mathbb{F}_q)$.*

Proof Let π_q be the q-power Frobenius map on E. For $P \in E(\overline{\mathbb{F}}_q)$ we have $\pi_q(P) = P$ if and only if $P \in E(\mathbb{F}_q)$. Hence, $E(\mathbb{F}_q) = \ker(\pi_q - 1)$. Since $\pi_q - 1$ is separable it follows that $\#E(\mathbb{F}_q) = \deg(\pi_q - 1)$.

Now, returning to the problem of the Theorem, write π_q and $\widehat{\pi}_q$ for the q-power Frobenius maps on E and $\widetilde{E}$ respectively. Since ϕ is defined over $\mathbb{F}_q$ it follows that $\widehat{\pi}_q \circ \phi = \phi \circ \pi_q$. Hence, $(\widehat{\pi}_q - 1) \circ \phi = \phi \circ (\pi_q - 1)$ and so (applying Exercise 8.1.12 twice) $\deg(\widehat{\pi}_q - 1) = \deg(\pi_q - 1)$. The result follows since $\#E(\mathbb{F}_q) = \deg(\pi_q - 1)$ and $\#\widetilde{E}(\mathbb{F}_q) = \deg(\widehat{\pi}_q - 1)$. □

The converse (namely, if E and $\widetilde{E}$ are elliptic curves over $\mathbb{F}_q$ and $\#E(\mathbb{F}_q) = \#\widetilde{E}(\mathbb{F}_q)$ then there is an isogeny from E to $\widetilde{E}$ over $\mathbb{F}_q$) is Tate's isogeny theorem [540]. This can be proved for elliptic curves using the theory of complex multiplication (see Remark 25.3.10).

We now give a refinement of Lemma 9.6.12. This result shows that a separable isogeny is determined by $\phi_1(x)$ when $\mathrm{char}(\Bbbk) \neq 2$.

Theorem 9.7.5 *Let the notation be as in Lemma 9.6.12. Let $\phi : E \to \widetilde{E}$ be a separable isogeny over $\Bbbk$. Then ϕ may be expressed by a rational function in the form*

$$\phi(x, y) = (\phi_1(x), cy\phi_1(x)' + \phi_3(x))$$

where $\phi_1(x)' = d\phi_1(x)/dx$ is the (formal) derivative of the rational function $\phi_1(x)$, where $c \in \overline{\Bbbk}^*$ is a non-zero constant, and where

$$2\phi_3(x) = -\widetilde{a}_1\phi_1(x) - \widetilde{a}_3 + c(a_1x + a_3)\phi_1(x)'.$$

Proof Let $\omega_E = dx/(2y + a_1x + a_3)$ be the invariant differential on E and $\omega_{\widetilde{E}} = dX/(2Y + \widetilde{a}_1X + \widetilde{a}_3)$ be the invariant differential on $\widetilde{E}$. Since ϕ is separable, then $\phi^*(\omega_{\widetilde{E}})$ is non-zero. Furthermore, since ϕ is unramified, Lemma 8.5.36 implies that $\mathrm{div}(\phi^*(\omega_{\widetilde{E}})) = \phi^*(\mathrm{div}(\omega_{\widetilde{E}})) = 0$. Hence, $\phi^*(\omega_{\widetilde{E}})$ is a multiple of ω_E and so

$$dx/(2y + a_1x + a_3) = c\phi^*(dX/(2Y + \widetilde{a}_1X + \widetilde{a}_3))$$

for some non-zero constant $c \in \overline{\Bbbk}$.

By Lemma 9.6.12, $X = \phi_1(x)$, $Y = y\phi_2(x) + \phi_3(x)$ and

$$2\phi_3(x) = -\widetilde{a}_1\phi_1(x) - \widetilde{a}_3 + (a_1x + a_3)\phi_2(x).$$

Now, since $dX/dx = \phi_1(x)'$

$$\phi^*(dX/(2Y + \widetilde{a}_1X + \widetilde{a}_3)) = \phi_1(x)'dx/(2(y\phi_2(x) + \phi_3(x)) + \widetilde{a}_1\phi_1(x) + \widetilde{a}_3).$$

Hence, substituting for $\phi_3(x)$

$$\begin{aligned}\phi^*(dX/(2Y + \widetilde{a}_1X + \widetilde{a}_3)) &= \phi_1(x)'dx/((2y + a_1x + a_3)\phi_2(x)) \\ &= (\phi_1(x)'/\phi_2(x))dx/(2y + a_1x + a_3).\end{aligned}$$

It follows that $\phi_2(x) = c\phi_1(x)'$ for some $c \in \overline{\Bbbk}^*$, which proves the result. □

In Section 25.1.1 we will make use of Theorem 9.7.5 in the case $c = 1$.

Exercise 9.7.6 Let the notation be as in Theorem 9.7.5 and suppose $\mathrm{char}(\Bbbk) = 2$. Show that there are only two possible values for the rational function $\phi_3(x)$.

9.8 Multiplication by n and division polynomials

Corollary 9.6.28 showed the fundamental fact that $\deg([m]) = m^2$ and so there are at most m^2 points of order dividing m on an elliptic curve. There are several other explanations for this fact. One explanation is to consider elliptic curves over $\mathbb{C}$: as a Riemann surface, they are a complex torus $\mathbb{C}/L$ where L is a rank 2 lattice (see Chapter 5 of Silverman [505], especially Proposition 5.4) and it follows that there are m^2 points of order m (this argument generalises immediately to Abelian varieties).

It follows from Corollary 9.6.28 that $\#E[m] \le m^2$, and elementary group theory implies $\#E[m]$ is therefore a divisor of m^2. Theorem 9.8.1 follows. A more precise version of this result is Theorem 9.10.13.

Theorem 9.8.1 *Let E be an elliptic curve over a finite field $\mathbb{F}_q$. Then $E(\mathbb{F}_q)$ is isomorphic as a group to a product of cyclic groups of order n_1 and n_2 such that $n_1 \mid n_2$.*

Proof (Sketch) Since $E(\mathbb{F}_q)$ is a finite Abelian group we apply the classification of finite Abelian groups (e.g., Theorem II.2.1 of [271]). Then use the fact that there are at most m^2 points in $E(\mathbb{F}_q)$ of order m for every $m \in \mathbb{N}$. □

Since $\#E[m] \le m^2$ (and, by Corollary 9.7.3, equal to m^2 when m is coprime to the characteristic) it is natural to seek polynomials whose roots give the (affine) points of order dividing m. We already saw such polynomials in Exercise 9.6.10 for the case $m = 2$ (and this gave an alternative proof that, in general, there are three points (x, y) over $\overline{\Bbbk}$ of order 2 on an elliptic curve; namely, the points $(x, 0)$ where x is a root of the polynomial in equation (9.9)). Since $[m]P = \mathcal{O}_E$ if and only if $[m](-P) = \mathcal{O}_E$ one might expect to use polynomials in $\Bbbk[x]$, but when m is even it turns out to be more convenient to have polynomials that feature the variable y (one reason being that this leads to polynomials of lower degree). When m is odd the polynomials will be univariate and of degree $(m^2 - 1)/2$ as expected. We now determine these polynomials, first for the cases $m = 3$ and $m = 4$.

Exercise 9.8.2 Let $E : y^2 = x^3 + a_2x^2 + a_4x + a_6$ be an elliptic curve over $\Bbbk$ (with $\mathrm{char}(\Bbbk) \ne 2$). Show that if $\mathrm{char}(\Bbbk) = 3$, $a_2 = 0$ and $a_4 \ne 0$ then there is no point (x, y) of order 3. Show that if $\mathrm{char}(\Bbbk) = 3$ and $a_2 \ne 0$ then (x, y) has order 3 if and only if $x^3 = a_6 - a_4^2/(4a_2)$. Hence, if $\mathrm{char}(\Bbbk) = 3$ then $\#E[3] \in \{1, 3\}$.

Show that if $\mathrm{char}(\Bbbk) \ne 3$ then (x, y) has order 3 if and only if

$$3x^4 + 4a_2x^3 + 6a_4x^2 + 12a_6x + (4a_2a_6 - a_4^2) = 0.$$

Exercise 9.8.3 Let $E : y^2 = x^3 + a_4x + a_6$ be an elliptic curve over $\Bbbk$ with $\mathrm{char}(\Bbbk) \ne 2$. Show that if $P = (x, y) \in E(\overline{\Bbbk})$ satisfies $P \in E[4]$ and $P \notin E[2]$ then $[2]P$ is of the form $(x_2, 0)$ for some $x_2 \in \Bbbk$. Hence, show that x satisfies

$$x^6 + 5a_4x^4 + 20a_6x^3 - 5a_4^2x^2 - 4a_4a_6x - (a_4^3 + 8a_6^2).$$

We now state the polynomials whose roots give affine points of order dividing m for the case of elliptic curves in short Weierstrass form. The corresponding polynomials for elliptic curves over fields of characteristic 2 are given in Section 4.4.5.a of [16] and Section III.4.2 of [61]. Division polynomials for elliptic curves in general Weierstrass form are discussed in Section III.4 of [61].

Definition 9.8.4 Let $E : y^2 = x^3 + a_4 x + a_6$ be an elliptic curve over $\mathbb{k}$ with $\mathrm{char}(\mathbb{k}) \neq 2$. The **division polynomials** are defined by

$$\begin{aligned}
\psi_1(x, y) &= 1 \\
\psi_2(x, y) &= 2y \\
\psi_3(x, y) &= 3x^4 + 6a_4x^2 + 12a_6x - a_4^2 \\
\psi_4(x, y) &= 4y(x^6 + 5a_4x^4 + 20a_6x^3 - 5a_4^2x^2 - 4a_4a_6x - (a_4^3 + 8a_6^2)) \\
\psi_{2m+1}(x, y) &= \psi_{m+2}(x, y)\psi_m(x, y)^3 - \psi_{m-1}(x, y)\psi_{m+1}(x, y)^3, \ (m \geq 2) \\
\psi_{2m}(x, y) &= \tfrac{1}{2y}\psi_m(x, y)(\psi_{m+2}(x, y), \psi_{m-1}(x, y)^2 \\
&\quad -\psi_{m-2}(x, y)\psi_{m+1}(x, y)^2), \ (m \geq 3).
\end{aligned}$$

Lemma 9.8.5 *Let E be an elliptic curve in short Weierstrass form over $\mathbb{k}$ with* $\mathrm{char}(\mathbb{k}) \neq 2$. *Let $m \in \mathbb{N}$. Then $\psi_m(x, y) \in \mathbb{k}[x, y]$. If m is odd then $\psi_m(x, y)$ is a polynomial in x only and $\psi_m(x, y) = mx^{(m^2-1)/2} + \cdots \in \mathbb{k}[x]$. If m is even then $\psi_m(x, y) = yh(x)$ where $h(x) = mx^{(m^2-4)/2} + \cdots \in \mathbb{k}[x]$.*

Proof The case $m = 2$ is trivial and the cases $m = 3$ and 4 were done in Exercises 9.8.2 and 9.8.3. The rest are easily proved by induction. □

Theorem 9.8.6 *Let E be an elliptic curve in short Weierstrass form over $\mathbb{k}$ with* $\mathrm{char}(\mathbb{k}) \neq 2, 3$. *Let $m \in \mathbb{N}$ and $\psi_m(x, y)$ as above. Then $P = (x_P, y_P) \in E(\overline{\mathbb{k}})$ satisfies $[m]P = \mathcal{O}_E$ if and only if $\psi_m(x_P, y_P) = 0$. Furthermore, there are polynomials $A_m(x) \in \mathbb{k}[x]$ and $B_m(x, y) \in \mathbb{k}[x, y]$ such that*

$$[m](x, y) = \left(\frac{A_m(x)}{\psi_m(x, y)^2}, \frac{B_m(x, y)}{\psi_m(x, y)^3} \right).$$

Proof This can be proved in various ways: Section 9.5 of Washington [560] gives a proof for elliptic curves over $\mathbb{C}$ and then deduces the result for general fields of characteristic not equal to 2, Charlap and Robbins [119] give a proof (Sections 7 to 9) using considerations about divisors and functions, other sources (such as Exercise 3.7 of [505]) suggest a (tedious) verification by induction. □

9.9 Endomorphism structure

The aim of this section is to discuss the structure of the ring $\mathrm{End}_{\mathbb{k}}(E)$. Note that $\mathbb{Z} \subseteq \mathrm{End}_{\mathbb{k}}(E)$ and that, by Lemma 9.6.11, $\mathrm{End}_{\mathbb{k}}(E)$ is a torsion-free $\mathbb{Z}$-module. For an isogeny $\phi : E \to E$ and an integer $m \in \mathbb{Z}$ we write $m\phi$ for the isogeny $[m] \circ \phi$.

To understand the endomorphism rings of elliptic curves one introduces the **Tate module** $T_l(E)$. This is defined, for any prime $l \neq \mathrm{char}(\mathbb{k})$, to be the inverse limit of the groups $E[l^i]$

(this is the same process as used to construct the p-adic ($= l$-adic) numbers $\mathbb{Z}_l$ as the inverse limit of the rings $\mathbb{Z}/l^i\mathbb{Z}$). More precisely, for each $i \in \mathbb{N}$ fix a pair $\{P_{i,1}, P_{i,2}\}$ of generators for $E[l^i]$ such that $P_{i-1,j} = [l]P_{i,j}$ for $i > 1$ and $j \in \{1, 2\}$. Via this basis, we can identify $E[l^i]$ with $(\mathbb{Z}/l^i\mathbb{Z})^2$. Indeed, this is an isomorphism of $(\mathbb{Z}/l^i\mathbb{Z})$-modules. It follows that $T_l(E)$ is a $\mathbb{Z}_l$-module that is isomorphic to $\mathbb{Z}_l^2$ as a $\mathbb{Z}_l$-module. Hence, the set $\mathrm{End}_{\mathbb{Z}_l}(T_l(E))$ of $\mathbb{Z}_l$-linear maps from $T_l(E)$ to itself is isomorphic as a $\mathbb{Z}_l$-module to $M_2(\mathbb{Z}_l)$. We refer to Section III.7 of Silverman [505] for the details.

An isogeny $\phi : E \to \widetilde{E}$ gives rise to a linear map from $E[l^i]$ to $\widetilde{E}[l^i]$ for each i. Writing $\phi(P_{i,1}) = [a]\widetilde{P}_{i,1} + [b]\widetilde{P}_{i,2}$ and $\phi(P_{i,2}) = [c]\widetilde{P}_{i,1} + [d]\widetilde{P}_{i,2}$ (where $\{\widetilde{P}_{i,1}, \widetilde{P}_{i,2}\}$ is a basis for $\widetilde{E}[l^i]$) we can represent ϕ as a matrix $\left(\begin{smallmatrix} a & b \\ c & d \end{smallmatrix}\right) \in M_2(\mathbb{Z}/l^i\mathbb{Z})$. It follows that ϕ corresponds to an element $\phi_l \in M_2(\mathbb{Z}_l)$.

Write $\mathrm{Hom}_{\mathbb{Z}_l}(T_l(E_1), T_l(E_2))$ for the set of $\mathbb{Z}_l$-module homomorphisms from $T_l(E_1)$ to $T_l(E_2)$. Since $T_l(E)$ is isomorphic to $M_2(\mathbb{Z}_l)$ it follows that $\mathrm{Hom}_{\mathbb{Z}_l}(T_l(E_1), T_l(E_2))$ is a $\mathbb{Z}_l$-module of rank 4. An important result is that

$$\mathrm{Hom}_{\Bbbk}(E_1, E_2) \otimes \mathbb{Z}_l \longrightarrow \mathrm{Hom}_{\mathbb{Z}_l}(T_l(E_1), T_l(E_2))$$

is injective (Theorem III.7.4 of [505]). It follows that $\mathrm{Hom}_{\Bbbk}(E_1, E_2)$ is a $\mathbb{Z}$-module of rank at most 4.

The map $\phi \mapsto \widehat{\phi}$ is an involution in $\mathrm{End}_{\Bbbk}(E)$ and $\phi \circ \widehat{\phi} = [d]$ where $d > 0$. This constrains what sort of ring $\mathrm{End}_{\Bbbk}(E)$ can be (Silverman [505] Theorem III.9.3). The result is as follows (for the definitions of orders in quadratic fields see Section A.12, and for quaternion algebras see Vignéras [558]).

Theorem 9.9.1 *Let E be an elliptic curve over a field $\Bbbk$. Then $\mathrm{End}_{\Bbbk}(E)$ is either $\mathbb{Z}$, an order in an imaginary quadratic field, or an order in a definite quaternion algebra.*

Proof See Corollary III.9.4 of [505]. □

When $\Bbbk$ is a finite field then the case $\mathrm{End}_{\overline{\Bbbk}}(E) = \mathbb{Z}$ is impossible (see Theorem V.3.1 of [505]).

Example 9.9.2 Let $E : y^2 = x^3 + x$ over $\mathbb{F}_p$ where $p \equiv 3 \pmod 4$ is prime. Then $\xi(x, y) = (-x, iy)$ is an isogeny where $i \in \mathbb{F}_{p^2}$ satisfies $i^2 = -1$. One can verify that $\xi^2 = \xi \circ \xi = [-1]$. One can show that $\#E(\mathbb{F}_p) = p + 1$ (Exercise 9.10.5) and then Theorem 9.10.3 implies that the Frobenius map $\pi_p(x, y) = (x^p, y^p)$ satisfies $\pi_p^2 = [-p]$. Finally, we have $\xi \circ \pi_p(x, y) = (-x^p, iy^p) = -\pi_p \circ \xi(x, y)$. Hence, $\mathrm{End}_{\overline{\mathbb{F}}_p}(E)$ is isomorphic to a subring of the quaternion algebra (be warned that we are recycling the symbol i here) $\mathbb{Q}[i, j]$ with $i^2 = -1$, $j^2 = -p$, $ij = -ji$. Note that $\mathrm{End}_{\mathbb{F}_p}(E)$ is isomorphic to an order, containing $\mathbb{Z}[\sqrt{-p}]$, in the ring of integers of the imaginary quadratic field $\mathbb{Q}(\sqrt{-p})$.

Every endomorphism on an elliptic curve satisfies a quadratic characteristic polynomial with integer coefficients.

Theorem 9.9.3 *Let E be an elliptic curve over $\Bbbk$ and $\phi \in \mathrm{End}_{\Bbbk}(E)$ be a non-zero isogeny. Let $d = \deg(\phi)$. Then there is an integer t such that $\phi^2 - t\phi + d = 0$ in $\mathrm{End}_{\Bbbk}(E)$. In other* *words, for all $P \in E(\overline{\Bbbk})$,*

$$\phi(\phi(P)) - [t]\phi(P) + [d]P = \mathcal{O}_E.$$

Proof (Sketch) Choose an auxiliary prime $l \neq \mathrm{char}(\Bbbk)$. Then ϕ acts on the Tate module $T_l(E)$ and so corresponds to a matrix $M \in \mathrm{Hom}_{\mathbb{Z}_l}(T_l(E), T_l(E))$. Such a matrix has a determinant d and a trace t. The trick is to show that $d = \deg(\phi)$ and $t = 1 + \deg(\phi) - \deg(1 - \phi)$ (which are standard facts for 2×2 matrices when deg is replaced by det). These statements are independent of l. Proposition V.2.3 of Silverman [505] gives the details (this proof uses the Weil pairing). A slightly simpler proof is given in Lemma 24.4 of [114]. □

Definition 9.9.4 The integer t in Theorem 9.9.3 is called the **trace** of the endomorphism.

Exercise 9.9.5 Show that if $\phi \in \mathrm{End}_{\Bbbk}(E)$ satisfies the equation $T^2 - tT + d = 0$ then so does $\widehat{\phi}$.

Lemma 9.9.6 *Suppose $\phi \in \mathrm{End}_{\overline{\Bbbk}}(E)$ has characteristic polynomial $P(T) = T^2 - tT + d \in \mathbb{Z}[T]$. Let $\alpha, \beta \in \mathbb{C}$ be the roots of $P(T)$. Then, for $n \in \mathbb{N}$, ϕ^n satisfies the polynomial $(T - \alpha^n)(T - \beta^n) \in \mathbb{Z}[T]$.*

Proof This is a standard result: let M be a matrix representing ϕ (or at least, representing the action of ϕ on the Tate module for some l) in Jordan form $M = \left(\begin{smallmatrix} \alpha & \gamma \\ 0 & \beta \end{smallmatrix}\right)$. Then M^n has Jordan form $\left(\begin{smallmatrix} \alpha^n & * \\ 0 & \beta^n \end{smallmatrix}\right)$ and the result follows by the previous statements. □

9.10 Frobenius map

We have seen that the q-power Frobenius on an elliptic curve over $\mathbb{F}_q$ is a non-zero isogeny of degree q (Corollary 9.6.15) and that isogenies on elliptic curves satisfy a quadratic characteristic polynomial. Hence, there is an integer t such that

$$\pi_q^2 - t\pi_q + q = 0. \tag{9.11}$$

Definition 9.10.1 The integer t in equation (9.11) is called the **trace of Frobenius**. The polynomial $P(T) = T^2 - tT + q$ is the **characteristic polynomial of Frobenius**.

Note that $\mathrm{End}_{\mathbb{F}_q}(E)$ always contains the order $\mathbb{Z}[\pi_q]$, which is an order of discriminant $t^2 - 4q$.

Example 9.10.2 Equation (9.11) implies

$$([t] - \pi_q) \circ \pi_q = [q]$$

and so we have $\widehat{\pi_q} = [t] - \pi_q$.

Theorem 9.10.3 *Let E be an elliptic curve over $\mathbb{F}_q$ and let $P(T)$ be the characteristic polynomial of Frobenius. Then $\#E(\mathbb{F}_q) = P(1)$.*

Proof We have $E(\mathbb{F}_q) = \ker(\pi_q - 1)$ and, since $\pi_q - 1$ is separable, $\#E(\mathbb{F}_q) = \deg(\pi_q - 1)$. Now, $P(1) = 1 + q - t$ where, as noted in the proof of Theorem 9.9.3, $t = 1 + \deg(\pi_q) - \deg(1 - \pi_q)$. □

Exercise 9.10.4 Let $p \equiv 2 \pmod 3$. Show that the elliptic curve $E : y^2 = x^3 + a_6$ for $a_6 \in \mathbb{F}_p^*$ has $p + 1$ points over $\mathbb{F}_p$.
[Hint: Rearrange the equation.]

Exercise 9.10.5 Let $p \equiv 3 \pmod 4$ and $a_4 \in \mathbb{F}_p^*$. Show that $E : y^2 = x^3 + a_4 x$ over $\mathbb{F}_p$ has $\#E(\mathbb{F}_p) = p + 1$.
[Hint: Write the right-hand side as $x(x^2 + a_4)$ and use the fact that $(\frac{-1}{p}) = -1$.]

Theorem 9.10.6 *(Hasse) Let E be an elliptic curve over $\mathbb{F}_q$ and denote by t the trace of the q-power Frobenius map. Then $|t| \le 2\sqrt{q}$.*

Proof (Sketch) The idea is to use the fact that $\deg : \mathrm{End}(E) \to \mathbb{Z}$ is a positive definite quadratic form. See Theorem V.1.1 of [505], Theorem 4.2 of [560], Theorem 1 of Chapter 25 of [114] or Theorem 13.4 of [119]. □

In other words, the number of points on an elliptic curve over $\mathbb{F}_q$ lies in the **Hasse interval** $[q + 1 - 2\sqrt{q}, q + 1 + 2\sqrt{q}]$.

Corollary 9.10.7 *Let E be an elliptic curve over $\mathbb{F}_q$ and let $P(T)$ be the characteristic polynomial of Frobenius. Let $\alpha, \beta \in \mathbb{C}$ be such that $P(T) = (T - \alpha)(T - \beta)$. Then $\beta = q/\alpha = \overline{\alpha}$ and $|\alpha| = |\beta| = \sqrt{q}$.*

Proof It follows from the proof of Theorem 9.10.6 that if $P(T) \in \mathbb{Z}[T]$ has a real root then it is a repeated root (otherwise, the quadratic form is not positive definite). Obviously, if the root α is not real then $\beta = \overline{\alpha}$. Since the constant coefficient of $P(T)$ is q it follows that $q = \alpha\beta = \alpha\overline{\alpha} = |\alpha|^2$ and similarly for β. □

The case of repeated roots of $P(T)$ only happens when $\alpha = \pm\sqrt{q} \in \mathbb{Z}$ and $P(T) = (T \pm \sqrt{q})^2$. The condition $|\alpha| = |\beta| = \sqrt{q}$ is known as the **Riemann hypothesis for elliptic curves**. This concept has been generalised to general varieties over finite fields as part of the Weil conjectures (proved by Deligne).

Corollary 9.10.8 *Let E be an elliptic curve over $\mathbb{F}_q$ and let $P(T) = (T - \alpha)(T - \beta)$ be the characteristic polynomial of Frobenius. Let $n \in \mathbb{N}$. Then $\#E(\mathbb{F}_{q^n}) = (1 - \alpha^n)(1 - \beta^n)$.*

Proof We have $E(\mathbb{F}_{q^n}) = \ker(\pi_{q^n} - 1) = \ker(\pi_q^n - 1)$. The result follows from Lemma 9.9.6. □

Corollary 9.10.8 shows that for practical calculations we can identify the isogeny π_q with a complex number α that is one of the roots of $P(T)$. The name "complex multiplication" for endomorphisms of elliptic curves that are not in $\mathbb{Z}$ comes from this identification. When working with elliptic curves over $\mathbb{C}$ the analogy is even stronger, see Theorem 5.5 of [505].

Exercise 9.10.9 Let E be an elliptic curve over $\mathbb{F}_q$. Write $\#E(\mathbb{F}_{q^n}) = q^n - t_n + 1$ for $n \in \mathbb{N}$. Show that for $i, j \in \mathbb{N}$ with $i < j$ we have $t_i t_j = t_{i+j} + q^i t_{j-i}$. Some special cases are

$$t_{2n} = t_n^2 - 2q^n, \quad t_{n+1} = t_n t_1 - q t_{n-1}.$$

Hence, give an algorithm to efficiently compute t_n for any value n, given q and t_1.

Exercise 9.10.10 Let $E_a : y^2 + xy = x^3 + ax^2 + 1$ over $\mathbb{F}_2$ where $a \in \{0, 1\}$. Show that $\#E_a(\mathbb{F}_2) = 2 + (-1)^a + 1$ so $P(T) = T^2 + (-1)^a T + 2$. These curves are called **Koblitz curves** (Koblitz called them **anomalous binary curves**). Show that if n is composite then $\#E_a(\mathbb{F}_{2^n})$ is not of the form $2r$ or $4r$ where r is prime. Hence, find all $3 < n < 200$ such that $\#E_0(\mathbb{F}_{2^n}) = 2r$ or $\#E_1(\mathbb{F}_{2^n}) = 4r$ where r is prime.

We have seen that the number of points on an elliptic curve over a finite field lies in the Hasse interval. An important result of Waterhouse [561] specifies exactly which group orders arise.

Theorem 9.10.11 *(Waterhouse) Let $q = p^m$ and let $t \in \mathbb{Z}$ be such that $|t| \le 2\sqrt{q}$. Then there is an elliptic curve over $\mathbb{F}_q$ with $\#E(\mathbb{F}_q) = q - t + 1$ if and only if one of the following conditions holds:*

1. $\gcd(t, p) = 1$;
2. *m is even and $t = \pm 2\sqrt{q}$;*
3. *m is even, $p \not\equiv 1 \pmod 3$ and $t = \pm\sqrt{q}$;*
4. *m is odd, $p = 2, 3$ and $t = \pm p^{(m+1)/2}$;*
5. *Either m is odd or (m is even and $p \not\equiv 1 \pmod 4$) and $t = 0$.*

Proof The proof given by Waterhouse relies on Honda–Tate theory; one shows that the above cases give precisely the polynomials $T^2 - tT + q$ with roots being Weil numbers. See Theorem 4.1 of [561]. □

In the cases $\gcd(t, p) \ne 1$ (i.e., $p \mid t$), the elliptic curve is said to be **supersingular**. This case is discussed further in Section 9.11.

Example 9.10.12 Let $p \equiv 3 \pmod 4$ be prime, let $g \in \mathbb{F}_{p^2}$ be a primitive root and $E : y^2 = x^3 + g^2 x$. Let $u = 1/\sqrt{g} \in \mathbb{F}_{p^4}$. Consider the map $\phi(x, y) = (u^2 x, u^3 y)$ that maps E to $\widetilde{E} : Y^2 = X^3 + (u^4 g^2)X = X^3 + X$. By Exercise 9.10.5, $\#\widetilde{E}(\mathbb{F}_p) = p + 1$ and the p-power Frobenius map $\widetilde{\pi}_p$ on $\widetilde{E}$ satisfies $(\widetilde{\pi}_p)^2 = -p$.

Define $\psi \in \mathrm{End}_{\overline{\mathbb{F}}_q}(E)$ by $\psi = \phi^{-1} \circ \widetilde{\pi}_p \circ \phi$. Then $\psi(x, y) = (w_1 x^p, w_2 y^p)$ where $w_1 = u^{2p}/u^2$ and $w_2 = u^{3p}/u^3$. One can verify that $w_1, w_2 \in \mathbb{F}_{p^2}$ (just show that $w_i^{p^2} = w_i$) and that $w_1^{p+1} = 1$ and $w_2^{p+1} = -1$. Finally, one has $\psi(\psi(x, y)) = \psi(w_1 x^p, w_2 y^p) = (w_1^{p+1} x^{p^2}, w_2^{p+1} y^{p^2}) = (x^{p^2}, -y^{p^2}) = -\pi_{p^2}(x, y)$ on E. On the other hand, by definition

$$\psi^2 = \phi^{-1} \circ (\widetilde{\pi}_p)^2 \circ \phi = \phi^{-1} \circ [-p] \circ \phi = [-p]$$

on E. Hence, we have shown that $\pi_{p^2} = [p]$ on E. The characteristic polynomial of π_{p^2} is therefore $(T - p)^2$ and so $\#E(\mathbb{F}_{p^2}) = p^2 - 2p + 1$.

Since $\pi_{p^2} \in \mathbb{Z}$ in $\mathrm{End}_{\overline{\Bbbk}}(E)$ the quaternion algebra structure comes from other endomorphisms. We already met $\psi \in \mathrm{End}_{\mathbb{F}_{p^2}}(E)$ such that $\psi^2 = -p$. The endomorphism ring also contains the map $\xi(x, y) = (-x, iy)$ where $i \in \mathbb{F}_{p^2}$ satisfies $i^2 = -1$. One can verify that $\xi^2 = -1$ and $\xi\psi = -\psi\xi$ (since $i^p = -i$ as $p \equiv 3 \pmod 4$); as was seen already in Example 9.9.2.

We know from Theorem 9.8.1 that the group structure of an elliptic curve over a finite field $\mathbb{F}_q$ is of the form $\mathbb{Z}/n_1\mathbb{Z} \times \mathbb{Z}/n_2\mathbb{Z}$ for some integers n_1, n_2 such that $n_1 \mid n_2$. It follows from the Weil pairing (see Exercise 26.2.5 or Section 3.8 of [505]) that $n_1 \mid (q - 1)$.

The following result gives the group structures of elliptic curves.[1]

Theorem 9.10.13 *Let $q = p^m$, let $t \in \mathbb{Z}$ be such that $|t| \le 2\sqrt{q}$ and let $N = q - t + 1$ be a possible group order for an elliptic curve as in Theorem 9.10.11. Write $N = \prod_l l^{h_l}$ for the prime factorisation of N. Then the possible group structures of elliptic curves over $\mathbb{F}_q$ with N points are (i.e., only these cases are possible, and every case does arise for every q)*

$$\mathbb{Z}/p^{h_p}\mathbb{Z} \times \prod_{l \neq p} (\mathbb{Z}/l^{a_l}\mathbb{Z} \times \mathbb{Z}/l^{h_l - a_l}\mathbb{Z})$$

where:

1. *if $\gcd(t, p) = 1$ then $0 \le a_l \le \min\{v_l(q - 1), \lfloor h_l/2 \rfloor\}$ where $v_l(q - 1)$ denotes the integer b such that $l^b \| (q - 1)$,*
2. *if $t = \pm 2\sqrt{q}$ then $a_l = h_l/2$ (i.e., the group is $(\mathbb{Z}/(\sqrt{q} \pm 1)\mathbb{Z})^2$),*
3. *if $t = \pm\sqrt{q}$ or $t = \pm p^{(m+1)/2}$ then the group is cyclic (i.e., all $a_l = 0$),*
4. *if $t = 0$ then either the group is cyclic (i.e., all $a_l = 0$) or is $\mathbb{Z}/2\mathbb{Z} \times \mathbb{Z}/((q + 1)/2)\mathbb{Z}$ (i.e., all $a_l = 0$ except $a_2 = 1$).*

Proof See Voloch [559] or Theorem 3 of Rück [454] (note that it is necessary to prove that Rück's conditions imply those written above by considering possible divisors $d \mid (q - 1)$ and $d \mid (q - t + 1)$ in the supersingular cases). □

Exercise 9.10.14 Let q be a prime power, $\gcd(t, q) = 1$, and $N = q + 1 - t$ a possible value for $\#E(\mathbb{F}_q)$. Show that there exists an elliptic curve over $\mathbb{F}_q$ with N points and which is cyclic as a group.

Another useful result, which relates group structures and properties of the endomorphism ring, is Theorem 9.10.16. Exercise 9.10.15 shows that the final condition makes sense.

Exercise 9.10.15 Let E be an elliptic curve over $\mathbb{F}_q$ and let $t = q + 1 - \#E(\mathbb{F}_q)$. Show that if $n^2 \mid (q + 1 - t)$ and $n \mid (q - 1)$ then $n^2 \mid (t^2 - 4q)$.

[1] This result has been discovered by several authors. Schoof determined the group structures of supersingular elliptic curves in his thesis. The general statement was given by Tsfasman in 1985, Rück in 1987 and Voloch in 1988.

Theorem 9.10.16 *Let p be a prime, $q = p^m$, E an elliptic curve over $\mathbb{F}_q$ and $t = q + 1 - \#E(\mathbb{F}_q)$. Let $n \in \mathbb{N}$ be such that $p \nmid n$. Then $E[n] \subseteq E(\mathbb{F}_q)$ if and only if $n^2 \mid (q + 1 - t)$, $n \mid (q - 1)$, and (either $t = \pm 2\sqrt{q}$ (equivalently, $\pi_q \in \mathbb{Z}$) or $\mathrm{End}_{\mathbb{F}_q}(E)$ contains the order of discriminant $(t^2 - 4q)/n^2$).*

Proof If the kernel of $\pi_q - 1$ contains the kernel of $[n]$ then, by Theorem 9.6.18, there is an isogeny $\psi \in \mathrm{End}_{\mathbb{F}_q}(E)$ such that $\pi_q - 1 = \psi \circ [n]$. We write $\psi = (\pi_q - 1)/n$. The result follows easily; see Proposition 3.7 of Schoof [476] for the details. □

Exercise 9.10.17 Let E be an elliptic curve over $\mathbb{F}_q$ with[2] $\gcd(q, t) = 1$, where $\#E(\mathbb{F}_q) = q + 1 - t$. Deduce from Theorem 9.10.16 that if $\mathrm{End}_{\overline{\mathbb{F}}_q}(E) = \mathbb{Z}[\pi_q]$ then $E(\mathbb{F}_q)$ is a cyclic group.

9.10.1 Complex multiplication

A lot of information about the numbers of points on elliptic curves arises from the theory of complex multiplication. We do not have space to develop this theory in detail. Some crucial tools are the lifting and reduction theorems of Deuring (see Sections 13.4 and 13.5 of Lang [328] or Chapter 10 of Washington [560]). We summarise some of the most important ideas in the following theorem.

Theorem 9.10.18 *Let $\mathcal{O}$ be an order in an imaginary quadratic field K. Then there is a number field L containing K (called the ring class field) and an elliptic curve E over L with $\mathrm{End}_{\overline{L}}(E) \cong \mathcal{O}$.*

Let p be a rational prime that splits completely in L, and let $\wp$ be a prime of $\mathcal{O}_L$ above p (so that $\mathcal{O}_L/\wp \cong \mathbb{F}_p$). If E has good reduction modulo $\wp$ (this holds if $\wp$ does not divide the discriminant of E) write $\overline{E}$ for the elliptic curve over $\mathbb{F}_p$ obtained as the reduction of E modulo $\wp$. Then $\mathrm{End}_{\overline{\mathbb{F}}_p}(\overline{E}) \cong \mathcal{O}$ and there is an element $\pi \in \mathcal{O}$ such that $p = \pi\overline{\pi}$ (where the overline denotes complex conjugation). Furthermore

$$\#\overline{E}(\mathbb{F}_p) = p + 1 - (\pi + \overline{\pi}). \tag{9.12}$$

Conversely, every elliptic curve $\overline{E}$ over $\mathbb{F}_p$ such that $\mathrm{End}_{\overline{\mathbb{F}}_p}(\overline{E}) \cong \mathcal{O}$ arises in this way as a reduction modulo $\wp$ of an elliptic curve over L.

Proof This is Theorem 14.16 of Cox [145]; we refer to the books [145, 328] for much more information about complex multiplication and elliptic curves. □

Remark 9.10.19 An important consequence of the theory of complex multiplication is that the weighted number of $\mathbb{F}_q$-isomorphism classes of elliptic curves over $\mathbb{F}_q$ with number of points equal to $q + 1 - t$ is the Hurwitz class number[3] $H(t^2 - 4q)$ (see Theorem 14.18 of Cox [145], Section 1.5 of Lenstra [339] or Schoof [476]). The Hurwitz class number is

[2] In fact, if $\gcd(q, t) \neq 1$ then the condition $\mathrm{End}_{\overline{\mathbb{F}}_q}(E) = \mathbb{Z}[\pi_q]$ never holds.

[3] Lenstra and Schoof call it the Kronecker class number.

the sum of the (weighted) class numbers of the orders containing the order of discriminant $t^2 - 4q$ (see the references mentioned or Section 5.3.2 of Cohen [127]).

These results imply that the number of elliptic curves over $\mathbb{F}_q$ with $q + 1 - t$ points is $O(u \log(u) \log(\log(u)))$, where $u = \sqrt{4q - t^2}$. The bound $h(-D) < \sqrt{D} \log(D)$ for fundamental discriminants is Exercise 5.27 of Cohen [127]; the case of general discriminants was discussed by Lenstra [339] and the best result is due to McKee [372].

Example 9.10.20 Let $p \equiv 1 \pmod 4$ be prime and let $a_4 \in \mathbb{Z}$ be such that $p \nmid a_4$. Let $E : y^2 = x^3 + a_4 x$ be an elliptic curve over $\mathbb{Q}$ and denote by $\overline{E}$ the elliptic curve over $\mathbb{F}_p$ obtained as the reduction of E modulo p. We will determine $\#\overline{E}(\mathbb{F}_p)$.

The curve E has the endomorphism $\psi(x, y) = (-x, iy)$ (where $i \in \mathbb{F}_{p^2}$ satisfies $i^2 = -1$) satisfying $\psi^2 = [-1]$ and so $\mathrm{End}_{\overline{\mathbb{Q}}}(E)$ contains $\mathbb{Z}[\psi] \cong \mathbb{Z}[i]$. Since $\mathbb{Z}[i]$ is a maximal order it follows that $\mathrm{End}_{\overline{\mathbb{Q}}}(E) = \mathbb{Z}[i]$.

Note that every prime $p \equiv 1 \pmod 4$ can be written as $p = a^2 + b^2$ for $a, b \in \mathbb{Z}$ (see Theorem 1.2 of Cox [145]). Note that there are eight choices for the pair (a, b) in $p = a^2 + b^2$, namely $(\pm a, \pm b)$, $(\pm b, \pm a)$ with all choices of sign independent (note that $a \neq b$ since p is odd).

In other words, $p = (a + bi)(a - bi)$ where $i^2 = -1$. By Theorem 9.10.18 the reduction modulo p of E has $\#\overline{E}(\mathbb{F}_p) = p + 1 - (\pi + \overline{\pi})$ where $\pi\overline{\pi} = p$. Hence, $\pi = a + bi$ for one of the pairs (a, b) and $\#\overline{E}(\mathbb{F}_p) = p + 1 - t$ where

$$t \in \{2a, -2a, 2b, -2b\}.$$

Section 4.4 of Washington [560] gives much more detail about this case.

In practice, one uses the Cornacchia algorithm to compute the integers a and b such that $p = a^2 + b^2$ and so it is efficient to compute $\#E(\mathbb{F}_p)$ for elliptic curves of this form for very large primes p. This idea can be extended to many other curves and is known as the **complex multiplication method** or **CM method**.

Exercise 9.10.21 Determine the number of points on $E : y^2 = x^3 + a_4 x$ modulo $p = 1429 = 23^2 + 30^2$ for $a_4 = 1, 2, 3, 4$.

Exercise 9.10.22 Let p be an odd prime such that $p \equiv 1 \pmod 3$. Then there exist integers a, b such that $p = a^2 + ab + b^2$ (see Chapter 1 of [145] and note that $p = x^2 + 3y^2$ implies $p = (x - y)^2 + (x - y)(2y) + (2y)^2$). Show that the number of points on $y^2 = x^3 + a_6$ over $\mathbb{F}_p$ is $p + 1 - t$ where

$$t \in \{\pm(2a + b), \pm(2b + a), \pm(b - a)\}.$$

Example 9.10.23 The six values $a_6 = 1, 2, 3, 4, 5, 6$ all give distinct values for $\#E(\mathbb{F}_7)$ for the curve $E : y^2 = x^3 + a_6$, namely 12, 9, 13, 3, 7, 4 respectively.

9.10.2 Counting points on elliptic curves

A computational problem of fundamental importance is to compute $\#E(\mathbb{F}_q)$ where E is an elliptic curve over a finite field $\mathbb{F}_q$. Due to lack of space we are unable to give a full treatment of this topic.

We know that $\#E(\mathbb{F}_q)$ lies in the Hasse interval $[q+1-2\sqrt{q}, q+1+2\sqrt{q}]$. In many cases, to determine $\#E(\mathbb{F}_q)$ it suffices to determine the order n of a random point $P \in E(\mathbb{F}_q)$. Determining all multiples of n that lie in the Hasse interval for a point in $E(\mathbb{F}_q)$ can be done using the baby-step–giant-step algorithm in $\tilde{O}(q^{1/4})$ bit operations (see Exercise 13.3.11). If there is only one multiple of n in the Hasse interval then we have determined $\#E(\mathbb{F}_q)$. This process will not determine $\#E(\mathbb{F}_q)$ uniquely if $n \le 4\sqrt{q}$. Mestre suggested determining the order of points on both $E(\mathbb{F}_q)$ and its quadratic twist. This leads to a randomised algorithm to compute $\#E(\mathbb{F}_q)$ in $\tilde{O}(q^{1/4})$ bit operations. We refer to Section 3 of Schoof [477] for details.

A polynomial-time algorithm to compute $\#E(\mathbb{F}_q)$ was given by Schoof [475, 477]. Improvements have been given by numerous authors, especially Atkin and Elkies. The crucial idea is to use equation (9.11). Indeed, the basis of Schoof's algorithm is that if P is a point of small prime order l then one can compute $t \pmod{l}$ by solving the (easy) discrete logarithm problem

$$\pi_q(\pi_q(P)) + [q]P = [t \pmod{l}]\pi_q(P).$$

One finds a point P of order l using the division polynomials (in fact, Schoof never writes down an explicit P, but rather works with a "generic" point of order l by performing polynomial arithmetic modulo $\psi_l(x, y)$). Repeating this idea for different small primes l and applying the Chinese remainder theorem gives t. We refer to [477], Chapters VI and VII of [60], Chapter VI of [61] and Chapter 17 of [16] for details and references.

Exercise 9.10.24 Let $E : y^2 = F(x)$ over $\mathbb{F}_q$. Show that one can determine $t \pmod{2}$ by considering the number of roots of $F(x)$ in $\mathbb{F}_q$.

There are a number of point counting algorithms using p-adic ideas. We do not have space to discuss these algorithms. See Chapter VI of [61] and Chapter IV of [16] for details and references.

9.11 Supersingular elliptic curves

This section is about a particular class of elliptic curves over finite fields that have quite different properties to the general case. For many cryptographic applications these elliptic curves are avoided, though in pairing-based cryptography they have some desirable properties.

Exercise 9.11.1 Let $q = p^m$ where p is prime and let E be an elliptic curve over $\mathbb{F}_q$. Show using Exercise 9.10.9 that if $\#E(\mathbb{F}_q) \equiv 1 \pmod{p}$ then $\#E(\mathbb{F}_{q^n}) \equiv 1 \pmod{p}$ for all $n \in \mathbb{N}$. Hence, show that $E[p] = \{\mathcal{O}_E\}$ for such an elliptic curve.

Theorem 9.11.2 *Let E be an elliptic curve over $\mathbb{F}_{p^m}$ where p is prime. The following are equivalent:*

1. $\#E(\mathbb{F}_{p^m}) = p^m + 1 - t$ *where* $p \mid t$*;*
2. $E[p] = \{\mathcal{O}_E\}$*;*
3. $\mathrm{End}_{\overline{\mathbb{F}}_p}(E)$ *is not commutative (hence, by Theorem 9.9.1, it is an order in a quaternion algebra);*
4. *The characteristic polynomial of Frobenius* $P(T) = T^2 - tT + p^m$ *factors over* $\mathbb{C}$ *with roots* α_1, α_2 *such that* $\alpha_i/\sqrt{p^m}$ *are roots of unity. (Recall that a root of unity is a complex number* z *such that there is some* $n \in \mathbb{N}$ *with* $z^n = 1$*.)*

Proof The equivalence of Properties 1, 2 and 3 is shown in Theorem 3.1 of Silverman [505]. Property 4 is shown in Proposition 13.6.2 of Husemöller [272]. □

Definition 9.11.3 An elliptic curve E over $\mathbb{F}_{p^m}$ is **supersingular** if it satisfies any of the conditions of Theorem 9.11.2. An elliptic curve is **ordinary** if it does not satisfy any of the conditions of Theorem 9.11.2.

We stress that a supersingular curve is not singular as a curve.

Example 9.11.4 Let $p \equiv 2 \pmod 3$ be prime and let $a_6 \in \mathbb{F}_p^*$. The elliptic curve $E : y^2 = x^3 + a_6$ is supersingular since, by Exercise 9.10.4, it has $p + 1$ points. Another way to show supersingularity for this curve is to use the endomorphism $\rho(x, y) = (\zeta_3 x, y)$ as in Exercise 9.6.25 (where $\zeta_3 \in \mathbb{F}_{p^2}$ is such that $\zeta_3^2 + \zeta_3 + 1 = 0$). Since ρ does not commute with the p-power Frobenius map π_p (specifically, $\pi_p\rho = \rho^2\pi_p$ since $\zeta_3 \notin \mathbb{F}_p$) the endomorphism ring is not commutative.

To determine the quaternion algebra, one can proceed as follows. First, show that ρ satisfies the characteristic polynomial $T^2 + T + 1 = 0$ (since $\rho^3(P) = P$ for all $P \in E(\overline{\mathbb{F}}_p)$). Then consider the isogeny $\phi = [1] - \rho$, which has dual $\widehat{\phi} = [1] - \rho^2$. The degree d of ϕ satisfies $[d] = \phi\widehat{\phi} = (1-\rho)(1-\rho^2) = 1 - \rho - \rho^2 + 1 = 3$. Hence, ϕ has degree 3. The trace of ϕ is $t = 1 + \deg(\phi) - \deg(1-\phi) = 1 + 3 - \deg(\rho) = 3$. One can show that $(\rho\phi)^2 = [-3]$ and so the quaternion algebra is $\mathbb{Q}[i, j]$ with $i^2 = -3$ and $j^2 = -p$.

Example 9.11.5 Let $p \equiv 3 \pmod 4$ be prime and $a_4 \in \mathbb{F}_p^*$. Exercise 9.10.5 implies that $E : y^2 = x^3 + a_4 x$ is supersingular. An alternative proof of supersingularity follows from Example 9.9.2; since $\xi(x, y) = (-x, iy)$ does not commute with the p-power Frobenius.

Example 9.11.6 Let $\mathbb{F}_q$ be a finite field of characteristic 2 and $F(x) \in \Bbbk[x]$ a monic polynomial of degree 3. Then $E : y^2 + y = F(x)$ is supersingular. This follows from the fact that $(x, y) \in E(\mathbb{F}_{q^n})$ if and only if $(x, y + 1) \in E(\mathbb{F}_{q^n})$ and hence $\#E(\mathbb{F}_{q^n})$ is odd for all n. It follows that there are no points of order 2 on $E(\overline{\mathbb{F}}_2)$ and so E is supersingular.

Exercise 9.11.7 Use Waterhouse's theorem to show that, for every prime p and $m \in \mathbb{N}$, there exists a supersingular curve over $\mathbb{F}_{p^m}$.

Bröker [100] has given an algorithm to construct supersingular elliptic curves over finite fields using the CM method. The method has expected polynomial-time, assuming a generalisation of the Riemann hypothesis is true.

Property 4 of Theorem 9.11.2 implies that if E is a supersingular curve then $\pi_q^m = [p^M]$ for some $m, M \in \mathbb{N}$. In other words, $\pi_q^m \in \mathbb{Z}$. In examples we have seen $\pi^2 = [-q]$. A natural question is how large the integer m can be.

Lemma 9.11.8 *Let E be a supersingular elliptic curve over $\mathbb{F}_q$ and let $P(T)$ be the characteristic polynomial of Frobenius. Then every non-square factor of $\frac{1}{q}P(T\sqrt{q})$ divides $\Phi_m(T^2)$ in $\mathbb{R}[x]$ for some $m \in \{1, 2, 3, 4, 6\}$, where $\Phi_m(x)$ is the mth cyclotomic polynomial (see Section 6.1).*

Proof Waterhouse's theorem gives the possible values for the characteristic polynomial $P(T) = T^2 - tT + q$ of Frobenius. The possible values for t are $0, \pm\sqrt{q}, \pm 2\sqrt{q}, \pm\sqrt{2q}$ (when q is a power of 2) or $\pm\sqrt{3q}$ (when q is a power of 3).

By part 4 of Theorem 9.11.2, every root α of $P(T)$ is such that $\alpha/\sqrt{q}$ is a root of unity. If $P(T) = (T - \alpha)(T - \beta)$ then

$$(T - \alpha/\sqrt{q})(T - \beta/\sqrt{q}) = \tfrac{1}{q}P(T\sqrt{q}).$$

So, write $Q(T) = P(T\sqrt{q})/q \in \mathbb{R}[T]$. The first three values for t in the above list give $Q(T)$ equal to $T^2 + 1$, $T^2 \pm T + 1$ and $T^2 \pm 2T + 1 = (T \pm 1)^2$ respectively. The result clearly holds in these cases (the condition about "non-square factors" is needed since $(T \pm 1)$ divides $\Phi_1(T^2) = (T - 1)(T + 1)$, but $(T \pm 1)^2$ does not divide any cyclotomic polynomial.

We now deal with the remaining two cases. Let $t = \pm 2^{(m+1)/2}$ where $q = 2^m$. Then $Q(T) = T^2 \pm \sqrt{2}T + 1$ and we have

$$(T^2 + \sqrt{2}T + 1)(T^2 - \sqrt{2}T + 1) = T^4 + 1 = \Phi_4(T^2).$$

Similarly, when $t = \pm 3^{(m+1)/2}$ and $q = 3^m$ then $Q(T) = T^2 \pm \sqrt{3}T + 1$ and

$$(T^2 + \sqrt{3}T + 1)(T^2 - \sqrt{3}T + 1) = T^4 - T^2 + 1 = \Phi_6(T^2).$$ □

Corollary 9.11.9 *Let E be a supersingular elliptic curve over $\mathbb{F}_q$. Then there is an integer $m \in \{1, 2, 3, 4, 6\}$ such that $\pi_q^m \in \mathbb{Z}$ and the exponent of the group $E(\mathbb{F}_q)$ divides $(q^m - 1)$. Furthermore, the cases $m = 3, 4, 6$ only occur when q is a square, a power of 2, or a power of 3 respectively.*

Exercise 9.11.10 Prove Corollary 9.11.9.

In general, the endomorphism ring of a supersingular elliptic curve is generated over $\mathbb{Z}$ by the Frobenius map and some "complex multiplication" isogeny. However, as seen in Example 9.10.12, the Frobenius can lie in $\mathbb{Z}$, in which case two independent "complex multiplications" are needed (though, as in Example 9.10.12, one of them will be very closely related to a Frobenius map on a related elliptic curve).

It is known that the endomorphism ring $\mathrm{End}_{\overline{\Bbbk}}(E)$ of a supersingular elliptic curve E over $\Bbbk$ is a **maximal order** in a quaternion algebra (see Theorem 4.2 of Waterhouse [561]) and that the quaternion algebra is ramified at exactly p and ∞. Indeed, [561] shows that when $t = \pm 2\sqrt{q}$ then all endomorphisms are defined over $\mathbb{F}_q$ and every maximal order arises. In other cases not all endomorphisms are defined over $\mathbb{F}_q$ and the maximal order is an order that contains π_q and is maximal at p (i.e., the index is not divisible by p).

We now present some results on the number of supersingular curves over finite fields.

Theorem 9.11.11 *Let $\mathbb{F}_q$ be a field of characteristic p and $E/\mathbb{F}_q$ a supersingular elliptic curve. Then $j(E) \in \mathbb{F}_{p^2}$. Furthermore:*

1. *The number of $\overline{\mathbb{F}}_q$-isomorphism classes of supersingular elliptic curves over $\mathbb{F}_{p^2}$ is 1 if $p = 2, 3$ and $\lfloor p/12 \rfloor + \epsilon_p$ where $\epsilon_p = 0, 1, 1, 2$ respectively if $p \equiv 1, 5, 7, 11 \pmod{12}$.*
2. *The number of $\overline{\mathbb{F}}_q$-isomorphism classes of supersingular elliptic curves over $\mathbb{F}_p$ is 1 if $p = 2, 3$ and is equal to the Hurwitz class number $H(-4p)$ if $p > 3$. Furthermore*

$$H(-4p) = \begin{cases} \frac{1}{2}h(-4p) & \text{if } p \equiv 1 \pmod{4}, \\ h(-p) & \text{if } p \equiv 7 \pmod{8}, \\ 2h(-p) & \text{if } p \equiv 3 \pmod{8} \end{cases}$$

where $h(d)$ is the usual ideal class number of the quadratic field $\mathbb{Q}(\sqrt{d})$.

Proof The claim that $j(E) \in \mathbb{F}_{p^2}$ is Theorem 3.1(iii) of [505] or Theorem 5.6 of [272]. The formula for the number of supersingular j-invariants in $\mathbb{F}_{p^2}$ is Theorem 4.1(c) of [505] or Section 13.4 of [272]. The statement about the number of supersingular j-invariants in $\mathbb{F}_p$ is given in Theorem 14.18 of Cox [145] (the supersingular case is handled on page 322). The precise formula for $H(-4p)$ is equation (1.11) of Gross [244]. (Gross also explains the relation between isomorphism classes of supersingular curves and Brandt matrices.) □

Lemma 9.11.12 *Let E_1, E_2 be elliptic curves over $\mathbb{F}_q$. Show that if E_1 and E_2 are ordinary, $\#E_1(\mathbb{F}_q) = \#E_2(\mathbb{F}_q)$ and $j(E_1) = j(E_2)$ then they are isomorphic over $\mathbb{F}_q$.*

Proof (Sketch) Since $j(E_1) = j(E_2)$ the curves are isomorphic over $\overline{\mathbb{F}}_q$. If $\#E_1(\mathbb{F}_q) = q + 1 - t$ and E_2 is not isomorphic to E_1 over $\mathbb{F}_q$, then E_2 is a non-trivial twist of E_1. If $j(E_1) \neq 0, 1728$ then $\#E_2(\mathbb{F}_q) = q + 1 + t \neq \#E_1(\mathbb{F}_q)$, since $t \neq 0$ (this is where we use the fact that E_1 is ordinary). In the cases $j(E_1) = 0, 1728$, one needs to use the formulae of Example 9.10.20 and Exercise 9.10.22 and show that these group orders are distinct when $t \neq 0$.

An alternative proof, using less elementary methods, is given in Proposition 14.19 (page 321) of Cox [145]. □

Exercise 9.11.13 Give an example of supersingular curves E_1, E_2 over $\mathbb{F}_p$ such that $j(E_1) = j(E_2)$, $\#E_1(\mathbb{F}_p) = \#E_2(\mathbb{F}_p)$ and E_1 is not isomorphic to E_2 over $\mathbb{F}_p$.

9.12 Alternative models for elliptic curves

We have introduced elliptic curves using Weierstrass equations, but there are many different models and some of them have computational advantages. We present the Montgomery model and the twisted Edwards model. A mathematically important model, which we do not discuss directly, is the intersection of two quadratic surfaces; see Section 2.5 of Washington [560] for details. It is not the purpose of this book to give an implementation guide, so we refrain from providing the optimised addition algorithms. Readers are advised to consult Sections 13.2 and 13.3 of [16] or the *Explicit Formulas Database* [49].

9.12.1 Montgomery model

This model, for elliptic curves over fields of odd characteristic, was introduced by Montgomery [392] in the context of efficient elliptic curve factoring using $(x : z)$ coordinates. It is a very convenient model for arithmetic in (a projective representation of) the algebraic group quotient $E(\Bbbk)$ modulo the equivalence relation $P \equiv -P$. Versions of the Montgomery model have been given in characteristic 2, but they are not so successful; we refer to Stam [519] for a survey.

Definition 9.12.1 Let $\Bbbk$ be a field such that $\mathrm{char}(\Bbbk) \neq 2$. Let $A, B \in \Bbbk$, $B \neq 0$. The **Montgomery model** is

$$By^2 = x^3 + Ax^2 + x. \tag{9.13}$$

According to Definition 7.2.8, when $B \neq 1$, the Montgomery model is not an elliptic curve. However, the theory all goes through in the more general case, and so we refer to curves in Montgomery model as elliptic curves.

Exercise 9.12.2 Show that the Montgomery model is non-singular if and only if $B(A^2 - 4) \neq 0$.

Exercise 9.12.3 Show that there is a unique point at infinity on the Montgomery model of an elliptic curve. Show that this point is not singular, and is always $\Bbbk$-rational.

Lemma 9.12.4 *Let $\Bbbk$ be a field such that* $\mathrm{char}(\Bbbk) \neq 2$. *Let $E : y^2 = x^3 + a_2x^2 + a_4x + a_6$ be an elliptic curve over $\Bbbk$ in Weierstrass form. There is an isomorphism over $\Bbbk$ from E to a Montgomery model if and only if $F(x) = x^3 + a_2x^2 + a_4x + a_6$ has a root $x_P \in \Bbbk$ such that $(3x_P^2 + 2a_2x_P + a_4)$ is a square in $\Bbbk$. This isomorphism maps $\mathcal{O}_E$ to the point at infinity on the Montgomery model and is a group homomorphism.*

Proof Let $P = (x_P, 0) \in E(\Bbbk)$. First, move P to $(0, 0)$ by the change of variable $X = x - x_P$. The map $(x, y) \mapsto (x - x_P, y)$ is an isomorphism to $y^2 = X^3 + a_2'X^2 + a_4'X$ where $a_2' = 3x_P + a_2$ and $a_4' = 3x_P^2 + 2a_2x_P + a_4$. Let $w = \sqrt{a_4'}$, which lies in $\Bbbk$ by the assumption of the Lemma. Consider the isomorphism $(X, y) \mapsto (U, V) = (X/w, y/w)$ that maps to

$$(1/w)V^2 = U^3 + (a_2'/w)U^2 + U.$$

Taking $A = a_2'/w$, $B = 1/w \in \Bbbk$ gives the result.

Conversely, suppose $By^2 = x^3 + Ax^2 + x$ is a Montgomery model of an elliptic curve over $\Bbbk$. Multiplying though by B^3 gives $(B^2y)^2 = (Bx)^3 + AB(Bx)^2 + B^2(Bx)$ and so $(U, V) = (Bx, B^2y)$ satisfies the Weierstrass equation $V^2 = U^3 + ABU^2 + B^2U$. Taking $a_2 = AB, a_4 = B^2$ and $a_6 = 0$ one can check that the conditions in the statement of the Lemma hold (the polynomial $F(x)$ has the root 0, and $a_4' = B^2$ is a square).

The maps extend to the projective curves and map $(0 : 1 : 0)$ to $(0 : 1 : 0)$. The fact that they are group homomorphisms follows from a generalisation of Theorem 9.2.1. □

When the conditions of Lemma 9.12.4 hold we say that the elliptic curve E can be written in Montgomery model. Throughout this section, when we refer to an elliptic curve E in Montgomery model, we assume that E is specified by an affine equation as in equation (9.13).

Lemma 9.12.5 *Let $P_1 = (x_1, y_2)$, $P_2 = (x_2, y_2)$ be points on the elliptic curve $By^2 = x^3 + Ax^2 + x$ such that $x_1 \neq x_2$ and $x_1x_2 \neq 0$. Then $P_1 + P_2 = (x_3, y_3)$ where*

$$x_3 = B(x_2y_1 - x_1y_2)^2/(x_1x_2(x_2 - x_1)^2).$$

Writing $P_1 - P_2 = (x_4, y_4)$ one finds

$$x_3x_4 = (x_1x_2 - 1)^2/(x_1 - x_2)^2.$$

For the case $P_2 = P_1$ we have $[2](x_1, y_1) = (x_3, y_3)$ where

$$x_3 = (x_1^2 - 1)^2/(4x_1(x_1^2 + Ax_1 + 1)).$$

Proof The standard addition formula gives $x_3 = B((y_2 - y_1)/(x_2 - x_1))^2 - (A + x_1 + x_2)$, which yields

$$\begin{aligned} x_3(x_2 - x_1)^2 &= By_1^2 + By_2^2 - 2By_1y_2 - (A + x_1 + x_2)(x_2 - x_1)^2 \\ &= -2By_1y_2 + 2Ax_1x_2 + x_1^2x_2 + x_1x_2^2 + x_1 + x_2 \\ &= \tfrac{x_2}{x_1}By_1^2 + \tfrac{x_1}{x_2}By_2^2 - 2By_1y_2 \\ &= B(x_2y_1 - x_1y_2)^2/(x_1x_2). \end{aligned}$$

Replacing P_2 by $-P_2$ gives $P_1 - P_2 = (x_4, y_4)$ with $x_4(x_2 - x_1)^2 = B(x_2y_1 + x_1y_2)^2/(x_1x_2)$. Multiplying the two equations gives

$$\begin{aligned} x_3x_4(x_2 - x_1)^4 &= B^2(x_2y_1 - x_1y_2)^2(x_2y_1 + x_1y_2)^2/(x_1x_2)^2 \\ &= \left(\frac{x_2By_1^2}{x_1} - \frac{x_1By_2^2}{x_2}\right)^2 \\ &= (x_1x_2(x_1 - x_2) + (x_2 - x_1))^2 \end{aligned}$$

from which we deduce that $x_3x_4(x_2 - x_1)^2 = (x_1x_2 - 1)^2$. In the case $P_1 = P_2$, we have $x_3 4By_1^2 = (3x_1^2 + 2Ax_1 + 1)^2 - (A + 2x_1)4By_1^2$, which implies $4x_1x_3(x_1^2 + Ax_1 + 1) = (x_1^2 - 1)^2$. □

In other words, one can compute the x-coordinate of $[2]P$ using only the x-coordinate of P. Similarly, given the x-coordinates of P_1, P_2 and $P_1 - P_2$ (i.e., x_1, x_2 and x_4) one can compute the x-coordinate of $P_1 + P_2$. The next exercise shows how to do this projectively.

Exercise 9.12.6 Let $P = (x_P, y_P) \in E(\mathbb{F}_q)$ be a point on an elliptic curve given in a Montgomery model. Define $X_1 = x_P$, $Z_1 = 1$, $X_2 = (X_1^2 - 1)^2$, $Z_2 = 4x_1(x_1^2 + Ax_1 + 1)$. Given (X_n, Z_n), (X_m, Z_m), (X_{m-n}, Z_{m-n}) define

$$X_{n+m} = Z_{m-n}(X_nX_m - Z_nZ_m)^2$$
$$Z_{n+m} = X_{m-n}(X_nZ_m - X_mZ_n)^2$$

and

$$X_{2n} = (X_n^2 - Z_n^2)^2$$
$$Z_{2n} = 4X_nZ_n(X_n^2 + AX_nZ_n + Z_n^2).$$

Show that the x-coordinate of $[m]P$ is X_m/Z_m.

Exercise 9.12.7★ Write a "double and add" algorithm to compute the x-coordinate of $[n]P$ using the projective Montgomery addition formula. Give alternative versions of the Montgomery addition formulae that show that each iteration of your algorithm requires only 7 multiplications and 4 squarings in $\mathbb{F}_q$.

The most efficient formulae for exponentiation using a ladder algorithm on Montgomery curves are given in Section 6.2 of Gaudry and Lubicz [227] (also see [49]).

Exercise 9.12.8 Let $E : By^2 = x(x^2 + a_2x + a_4)$ be an elliptic curve over $\Bbbk$ (where char$(\Bbbk) \neq 2$). Show that the solutions $(x, y) \in E(\overline{\Bbbk})$ to $[2](x, y) = (0, 0)$ are the points $(\sqrt{a_4}, \pm\sqrt{a_4(a_2 + 2\sqrt{a_4})/B})$ and $(-\sqrt{a_4}, \pm\sqrt{a_4(a_2 - 2\sqrt{a_4})/B})$.

Lemma 9.12.9 *(Suyama) If E is an elliptic curve given by a Montgomery model then $4 \mid \#E(\mathbb{F}_q)$.*

Proof If $A^2 - 4 = (A - 2)(A + 2)$ is a square then the full 2-torsion is over $\mathbb{F}_q$. If $(A - 2)(A + 2)$ is not a square then one of $(A \pm 2)$ is a square in $\mathbb{F}_q$ and the other is not. If $B(A + 2)$ is a square then $(1, \sqrt{(A + 2)/B})$ is defined over $\mathbb{F}_q$ and, by Exercise 9.12.8, has order 4. Similarly, if $B(A - 2)$ is a square then $(-1, \sqrt{(A - 2)/B})$ is defined over $\mathbb{F}_q$ and has order 4. □

Let $E : By^2 = x^3 + Ax^2 + x$ be an elliptic curve over $\Bbbk$ in Montgomery model. If $u \in \Bbbk^*$ then E is isomorphic to $E^{(u)} : (uB)Y^2 = X^3 + AX^2 + X$ where the corresponding isomorphism $\phi : E \to E^{(u)}$ is $\phi(x, y) = (x, y/\sqrt{u})$. If u is not a square in $\Bbbk$ then ϕ is not defined over $\Bbbk$ and so $E^{(u)}$ is the **quadratic twist** of E.

Exercise 9.12.10 Show that every elliptic curve E in Montgomery model over a finite field $\mathbb{F}_q$ is such that either E or its quadratic twist $E^{(d)}$ has a point of order 4.

Theorem 9.12.11 *Let E be an elliptic curve over $\mathbb{F}_q$ (char$(\mathbb{F}_q) \neq 2$) such that $4 \mid \#E(\mathbb{F}_q)$. Then E is either isomorphic or 2-isogenous over $\mathbb{F}_q$ to an elliptic curve in Montgomery model.*

Proof Suppose $P \in E(\mathbb{F}_q)$ has order 4. Write $P_0 = [2]P$ and change coordinates so that $P_0 = (0, 0)$. By Exercise 9.12.8 it follows that a_4 is a square in $\mathbb{F}_q$ and so by Lemma 9.12.4 is isomorphic to an elliptic curve in Montgomery model.

Suppose now that there is no point of order 4 in $E(\mathbb{F}_q)$. Then $\#E(\mathbb{F}_q)[2] = 4$ and so all points of order 2 are defined over $\mathbb{F}_q$. In other words, one can write E as $y^2 = x(x-a)(x-b) = x(x^2 - (a+b)x + ab)$ where $a, b \in \mathbb{F}_q$. Now take the 2-isogeny as in Example 9.6.9. This maps E to $E' : Y^2 = X(X^2 + 2(a+b)X + (a-b)^2)$. By Lemma 9.12.4 it follows that E' is isomorphic to an elliptic curve in Montgomery model. □

We have already seen the quadratic twist of a Montgomery model. It is natural to consider whether there are other twists.

Theorem 9.12.12 *Let $q = p^n$ where $p > 3$ is prime. If $E/\mathbb{F}_q$ is an ordinary elliptic curve admitting a Montgomery model then only one non-trivial twist also admits a Montgomery model. Furthermore, this twist is the quadratic twist.*

Proof When $j(E) \neq 0, 1728$ then the quadratic twist is the only non-trivial twist, so there is nothing to prove. So we consider $j(E) = 1728$ and $j(E) = 0$. The crucial observation will be that the other twists E' do not satisfy $4 \mid \#E'(\mathbb{F}_q)$.

By Example 9.10.20, if $j(E) = 1728$ then $q \equiv 1 \pmod 4$, $q = a^2 + b^2$ for some $a, b \in \mathbb{Z}$, and the group orders are $q + 1 \pm 2a$ and $q + 1 \pm 2b$. Note that, without loss of generality, the solution (a, b) to $q = a^2 + b^2$ is such that a is odd and b is even. Then $2a \not\equiv 2b \pmod 4$ and so only one of $q + 1 + 2a$ and $q + 1 + 2b$ is divisible by 4. Since $q + 1 + 2a \equiv q + 1 - 2a \pmod 4$ (and similarly for the other case) it follows that only one pair of quadratic twists can be given in Montgomery model.

By Exercise 9.10.22, if $j(E) = 0$ then $q \equiv 1 \pmod 3$, $q = a^2 + ab + b^2$ for some $a, b \in \mathbb{Z}$, and the possible group orders are

$$q + 1 \pm (a - b), \quad q + 1 \pm (2a + b), \quad q + 1 \pm (2b + a).$$

Without loss of generality, a is odd and b may be either odd or even. If a and b are both odd then $2a - b$ and $2b - a$ are both odd and so $q + 1 \pm (a + b)$ is the only pair of group orders that are even. Similarly, if a is odd and b is even then $a + b$ and $2b + a$ are both odd and so $q + 1 \pm (2a + b)$ is the only pair of group orders that are even. This completes the proof. □

Example 9.12.13 The elliptic curve $y^2 = x^3 + a_4 x$ is isomorphic over $\overline{\mathbb{k}}$ to the curve $\sqrt{a_4} Y^2 = X^3 + X$ in Montgomery form via $(x, y) \mapsto (X, Y) = (x/\sqrt{a_4}, y/a_4)$.

The elliptic curve $y^2 = x^3 + a_6$ is isomorphic over $\overline{\Bbbk}$ to the curve

$$1/(\sqrt{3}(-a_6)^{1/3})Y^2 = X^3 + \sqrt{3}X^2 + X$$

in Montgomery model. To see this, consider the point $P = ((-a_6)^{1/3}, 0)$ and move it to $(0, 0)$ via $W = x - a_6^{1/3}$, giving $y^2 = W^3 + 3(-a_6)^{1/3}W^2 + 3(-a_6)^{2/3}W$.

9.12.2 Edwards model

Euler and Gauss considered the genus 1 curve $x^2 + y^2 = 1 - x^2y^2$ and described a group operation on its points. Edwards generalised this to a wide class of elliptic curves (we refer to [175] for details and historical discussion). Further extensions were proposed by Bernstein, Birkner, Joye, Lange, and Peters (see [46] and its references). Edwards curves have several important features: they give a complete group law on $E(\mathbb{F}_q)$ for some fields $\mathbb{F}_q$ (in other words, there is a single rational map $+ : E \times E \to E$ that computes addition for all[4] possible inputs in $E(\mathbb{F}_q) \times E(\mathbb{F}_q)$) and the addition formulae can be implemented extremely efficiently in some cases. Hence, this model for elliptic curves is very useful for many cryptographic applications.

Definition 9.12.14 Let $\Bbbk$ be a field such that $\mathrm{char}(\Bbbk) \neq 2$. Let $a, d \in \Bbbk$ satisfy $a \neq 0, d \neq 0, a \neq d$. The **twisted Edwards model** is

$$ax^2 + y^2 = 1 + dx^2y^2.$$

Exercise 9.12.15 Show that a curve in twisted Edwards model is non-singular as an affine curve. Show that if any of the conditions $a \neq 0, d \neq 0$ and $a \neq d$ are not satisfied then the affine curve has a singular point.

Bernstein, Lange and Farashahi [53] have also formulated an Edwards model for elliptic curves in characteristic 2.

The Weierstrass model of an elliptic curve over $\Bbbk$ (where $\mathrm{char}(\Bbbk) \neq 2$) is of the form $y^2 = F(x)$ and it would be natural to write the twisted Edwards model in the form $y^2 = (1 - ax^2)/(1 - dx^2)$. A natural formulation of the group law would be such that the inverse of a point (x, y) is $(x, -y)$, however this leads to having identity element $(x, y) = (1/\sqrt{a}, 0)$. Instead, for historical reasons and to make the identity $\Bbbk$-rational, it is traditional to think of the curve as

$$x^2 = (1 - y^2)/(a - dy^2).$$

The identity element is then $(0, 1)$ and the inverse of (x, y) is $(-x, y)$.

[4] Note that this is a stronger statement than the unified group law of Exercise 9.1.1 as the group law on (twisted) Edwards curve also includes addition of a point with its inverse or the identity element. Also, the group law on (twisted) Edwards curves achieves this with no loss of efficiency, unlike Exercise 9.1.1. On the other hand, we should mention that the group law on (twisted) Edwards curves is never complete for the group $E(\overline{\mathbb{F}}_q)$.

The group operation on twisted Edwards models is

$$(x_1, y_1) + (x_2, y_2) = \left(\frac{x_1y_2 + x_2y_1}{1 + dx_1x_2y_1y_2}, \frac{y_1y_2 - ax_1x_2}{1 - dx_1x_2y_1y_2} \right). \quad (9.14)$$

This is shown to be a group law in [50, 46]. A geometric description of the Edwards group law on the singular curve is given by Arène, Lange, Naehrig and Ritzenthaler [12]. An inversion-free (i.e., projective) version and explicit formulae for efficient arithmetic are given in [46].

Exercise 9.12.16 Let E be a curve over $\Bbbk$ in twisted Edwards model. Show that $(0, -1) \in E(\Bbbk)$ has order 2 and that $(\pm 1/\sqrt{a}, 0) \in E(\overline{\Bbbk})$ have order 4.

Exercise 9.12.17 Determine the points at infinity on a curve in twisted Edwards model and show they are singular.

We now give a non-singular projective model for twisted Edwards models that allows us to view the points at infinity and determine their orders.

Lemma 9.12.18 *Let $\Bbbk$ be a field of characteristic not equal to 2. Let $a, d \in \Bbbk$ with $a, d \neq 0$. There are four points at infinity over $\overline{\Bbbk}$ on a twisted Edwards model over $\Bbbk$ and they all have order dividing 4.*

Proof (Sketch) The rational map $\phi(x, y) = (X_0 = xy, X_1 = x, X_2 = y, X_3 = 1)$ maps a twisted Edwards curve to the projective algebraic set

$$X = V(aX_1^2 + X_2^2 - X_3^2 - dX_0^2, X_1X_2 - X_0X_3) \subset \mathbb{P}^3.$$

It can be shown that X is irreducible and of dimension 1.

The points at infinity on the affine twisted Edwards model correspond to the points

$$(1 : \pm\sqrt{d/a} : 0 : 0) \quad \text{and} \quad (1 : 0 : \pm\sqrt{d} : 0)$$

with $X_3 = 0$. To see that the points at infinity on X are non-singular set $X_0 = 1$ and obtain the Jacobian matrix

$$\begin{pmatrix} 2aX_1 & 2X_2 & -2X_3 \\ X_2 & X_1 & -1 \end{pmatrix},$$

which is seen to have rank 2 when evaluated at the points $(\pm\sqrt{d/a}, 0, 0)$ and $(0, \pm\sqrt{d}, 0)$.

Let $(X_0 : X_1 : X_2 : X_3)$ and $(Z_0 : Z_1 : Z_2 : Z_3)$ be points on X and define the values

$$S_1 = (X_1Z_2 + Z_1X_2), \qquad S_2 = (X_2Z_2 - aX_1Z_1),$$
$$S_3 = (X_3Z_3 + dX_0Z_0), \quad S_4 = (X_3Z_3 - dX_0Z_0).$$

The group law formula on the affine twisted Edwards curve corresponds to the formula

$$(X_0 : X_1 : X_2 : X_3) + (Z_0 : Z_1 : Z_2 : Z_3) = (S_1S_2 : S_1S_4 : S_2S_3 : S_3S_4).$$

One can verify that $(0 : 0 : 1 : 1)$ is the identity by computing

$$(X_0 : X_1 : X_2 : X_3) + (0 : 0 : 1 : 1) = (X_1X_2 : X_1X_3 : X_2X_3 : X_3^2).$$

When $X_3 \neq 0$ one replaces the first coordinate X_1X_2 by X_0X_3 and divides by X_3 to get $(X_0 : X_1 : X_2 : X_3)$. When $X_3 = 0$ one multiplies through by X_0, replaces X_0X_3 by X_1X_2 everywhere, and divides by X_1X_2.

Similarly, one can verify that $(0 : 0 : -1 : 1)$ and $(1 : \pm\sqrt{d/a} : 0 : 0)$ have order 2, and $(1 : 0 : \pm\sqrt{d} : 0)$ have order 4. □

We now show that the Edwards group law is complete for points defined over $\Bbbk$ in certain cases.

Lemma 9.12.19 *Let $\Bbbk$ be a field,* char($\Bbbk$) $\neq 2$ *and let $a, d \in \Bbbk$ be such that $a \neq 0, d \neq 0, a \neq d$. Suppose a is a square in $\Bbbk^*$ and d is not a square in $\Bbbk^*$. Then the affine group law formula for twisted Edwards curves of equation (9.14) is defined for all points over $\Bbbk$.*

Proof Let $\epsilon = dx_1x_2y_1y_2$. Suppose, for contradiction, that $\epsilon = \pm 1$. Then $x_1, x_2, y_1, y_2 \neq 0$. One can show, by substituting $ax_2^2 + y_2^2 = 1 + dx_2^2y_2^2$, that

$$dx_1^2y_1^2(ax_2^2 + y_2^2) = ax_1^2 + y_1^2.$$

Adding $\pm 2\sqrt{a}\epsilon x_1y_1$ to both sides and inserting the definition of ϵ gives

$$(\sqrt{a}x_1 \pm \epsilon y_1)^2 = dx_1^2y_1^2(\sqrt{a}x_2 \pm y_2)^2.$$

Hence, if either $\sqrt{a}x_2 + y_2 \neq 0$ or $\sqrt{a}x_2 - y_2 \neq 0$ then one can deduce that d is a square in $\Bbbk^*$. On the other hand, if $\sqrt{a}x_2 + y_2 = \sqrt{a}x_2 - y_2 = 0$ one deduces that $x_2 = 0$. Both cases are a contradiction. □

It turns out that twisted Edwards curves and Montgomery curves cover exactly the same $\Bbbk$-isomorphism classes of elliptic curves.

Lemma 9.12.20 *Let $M : By^2 = x^3 + Ax^2 + x$ be a Montgomery model for an elliptic curve over $\Bbbk$ (so $B \neq 0$ and $A^2 \neq 4$). Define $a = (A+2)/B$ and $d = (A-2)/B$. Then $a \neq 0, d \neq 0$ and $a \neq d$. The map $(x, y) \mapsto (X = x/y, Y = (x-1)/(x+1))$ is a birational map over $\Bbbk$ from M to the twisted Edwards curve*

$$E : aX^2 + Y^2 = 1 + dX^2Y^2.$$

Conversely, if E is as above then define $A = 2(a+d)/(a-d)$ and $B = 4/(a-d)$. Then $(X, Y) \mapsto (x = (1+Y)/(1-Y), y = (1+Y)/(X(1-Y)))$ is a birational map over $\Bbbk$ from E to M.

Exercise 9.12.21 Prove Lemma 9.12.20.

The birational map in Lemma 9.12.20 is a group homomorphism. Indeed, the proofs of the group law in [50, 47] use this birational map to transfer the group law from the Montgomery model to the twisted Edwards model.

Exercise 9.12.22 Show that the birational map from Montgomery model to twisted Edwards model in Lemma 9.12.20 is undefined only for points P of order dividing 2 and

$P = (-1, \pm\sqrt{(A-2)/B})$ (which has order 4). Show that the map from Edwards model to Montgomery model is undefined only for points $P = (0, \pm 1)$ and points at infinity.

Exercise 9.12.23 Show that a non-trivial quadratic twist of the twisted Edwards model $ax^2 + y^2 = 1 + dx^2y^2$ over $\Bbbk$ is $aux^2 + y^2 = 1 + dux^2y^2$ where $u \in \Bbbk^*$ is a non-square.

Exercise 9.12.24 Show that if an elliptic curve E can be written in twisted Edwards model then the only non-trivial twist of E that can also be written in twisted Edwards model is the quadratic twist.

Example 9.12.25 The curve

$$x^2 + y^2 = 1 - x^2y^2$$

has an automorphism $\rho(x, y) = (ix, 1/y)$ (which fixes the identity point $(0, 1)$) for $i = \sqrt{-1}$. One has $\rho^2 = -1$. Hence, this curve corresponds to a twist of the Weierstrass curve $y^2 = x^3 + x$ having j-invariant 1728.

Example 9.12.26 Elliptic curves with CM by $D = -3$ (equivalently, j-invariant 0) can only be written in Edwards model if $\sqrt{3} \in \mathbb{F}_q$. Taking $d = (\sqrt{3}+2)/(\sqrt{3}-2)$ gives the Edwards curve

$$E : x^2 + y^2 = 1 + dx^2y^2,$$

which has j-invariant 0. We construct the automorphism corresponding to ζ_3 in stages. First we give the isomorphism $\phi : E \rightarrow M$ where $M : BY^2 = X^3 + AX^2 + X$ is the curve in Montgomery model with $A = 2(1+d)/(1-d)$ and $B = 4/(1-d)$. This map is $\phi(x, y) = ((1+y)/(1-y), (1+y)/(x(1-y)))$ as in Lemma 9.12.20. The action of ζ_3 on M is given by

$$\zeta(X, Y) = (\zeta_3 X + (1-\zeta_3)/\sqrt{3}, Y).$$

Then we apply $\phi^{-1}(X, Y) = (X/Y, (X-1)/(X+1))$.

9.13 Statistical properties of elliptic curves over finite fields

There are a number of questions, relevant for cryptography, about the set of all elliptic curves over $\mathbb{F}_q$.

The theory of complex multiplication states that if $|t| < 2\sqrt{q}$ and $\gcd(t, q) = 1$ then the number of isomorphism classes of elliptic curves E over $\mathbb{F}_q$ with $\#E(\mathbb{F}_q) = q + 1 - t$ is given by the Hurwitz class number $H(t^2 - 4q)$. Theorem 9.11.11 gave a similar result for the supersingular case. As noted in Section 9.10.1, this means that the number of $\mathbb{F}_q$-isomorphism classes of elliptic curves over $\mathbb{F}_q$ with $q + 1 - t$ points is $O(D\log(D)\log(\log(D)))$, where $D = \sqrt{4q - t^2}$. We now give Lenstra's bounds on the number of $\mathbb{F}_q$-isomorphism classes of elliptic curves with group orders in a subset of the Hasse interval.

Since the number of elliptic curves in short Weierstrass form (assuming now that $2 \nmid q$) that are $\mathbb{F}_q$-isomorphic to a given curve E is $(q-1)/\#\mathrm{Aut}(E)$, it is traditional to count the number of $\mathbb{F}_q$-isomorphism classes weighted by #Aut(E) (see Section 1.4 of Lenstra [339] for discussion and precise definitions). In other words, each $\mathbb{F}_q$-isomorphism class of elliptic curves with $j(E) = 0$ or $j(E) = 1728$ contributes less than one to the total. This makes essentially no difference to the asymptotic statements in Theorem 9.13.1. The weighted sum of all $\mathbb{F}_p$-isomorphism classes of elliptic curves over $\mathbb{F}_p$ is p.

Theorem 9.13.1 *(Proposition 1.9 of Lenstra [339] with the improvement of Theorem 2 of McKee [372]) There exists a constant $C_1 \in \mathbb{R}_{>0}$ such that, for any prime $p > 3$ and any $S \subset [p+1-2\sqrt{p}, p+1+2\sqrt{p}] \cap \mathbb{Z}$, the weighted sum of $\mathbb{F}_p$-isomorphism classes of elliptic curves $E/\mathbb{F}_p$ with $\#E(\mathbb{F}_p) \in S$ is at most $C_1 \#S \sqrt{p} \log(p) \log(\log(p))$.*

There exists a constant $C_2 \in \mathbb{R}_{>0}$ such that, for any prime $p > 3$ and any $S \subset [p+1-\sqrt{p}, p+1+\sqrt{p}] \cap \mathbb{Z}$, the weighted sum of $\mathbb{F}_p$-isomorphism classes of elliptic curves $E/\mathbb{F}_p$ with $\#E(\mathbb{F}_p) \in S$ is at least $C_2(\#S-2)\sqrt{p}/\log(p)$.

Lenstra also gave a result about divisibility of the group order by small primes.

Theorem 9.13.2 *(Proposition 1.14 of [339]) Let $p > 3$ and $l \neq p$ be primes. Then the weighted sum of all elliptic curves E over $\mathbb{F}_p$ such that $l \mid \#E(\mathbb{F}_p)$ is $p/(l-1) + O(l\sqrt{p})$ if $p \not\equiv 1 \pmod{l}$ and $pl/(l^2-1) + O(l\sqrt{p})$ if $p \equiv 1 \pmod{l}$. (Here the constants in the O are independent of l and p.)*

This result was generalised by Howe [267] to count curves with $N \mid \#E(\mathbb{F}_q)$ where N is not prime.

For cryptography it is important to determine the probability that a randomly chosen elliptic curve over $\mathbb{F}_q$ (i.e., choosing coefficients $a_4, a_6 \in F_q$ uniformly at random) is prime. A conjectural result was given by Galbraith and McKee [208].

Conjecture 9.13.3 *Let P_1 be the probability that a number within $2\sqrt{p}$ of $p+1$ is prime. Then the probability that an elliptic curve over $\mathbb{F}_p$ (p prime) has a prime number of points is asymptotic to $c_p P_1$ as $p \to \infty$, where*

$$c_p = \frac{2}{3} \prod_{l>2} \left(1 - \frac{1}{(l-1)^2}\right) \prod_{l \mid (p-1), l>2} \left(1 + \frac{1}{(l+1)(l-2)}\right).$$

Here the products are over all primes l satisfying the stated conditions.

Galbraith and McKee also give a precise conjecture for the probability that a random elliptic curve E over $\mathbb{F}_p$ has $\#E(\mathbb{F}_p) = kr$ where r is prime and $k \in \mathbb{N}$ is small.

Related problems have also been considered. For example, Koblitz [310] studies the probability that $\#E(\mathbb{F}_p)$ is prime for a fixed elliptic curve E over $\mathbb{Q}$ as p varies. A similar situation arises in the Sato-Tate distribution; namely, the distribution on $[-1, 1]$ arising from $(\#E(\mathbb{F}_p) - (p+1))/(2\sqrt{p})$ for a fixed elliptic curve E over $\mathbb{Q}$ as p varies. We refer to Murty and Shparlinksi [401] for a survey of other results in this area (including discussion of the Lang–Trotter conjecture).

9.14 Elliptic curves over rings

The elliptic curve factoring method (and some other theoretical applications in cryptography) use elliptic curves over the ring $\mathbb{Z}/N\mathbb{Z}$. When $N = \prod_{i=1}^{k} p_i$ is square-free[5] one can use the Chinese remainder theorem to interpret a triple (x, y, z) such that $y^2z + a_1xyz + a_3yz^2 \equiv x^3 + a_2x^2z + a_4xz^2 + a_6z^3 \pmod{N}$ as an element of the direct sum $\oplus_{i=1}^{k} E(\mathbb{F}_{p_i})$ of groups of elliptic curves over fields. It is essential to use the projective representation, since there can be points that are the point at infinity modulo p_1 but not the point at infinity modulo p_2 (in other words, $p_1 \mid z$ but $p_2 \nmid z$). Considering triples (x, y, z) such that $\gcd(x, y, z) = 1$ (otherwise, the point modulo some prime is $(0, 0, 0)$) up to multiplication by elements in $(\mathbb{Z}/N\mathbb{Z})^*$ leads to a projective elliptic curve point in $E(\mathbb{Z}/N\mathbb{Z})$. The usual formulae for the group operations can be used modulo N and, when they are defined, give a group law. We refer to Section 2.11 of Washington [560] for a detailed discussion, including a set of formulae for all cases of the group law. For a more theoretical discussion we refer to Lenstra [339, 340].

[5] The non-square-free case is more subtle. We do not discuss it.

10
Hyperelliptic curves

Hyperelliptic curves are a natural generalisation of elliptic curves, and it was suggested by Koblitz [298] that they might be useful for public key cryptography. Note that there is not a group law on the points of a hyperelliptic curve; instead, we use the divisor class group of the curve. The main goals of this chapter are to explain the geometry of hyperelliptic curves, to describe Cantor's algorithm [105] (and variants) to compute in the divisor class group of hyperelliptic curves and then to state some basic properties of the divisor class group.

Definition 10.0.1 Let $\Bbbk$ be a perfect field. Let $H(x), F(x) \in \Bbbk[x]$ (we stress that $H(x)$ and $F(x)$ are not assumed to be monic). An affine algebraic set of the form $C : y^2 + H(x)y = F(x)$ is called a **hyperelliptic equation**. The **hyperelliptic involution** $\iota : C \to C$ is defined by $\iota(x, y) = (x, -y - H(x))$.

Exercise 10.0.2 Let C be a hyperelliptic equation over $\Bbbk$. Show that if $P \in C(\Bbbk)$ then $\iota(P) \in C(\Bbbk)$.

When the projective closure of the algebraic set C in Definition 10.0.1 is irreducible, dimension 1, non-singular and of genus $g \geq 2$, then we will call it a hyperelliptic curve. By definition, a curve is projective and non-singular. We will give conditions for when a hyperelliptic equation is non-singular. Exercise 10.1.15 will give a projective non-singular model, but, in practice, one can work with the affine hyperelliptic equation. To "see" the points at infinity we will move them to points on a related affine equation, namely, the curve $C^\dagger$ of equation (10.2).

The genus has already been defined (see Definition 8.4.7) as a measure of the complexity of a curve. The treatment of the genus in this chapter is very "explicit". We will give precise conditions (Lemmas 10.1.6 and 10.1.8) that explain when the degree of a hyperelliptic equation is minimal. From this minimal degree we define the genus. In contrast, the approach of most other authors is to use the Riemann–Roch theorem.

We remark that one can also consider the algebraic group quotient $\mathrm{Pic}^0_{\mathbb{F}_q}(C)/[-1]$ of equivalence classes $\{D, -D\}$ where D is a reduced divisor. For genus 2 curves this object can be described as a variety, called the **Kummer surface**. It is beyond the scope of this book to give the details of this case. We refer to Chapter 3 of Cassels and Flynn [115] for background. Gaudry [224] and Gaudry and Lubicz [227] have given fast algorithms for computing with this algebraic group quotient.

10.1 Non-singular models for hyperelliptic curves

Consider the singular points on the affine curve $C(x, y) = y^2 + H(x)y - F(x) = 0$. The partial derivatives are $\partial C(x, y)/\partial y = 2y + H(x)$ and $\partial C(x, y)/\partial x = H'(x)y - F'(x)$, so a singular point in particular satisfies $2F'(x) + H(x)H'(x) = 0$. If $H(x) = 0$ and if the characteristic of $\Bbbk$ is not 2 then C is non-singular over $\overline{\Bbbk}$ if and only if $F(x)$ has no repeated root in $\overline{\Bbbk}$.

Exercise 10.1.1 Show that the curve $y^2 + H(x)y = F(x)$ over $\Bbbk$ has no affine singular points if and only if one of the following conditions hold.

1. $\mathrm{char}(\Bbbk) = 2$ and $H(x)$ is a non-zero constant.
2. $\mathrm{char}(\Bbbk) = 2$, $H(x)$ is a non-zero polynomial and $\gcd(H(x), F'(x)^2 - F(x)H'(x)^2) = 1$.
3. $\mathrm{char}(\Bbbk) \neq 2$, $H(x) = 0$ and $\gcd(F(x), F'(x)) = 1$.
4. $\mathrm{char}(\Bbbk) \neq 2$, $H(x) \neq 0$ and $\gcd(H(x)^2 + 4F(x), 2F'(x) + H(x)H'(x)) = 1$ (this applies even when $H(x) = 0$ or $H'(x) = 0$).

We will now give a simple condition for when a hyperelliptic equation is geometrically irreducible and of dimension 1. The proof also applies in many other cases. For the remaining cases, one has to test irreducibility directly.

Lemma 10.1.2 *Let $C(x, y) = y^2 + H(x)y - F(x)$ over $\Bbbk$ be a hyperelliptic equation. Suppose that $\deg(F(x))$ is odd. Suppose also that there is no point $P = (x_P, y_P) \in C(\overline{\Bbbk})$ such that $(\partial C(x, y)/\partial x)(P) = (\partial C(x, y)/\partial y)(P) = 0$. Then the affine algebraic set $V(C(x, y))$ is geometrically irreducible. The dimension of $V(C(x, y))$ is 1.*

Proof From Theorem 5.3.8, $C(x, y) = 0$ is $\overline{\Bbbk}$-reducible if and only if $C(x, y)$ factors over $\overline{\Bbbk}[x, y]$. By considering $C(x, y)$ as an element of $\overline{\Bbbk}(x)[y]$ it follows that such a factorisation must be of the form $C(x, y) = (y - a(x))(y - b(x))$ with $a(x), b(x) \in \overline{\Bbbk}[x]$. Since $\deg(F)$ is odd it follows that $\deg(a(x)) \neq \deg(b(x))$ and that at least one of $a(x)$ and $b(x)$ is non-constant. Hence, $a(x) - b(x)$ is a non-constant polynomial, so let $x_P \in \overline{\Bbbk}$ be a root of $a(x) - b(x)$ and set $y_P = a(x_P) = b(x_P)$ so that $(x_P, y_P) \in C(\overline{\Bbbk})$. It is then easy to check that both partial derivatives vanish at P. Hence, under the conditions of the Lemma, $V(C(x, y))$ is $\overline{\Bbbk}$-irreducible and so is an affine variety.

Now that $V(C(x, y))$ is known to be a variety we can consider the dimension. The function field of the affine variety is $\Bbbk(x)(y)$, which is a quadratic algebraic extension of $\Bbbk(x)$ and so has transcendence degree 1. Hence, the dimension of is 1. □

Let $H(x), F(x) \in \Bbbk[x]$ be such that $y^2 + H(x)y = F(x)$ is a non-singular affine curve. Define $D = \max\{\deg(F(x)), \deg(H(x)) + 1\}$. The projective closure of C in $\mathbb{P}^2$ is given by

$$y^2 z^{D-2} + z^{D-1} H(x/z) y = z^D F(x/z). \tag{10.1}$$

Exercise 10.1.3 Show that if $D > 2$ then there are at most two points at infinity on the curve of equation (10.1). Show further that if $D > 3$ and $\deg(F) > \deg(H) + 1$ then there is a unique point $(0 : 1 : 0)$ at infinity, which is a singular point.

In Definition 10.1.10 we will define the genus of a hyperelliptic curve in terms of the degree of the hyperelliptic equation. To do this, it will be necessary to have conditions that ensure that this degree is minimal. Example 10.1.4 and Exercise 10.1.5 show how a hyperelliptic equation that is a variety can be isomorphic to an equation of significantly lower degree (remember that isomorphism is only defined for varieties).

Example 10.1.4 The curve $y^2 + xy = x^{200} + x^{101} + x^3 + 1$ over $\mathbb{F}_2$ is isomorphic over $\mathbb{F}_2$ to the curve $Y^2 + xY = x^3 + 1$ via the map $(x, y) \mapsto (x, Y + x^{100})$.

Exercise 10.1.5 Let $\Bbbk$ be any field. Show that the affine algebraic variety $y^2 + (1 - 2x^3)y = -x^6 + x^3 + x + 1$ is isomorphic to a variety having an equation of total degree 2. Show that the resulting curve has genus 0.

Lemma 10.1.6 *Let $\Bbbk$ be a perfect field of characteristic 2 and $h(x), f(x) \in \Bbbk[x]$. Suppose the hyperelliptic equation $C : y^2 + h(x)y = f(x)$ is a variety. Then it is isomorphic over $\Bbbk$ to $Y^2 + H(x)Y = F(x)$ where one of the following conditions hold:*

1. $\deg(F(x)) > 2\deg(H(x))$ *and* $\deg(F(x))$ *is odd;*
2. $\deg(F(x)) = 2\deg(H(x)) = 2d$ *and the equation $u^2 + H_d u + F_{2d}$ has no solution in $\Bbbk$ (where $H(x) = H_d x^d + H_{d-1}x^{d-1} + \cdots + H_0$ and $F(x) = F_{2d}x^{2d} + \cdots + F_0$);*
3. $\deg(F(x)) < \deg(H(x))$.

Proof Let $d_H = \deg(H(x))$ and $d_F = \deg(F(x))$. The change of variables $y = Y + cx^i$ transforms $y^2 + H(x)y = F(x)$ to $Y^2 + H(x)Y = F(x) + c^2x^{2i} + H(x)cx^i$. Hence, if $\deg(F(x)) > 2\deg(H(x))$ and $\deg(F(x))$ is even then one can remove the leading coefficient by choosing $i = \deg(F(x))/2$ and $c = \sqrt{F_{2i}}$ (remember that $\mathrm{char}(\Bbbk) = 2$ and $\Bbbk$ is perfect so $c \in \Bbbk$). Similarly, if $\deg(H(x)) \leq j = \deg(F(x)) < 2\deg(H(x))$ then one can remove the leading coefficient $F_j x^i$ from F by taking $i = j - \deg(H(x))$ and $c = F_j / H_{d_H}$. Repeating these processes yields the first and third claims. The second case follows easily. □

Note that in the second case in Lemma 10.1.6 one can lower the degree using a $\overline{\Bbbk}$-isomorphism. Hence, geometrically (i.e., over $\overline{\Bbbk}$) one can assume that a hyperelliptic equation is of the form of case 1 or 3.

Example 10.1.7 The affine curve $y^2 + x^3 y = x^6 + x + 1$ is isomorphic over $\mathbb{F}_{2^2}$ to $Y^2 + x^3 Y = x + 1$ via $Y = y + ux^3$ where $u \in \mathbb{F}_{2^2}$ satisfies $u^2 + u = 1$. (Indeed, these curves are quadratic twists; see Definition 10.2.2.)

Lemma 10.1.8 *Let* $\Bbbk$ *be a field such that* $\mathrm{char}(\Bbbk) \neq 2$. *Every hyperelliptic curve over* $\Bbbk$ *is isomorphic over* $\Bbbk$ *to an equation of the form* $y^2 + (H_d x^d + \cdots + H_0)y = F_{2d}x^{2d} + F_{2d-1}x^{2d-1} + \cdots + F_0$ *where either:*

1. $H_d = 0$ *and* $(F_{2d} \neq 0$ *or* $F_{2d-1} \neq 0)$;
2. $H_d \neq 0$ *and* $(F_{2d} \neq -(H_d/2)^2$ *or* $F_{2d-1} \neq -H_d H_{d-1}/2)$.

Proof If $H_d = F_{2d} = F_{2d-1} = 0$ then just replace d by $d - 1$. If $H_d \neq 0$ and both $F_{2d} = -(H_d/2)^2$ and $F_{2d-1} = -H_d H_{d-1}/2$ then the morphism $(x, y) \mapsto (x, Y = y + \frac{H_d}{2}x^d)$ maps the hyperelliptic equation to

$$(Y - \tfrac{H_d}{2}x^d)^2 + (H_d x^d + \cdots + H_0)(Y - \tfrac{H_d}{2}x^d) - (F_{2d}x^{2d} + F_{2d-1}x^{2d-1} + \cdots + F_0).$$

This can be shown to have the form

$$Y^2 + h(x)Y = f(x)$$

with $\deg(h(x)) \leq d - 1$ and $\deg(f(x)) \leq 2d - 2$. (This is what happened in Exercise 10.1.5.) □

Exercise 10.1.9 Show that the hyperelliptic curve $y^2 + (2x^3 + 1)y = -x^6 + x^5 + x + 1$ is isomorphic to $Y^2 + Y = x^5 + x^3 + x + 1$.

10.1.1 Projective models for hyperelliptic curves

For the rest of the chapter we will assume that our hyperelliptic equations are $\overline{\Bbbk}$-irreducible and non-singular as affine algebraic sets. We also assume that when $\mathrm{char}(\Bbbk) = 2$ one of the conditions of Lemma 10.1.6 holds and when $\mathrm{char}(\Bbbk) \neq 2$ one of the conditions of Lemma 10.1.8 holds. The interpretation of $\deg(H(x))$ and $\deg(F(x))$ in terms of the genus of the curve will be discussed in Section 10.1.3.

Suppose $y^2 + H(x)y = F(x)$ is a non-singular affine hyperelliptic equation for a projective non-singular curve C. Write H_j for the coefficients of $H(x)$ and F_j for the coefficients of $F(x)$. Define $d_H = \deg(H(x))$ and $d_F = \deg(F(x))$. Let $d = \max\{d_H, \lceil d_F/2 \rceil\}$ and suppose $d > 0$. Set $H_d = \cdots = H_{d_H+1} = 0$ and $F_{2d} = \cdots = F_{d_F+1} = 0$ if necessary.

The rational map $(Z, Y) = \rho(x, y) = (1/x, y/x^d)$ maps C to the affine algebraic set

$$C^\dagger : Y^2 + (H_d + H_{d-1}Z + \cdots H_0 Z^d)Y = F_{2d} + F_{2d-1}Z + \cdots F_0 Z^{2d}. \qquad (10.2)$$

It is easy to check that $C^\dagger$ is geometrically irreducible, since C is. Now, all affine points (z, y) on $C^\dagger$ with $z \neq 0$ correspond to affine points on C and so are non-singular (since non-singularity is a local property; see Remark 7.1.3). To show that $C^\dagger$ is non-singular it is sufficient to consider the points $(z, y) = (0, \alpha)$ where $\alpha^2 + H_d\alpha - F_{2d} = 0$.

The partial derivatives evaluated at $(0, \alpha)$ are $2\alpha + H_d$ and $H_{d-1}\alpha - F_{2d-1}$. When $\mathrm{char}(\Bbbk) \neq 2$ the point being singular would imply $H_d = -2\alpha$ in which case $F_{2d} = \alpha^2 + H_d\alpha = -\alpha^2 = -(H_d/2)^2$ and $F_{2d-1} = H_{d-1}\alpha = -H_d H_{d-1}/2$. One easily sees that these equations contradict the conditions of Lemma 10.1.8.

Hence, $C^\dagger$ is a hyperelliptic curve and $\rho : C \to C^\dagger$ is a birational map. It follows that ρ induces a morphism between the corresponding projective curves. The point(s) $(0, \alpha)$ are the images of the point(s) at infinity on C. Hence, we can use $C^\dagger$ to visualise the points at infinity on C.

Up to now the phrase "hyperelliptic curve" has meant a projective non-singular curve of genus $g \geq 2$ that has an affine model as a hyperelliptic equation. Definition 10.1.10 gives an equivalent formulation that will be used throughout the book. Technically, this is an abuse of notation since C is not projective.

Definition 10.1.10 Let $\Bbbk$ be a perfect field. Let $H(x), F(x) \in \Bbbk[x]$ be such that:

- $\deg(H(x)) \geq 3$ or $\deg(F(x)) \geq 5$;
- the affine hyperelliptic equation $C : y^2 + H(x)y = F(x)$ is $\overline{\Bbbk}$-irreducible and non-singular;
- the conditions of Lemma 10.1.6 and Lemma 10.1.8 hold.

Then C is called a **hyperelliptic curve**. The **genus** of the hyperelliptic curve is $g = \max\{\deg(H(x)) - 1, \lfloor \deg(F(x)) - 1)/2 \rfloor\}$ (see Section 10.1.3 for justification of this).

It looks like Definition 10.1.10 excludes some potentially interesting equations (such as $y^2 + H(x)y = F(x)$ where $\deg(F(x)) = 4$ and $\deg(H(x)) = 2$). In fact, it can be shown that all the algebraic sets excluded by the definition are either $\overline{\Bbbk}$-reducible, singular over $\overline{\Bbbk}$ or birational over $\overline{\Bbbk}$ to a curve of genus 0 or 1 over $\overline{\Bbbk}$.

The equation $\alpha^2 + H_d\alpha - F_{2d} = 0$ can have a $\Bbbk$-rational repeated root, two roots in $\Bbbk$, or two conjugate roots in $\overline{\Bbbk}$. It follows that there are three possible behaviours at infinity: a single $\Bbbk$-rational point, two distinct $\Bbbk$-rational points and a pair of distinct points defined over a quadratic extension of $\Bbbk$ (which are Galois conjugates). These three cases correspond to the fact that the place at infinity in $\Bbbk[x]$ is ramified, split or inert respectively in the field extension $\Bbbk(C)/\Bbbk(x)$. A natural terminology for the three types of behaviour at infinity is therefore to call them ramified, split and inert.

Definition 10.1.11 Let C be a hyperelliptic curve and let $C^\dagger$ be as in equation (10.2). Let $\rho : C \to C^\dagger$ be as above. Let α^+, α^- be the roots in $\overline{\Bbbk}$ of the polynomial $\alpha^2 + H_d\alpha - F_{2d}$. We write ∞^+ for the **point at infinity** on C such that $\rho(\infty^+) = (0, \alpha^+)$ and ∞^- for the point such that $\rho(\infty^-) = (0, \alpha^-)$.

If $\alpha^+ = \alpha^-$ then C is called a **ramified model of a hyperelliptic curve**. If there are two distinct points at infinity with $\alpha^+, \alpha^- \in \Bbbk$ then C is called a **split model of a hyperelliptic curve** and if $\alpha^+, \alpha^- \notin \Bbbk$ then C is an **inert model of a hyperelliptic curve**.

One finds in the literature the names **imaginary hyperelliptic curve** (respectively, **real hyperelliptic curve**) for ramified model and split model respectively. Exercise 10.1.13

classifies ramified hyperelliptic models. Exercise 10.1.14 shows that if $C(\Bbbk) \neq \varnothing$ then one may transform C into a ramified or split model. Hence, when working over finite fields, it is not usually necessary to deal with curves having an inert model.

Exercise 10.1.12 With notation as in Definition 10.1.11 show that $\iota(\infty^+) = \infty^-$.

Exercise 10.1.13 Let $C : y^2 + H(x)y = F(x)$ be a hyperelliptic curve over $\Bbbk$ satisfying all the conditions above. Let $d = \max\{\deg(H(x)), \lceil \deg(F(x))/2 \rceil\}$. Show that this is a ramified model if and only if $(\deg(H(x)) < d$ and $\deg(F(x)) = 2d - 1)$ or $(\mathrm{char}(\Bbbk) \neq 2$, $\deg(F(x)) = 2d$, $\deg(H(x)) = d$ and $F_{2d} = -(H_d/2)^2)$.

Exercise 10.1.14 Let $C : y^2 + H(x)y = F(x)$ be a hyperelliptic curve over $\Bbbk$ and let $P \in C(\Bbbk)$. Define the rational map

$$\rho_P(x, y) = (1/(x - x_P), y/(x - x_P)^d).$$

Then $\rho_P : C \to C'$ where C' is also a hyperelliptic curve. Show that ρ_P is just the translation map $P \mapsto (0, y_P)$ followed by the map ρ and so is an isomorphism from C to C'.

Show that if $P = \iota(P)$ then C is birational over $\Bbbk$ (using ρ_P) to a hyperelliptic curve with ramified model. Show that if $P \neq \iota(P)$ then C is birational over $\Bbbk$ to a hyperelliptic curve with split model.

We now indicate a different projective model for hyperelliptic curves.

Exercise 10.1.15 Let the notation and conditions be as above. Assume $C : y^2 + H(x)y = F(x)$ is irreducible and non-singular as an affine curve. Let $Y, X_d, X_{d-1}, \ldots, X_1, X_0$ be coordinates for $\mathbb{P}^{d+1}$ (one interprets $X_i = x^i$). The **projective hyperelliptic equation** is the projective algebraic set in $\mathbb{P}^{d+1}$ given by

$$\begin{aligned} Y^2 + (H_d X_d + H_{d-1}X_{d-1} + \cdots + H_0 X_0)Y &= F_{2d}X_d^2 + F_{2d-1}X_d X_{d-1} + \cdots \\ &\quad + F_1 X_1 X_0 + F_0 X_0^2, \\ X_i^2 &= X_{i-1}X_{i+1}, \qquad \text{for } 1 \le i \le d-1, \\ X_d X_i &= X_{\lceil (d+i)/2 \rceil} X_{\lfloor (d+i)/2 \rfloor}, \qquad \text{for } 0 \le i \le d-2. \end{aligned} \tag{10.3}$$

1. Give a birational map (assuming for the moment that the above model is a variety) between the affine algebraic set C and the model of equation (10.3).
2. Show that the hyperelliptic involution ι extends to equation (10.3) as

$$\iota(Y : X_d : \cdots : X_0) = (-Y - H_d X_d - H_{d-1}X_{d-1} - \cdots - H_0 X_0 : X_d : \cdots : X_0).$$

3. Show that the points at infinity on equation (10.3) satisfy $X_0 = X_1 = X_2 = \cdots = X_{d-1} = 0$ and $Y^2 + H_d X_d Y - F_{2d} X_d^2 = 0$. Show that if $F_{2d} = H_d = 0$ then there is a single point at infinity.
4. Show that if the conditions of Lemma 10.1.6 or Lemma 10.1.8 hold then equation (10.3) is non-singular at infinity.
5. Show that equation (10.3) is a variety.

10.1.2 Uniformisers on hyperelliptic curves

The aim of this section is to determine uniformisers for all points on hyperelliptic curves. We begin in Lemma 10.1.16 by determining uniformisers for the affine points of a hyperelliptic curve.

Lemma 10.1.16 *Let $P = (x_P, y_P) \in C(\Bbbk)$ be a point on a hyperelliptic curve. If $P = \iota(P)$ then $(y - y_P)$ is a uniformiser at P (and $v_P(x - x_P) = 2$). If $P \neq \iota(P)$ then $(x - x_P)$ is a uniformiser at P.*

Proof We have

$$\begin{aligned}(y - y_P)(y + y_P + H(x_P)) &= y^2 + H(x_P)y - (y_P^2 + H(x_P)y_P) \\ &= F(x) + y(H(x_P) - H(x)) - F(x_P).\end{aligned}$$

Now, use the general fact for any polynomial that $F(x) = F(x_P) + (x - x_P)F'(x_P) \pmod{(x - x_P)^2}$. Hence, the above expression is congruent modulo $(x - x_P)^2$ to

$$(x - x_P)(F'(x_P) - yH'(x_P)) \pmod{(x - x_P)^2}.$$

When $P = \iota(P)$ then $(y - y_P)(y + (y_P + H(x_P))) = (y - y_P)^2$. Note also that $F'(x_P) - y_P H'(x_P)$ is not zero since $2y_P + H(x_P) = 0$ and yet C is not singular. Writing $G(x, y) = (y - y_P)^2/(x - x_P) \in \Bbbk[x, y]$ we have $G(x_P, y_P) \neq 0$ and

$$x - x_P = (y - y_P)^2 \frac{1}{G(x, y)}.$$

Hence, a uniformiser at P is $(y - y_P)$ and $v_P(x - x_P) = 2$.

For the case $P \neq \iota(P)$ note that $v_P(y - y_P) > 0$ and $v_P(y + y_P + H(x_P)) = 0$. It follows that $v_P(y - y_P) \geq v_P(x - x_P)$. □

We now consider uniformisers at infinity on a hyperelliptic curve C over $\Bbbk$. The easiest way to proceed is to use the curve $C^\dagger$ of equation (10.2).

Lemma 10.1.17 *Let C be a hyperelliptic curve and let $\rho : C \to C^\dagger$ be as in equation (10.2). Let $P = \rho(\infty^+) = (0, \alpha^+) \in C^\dagger(\Bbbk)$. If $\iota(\infty^+) = \infty^+$ (i.e., if there is one point at infinity) then $Y - \alpha^+$ is a uniformiser at P on $C^\dagger$ and so $(y/x^d) - \alpha^+$ is a uniformiser at ∞^+ on C. If $\iota(\infty^+) \neq \infty^+$ then Z is a uniformiser at P on $C^\dagger$ (i.e., $1/x$ is a uniformiser at ∞^+ on C).*

Proof Note that if $\iota(\infty^+) = \infty^+$ then $\iota(P) = P$ and if $\iota(\infty^+) \neq \infty^+$ then $\iota(P) \neq P$. It immediately follows from Lemma 10.1.16 that $Y - \alpha^+$ or Z is a uniformiser at P on $C^\dagger$. Lemma 8.1.13, Exercise 8.2.8 and Lemma 8.2.6 show that for any $f \in \Bbbk(C^\dagger)$ and $P \in C(\Bbbk)$, $v_P(f \circ \rho) = v_{\rho(P)}(f)$. Hence, uniformisers at infinity on C are $(Y - \alpha^+) \circ \rho = (y/x^d) - \alpha^+$ or $Z \circ \rho = 1/x$. □

Exercise 10.1.18 Let C be a hyperelliptic curve in ramified model. Show that $v_\infty(x) = -2$. Show that if the curve has equation $y^2 = F(x)$ where $\deg(F(x)) = 2g+1$ then x^g/y is an alternative uniformiser at infinity.

Now suppose C is given as a split or inert model. Show that $v_{\infty^+}(x) = v_{\infty^-}(x) = -1$.

Exercise 10.1.19 Let C be a hyperelliptic curve (ramified, split or inert). If $u(x) = (x - x_0)$ is a function on C and $P_0 = (x_0, y_0) \in C(\overline{\Bbbk})$ then $\operatorname{div}(u(x)) = (P_0) + (\iota(P_0)) - (\infty^+) - (\infty^-)$.

Exercise 10.1.20 Let C be a hyperelliptic curve of genus g. Show that if C is in ramified model then $v_\infty(y) = -(2g+1)$ and if C is in split model then $v_{\infty^+}(y) = v_{\infty^-}(y) = -(g+1)$.

Exercise 10.1.21 Let C be a hyperelliptic curve. Let $A(x), B(x) \in \Bbbk[x]$ and let $P = (x_P, y_P) \in C(\overline{\Bbbk})$ be a point on the affine curve. Show that $v_P(A(x) - yB(x))$ is equal to e where $(x - x_P)^e \| (A(x)^2 + H(x)A(x)B(x) - F(x)B(x)^2)$.

We now describe a polynomial that will be crucial for arithmetic on hyperelliptic curves with a split model. Essentially, $G^+(x)$ is a function that cancels the pole of y at ∞^+. This leads to another choice of uniformiser at ∞^+ for these models.

Exercise 10.1.22 Let $C : y^2 + H(x)y = F(x)$ be a hyperelliptic curve with split model over $\Bbbk$ of genus g. Let $\alpha^+, \alpha^- \in \Bbbk$ be the roots of $Y^2 + H_d Y - F_{2d}$. Show that there exists a polynomial $G^+(x) = \alpha^+ x^d + \cdots \in \Bbbk[x]$ of degree $d = g+1$ such that $\deg(G^+(x)^2 + H(x)G^+(x) - F(x)) \le d - 1 = g$. Similarly, show that there is a polynomial $G^-(x) = \alpha^- x^d + \cdots$ such that $\deg(G^-(x)^2 + H(x)G^-(x) - F(x)) \le d - 1 = g$. Indeed, show that $G^-(x) = -G^+(x) - H(x)$.

Exercise 10.1.23 Let $C : y^2 + H(x)y = F(x)$ be a hyperelliptic curve with split model over $\Bbbk$ of genus g and let $G^+(x)$ be as in Exercise 10.1.22. Show that $v_{\infty^+}(y - G^+(x)) \ge 1$.

10.1.3 The genus of a hyperelliptic curve

In Lemma 10.1.6 and Lemma 10.1.8 we showed that some hyperelliptic equations $y^2 + h(x)y = f(x)$ are birational to hyperelliptic equations $y^2 + H(x)y = F(x)$ with $\deg(F(x)) < \deg(f(x))$ or $\deg(H(x)) < \deg(h(x))$. Hence, it is natural to suppose that the geometry of the curve C imposes a lower bound on the degrees of the polynomials $H(x)$ and $F(x)$ in its curve equation. The right measure of the complexity of the geometry is the genus.

Indeed, the Riemann–Roch theorem implies that if C is a hyperelliptic curve over $\Bbbk$ of genus g and there is a function $x \in \Bbbk(C)$ of degree 2 then C is birational over $\Bbbk$ to an equation of the form $y^2 + H(x)y = F(x)$ with $\deg(H(x)) \le g+1$ and $\deg(F(x)) \le 2g+2$. Furthermore, the Hurwitz genus formula shows that if $y^2 + H(x)y = F(x)$ is non-singular and with degrees reduced as in Lemma 10.1.6 and Lemma 10.1.8 then the genus is $\max\{\deg(H(x)) - 1, \lceil \deg(F(x))/2 - 1 \rceil\}$. (Theorem 8.7.3, as it is stated, cannot be applied for hyperelliptic curves in characteristic 2, but a more general version of the Hurwitz genus formula proves the above statement about the genus.) Hence, writing $d = g + 1$, the conditions of Lemma 10.1.6 and Lemma 10.1.8 together with

$$\deg(H(x)) = d \quad \text{or} \quad 2d - 1 \leq \deg(F(x)) \leq 2d \tag{10.4}$$

are equivalent to the curve $y^2 + H(x)y = F(x)$ having genus g.

It is not necessary for us to prove the Riemann–Roch theorem or the Hurwitz genus formula. Our discussion of Cantor reduction (see Lemma 10.3.20 and Lemma 10.4.6) will directly prove a special case of the Riemann–Roch theorem for hyperelliptic curves, namely that every divisor class contains a representative corresponding to an effective divisor of degree at most $g = d - 1$.

The reader should interpret the phrase "hyperelliptic curve of genus g" as meaning the conditions of Lemma 10.1.6 and Lemma 10.1.8 together with equation (10.4) on the degrees of $H(x)$ and $F(x)$ hold.

10.2 Isomorphisms, automorphisms and twists

We consider maps between hyperelliptic curves in this section. We are generally interested in isomorphisms over $\overline{\Bbbk}$ rather than just $\Bbbk$.

In the elliptic curve case (see Section 9.3) there was no loss of generality by assuming that isomorphisms fix infinity (since any isomorphism can be composed with a translation map). Since the points on a hyperelliptic curve do not, in general, form a group, one can no longer make this assumption. Nevertheless, many researchers have restricted attention to the special case of maps between curves that map points at infinity (with respect to an affine model of the domain curve) to points at infinity on the image curve. Theorem 10.2.1 classifies this special case.

In this chapter, and in the literature as a whole, isomorphisms are usually not assumed to fix infinity. For example, the isomorphism ρ_P defined earlier in Exercise 10.1.14 does not fix infinity. Isomorphisms that map points at infinity to points at infinity map ramified models to ramified models and unramified models to unramified models.

Theorem 10.2.1 *Let $C_1 : y_1^2 + H_1(x_1)y_1 = F_1(x_1)$ and $C_2 : y_2^2 + H_2(x_2)y_2 = F_2(x_2)$ be hyperelliptic curves over $\Bbbk$ of genus g. Then every isomorphism $\phi : C_1 \to C_2$ over $\Bbbk$ that maps points at infinity of C_1 to points at infinity of C_2 is of the form*

$$\phi(x_1, y_1) = (ux_1 + r, wy_1 + t(x_1))$$

where $u, w, r \in \Bbbk$ and $t \in \Bbbk[x_1]$. If C_1 and C_2 have ramified models then $\deg(t) \leq g$. *If C_1 and C_2 have split or inert models then* $\deg(t) \leq g + 1$, *and the leading coefficient of $t(x_1)$ is not equal to the leading coefficient of $-wG^+(x_1)$ or $-wG^-(x_1)$ (where G^+ and G^- are as in Exercise 10.1.22).*

Proof (Sketch) The proof is essentially the same as the proof of Proposition 3.1(b) of Silverman [505]; one can also find the ramified case in Proposition 1.2 of Lockhart [353]. One notes that the valuations at infinity of x_1 and x_2 have to agree, and similarly for y_1 and y_2. It follows that x_2 lies in the same Riemann–Roch spaces as x_1 and similarly for y_2 and y_1. The result follows (the final conditions are simply that the valuations at infinity of y_1 and y_2 must agree, so we are prohibited from setting $y_2 = w(y_1 + t(x))$ such that it lowers the valuation of y_2). □

We now introduce quadratic twists in the special case of finite fields. As mentioned in Example 9.5.2, when working in characteristic zero there are infinitely many quadratic twists.

Definition 10.2.2 Let $C : y^2 = F(x)$ be a hyperelliptic curve over a finite field $\mathbb{k}$ where $\mathrm{char}(\mathbb{k}) \neq 2$. Let $u \in \mathbb{k}^*$ be a non-square (i.e., there is no $v \in \mathbb{k}^*$ such that $u = v^2$) and define $C^{(u)} : y^2 = uF(x)$.

Let $C : y^2 + H(x)y = F(x)$ be a hyperelliptic curve over a finite field $\mathbb{k}$ where $\mathrm{char}(\mathbb{k}) = 2$. Let $u \in \mathbb{k}$ be such that $\mathrm{Tr}_{\mathbb{k}/\mathbb{F}_2}(u) = 1$. Define $C^{(u)} : y^2 + H(x)y = F(x) + uH(x)^2$.

In both cases, the $\mathbb{k}$-isomorphism class of the curve $C^{(u)}$ is called the non-trivial **quadratic twist** of C.

Exercise 10.2.3 Show that the quadratic twist is well-defined when $\mathbb{k}$ is a finite field. In other words, show that in the case $\mathrm{char}(\mathbb{k}) \neq 2$ if u and u' are two different non-squares in $\mathbb{k}^*$ then the corresponding curves $C^{(u)}$ and $C^{(u')}$ as in Definition 10.2.2 are isomorphic over $\mathbb{k}$. Similarly, if $\mathrm{char}(\mathbb{k}) = 2$ and for two different choices of trace one elements $u, u' \in \mathbb{k}$ show that the corresponding curves $C^{(u)}$ and $C^{(u')}$ are isomorphic over $\mathbb{k}$.

Exercise 10.2.4 Let C be a hyperelliptic curve over a finite field $\mathbb{k}$ and let $C^{(u)}$ be a non-trivial quadratic twist. Show that $\#C(\mathbb{F}_q) + \#C^{(u)}(\mathbb{F}_q) = 2(q + 1)$.

We now consider automorphisms. Define $\mathrm{Aut}(C)$ to be the set of all isomorphisms $\phi : C \to C$ over $\overline{\mathbb{k}}$. As usual, $\mathrm{Aut}(C)$ is a group under composition.

Example 10.2.5 Let $p > 2$ be a prime and $C : y^2 = x^p - x$ over $\mathbb{F}_p$. For $a \in \mathbb{F}_p^*, b \in \mathbb{F}_p$ one has isomorphisms

$$\phi_a(x, y) = (ax, \pm\sqrt{a}y) \quad \text{and} \quad \psi_{b,\pm}(x, y) = (x + b, \pm y)$$

from C to itself (in both cases they fix the point at infinity). Hence, the subgroup of $\mathrm{Aut}(C)$ consisting of maps that fix infinity is a group of at least $2p(p - 1)$ elements.

There is also the birational map $\rho(x, y) = (-1/x, y/x^{(p+1)/2})$ that corresponds to an isomorphism $\rho : C \to C$ on the projective curve. This morphism does not fix infinity. Since all the compositions $\psi_{b',\pm} \circ \rho \circ \psi_{b,\pm} \circ \phi_a$ are distinct one has $2p^2(p - 1)$ isomorphisms of this form. Hence, $\mathrm{Aut}(C)$ has size at least $2p(p - 1) + 2p^2(p - 1) = 2p(p + 1)(p - 1)$.

Exercise 10.2.6 Let $p > 2$ be a prime and $C : y^2 = x^p - x + 1$ over $\mathbb{F}_p$. Show that the subgroup of $\mathrm{Aut}(C)$ consisting of automorphisms that fix infinity has order $2p$.

Exercise 10.2.7 Let $p > 2$ be a prime and $C : y^2 = x^n + 1$ over $\mathbb{F}_p$ with $n \neq p$ (when $n = p$ the equation is singular). Show that the subgroup of Aut(C) consisting of automorphisms that fix infinity has order $2n$.

Exercise 10.2.8 Let $p \equiv 1 \pmod 8$ and let $C : y^2 = x^5 + Ax$ over $\mathbb{F}_p$. Write $\zeta_8 \in \overline{\mathbb{F}}_p$ for a primitive 8th root of unity. Show that $\zeta_8 \in \mathbb{F}_{p^4}$. Show that $\psi(x, y) = (\zeta_8^2 x, \zeta_8 y)$ is an automorphism of C. Show that $\psi^4 = \iota$.

10.3 Effective affine divisors on hyperelliptic curves

This section is about how to represent effective divisors on affine hyperelliptic curves, and algorithms to compute with them. A convenient way to represent divisors is using Mumford representation, and this is only possible if the divisor is semi-reduced.

Definition 10.3.1 Let C be a hyperelliptic curve over $\Bbbk$ and denote by $C \cap \mathbb{A}^2$ the affine curve. An **effective affine divisor** on C is

$$D = \sum_{P \in (C \cap \mathbb{A}^2)(\overline{\Bbbk})} n_P(P)$$

where $n_P \geq 0$ (and, as always, $n_P \neq 0$ for only finitely many P). A divisor on C is **semi-reduced** if it is an effective affine divisor and for all $P \in (C \cap \mathbb{A}^2)(\overline{\Bbbk})$ we have:

1. If $P = \iota(P)$ then $n_P = 1$.
2. If $P \neq \iota(P)$ then $n_P > 0$ implies $n_{\iota(P)} = 0$.

We slightly adjust the notion of equivalence for divisors on $C \cap \mathbb{A}^2$.

Definition 10.3.2 Let C be a hyperelliptic curve over a field $\Bbbk$ and let $f \in \Bbbk(C)$. We define

$$\operatorname{div}(f) \cap \mathbb{A}^2 = \sum_{P \in (C \cap \mathbb{A}^2)(\overline{\Bbbk})} v_P(f)(P).$$

Two divisors D, D' on $C \cap \mathbb{A}^2$ are **equivalent**, written $D \equiv D'$, if there is some function $f \in \overline{\Bbbk}(C)$ such that $D = D' + \operatorname{div}(f) \cap \mathbb{A}^2$.

Lemma 10.3.3 *Let C be a hyperelliptic curve. Every divisor on $C \cap \mathbb{A}^2$ is equivalent to a semi-reduced divisor.*

Proof Let $D = \sum_{P \in C \cap \mathbb{A}^2} n_P(P)$. By Exercise 10.1.19 the function $x - x_P$ has divisor $(P) + (\iota(P))$ on $C \cap \mathbb{A}^2$. If $n_P < 0$ for some $P \in (C \cap \mathbb{A}^2)(\overline{\Bbbk})$ then, by adding an appropriate multiple of $\operatorname{div}(x - x_P)$, one can arrange that $n_P = 0$ (this will increase $n_{\iota(P)}$). Similarly, if $n_P > 0$ and $n_{\iota(P)} > 0$ (or if $P = \iota(P)$ and $n_P \geq 2$) then subtracting a multiple of $\operatorname{div}(x - x_P)$ lowers the values of n_P and $n_{\iota(P)}$. Repeating this process yields a semi-reduced divisor. □

Example 10.3.4 Let $P_1 = (x_1, y_1)$ and $P_2 = (x_2, y_2)$ be points on a hyperelliptic curve C such that $x_1 \neq x_2$. Let $D = -(P_1) + 2(P_2) + (\iota(P_2))$. Then D is not semi-reduced. One has

$$D + \operatorname{div}(x - x_1) = D + (P_1) + (\iota(P_1)) = (\iota(P_1)) + 2(P_2) + (\iota(P_2)),$$

which is still not semi-reduced. Subtracting $\operatorname{div}(x - x_2)$ from the above gives

$$D + \operatorname{div}((x - x_1)/(x - x_2)) = (\iota(P_1)) + (P_2),$$

which is semi-reduced.

10.3.1 Mumford representation of semi-reduced divisors

Mumford [399] introduced[1] a representation for semi-reduced divisors. The condition that the divisor is semi-reduced is crucial: if points $P = (x_P, y_P)$ and (x_P, y'_P) with $y_P \neq y'_P$ both appear in the support of the divisor then no polynomial $v(x)$ can satisfy both $v(x_P) = y_P$ and $v(x_P) = y'_P$.

Lemma 10.3.5 *Let $D = \sum_{i=1}^{l} e_i(x_i, y_i)$ be a non-zero semi-reduced divisor on a hyperelliptic curve $C : y^2 + H(x)y = F(x)$ (hence, D is affine and effective). Define*

$$u(x) = \prod_{i=1}^{l} (x - x_i)^{e_i} \in \overline{\Bbbk}[x].$$

Then there is a unique polynomial $v(x) \in \overline{\Bbbk}[x]$ such that $\deg(v(x)) < \deg(u(x))$, $v(x_i) = y_i$ *for all $1 \le i \le l$, and*

$$v(x)^2 + H(x)v(x) - F(x) \equiv 0 \pmod{u(x)}. \tag{10.5}$$

In particular, $v(x) = 0$ if and only if $u(x) \mid F(x)$.

Proof Since D is semi-reduced there is no conflict in satisfying the condition $v(x_i) = y_i$. If all $e_i = 1$ then the result is trivial. For each i such that $e_i > 1$ write $v(x) = y_i + (x - x_i)W(x)$ for some polynomial $W(x)$. We compute $v(x) \pmod{(x - x_i)^{e_i}}$ so it satisfies $v(x)^2 + H(x)v(x) - F(x) \equiv 0 \pmod{(x - x_i)^{e_i}}$ by Hensel lifting (see Section 2.13) as follows: if $v(x)^2 + H(x)v(x) - F(x) = (x - x_i)^j G_j(x)$ then set $v^\dagger(x) = v(x) + w(x - x_i)^j$ where w is an indeterminate and note that

$$v^\dagger(x)^2 + H(x)v^\dagger(x) - F(x) \equiv (x - x_i)^j (G_j(x) + 2v(x)w + H(x)w) \pmod{(x - x_i)^{j+1}}.$$

It suffices to find w such that this is zero, in other words, solve $G_j(x_i) + w(2y_i + H(x_i)) = 0$. Since D is semi-reduced, we know $2y_i + H(x_i) \neq 0$ (since $P = \iota(P)$ implies $n_P = 1$). The result follows by the Chinese remainder theorem. □

[1] Mumford remarks on pages 3–17 of [399] that a special case of these polynomials arises in the work of Jacobi. However, Jacobi only gives a representation for semi-reduced divisors with g points in their support, rather than arbitrary semi-reduced divisors.

Definition 10.3.6 Let D be a non-zero semi-reduced divisor. The polynomials $(u(x), v(x))$ of Lemma 10.3.5 are the **Mumford representation** of D. If $D = 0$ then take $u(x) = 1$ and $v(x) = 0$. A pair of polynomials $u(x), v(x) \in \overline{\Bbbk}[x]$ is called a **Mumford representation** if $u(x)$ is monic, $\deg(v(x)) < \deg(u(x))$ and if equation (10.5) holds.

We have shown that every semi-reduced divisor D has a Mumford representation and that the polynomials satisfying the conditions in Definition 10.3.6 are unique. We now show that one can easily recover an affine divisor D from the pair $(u(x), v(x))$: write $u(x) = \prod_{i=1}^{l}(x - x_i)^{e_i}$ and let $D = \sum_{i=1}^{l} e_i(x_i, v(x_i))$.

Exercise 10.3.7 Show that the processes of associating a Mumford representation to a divisor and associating a divisor to a Mumford representation are inverse to each other. More precisely, let D be a semi-reduced divisor on a hyperelliptic curve. Show that if one represents D in Mumford representation, and then obtains a corresponding divisor D' as explained above, then $D' = D$.

Exercise 10.3.8 Let $u(x), v(x) \in \Bbbk[x]$ be such that equation (10.5) holds. Let D be the corresponding semi-reduced divisor. Show that

$$D = \sum_{P \in (C \cap \mathbb{A}^2)(\overline{\Bbbk})} \min\{v_P(u(x)), v_P(y - v(x))\}(P).$$

This is called the **greatest common divisor** of $\operatorname{div}(u(x))$ and $\operatorname{div}(y - v(x))$ and is denoted $\operatorname{div}(u(x), y - v(x))$.

Exercise 10.3.9 Let $(u_1(x), v_1(x))$ and $(u_2(x), v_2(x))$ be the Mumford representations of two semi-reduced divisors D_1 and D_2. Show that if $\gcd(u_1(x), u_2(x)) = 1$ then $\operatorname{Supp}(D_1) \cap \operatorname{Supp}(D_2) = \varnothing$.

Lemma 10.3.10 *Let C be a hyperelliptic curve over $\Bbbk$ and let D be a semi-reduced divisor on C with Mumford representation $(u(x), v(x))$. Let $\sigma \in \operatorname{Gal}(\overline{\Bbbk}/\Bbbk)$.*

1. *$\sigma(D)$ is semi-reduced.*
2. *The Mumford representation of $\sigma(D)$ is $(\sigma(u(x)), \sigma(v(x)))$.*
3. *D is defined over $\Bbbk$ if and only if $u(x), v(x) \in \Bbbk[x]$.*

Exercise 10.3.11 Prove Lemma 10.3.10.

Exercise 10.3.8 shows that the Mumford representation of a semi-reduced divisor D is natural from the point of view of principal divisors. This explains why condition (10.5) is the natural definition for the Mumford representation. There are two other ways to understand condition (10.5). First, the divisor D corresponds to an ideal in the ideal class group of the affine coordinate ring $\Bbbk[x, y]$, and condition (10.5) shows this ideal is equal to the $\Bbbk[x, y]$-ideal $(u(x), y - v(x))$. Second, from a purely algorithmic point of view, condition (10.5) is needed to make the Cantor reduction algorithm work (see Section 10.3.3).

A divisor class contains infinitely many divisors whose affine part is semi-reduced. Later we will define a reduced divisor to be one whose degree is sufficiently small. One can then consider whether there is a unique such representative of the divisor class. This issue will be considered in Lemma 10.3.24 below.

Exercise 10.3.12 is relevant for the index calculus algorithms on hyperelliptic curves and it is convenient to place it here.

Exercise 10.3.12 A semi-reduced divisor D defined over $\Bbbk$ with Mumford representation $(u(x), v(x))$ is said to be a **prime divisor** if the polynomial $u(x)$ is irreducible over $\Bbbk$. Show that if D is not a prime divisor, then D can be efficiently expressed as a sum of prime divisors by factoring $u(x)$. More precisely, show that if $u(x) = \prod u_i(x)^{c_i}$ is the complete factorisation of $u(x)$ over $\Bbbk$, then $D = \sum c_i \operatorname{div}(u_i(x), y - v_i(x))$ where $v_i(x) = v(x) \bmod u_i(x)$.

10.3.2 Addition and semi-reduction of divisors in Mumford representation

We now present Cantor's algorithm [111][2] for addition of semi-reduced divisors on a hyperelliptic curve C. As above, we take a purely geometric point of view. An alternative, and perhaps more natural, interpretation of Cantor's algorithm is multiplication of ideals in $\Bbbk[x, y] \subset \Bbbk(C)$.

Given two semi-reduced divisors D_1 and D_2 with Mumford representation $(u_1(x), v_1(x))$ and $(u_2(x), v_2(x))$ we want to compute the Mumford representation $(u_3(x), v_3(x))$ of the sum $D_1 + D_2$. Note that we are not yet considering reduction of divisors in the divisor class group. There are two issues that make addition not completely trivial. First, if P is in the support of D_1 and $\iota(P)$ is in the support of D_2 then we remove a suitable multiple of $(P) + (\iota(P))$ from $D_1 + D_2$. Second, we must ensure that the Mumford representation takes multiplicities into account (i.e., so that equation (10.5) holds for $(u_3(x), v_3(x))$).

Example 10.3.13 Let $P = (x_P, y_P)$ on $y^2 + H(x)y = F(x)$ be such that $P \neq \iota(P)$. Let $D_1 = D_2 = (P)$ so that $u_1(x) = u_2(x) = (x - x_P)$ and $v_1(x) = v_2(x) = y_P$. Then $D_1 + D_2 = 2(P)$. The Mumford representation for this divisor has $u_3(x) = (x - x_P)^2$ and $v(x) = y_P + w(x - x_P)$ for some $w \in \overline{\Bbbk}$. To satisfy equation (10.5) one finds that

$$y_P^2 + 2y_P w(x - x_P) + H(x)y_P + wH(x)(x - x_P) - F(x) \equiv 0 \pmod{(x - x_P)^2}.$$

Writing $F(x) \equiv F(x_P) + F'(x_P)(x - x_P) \pmod{(x - x_P)^2}$ and $H(x) \equiv H(x_P) + H'(x_P)(x - x_P) \pmod{(x - x_P)^2}$ gives

$$w = \frac{F'(x_P) - y_P H'(x_P)}{2y_P + H(x_P)},$$

which is defined since $P \neq \iota(P)$.

To help motivate the formula for $v_3(x)$ in Theorem 10.3.14 we now make some observations. First, note that the equation

$$1 = s_1(x)(x - x_P) + s_3(x)(2y_P + H(x))$$

[2] The generalisation of Cantor's algorithm to all hyperelliptic curves was given by Koblitz [311].

has the solution

$$s_3(x) = \frac{1}{2y_P + H(x_P)} \quad \text{and} \quad s_1(x) = -s_3(x)(H'(x_P) + (x - x_P)G(x))$$

where $G(x) = (H(x) - H(x_P) - H'(x_P)(x - x_P))/(x - x_P)^2$. Furthermore, note that

$$v(x) \equiv s_1(x)(x - x_P)y_P + s_3(x)(y_P^2 + F(x)) \pmod{(x - x_P)^2}.$$

The core of Cantor's addition and semi-reduction algorithm is to decide which functions $(x - x_P)$ are needed (and to which powers) to obtain a semi-reduced divisor equivalent to $D_1 + D_2$. The **crucial observation** is that if P is in the support of D_1 and $\iota(P)$ is in the support of D_2 then $(x - x_P) \mid u_1(x)$, $(x - x_P) \mid u_2(x)$ and $v_1(x_P) = -v_2(x_P) - H(x_P)$ and so $(x - x_P) \mid (v_1(x) + v_2(x) + H(x))$. The exact formulae are given in Theorem 10.3.14. The process is called **Cantor's addition algorithm** or **Cantor's composition algorithm**.

Theorem 10.3.14 *Let $(u_1(x), v_1(x))$ and $(u_2(x), v_2(x))$ be Mumford representations of two semi-reduced divisors D_1 and D_2. Let $s(x) = \gcd(u_1(x), u_2(x), v_1(x) + v_2(x) + H(x))$ and let $s_1(x), s_2(x), s_3(x) \in \Bbbk[x]$ be such that*

$$s(x) = s_1(x)u_1(x) + s_2(x)u_2(x) + s_3(x)(v_1(x) + v_2(x) + H(x)).$$

Define $u_3(x) = u_1(x)u_2(x)/s(x)^2$ and

$$v_3(x) = (s_1(x)u_1(x)v_2(x) + s_2(x)u_2(x)v_1(x) + s_3(x)(v_1(x)v_2(x) + F(x)))/s(x). \quad (10.6)$$

Then $u_3(x), v_3(x) \in \Bbbk[x]$ and the Mumford representation of the semi-reduced divisor D equivalent to $D_1 + D_2$ is $(u_3(x), v_3(x))$.

Proof Let $D = D_1 + D_2 - \mathrm{div}(s(x)) \cap \mathbb{A}^2$ so that D is equivalent to $D_1 + D_2$. By the "crucial observation" above, $s(x)$ has a root x_P for some point $P = (x_P, y_P)$ on the curve if and only if P and $\iota(P)$ lie in the supports of D_1 and D_2. Taking multiplicities into account, it follows that D is semi-reduced.

It is immediate that $s(x)^2 \mid u_1(x)u_2(x)$ and so $u_3(x) \in \Bbbk[x]$. It is also immediate that $u_3(x)$ is the correct first component of the Mumford representation of D.

To show $v_3(x) \in \Bbbk[x]$ re-write $v_3(x)$ as

$$v_3 = \frac{v_2(s - s_2u_2 - s_3(v_1 + v_2 + H)) + s_2u_2v_1 + s_3(v_1v_2 + F)}{s} \quad (10.7)$$

$$= v_2 + s_2(v_1 - v_2)(u_2/s) + s_3(F - v_2H - v_2^2)/s. \quad (10.8)$$

Since $s(x) \mid u_2(x)$ and $u_2(x) \mid (F - v_2H - v_2^2)$ the result follows.

We now need the equation

$$(v_1 + v_2 + H)(v_3 - y) \equiv (y - v_1)(y - v_2) \pmod{u_3}. \quad (10.9)$$

This is proved by inserting the definition of v_3 from equation (10.6) to get

$$(v_1 + v_2 + H)(v_3 - y) \equiv -(v_1 + v_2 + H)y + (s_1u_1(v_2^2 + Hv_2 - F) + s_2u_2(v_1^2 + Hv_1 - F) + (v_1v_2 + F)(s_1u_1 + s_2u_2 + s_3(v_1 + v_2 + H))/s \pmod{u_3(x)}.$$

Then using $(y - v_1)(y - v_2) = F - (v_1 + v_2 + H)y + v_1 v_2$ and $u_i \mid (v_i^2 + Hv_i - F)$ for $i = 1, 2$ proves equation (10.9).

Finally, it remains to prove that equation (10.5) holds. We do this by showing that

$$v_P(v(x)^2 + H(x)v(x) - F(x)) \geq v_P(u(x))$$

for all $P = (x_P, y_P) \in \operatorname{Supp}(D)$. Suppose first that $P \neq \iota(P)$ and that $(x - x_P)^e \| u_3(x)$. Then it is sufficient to show that $v_P(y - v_3(x)) \geq e$. This will follow from equation (10.9). First note that $v_P(y - v_3) = v_P((v_1 + v_2 + H)(v_3 - y))$ and that this is at least $\min\{v_P(u_3), v_P((y - v_1)(y - v_2))\}$. Then $v_P(y - v_1) + v_P(y - v_2) \geq v_P(u_1(x)) + v_P(u_2(x)) \geq e$.

Now for the case $P = \iota(P) \in \operatorname{Supp}(D)$. Recall that such points only occur in semi-reduced divisors with multiplicity 1. Since $u_3(x)$ is of minimal degree we know $(x - x_P) \| u_3(x)$. It suffices to show that $v_3(x_P) = y_P$, but this follows from equation (10.8). Without loss of generality, $P \in \operatorname{Supp}(D_2)$ and $P \notin \operatorname{Supp}(D_1)$ (if $P \in \operatorname{Supp}(D_i)$ for both $i = 1, 2$ then $P \notin \operatorname{Supp}(D)$) so $(x - x_P) \nmid s(x)$, $v_2(x_P) = y_P$ and $(u_2/s)(x_P) = 0$. Hence, $v_3(x_P) = v_2(x_P) + 0 = y_P$. □

Exercise 10.3.15 Let $C : y^2 + (x^2 + 2x + 10)y = x^5 + x + 1$ over $\mathbb{F}_{11}$. Let $D_1 = (0, 4) + (6, 4)$ and $D_2 = (0, 4) + (1, 1)$. Determine the Mumford representation of $D_1, D_2, 2D_1, D_1 + D_2$.

We remark that, in practical implementation, one almost always has $\gcd(u_1(x), u_2(x)) = 1$ and so $s(x) = 1$ and the addition algorithm can be simplified. Indeed, it is possible to give explicit formulae for the general cases in the addition algorithm for curves of small genus, we refer to Sections 14.4, 14.5 and 14.6 of [16].

Exercise 10.3.16 Show that the Cantor addition algorithm for semi-reduced divisors of degree $\leq m$ has complexity $O(m^2 M(\log(q))$ bit operations.

10.3.3 Reduction of divisors in Mumford representation

Suppose we have an affine effective divisor D with Mumford representation $(u(x), v(x))$. We wish to obtain an equivalent divisor (affine and effective) whose Mumford representation has $\deg(u(x))$ of low degree. We will show in Theorem 10.3.21 and Lemma 10.4.6 that one can ensure $\deg(u(x)) \leq g$, where g is the genus; we will call such divisors reduced. The idea is to consider

$$u^\dagger(x) = \operatorname{monic}((v(x)^2 + H(x)v(x) - F(x))/u(x)),\ v^\dagger(x) = -v(x) - H(x) \pmod{u^\dagger(x)} \tag{10.10}$$

where $\operatorname{monic}\left(u_0 + u_1 x + \cdots + u_k x^k\right)$ for $u_k \neq 0$ is defined to be $(u_0/u_k) + (u_1/u_k)x + \cdots + x^k$. Obtaining $(u^\dagger(x), v^\dagger(x))$ from $(u(x), v(x))$ is a **Cantor reduction step**. This operation appears in the classical reduction theory of binary quadratic forms.

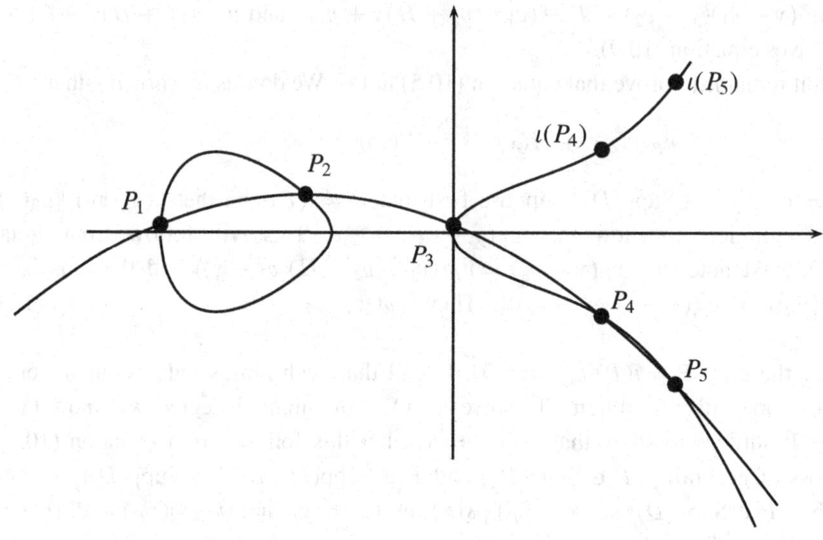

Figure 10.1 Cantor reduction on a hyperelliptic curve.

Lemma 10.3.17 *Let D be an affine effective divisor on a hyperelliptic curve C with Mumford representation $(u(x), v(x))$. Define $(u^\dagger(x), v^\dagger(x))$ as in equation (10.10). Then $(u^\dagger(x), v^\dagger(x))$ is the Mumford representation of a semi-reduced divisor $D^\dagger$ and $D^\dagger \equiv D$ on $C \cap \mathbb{A}^2$.*

Proof One checks that $(u^\dagger(x), v^\dagger(x))$ satisfies condition (10.5) and so there is an associated semi-reduced divisor $D^\dagger$.

Write $D = (P_1) + \cdots + (P_n)$ (where the same point can appear more than once). Then $\mathrm{div}(y - v(x)) \cap \mathbb{A}^2 = (P_1) + \cdots + (P_n) + (P_{n+1}) + \cdots + (P_{n+m})$ for some points $P_{n+1}, \ldots, P_{n+m}$ (not necessarily distinct from the earlier n points, or from each other) and $\mathrm{div}(v(x)^2 + H(x)v(x) - F(x)) \cap \mathbb{A}^2 = \mathrm{div}((y - v(x))(-y - H(x) - v(x))) \cap \mathbb{A}^2 = (P_1) + (\iota(P_1)) + \cdots + (P_{n+m}) + (\iota(P_{n+m}))$. Now, $\mathrm{div}(u^\dagger(x)) = (P_{n+1}) + (\iota(P_{n+1})) + \cdots + (P_{n+m}) + (\iota(P_{n+m}))$. It follows that $D^\dagger = (\iota(P_{n+1})) + \cdots + (\iota(P_{n+m}))$ and that $D = D^\dagger + \mathrm{div}(y - v(x)) \cap \mathbb{A}^2 - \mathrm{div}(u^\dagger(x)) \cap \mathbb{A}^2$. □

Example 10.3.18 Consider

$$C : y^2 = F(x) = x^5 + 2x^4 - 8x^3 + 10x^2 + 40x + 1$$

over $\mathbb{Q}$. Let $P_1 = (-4, 1)$, $P_2 = (-2, 5)$, $P_3 = (0, 1)$ and $D = (P_1) + (P_2) + (P_3)$. The Mumford representation of D is $(u(x), v(x)) = (x(x + 2)(x + 4), -x^2 - 4x + 1)$, which is easily checked by noting that $v(x_{P_i}) = y_{P_i}$ for $1 \le i \le 3$.

To reduce D one sets $u^\dagger(x) = \mathrm{monic}\left((v(x)^2 - F(x))/u(x)\right) = \mathrm{monic}(-x^2 + 5x - 6) = (x - 3)(x - 2)$ and $v^\dagger(x) = -v(x) \pmod{u^\dagger(x)} = 9x - 7$.

One can check that $\mathrm{div}(y - v(x)) = (P_1) + (P_2) + (P_3) + (P_4) + (P_5)$ where $P_4 = (2, -11)$ and $P_5 = (3, -20)$, that $\mathrm{div}(u^\dagger(x)) = (P_4) + (\iota(P_4)) + (P_5) + (\iota(P_5))$ and that $D \equiv \mathrm{div}(u^\dagger(x), y - v^\dagger(x)) \cap \mathbb{A}^2 = (\iota(P_4)) + (\iota(P_5))$. See Figure 10.1 for an illustration.

Exercise 10.3.19 Show that the straight lines $l(x, y)$ and $v(x)$ in the elliptic curve addition law (Definition 7.9.1) correspond to the polynomials $y - v(x)$ and $u^\dagger(x)$ (beware of the double meaning of $v(x)$ here) in a Cantor reduction step.

Lemma 10.3.20 *Let $C : y^2 + H(x)y = F(x)$ and let $(u(x), v(x))$ be the Mumford representation of a semi-reduced divisor D. Write $d_H = \deg(H(x))$, $d_F = \deg(F(x))$, $d_u = \deg(u(x))$ and $d_v = \deg(v(x))$. Let $d = \max\{d_H, \lceil d_F/2 \rceil\}$. Let $(u^\dagger(x), v^\dagger(x))$ be the polynomials arising from a Cantor reduction step.*

1. *If $d_v \geq d$ then $\deg(u^\dagger(x)) \leq d_u - 2$.*
2. *If $d_F \leq 2d - 1$ and $d_u \geq d > d_v$ then $\deg(u^\dagger(x)) \leq d - 1$ (this holds even if $d_H = d$).*
3. *If $d_F = 2d$ and $d_u > d > d_v$ then $\deg(u^\dagger(x)) \leq d - 1$.*

Proof Note that $d_u > d_v$. If $d_v \geq d$ then

$$\begin{aligned}\deg(v(x)^2 + H(x)v(x) - F(x)) &\leq \max\{2d_v, d_H + d_v, d_F\}\\ &\leq \max\{2(d_u - 1), d + (d_u - 1), 2d\}.\end{aligned}$$

Hence, $\deg(u^\dagger(x)) = \deg(v^2 + Hv - F) - d_u \leq \max\{d_u - 2, d - 1, 2d - d_u\} = d_u - 2$.

If $d_F \leq 2d - 1$ and $d_u \geq d > d_v$ then, by a similar argument, $\deg(u^\dagger(x)) \leq 2d - 1 - d_u \leq d - 1$. Finally, if $d_F = 2d$ and $d_u > d > d_v$ then $\deg(v^2 + Hv + F) = 2d$ and $\deg(u^\dagger) = 2d - d_u < d$. □

Theorem 10.3.21 *Suppose $C : y^2 + H(x)y = F(x)$ is a hyperelliptic curve of genus g with $\deg(F(x)) \leq 2g + 1$. Then every semi-reduced divisor is equivalent to a semi-reduced divisor of degree at most g.*

Proof Perform Cantor reduction steps repeatedly. By Lemma 10.3.20 the desired condition will eventually hold. □

Theorem 10.3.21 is an "explicit Riemann–Roch theorem" for hyperelliptic curves with a single point at infinity (also for hyperelliptic curves $y^2 + H(x)y = F(x)$ with two points at infinity but $\deg(F(x)) \leq 2g + 1$) as it shows that every divisor class contains a representative as an affine effective divisor of degree at most g. The general result is completed in Lemma 10.4.6 below.

Definition 10.3.22 Let $C : y^2 + H(x)y = F(x)$ be a hyperelliptic curve of genus g. A semi-reduced divisor on C is **reduced** if its degree is at most g.

Exercise 10.3.23 Let $C : y^2 + H(x)y = F(x)$ be a hyperelliptic curve with $d = \max\{\deg(H), \lceil \deg(F)/2 \rceil\}$. Let $(u(x), v(x))$ be the Mumford representation of a divisor with $\deg(v(x)) < \deg(u(x)) < d$. Show that if $\deg(F(x)) \geq 2d - 1$ and one performs a Cantor reduction step on $(u(x), v(x))$ then the resulting polynomials $(u^\dagger(x), v^\dagger(x))$ are such that $\deg(u^\dagger(x)) \geq d$.

When $\deg(F) = 2d$ then Lemma 10.3.20 is not sufficient to prove an analogue of Theorem 10.3.21. However, one can at least reduce to a divisor of degree $d = g + 1$. It is notable that performing a Cantor reduction step on a divisor of degree d in this case usually yields another divisor of degree d. This phenomena will be discussed in detail in Section 10.4.2.

We now consider the uniqueness of the reduced divisor of Theorem 10.3.21. Lemma 10.3.24 below shows that non-uniqueness can only arise with split or inert models. It follows that there is a unique reduced divisor in every divisor class for hyperelliptic curves with ramified model. For hyperelliptic curves with split or inert model there is not necessarily a unique reduced divisor.

Lemma 10.3.24 *Let $y^2 + H(x)y = F(x)$ be a hyperelliptic curve over $\mathbb{k}$ of genus g. Let $d_H = \deg(H(x))$ and $d_F = \deg(F(x))$. Let D_1 and D_2 be semi-reduced divisors of degree at most g. Assume that $D_1 \neq D_2$ but $D_1 \equiv D_2$. Then $d_F = 2g + 2$ or $d_H = g + 1$.*

Proof First note that $d_H \leq g + 1$ and $d_F \leq 2g + 2$. Let $D_3' = D_1 + \iota_*(D_2)$ so that $D_3' \equiv D_1 - D_2 \equiv 0$ as an affine divisor. Let D_3 be the semi-reduced divisor equivalent to D_3' (i.e., by removing all occurences $(P) + (\iota(P))$). Note that the degree of D_3 is at most $2g$ and that $D_3 \neq 0$. Since $D_3 \equiv 0$ and D_3 is an effective affine divisor we have $D_3 = \operatorname{div}(G(x, y))$ on $C \cap \mathbb{A}^2$ for some non-zero polynomial $G(x, y)$. Without loss of generality, $G(x, y) = a(x) - b(x)y$. Furthermore, $b(x) \neq 0$ (since $\operatorname{div}(a(x))$ is not semi-reduced for any non-constant polynomial $a(x)$).

Exercise 10.1.21 shows that the degree of $\operatorname{div}(a(x) - b(x)y)$ on $C \cap \mathbb{A}^2$ is the degree of $a(x)^2 + H(x)a(x)b(x) - F(x)b(x)^2$. We need this degree to be at most $2g$. This is easily achieved if $d_F \leq 2g$ (in which case $d_H = g + 1$ for the curve to have genus g). However, if $2g + 1 \leq d_F \leq 2g + 2$ then we need either $\deg(a(x)^2) = \deg(F(x)b(x)^2)$ or $\deg(H(x)a(x)b(x)) = \deg(F(x)b(x)^2)$. The former case is only possible if d_F is even (i.e., $d_F = 2g + 2$). If $d_F = 2g + 1$ and $d_H \leq g$ then the latter case implies $\deg(a(x)) \geq g + 1 + \deg(b(x))$ and so $\deg(a(x)^2) > \deg(F(x)b(x)^2)$ and $\deg(G(x, y)) > 2g$. □

For hyperelliptic curves of fixed (small) genus it is possible to give explicit formulae for the general cases of the composition and reduction algorithms. For genus 2 curves this was done by Harley [251] (the basic idea is to formally solve for $u^\dagger(x)$ such that $u^\dagger(x)u(x) = \operatorname{monic}(v(x)^2 + H(x)v(x) - F(x))$ as in equation (10.10)). For extensive discussion and details (and also for non-affine coordinate systems for efficient hyperelliptic arithmetic) we refer to Sections 14.4, 14.5 and 14.6 of [16].

10.4 Addition in the divisor class group

We now show how Cantor's addition and reduction algorithms for divisors on the affine curve can be used to perform arithmetic in the divisor class group of the projective curve. A first remark is that Lemma 10.3.3 implies that every degree zero divisor class on a hyperelliptic curve has a representative of the form $D + n^+(\infty^+) + n^-(\infty^-)$ where D is a semi-reduced (hence, affine and effective) divisor and $n^+, n^- \in \mathbb{Z}$ (necessarily, $\deg(D) + n^+ + n^- = 0$).

10.4.1 Addition of divisor classes on ramified models

On a hyperelliptic curve with ramified model there is only a single point at infinity. We will show in this section that, for such curves, one can compute in the divisor class group using only affine divisors.

We use the Cantor algorithms for addition, semi-reduction and reduction. In general, if one has a semi-reduced divisor D then, by case 1 of Lemma 10.3.20, a reduction step reduces the degree of D by 2. Hence, at most $\deg(D)/2$ reduction steps are possible.

Theorem 10.4.1 *Let C be a hyperelliptic curve with ramified model. Then every degree zero divisor class on C has a unique representative of the form $D - n(\infty)$ where D is semi-reduced and where $0 \le n \le g$.*

Proof Theorem 10.3.21 showed that every affine divisor is equivalent to a semi-reduced divisor D such that $0 \le \deg(D) \le g$. This corresponds to the degree zero divisor $D - n(\infty)$ where $n = \deg(D)$. Uniqueness was proved in Lemma 10.3.24. □

A degree zero divisor of the form $D - n(\infty)$ where D is a semi-reduced divisor of degree n and $0 \le n \le g$ is called **reduced**. We represent D using Mumford representation as $(u(x), v(x))$ and we know that the polynomials $u(x)$ and $v(x)$ are unique. The divisor class is defined over $\Bbbk$ if and only if the corresponding polynomials $u(x), v(x) \in \Bbbk[x]$. Addition of divisors is performed using Cantor's composition and reduction algorithms as above.

Exercise 10.4.2 Let $C : y^2 + H(x)y = F(x)$ be a ramified model of a hyperelliptic curve over $\mathbb{F}_q$. Show that the inverse (also called the negative) of a divisor class on C represented as $(u(x), v(x))$ is $(u(x), -v(x) - (H(x) \pmod{u(x)}))$.

Exercise 10.4.3 Let C be a hyperelliptic curve over $\Bbbk$ of genus g with ramified model. Let D_1 and D_2 be reduced divisors on C. Show that one can compute a reduced divisor representing $D_1 + D_2$ in $O(g^3)$ operations in $\Bbbk$. Show that one can compute $[n]D_1$ in $O(\log(n)g^3)$ operations in $\Bbbk$ (here $[n]D_1$ means the n-fold addition $D_1 + D_1 + \cdots + D_1$).

When the genus is 2 (i.e., $d = 3$) and one adds two reduced divisors (i.e., effective divisors of degree ≤ 2) then the sum is an effective divisor of degree at most 4 and so only one reduction operation is needed to compute the reduced divisor. Similarly, for curves of any genus, at most one reduction operation is needed to compute a reduced divisor equivalent to $D + (P)$ where D is a reduced divisor (such ideas were used by Katagi, Akishita, Kitamura and Takagi [299, 298] to speed up cryptosystems using hyperelliptic curves).

For larger genus there are several variants of the divisor reduction algorithm. In Section 4 of [111], Cantor gives a method that uses higher degree polynomials than $y - v(x)$ and requires fewer reduction steps. In Section VII.2.1 of [61], Gaudry presents a reduction algorithm, essentially due to Lagrange, that is useful when $g \ge 3$. The NUCOMP algorithm (originally proposed by Shanks in the number field setting) is another useful alternative.

We refer to Jacobson and van der Poorten [289] and Section VII.2.2 of [61] for details. It seems that NUCOMP should be used once the genus of the curve exceeds 10 (and possibly even for $g \geq 7$).

Exercise 10.4.4 Let C be a hyperelliptic curve of genus 2 over a field $\mathbb{k}$ with a ramified model. Show that every $\mathbb{k}$-rational divisor class has a unique representative of one of the following four forms:

1. $(P) - (\infty)$ where $P \in C(\mathbb{k})$, including $P = \infty$. Here $u(x) = (x - x_P)$ or $u(x) = 1$.
2. $2(P) - 2(\infty)$ where $P \in C(\mathbb{k})$, excluding points P such that $P = \iota(P)$. Here $u(x) = (x - x_P)^2$.
3. $(P) + (Q) - 2(\infty)$ where $P, Q \in C(\mathbb{k})$ are such that $P, Q \neq \infty$, $P \neq Q$, $P \neq \iota(Q)$. Here $u(x) = (x - x_P)(x - x_Q)$.
4. $(P) + (\sigma(P)) - 2(\infty)$ where $P \in C(\mathbb{K}) - C(\mathbb{k})$ for any quadratic field extension $\mathbb{K}/\mathbb{k}$, $\mathrm{Gal}(\mathbb{K}/\mathbb{k}) = \langle \sigma \rangle$ and $\sigma(P) \notin \{P, \iota(P)\}$. Here $u(x)$ is an irreducible quadratic in $\mathbb{k}[x]$.

Exercise 10.4.5 can come in handy when computing pairings on hyperelliptic curves.

Exercise 10.4.5 Let $D_1 = \mathrm{div}(u_1(x), y - v_1(x)) \cap \mathbb{A}^2$ and $D_2 = \mathrm{div}(u_2(x), y - v_2(x)) \cap \mathbb{A}^2$ be semi-reduced divisors on a hyperelliptic curve with ramified model over $\mathbb{k}$. Write $d_1 = \deg(u_1(x))$ and $d_2 = \deg(u_2(x))$. Let $D_3 = \mathrm{div}(u_3(x), y - v_3(x)) \cap \mathbb{A}^2$ be a semi-reduced divisor of degree d_3 such that $D_3 - d_3(\infty) \equiv D_1 - d_1(\infty) + D_2 - d_2(\infty)$. Show that if $d_2 = d_3$ then $D_1 - d_1(\infty) \equiv D_3 - D_2$.

10.4.2 Addition of divisor classes on split models

This section is rather detailed and can safely be ignored by most readers. It presents results of Paulus and Rück [429] and Galbraith, Harrison and Mireles [202].

Let C be a hyperelliptic curve of genus g over $\mathbb{k}$ with a split model. We have already observed that every degree zero divisor class has a representative of the form $D + n^+(\infty^+) + n^-(\infty^-)$ where D is semi-reduced and $n^+, n^- \in \mathbb{Z}$. Lemma 10.3.20 has shown that we may assume $0 \leq \deg(D) \leq g + 1$. One could consider the divisor to be reduced if this is the case, but this would not be optimal.

The Riemann–Roch theorem implies we should be able to take $\deg(D) \leq g$ but Cantor reduction becomes "stuck" if the input divisor has degree $g + 1$. The following simple trick allows us to reduce to semi-reduced divisors of degree g (and this essentially completes the proof of the "Riemann–Roch theorem" for these curves). Recall the polynomial $G^+(x)$ of degree $d = g + 1$ from Exercise 10.1.22.

Lemma 10.4.6 *Let $y^2 + H(x)y = F(x)$ be a hyperelliptic curve of genus g over $\mathbb{k}$ with split model. Let $u(x), v(x)$ be a Mumford representation such that* $\deg(u(x)) = g + 1$. *Define*

$$v^{\ddagger}(x) = G^+(x) + (v(x) - G^+(x) \pmod{u(x)}) \in \mathbb{k}[x]$$

where we mean that $v(x) - G^+(x)$ is reduced to a polynomial of degree at most $\deg(u(x)) - 1 = g$. *Define*

$$u^\dagger(x) = \text{monic}\left(\frac{v^\ddagger(x)^2 + H(x)v^\ddagger(x) - F(x)}{u(x)}\right) \quad \textit{and}$$
$$v^\dagger(x) = -v^\ddagger(x) - H(x) \pmod{u^\dagger(x)}. \tag{10.11}$$

Then $\deg(u^\dagger(x)) \leq g$ *and*

$$\operatorname{div}(u(x), y - v(x)) \cap \mathbb{A}^2 = \operatorname{div}(u^\dagger(x), y - v^\dagger(x)) \cap \mathbb{A}^2 - \operatorname{div}(u^\dagger(x)) \cap \mathbb{A}^2 + \operatorname{div}(y - v^\ddagger(x)) \cap \mathbb{A}^2. \tag{10.12}$$

Proof Note that $v^\ddagger(x) \equiv v(x) \pmod{u(x)}$ and so $v^\ddagger(x)^2 + H(x)v^\ddagger(x) - F(x) \equiv 0 \pmod{u(x)}$; hence, $u^\dagger(x)$ is a polynomial. The crucial observation is that $\deg(v^\ddagger(x)) = \deg(G^+(x)) = d = g + 1$ and so the leading coefficient of $v^\ddagger(x)$ agrees with that of $G^+(x)$. Hence, $\deg(v^\ddagger(x)^2 + H(x)v^\ddagger(x) - F(x)) \leq 2d - 1 = 2g + 1$ and so $\deg(u^\dagger(x)) \leq 2d - 1 - d = d - 1 = g$ as claimed. To show equation (10.12) it is sufficient to write $u(x)u^\dagger(x) = \prod_{i=1}^{l}(x - x_i)^{e_i}$ and to note that

$$\begin{aligned}\operatorname{div}(y - v^\ddagger(x)) \cap \mathbb{A}^2 &= \sum_{i=1}^{l} e_i(x_i, v^\ddagger(y_i)) \\ &= \operatorname{div}(u(x), y - v(x)) \cap \mathbb{A}^2 + \operatorname{div}(u^\dagger(x), y + H(x) + v^\dagger(x)) \cap \mathbb{A}^2\end{aligned}$$

and that $\operatorname{div}(u^\dagger(x)) = \operatorname{div}(u^\dagger(x), y - v^\dagger(x)) + \operatorname{div}(u^\dagger(x), y + v^\dagger(x) + H(x))$. □

Example 10.4.7 Let $C : y^2 = F(x) = x^6 + 6 = (x-1)(x+1)(x-2)(x+2)(x-3)(x+3)$ over $\mathbb{F}_7$. Then $G^+(x) = x^3$. Consider the divisor $D = (1, 0) + (-1, 0) + (2, 0)$ with Mumford representation $(u(x), v(x)) = ((x-1)(x+1)(x-2), 0)$. Performing standard Cantor reduction gives $u^\dagger(x) = F(x)/u(x) = (x+2)(x-3)(x+3)$, which corresponds to the trivial divisor equivalence $D \equiv (-2, 0) + (3, 0) + (-3, 0)$. Instead, we take $v^\ddagger = G^+(x) + (-G^+(x) \pmod{u(x)}) = x^3 + (-x^3 + u(x)) = u(x)$. Then $u^\dagger(x) = \text{monic}\left((v^\ddagger(x)^2 - F(x))/u(x)\right) = x^2 + 5x + 2$ and $v^\dagger(x) = 3x + 5$. The divisor $\operatorname{div}(u^\dagger(x), y - v^\dagger(x)) \cap \mathbb{A}^2$ is a sum $(P) + (\sigma(P))$ where $P \in C(\mathbb{F}_{7^2}) - C(\mathbb{F}_7)$ and σ is the non-trivial element of $\operatorname{Gal}(\mathbb{F}_{7^2}/\mathbb{F}_7)$.

The operation $(u(x), v(x)) \mapsto (u^\dagger(x), v^\dagger(x))$ of equation (10.11) is called **composition and reduction at infinity**; the motivation for this is given in equation (10.17) below. Some authors call it a **baby step**. This operation can be performed even when $\deg(u(x)) < d$, and we analyse it in the general case in Lemma 10.4.14.

Exercise 10.4.8 Let the notation be as in Lemma 10.4.6. Let $d_u = \deg(u(x))$ so that $v^\ddagger(x)$ agrees with $G^+(x)$ for the leading $d - d_u + 1$ coefficients and so $m = \deg(v^\ddagger(x)^2 +$

$H(x)v^\ddagger(x) - F(x)) \le d + d_u - 1$. Let $d_{u^\dagger} = \deg(u^\dagger(x))$ so that $m = d_u + d_{u^\dagger}$. Show that $v_{\infty^-}(y - v^\ddagger(x)) = -d$, $\mathrm{div}(y - v^\ddagger(x)) =$

$$\mathrm{div}(u(x), y - v(x)) \cap \mathbb{A}^2 + \mathrm{div}(u^\dagger(x), y + H(x) + v^\dagger(x)) \cap \mathbb{A}^2 - (d_u + d_{u^\dagger} - d)(\infty^+) - d(\infty^-) \quad (10.13)$$

and $v_{\infty^+}(y - v^\ddagger(x)) = -(d_u + d_{u^\dagger} - d)$.

We now discuss how to represent divisor classes. An obvious choice is to represent classes as $D - d(\infty^+)$ where D is an affine effective divisor of degree d (see Paulus and Rück [429] for a full discussion of this case). A more natural representation, as pointed out by Galbraith, Harrison and Mireles [202], is to use balanced representations at infinity: in other words, when g is even, to represent divisor classes as $D - (g/2)((\infty^+) + (\infty^-))$ where D is an effective divisor of degree g.

Definition 10.4.9 Let C be a hyperelliptic curve of genus g over $\Bbbk$ with split model. If g is even then define $D_\infty = \frac{g}{2}((\infty^+) + (\infty^-))$. If g is odd then define $D_\infty = \frac{(g+1)}{2}(\infty^+) + \frac{(g-1)}{2}(\infty^-)$.

Let $u(x), v(x) \in \Bbbk[x]$ be the Mumford representation of a semi-reduced divisor $D = \mathrm{div}(u(x), y - v(x)) \cap \mathbb{A}^2$ and $n \in \mathbb{Z}$. Then $\mathrm{div}(u(x), v(x), n)$ denotes the degree zero divisor

$$D + n(\infty^+) + (g - \deg(u(x)) - n)(\infty^-) - D_\infty.$$

If $0 \le \deg(u(x)) \le g$ and $0 \le n \le g - \deg(u(x))$ then such a divisor is called **reduced**.

Uniqueness of this representation is shown in Theorem 10.4.19. When g is odd then one could also represent divisor classes using $D_\infty = (g + 1)/2((\infty^+) + (\infty^-))$. This is applicable in the inert case too. A problem is that this would lead to polynomials of higher degree than necessary in the Mumford representation, and divisor class representatives would no longer necessarily be unique.

It is important to realise that $u(x)$ and $v(x)$ are only used to specify the affine divisor. The values of $v_{\infty^+}(y - v(x))$ and $v_{\infty^-}(y - v(x))$ have no direct influence over the degree zero divisor under consideration. Note also that we allow $n \in \mathbb{Z}$ in Definition 10.4.9 in general, but reduced divisors must have $n \in \mathbb{Z}_{\ge 0}$.

For hyperelliptic curves with split model, $\infty^+, \infty^- \in \Bbbk$ and so a divisor $(u(x), v(x), n)$ is defined over $\Bbbk$ if and only if $u(x), v(x) \in \Bbbk[x]$. Note that when the genus is even then D_∞ is $\Bbbk$-rational even when the model is inert, though in this case a divisor $(u(x), v(x), n)$ with $n \ne 0$ is not defined over $\Bbbk$ if $u(x), v(x) \in \Bbbk[x]$.

We may now consider Cantor's addition algorithm in this setting.

Lemma 10.4.10 *Let C be a hyperelliptic curve over $\Bbbk$ of genus g with split model. Let $\mathrm{div}(u_1(x), v_1(x), n_1)$ and $\mathrm{div}(u_2(x), v_2(x), n_2)$ be degree zero divisors as above. Write $D_i = \mathrm{div}(u_i(x), y - v_i(x)) \cap \mathbb{A}^2$ for $i = 1, 2$ and let $D_3 = \mathrm{div}(u_3(x), y - v_3(x)) \cap \mathbb{A}^2$ be the semi-reduced divisor equivalent to $D_1 + D_2$, and $s(x)$ such that $D_1 + D_2 = D_3 +$*

$\mathrm{div}(s(x)) \cap \mathbb{A}^2$. *Let* $m = g/2$ *when* g *is even and* $m = (g+1)/2$ *otherwise. Then*

$$\mathrm{div}(u_1, v_1, n_1) + \mathrm{div}(u_2, v_2, n_2) \equiv \mathrm{div}(u_3, v_3, n_1 + n_2 + \deg(s) - m). \qquad (10.14)$$

Proof We will show that

$$\mathrm{div}(u_1, v_1, n_1) + \mathrm{div}(u_2, v_2, n_2) = \mathrm{div}(u_3, v_3, n_1 + n_2 + \deg(s) - m) + \mathrm{div}(s(x)).$$

The left-hand side is

$$D_1 + D_2 + (n_1 + n_2 - m)(\infty^+) + (3m - \deg(u_1) - \deg(u_2) - n_1 - n_2)(\infty^-) - D_\infty. \qquad (10.15)$$

Replacing $D_1 + D_2$ by $D_3 + \mathrm{div}(s(x)) \cap \mathbb{A}^2$ has no effect on the coefficients of ∞^+ or ∞^-, but since we actually need $\mathrm{div}(s(x))$ on the whole of C we have $D_1 + D_2 = D_3 + \mathrm{div}(s(x)) + \deg(s(x))((\infty^+) + (\infty^-))$. Writing $\mathrm{div}(u_3, v_3, n_3) = \mathrm{div}(u_3, y - v_3) \cap \mathbb{A}^2 + n_3(\infty^+) + (g - \deg(u_3) - n_3)(\infty^-) - D_\infty$ gives $n_3 = n_1 + n_2 + \deg(s(x)) - m$ as required.

Note that $\deg(u_3) + \deg(s) = \deg(u_1) + \deg(u_2)$, so the coefficient of ∞^- in equation (10.15) is also correct (as it must be). □

We now discuss reduction of divisors on a hyperelliptic curve with a split model. We first consider the basic Cantor reduction step. There are two relevant cases for split models (namely, the first and third cases in Lemma 10.3.20) that we handle as Lemma 10.4.11 and Exercise 10.4.12.

Lemma 10.4.11 *Let* $C : y^2 + H(x)y = F(x)$ *where* $\deg(F(x)) = 2d = 2g + 2$ *be a hyperelliptic curve over* $\Bbbk$ *of genus* g *with split model. Let* $\mathrm{div}(u(x), v(x), n)$ *be a degree zero divisor as in Definition 10.4.9. Let* $(u^\dagger(x), v^\dagger(x))$ *be the polynomials arising from a Cantor reduction step (i.e.,* $u^\dagger(x)$ *and* $v^\dagger(x)$ *are given by equation (10.10)). If* $\deg(v(x)) \geq d = g + 1$ *then set* $n^\dagger = n + \deg(v(x)) - \deg(u^\dagger(x)) = n + (\deg(u(x)) - \deg(u^\dagger(x)))/2$ *and if* $\deg(v(x)) < g + 1 < \deg(u(x))$ *then set* $n^\dagger = n + g + 1 - \deg(u^\dagger(x))$*. Then*

$$\mathrm{div}(u, v, n) = \mathrm{div}(u^\dagger, v^\dagger, n^\dagger) + \mathrm{div}(y - v(x)) - \mathrm{div}(u^\dagger(x)) \qquad (10.16)$$

and $\mathrm{div}(u, v, n) \equiv \mathrm{div}(u^\dagger, v^\dagger, n^\dagger)$.

Proof If $\deg(v(x)) \geq d$ then $\deg(u(x)) + \deg(u^\dagger(x)) = 2\deg(v(x))$ and $v_{\infty^+}(y - v(x)) = v_{\infty^-}(y - v(x)) = -\deg(v(x))$. For equation (10.16) to be satisfied we require

$$n = n^\dagger + v_{\infty^+}(y - v(x)) - v_{\infty^+}(u^\dagger(x))$$

and the formula for $n^\dagger$ follows (the coefficients of ∞^- must also be correct, as the divisors all have degree zero).

In the second case of reduction, we have $\deg(v(x)) < d < \deg(u(x))$ and hence $\deg(u(x)) + \deg(u^\dagger(x)) = 2d$ and $v_{\infty^+}(y - v(x)) = v_{\infty^-}(y - v(x)) = -d$. The formula for $n^\dagger$ follows as in the first case. □

Exercise 10.4.12 Let $C : y^2 + H(x)y = F(x)$ where $\deg(F(x)) < 2d = 2g + 2$ be a hyperelliptic curve over $\Bbbk$ of genus g with split model. Let $\operatorname{div}(u(x), v(x), n)$ be a degree zero divisor as in Definition 10.4.9 such that $d \le \deg(u(x))$. Let $(u^\dagger(x), v^\dagger(x))$ be the polynomials arising from a Cantor reduction step. Show that $\operatorname{div}(u, v, n) \equiv \operatorname{div}(u^\dagger, v^\dagger, n^\dagger)$ where if $\deg(v(x)) < d$ then $n^\dagger = n + g + 1 - \deg(u^\dagger(x))$ and if $\deg(v(x)) \ge d$ then $n^\dagger = n + \deg(v(x)) - \deg(u^\dagger(x))$.

Example 10.4.13 Let $C : y^2 = x^6 + 3$ over $\mathbb{F}_7$. Let $D_1 = \operatorname{div}((x-1)(x-2), 2, 0) = (1, 2) + (2, 2) - D_\infty$ and $D_2 = \operatorname{div}((x-3)(x-4), 2, 0) = (3, 2) + (4, 2) - D_\infty$. Cantor addition gives $D_1 + D_2 = D_3 = \operatorname{div}((x-1)(x-2)(x-3)(x-4), 2, -1)$, which is not a reduced divisor. Applying Cantor reduction to D_3 results in $u^\dagger(x) = (x-5)(x-6)$ and $v^\dagger(x) = -2$ and $n^\dagger = n_3 + (g+1) - \deg(u^\dagger(x)) = -1 + 3 - 2 = 0$. Hence, we have $D_3 \equiv \operatorname{div}((x-5)(x-6), -2, 0)$, which is a reduced divisor.

We now explain the behaviour of a composition at infinity and reduction step.

Lemma 10.4.14 *Let $C : y^2 + H(x)y = F(x)$ where $\deg(F(x)) = 2d = 2g + 2$ be a hyperelliptic curve over $\Bbbk$ of genus g with split model. Let $\operatorname{div}(u(x), v(x), n)$ be a degree zero divisor as in Definition 10.4.9 such that $1 \le \deg(u(x)) \le g + 1$. Let $v^\ddagger(x)$, $u^\dagger(x)$ and $v^\dagger(x)$ be as in Lemma 10.4.6. Let $n^\dagger = n + \deg(u(x)) - (g+1)$ and $D^\dagger = \operatorname{div}(u^\dagger(x), v^\dagger(x), n^\dagger)$. Then*

$$D = D^\dagger + \operatorname{div}(y - v^\ddagger(x)) - \operatorname{div}(u^\dagger(x)).$$

If one uses $G^-(x)$ in Lemma 10.4.6 then $n^\dagger = n + g + 1 - \deg(u^\dagger(x))$.

It follows that if $\deg(u(x)) = g + 1$ then $\operatorname{div}(u, y - v) \cap \mathbb{A}^2 \equiv \operatorname{div}(u^\dagger, y - v^\dagger) \cap \mathbb{A}^2$ and there is no adjustment at infinity (the point of the operation in this case is to lower the degree from $\deg(u(x)) = g + 1$ to $\deg(u^\dagger(x)) \le g$). But if, for example, $\deg(u(x)) = \deg(u^\dagger(x)) = g$ then we have

$$\operatorname{div}(u, y - v) \cap \mathbb{A}^2 - D_\infty \equiv \operatorname{div}(u^\dagger, y - v^\dagger) \cap \mathbb{A}^2 + (\infty^+) - (\infty^-) - D_\infty \qquad (10.17)$$

and so the operation corresponds to addition of D with the degree zero divisor $(\infty^-) - (\infty^+)$. This justifies the name "composition at infinity". To add $(\infty^+) - (\infty^-)$ one should use $G^-(x)$ instead of $G^+(x)$ in Lemma 10.4.6.

Exercise 10.4.15 Prove Lemma 10.4.14.

We can finally put everything together and obtain the main result about reduced divisors on hyperelliptic curves with split model.

Theorem 10.4.16 *Let C be a hyperelliptic curve over $\Bbbk$ of genus g with split model. Then every divisor class contains a reduced divisor as in Definition 10.4.9.*

Proof We have shown the existence of a divisor in the divisor class with semi-reduced affine part, and hence of the form $(u(x), v(x), n)$ with $n \in \mathbb{Z}$. Cantor reduction and composition and reduction at infinity show that we can assume $\deg(u(x)) \le g$. Finally, to show that one may assume $0 \le n \le g - \deg(u(x))$ note that Lemma 10.4.14 maps n to $n^\dagger = n + (g + 1) - \deg(u(x))$. Hence, if $n > g - \deg(u(x))$ then $n > n^\dagger \ge 0$ and continuing the process gives a reduced divisor. On the other hand, if $n < 0$ then using $G^-(x)$ instead one has $n^\dagger = n + g + 1 - \deg(u^\dagger(x)) \le g - \deg(u^\dagger(x))$. □

Exercise 10.4.17 Let $C : y^2 + H(x)y = F(x)$ be a hyperelliptic curve of genus g over $\mathbb{F}_q$ with split model. If g is even, show that the inverse of $\mathrm{div}(u(x), v(x), n)$ is $\mathrm{div}(u(x), -v(x) - (H(x) \pmod{u(x)}), g - \deg(u(x)) - n)$. If g is odd then show that computing the inverse of a divisor may require performing composition and reduction at infinity.

Example 10.4.18 Let $C : y^2 = x^6 + x + 1$ over $\mathbb{F}_{37}$. Then $d = 3$ and $G^+(x) = x^3$. Let $D = (1, 22) + (2, 17) + (\infty^+) - (\infty^-) - D_\infty$, which is represented as $\mathrm{div}(u(x), v(x), 1)$ where $u(x) = (x-1)(x-2) = x^2 + 34x + 2$ and $v(x) = 32x + 27$. This divisor is not reduced. Then $v^\ddagger(x) = x^3 + 25x + 33$ and $\deg(v^\ddagger(x)^2 - F(x)) = 4$. Indeed, $v^\ddagger(x)^2 - F(x) = 13u(x)u^\dagger(x)$ where $u^\dagger(x) = x^2 + 28x + 2$. It follows that $v^\dagger(x) = 7x + 22$ and that

$$\mathrm{div}(u(x), v(x), 1) \equiv \mathrm{div}(u^\dagger(x), v^\dagger(x), 0),$$

which is reduced.

Explicit formulae for all these operations for genus 2 curves of the form $y^2 = x^6 + F_4x^4 + F_3x^3 + F_2x^2 + F_1x + F_0$ have been given by Erickson, Jacobson, Shang, Shen and Stein [184].

Uniqueness of the representation

We have shown that every divisor class for hyperelliptic curves with a split model contains a reduced divisor. We now discuss the uniqueness of this reduced divisor, following Paulus and Rück [429].

Theorem 10.4.19 *Let C be a hyperelliptic curve over $\Bbbk$ of genus g with split model. Then every divisor class has a unique representative of the form*

$$D + n(\infty^+) + (g - \deg(D) - n)(\infty^-) - D_\infty$$

where D is a semi-reduced divisor (hence, affine and effective) and $0 \le n \le g - \deg(D)$.

Proof Existence has already been proved using the reduction algorithms above, so it suffices to prove uniqueness. Hence, suppose

$$\begin{aligned} &D_1 + n_1(\infty^+) + (g - \deg(D_1) - n_1)(\infty^-) - D_\infty \\ &\equiv D_2 + n_2(\infty^+) + (g - \deg(D_2) - n_2)(\infty^-) - D_\infty \end{aligned}$$

with all terms satisfying the conditions of the theorem. Then, taking the difference and adding $\mathrm{div}(u_2(x)) = D_2 + \iota(D_2) - \deg(D_2)((\infty^+) + (\infty^-))$, there is a function $f(x, y)$

such that

$$\mathrm{div}(f(x, y)) = D_1 + \iota(D_2) - (n_2 + \deg(D_2) - n_1)(\infty^+) - (n_1 + \deg(D_1) - n_2)(\infty^-).$$

Since $f(x, y)$ has poles only at infinity it follows that $f(x, y) = a(x) + yb(x)$ where $a(x), b(x) \in \Bbbk[x]$. Now, $0 \le n_i \le n_i + \deg(D_i) \le g$ and so $-g \le v_{\infty^+}(f(x, y)) = -(n_2 + \deg(D_2) - n_1) \le g$ and $-g \le v_{\infty^-}(f(x, y)) = -(n_1 + \deg(D_1) - n_2) \le g$. But $v_{\infty^+}(y) = v_{\infty^-}(y) = -(g + 1)$ and so $b(x) = 0$ and $f(x, y) = a(x)$. But $\mathrm{div}(a(x)) = D + \iota(D) - \deg(a(x))((\infty^+) + (\infty^-))$ and so $D_1 = D_2$, $n_1 + \deg(D_1) - n_2 = n_2 + \deg(D_2) - n_1$ and $n_1 = n_2$. □

Exercise 10.4.20 Let C be a hyperelliptic curve over $\Bbbk$ of genus $g = d - 1$ with split model. Show that $(\infty^+) - (\infty^-)$ is not a principal divisor and that this divisor is represented as $(1, 0, \lceil g/2 \rceil + 1)$.

10.5 Jacobians, Abelian varieties and isogenies

As mentioned in Section 7.8, we can consider $\mathrm{Pic}^0_{\Bbbk}(C)$ as an algebraic group, by considering the **Jacobian variety** J_C of the curve. The fact that the divisor class group is an algebraic group is not immediate from our description of the group operation as an algorithm (rather than a formula).

Indeed, J_C is an Abelian variety (namely, a projective algebraic group). The dimension of the variety J_C is equal to the genus of C. Unfortunately, we do not have space to introduce the theory of Abelian varieties and Jacobians in this book. We remark that the Mumford representation directly gives an affine part of the Jacobian variety of a hyperelliptic curve (see Propositions 1.2 and 1.3 of Mumford [399] for the details).

An explicit description of the Jacobian variety of a curve of genus 2 has been given by Flynn; we refer to Chapter 2 of Cassels and Flynn [115] for details, references and further discussion.

There are several important concepts in the theory of Abelian varieties that are not able to be expressed in terms of divisor class groups.[3] Hence, our treatment of hyperelliptic curves will not be as extensive as the case of elliptic curves. In particular, we do not give a rigorous discussion of isogenies (i.e., morphisms of varieties that are group homomorphisms with finite kernel) for Abelian varieties of dimension $g > 1$. However, we do mention one important result. The Poincaré reducibility theorem (see Theorem 1 of Section 19 (page 173) of Mumford [398]) states that if A is an Abelian variety over $\Bbbk$ and B is an Abelian subvariety of A (i.e., B is a subset of A that is an Abelian variety over $\Bbbk$) then there is an Abelian subvariety $B' \subseteq A$ over $\Bbbk$ such that $B \cap B'$ is finite and $B + B' = A$. It follows that A is isogenous over $\Bbbk$ to $B \times B'$. If an Abelian variety A over $\Bbbk$ has no Abelian subvarieties over $\Bbbk$ then we call it **simple**. An Abelian variety is **absolutely simple** if is has no Abelian subvarieties over $\overline{\Bbbk}$.

[3] There are two reasons for this: first, the divisor class group is merely an abstract group and so does not have the geometric structure necessary for some of these concepts; second, not every Abelian variety is a Jacobian variety.

Despite not discussing isogenies in full generality, it is possible to discuss isogenies that arise from maps between curves purely in terms of divisor class groups. We now give some examples, but first introduce a natural notation.

Definition 10.5.1 Let C be a curve over a field $\Bbbk$ and let $n \in \mathbb{N}$. For $D \in \mathrm{Pic}^0_{\Bbbk}(C)$ define

$$[n]D = D + \cdots + D \quad (n \text{ times}).$$

Indeed, we usually assume that $[n]D$ is a reduced divisor representing the divisor class nD. Define

$$\mathrm{Pic}^0_{\Bbbk}(C)[n] = \{D \in \mathrm{Pic}^0_{\Bbbk}(C) : [n]D = 0\}.$$

Recall from Corollary 8.3.10 that if $\phi : C_1 \to C_2$ is a non-constant rational map (and hence a non-constant morphism) over $\Bbbk$ between two curves then there are corresponding group homomorphisms $\phi^* : \mathrm{Pic}^0_{\Bbbk}(C_2) \to \mathrm{Pic}^0_{\Bbbk}(C_1)$ and $\phi_* : \mathrm{Pic}^0_{\Bbbk}(C_1) \to \mathrm{Pic}^0_{\Bbbk}(C_2)$. Furthermore, by part 5 of Theorem 8.3.8 we have $\phi_*\phi^*(D) = [\deg(\phi)]D$ on $\mathrm{Pic}^0_{\Bbbk}(C_2)$.

In the special case of a non-constant rational map $\phi : C \to E$ over $\Bbbk$ where E is an elliptic curve we can compose with the Abel–Jacobi map $E \to \mathrm{Pic}^0_{\Bbbk}(E)$ of Theorem 7.9.8 given by $P \mapsto (P) - (\mathcal{O}_E)$ to obtain group homomorphisms that we call $\phi^* : E \to \mathrm{Pic}^0_{\Bbbk}(C)$ and $\phi_* : \mathrm{Pic}^0_{\Bbbk}(C) \to E$.

Exercise 10.5.2 Let $\phi : C \to E$ be a non-constant rational map over $\Bbbk$ where E is an elliptic curve over $\Bbbk$. Let $\phi^* : E \to \mathrm{Pic}^0_{\Bbbk}(C)$ and $\phi_* : \mathrm{Pic}^0_{\Bbbk}(C) \to E$ be the group homomorphisms as above. Show that ϕ_* is surjective as a map from $\mathrm{Pic}^0_{\overline{\Bbbk}}(C)$ to $E(\overline{\Bbbk})$ and that the kernel of ϕ^* is contained in $E[\deg(\phi)]$.

If C is a curve of genus 2 and there are two non-constant rational maps $\phi_i : C \to E_i$ over $\overline{\Bbbk}$ for elliptic curves E_1, E_2 then one naturally has a group homomorphism $\phi_{1,*} \times \phi_{2,*} : \mathrm{Pic}^0_{\overline{\Bbbk}}(C) \to E_1(\overline{\Bbbk}) \times E_2(\overline{\Bbbk})$. If $\ker(\phi_{1,*}) \cap \ker(\phi_{2,*})$ is finite then it follows from the theory of Abelian varieties that the Jacobian variety J_C is isogenous to the product $E_1 \times E_2$ of the elliptic curves and one says that J_C is a **split Jacobian**.

Example 10.5.3 Let $C : y^2 = x^6 + 2x^2 + 1$ be a genus 2 curve over $\mathbb{F}_{11}$. Consider the rational maps

$$\phi_1 : C \to E_1 : Y^2 = X^3 + 2X + 1$$

given by $\phi_1(x, y) = (x^2, y)$ and

$$\phi_2 : C \to E_2 : Y^2 = X^3 + 2X^2 + 1$$

given by $\phi_2(x, y) = (1/x^2, y/x^3)$. The two elliptic curves E_1 and E_2 are neither isomorphic or isogenous. One has $\#E_1(\mathbb{F}_{11}) = 16$, $\#E_2(\mathbb{F}_{11}) = 14$ and $\#\mathrm{Pic}^0_{\mathbb{F}_{11}}(C) = 14 \cdot 16$.

It can be shown (this is not trivial) that $\ker(\phi_{1,*}) \cap \ker(\phi_{2,*})$ is finite. Further, since $\deg(\phi_1) = \deg(\phi_2) = 2$ it can be shown that the kernel of $\phi_{1,*} \times \phi_{2,*}$ is contained in $\mathrm{Pic}^0_{\overline{\Bbbk}}(C)[2]$.

The Jacobian of a curve satisfies the following universal property. Let $\phi : C \to A$ be a morphism where A is an Abelian variety. Let $P_0 \in C(\overline{\Bbbk})$ be such that $\phi(P_0) = 0$ and consider the Abel–Jacobi map $\psi : C \to J_C$ (corresponding to $P \mapsto (P) - (P_0)$). Then there is a homomorphism of Abelian varieties $\phi' : J_C \to A$ such that $\phi = \phi' \circ \psi$. Exercise 10.5.4 gives a special case of this universal property.

Exercise 10.5.4 Let $C : y^2 = x^6 + a_2x^4 + a_4x^2 + a_6$ over $\Bbbk$ where $\mathrm{char}(\Bbbk) \neq 2$, and let $\phi(x, y) = (x^2, y)$ be non-constant rational map $\phi : C \to E$ over $\Bbbk$ where E is an elliptic curve. Let $P_0 \in C(\overline{\Bbbk})$ be such that $\phi(P_0) = \mathcal{O}_E$. Show that the composition

$$C(\overline{\Bbbk}) \to \mathrm{Pic}^0_{\overline{\Bbbk}}(C) \to E(\overline{\Bbbk})$$

where the first map is the Abel–Jacobi map $P \mapsto (P) - (P_0)$ and the second map, ϕ_*, is just the original map ϕ.

There is a vast literature on split Jacobians and we are unable to give a full survey. We refer to Sections 4, 5 and 6 of Kuhn [321] or Chapter 14 of Cassels and Flynn [115] for further examples.

10.6 Elements of order n

We now bound the size of the set of elements of order dividing n in the divisor class group of a curve. As with many other results in this chapter, the best approach is via the theory of Abelian varieties. We state Theorem 10.6.1 for general curves, but without proof. The result is immediate for Abelian varieties over $\mathbb{C}$, as they are isomorphic to $\mathbb{C}^g/L$ where L is a rank $2g$ lattice. The elements of order n in $\mathbb{C}^g/L$ are given by the n^{2g} points in $\frac{1}{n}L/L$.

Theorem 10.6.1 *Let C be a curve of genus g over $\Bbbk$ and let $n \in \mathbb{N}$. If $\mathrm{char}(\Bbbk) = 0$ or $\gcd(n, \mathrm{char}(\Bbbk)) = 1$ then $\#\mathrm{Pic}^0_{\overline{\Bbbk}}(C)[n] = n^{2g}$. If $\mathrm{char}(\Bbbk) = p > 0$ then $\#\mathrm{Pic}^0_{\overline{\Bbbk}}(C)[p] = p^e$ where $0 \leq e \leq g$.*

Proof See Theorem 4 of Section 7 of Mumford [398]. □

10.7 Hyperelliptic curves over finite fields

There are a finite number of points on a curve C of genus g over a finite field $\mathbb{F}_q$. There are also finitely many possible values for the Mumford representation of a reduced divisor on a hyperelliptic curve over a finite field. Hence, the divisor class group $\mathrm{Pic}^0_{\mathbb{F}_q}(C)$ of a curve over a finite field is a finite group. Since the affine part of a reduced divisor is a sum of at most g points (possibly defined over a field extension of degree bounded by g) it is not surprising that there is a connection between $\{\#C(\mathbb{F}_{q^i}) : 1 \leq i \leq g\}$ and $\#\mathrm{Pic}^0_{\mathbb{F}_q}(C)$. Indeed, there is also a connection between $\{\#\mathrm{Pic}^0_{\mathbb{F}_{q^i}}(C) : 1 \leq i \leq g\}$ and $\#C(\mathbb{F}_q)$. The aim of this section is to describe these connections. We also give some important bounds on these numbers (analogous to the Hasse bound for elliptic curves). Most results are presented for general curves (i.e., not only hyperelliptic curves).

One of the most important results in the theory of curves over finite fields is the following theorem of Hasse and Weil. The condition that the roots of $L(t)$ have absolute value $\sqrt{q}$ can be interpreted as an analogue of the Riemann hypothesis. This result gives precise bounds on the number of points on curves and divisor class groups over finite fields.

Theorem 10.7.1 *(Hasse–Weil) Let C be a curve of genus g over $\mathbb{F}_q$. There exists a polynomial $L(t) \in \mathbb{Z}[t]$ of degree $2g$ with the following properties:*

1. *$L(1) = \#\mathrm{Pic}^0_{\mathbb{F}_q}(C)$.*
2. *One can write $L(t) = \prod_{i=1}^{2g}(1 - \alpha_i t)$ with $\alpha_i \in \mathbb{C}$ such that $\alpha_{g+i} = \overline{\alpha_i}$ (this is complex conjugation) and $|\alpha_i| = \sqrt{q}$ for $1 \le i \le g$.*
3. *$L(t) = q^g t^{2g} L(1/(qt))$ and so*
$$L(t) = 1 + a_1 t + \cdots + a_{g-1}t^{g-1} + a_g t^g + q a_{g-1} t^{g+1} + \cdots + q^{g-1} a_1 t^{2g-1} + q^g t^{2g}.$$
4. *For $n \in \mathbb{N}$ define $L_n(t) = \prod_{i=1}^{2g}(1 - \alpha_i^n t)$. Then $\#\mathrm{Pic}^0_{\mathbb{F}_{q^n}}(C) = L_n(1)$.*

Proof The polynomial $L(t)$ is the numerator of the zeta function of C. For details see Section V.1 of Stichtenoth [529], especially Theorem V.1.15. The proof that $|\alpha_i| = \sqrt{q}$ for all $1 \le i \le 2g$ is Theorem V.2.1 of Stichtenoth [529].

A proof of some parts of this result in a special case is given in Exercise 10.7.14. □

Exercise 10.7.2 Show that part 3 of Theorem 10.7.1 follows immediately from part 2.

Definition 10.7.3 The polynomial $L(t)$ of Theorem 10.7.1 is called the ***L*-polynomial** of the curve C over $\mathbb{F}_q$.

Theorem 10.7.4 *(Schmidt) Let C be a curve of genus g over $\mathbb{F}_q$. There there exists a divisor D on C of degree 1 that is defined over $\mathbb{F}_q$.*

We stress that this result does not prove that C has a point defined over $\mathbb{F}_q$ (though when q is large compared with the genus, existence of a point in $C(\mathbb{F}_q)$ will follow by the Weil bounds). The result implies that even a curve with no points defined over $\mathbb{F}_q$ does have a divisor of degree 1 (hence, not an effective divisor) that is defined over $\mathbb{F}_q$.

Proof See Corollary V.1.11 of Stichtenoth [529]. □

We now describe the precise connection between the roots α_i of the polynomial $L(t)$ (corresponding to $\mathrm{Pic}^0_{\mathbb{F}_q}(C)$) and $\#C(\mathbb{F}_{q^n})$ for $n \in \mathbb{N}$.

Theorem 10.7.5 *Let C be a curve of genus g over $\mathbb{F}_q$ and let $\alpha_i \in \mathbb{C}$ for $1 \le i \le 2g$ be as in Theorem 10.7.1. Let $n \in \mathbb{N}$. Then*

$$\#C(\mathbb{F}_{q^n}) = q^n + 1 - \sum_{i=1}^{2g} \alpha_i^n. \tag{10.18}$$

Proof See Corollary V.1.16 of Stichtenoth [529]. □

Equation (10.18) can be read in two ways. On the one hand, it shows that given $L(t)$ one can determine $\#C(\mathbb{F}_{q^n})$. On the other hand, it shows that if one knows $\#C(\mathbb{F}_{q^n})$ for $1 \le n \le g$ then one has g non-linear equations in the g variables $\alpha_1, \dots, \alpha_g$ (there are only g variables since $\alpha_{i+g} = q/\alpha_i$ for $1 \le i \le g$). The following result shows that one can therefore deduce the coefficients $a_1, \dots, a_g$ giving the polynomial $L(t)$.

Lemma 10.7.6 *(Newton's identities) Let $\alpha_1, \dots, \alpha_{2g} \in \mathbb{C}$ and define $t_n = \sum_{i=1}^{2g} \alpha_i^n$. Let $a_1, \dots, a_{2g}$ be such that $\prod_{i=1}^{2g}(x - \alpha_i) = x^{2g} + a_1 x^{2g-1} + \cdots + a_{2g}$. Then, for $1 \le n \le 2g$,*

$$na_n = -t_n - \sum_{i=1}^{n-1} a_{n-i} t_i.$$

In particular, $a_1 = -t_1$ and $a_2 = (t_1^2 - t_2)/2$.

Exercise 10.7.7★ Prove Lemma 10.7.6.

Exercise 10.7.8 Suppose C is a genus 3 curve over $\mathbb{F}_7$ such that $\#C(\mathbb{F}_7) = 8, \#C(\mathbb{F}_{7^2}) = 92, \#C(\mathbb{F}_{7^3}) = 344$. Determine $L(t)$ and hence $\#\mathrm{Pic}^0_{\mathbb{F}_7}(C)$. (One can take $y^2 = x^7 + x + 1$ for C.)

Exercise 10.7.9 **(Weil bounds)** Let C be a curve of genus g over $\mathbb{F}_q$. Use Theorem 10.7.1 and Theorem 10.7.5 to show that

$$|\#C(\mathbb{F}_{q^n}) - (q^n + 1)| \le 2g\sqrt{q^n}$$

and

$$(\sqrt{q^n} - 1)^{2g} \le \#\mathrm{Pic}^0_{\mathbb{F}_{q^n}}(C) \le (\sqrt{q^n} + 1)^{2g}.$$

More precise bounds on $\#C(\mathbb{F}_q)$ are known; we refer to Section V.3 of Stichtenoth [529] for discussion and references.

Consider the q-power Frobenius map $\pi : C \to C$ given by $\pi(x, y) = (x^q, y^q)$. This map induces a morphism $\pi : J_C \to J_C$ (indeed, an isogeny of Abelian varieties) where J_C is the Jacobian variety of C. By considering the action of π on the Tate module (the Tate module of an Abelian variety is defined in the analogous way to elliptic curves, see Section 19 of Mumford [398]) it can be shown that π satisfies a characteristic equation given by a monic polynomial $P(T) \in \mathbb{Z}[T]$ of degree $2g$. It can further be shown that $P(T) = T^{2g} L(1/T)$ (we refer to Section 21 of [398], especially the subsection entitled "Application II: The Riemann Hypothesis").

Definition 10.7.10 Let C be a curve over $\mathbb{F}_q$. The **characteristic polynomial of Frobenius** is the polynomial $P(T) = T^{2g} L(1/T)$.

The Frobenius map $\pi : C \to C$ also induces the map $\pi_* : \mathrm{Pic}^0_{\overline{\mathbb{F}}_q}(C) \to \mathrm{Pic}^0_{\overline{\mathbb{F}}_q}(C)$, and we abuse notation by calling it π as well. If D is any divisor representing a divisor class in

$\mathrm{Pic}^0_{\mathbb{F}_q}(C)$ then $P(\pi)D \equiv 0$. In other words, if $P(T) = T^{2g} + a_1 T^{2g-1} + \cdots + a_1 q^{g-1} T + q^g$ then

$$\pi^{2g}(D) + [a_1]\pi^{2g-1}(D) + \cdots + [a_1 q^{g-1}]\pi(D) + [q^g]D \equiv 0 \qquad (10.19)$$

where the notation $[n]D$ is from Definition 10.5.1.

Exercise 10.7.11 Let C be a curve over $\mathbb{F}_q$ and D a reduced divisor on C over $\overline{\mathbb{F}}_q$ with Mumford representation $(u(x), v(x))$. Let π be the q-power Frobenius map on C. For a polynomial $u(x) = \sum_{i=0}^{d} u_i x^i$ define $u^{(q)}(x) = \sum_{i=0}^{d} u_i^q x^i$. Show that the Mumford representation of $\pi_*(D)$ is $(u^{(q)}(x), v^{(q)}(x))$.

Example 10.7.12 (Koblitz [298]) Let $a \in \{0, 1\}$ and consider the genus 2 curve $C_a : y^2 + xy = x^5 + ax^2 + 1$ over $\mathbb{F}_2$. One can verify that $\#C_0(\mathbb{F}_2) = 4$, $\#C_1(\mathbb{F}_2) = 2$ and $\#C_0(\mathbb{F}_{2^2}) = \#C_1(\mathbb{F}_{2^2}) = 4$. Hence, the characteristic polynomial of Frobenius is $P(T) = T^4 + (-1)^a T^3 + 2(-1)^a T + 4$. One can determine $\#\mathrm{Pic}^0_{\mathbb{F}_{2^n}}(C_a)$ for any $n \in \mathbb{N}$. If n is composite and $m \mid n$ one has $\#\mathrm{Pic}^0_{\mathbb{F}_{2^m}}(C_a) \mid \#\mathrm{Pic}^0_{\mathbb{F}_{2^n}}(C_a)$. For cryptographic applications one would like $\#\mathrm{Pic}^0_{\mathbb{F}_{2^n}}(C_a)/\#\mathrm{Pic}^0_{\mathbb{F}_2}(C_a)$ to be prime, so restrict attention to primes values for n. For example, taking $n = 113$ and $a = 1$ gives group order $2 \cdot r$ where $r = 539 \cdots 381$ is a 225-bit prime.

If $D \in \mathrm{Pic}^0_{\mathbb{F}_{2^n}}(C_1)$ then $\pi^4(D) - \pi^3(D) - [2]\pi(D) + [4]D \equiv 0$ where π is the map induced on $\mathrm{Pic}^0_{\mathbb{F}_{2^n}}(C_1)$ from the 2-power Frobenius map $\pi(x, y) = (x^2, y^2)$ on C.

A major result, whose proof is beyond the scope of this book, is Tate's isogeny theorem.

Theorem 10.7.13 *(Tate) Let A and B be Abelian varieties over a field $\mathbb{F}_q$. Then A is $\mathbb{F}_q$-isogenous to B if and only if $P_A(T) = P_B(T)$. Similarly, A is $\mathbb{F}_q$-isogenous to an Abelian subvariety of B if and only if $P_A(T) \mid P_B(T)$.*

Proof See [540]. □

Exercise 10.7.14 gives a direct proof of Theorems 10.7.1 and 10.7.5 for genus 2 curves with ramified model.

Exercise 10.7.14★ Let q be an odd prime power. Let $F(x) \in \mathbb{F}_q[x]$ be square-free and of degree 5. Then $C : y^2 = F(x)$ is a hyperelliptic curve over $\mathbb{F}_q$ of genus 2 with a ramified model. For $n = 1, 2$ let $N_n = \#C(\mathbb{F}_{q^n})$ and define $t_n = q^n + 1 - N_n$ so that $N_n = q^n + 1 - t_n$. Define $a_1 = -t_1$ and $a_2 = (t_1^2 - t_2)/2$. Show, using direct calculation and Exercise 10.4.4, that $\mathrm{Pic}^0_{\mathbb{F}_q}(C)$ has order $q^2 + a_1(q + 1) + a_2 + 1$.

10.8 Supersingular curves

Recall from Theorem 10.6.1 that if C is a curve of genus g over a field $\Bbbk$ of characteristic p then $\#\mathrm{Pic}^0_{\Bbbk}(C)[p] \le p^g$.

Definition 10.8.1 Let $\Bbbk$ be a field such that $\mathrm{char}(\Bbbk) = p > 0$ and let C be a curve of genus g over $\Bbbk$. The **p-rank** of C is the integer $0 \le r \le g$ such that $\#\mathrm{Pic}^0_{\overline{\Bbbk}}(C)[p] = p^r$.

An Abelian variety of dimension g over $\mathbb{F}_q$ is defined to be **supersingular** if it is isogenous over $\overline{\mathbb{F}}_q$ to E^g where E is a supersingular elliptic curve over $\overline{\mathbb{F}}_q$. A curve C over $\mathbb{F}_q$ is **supersingular** if J_C is a supersingular Abelian variety. It follows that the p-rank of a supersingular Abelian variety over $\mathbb{F}_{p^n}$ is zero. The converse is not true (i.e., p-rank zero does not imply supersingular) when the dimension is 3 or more; see Example 10.8.8). If the p-rank of a dimension g Abelian variety A over $\mathbb{F}_{p^n}$ is g then A is said to be **ordinary**.

Lemma 10.8.2 *Suppose A is a supersingular Abelian variety over $\mathbb{F}_q$ and write $P_A(T)$ for the characteristic polynomial of Frobenius on A. The roots α of $P_A(T)$ are such that $\alpha/\sqrt{q}$ is a root of unity.*

Proof Since the isogeny to E^g is defined over some finite extension $\mathbb{F}_{q^n}$ it follows from part 4 of Theorem 9.11.2 that $\alpha^n/\sqrt{q^n}$ is a root of unity. Hence, $\alpha/\sqrt{q}$ is a root of unity. □

The converse of Lemma 10.8.2 follows from the Tate isogeny theorem.

Example 10.8.3 Let $C : y^2 + y = x^5$ over $\mathbb{F}_2$. One can check that $\#C(\mathbb{F}_2) = 3$ and $\#C(\mathbb{F}_{2^2}) = 5$ and so the characteristic polynomial of the 2-power Frobenius is $P(T) = T^4 + 4 = (T^2 + 2T + 2)(T^2 - 2T + 2)$. It follows from Theorem 10.7.13 (Tate's isogeny theorem) that J_C is isogenous to $E_1 \times E_2$ where E_1 and E_2 are supersingular curves over $\mathbb{F}_2$. The characteristic polynomial of the 2^2-power Frobenius can be shown to be $T^4 + 8T^2 + 16 = (T^2 + 4)^2$ and it follows that J_C is isogenous over $\mathbb{F}_{2^2}$ to the square of a supersingular elliptic curve. Hence, C is a supersingular curve.

Note that the endomorphism ring of J_C is non-commutative since the map $\phi(x, y) = (\zeta_5 x, y)$, where $\zeta_5 \in \mathbb{F}_{2^4}$ is a root of $z^4 + z^3 + z^2 + z + 1 = 0$, does not commute with the 2-power Frobenius map.

Exercise 10.8.4★ Show that if C is a supersingular curve over $\mathbb{F}_q$ of genus 2 then $\#\mathrm{Pic}^0_{\mathbb{F}_q}(C) \mid (q^k - 1)$ for some $1 \le k \le 12$.

The following result shows that computing the p-rank and determining supersingularity are easy when $P(T)$ is known.

Theorem 10.8.5 *Let A be an Abelian variety of dimension g over $\mathbb{F}_{p^n}$ with characteristic polynomial of Frobenius $P(T) = T^{2g} + a_1 T^{2g-1} + \cdots + a_g T^g + \cdots + p^{ng}$.*

1. *The p-rank of A is the smallest integer $0 \le r \le g$ such that $p \mid a_i$ for all $1 \le i \le g - r$. (In other words, the p-rank is zero if $p \mid a_i$ for all $1 \le i \le g$ and the p-rank is g if $p \nmid a_1$.)*
2. *A is supersingular if and only if*

$$p^{\lceil in/2 \rceil} \mid a_i \quad \text{for all} \quad 1 \le i \le g.$$

Proof Part 1 is Satz 1 of Stichtenoth [528]. Part 2 is Proposition 1 of Stichtenoth and Xing [530]. □

We refer to Yui [572] for a survey of the Cartier–Manin matrix and related criteria for the p-rank.

Exercise 10.8.6 Let A be an Abelian variety of dimension 2 over $\mathbb{F}_p$ that has p-rank zero. Show that A is supersingular.

In fact, the result of Exercise 10.8.6 holds when $\mathbb{F}_p$ is replaced by any finite field; see page 9 of Li and Oort [348].

Exercise 10.8.7 Let $C : y^2 + y = F(x)$ over $\mathbb{F}_{2^n}$ where $\deg(F(x)) = 5$ be a genus 2 hyperelliptic curve. Show that C has 2-rank zero (and hence is supersingular).

Example 10.8.8 shows that, once the genus is at least 3, p-rank zero does not imply supersingularity.

Example 10.8.8 Define $C : y^2 + y = x^7$ over $\mathbb{F}_2$. Then $P(T) = T^6 - 2T^3 + 2^3$ and so by Theorem 10.8.5 the 2-rank of C is zero but C is not supersingular.

Example 10.8.9 (Hasse/Hasse-Davenport/Duursma [171]) Let $p > 2$ be prime and $C : y^2 = x^p - x + 1$ over $\mathbb{F}_p$. One can verify that C is non-singular and the genus of C is $(p-1)/2$. It is shown in [171] that, over $\mathbb{F}_{p^2}$, $L(T) = \Phi_p((\frac{-1}{p})pT)$ where $\Phi_p(T)$ is the p-th cyclotomic polynomial. It follows that the roots of $P(T)$ are roots of unity and so C is supersingular.

PART III

EXPONENTIATION, FACTORING AND DISCRETE LOGARITHMS

11

Basic algorithms for algebraic groups

In Section 4.1 a number of basic computational tasks for an algebraic group G were listed. Some of these topics have been discussed already, especially providing efficient group operations and compact representations for group elements. But some other topics (such as efficient exponentiation, generating random elements in G and hashing from or into G) require further attention. The goal of this chapter is to briefly give some details about these tasks for the algebraic groups of most interest in the book.

The main goal of the chapter is to discuss exponentiation and multi-exponentiation. These operations are crucial for efficient discrete logarithm cryptography and there are a number of techniques available for specific groups that give performance improvements.

It is beyond the scope of this book to present a recipe for the best possible exponentiation algorithm in a specific application. Instead, our focus is on explaining the mathematical ideas that are used. For an "implementors guide" in the case of elliptic curves we refer to Bernstein and Lange [51].

Let G be a group (written in multiplicative notation). Given $g \in G$ and $a \in \mathbb{N}$ we wish to compute g^a. We assume in this chapter that a is a randomly chosen integer of size approximately the same as the order of g, and so a varies between executions of the exponentiation algorithm. If g does not change between executions of the algorithm then we call it a **fixed base** and otherwise it is a **variable base**.

As mentioned in Section 2.8, there is a significant difference between the cases where g is fixed (and one is computing g^a repeatedly for different values of a) and the case where both g and a vary. Section 2.8 already briefly mentioned addition chains and sliding window methods. The literature on addition chains is enormous and we do not delve further into this topic. Window methods date back to Brauer in 1939 and sliding windows to Thurber; we refer to Bernstein's excellent survey [41] for historical details. Other references for fast exponentiation are Chapters 9 and 15 of [16], Chapter 3 of [248] and Sections 14.6 and 14.7 of [376].

11.1 Efficient exponentiation using signed exponents

In certain algebraic groups, computing the inverse of a group element is much more efficient than a general group operation. For example, Exercise 6.3.1 and Lemma 6.3.12 show that inversion in $G_{q,2}$ and $\mathbb{T}_2(\mathbb{F}_q)$ is easy. Similarly, inversion in elliptic and hyperelliptic curve groups is easy (see Section 9.1 and Exercises 10.4.2 and 10.4.17). Hence, one can exploit inversion when computing exponentiation and it is desirable to consider signed expansions for exponents.

Signed expansions and addition–subtraction chains have a long history.[1] Morain and Olivos [394] realised that, since inversion is easy for elliptic curve groups, signed expansions are natural in this context.

11.1.1 Non-adjacent form

We now discuss the non-adjacent form (NAF) of an integer a. This is the best signed expansion in the sense that it has the minimal number of non-zero coefficients and can be computed efficiently. The non-adjacent form was discussed by Reitwiesner [449] (where it is called "property M"). Reitwiesner proved that the NAF is unique, has minimal weight among binary expansions with coefficients in $\{-1, 0, 1\}$ and he gave an algorithm to compute the NAF of an integer. These results have been re-discovered and simplified numerous times (we refer to Section IV.2.4 of [60] and Section 5 of [490] for references).

Definition 11.1.1 Let $a \in \mathbb{N}$. A representation $a = \sum_{i=0}^{l} a_i 2^i$ is a **non-adjacent form** or **NAF** if $a_i \in \{-1, 0, 1\}$ for all $0 \le i \le l$ and $a_i a_{i+1} = 0$ for all $0 \le i < l$. If $a_l \neq 0$ then the **length** of the NAF is $l + 1$.

One can transform an integer a into NAF representation using Algorithm 6. This is a "right-to-left" algorithm in the sense that it processes the least significant bits first. We define the operator $a \pmod{s\ 2m}$ to be reduction of a modulo $2m$ to the range $\{-m + 1, \ldots, -1, 0, 1, \ldots, m\}$. In particular, if a is odd then a (mods 4) $\in \{-1, 1\}$. An alternative right-to-left algorithm is given in the proof of Theorem 11.1.12.

Algorithm 6 Convert an integer to non-adjacent form

INPUT: $a \in \mathbb{N}$
OUTPUT: $(a_l \ldots a_0)$

```
1: i = 0
2: while a ≠ 0 do
3:     if a even then
4:         a_i = 0
5:     else
6:         a_i = a (mods 4)
7:     end if
8:     a = (a − a_i)/2
9:     i = i + 1
10: end while
11: return a_l ... a_0
```

[1] Reitwiesner's long paper [449] suggests signed expansions as a way to achieve faster arithmetic (e.g., multiplication and division), but does not discuss exponentiation. Brickell, in 1982, seems to have been the first to suggest using negative powers of g to speed up the computation of g^a; this was in the context of computing $m^e \pmod{N}$ in RSA and required the precomputation of $m^{-1} \pmod{N}$.

Exercise 11.1.2 Prove that Algorithm 6 outputs a NAF.

Example 11.1.3 We compute the NAF representation of $a = 91$. Since $91 \equiv 3 \pmod 4$ the first digit is -1, which we denote as $\bar{1}$. Note that $2^2 \| 92$ so the next digit is 0. Now, $92/4 = 23 \equiv 3 \pmod 4$ and the next digit is $\bar{1}$. Since $2^3 \| 24$ the next 2 digits are 0. Continuing one finds the expansion to be $10\bar{1}00\bar{1}0\bar{1}$.

Lemma 11.1.4 shows that a simple way to compute a NAF of an integer a is to compute the binary representation of $3a$, subtract the binary representation of a (writing the result in signed binary expansion; in other words, performing the subtraction without carries) and discard the least significant bit. We write this as $((3a) - a)/2$.

Lemma 11.1.4 *Let $a \in \mathbb{N}$. Then the signed binary expansion $((3a) - (a))/2$ is in non-adjacent form.*

Proof Write a in binary as $(a_l \dots a_0)_2$ and write $3a$ in binary as $(b_{l+2} \dots b_0)_2$. Set $a_{-1} = a_{l+1} = a_{l+2} = 0$ and $c_{-1} = 0$. Then $b_i = a_i + a_{i-1} + c_{i-1} - 2c_i$ where $c_i = \lfloor (a_i + a_{i-1} + c_{i-1})/2 \rfloor \in \{0, 1\}$ is the carry from the ith addition.

Now consider the signed expansion $s_i = b_i - a_i \in \{-1, 0, 1\}$. In other words, $s_i = a_{i-1} + c_{i-1} - 2c_i$. Clearly, $b_0 = a_0$ and so $s_0 = 0$. We show that $s_i s_{i+1} = 0$ for $1 \le i \le l + 1$. Suppose i is such that $s_i \ne 0$. Since $a_{i-1} + c_{i-1} \in \{0, 1, 2\}$ and $a_{i-1} + c_{i-1} \equiv s_i \pmod 2$ it follows that $a_{i-1} + c_{i-1} = 1$. This then implies that $c_i = \lfloor (1 + a_i)/2 \rfloor = a_i$. Hence, $c_{i+1} = a_i$ and $s_{i+1} = a_i + c_i - 2c_{i+1} = 0$. □

Example 11.1.5 Taking $a = 91$ again, we have $3 \cdot 91 = 273 = (100010001)_2$. Computing $(3a) - a$ is

$$100010001 - 1011011 = 10\bar{1}00\bar{1}01\bar{1}0.$$

Exercise 11.1.6 Compute NAFs for $a = 100, 201, 302$ and 403.

We now state and prove some properties of NAFs.

Exercise 11.1.7 Show that if $a_l \dots a_0$ is a NAF of a then $(-a_l) \dots (-a_0)$ is a NAF of $-a$.

Lemma 11.1.8 *The NAF representation of $a \in \mathbb{Z}$ is unique.*

Proof Without loss of generality we may assume $a > 1$. Let $a \in \mathbb{N}$ be the smallest positive integer such that a has two (or more) distinct representations as a NAF, call them $\sum_{i=0}^{l} a_i 2^i$ and $\sum_{i=0}^{l'} a_i' 2^i$. If a is even then $a_0 = a_0' = 0$ and so we have two distinct NAF representations of $a/2$, which contradicts the minimality of a. If a is odd then $a \equiv \pm 1 \pmod 4$ and so $a_0 = a_0'$ and $a_1 = a_1' = 0$. Hence, we obtain two distinct NAF representations of $(a - a_0)/4 < a$, which again contradicts the minimality of a. □

Exercise 11.1.9★ Let $a \in \mathbb{N}$. Show that a has a length $l + 1$ NAF representation if and only if $2^l - d_l \le a \le 2^l + d_l$ where

$$d_l = \begin{cases} (2^l - 2)/3 & \text{if } l \text{ is odd} \\ (2^l - 1)/3 & \text{if } l \text{ is even.} \end{cases}$$

Also show that if $a > 0$ then $a_l = 1$ and prove that the length of a NAF is at most one more than the length of the binary expansion of a.

Definition 11.1.10 Let $\mathcal{D} \subset \mathbb{Z}$ be such that $0 \in \mathcal{D}$. The **weight** of a representation $a = \sum_{i=0}^{l} a_i 2^i$ where $a_i \in \mathcal{D}$ is the number of values $0 \le i \le l$ such that $a_i \ne 0$. The weight of a is denoted weight(a). The **density** of the representation is weight$(a)/(l+1)$.

Exercise 11.1.11 Show that if $a \in \mathbb{N}$ is uniformly chosen in $2^l \le a < 2^{l+1}$ and represented using the standard binary expansion then the expected value of the weight is $(l+1)/2$ and therefore the expected value of the density is $1/2$.

Theorem 11.1.12 *The NAF of an integer $a \in \mathbb{N}$ has minimal weight among all signed expansions $a = \sum_{i=0}^{l} a_i 2^i$ where $a_i \in \{-1, 0, 1\}$.*

Proof Let $a = \sum_{i=0}^{l} a_i 2^i$ where $a_i \in \{-1, 0, 1\}$ be any signed expansion of a. Perform the following string re-writing process from right to left (i.e., starting with a_0). If $a_i = 0$ or $a_{i+1} = 0$ then do nothing. Otherwise (i.e., $a_i \ne 0$ and $a_{i+1} \ne 0$) there exists an integer $k \ge 1$ such that the sequence $a_{i+k} a_{i+k-1} \ldots a_i a_{i-1}$ is of the form

$$01\ldots10, \qquad \bar{1}1\ldots10, \qquad 1\bar{1}\ldots\bar{1}0, \qquad 0\bar{1}\ldots\bar{1}0.$$

In each case, replace the pattern with the following

$$10\ldots0\bar{1}0, \qquad 0\ldots0\bar{1}0, \qquad 0\ldots010, \qquad \bar{1}0\ldots010.$$

In each case, the resulting substring has weight less than or equal to the weight of the previous substring and is in non-adjacent form (at least up to $a_{i+k-1}a_{i+k} = 0$). Continuing the process therefore yields a NAF expansion of a of weight less than or equal to the weight of the original signed expansion. □

Example 11.1.13 We re-compute the NAF representation of $a = 91$, using the method in the proof of Theorem 11.1.12. First, note that the binary expansion of 91 is 1011011. One replaces the initial 011 by $10\bar{1}$ to get $101110\bar{1}$. One then replaces 0111 by $100\bar{1}$. Continuing one determines the NAF of 91 to be $10\bar{1}00\bar{1}0\bar{1}$.

We have established that the NAF of an integer a is unique, has minimal weight and has length at most one bit more than the binary expansion of a. Finally, we sketch a probabilistic argument that shows that the density of a NAF is expected to be $1/3$.

Lemma 11.1.14 *Let $l \in \mathbb{N}$. Define d_l to be the expected value of the density of the NAF representation of uniformly chosen integers $2^{l+1}/3 < a < 2^{l+2}/3$. Then d_l tends to $1/3$ as l goes to infinity.*

Proof (Sketch) Write $a = \sum_{i=0}^{l} a_i 2^l$ for the NAF representation of $a \in \mathbb{N}$. Note that Algorithm 6 has the property that if $a_i \ne 0$ then $a_{i+1} = 0$, but the value of a_{i+2} is independent of the previous operations of the algorithm. Hence, if the bits of a are considered to be chosen uniformly at random then the probability that $a_{i+2} \ne 0$ is $1/2$. Similarly, the probability that $a_{i+2} = 0$ but $a_{i+3} \neq 0$ is $1/4$, and so on. Hence, the expected number of zeroes after the non-zero a_i is

$$E = 1 \cdot \tfrac{1}{2} + 2 \cdot \tfrac{1}{4} + 3 \cdot \tfrac{1}{8} + \cdots = \sum_{i=1}^{\infty} \frac{i}{2^i}$$

(at least approximately, since the expansion is not infinite).

Now

$$E = 2E - E = \sum_{i=1}^{\infty} \frac{i}{2^{i-1}} - \sum_{i=1}^{\infty} \frac{i}{2^i} = 1 + \sum_{j=1}^{\infty} \frac{1}{2^j} = 2.$$

Hence, on average, there are two zeroes between adjacent non-zero coefficients and the density tends to $1/3$. □

Exercise 11.1.15 Prove that the number of distinct NAFs of length k is $(2^{k+2} - (-1)^k)/3$.

Exercise 11.1.16★ Write down an algorithm to list all NAFs of length k.

For some applications it is desired to compute a low-density signed expansion from left to right; Joye and Yen [287] give an algorithm to do this.

Generalisations of the NAF are the width-w non-adjacent form (w-NAF) and fractional window NAFs. We refer to Miyaji, Ono and Cohen [386], Section IV.2.5 of Blake, Seroussi and Smart [60] and Section 3.2 of Solinas [517] for the former, and Möller [388, 389] for the latter. For fast exponentiation, one can perform sliding windows over fractional NAFs or sliding fractional windows over NAFs.

11.2 Multi-exponentiation

An n-dimensional **multi-exponentiation** (also called **simultaneous multiple exponentiation**) is the problem of computing a product $g_1^{a_1} \cdots g_n^{a_n}$. The question of how efficiently this can be done was asked by Richard Bellman as problem 5125 of volume 70, number 6 of the *American Mathematical Monthly* in 1963. A solution was given by E. G. Straus[2] in [534]; the idea was re-discovered by Shamir and is often attributed to him. We only give a brief discussion of this topic and refer to Section 9.1.5 of [16] and Section 3.3.3 of [248] for further details.

Algorithm 7 computes an n-dimensional multi-exponentiation. We write $a_{i,j}$ for the jth bit of a_i (where, as usual in this book, the least significant bit is $a_{i,0}$); if $j > \log_2(a_i)$ then $a_{i,j} = 0$. The main idea is to use a single accumulating variable (in this case called h) and to perform only one squaring. If a value g_i does not change between executions of the algorithm then we call it a **fixed base** and otherwise it is a **variable base** (the precomputation can be improved when some of the g_i are fixed). We assume that the integers a_i all vary.

[2] Straus had the remarkable ability to solve crossword puzzles in English (his third language) using only the horizontal clues; see his commemorative issue of the *Pacific Journal of Mathematics*.

Algorithm 7 Basic multi-exponentiation

INPUT: $g_1, \dots, g_n \in G, a_1, \dots, a_n \in \mathbb{N}$

OUTPUT: $\prod_{i=1}^{n} g_i^{a_i}$

1: Precompute all $u_{b_1,\dots,b_n} = \prod_{i=1}^{n} g_i^{b_i}$ for $b_i \in \{0, 1\}$
2: Set $l = \max_{1 \le i \le n}\{\lfloor \log_2(a_i) \rfloor\}$
3: $h = u_{a_{1,l},\dots,a_{n,l}}$
4: $j = l - 1$
5: **while** $j \ge 0$ **do**
6: $\quad h = h^2$
7: $\quad h = h u_{a_{1,j},\dots,a_{n,j}}$
8: $\quad j = j - 1$
9: **end while**
10: **return** h

Example 11.2.1 One can compute $g_1^7 g_2^5$ by setting $h = u_{1,1} = g_1 g_2$ and then computing $h = h^2 = g_1^2 g_2^2$, $h = h u_{1,0} = g_1^3 g_2^2$, $h = h^2 = g_1^6 g_2^4$, $h = h u_{1,1} = g_1^7 g_2^5$.

Exercise 11.2.2 Show that one can perform the precomputation in Algorithm 7 in $2^n - n - 1$ multiplications. Show that the main loop of Algorithm 7 performs l squarings and l multiplications.

Exercise 11.2.3 (Yen, Laih and Lenstra [571]) Show that, by performing further precomputation, one can obtain a sliding window multi-exponentiation algorithm that still requires l squarings in the main loop, but fewer multiplications. Determine the precomputation cost.

An alternative approach[3] to multi-exponentiation is called **interleaving**. The basic idea is to replace line 7 in Algorithm 7 by

$$\textbf{for } i = 1 \textbf{ to } n \textbf{ do } h = h g_i^{a_{i,j}} \textbf{ end for}$$

and to omit the precomputation. This version is usually less efficient than Algorithm 7 unless n is rather large. However, the benefit of interleaving comes when using sliding windows: since the precomputation cost and storage requirements for the method in Exercise 11.2.3 are so high, it is often much more practical to use a sliding window version in the setting of interleaving. We refer to [387] and Section 3.3.3 of [248] for further discussion of this method.

Exercise 11.2.4 Write pseudocode for multi-exponentiation using interleaving and sliding windows.

Another approach, when signed expansions are being used, is to find a representation for the exponents $a_1, \dots, a_n$ so that the jth component of the representations of all a_i is

[3] Independently discovered by Möller [387] and Gallant, Lambert and Vanstone [216].

simultaneously zero relatively often. Such a method was developed by Solinas [518] and is called a **joint sparse form**. We refer to Section 9.1.5 of [16] and Section 3.3.3 of [248].

Multi-exponentiation for algebraic group quotients

In algebraic group quotients, multiplication is not well-defined and so extra information is needed to be able to compute $\prod_{i=1}^{n} g_i^{a_i}$. A large survey of exponentiation algorithms and multi-exponentiation algorithms for algebraic group quotients is given in Chapter 3 of Stam's thesis [520]. In particular, he gives the Montgomery Euclidean ladder in Section 3.3 (also see Section 4.3 of [519]). Due to lack of space we do not discuss this topic further.

11.3 Efficient exponentiation in specific algebraic groups

We now discuss some exponentiation methods that exploit specific features of algebraic groups.

11.3.1 Alternative basic operations

So far, all exponentiation algorithms have been based on squaring (and hence have used representations of integers to the base 2). We now briefly mention some alternatives to squaring as the basic operation. First, we discuss halving and tripling. Frobenius expansions will be discussed in Section 11.3.2.

When one has several possible basic operations then one can consider **multi-base representations** of integers for exponentiation. These ideas were first proposed by Dimitrov, Jullien and Miller [167], but we do not consider them further in this book.

Point halving on elliptic curves

This idea, independently discovered by Knudsen [307] and Schroeppel, applies to subgroups of odd order in ordinary elliptic curves over finite fields of characteristic 2. The formulae for point halving were given in Exercise 9.1.4: given $P = (x_P, y_P) \in E(\mathbb{F}_{2^n})$ one finds $Q = (x_Q, y_Q) \in E(\mathbb{F}_{2^n})$ such that $[2]Q = P$ by solving $\lambda_Q^2 + \lambda_Q = x_P + a_2$. For either solution let $x_Q = \sqrt{x_P(\lambda_Q + 1) + y_P} = \sqrt{x_P(\lambda_P + \lambda_Q + x_P + 1)}$ and $y_Q = x_Q(\lambda_Q + x_Q)$. One must ensure that the resulting point Q has odd order. When $2 \| \#E(\mathbb{F}_{q^n})$ this is easy as, by Exercise 9.1.4, it is sufficient to check that $\mathrm{Tr}_{\mathbb{F}_{2^n}/\mathbb{F}_2}(x_Q) = \mathrm{Tr}_{\mathbb{F}_{2^n}/\mathbb{F}_2}(a_2)$. In practice, it is more convenient to check whether $\mathrm{Tr}_{\mathbb{F}_{2^n}/\mathbb{F}_2}(x_Q^2) = \mathrm{Tr}_{\mathbb{F}_{2^n}/\mathbb{F}_2}(a_2^2)$.

Exercise 11.3.1 Write down the point halving algorithm.

Knudsen suggests representing points using the pair (x_P, λ_P) instead of (x_P, y_P). In any case, this can be done internally in the exponentiation algorithm. When $\mathbb{F}_{2^n}$ is represented using a normal basis over $\mathbb{F}_2$ then halving can be more efficient than doubling on such an elliptic curve. One can therefore use expansions of integers to the "base 2^{-1}" for efficient exponentiation. We refer to Section 13.3.5 of [16] and [307] for the details.

Tripling

Doche, Icart and Kohel [169] suggested to speed up the computation of $[m]P$ on E for small m by splitting it as $\hat{\phi} \circ \phi$ where $\phi : E \to E'$ is an isogeny of degree m. We refer to Exercise 9.6.30 for an example of this in the case $m = 3$, and to [169] for the details in general.

11.3.2 Frobenius expansions

Koblitz (in Section 5 of [309]) presented a very efficient doubling formula for $E : y^2 + y = x^3$ over $\mathbb{F}_2$ (see Exercise 9.1.3). Defining $\pi(x, y) = (x^2, y^2)$ one can write this as $[2]P = -\pi^2(P)$ for all $P \in E(\mathbb{F}_{2^m})$ for any integer m. We assume throughout this section that finite fields $\mathbb{F}_{p^m}$ are represented using a normal basis so that raising to the power p is very fast. Menezes and Vanstone [374] and Koblitz [311, 312] explored further how to speed up arithmetic on curves over small fields. However, the curves used in [309, 374, 311] are supersingular and so are less commonly used for cryptography.

The Frobenius map can be used to speed up elliptic curve exponentiation on more general curves. For cryptographic applications we assume that E is an elliptic curve over $\mathbb{F}_q$ such that $\#E(\mathbb{F}_{q^m})$ has a large prime divisor r for some $m > 1$.[4] Let π be the q-power Frobenius map on E. The trick is to replace an integer a with a sequence $a_0, \ldots, a_l$ of "small" integers such that

$$[a]P = \sum_{i=0}^{l} [a_i]\pi^i(P)$$

for the point $P \in E(\mathbb{F}_{q^m})$ of interest.

Definition 11.3.2 Let E be an elliptic curve over $\mathbb{F}_q$ and let π be the q-power Frobenius map. Let $S \subset \mathbb{Z}$ be a finite set such that $0 \in S$ (the set S is usually obvious from the context). A **Frobenius expansion** with **digit set** S is an endomorphism of the form

$$\sum_{i=0}^{l} [a_i]\pi^i$$

where $a_i \in S$ and $a_l \neq 0$. The **length** of a Frobenius expansion is $l + 1$. The **weight** of a Frobenius expansion is the number of non-zero a_i.

Many papers write τ for the Frobenius map and speak of τ-adic expansions, but we will call them **π-adic expansions** in this book.

Example 11.3.3 Let $E : y^2 + xy = x^3 + ax^2 + 1$ over $\mathbb{F}_2$ where $a \in \{0, 1\}$. Consider the group $E(\mathbb{F}_{2^m})$ and write $\pi(x, y) = (x^2, y^2)$. From Exercise 9.10.10 we know that $\pi^2 + (-1)^a\pi + 2 = 0$. Hence, one can replace the computation $[2]P$ by $-\pi(\pi(P)) - (-1)^a\pi(P)$. At first sight, there is no improvement here (we have replaced a doubling with

[4] Note that, for any fixed elliptic curve E over $\mathbb{F}_q$ and any fixed $c \in \mathbb{R}_{>0}$, it is not known if there are infinitely many $m \in \mathbb{N}$ such that $\#E(\mathbb{F}_{q^m})$ has a prime factor r such that $r > cq^m$. However, in practice one finds a sufficient quantity of suitable examples.

an elliptic curve addition). However, the idea is to represent an integer by a polynomial in π. For example, one can verify that

$$-T^5 + T^3 + 1 \equiv 9 \pmod{T^2 + T + 2}$$

and so one can compute $[9]P$ (normally taking 3 doublings and an addition) as $-\pi^5(P) + \pi^3(P) + P$ using only two elliptic curve additions.

This idea can be extended to any algebraic group G (in particular, elliptic curves, divisor class groups of hyperelliptic curves and algebraic tori) that is defined over a field $\mathbb{F}_q$ but for which one works in the group $G(\mathbb{F}_{q^m})$.

Exercise 11.3.4 Give an algorithm to compute $[a]P$ when $a = \sum_{i=0}^{l} a_i \pi^i(P)$ is a Frobenius expansion. What is the cost of the algorithm?

Definition 11.3.5 Let $S \subseteq \mathbb{Z}$ such that $0 \in S$ and if $a \in S$ then $-a \in S$. Let $a = \sum_{i=0}^{l} a_i \pi^i$ be a Frobenius expansion with $a_i \in S$. Then a is in **non-adjacent form** if $a_i a_{i+1} = 0$ for all $0 \le i < l$. Such an expansion is also called a **π-NAF**.

An important task is to convert an integer n into a Frobenius expansion in non-adjacent form. In fact, to get short expansions we will need to convert a general element $n_0 + n_1\pi$ to a π-NAF, so we study the more general problem. The crucial result is the following.

Lemma 11.3.6 *Let π satisfy $\pi^2 - t\pi + q = 0$. An element $n_0 + n_1\pi$ in $\mathbb{Z}[\pi]$ is divisible by π if and only if $q \mid n_0$. In this case*

$$(n_0 + n_1\pi)/\pi = (n_1 + tn_0/q) + \pi(-n_0/q). \tag{11.1}$$

Similarly, it is divisible by π^2 if and only if $q \mid n_0$ and $qn_1 \equiv -tn_0 \pmod{q^2}$.

Proof Note that $\pi^2 = t\pi - q$. Since

$$\pi(m_0 + m_1\pi) = -qm_1 + \pi(m_0 + tm_1)$$

it follows that $n_0 + n_1\pi = \pi(m_0 + m_1\pi)$ if and only if

$$n_0 = -qm_1, \quad \text{and} \quad n_1 = m_0 + tm_1.$$

Writing $m_1 = -n_0/q$ and $m_0 = n_1 - tm_1$ yields equation (11.1).

Repeating the argument, one can divide the element in equation (11.1) by π if and only if $q \mid (n_1 + tn_0/q)$. The result follows. □

The idea of the algorithm for computing a π-NAF, given $n_0 + n_1\pi$, is to add a suitable integer so that the result is divisible by π^2, divide by π^2 and then repeat. This approach is only really practical when $q = 2$, so we restrict to this case.

Lemma 11.3.7 *Let $\pi^2 - t\pi + 2 = 0$ where $t = \pm 1$. Then $n_0 + n_1\pi$ is either divisible by π or else there is some $\epsilon = \pm 1$ such that*

$$(n_0 + \epsilon) + n_1\pi \equiv 0 \pmod{\pi^2}.$$

Indeed, $\epsilon = (n_0 + 2n_1 \pmod 4) - 2$ if one defines $n_0 + 2n_1 \pmod 4 \in \{1, 3\}$.

Proof If $\pi \nmid (n_0 + n_1\pi)$ then n_0 is odd and so $n_0 \pm 1$ is even. One can choose the sign such that $2n_1 \equiv -(n_0 \pm 1) \pmod 4$, in which case the result follows. □

The right-to-left algorithm[5] to generate a π-NAF is then immediate (see Algorithm 8; this algorithm computes $u = -\epsilon$ in the notation of Lemma 11.3.7). To show that the algorithm terminates we introduce the norm map: for any $a, b \in \mathbb{R}$ define $\mathrm{N}(a + b\pi) = (a + b\pi)(a + b\bar{\pi}) = a^2 + tab + qb^2$ where $\pi, \bar{\pi} \in \mathbb{C}$ are the roots of the polynomial $x^2 - tx + q = 0$. This map agrees with the norm map with respect to the quadratic field extension $\mathbb{Q}(\pi)/\mathbb{Q}$ and so is multiplicative. Note also that $\mathrm{N}(a + b\pi) \geq 0$ and equals zero only when $a = b = 0$. Meier and Staffelbach [373] note that if $n_0 + n_1\pi$ is divisible by π then $\mathrm{N}((n_0 + n_1\pi)/\pi) = \frac{1}{2}\mathrm{N}(n_0 + n_1\pi)$. This suggests that the length of the π-NAF will grow like $\log_2(\mathrm{N}(n_0 + n_1\pi))$. The case $\mathrm{N}((n_0 \pm 1 + n_1\pi)/\pi)$ needs more care. Lemma 3 of Meier and Staffelbach [373] states that if $\mathrm{N}(n_0 + n_1\pi) < 2^n$ then there is a corresponding Frobenius expansion[6] of length at most n. Theorem 2 of Solinas gives a formula for the norm in terms of the length $k = l + 1$ of the corresponding π-NAF, from which he deduces (equation (53) of [517])

$$\log_2(\mathrm{N}(n_0 + n_1\pi)) - 0.55 < k < \log_2(\mathrm{N}(n_0 + n_1\pi)) + 3.52$$

when $k \geq 30$.

Algorithm 8 Convert $n_0 + n_1\pi$ to non-adjacent form

INPUT: $n_0, n_1 \in \mathbb{Z}$
OUTPUT: $a_0, \ldots, a_l \in \{-1, 0, 1\}$
1: **while** $n_0 \neq 0$ and $n_1 \neq 0$ **do**
2:     **if** n_0 odd **then**
3:         $u = 2 - (n_0 + 2n_1 \pmod 4)$
4:         $n_0 = n_0 - u$
5:     **else**
6:         $u = 0$
7:     **end if**
8:     Output u
9:     $(n_0, n_1) = (n_1 + tn_0/2, -n_0/2)$
10: **end while**

Example 11.3.8 Suppose $\pi^2 + \pi + 2 = 0$. To convert $-1 + \pi$ to a π-NAF one writes $n_0 = -1$ and $n_1 = 1$. Let $u = 2 - (n_0 + 2n_1 \pmod 4) = 2 - (1) = 1$. Output 1 and set $n_0 = n_0 - 1$ to get $-2 + \pi$. Dividing by π yields $2 + \pi$. One can divide by π again (output 0 first) to get $-\pi$, output 0 and divide by π again to get -1. The π-NAF is therefore $1 - \pi^3$.

[5] Solinas states that this algorithm was joint work with R. Reiter.
[6] Meier and Staffelbach do not consider Frobenius expansions in non-adjacent form.

To see this directly using the equation $\pi^2 + \pi + 2 = 0$ write

$$-1 + \pi + (\pi^2 + \pi + 2)(1 - \pi) = 1 - \pi^3.$$

Exercise 11.3.9 Verify that Algorithm 8 does output a π-NAF.

Exercise 11.3.10 Let $\pi^2 - \pi + 2 = 0$. Use Algorithm 8 to convert $107 + 126\pi$ into non-adjacent form.

Exercise 11.3.11 Show, using the same methods as Lemma 11.1.14, that the average density of a π-NAF tends to $1/3$ when $q = 2$.

Reducing the length of Frobenius expansions

As we have seen, $\mathrm{N}(n + 0\pi) = n^2$, while the norm only decreases by a factor of roughly 2 each time we divide by π. Hence, the Frobenius expansions output by Algorithm 8 on input n have length roughly $2\log_2(n)$. Since the density is $1/3$ it follows that the weight of the Frobenius expansions is roughly $\frac{2}{3}\log_2(n)$. Exponentiation using Frobenius expansions is therefore faster than using the square-and-multiply algorithm, even with sliding windows (since the latter method always needs $\log_2(n)$ doublings and also some additions). However, it is a pity that the expansions are so long. It is natural to seek shorter expansions that still have the same density. The crucial observation, due to Meier and Staffelbach [373], is that Algorithm 8 outputs a Frobenius expansion $\sum_i [a_i]\pi^i$ that acts the same as $[n]$ on all points in $E(\overline{\mathbb{F}}_q)$, whereas, for a given application, one only needs a Frobenius expansion that acts the same as $[n]$ on the specific subgroup $\langle P \rangle$ of prime order r.

Definition 11.3.12 Let E be an elliptic curve over $\mathbb{F}_p$, $P \in E(\mathbb{F}_{p^m})$, and let π be the p-power Frobenius map on E. We say that two Frobenius expansions $a(\pi), b(\pi) \in \mathbb{Z}[\pi]$ are **equivalent** with respect to P if

$$a(\pi)(P) = b(\pi)(P).$$

Exercise 11.3.13 Let the notation be as in Definition 11.3.12. Show that if $a(\pi) \equiv b(\pi) \pmod{\pi^m - 1}$ then $a(\pi)$ and $b(\pi)$ are equivalent with respect to P.

Show that if $Q \in \langle P \rangle$ and $a(\pi)$ and $b(\pi)$ are equivalent with respect to P then $a(\pi)$ and $b(\pi)$ are equivalent with respect to Q.

A simple idea is to replace all powers π^i by $\pi^{i \pmod m}$. This will reduce the length of a Frobenius expansion, but it does not significantly change the weight (and hence the cost of exponentiation does not change).

The goal is therefore to find an element $n_0 + n_1\pi$ of "small" norm that is equivalent to $[n]$ with respect to P. Then one applies Algorithm 8 to the pair (n_0, n_1), not to $(n, 0)$. There are two simple ways to find an element of small norm, both of which apply the Babai rounding method (see Section 18.2) in a suitable lattice. They differ in how one expresses the fact that $(n_0 + n_1\pi)P = [n]P$ for the point P of interest.

- Division with remainder in $\mathbb{Z}[\pi]$.
This method was proposed by Meier and Staffelbach and is also used by Solinas (Section 5.1 of [517]). Since $(\pi^m - 1)(P) = \mathcal{O}_E$ when $P \in E(\mathbb{F}_{q^m})$ one wants to determine the remainder of dividing n by $(\pi^m - 1)$. The method is to consider the element $\gamma = n/(\pi^m - 1) \in \mathbb{R}[\pi]/(\pi^2 - t\pi + q)$ and find a close vector to it (using Babai rounding) in the lattice $\mathbb{Z}[\pi]$. In other words, write $\gamma = \gamma_0 + \gamma_1\pi$ with $\gamma_0, \gamma_1 \in \mathbb{R}$ and round them to the nearest integers g_0, g_1 (in the special case of $\pi^2 \pm \pi + 2 = 0$, there is an exact description of a fundamental domain for the lattice that can be used to "correct" the Babai rounding method if it does not reach the closest lattice element). Lemma 3 of [373] and Proposition 57 of [517] state that $\mathrm{N}(\gamma - (g_0 + g_1\pi)) \le 4/7$. One can then define

$$n_0 + n_1\pi = n - (g_0 + g_1\pi)(\pi^m - 1) \pmod{\pi^2 - t\pi + q}.$$

- The Gallant–Lambert–Vanstone method [216].
This method appears in a different context (see Section 11.3.3), but it is also suitable for the present application. We assume that $P \in E(\mathbb{F}_{q^m})$ has prime order r where $r \| \#E(\mathbb{F}_{q^m})$. Since $\pi(P) \in E(\mathbb{F}_{q^m})$ has order r it follows that $\pi(P) = [\lambda]P$ for some $\lambda \in \mathbb{Z}/r\mathbb{Z}$. The problem is therefore to find small integers n_0 and n_1 such that

$$n_0 + n_1\lambda \equiv n \pmod{r}.$$

One defines the **GLV lattice**

$$L = \{(x_0, x_1) \in \mathbb{Z}^2 : x_0 + x_1\lambda \equiv 0 \pmod{r}\}.$$

A basis for L is given in Exercise 11.3.22. The idea is to find a lattice vector $(n_0', n_1') \in L$ close to $(n, 0)$. Then $|n_1'|$ is "small" and $|n - n_0'|$ is "small". Define $n_0 = n - n_0'$ and $n_1 = -n_1'$ so that

$$n_0 + n_1\lambda \equiv n \pmod{r}$$

as required.

We can compute a reduced basis for the lattice and then use Babai rounding to solve the closest vector problem (CVP). Note that the reduced basis can be precomputed. Since the dimension is two, one can use the Gauss lattice reduction algorithm (see Section 17.1). Alternatively, one can use Euclid's algorithm to compute (n_0, n_1) directly (as discussed in Section 17.1.1, Euclid's algorithm is closely related to the Gauss algorithm).

Example 11.3.14 The elliptic curve $E : y^2 + xy = x^3 + x^2 + 1$ over $\mathbb{F}_{2^{19}}$ has $2r$ points where $r = 262543$ is prime. Let $\pi(x, y) = (x^2, y^2)$. Then $\pi^2 - \pi + 2 = 0$. Let $n = 123456$. We want to write n as $n_0 + n_1\pi$ on the subgroup of $E(\mathbb{F}_{2^{19}})$ of order r.

For the "division with remainder in $\mathbb{Z}[\pi]$" method we first use Lucas sequences (as in Exercise 9.10.9) to determine that

$$\pi^{19} - 1 = -(171 + 457\pi)$$

(one can think of this as equality of complex numbers where π is a root of $x^2 - x + 2$, or as congruence of polynomials modulo $\pi^2 - \pi + 2$). It is convenient to change the sign (the method works in both cases). The norm of $171 + 457\pi$ is $\#E(\mathbb{F}_{2^{19}}) = 2r = 525086$. and so

$$\frac{n}{-\pi^{19}+1} = \frac{n(171+457\bar{\pi})}{525086} \approx 147.653 - 107.448\pi.$$

(since $\bar{\pi} = 1 - \pi$). Rounding gives $148 - 107\pi$ and

$$n - (148 - 107\pi)(171 + 457\pi) \equiv 350 - 440\pi \pmod{\pi^2 - \pi + 2}.$$

This is a short representative for n, but its norm is larger than $8r/7$, which is not optimal. Section 5.1 of Solinas [517] shows how to choose a related element of smaller norm. In this case, the correct choice of rounding is $147 - 107\pi$ giving

$$n - (147 - 107\pi)(171 + 457\pi) \equiv 521 + 17\pi \pmod{\pi^2 - \pi + 2},$$

which has norm less than $8r/7$.

Now for the Gallant–Lambert–Vanstone method. We compute $\gcd(x^{19} - 1, x^2 - x + 2) = (x - \lambda)$ in $\mathbb{F}_r[x]$, where $\lambda = 84450$. The lattice with (row) basis

$$\begin{pmatrix} r & 0 \\ -\lambda & 1 \end{pmatrix}$$

has LLL (or Gauss) reduced basis

$$B = \begin{pmatrix} 171 & 457 \\ 457 & -314 \end{pmatrix}.$$

Writing $\underline{b}_1, \underline{b}_2$ for the rows of the reduced matrix one finds $(n, 0) \approx 147.65\underline{b}_1 + 214.90\underline{b}_2$. One computes

$$(n, 0) - (148, 215)\, B = (-107, -126)\,.$$

One can verify that $-107 - 126\lambda \equiv n \pmod{r}$. (Exercise 11.3.16 shows how to get this element using remainders in $\mathbb{Z}[\pi]$.) The corresponding Frobenius expansion can be obtained from the solution to Exercise 11.3.10.

Exercise 11.3.15 Prove that both the above methods yield an element $n_0 + n_1\pi \in \mathbb{Z}[\pi]$ that is equivalent to n.

Exercise 11.3.16 Show that if $P \in E(\mathbb{F}_{q^m})$ but $P \notin E(\mathbb{F}_q)$ then instead of computing the remainder in $\mathbb{Z}[\pi]$ modulo the polynomial $(\pi^m - 1)$ one can use $(\pi^m - 1)/(\pi - 1)$. Repeat Example 11.3.14 using this polynomial.

In practice, it is unnecessary to determine the minimal solution (n_0, n_1) as long as n_0 and n_1 have bit-length roughly $\frac{1}{2}\log_2(r)$ (where the point P has order r). We also stress that computing the q-power Frobenius map π is assumed to be very fast, so the main task is to minimise the *weight* of the representation, not its length.

Remark 11.3.17 In cryptographic protocols, one is often computing $[a]P$, where a is a randomly chosen integer modulo r. Rather than choosing a random integer a and then converting to a Frobenius expansion, one could choose a random Frobenius expansion of given weight and length (this trick appears in Section 6 of [312] where it is attributed to H. W. Lenstra Jr.).

We have analysed π-NAFs in the case $q = 2$. Müller [397] gives an algorithm to compute Frobenius expansions for elliptic curves over $\mathbb{F}_{2^e}$ with $e > 1$ (but still small). The coefficients of the expansion lie in $\{-2^{e-1}, \dots, 2^{e-1}\}$. Smart [512] gives an algorithm for odd q, with a similar coefficient set; see Exercise 11.3.18. Lange [330] generalises to hyperelliptic curves. In all cases, the output is not necessarily in non-adjacent form; to obtain a π-NAF in these cases seems to require much larger digit sets. In any case, the asymptotic density of π-NAFs with large digit set is not significantly smaller than $1/2$ and this can easily be bettered using window methods (see Exercise 11.3.19).

Exercise 11.3.18 Let $q > 2$. Show that Algorithm 8 can be generalised (not to compute a π-NAF, but just a π-adic expansion) by taking digit set $\{-\lfloor q/2 \rfloor, \dots, -1, 0, 1, \dots, \lfloor q/2 \rfloor\}$ (or this set with $-\lfloor q/2 \rfloor$ removed when $q > 2$ is even).

Exercise 11.3.19 Let E be an elliptic curve over a field $\mathbb{F}_q$, let π be the q-power Frobenius map, and let $P \in E(\mathbb{F}_{q^m})$. Let $S = \{-(q-1)/2, \dots, -1, 0, 1, \dots, (q-1)/2\}$ if q is odd and $S = \{-(q-2)/2, \dots, -1, 0, 1, \dots, q/2\}$ if q is even.

Suppose one has a Frobenius expansion

$$a(\pi) = \sum_{j=0}^{l} [a_j]\pi^j$$

with $a_j \in S$. Let $w \in \mathbb{N}$. Give a sliding window method to compute $[a(\pi)]P$ using windows of length w. Give an upper bound on the cost of this algorithm (including precomputation) ignoring the cost of evaluating π.

Exercise 11.3.20 (Brumley and Järvinen [105]) Let E be an elliptic curve over $\mathbb{F}_q$, π be the q-power Frobenius and $P \in E(\mathbb{F}_{q^m})$ have prime order r where $r \| \#E(\mathbb{F}_{q^m})$. Given a Frobenius expansion $a(\pi) = \sum_i [a_i]\pi^i$, show how to efficiently compute $a \in \mathbb{Z}$ such that $a(\pi)(P) = [a]P$.

For a complete presentation of Frobenius expansions, and further references, we refer to Section 15.1 of [16]. For multi-exponentiation using Frobenius expansions there is also a π-adic joint sparse form. We refer to Section 15.1.1.e of [16] for details.

11.3.3 GLV method

This method is due to Gallant, Lambert and Vanstone [216] for elliptic curves and Stam and Lenstra (see Section 4.4 of [521]) for tori.[7] The idea is to use an "efficiently

[7] A patent on the method was filed by Gallant, Lambert and Vanstone in 1999.

computable" (see below for a clarification of this term) group homomorphism ψ and replace the computation g^a in a group of order r by the multi-exponentiation $g^{a_0}\psi(g)^{a_1}$ for suitable integers a_0 and a_1 such that $|a_0|, |a_1| \approx \sqrt{r}$. Typical choices for ψ are an automorphism of an elliptic curve or the Frobenius map on an elliptic curve or torus over an extension field.

More precisely, let $g \in G(\mathbb{F}_q)$ be an element of prime order r in an algebraic group and let ψ be a group homomorphism such that $\psi(g) \in \langle g \rangle$ (this is automatic if $\psi : G(\mathbb{F}_q) \to G(\mathbb{F}_q)$ and $r \| \#G(\mathbb{F}_q)$). Then $\psi(g) = g^\lambda$ for some $\lambda \in \mathbb{Z}/r\mathbb{Z}$. The meaning of "efficiently computable" is essentially that computing $\psi(g)$ is much faster than computing g^λ using exponentiation algorithms. Hence, we require that λ and $r - \lambda$ are not small; in particular, the map $\psi(P) = -P$ on an elliptic curve is not interesting for this application.

Example 11.3.21 Consider $\mathbb{T}_2(\mathbb{F}_{p^2})$, with elements represented as in Definition 6.3.7 so that $u \in \mathbb{F}_{p^2}$ corresponds to $g = (u+\theta)/(u+\overline{\theta}) \in \mathbb{F}_{p^4}$. It follows by Lemma 6.3.12 that $A - u^p$ (where $\theta^2 + A\theta + B = 0$) corresponds to $g^p = (u^p + \overline{\theta})/(u^p + \theta) = \left((u^p+\theta)(u^p+\overline{\theta})\right)^{-1}$. Since computing u^p for $u \in \mathbb{F}_{p^2}$ is easy, the map $\psi(u) = A - u^p$ is a useful efficiently computable group homomorphism with respect to the torus group operation.

One can also perform exponentiation in $G_{q^2,2} \subseteq \mathbb{F}_{p^4}^*$ using Frobenius. Given an exponent a such that $1 \le a < p^2$ one lets a_0 and a_1 be the coefficients in the base-p representation of a and computes g^a as $g^{a_0}(g^p)^{a_1}$. Note that g^p is efficient to compute as it is a linear map on the four-dimensional vector space $\mathbb{F}_{p^4}$ over $\mathbb{F}_p$.

Other examples include the automorphism ζ_3 on $y^2 = x^3 + B$ in Example 9.4.2 and the automorphisms in Exercises 9.4.5 and 10.2.8. Computing the eigenvalue λ for ψ is usually easy in practice: for elliptic curves λ is a root of the characteristic polynomial of ψ modulo r.

In some applications (e.g., $\mathbb{T}_2(\mathbb{F}_{p^3})$ or some automorphisms on genus 2 curves such as the one in Exercise 10.2.8) one can replace g^a by $g^{a_0}\psi(g)^{a_1}\cdots\psi^{l-1}(g)^{a_{l-1}}$ for some $l > 2$. We call this the l-dimensional GLV method. We stress that l cannot be chosen arbitrarily; in Example 11.3.21 the map ψ^2 is the identity map and so is not useful.

In the previous section we sketched, for elliptic curves, the GLV method to represent an integer as a short Frobenius expansion with relatively small coefficients. One can do the same for any endomorphism ψ as long as $\psi(P) = [\lambda]P$ (or $\psi(g) = g^\lambda$ in multiplicative notation). The **GLV lattice** is

$$L = \{(x_0, \ldots, x_l) \in \mathbb{Z}^{l+1} : x_0 + x_1\lambda + \cdots + x_l\lambda^l \equiv 0 \pmod{r}\}.$$

A basis for L is given in Exercise 11.3.22. As explained earlier, to convert an integer a into GLV representation one finds a lattice vector $(a_0', a_1', \ldots, a_l') \in L$ close to $(a, 0, \ldots, 0)$ (using Babai rounding) then sets $a_0 = a - a_0'$ and $a_i = -a_i'$ for $1 \le i \le l$.

Exercise 11.3.22 Show that

$$\begin{pmatrix} r & 0 & 0 & \cdots & 0 \\ -\lambda & 1 & 0 & \cdots & 0 \\ -\lambda^2 & 0 & 1 & \cdots & 0 \\ \vdots & & & & \vdots \\ -\lambda^l & 0 & 0 & \cdots & 1 \end{pmatrix}$$

is a basis for the GLV lattice L.

Exercise 11.3.23 Show how to compute the coefficients $a_0, \ldots, a_l$ for the GLV method using Babai rounding.

Exercise 11.3.24 gives a construction of homomorphisms for the GLV method that apply to a large class of curves. We refer to Galbraith, Lin and Scott [207] for implementation results that show the benefit of using this construction.

Exercise 11.3.24 (Iijima, Matsuo, Chao and Tsujii [274]) Let $p > 3$ be a prime and let $E : y^2 = x^3 + a_4x + a_6$ be an ordinary elliptic curve over $\mathbb{F}_p$ with $p + 1 - t$ points (note that $t \neq 0$). Let $u \in \mathbb{F}_{p^2}^*$ be a non-square and define $E' : Y^2 = X^3 + u^2a_4X + u^3a_6$ over $\mathbb{F}_{p^2}$. Show that E' is the quadratic twist of $E(\mathbb{F}_{p^2})$ and that $\#E'(\mathbb{F}_{p^2}) = (p-1)^2 + t^2$. Let $\phi : E \to E'$ be the isomorphism $\phi(x, y) = (ux, u^{3/2}y)$ defined over $\mathbb{F}_{p^4}$.

Let $\pi(x, y) = (x^p, y^p)$ and define

$$\psi = \phi \circ \pi \circ \phi^{-1}.$$

Show that $\psi : E' \to E'$ is an endomorphism of E' that is defined over $\mathbb{F}_{p^2}$. Show that $\psi^2 = [-1]$.

Let $r \mid \#E'(\mathbb{F}_{p^2})$ be a prime such that $r > 2p$ and $r^2 \nmid \#E'(\mathbb{F}_{p^2})$. Let $P \in E'(\mathbb{F}_{p^2})$ have order r. Show that $\psi^2(P) - [t]\psi(P) + [p]P = \mathcal{O}_{E'}$. Hence, deduce that $\psi(P) = [\lambda]P$ where $\lambda = t^{-1}(p-1) \pmod{r}$. Note that it is possible for $\#E'(\mathbb{F}_{p^2})$ to be prime, since E' is not defined over $\mathbb{F}_p$.

As in Remark 11.3.17, for some applications one might be able to choose a random GLV expansion directly, rather than choosing a random integer and converting it to GLV representation.

There is a large literature on the GLV method, including several different algorithms to compute the integers $a_0, \ldots, a_l$. As noted earlier, reducing the bit-length of the a_i by one or two bits has very little effect on the overall running time. Instead, the weight of the entries $a_0, \ldots, a_l$ is more critical. We refer to Sections 15.2.1 and 15.2.2 of [16] for further details and examples of the GLV method. Section 15.2.3 of [16] discusses combining the GLV method with Frobenius expansions.

11.4 Sampling from algebraic groups

A natural problem, given an algebraic group G over a finite field $\mathbb{F}_q$, is to generate a "random" element of G. By "random" we usually mean uniformly at random from $G(\mathbb{F}_q)$, although sometimes it may be appropriate to weaken this condition. The first problem is to generate a random integer in $[0, p-1]$ or $[1, p-1]$. Examples 11.4.1 and 11.4.2 give two simple approaches. Chapter 7 of Sidorenko [503] is a convenient survey.

Example 11.4.1 One way to generate a random integer in $[0, p-1]$ is to generate random binary strings x of length k (where $2^{k-1} < p \le 2^k$) and only output those satisfying $0 \le x \le p-1$.

Example 11.4.2 Another method is to generate a binary string that is longer than p and then return this value reduced modulo p. We refer to Section 7.4 of Shoup [497] for a detailed analysis of this method (briefly, if $2^{k-1} < p \le 2^k$ and one generates a $k+l$ bit string then the statistical difference of the output from uniform is $1/2^l$). Section 7.5 of [497] discusses how to generate a random k-bit prime and Section 7.7 of [497] discusses how to generate a random integer of known factorisation.

Exercise 11.4.3 Show that the expected number of trials of the algorithm in Example 11.4.1 is less than 2.

Exercise 11.4.4 Give an algorithm to generate an element of $\mathbb{F}_{p^n}^*$ uniformly at random, assuming that generating random integers modulo p is easy.

Appendix B.2.4 of Katz and Lindell [300] gives a thorough discussion of sampling randomly in $(\mathbb{Z}/N\mathbb{Z})^*$ and $\mathbb{F}_p^*$.

Algorithm 5 shows how to compute a generator for $\mathbb{F}_p^*$, when the factorisation of $p-1$ is known. Generalising this algorithm to $\mathbb{F}_{p^n}^*$, when the factorisation of $p^n - 1$ is known, is straightforward. In practice, one often works in a subgroup $G \subseteq \mathbb{F}_q^*$ of prime order r. To sample uniformly from G, one can generate a uniform element in $\mathbb{F}_q^*$ and then raise to the power $(q-1)/r$. This exponentiation can be accelerated using any of the techniques discussed earlier in this chapter.

Exercise 11.4.5 Let q be a prime power such that the factorisation of $q-1$ is known. Give an algorithm to determine the order of an element $g \in \mathbb{F}_q^*$.

Exercise 11.4.6 Let q be a prime power. Let G be a subgroup of $\mathbb{F}_q^*$ such that the factorisation of the order of G is known. Give an algorithm to compute a generator of G.

An alternative approach to the sampling problem for a finite Abelian group G is given in Exercise 11.4.7 (these ideas will also be used in Exercise 15.5.2). However, this method is often not secure for applications in discrete logarithm cryptography. The reason is that one usually wants to sample group elements at random such that no information about their discrete logarithm is known, whereas the construction in Exercise 11.4.7 (especially when used to define a hash function) may give an attacker a way to break the cryptosystem.

Exercise 11.4.7 Let G be a finite Abelian group. Let $g_1, \dots, g_k$ be fixed elements that generate G. Let $m_1, \dots, m_k$ be the orders of $g_1, \dots, g_k$ respectively. Then one can generate an element of G at random by choosing integers $a_1, \dots, a_k$ uniformly at random such that $0 \le a_i < m_i$ for $1 \le i \le k$ and computing

$$\prod_{i=1}^{k} g_i^{a_i}. \tag{11.2}$$

Show that this process does sample from G with uniform distribution.

11.4.1 Sampling from tori

We now mention further techniques to speed up sampling from subgroups of $\mathbb{F}_q^*$.

Example 11.4.8 Lemma 6.3.4 shows that $\mathbb{T}_2(\mathbb{F}_q)$ and $G_{2,q} \subseteq \mathbb{F}_{q^2}^*$ are in one-to-one correspondence with the set

$$\{1\} \cup \{(a+\theta)/(a+\overline{\theta}) : a \in \mathbb{F}_q\}.$$

Hence, one can sample from $\mathbb{T}_2(\mathbb{F}_q)$ or $G_{2,q}$ as follows: choose uniformly $0 \le a \le p$ and, if $a = p$, output 1, otherwise output $(a+\theta)/(a+\overline{\theta})$.

Generating elements of $\mathbb{T}_6(\mathbb{F}_q)$ or $G_{6,q}$ uniformly at random is less simple since the compression map does not map to the whole of $\mathbb{A}^2(\mathbb{F}_q)$. Indeed, the group $G_{6,q}$ has $q^2 - q + 1 < q^2$ elements. Example 6.4.4 showed, in the case $q \equiv 2, 5 \pmod 9$, how to map

$$A = \{(a,b) \in \mathbb{A}^2(\mathbb{F}_q) : a^2 - ab + b^2 \ne 1\} \tag{11.3}$$

to a subset of $\mathbb{T}_6(\mathbb{F}_q)$.

Exercise 11.4.9 Let $q \equiv 2, 5 \pmod 9$. Give an algorithm to generate points in the set A of equation (11.3) uniformly at random. Hence, show how to efficiently choose random elements of a large subset of $\mathbb{T}_6(\mathbb{F}_q)$ or $G_{6,q}$ uniformly at random.

11.4.2 Sampling from elliptic curves

Let $E : y^2 = x^3 + a_4x + a_6$ be an elliptic curve over $\mathbb{F}_q$ where q is not a power of 2. To generate points in $E(\mathbb{F}_q)$ one can proceed as follows: choose a random $x \in \mathbb{F}_q$, test whether $x^3 + a_4x + a_6$ is a square in $\mathbb{F}_q$, if not then repeat, otherwise take square roots to get y and output (uniformly) one of $\pm y$. It is not surprising that an algorithm to generate random points is randomised, but something that did not arise previously is that this algorithm uses a randomised subroutine (i.e., to compute square roots efficiently) and may need to be repeated several times before it succeeds (i.e., it is a Las Vegas algorithm). Hence, only an expected run time for the algorithm can be determined.

A more serious problem is that the output is not uniform. For example, $\mathcal{O}_E$ is never output, and points $(x, 0)$ occur with probability twice the probability of (x, y) with $y \ne 0$.

A solution for affine points on elliptic curves is to reject P with probability $1/2$ if P is of the form $(x, 0)$. For a detailed analysis and generalisation of this algorithm see von zur Gathen, Shparlinski and Sinclair [222].

Exercise 11.4.10★ Give a randomised algorithm to sample elliptic curve points uniformly at random. Determine the expected running time of this algorithm.

Deterministic sampling of elliptic curve points

The above methods are randomised, not just due to the randomness that naturally arises when sampling, but also because of the use of randomised algorithms for solving quadratic equations, and because not every x in the field is an x-coordinate of an elliptic curve point. It is of interest to minimise the reliance on randomness, especially when using the above ideas to construct a hash function (otherwise, there may be timing attacks). We first give an easy example.

Exercise 11.4.11 (Boneh and Franklin [75]) Let $p \equiv 2 \pmod 3$ be prime. Consider the supersingular elliptic curve $E : y^2 = x^3 + a_6$ over $\mathbb{F}_p$. One can sample points uniformly in $E(\mathbb{F}_p) - \{\mathcal{O}_E\}$ by uniformly choosing $y \in \mathbb{F}_p$ and setting $x = (y^2 - a_6)^{1/3} \pmod p$. The cube root is computed efficiently by exponentiation to the power $(2p-1)/3 \equiv 3^{-1} \pmod{p-1}$.

The first general results on deterministic methods to find points on curves over finite fields $\Bbbk$ are due to Schinzel and Skałba [462]. Given $a_6 \in \Bbbk$ (the case $\mathrm{char}(\Bbbk) = 2$ is not interesting since the curve is singular, and the case $\mathrm{char}(\Bbbk) = 3$ is easy since taking cube roots is easy, so assume $\mathrm{char}(\Bbbk) \neq 2, 3$), they give a formula, in terms of a_6, for four values $y_1, \dots, y_4$ such that the equation $x^3 + a_6 = y_i^2$ has a solution $x \in \Bbbk$ for some $1 \le i \le 4$. This method therefore produces at most 12 points on any given curve.

Skałba [510] gave results for general curves $y^2 = F(x)$ where $F(x) = x^3 + a_4 x + a_6$. This method can give more than a fixed number of points for any given curve. More precisely, Skałba gives explicit rational functions $X_i(t) \in \mathbb{Q}(t)$ for $1 \le i \le 3$ such that there is a rational function $U(t) \in \mathbb{Q}(t)$ and such that

$$F(X_1(t^2))F(X_2(t^2))F(X_3(t^2)) = U(t)^2.$$

In other words, Skałba gives a rational map from $\mathbb{A}^1$ to the variety $F(x_1)F(x_2)F(x_3) = u^2$. Evaluating at $t \in \mathbb{F}_p$, where $p > 3$ is prime, it follows that at least one of the $F(X_i(t^2))$ is a square in $\mathbb{F}_p^*$. One can therefore find a point on E by taking square roots. Note that efficient algorithms for computing square roots modulo p are randomised in general. Skałba avoids this problem by assuming that the required quadratic non-residue has been precomputed.

Shallue and van de Woestijne [491] improve upon Skałba's algorithm in several ways. First, and most significantly, they show that a deterministic sampling algorithm does not require a quadratic non-residue modulo p. They achieve this by cleverly using all three values $F(X_1(t^2))$, $F(X_2(t^2))$ and $F(X_3(t^2))$. In addition, they give a simpler rational map (see Exercise 11.4.12) from $\mathbb{A}^1$ to the variety $F(x_1)F(x_2)F(x_3) = u^2$, and handle the characteristic 2 and 3 cases.

Exercise 11.4.12 (Shallue and van de Woestijne [491]) Let $F(x) = x^3 + Ax + B$ and $H(u, v) = u^2 + uv + v^2 + A(u + v) + B$. Let $V : F(x_1)F(x_2)F(x_3) = u^2$ and let $S : y^2 H(u, v) = -F(u)$. Show that the map $\psi(u, v, y) \to (v, -A - u - v, u + y^2, F(u + y^2)H(u, v)/y)$ is a rational map from S to V. Let $p > 3$ be prime. Fix $u \in \mathbb{F}_p$ such that $F(u) \neq 0$ and $3u^2 + 2Au + 4B - A^2 \neq 0$. Show that the surface S for this fixed value of u is

$$[y(v + u/2 + A/2)]^2 + [3u^2/4 + Au/2 + B - A^2/4]y^2 = -F(u).$$

Hence, show there is a rational map from $\mathbb{A}^1$ to S and hence a rational map from $\mathbb{A}^1$ to V.

It is worth noting that there can be no rational map $\phi : \mathbb{P}^1 \to C$ when C is a curve of genus at least 1. This follows from the Hurwitz genus formula: if the map has degree d then we have $-2 = 2g(\mathbb{P}^1) - 2 = d(2g(C) - 2) + R \geq 0$ where R is a positive integer counting the ramification, which is a contradiction. The above maps do not contradict this fact. They are not rational maps from $\mathbb{P}^1$ (or $\mathbb{A}^1$) to an elliptic curve; there is always one part of the function (such as computing a square-root or cube-root) that is not a rational map.

Icart [273] has given a simpler map for elliptic curves $y^2 = x^3 + Ax + B$ over $\mathbb{F}_q$ when $q \equiv 2 \pmod 3$. Let $u \in \mathbb{F}_q$. Define

$$v = (3A - u^4)/(6u), \quad x = \left(v^2 - B - u^6/27\right)^{1/3} + u^2/3 \quad \text{and} \quad y = ux + v \qquad (11.4)$$

where the cube root is computed by exponentiating to the power $(2q - 1)/3 \equiv 3^{-1} \pmod{(q - 1)}$.

Exercise 11.4.13 Verify that the point (x, y) of equation (11.4) is a point on $E : y^2 = x^3 + Ax + B$ over $\mathbb{F}_q$. Show that, given a point (x, y) on an elliptic curve E over $\mathbb{F}_q$ as above, one can efficiently compute $u \in \mathbb{F}_q$, if it exists, such that the process of equation (11.4) gives the point (x, y).

In most cryptographic applications, we are interested in sampling from subgroups of $E(\mathbb{F}_q)$ of prime order r. As mentioned earlier, the simplest way to transform elements sampled randomly in $E(\mathbb{F}_q)$ into random elements of the subgroup is to exponentiate to the power $\#E(\mathbb{F}_q)/r$ (assuming that $r \| \#E(\mathbb{F}_q)$).

11.4.3 Hashing to algebraic groups

Recall that sampling from algebraic groups is the task of selecting group elements uniformly at random. On the other hand, a hash function $H : \{0, 1\}^l \to G(\mathbb{F}_q)$ is a deterministic algorithm that takes an input $\mathsf{m} \in \{0, 1\}^l$ and outputs a group element. It is required that the output distribution of H, corresponding to the uniform distribution of the message space, is close to uniform in the group $G(\mathbb{F}_q)$. The basic idea is to use m as the randomness required by the sampling algorithm.

Recall that a hash function is also usually required to satisfy some security requirements such as collision resistance. This is usually achieved by first applying a collision resistant hash function $H' : \{0,1\}^l \to \{0,1\}^l$ and setting $\mathsf{m}' = H'(\mathsf{m})$. In this section we are only concerned with the problem of using m' as input to a sampling algorithm.

The first case to consider is hashing to $\mathbb{F}_p^*$. If $p > 2^l + 1$ then we are in trouble since one cannot get uniform coverage of a set of size $p-1$ using fewer than $p-1$ elements. This shows that we always need $l > \log_2(\#G(\mathbb{F}_q))$ (though, in some applications, it might be possible to still have a useful cryptographic system even when the image of the hash function is a subset of the group).

Example 11.4.14 Suppose $2^l > p$ and $\mathsf{m} \in \{0,1\}^l$. The method of Example 11.4.2 gives output close to uniform (at least, if $l - \log_2(p)$ is reasonably large).

Exercise 11.4.15 Let $q = p^n < 2^l$. Give a hash function $H : \{0,1\}^l \to \mathbb{F}_q$.

It is relatively straightforward to turn the algorithms of Example 11.4.8 and Exercise 11.4.9 into hash functions. In the elliptic curve case, there is a growing literature on transforming a sampling algorithm into a hash function. We do not give the details.

11.4.4 Hashing from algebraic groups

In some applications, it is also necessary to have a hash function $H : G(\mathbb{F}_q) \to \{0,1\}^l$ where G is an algebraic group. Motivation for this problem is given in the discussion of key derivation functions in Section 20.2.3. A framework for problems of this type is **randomness extraction**. It is beyond the scope of this book to give a presentation of this topic, but some related results are given in Sections 21.7 and 21.6.

11.5 Determining group structure and computing generators for elliptic curves

Since $\mathbb{F}_q^*$ is cyclic, it follows that all subgroups of finite fields and tori are cyclic. However, elliptic curves and divisor class groups of hyperelliptic curves can be non-cyclic. Determining the group structure and a set of generators for an algebraic group $G(\mathbb{F}_q)$ can be necessary for some applications. It is important to remark that solutions to these problems are not expected to exist if the order $N = \#G(\mathbb{F}_q)$ is not known or if the factorisation of N is not known.

Let E be an elliptic curve over $\mathbb{F}_q$ and let $N = \#E(\mathbb{F}_q)$. If N has no square factors then $E(\mathbb{F}_q)$ is isomorphic as a group to $\mathbb{Z}/N\mathbb{Z}$. If $r^2 \| N$ then there could be a point of order r^2 or two "independent" points of order r (i.e., $E(\mathbb{F}_q)$ has a non-cyclic subgroup of order r^2 but exponent r).

The Weil pairing (see Section 26.2) can be used to determine the group structure of an elliptic curve. Let r be a prime and $P, Q \in E(\mathbb{F}_q)$ of order r. The key fact is that the

Weil pairing is alternating and so $e_r(P, P) = 1$. It follows from the non-degeneracy of the pairing that $e_r(P, Q) = 1$ if and only if $Q \in \langle P \rangle$. The Weil pairing also shows that one can only have two independent points when r divides $(q - 1)$.

Given the factorisation of $\gcd(q - 1, \#E(\mathbb{F}_q))$, the group structure can be determined using a randomised algorithm due to Miller [383, 385]. We present this algorithm in Algorithm 9. Note that the algorithm of Theorem 2.15.10 is used in lines 7 and 10. The expected running time is polynomial, but we refer to Miller [385] for the details.

Algorithm 9 Miller's algorithm for group structure

INPUT: $E/\mathbb{F}_q$, $N_0, N_1 \in \mathbb{N}$ and the factorisation of N_0, where $\#E(\mathbb{F}_q) = N_0 N_1$, $\gcd(N_1, q - 1) = 1$ and all primes dividing N_0 divide $q - 1$

OUTPUT: Integers m and n such that $E(\mathbb{F}_q) \cong (\mathbb{Z}/m\mathbb{Z}) \times (\mathbb{Z}/n\mathbb{Z})$ as a group

1: Write $N_0 = \prod_{i=1}^{k} l_i^{e_i}$ where $l_1, \ldots, l_k$ are distinct primes
2: For all $1 \le i \le k$ such that $e_i = 1$ set $N_0 = N_0/l_i$, $N_1 = N_1 l_i$
3: $m = 1, n = 1$
4: **while** $mn \neq N_0$ **do**
5: Choose random points $P', Q' \in E(\mathbb{F}_q)$
6: $P = [N_1]P'$, $Q = [N_1]Q'$
7: Find the exact orders m' and n' of P and Q
8: $n = \mathrm{lcm}(m', n')$
9: $\alpha = e_n(P, Q)$
10: Let m be the exact order of α in $\mu_n = \{z \in \mathbb{F}_q^* : z^n = 1\}$
11: **end while**
12: **return** m and nN_1

Exercise 11.5.1 Show that Algorithm 9 is correct.

Exercise 11.5.2 Modify Algorithm 9 so that it outputs generators for $E(\mathbb{F}_q)$.

Kohel and Shparlinski [317] give a deterministic algorithm to compute the group structure and to find generators for $E(\mathbb{F}_q)$. Their algorithm requires $O(q^{1/2+\epsilon})$ bit operations.

11.6 Testing subgroup membership

In many cryptographic protocols, it is necessary to verify that the elements received really do correspond to group elements with the right properties. There are a variety of attacks that can be performed otherwise, some of which are briefly mentioned in Section 20.4.2.

The first issue is whether a binary string corresponds to an element of the "parent group" $G(\mathbb{F}_q)$. This is usually easy to check when $G(\mathbb{F}_q) = \mathbb{F}_q^*$. In the case of elliptic curves, one must parse the bitstring as a point (x, y) and determine that (x, y) does satisfy the curve equation.

The more difficult problem is testing whether a group element g lies in the desired subgroup. For example, if $r \| \#G(\mathbb{F}_q)$ and we are given a group element g, to ensure that g lies in the unique subgroup of order r one can compute g^r and check if this is the identity. Efficient exponentiation algorithms can be used, but the computational cost is still significant. In some situations, one can more efficiently test subgroup membership. One notable case is when $\#G(\mathbb{F}_q)$ is prime; this is one reason why elliptic curves of prime order are so convenient for cryptography.

Exercise 11.6.1 (King) Let E be an elliptic curve over $\mathbb{F}_q$ such that $\#E(\mathbb{F}_q) = 2^m r$ where m is small and r is prime. Show how to use point halving (see Exercise 9.1.4) to efficiently determine whether a point $P \in E(\mathbb{F}_q)$ has order dividing r.

An alternative way to prevent attacks due to elements of incorrect group order is to "force" all group elements to lie in the required subgroup by exponentiating to a **cofactor** (such as $\#G(\mathbb{F}_q)/r$). When the cofactor is small, this can be a more efficient way to deal with the problem than testing subgroup membership, though one must ensure the cryptographic system can function correctly in this setting.

With algebraic group quotients represented using traces (i.e., LUC and XTR) one represents a finite field element using a trace. This value corresponds to a valid element of the extension field only if certain conditions hold. In the case of LUC, we represent $g \in G_{2,p}$, where p is prime, by the trace $V = \mathrm{Tr}(g)$. A value V corresponds to an element of $G_{2,q}$ if and only if the quadratic polynomial $(x - g)(x - g^p) = x^2 - Vx + 1$ is irreducible (in other words, if $(\frac{V^2-4}{p}) = -1$). Similarly, in XTR one needs to check whether the polynomial $x^3 - tx^2 + t^p x - 1$ is irreducible; Lenstra and Verheul [338] have given efficient algorithms to do this. Section 4 of [338] also discusses subgroup attacks in the context of XTR and countermeasures in this context.

12

Primality testing and integer factorisation using algebraic groups

There are numerous books about primality testing and integer factorisation, of which the most notable is Crandall and Pomerance [150]. There is no need to reproduce all the details of these topics. Hence, the purpose of this chapter is simply to sketch a few basic ideas that will be used later. In particular, we describe methods for primality testing and integer factorisation that exploit the structure of algebraic groups.

Definition 12.0.1 A **primality test** is a randomised algorithm that, on input $N \in \mathbb{N}$, outputs a single bit b such that if N is prime then $b = 1$. A composite integer that passes a primality test is called a **pseudoprime**. An algorithm **splits** $N \in \mathbb{N}$ if it outputs a pair (a, b) of integers such that $1 < a, b < N$ and $N = ab$.

12.1 Primality testing

The simplest primality test is **trial division** (namely, testing whether N is divisible by any integer up to $\sqrt{N}$). This algorithm is not useful for factoring numbers chosen for cryptography, but the first step of most general-purpose factoring algorithms is to run trial division to remove all 'small' prime factors of N before trying more elaborate methods. Hence, for the remainder of this section we may assume that N is odd (and usually that it is not divisible by any primes less than, say, 10^6).

12.1.1 Fermat test

Let $N \in \mathbb{N}$. If N is prime then the algebraic group $G_m(\mathbb{Z}/N\mathbb{Z}) = (\mathbb{Z}/N\mathbb{Z})^*$ over the ring $\mathbb{Z}/N\mathbb{Z}$ has $N - 1$ elements. In other words, if a is an integer such that $\gcd(a, N) = 1$ and

$$a^{N-1} \not\equiv 1 \pmod{N}$$

then N is not prime. Such a number a is called a **compositeness witness** for N. The hope is that if N is not prime then the order of the group $G_m(\mathbb{Z}/N\mathbb{Z})$ is not a divisor of $N - 1$ and so a compositeness witness exists. Hence, the Fermat test is to choose random $1 < a < N$ and compute $a^{N-1} \pmod{N}$.

As is well-known, there are composite numbers N that are pseudoprimes for the Fermat test.

Definition 12.1.1 An integer $N \in \mathbb{N}$ is a **Carmichael number** if N is composite and

$$a^{N-1} \equiv 1 \pmod{N}$$

for all $a \in \mathbb{N}$ such that $\gcd(a, N) = 1$.

If $N = \prod_{i=1}^{l} p_i^{e_i}$ is composite then $G_m(\mathbb{Z}/N\mathbb{Z}) \cong \prod_{i=1}^{l} G_m(\mathbb{Z}/p_i^{e_i}\mathbb{Z})$ and has order $\varphi(N)$ and exponent $\lambda(N) = \mathrm{lcm}\{p_i^{e_i-1}(p_i - 1) : 1 \le i \le l\}$.

Exercise 12.1.2 Show that all Carmichael numbers are odd. Show that N is a Carmichael number if and only if $\lambda(N) \mid (N-1)$. Show that a composite number $N \in \mathbb{N}$ is a Carmichael number if and only if $N = \prod_{i=1}^{l} p_i$ is a product of distinct primes such that $(p_i - 1) \mid (N - 1)$ for $i = 1, \dots, l$.

Exercise 12.1.3 Show that $561 = 3 \cdot 11 \cdot 17$ is a Carmichael number.

It was shown by Alford, Granville and Pomerance in 1992 that there are infinitely many Carmichael numbers.

It is natural to replace $G_m(\mathbb{Z}/N\mathbb{Z})$ with any algebraic group or algebraic group quotient, such as the torus $\mathbb{T}_2$, the algebraic group quotient corresponding to Lucas sequences (this gives rise to the $p + 1$ test) or an elliptic curve of predictable group order.

Exercise 12.1.4 Design a primality test based on the algebraic group $\mathbb{T}_2(\mathbb{Z}/N\mathbb{Z})$, which has order $N + 1$ if N is prime. Also show to use Lucas sequences to test N for primality using the algebraic group quotient.

Exercise 12.1.5 Design a primality test for integers $N \equiv 3 \pmod{4}$ based on the algebraic group $E(\mathbb{Z}/N\mathbb{Z})$ where E is a suitably chosen supersingular elliptic curve.

Exercise 12.1.6 Design a primality test for integers $N \equiv 1 \pmod{4}$ based on the algebraic group $E(\mathbb{Z}/N\mathbb{Z})$ where E is a suitably chosen elliptic curve.

12.1.2 The Miller–Rabin test

This primality test is also called the Selfridge–Miller–Rabin test or the strong prime test. It is a refinement of the Fermat test, and works very well in practice. Rather than changing the algebraic group, the idea is to make better use of the available information. It is based on the following trivial lemma, which is false if p is replaced by a composite number N (except for $N = p^a$ where p is odd).

Lemma 12.1.7 *Let p be prime. If $x^2 \equiv 1 \pmod{p}$ then $x \equiv \pm 1 \pmod{p}$.*

For the Miller–Rabin test write $N - 1 = 2^b m$ where m is odd and consider the sequence $a_0 = a^m \pmod{N}$, $a_1 = a_0^2 = a^{2m} \pmod{N}$, $\dots$, $a_b = a_{b-1}^2 = a^{N-1} \pmod{N}$ where $\gcd(a, N) = 1$. If N is prime then this sequence must have the form $(*, *, \dots, *, -1, 1, \dots, 1)$ or $(-1, 1, \dots, 1)$ or $(1, \dots, 1)$ (where $*$ denotes numbers whose values are not relevant). Any deviation from this form means that the number N is composite.

An integer N is called a **base-a probable prime** if the Miller–Rabin sequence has the good form and is called a **base-a pseudoprime** if it is a base-a probable prime that is actually composite.

Exercise 12.1.8 Let $N = 561$. Note that $\gcd(2, N) = 1$ and $2^{N-1} \equiv 1 \pmod{N}$. Show that the Miller–Rabin method with $a = 2$ demonstrates that N is composite. Show that this failure allows one to immediately split N.

Theorem 12.1.9 *Let $n > 9$ be an odd composite integer. Then N is a base-a pseudoprime for at most $\varphi(N)/4$ bases between 1 and N.*

Proof See Theorem 3.5.4 of [150] or Theorem 10.6 of Shoup [497]. □

Hence, if a number N passes several Miller–Rabin tests for several randomly chosen bases a then one can believe that with high probability N is prime (Section 5.4.2 of Stinson [532] gives a careful analysis of the probability of success of a closely related algorithm using Bayes' theorem). Such an integer is called a **probable prime**. In practice, one chooses $O(\log(N))$ random bases a and runs the Miller–Rabin test for each. The total complexity is therefore $O(\log(N)^4)$ bit operations (which can be improved to $O(\log(N)^2 M(\log(N)))$ where $M(m)$ is the cost of multiplying two m-bit-integers).

12.1.3 Primality proving

Agrawal, Kayal and Saxena (AKS) discovered a deterministic algorithm that runs in polynomial-time and determines whether or not N is prime. We refer to Section 4.5 of [150] for details. The original AKS test has been improved significantly. A variant due to Bernstein requires $O(\log(N)^{4+o(1)})$ bit operations using fast arithmetic (see Section 4.5.4 of [150]).

There is also a large literature on primality proving using Gauss and Jacobi sums and using elliptic curves. We refer to Sections 4.4 and 7.6 of [150].

In practice, the Miller–Rabin test is still widely used for cryptographic applications.

12.2 Generating random primes

Definition 12.2.1 Let $X \in \mathbb{N}$, then $\pi(X)$ is defined to be the number of primes $1 < p < X$.

The famous **prime number theorem** states that $\pi(X)$ is asymptotically equal to $X/\log(X)$ (as always log denotes the natural logarithm). In other words, primes are rather common among the integers. If one choose a random integer $1 < p < X$ then the probability that p is prime is therefore about $1/\log(X)$ (equivalently, about $\log(X)$ trials are required to find a prime between 1 and X). In practice, this probability increases significantly if one choose p to be odd and not divisible by 3.

Theorem 12.2.2 *Random (probable) prime numbers of a given size X can be generated using the Miller–Rabin algorithm in expected $O(\log(X)^5)$ bit operations (or $O(\log(X)^3 M(\log(X)))$ using fast arithmetic).*

Exercise 12.2.3 For certain cryptosystems based on the discrete logarithm problem it is required to produce a k_1-bit prime p such that $p-1$ has a k_2-bit prime factor q. Give a method that takes integers $k_2 < k_1$ and outputs p and q such that p is a k_1-bit prime, q is a k_2-bit prime and $q \mid (p-1)$.

Exercise 12.2.4 For certain cryptosystems based on the discrete logarithm problem (see Chapter 6) it is required to produce a k_1-bit prime p such that $\Phi_k(p)$ has a k_2-bit prime factor q (where $\Phi_k(x)$ is the kth cyclotomic polynomial). Give a method that takes integers k, k_1, k_2 such that $k_2 < \varphi(k)k_1$ and outputs p and q such that p is a k_1-bit prime, q is a k_2-bit prime and $q \mid \Phi_k(p)$.

Exercise 12.2.5 A **strong prime** is defined to be a prime p such that $q = (p-1)/2$ is prime, $(p+1)/2$ is prime and $(q-1)/2$ is prime (it is conjectured that infinitely many such primes exist). Some RSA systems require the RSA moduli to be a product of strong primes. Give an algorithm to generate strong primes.

12.2.1 Primality certificates

For cryptographic applications it may be required to provide a **primality certificate**. This is a mathematical proof that can be checked in polynomial-time and that establishes the primality of a number n. Pratt [440] showed that there exists a short primality certificate for every prime. Primality certificates are not so important since the discovery of the AKS test, but primes together with certificates can be generated (and the certificates verified) more quickly than using the AKS test, so this subject could still be of interest. We refer to Section 4.1.3 of Crandall and Pomerance [150] and Maurer [363] for further details.

One basic tool for primality certificates is Lucas' converse of Fermat's little theorem.

Theorem 12.2.6 *(Lucas) Let $N \in \mathbb{N}$. If there is an integer a such that* $\gcd(a, N) = 1$, $a^{N-1} \equiv 1 \pmod{N}$ *and* $a^{(N-1)/l} \not\equiv 1 \pmod{N}$ *for all primes $l \mid (N-1)$ then N is prime.*

Exercise 12.2.7 Prove Theorem 12.2.6.

In practice, one can weaken the hypothesis of Theorem 12.2.6.

Theorem 12.2.8 *(Pocklington) Suppose $N - 1 = FR$ where the complete factorisation of F is known. Suppose there is an integer a such that $a^{N-1} \equiv 1 \pmod{N}$ and*

$$a^{(N-1)/q} \not\equiv 1 \pmod{N}$$

for every prime $q \mid F$. Then every prime factor of N is congruent to 1 modulo F. Hence, if $F \geq \sqrt{N}$ then N is prime.

Exercise 12.2.9 Prove Theorem 12.2.8.

Exercise 12.2.10 A **Sophie Germain prime** (in cryptography the name **safe prime** is commonly used) is a prime p such that $(p-1)/2$ is also prime. It is conjectured that there are infinitely many Sophie Germain primes. Give a method to generate a k-bit Sophie Germain prime together with a certificate of primality, such that the output is close to uniform over the set of all k-bit Sophie Germain primes.

12.3 The $p-1$ factoring method

First we recall the notion of a smooth integer. These are discussed in more detail in Section 15.1.

Definition 12.3.1 Let $N = \prod_{i=1}^{r} p_i^{e_i} \in \mathbb{N}$ (where we assume the p_i are distinct primes and $e_i \geq 1$) and let $B \in \mathbb{N}$. Then N is **B-smooth** if all $p_i \leq B$ and N is **B-power smooth** (or **strongly B-smooth**) if all $p_i^{e_i} \leq B$.

Example 12.3.2 $528 = 2^4 \cdot 3 \cdot 11$ is 14-smooth but is not 14-power smooth.

The $p-1$ method was published by Pollard [435].[1] The idea is to suppose that N has prime factors p and q where $p-1$ is B-power smooth but $q-1$ is not B-power smooth. Then if $1 < a < N$ is randomly chosen we have $a^{B!} \equiv 1 \pmod{p}$ and, with high probability, $a^{B!} \not\equiv 1 \pmod{q}$. Hence, $\gcd(a^{B!} - 1, N)$ splits N. Algorithm 10 gives the Pollard $p-1$ algorithm.

Example 12.3.3 Let $N = 124639$ and let $B = 8$. Choose $a = 2$. One can check that

$$\gcd(a^{B!} \pmod{N} - 1, N) = 113$$

from which one deduces that $N = 113 \cdot 1103$.

This example worked because the prime $p = 113$ satisfies $p - 1 = 2^4 \cdot 7 \mid 8!$ and so $2^{8!} \equiv 1 \pmod{p}$ while the other prime satisfies $q - 1 = 2 \cdot 19 \cdot 29$, which is not 8-smooth.

Of course, the "factor" returned from the gcd may be 1 or N. If the factor is not 1 or N then we have split N as $N = ab$. We now test each factor for primality and attempt to split any composite factors further.

Algorithm 10 Pollard $p-1$ algorithm

INPUT: $N \in \mathbb{N}$
OUTPUT: Factor of N

```
1: Choose a suitable value for B
2: Choose a random 1 < a < N
3: b = a
4: for i = 2 to B do
5:     b = b^i (mod N)
6: end for
7: return gcd(b − 1, N)
```

Exercise 12.3.4 Factor $N = 10028219737$ using the $p-1$ method.

Lemma 12.3.5 *The complexity of Algorithm 10 is $O(B \log(B) M(\log(N)))$ bit operations.*

[1] According to [567], the first stage of the method was also known to D. N. Lehmer and D. H. Lehmer, though they never published it.

Proof The main loop is repeated B times and contains an exponentiation modulo N to a power $i < B$. The cost of the exponentiation is $O(\log(B)M(\log(N)))$ bit operations. □

The algorithm is therefore exponential in B and so is only practical if B is relatively small. If $B = O(\log(N)^i)$ then the algorithm is polynomial-time. Unfortunately, the algorithm only splits numbers of a special form (namely, those for which there is a factor p such that $p-1$ is very smooth).

Exercise 12.3.6 Show that searching only over prime power values for i in Algorithm 10 lowers the complexity to $O(BM(\log(N)))$ bit operations.

It is usual to have a **second stage** or **continuation** to the Pollard $p-1$ method. Suppose that Algorithm 10 terminates with $\gcd(b-1, N) = 1$. If there is a prime $p \mid N$ such that $p-1 = SQ$ where S is B-smooth and Q is prime then the order of b modulo p is Q. One will therefore expect to split N by computing $\gcd(b^Q \pmod N) - 1, N)$. The second stage is to find Q if it is not too big. One therefore chooses a bound $B' > B$ and wants to compute $\gcd(b^Q \pmod N) - 1, N)$ for all primes $B < Q \le B'$.

We give two methods to do this: the standard continuation (Exercise 12.3.7) has the same complexity as the first stage of the $p-1$ method, but the constants are much better; the FFT continuation (Exercise 12.3.8) has better complexity and shows that if sufficient storage is available then one can take B' to be considerably bigger than B. Further improvements are given in Sections 4.1 and 4.2 of Montgomery [392].

Exercise 12.3.7 (Standard continuation) Show that one can compute $\gcd(b^Q \pmod N) - 1, N)$ for all primes $B < Q \le B'$ in $O((B' - B)M(\log(N)))$ bit operations.

Exercise 12.3.8 (Pollard's FFT continuation) Let $w = \lceil \sqrt{B' - B} \rceil$. We will exploit the fact that $Q = B + vw - u$ for some $0 \le u < w$ and some $1 \le v \le w$ (this is very similar to the baby-step–giant-step algorithm; see Section 13.3). Let $P(x) = \prod_{i=0}^{w-1}(x - b^i) \pmod N$, computed as in Section 2.16. Now compute $\gcd(P(g^{B+vw}) \pmod N), N)$ for $v = 1, 2, \ldots, w$. For the correct value v we have

$$P(g^{B+vw}) = \prod_i (g^{B+vw} - g^i) = (g^{B+vw} - g^u)\prod_{i \ne u}(g^{B+vw} - g^i)$$

$$= g^u(g^{B+vw-u} - 1)\prod_{i \ne u}(g^{B+vw} - g^i).$$

Since $g^{B+vw-u} = g^Q \equiv 1 \pmod p$ then $\gcd(P(g^{B+vw}) \pmod N), N)$ is divisible by p. Show that the time complexity of this continutation is $O(M(w)\log(w)M(\log(N))$, which asymptotically is $O(\sqrt{B'}\log(B')^2\log(\log(B'))M(\log(N)))$ bit operations. Show that the storage required is $O(w\log(w)) = O(\sqrt{B'}\log(B'))$ bits.

Exercise 12.3.9 The $p+1$ factoring method uses the same idea as the $p-1$ method but in the algebraic group $\mathbb{T}_2$ or the algebraic group quotient corresponding to Lucas sequences. Write down the details of the $p+1$ factoring method using Lucas sequences.

12.4 Elliptic curve method

Let N be an integer to be factored and let $p \mid N$ be prime. One can view Pollard's $p - 1$ method as using an auxiliary group (namely, $G_m(\mathbb{F}_p)$) that may have smooth order. The idea is then to obtain an element modulo N (namely, $a^{B!}$), that is congruent modulo p (but not modulo some other prime $q \mid N$) to the identity element of the auxiliary group.

Lenstra's idea was to replace the group G_m in the Pollard $p - 1$ method with the group of points on an elliptic curve. The motivation was that even if $p - 1$ is not smooth it is reasonable to expect that there is an elliptic curve E over $\mathbb{F}_p$ such that $\#E(\mathbb{F}_p)$ is rather smooth. Furthermore, since there are lots of different elliptic curves over the field $\mathbb{F}_p$ we have a chance to split N by trying the method with lots of different elliptic curves. We refer to Section 9.14 for some remarks on elliptic curves modulo N.

If E is a "randomly chosen" elliptic curve modulo N with a point P on E modulo N then one hopes that the point $Q = [B!]P$ is congruent modulo p (but not modulo some other prime q) to the identity element. One constructs E and P together, for example choosing $1 < x_P, y_P, a_4 < N$ and setting $a_6 = y_P^2 - x_P^3 - a_4 x_P \pmod{N}$. If one computes $Q = (x : y : z)$ using inversion-free arithmetic and projective coordinates (as in Exercise 9.1.5) then $Q \equiv \mathcal{O}_E \pmod{p}$ is equivalent to $p \mid z$. Here we are performing elliptic curve arithmetic over the ring $\mathbb{Z}/N\mathbb{Z}$ (see Section 9.14).

The resulting algorithm is known as the **elliptic curve method** or **ECM** and it is very widely used, both as a general-purpose factoring algorithm in computer algebra packages, and as a subroutine of the number field sieve. An important consequence of Lenstra's suggestion of replacing the group $\mathbb{F}_p^*$ by $E(\mathbb{F}_p)$ is that it motivated Miller and Koblitz to suggest using $E(\mathbb{F}_p)$ instead of $\mathbb{F}_p^*$ for public key cryptography.

Algorithm 11 gives a sketch of one round of the ECM algorithm. If the algorithm fails then one should repeat it, possibly increasing the size of B. Note that it can be more efficient to compute $[B!]P$ as a single exponentiation rather than a loop as in line 5 of Algorithm 11; see [47].

Algorithm 11 Elliptic curve factoring algorithm

INPUT: $N \in \mathbb{N}$
OUTPUT: Factor of N

1: Choose a suitable value for B
2: Choose random elements $0 \le x, y, a_4 < N$
3: Set $a_6 = y^2 - x^3 - a_4 x \pmod{N}$
4: Set $P = (x : y : 1)$
5: **for** $i = 2$ to B **do**
6:     Compute $P = [i]P$
7: **end for**
8: **return** $\gcd(N, z)$ where $P = (x : y : z)$

Exercise 12.4.1 Show that the complexity of Algorithm 11 is $O(B\log(B)M(\log(N)))$ bit operations.

Exercise 12.4.2 Show that the complexity of Algorithm 11 can be lowered to $O(BM(\log(N)))$ bit operations using the method of Exercise 12.3.6.

Many of the techniques used to improve the Pollard $p-1$ method (such as the standard continuation, though not Pollard's FFT continuation) also apply directly to the elliptic curve method. We refer to Section 7.4 of [150] for details. One can also employ all known techniques to speed up elliptic curve arithmetic. Indeed, the Montgomery model for elliptic curves (Section 9.12.1) was discovered in the context of ECM rather than ECC (elliptic curve cryptography).

In practice, we repeat the algorithm a number of times for random choices of B, x, y and a_4. The difficult problems are to determine a good choice for B and to analyse the probability of success. We discuss these issues in Section 15.3 where we state Lenstra's conjecture that the elliptic curve method factors integers in subexponential time.

12.5 Pollard–Strassen method

Pollard [435] and, independently, Strassen gave a deterministic algorithm to factor an integer N in $\tilde{O}(N^{1/4})$ bit operations. It is based on the idea[2] of Section 2.16; namely, that one can evaluate a polynomial of degree n in $(\mathbb{Z}/N\mathbb{Z})[x]$ at n values in $O(M(n)\log(n))$ operations in $\mathbb{Z}/N\mathbb{Z}$. A different factoring algorithm with this complexity is given in Exercise 19.4.7.

The trick to let $B = \lceil N^{1/4} \rceil$, $F(x) = x(x-1)\cdots(x-B+1)$ (which has degree B) and to compute $F(jB) \pmod{N}$ for $1 \le j \le B$. Computing these values requires $O(M(B)\log(B)M(\log(N))) = O(N^{1/4}\log(N)^3\log(\log(N))^2\log(\log(\log(N))))$ bit operations. Once this list of values has been computed, one computes $\gcd(N, F(jB) \pmod{N})$ until one finds a value that is not 1. This will happen, for some j, since the smallest prime factor of N is of the form $jB - i$ for some $1 \le j \le B$ and some $0 \le i < B$. Note that $M = \gcd(N, F(jB) \pmod{N})$ may not be prime, but one can find the prime factors of it in $\tilde{O}(N^{1/4})$ bit operations by computing $\gcd(M, jB - i)$ for that value of j and all $0 \le i < B$. Indeed, one can find all prime factors of N that are less than $N^{1/2}$ (and hence factor N completely) using this method. The overall complexity is $\tilde{O}(N^{1/4})$ bit operations.

Exercise 12.5.1★ Show that one can determine all primes p such that $p^2 \mid N$ in $\tilde{O}(N^{1/6})$ bit operations.

[2] Despite the title of this chapter, the Pollard–Strassen algorithm does not use algebraic groups or any group-theoretic property of the integers modulo N.

13

Basic discrete logarithm algorithms

This chapter is about algorithms to solve the discrete logarithm problem (DLP) and some variants of it. We focus mainly on deterministic methods that work in any group; later chapters will present the Pollard rho and kangaroo methods, and index calculus algorithms. In this chapter, we also present the concept of generic algorithms and prove lower bounds on the running time of a generic algorithm for the DLP. The starting point is the following definition (already given as Definition 2.1.1).

Definition 13.0.1 Let G be a group written in multiplicative notation. The **discrete logarithm problem (DLP)** is: given $g, h \in G$ find a, if it exists, such that $h = g^a$. We sometimes denote a by $\log_g(h)$.

As discussed after Definition 2.1.1, we intentionally do not specify a distribution on g or h or a above, although it is common to assume that g is sampled uniformly at random in G, and a is sampled uniformly from $\{1, \ldots, \#G\}$.

Typically, G will be an algebraic group over a finite field $\mathbb{F}_q$ and the order of g will be known. If one is considering cryptography in an algebraic group quotient then we assume that the DLP has been lifted to the covering group G. A solution to the DLP exists if and only if $h \in \langle g \rangle$ (i.e., h lies in the subgroup generated by g). We have discussed methods to test this in Section 11.6.

Exercise 13.0.2 Consider the discrete logarithm problem in the group of integers modulo p under **addition**. Show that the discrete logarithm problem in this case can be solved in polynomial-time.

Exercise 13.0.2 shows there are groups for which the DLP is easy. The focus in this book is on algebraic groups for which the DLP seems to be hard.

Exercise 13.0.3 Let N be composite. Define the discrete logarithm problem DLP-MOD-N in the multiplicative group of integers modulo N. Show that FACTOR $\leq_R$ DLP-MOD-N.

Exercise 13.0.3 gives some evidence that cryptosystems based on the DLP should be at least as secure as cryptosystems based on factoring.

13.1 Exhaustive search

The simplest algorithm for the DLP is to sequentially compute g^a for $0 \le a < r$ and test equality of each value with h. This requires at most $r - 2$ group operations and r comparisons.

Exercise 13.1.1 Write pseudocode for the exhaustive search algorithm for the DLP and verify the claims about the worst-case number of group operations and comparisons.

If the cost of testing equality of group elements is $O(1)$ group operations then the worst-case running time of the algorithm is $O(r)$ group operations. It is natural to assume that testing equality is always $O(1)$ group operations, and this will always be true for the algebraic groups considered in this book. However, as Exercise 13.1.2 shows, such an assumption is not entirely trivial.

Exercise 13.1.2 Suppose projective coordinates are used for elliptic curves $E(\mathbb{F}_q)$ to speed up the group operations in the exhaustive search algorithm. Show that testing equality between a point in projective coordinates and a point in affine or projective coordinates requires at least one multiplication in $\mathbb{F}_q$ (and so this cost is not linear). Show that, nevertheless, the cost of testing equality is less than the cost of a group operation.

For the rest of this chapter we assume that groups are represented in a compact way and that operations involving the representation of the group (e.g., testing equality) all cost less than the cost of one group operation. This assumption is satisfied for all the algebraic groups studied in this book.

13.2 The Pohlig–Hellman method

Let g have order N and let $h = g^a$ so that h lies in the cyclic group generated by g. Suppose $N = \prod_{i=1}^{n} l_i^{e_i}$. The idea of the Pohlig–Hellman[1] method [432] is to compute a modulo the prime powers $l_i^{e_i}$ and then recover the solution using the Chinese remainder theorem. The main ingredient is the following group homomorphism, which reduces the discrete logarithm problem to subgroups of prime power order.

Lemma 13.2.1 *Suppose g has order N and $l^e \mid N$. The function*

$$\Phi_{l^e}(g) = g^{N/l^e}$$

is a group homomorphism from $\langle g \rangle$ to the unique cyclic subgroup of $\langle g \rangle$ of order l^e. Hence, if $h = g^a$ then

$$\Phi_{l^e}(h) = \Phi_{l^e}(g)^{a \pmod{l^e}}.$$

Exercise 13.2.2 Prove Lemma 13.2.1.

[1] The paper [432] is authored by Pohlig and Hellman and so the method is usually referred to by this name, although Silver, Schroeppel, Block and Nechaev also discovered it.

Using Φ_{l^e} one can reduce the DLP to subgroups of prime power order. To reduce the problem to subgroups of prime order one can the following: suppose g_0 has order l^e and $h_0 = g_0^a$ then we can write $a = a_0 + a_1 l + \cdots a_{e-1} l^{e-1}$ where $0 \le a_i < l$. Let $g_1 = g_0^{l^{e-1}}$. Raising to the power l^{e-1} gives

$$h_0^{l^{e-1}} = g_1^{a_0}$$

from which one can find a_0 by trying all possibilities (or using baby-step–giant-step or other methods).

To compute a_1 we define $h_1 = h_0 g_0^{-a_0}$ so that

$$h_1 = g_0^{a_1 l + a_2 l^2 + \cdots a_{e-1} l^{e-1}}.$$

Then a_1 is obtained by solving

$$h_1^{l^{e-2}} = g_1^{a_1}.$$

To obtain the next value we set $h_2 = h_1 g_0^{-l a_1}$ and repeat. Continuing gives the full solution modulo l^e. Once a is known modulo $l_i^{e_i}$ for all $l_i^{e_i} \| N$ one computes a using the Chinese remainder theorem. The full algorithm (in a slightly more efficient variant) is given in Algorithm 12.

Algorithm 12 Pohlig–Hellman algorithm

INPUT: $g, h = g^a, \{(l_i, e_i) : 1 \le i \le n\}$ such that order of g is $N = \prod_{i=1}^{n} l_i^{e_i}$
OUTPUT: a

```
1: Compute {g^{N/l_i^{f_i}}, h^{N/l_i^{f_i}} : 1 ≤ i ≤ n, 1 ≤ f_i ≤ e_i}
2: for i = 1 to n do
3:     a_i = 0
4:     for j = 1 to e_i do                 ▷ Reducing DLP of order l_i^{e_i} to cyclic groups
5:         Let g_0 = g^{N/l_i^j} and h_0 = h^{N/l_i^j}    ▷ These were already computed in line 1
6:         Compute u = g_0^{-a_i} and h_0 = h_0 u
7:         if h_0 ≠ 1 then
8:             Let g_0 = g^{N/l_i}, b = 1, T = g_0        ▷ Already computed in line 1
9:             while h_0 ≠ T do                           ▷ Exhaustive search
10:                b = b + 1, T = T g_0
11:            end while
12:            a_i = a_i + b l_i^{j-1}
13:        end if
14:    end for
15: end for
16: Use Chinese remainder theorem to compute a ≡ a_i (mod l_i^{e_i}) for 1 ≤ i ≤ n
17: return a
```

Example 13.2.3 Let $p = 19$, $g = 2$ and $h = 5$. The aim is to find an integer a such that $h \equiv g^a \pmod{p}$. Note that $p - 1 = 2 \cdot 3^2$. We first find a modulo 2. We have $(p-1)/2 = 9$ so define $g_0 = g^9 \equiv -1 \pmod{19}$ and $h_0 = h^9 \equiv 1 \pmod{19}$. It follows that $a \equiv 0 \pmod{2}$.

Now we find a modulo 9. Since $(p-1)/9 = 2$ we first compute $g_0 = g^2 \equiv 4 \pmod{19}$ and $h_0 \equiv h^2 \equiv 6 \pmod{19}$. To get information modulo 3 we compute

$$g_1 = g_0^3 \equiv 7 \pmod{19} \text{ and } h_0^3 \equiv 7 \pmod{19}.$$

It follows that $a \equiv 1 \pmod{3}$. To get information modulo 9 we remove the modulo 3 part by setting $h_1 = h_0/g_0 = 6/4 \equiv 11 \pmod{19}$. We now solve $h_1 \equiv g_1^{a_1} \pmod{19}$, which has the solution $a_1 \equiv 2 \pmod{3}$. It follows that $a \equiv 1 + 3 \cdot 2 \equiv 7 \pmod{9}$.

Finally, by the Chinese remainder theorem we obtain $a \equiv 16 \pmod{18}$.

Exercise 13.2.4 Let $p = 31$, $g = 3$ and $h = 22$. Solve the discrete logarithm problem of h to the base g using the Pohlig–Hellman method.

We recall that an integer is B-smooth if all its prime factors are at most B.

Theorem 13.2.5 *Let $g \in G$ have order N. Let $B \in \mathbb{N}$ be such that N is B-smooth Then Algorithm 12 solves the DLP in G using $O(\log(N)^2 + B\log(N))$ group operations.*[2]

Proof One can factor N using trial division in $O(BM(\log(N)))$ bit operations, where $M(n)$ is the cost of multiplying n-bit integers. We assume that $M(\log(N))$ is $O(1)$ group operations (this is true for all the algebraic groups of interest in this book). Hence, we may assume that the factorisation of N is known.

Computing all $\Phi_{l_i^{e_i}}(g)$ and $\Phi_{l_i^{e_i}}(h)$ can be done naively in $O(\log(N)^2)$ group operations, but we prefer to do it in $O(\log(N)\log\log(N))$ group operations using the method of Section 2.15.1.

Lines 5 to 13 run $\sum_{i=1}^n e_i = O(\log(N))$ times and, since each $l_i \geq 2$, we have $\sum_{i=1}^n e_i \leq \log_2(N)$. The computation of u in line 6 requires $O(e_i \log(l_i))$ group operations. Together this gives a bound of $O(\log(N)^2)$ group operations to the running time. (Note that when $N = 2^e$ then the cost of these lines is $e^2\log(2) = O(\log(N)^2)$ group operations.)

Solving each DLP in a cyclic group of order l_i using naive methods requires $O(l_i)$ group operations (this can be improved using the baby-step–giant-step method). There are $\leq \log_2(N)$ such computations to perform, giving $O(\log(N)B)$ group operations.

The final step is to use the Chinese remainder theorem to compute a, requiring $O(\log(N)M(\log(N)))$ bit operations, which is again assumed to cost at most $O(\log(N))$ group operations. □

Due to this method small primes give no added security in discrete logarithm systems. Hence, one generally uses elements of prime order r for cryptography.

[2] By this, we mean that the constant implicit in the $O(\cdot)$ is independent of B and N.

Exercise 13.2.6 Recall the Tonelli–Shanks algorithm for computing square roots modulo p from Section 2.9. A key step of the algorithm is to find a solution j to the equation $b = y^{2j} \pmod{p}$ where y has order 2^e. Write down the Pohlig–Hellman method to solve this problem. Show that the complexity is $O(\log(p)^2 M(\log(p)))$ bit operations.

Exercise 13.2.7 Let $B \in \mathbb{N}_{>3}$. Let $N = \prod_{i=1}^{n} x_i$ where $2 \le x_i \le B$. Prove that $\sum_{i=1}^{n} x_i \le B \log(N)/\log(B)$.

Hence, show that the Pohlig–Hellman method performs $O(\log(N)^2 + B\log(N)/\log(B))$ group operations.

Remark 13.2.8 As we will see, replacing exhaustive search by the baby-step–giant-step algorithm improves the complexity to $O(\log(N)^2 + \sqrt{B}\log(N)/\log(B))$ group operations (at the cost of more storage).

Algorithm 12 can be improved, when there is a prime power l^e dividing N with e large, by structuring it differently. Section 11.2.3 of Shoup [497] gives a method to compute the DLP in a group of order l^e in $O(e\sqrt{l} + e\log(e)\log(l))$ group operations (this is using baby-step–giant-step rather than exhaustive search). Algorithm 1 and Corollary 1 of Sutherland [537] give an algorithm that requires

$$O(e\sqrt{l} + e\log(l)\log(e)/\log(\log(e))) \tag{13.1}$$

group operations. Sutherland also considers non-cyclic groups.

If N is B-smooth then summing the improved complexity statements over the prime powers dividing N gives

$$O(\log(N)\sqrt{B}/\log(B) + \log(N)\log(\log(N))) \tag{13.2}$$

group operations for the DLP (it is not possible to have a denominator of $\log(\log(\log(N)))$ since not all the primes dividing N necessarily appear with high multiplicity).

13.3 Baby-step–giant-step (BSGS) method

This algorithm, usually credited to Shanks,[3] exploits an idea called the time/memory tradeoff. Suppose g has prime order r and that $h = g^a$ for some $0 \le a < r$. Let $m = \lceil \sqrt{r} \rceil$. Then there are integers a_0, a_1 such that $a = a_0 + m a_1$ and $0 \le a_0, a_1 < m$. It follows that

$$g^{a_0} = h(g^{-m})^{a_1}$$

and this observation leads to Algorithm 13. The algorithm requires storing a large list of values and it is important, in the second stage of the algorithm, to be able to efficiently determine whether or not an element lies in the list. There are a number of standard solutions to this problem including using binary trees, hash tables or sorting the list after line 7 of the algorithm (see, for example, parts II and III of [136] or Section 6.3 of [283]).

[3] Nechaev [405] states it was known to Gel'fond in 1962.

Algorithm 13 Baby-step–giant-step (BSGS) algorithm

INPUT: $g, h \in G$ of order r
OUTPUT: a such that $h = g^a$, or $\perp$

1: $m = \lceil \sqrt{r} \rceil$
2: Initialise an easily searched structure (such as a binary tree or a hash table) L
3: $x = 1$
4: **for** $i = 0$ to m **do** ▷ Compute baby steps
5: store (x, i) in L, easily searchable on the first coordinate
6: $x = xg$
7: **end for**
8: $u = g^{-m}$
9: $y = h, j = 0$
10: **while** $(y, \star) \notin L$ **do** ▷ Compute giant steps
11: $y = yu, j = j + 1$
12: **end while**
13: **if** $\exists (x, i) \in L$ such that $x = y$ **then**
14: **return** $i + mj$
15: **else**
16: **return** $\perp$
17: **end if**

Note that the BSGS algorithm is deterministic. The algorithm also solves the decision problem (is $h \in \langle g \rangle$?) though, as discussed in Section 11.6, there are usually faster solutions to the decision problem.

Theorem 13.3.1 *Let G be a group of order r. Suppose that elements of G are represented using $O(\log(r))$ bits and that the group operations can be performed in $O(\log(r)^2)$ bit operations. The BSGS algorithm for the DLP in G has running time $O(\sqrt{r}\log(r)^2)$ bit operations. The algorithm requires $O(\sqrt{r}\log(r))$ bits of storage.*

Proof The algorithm computes $\sqrt{r}$ group operations for the baby steps. The cost of inserting each group element into the easily searched structure is $O(\log(r)^2)$ bit operations, since comparisons require $O(\log(r))$ bit operations (this is where the assumption on the size of element representations appears). The structure requires $O(\sqrt{r}\log(r))$ bits of storage.

The computation of $u = g^{-m}$ in line 8 requires $O(\log(r))$ group operations.

The algorithm needs one group operation to compute each giant step. Searching the structure takes $O(\log(r)^2)$ bit operations. In the worst case, one has to compute m giant steps. The total running time is therefore $O(\sqrt{r}\log(r)^2)$ bit operations. □

The storage requirement of the BSGS algorithm quickly becomes prohibitive. For example, one can work with primes r such that $\sqrt{r}$ is more than the number of fundamental particles in the universe!

Remark 13.3.2 When solving the DLP it is natural to implement the group operations as efficiently as possible. For example, when using elliptic curves it would be tempting to use a projective representation for group elements (see Exercise 13.1.2). However this is not suitable for the BSGS method (or the rho and kangaroo methods) as one cannot efficiently detect a match $y \in L$ when there is a non-unique representation for the group element y.

Exercise 13.3.3 On average, the BSGS algorithm finds a match after half the giant steps have been performed. The average-case running time of the algorithm as presented is therefore approximately $1.5\sqrt{r}$ group operations. Show how to obtain an algorithm that requires, in the average case, approximately $\sqrt{2r}$ group operations and $\sqrt{r/2}$ group elements of storage.

Exercise 13.3.4 (Pollard [438]) A variant of the BSGS algorithm is to compute the baby steps and giant steps in parallel, storing the points together in a single structure. Show that if x and y are chosen uniformly in the interval $[0, r] \cap \mathbb{Z}$ then the expected value of $\max\{x, y\}$ is approximately $\frac{2}{3}r$. Hence, show that the average-case running time of this variant of the BSGS algorithm is $\frac{4}{3}\sqrt{r}$ group operations.

Exercise 13.3.5 Design a variant of the BSGS method that requires $O(r/M)$ group operations if the available storage is only for $M < \sqrt{r}$ group elements.

Exercise 13.3.6 (DLP in an interval) Suppose one is given g of order r in a group G and integers $0 \le b, w < r$. The DLP in an interval of length w is: given $h \in \langle g \rangle$ such that $h = g^a$ for some $b \le a < b + w$ to find a. Give a BSGS algorithm to find a in average-case $\sqrt{2w}$ group operations and $\sqrt{w/2}$ group elements of storage.

Exercise 13.3.7 Suppose one considers the DLP in a group G where computing the inverse g^{-1} is much faster than multiplication in the group. Show how to solve the DLP in an interval of length w using a BSGS algorithm in approximately $\sqrt{w}$ group operations in the average case.

Exercise 13.3.8 Suppose one is given $g, h \in G$ and $w, b, m \in \mathbb{N}$ such that $h = g^a$ for some integer a satisfying $0 \le a < w$ and $a \equiv b \pmod{m}$. Show how to reduce this problem to the problem of solving a DLP in an interval of length w/m.

Exercise 13.3.9 Let $g \in G$ have order $N = mr$ where r is prime and m is $\log(N)$-smooth. Suppose $h = g^x$ and w are given such that $0 \le x < w$. Show how one can compute x by combining the Pohlig–Hellman method and the BSGS algorithm in $O(\log(N)^2 + \sqrt{w/m})$ group operations.

Exercise 13.3.10 Suppose one is given $g, h \in G$ and $b_1, b_2, w \in \mathbb{Z}$ $(w > 0)$ such that $b_1 + w < b_2$ and $h = g^a$ for some integer a satisfying either $b_1 \le a < b_1 + w$ or $b_2 \le a < b_2 + w$. Give an efficient BSGS algorithm for this problem.

Exercise 13.3.11 Let $g \in G$ where the order of g and G are not known. Suppose one is given integers b, w such that the order of g lies in the interval $[b, b+w)$. Explain how to use the BSGS method to compute the order of g.

Exercise 13.3.12★ Suppose one is given an element g of order r and $h_1, \dots, h_n \in \langle g \rangle$. Show that one can solve the DLP of all n elements h_i to the base g in approximately $2\sqrt{nr}$ group operations (optimised for the worst case) or approximately $\sqrt{2nr}$ (optimised for the average case).

Exercise 13.3.13★ Suppose one is given $g \in G$ of order r, an integer w and an instance generator for the discrete logarithm problem that outputs $h = g^a \in G$ such that $0 \le a < w$ according to some known distribution on $\{0, 1, \dots, w-1\}$. Assume that the distribution is symmetric with mean value $w/2$. Determine the optimal BSGS algorithm to solve such a problem.

Exercise 13.3.14★ Suppose one is given $g, h \in G$ and $n \in \mathbb{N}$ such that $h = g^a$ where a has a representation as a non-adjacent form NAF (see Section 11.1.1) of length $n < \log_2(r)$. Give an efficient BSGS algorithm to find a. What is the running time?

13.4 Lower bound on complexity of generic algorithms for the DLP

This section presents a lower bound for the complexity of the discrete logarithm problem in groups of prime order for algorithms that do not exploit the representation of the group; such algorithms are called generic algorithms. The main challenge is to formally model such algorithms. Babai and Szemerédi [18] defined a black box group to be a group with elements represented (not necessarily uniquely) as binary strings and where multiplication, inversion and testing whether an element is the identity are all performed using oracles. Nechaev [405] used a different model (for which equality testing does not require an oracle query) and obtained $\Omega(\sqrt{r})$ time and space complexity.

Nechaev's paper concerns deterministic algorithms, and so his result does not cover the Pollard algorithms. Shoup [494] gave yet another model for generic algorithms (his model allows randomised algorithms) and proved $\Omega(\sqrt{r})$ time complexity for the DLP and some related problems. This lower bound is often called the **birthday bound** on the DLP.

Shoup's formulation has proven to be very popular with other authors and so we present it in detail. We also describe the model of generic algorithms by Maurer [364]. Further results in this area and extensions of the generic algorithm model (such as working with groups of composite order, working with groups endowed with pairings, providing access to decision oracles etc.) have been given by Maurer and Wolf [367], Maurer [364], Boneh and Boyen [71, 72], Boyen [90], Rupp, Leander, Bangerter, Dent and Sadeghi [456].

13.4.1 Shoup's model for generic algorithms

Fix a constant $t \in \mathbb{R}_{>0}$. When G is the group of points on an elliptic curve of prime order (and log means $\log_2$ as usual), one can take $t = 2$.

Definition 13.4.1 An **encoding** of a group G of order r is an injective function $\sigma : G \to \{0, 1\}^{\lceil t \log(r) \rceil}$.

A **generic algorithm** for a computational problem in a group G of order r is a probabilistic algorithm that takes as input r and $(\sigma(g_1), \ldots, \sigma(g_k))$ such that $g_1, \ldots, g_k \in G$ and returns a sequence $(a_1, \ldots, a_l, \sigma(h_1), \ldots, \sigma(h_m))$ for some $a_1, \ldots, a_l \in \mathbb{Z}/r\mathbb{Z}$ and $h_1, \ldots, h_m \in G$ (depending on the computational problem in question). The generic algorithm is given access to a perfect oracle O such that $O(\sigma(g_1), \sigma(g_2))$ returns $\sigma(g_1 g_2^{-1})$.

Note that one can obtain the encoding $\sigma(1)$ of the identity element by $O(\sigma(g_1), \sigma(g_1))$. One can then compute the encoding of g^{-1} from the encoding of g as $O(\sigma(1), \sigma(g))$. Defining $O'(\sigma(g_1), \sigma(g_2)) = O(\sigma(g_1), O(\sigma(1), \sigma(g_2)))$ gives an oracle for multiplication in G.

Example 13.4.2 A generic algorithm for the DLP in $\langle g \rangle$ where g has order r takes input $(r, \sigma(g), \sigma(h))$ and outputs a such that $h = g^a$. A generic algorithm for CDH (see Definition 20.2.1) takes input $(\sigma(g), \sigma(g^a), \sigma(g^b))$ and outputs $\sigma(g^{ab})$.

In Definition 13.4.1 we insisted that a generic algorithm take as input the order of the group, but this is not essential. Indeed, it is necessary to relax this condition if one wants to consider generic algorithms for, say, $(\mathbb{Z}/N\mathbb{Z})^*$ when N is an integer of unknown factorisation. To do this, one considers an encoding function to $\{0, 1\}^l$ and it follows that the order r of the group is at most 2^l. If the order is not given then one can consider a generic algorithm whose goal is to compute the order of a group. Theorem 2.3 and Corollary 2.4 of Sutherland [536] prove an $\Omega(r^{1/3})$ lower bound on the complexity of a generic algorithm to compute the order r of a group, given a bound M such that $\sqrt{M} < r < M$.

13.4.2 Maurer's model for generic algorithms

Maurer's formulation of generic algorithms [364] does not use any external representation of group elements (in particular, there are no randomly chosen encodings). Maurer considers a black box containing registers, specified by indices $i \in \mathbb{N}$, that store group elements. The model considers a set of operations and a set of relations. An oracle query $O(op, i_1, \ldots, i_{t+1})$ causes register i_{t+1} to be assigned the value of the t-ary operation op on the values in registers $i_1, \ldots, i_t$. Similarly, an oracle query $O(R, i_1, \ldots, i_t)$ returns the value of the t-ary relation R on the values in registers $i_1, \ldots, i_t$.

A **generic algorithm** in Maurer's model is an algorithm that takes as input the order of the group (as with Shoup's model, the order of the group can be omitted), makes oracle queries and outputs the value of some function of the registers (e.g., the value of one of the registers; Maurer calls such an algorithm an "extraction algorithm").

Example 13.4.3 To define a generic algorithm for the DLP in Maurer's model one imagines a black box that contains in the first register the value 1 (corresponding to g) and in the second register the value a (corresponding to $h = g^a$). Note that the black box is viewed as containing the additive group $\mathbb{Z}/r\mathbb{Z}$. The algorithm has access to an oracle $O(+, i, j, k)$ that assigns register k the sum of the elements in registers i and j, an oracle $O(-, i, j)$ that assigns register j the inverse of the element in register i and an oracle $O(=, i, j)$ that returns "true" if and only if registers i and j contain the same group element. The goal of the generic algorithm for the DLP is to output the value of the second register.

To implement the BSGS algorithm or Pollard rho algorithm in Maurer's model it is necessary to allow a further oracle that computes a well-ordering relation on the group elements.

We remark that the Shoup and Maurer models have been used to prove the security of cryptographic protocols against adversaries that behave like generic algorithms. Jager and Schwenk [275] have shown that both models are equivalent for this purpose.

13.4.3 The lower bound

We present the main result of this section using Shoup's model. A similar result can be obtained using Maurer's model (except that it is necessary to either ignore the cost of equality queries or else allow a total order relation on the registers).

We start with a result attributed by Shoup to Schwarz. In this section we only use the result when $k = 1$, but the more general case is used later in the book.

Lemma 13.4.4 *Let $F(x_1, \dots, x_k) \in \mathbb{F}_r[x_1, \dots, x_k]$ be a non-zero polynomial of total degree d. Then for $P = (P_1, \dots, P_k)$ chosen uniformly at random in $\mathbb{F}_r^k$ the probability that $F(P_1, \dots, P_k) = 0$ is at most d/r.*

Proof If $k = 1$ then the result is standard. We prove the result by induction on k. Write

$$F(x_1, \dots, x_k) = F_e(x_1, \dots, x_{k-1})x_k^e + F_{e-1}(x_1, \dots, x_{k-1})x_k^{e-1} + \cdots + F_0(x_1, \dots, x_{k-1})$$

where $F_i(x_1, \dots, x_{k-1}) \in \mathbb{F}_r[x_1, \dots, x_{k-1}]$ has total degree $\leq d - i$ for $0 \leq i \leq e$ and $e \leq d$. If $P = (P_1, \dots, P_{k-1}) \in \mathbb{F}_r^{k-1}$ is such that all $F_i(P) = 0$ then all r choices for P_k lead to a solution. The probability of this happening is at most $(d - e)/r$ (this is the probability that $F_e(P) = 0$). On the other hand, if some $F_i(P) \neq 0$ then there are at most e choices for P_k that give a root of the polynomial. The total probability is therefore $\leq (d - e)/r + e/r = d/r$. □

Theorem 13.4.5 *Let G be a cyclic group of prime order r. Let A be a generic algorithm for the DLP in G that makes at most m oracle queries. Then the probability, over uniformly chosen $a \in \mathbb{Z}/r\mathbb{Z}$ and uniformly chosen encoding function $\sigma : G \to \{0, 1\}^{\lceil t \log(r) \rceil}$, that $A(\sigma(g), \sigma(g^a)) = a$ is $O(m^2/r)$.*

Proof Instead of choosing a random encoding function in advance, the method of proof is to create the encodings "on the fly". The algorithm to produce the encodings is called the simulator. We also do not choose the instance of the DLP until the end of the game. The simulation will be perfect unless a certain bad event happens, and we will analyse the probability of this event.

Let $S = \{0, 1\}^{\lceil t \log(r) \rceil}$. The simulator begins by uniformly choosing two distinct σ_1, σ_2 in S and running $A(\sigma_1, \sigma_2)$. Algorithm A assumes that $\sigma_1 = \sigma(g)$ and $\sigma_2 = \sigma(h)$ for some $g, h \in G$ and some encoding function σ, but it is not necessary for the simulator to fix values for g and h.

It is necessary to ensure that the encodings are consistent with the group operations. This cannot be done perfectly without choosing g and h, but the following idea takes care of "trivial" consistency. The simulator maintains a list of pairs (σ_i, F_i) where $\sigma_i \in S$ and $F_i \in \mathbb{F}_r[x]$. The initial values are $(\sigma_1, 1)$ and (σ_2, x). Whenever A makes an oracle query on (σ_i, σ_j) the simulator computes $F = F_i - F_j$. If F appears as F_k in the list of pairs then the simulator replies with σ_k and does not change the list. Otherwise, a $\sigma \in S$ distinct from the previously used values is chosen uniformly at random, (σ, F) is added to the simulator's list and σ is returned to A.

After making at most m oracle queries, A outputs $b \in \mathbb{Z}/r\mathbb{Z}$. The simulator now chooses a uniformly at random in $\mathbb{Z}/r\mathbb{Z}$. Algorithm A wins if $b = a$.

Let the simulator's list contain precisely k polynomials $\{F_1(x), \ldots, F_k(x)\}$ for some $k \le m + 2$. Let E be the event that $F_i(a) = F_j(a)$ for some pair $1 \le i < j \le k$. The probability that A wins is

$$\Pr(A \text{ wins} \mid E) \Pr(E) + \Pr(A \text{ wins} \mid \neg E) \Pr(\neg E). \tag{13.3}$$

For each pair $1 \le i < j \le k$ the probability that $(F_i - F_j)(a) = 0$ is $1/r$ by Lemma 13.4.4. Hence, the probability of event E is at most $k(k-1)/2r = O(m^2/r)$. On the other hand, if event E does not occur then all A "knows" about a is that it lies in the set $\mathcal{X}$ of possible values for a for which $F_i(a) \ne F_j(a)$ for all $1 \le i < j \le k$. Let $N = \#\mathcal{X} \approx r - m^2/2$. Then $\Pr(\neg E) = N/r$ and $\Pr(A \text{ wins} \mid \neg E) = 1/N$.

Putting it all together, the probability that A wins is $O(m^2/r)$. □

Exercise 13.4.6 Prove Theorem 13.4.5 using Maurer's model for generic algorithms. [Hint: The basic method of proof is exactly the same. The difference is in formulation and analysis of the success probability.]

Corollary 13.4.7 *Let A be a generic algorithm for the DLP. If A succeeds with noticeable probability $1/\log(r)^c$ for some $c > 0$ then A must make $\Omega(\sqrt{r/\log(r)^c})$ oracle queries.*

13.5 Generalised discrete logarithm problems

A number of generalisations of the discrete logarithm problem have been proposed over the years. The motivation for such problems varies: sometimes the aim is to enable new cryptographic functionalities, other times the aim is to generate hard instances of the DLP more quickly than previous methods.

Definition 13.5.1 Let G be a finitely generated Abelian group. The **multi-dimensional discrete logarithm problem** or **representation problem**[4] is: given $g_1, g_2, \ldots, g_l, h \in G$ and $\mathcal{S}_1, \mathcal{S}_2, \ldots, \mathcal{S}_l \subseteq \mathbb{Z}$ to find $a_j \in \mathcal{S}_j$ for $1 \le j \le l$, if they exist, such that

$$h = g_1^{a_1} g_2^{a_2} \cdots g_l^{a_l}.$$

The **product discrete logarithm problem**[5] is: given $g, h \in G$ and $\mathcal{S}_1, \mathcal{S}_2, \ldots, \mathcal{S}_l \subseteq \mathbb{Z}$ to find $a_j \in \mathcal{S}_j$ for $1 \le j \le l$, if they exist, such that

$$h = g^{a_1 a_2 \cdots a_l}.$$

Remark 13.5.2 A natural variant of the product DLP is to compute only the product $a_1 a_2 \cdots a_l$ rather than the l-tuple $(a_1, \ldots, a_l)$. This is just the DLP with respect to a specific instance generator (see the discussion in Section 2.1.2). Precisely, consider an instance generator that, on input of a security parameter κ, outputs a group element g of prime order r and then chooses $a_j \in \mathcal{S}_j$ for $1 \le j \le l$ and computes $h = g^{a_1 a_2 \cdots a_l}$. The stated variant of the product DLP is the DLP with respect to this instance generator.

Note that the representation problem can be defined whether or not $G = \langle g_1, \ldots, g_l \rangle$ is cyclic. The solution to Exercise 13.5.4 applies in all cases. However, there may be other ways to tackle the non-cyclic case (e.g., exploiting efficiently computable group homomorphisms, see [214] for example), so the main interest is the case when G is cyclic of prime order r.

Example 13.5.3 The representation problem can arise when using the GLV method (see Section 11.3.3) with intentionally small coefficients. In this case, $g_2 = \psi(g_1)$, $\langle g_1, g_2 \rangle$ is a cyclic group of order r, and $h = g_1^{a_1} g_2^{a_2}$ where $0 \le a_1, a_2 < w \le \sqrt{r}$).

The number of possible choices for h in both the representation problem and product DLP is at most $\prod_{j=1}^{l} \#\mathcal{S}_j$ (it could be smaller if the same h can arise from many different combinations of $(a_1, \ldots, a_l)$). If l is even and $\#\mathcal{S}_j = \#\mathcal{S}_1$ for all j then there is an easy time/memory tradeoff algorithm requiring $O(\#\mathcal{S}_1^{l/2})$ group operations.

Exercise 13.5.4 Write down an efficient BSGS algorithm to solve the representation problem. What is the running time and storage requirement?

Exercise 13.5.5 Give an efficient BSGS algorithm to solve the product DLP. What is the running time and storage requirement?

[4] This computational problem seems to be first explicitly stated in the work of Brands [92] in the case $\mathcal{S}_i = \mathbb{Z}$.
[5] The idea of using product exponents for improved efficiency appears in Knuth [308] where it is called the "factor method".

It is natural to ask whether one can do better than the naive BSGS algorithms for these problems, at least for certain values of l. The following result shows that the answer in general turns out to be "no".

Lemma 13.5.6 *Assume l is even and $\#S_j = \#S_1$ for all $2 \le j \le l$. A generic algorithm for the representation problem with noticeable success probability $1/\log(\#S_1)^c$ needs $\Omega(\#S_1^{l/2}/\log(\#S_1)^{c/2})$ group operations.*

Proof Suppose A is a generic algorithm for the representation problem. Let G be a group of order r and let $g, h \in G$. Set $m = \lceil r^{1/l} \rceil$, $S_j = \{a \in \mathbb{Z} : 0 \le a < m\}$ and let $g_j = g^{m^j}$ for $0 \le j \le l-1$. If $h = g^a$ for some $a \in \mathbb{Z}$ then the base m-expansion $a_0 + a_1 m + \cdots + a_{l-1} m^{l-1}$ is such that

$$h = g^a = \prod_{j=0}^{l-1} g_j^{a_j}.$$

Hence, if A solves the representation problem then we have solved the DLP using a generic algorithm. Since we have shown that a generic algorithm for the DLP with success probability $1/\log(\#S_1)^c$ needs $\Omega(\sqrt{r/\log(\#S_1)^c})$ group operations, the result is proved. □

13.6 Low Hamming weight DLP

Recall that the **Hamming weight** of an integer is the number of ones in its binary expansion.

Definition 13.6.1 Let G be a group and let $g \in G$ have prime order r. The **low Hamming weight DLP** is: given $h \in \langle g \rangle$ and integers n, w to find a integer a (if it exists) whose binary expansion has length $\le n$ and Hamming weight $\le w$ such that $h = g^a$.

This definition makes sense even for $n > \log_2(r)$. For example, squaring is faster than multiplication in most representations of algebraic groups, so it could be more efficient to compute g^a by taking longer strings with fewer ones in their binary expansion.

Coppersmith developed a time/memory tradeoff algorithm to solve this problem. A thorough treatment of these ideas was given by Stinson in [531]. Without loss of generality, we assume that n and w are even (just add 1 to them if not).

The idea of the algorithm is to reduce solving $h = g^a$ where a has length n and Hamming weight w to solving $hg^{-a_2} = g^{a_1}$ where a_1 and a_2 have Hamming weight $w/2$. One does this by choosing a set $B \subset I = \{0, 1, \ldots, n-1\}$ of size $n/2$. The set B is the set of possible bit positions for the bits of a_1 and $(I - B)$ is the possible bit positions for the bits of a_2. The detailed algorithm is given in Algorithm 14. Note that one can compactly represent subsets $Y \subseteq I$ as n-bit strings.

Algorithm 14 Coppersmith's BSGS algorithm for the low Hamming weight DLP

INPUT: $g, h \in G$ of order r, n and w
OUTPUT: a of bit-length n and Hamming weight w such that $h = g^a$, or $\perp$

1: Choose $B \subset \{0, \ldots, n-1\}$ such that $\#B = n/2$
2: Initialise an easily searched structure (such as a binary tree, a heap, or a hash table) L
3: **for** $Y \subseteq B : \#Y = w/2$ **do**
4: Compute $b = \sum_{j \in Y} 2^j$ and $x = g^b$
5: store (x, Y) in L ordered according to first coordinate
6: **end for**
7: **for** $Y \subseteq (I - B) : \#Y = w/2$ **do**
8: Compute $b = \sum_{j \in Y} 2^j$ and $y = hg^{-b}$
9: **if** $y = x$ for some $(x, Y_1) \in L$ **then**
10: $a = \sum_{j \in Y \cup Y_1} 2^j$
11: **return** a
12: **end if**
13: **end for**
14: **return** $\perp$

Exercise 13.6.2 Write down an algorithm, to enumerate all $Y \subset B$ such that $\#Y = w/2$, which requires $O(\binom{n/2}{w/2} n)$ bit operations.

Lemma 13.6.3 *The running time of Algorithm 14 is* $O(\binom{n/2}{w/2})$ *group operations and the algorithm requires* $O(\binom{n/2}{w/2})$ *group elements of storage.*

Exercise 13.6.4 Prove Lemma 13.6.3.

Algorithm 14 is not guaranteed to succeed since the set B might not exactly correspond to a splitting of the bit positions of the integer a into two sets of Hamming weight $\leq w/2$. We now give a collection of subsets of I that is guaranteed to contain a suitable B.

Definition 13.6.5 Fix even integers n and w. Let $I = \{0, \ldots, n-1\}$. A **splitting system** is a set $\mathcal{B}$ of subsets of I of size $n/2$ such that for every $Y \subset I$ such that $\#Y = w$ there is a set $B \in \mathcal{B}$ such that $\#(B \cap Y) = w/2$.

Lemma 13.6.6 *For any even integers n and w there exists a splitting system* $\mathcal{B}$ *of size* $n/2$.

Proof For $0 \leq i \leq n-1$ define

$$B_i = \{i + j \pmod{n} : 0 \leq j \leq n/2 - 1\}$$

and let $\mathcal{B} = \{B_i : 0 \leq i \leq n/2 - 1\}$.

To show $\mathcal{B}$ is a splitting system, fix any $Y \subset I$ of size w. Define $v(i) = \#(Y \cap B_i) - \#(Y \cap (I - B_i)) \in \mathbb{Z}$ for $0 \leq i \leq n/2 - 1$. One can check that $v(i)$ is even, that $v(n/2) = -v(0)$ and that $v(i+1) - v(i) \in \{-2, 0, 2\}$. Hence, either $v(0) = 0$, or else the values

$\nu(i)$ change sign at least once as i goes from 0 to $n/2$. It follows that there exists some $0 \le i \le n/2$ such that $\nu(i) = 0$, in which case $\#(Y \cap B_i) = w/2$. □

One can run Algorithm 14 for all $n/2$ sets B in the splitting system $\mathcal{B}$ of Lemma 13.6.6. This gives a deterministic algorithm with running time $O(n\binom{n/2}{w/2})$ group operations. Stinson proposes different splitting systems giving a deterministic algorithm requiring $O(w^{3/2}\binom{n/2}{w/2})$ group operations. A more efficient randomised algorithm (originally proposed by Coppersmith) is to randomly choose sets B from the $\binom{n}{n/2}$ possible subsets of $\{0, \dots, n-1\}$ of size $n/2$. Theorem 13.6.9 determines the expected running time in this case.

Lemma 13.6.7 *Fix a set $Y \subset \{0, \dots, n-1\}$ such that $\#Y = w$. The probability that a randomly chosen $B \subseteq \{0, \dots, n-1\}$ having $\#B = n/2$ satisfies $\#(Y \cap B) = w/2$ is*

$$p_{Y,B} = \binom{w}{w/2}\binom{(n-w)}{(n-w)/2} \Big/ \binom{n}{n/2}.$$

Exercise 13.6.8 Prove Lemma 13.6.7.

Theorem 13.6.9 *The expected running time for the low Hamming weight DLP when running Algorithm 14 on randomly chosen sets B is $O(\sqrt{w}\binom{n/2}{w/2})$ exponentiations. The storage is $O(\binom{n/2}{w/2})$ group elements.*

Proof We expect to repeat the algorithm $1/p_{Y,B}$ times. One can show, using the fact $2^k/\sqrt{2k} \le \binom{k}{k/2} \le 2^k\sqrt{2/\pi k}$, that $1/p_{Y,B} \le c\sqrt{w}$ for some constant (see Stinson [531]). The result follows. □

Exercise 13.6.10 As with all BSGS methods, the bottleneck for this method is the storage requirement. Show how to modify the algorithm for the case where only M group elements of storage are available.

Exercise 13.6.11 Adapt Coppersmith's algorithm to the DLP for low weight signed expansions (e.g., NAFs, see Section 11.1.1).

All the algorithms in this section have large storage requirements. An approach due to van Oorschot and Wiener for solving such problems using less storage is presented in Section 14.8.1.

13.7 Low Hamming weight product exponents

Let G be an algebraic group (or algebraic group quotient) over $\mathbb{F}_p$ (p small) and let $g \in G(\mathbb{F}_{p^n})$ with $n > 1$. Let π_p be the p-power Frobenius on G, acting on G as $g \mapsto g^p$. Hoffstein and Silverman [262] proposed computing random powers of g efficiently by taking products of low Hamming weight Frobenius expansions.

In particular, for Koblitz elliptic curves (i.e., $p = 2$) they suggested using three sets and taking S_j for $1 \le j \le 3$ to be the set of Frobenius expansions of length n and weight 7. The BSGS algorithm in Section 13.5 applies to this problem, but the running time is not necessarily optimal since $\#\mathcal{S}_1\#\mathcal{S}_2 \neq \#\mathcal{S}_3$. Kim and Cheon [304] generalised the results of Section 13.6 to allow a more balanced time/memory tradeoff. This gives a small improvement to the running time.

Cheon and Kim [125] give a further improvement to the attack, which is similar to the use of equivalence classes in Pollard rho (see Section 14.4). They noted that the sets $\mathcal{S}_j$ in the Hoffstein–Silverman proposal have the property that for every $a \in \mathcal{S}_j$ there is some $a' \in \mathcal{S}_j$ such that $g^{a'} = \pi_p(g^a)$. In other words, π_p permutes $\mathcal{S}_j$ and each element $a \in \mathcal{S}_j$ lies in an orbit of size n under this permutation. Cheon and Kim define a unique representative of each orbit of π_p in $\mathcal{S}_j$ and show how to speed up the BSGS algorithm in this case by a factor of n.

Exercise 13.7.1★ Give the details of the Cheon–Kim algorithm. How many group operations does the algorithm perform when $n = 163$, and three sets with $w = 7$ are used?

14
Factoring and discrete logarithms using pseudorandom walks

This chapter is devoted to the rho and kangaroo methods for factoring and discrete logarithms (which were invented by Pollard) and some related algorithms. These methods use pseudorandom walks and require low storage (typically a polynomial amount of storage, rather than exponential as in the time/memory tradeoff). Although the rho factoring algorithm was developed earlier than the algorithms for discrete logarithms, the latter are much more important in practice.[1] Hence, we focus mainly on the algorithms for the discrete logarithm problem.

As in the previous chapter, we assume G is an algebraic group over a finite field $\mathbb{F}_q$ written in multiplicative notation. To solve the DLP in an algebraic group quotient using the methods in this chapter one would first lift the DLP to the covering group (though see Section 14.4 for a method to speed up the computation of the DLP in an algebraic group by essentially working in a quotient).

14.1 Birthday paradox

The algorithms in this chapter rely on results in probability theory. The first tool we need is the so-called "birthday paradox". This name comes from the following application, which surprises most people: among a set of 23 or more randomly chosen people, the probability that two of them share a birthday is greater than 0.5 (see Example 14.1.4).

Theorem 14.1.1 *Let S be a set of N elements. If elements are sampled uniformly at random from S then the expected number of samples to be taken before some element is sampled twice is less than $\sqrt{\pi N/2}+2 \approx 1.253\sqrt{N}$.*

The element that is sampled twice is variously known as a **repeat**, **match** or **collision**. For the rest of the chapter we will ignore the $+2$ and say that the expected number of samples is $\sqrt{\pi N/2}$.

[1] Pollard's paper [437] contains the remark "We are not aware of any particular need for such index calculations" (i.e., computing discrete logarithms) even though [437] cites the paper of Diffie and Hellman. Pollard worked on the topic before hearing of the cryptographic applications. Hence, Pollard's work is an excellent example of research pursued for its intrinsic interest, rather than motivated by practical applications.

Proof Let X be the random variable giving the number of elements selected from $\mathcal{S}$ (uniformly at random) before some element is selected twice. After l distinct elements have been selected then the probability that the next element selected is also distinct from the previous ones is $(1 - l/N)$. Hence, the probability $\Pr(X > l)$ is given by

$$p_{N,l} = 1(1 - 1/N)(1 - 2/N) \cdots (1 - (l-1)/N).$$

Note that $p_{N,l} = 0$ when $l \geq N$. We now use the standard fact that $1 - x \leq e^{-x}$ for $x \geq 0$. Hence

$$\begin{aligned} p_{N,l} \leq 1\, e^{-1/N} e^{-2/N} \cdots e^{-(l-1)/N} &= e^{-\sum_{j=0}^{l-1} j/N} \\ &= e^{-\frac{1}{2}(l-1)l/N} \\ &\leq e^{-(l-1)^2/2N}. \end{aligned}$$

By definition, the expected value of X is

$$\begin{aligned} \sum_{l=1}^{\infty} l \Pr(X = l) &= \sum_{l=1}^{\infty} l(\Pr(X > l-1) - \Pr(X > l)) \\ &= \sum_{l=0}^{\infty} (l + 1 - l) \Pr(X > l) \\ &= \sum_{l=0}^{\infty} \Pr(X > l) \\ &\leq 1 + \sum_{l=1}^{\infty} e^{-(l-1)^2/2N}. \end{aligned}$$

We estimate this sum using the integral

$$1 + \int_0^\infty e^{-x^2/2N} dx.$$

Since $e^{-x^2/2N}$ is monotonically decreasing and takes values in $[0, 1]$ the difference between the value of the sum and the value of the integral is at most 1. Making the change of variable $u = x/\sqrt{2N}$ gives

$$\sqrt{2N} \int_0^\infty e^{-u^2} du.$$

A standard result in analysis (see Section 11.7 of [305] or Section 4.4 of [569]) is that this integral is $\sqrt{\pi}/2$. Hence, the expected value for X is $\leq \sqrt{\pi N/2} + 2$. □

The proof only gives an upper bound on the probability of a collision after l trials. A lower bound of $e^{-l^2/2N - l^3/6N^2}$ for $N \geq 1000$ and $0 \leq l \leq 2N\log(N)$ is given in Wiener [563]; it is also shown that the expected value of the number of trials is $> \sqrt{\pi N/2} - 0.4$. A more precise analysis of the birthday paradox is given in Example II.10 of Flajolet and

Sedgewick [188] and Exercise 3.1.12 of Knuth [308]. The expected number of samples is $\sqrt{\pi N/2} + 2/3 + O(1/\sqrt{N})$.

We remind the reader of the meaning of expected value. Suppose the experiment of sampling elements of a set $\mathcal{S}$ of size N until a collision is found is repeated t times and each time we count the number l of elements sampled. Then the average of l over all trials tends to $\sqrt{\pi N/2}$ as t goes to infinity.

Exercise 14.1.2 Show that the number of elements that need to be selected from $\mathcal{S}$ to get a collision with probability 1/2 is $\sqrt{2\log(2)N} \approx 1.177\sqrt{N}$.

Exercise 14.1.3 One may be interested in the number of samples required when one is particularly unlucky. Determine the number of trials so that with probability 0.99 one has a collision. Repeat the exercise for probability 0.999.

The name "birthday paradox" arises from the following application of the result.

Example 14.1.4 In a room containing 23 or more randomly chosen people, the probability is greater than 0.5 that two people have the same birthday. This follows from $\sqrt{2\log(2)365} \approx 22.49$. Note also that $\sqrt{\pi 365/2} = 23.944\ldots$.

Finally, we mention that the expected number of samples from a set of size N until $k > 1$ collisions are found is approximately $\sqrt{2kN}$. A detailed proof of this fact is given by Kuhn and Struik as Theorem 1 of [320].

14.2 The Pollard rho method

Let g be a group element of prime order r and let $G = \langle g \rangle$. The discrete logarithm problem (DLP) is: given $h \in G$ to find a, if it exists, such that $h = g^a$. In this section we assume (as is usually the case in applications) that one has already determined that $h \in \langle g \rangle$.

The starting point of the rho algorithm is the observation that if one can find $a_i, b_i, a_j, b_j \in \mathbb{Z}/r\mathbb{Z}$ such that

$$g^{a_i} h^{b_i} = g^{a_j} h^{b_j} \qquad (14.1)$$

and $b_i \not\equiv b_j \pmod{r}$ then one can solve the DLP as

$$h = g^{(a_i - a_j)(b_j - b_i)^{-1} \pmod{r}}.$$

The basic idea is to generate pseudorandom sequences $x_i = g^{a_i} h^{b_i}$ of elements in G by iterating a suitable function $f : G \to G$. In other words, one chooses a starting value x_1 and defines the sequence by $x_{i+1} = f(x_i)$. A sequence $x_1, x_2, \ldots$ is called a **deterministic pseudorandom walk**. Since G is finite there is eventually a collision $x_i = x_j$ for some $1 \le i < j$ as in equation (14.1). This is presented as a collision between two elements in the same walk, but it could also be a collision between two elements in different walks. If the elements in the walks look like uniformly and independently chosen elements of G then, by the birthday paradox (Theorem 14.1.1), the expected value of j is $\sqrt{\pi r/2}$.

It is important that the function f be designed so that one can efficiently compute $a_i, b_i \in \mathbb{Z}/r\mathbb{Z}$ such that $x_i = g^{a_i} h^{b_i}$. The next step x_{i+1} depends only on the current step x_i and not on (a_i, b_i). The algorithms all exploit the fact that when a collision $x_i = x_j$ occurs then $x_{i+t} = x_{j+t}$ for all $t \in \mathbb{N}$. Pollard's original proposal used a cycle-finding method due to Floyd to find a self-collision in the sequence; we present this in Section 14.2.2. A better approach is to use distinguished points to find collisions; we present this in Section 14.2.4.

14.2.1 The pseudorandom walk

Pollard simulates a random function from G to itself as follows. The first step is to decompose G into n_S disjoint subsets (usually of roughly equal size) so that $G = \mathcal{S}_0 \cup \mathcal{S}_1 \cup \cdots \cup \mathcal{S}_{n_S-1}$. Traditional textbook presentations use $n_S = 3$ but, as explained in Section 14.2.5, it is better to take larger values for n_S; typical values in practice are 32, 256 or 2048.

The sets $\mathcal{S}_i$ are defined using a selection function $S : G \to \{0, \ldots, n_S - 1\}$ by $\mathcal{S}_i = \{g \in G : S(g) = i\}$. For example, in any computer implementation of G one represents an element $g \in G$ as a unique[2] binary string $b(g)$ and interpreting $b(g)$ as an integer one could define $S(g) = b(g) \pmod{n_S}$ (taking n_S to be a power of 2 makes this computation especially easy). To obtain different choices for S one could apply an $\mathbb{F}_2$-linear map L to the sequence of bits $b(g)$ so that $S(g) = L(b(g)) \pmod{n_S}$. These simple methods can be a poor choice in practice as they are not "sufficiently random". Some other ways to determine the partition are suggested in Section 2.3 of Teske [543] and Bai and Brent [23]. The strongest choice is to apply a hash function or randomness extractor to $b(g)$, though this may lead to an undesirable computational overhead.

Definition 14.2.1 The **rho walks** are defined as follows. Precompute $g_j = g^{u_j} h^{v_j}$ for $0 \le j \le n_S - 1$ where $0 \le u_j, v_j < r$ are chosen uniformly at random. Set $x_1 = g$. The **original rho walk** is

$$x_{i+1} = f(x_i) = \begin{cases} x_i^2 & \text{if } S(x_i) = 0 \\ x_i g_j & \text{if } S(x_i) = j,\ j \in \{1, \ldots, n_S - 1\}. \end{cases} \tag{14.2}$$

The **additive rho walk** is

$$x_{i+1} = f(x_i) = x_i g_{S(x_i)}. \tag{14.3}$$

An important feature of the walks is that each step requires only one group operation. Once the selection function S and the values u_j and v_j are chosen, the walk is deterministic. Even though these values may be chosen uniformly at random, the function f itself is not a random function as it has a compact description. Hence, the rho walks can only be described as pseudorandom. To analyse the algorithm we will consider the expectation of

[2] One often uses projective coordinates to speed up elliptic curve arithmetic, so it is natural to use projective coordinates when implementing these algorithms. But to define the pseudorandom walk one needs a unique representation for points, so projective coordinates are not appropriate. See Remark 13.3.2.

the running time over different choices for the pseudorandom walk. Many authors consider the expectation of the running time over all problem instances and random choices of the pseudorandom walk; they therefore write "expected running time" for what we are calling "average-case expected running time".

It is necessary to keep track of the decomposition

$$x_i = g^{a_i} h^{b_i}.$$

The values $a_i, b_i \in \mathbb{Z}/r\mathbb{Z}$ are obtained by setting $a_1 = 1, b_1 = 0$ and updating (for the original rho walk)

$$a_{i+1} = \begin{cases} 2a_i \pmod r & \text{if } S(x_i) = 0 \\ a_i + u_{S(x_i)} \pmod r & \text{if } S(x_i) > 0 \end{cases}$$
$$\text{and} \quad b_{i+1} = \begin{cases} 2b_i \pmod r & \text{if } S(x_i) = 0 \\ b_i + v_{S(x_i)} \pmod r & \text{if } S(x_i) > 0. \end{cases} \tag{14.4}$$

Putting everything together, we write

$$(x_{i+1}, a_{i+1}, b_{i+1}) = \mathsf{walk}(x_i, a_i, b_i)$$

for the random walk function. But it is important to remember that x_{i+1} only depends on x_i and not on (x_i, a_i, b_i).

Exercise 14.2.2 Give the analogue of equation (14.4) for the additive walk.

14.2.2 Pollard rho using Floyd cycle finding

We present the original version of Pollard rho. A single sequence $x_1, x_2, \dots$ of group elements is computed. Eventually, there is a collision $x_i = x_j$ with $1 \le i < j$. One pictures the walk as having a **tail** (which is the part $x_1, \dots, x_i$ of the walk that is not cyclic) followed by the **cycle** or **head** (which is the part $x_{i+1}, \dots, x_j$). Drawn appropriately, this resembles the shape of the greek letter ρ. The tail and cycle (or head) of such a random walk have expected length $\sqrt{\pi N/8}$ (see Flajolet and Odlyzko [187] for proofs of these, and many other, facts).

The goal is to find integers i and j such that $x_i = x_j$. It might seem that the only approach is to store all the x_i and, for each new value x_j, to check if it appears in the list. This approach would use more memory and time than the BSGS algorithm. If one were using a truly random walk then one would have to use this approach. The whole point of using a deterministic walk which eventually becomes cyclic is to enable better methods to find a collision.

Let l_t be the length of the **tail** of the "rho" and l_h be the length of the **cycle** of the "rho". In other words, the first collision is

$$x_{l_t + l_h} = x_{l_t}. \tag{14.5}$$

Floyd's cycle finding algorithm[3] is to compare x_i and x_{2i}. Lemma 14.2.3 shows that this will find a collision in at most $l_t + l_h$ steps. The crucial advantage of comparing x_{2i} and x_i is that it only requires storing two group elements. The rho algorithm with Floyd cycle finding is given in Algorithm 15.

Algorithm 15 The rho algorithm

INPUT: $g, h \in G$
OUTPUT: a such that $h = g^a$, or $\perp$

1: Choose randomly the function walk as explained above
2: $x_1 = g, a_1 = 1, b_1 = 0$
3: $(x_2, a_2, b_2) = \mathsf{walk}(x_1, a_1, b_1)$
4: **while** $(x_1 \neq x_2)$ **do**
5: $\quad (x_1, a_1, b_1) = \mathsf{walk}(x_1, a_1, b_1)$
6: $\quad (x_2, a_2, b_2) = \mathsf{walk}(\mathsf{walk}(x_2, a_2, b_2))$
7: **end while**
8: **if** $b_1 \equiv b_2 \pmod{r}$ **then**
9: $\quad$ **return** $\perp$
10: **else**
11: $\quad$ **return** $(a_2 - a_1)(b_1 - b_2)^{-1} \pmod{r}$
12: **end if**

Lemma 14.2.3 *Let the notation be as above. Then $x_{2i} = x_i$ if and only if $l_h \mid i$ and $i \geq l_t$. Further, there is some $l_t \leq i < l_t + l_h$ such that $x_{2i} = x_i$.*

Proof If $x_i = x_j$ then we must have $l_h \mid (i - j)$. Hence, the first statement of the Lemma is clear. The second statement follows since there is some multiple of l_h between l_t and $l_t + l_h$. □

Exercise 14.2.4 Let $p = 347, r = 173, g = 3, h = 11 \in \mathbb{F}_p^*$. Let $n_S = 3$. Determine l_t and l_h for the values $(u_1, v_1) = (1, 1)$, $(u_2, v_2) = (13, 17)$. What is the smallest of i for which $x_{2i} = x_i$?

Exercise 14.2.5 Repeat Exercise 14.2.4 for $g = 11, h = 3$ $(u_1, v_1) = (4, 7)$ and $(u_2, v_2) = (23, 5)$.

The smallest index i such that $x_{2i} = x_i$ is called the **epact**. The expected value of the epact is conjectured to be approximately $0.823\sqrt{\pi r/2}$.

Example 14.2.6 Let $p = 809$ and consider $g = 89$ which has prime order 101 in $\mathbb{F}_p^*$. Let $h = 799$ which lies in the subgroup generated by g.

Let $n_S = 4$. To define $S(g)$ write g in the range $1 \leq g < 809$, represent this integer in its usual binary expansion and then reduce modulo 4. Choose $(u_1, v_1) = (37, 34)$, $(u_2, v_2) =$

[3] Apparently this algorithm first appears in print in Knuth [308], but is credited there to Floyd.

(71, 69), $(u_3, v_3) = (76, 18)$ so that $g_1 = 343$, $g_2 = 676$, $g_3 = 627$. One computes the table of values (x_i, a_i, b_i) as follows:

i	x_i	a_i	b_i	$S(x_i)$
1	89	1	0	1
2	594	38	34	2
3	280	8	2	0
4	736	16	4	0
5	475	32	8	3
6	113	7	26	1
7	736	44	60	0

It follows that $l_t = 4$ and $l_h = 3$ and so the first collision detected by Floyd's method is $x_6 = x_{12}$. We leave as an exercise to verify that the discrete logarithm in this case is 50.

Exercise 14.2.7 Let $p = 569$ and let $g = 262$ and $h = 5$ which can be checked to have order 71 modulo p. Use the rho algorithm to compute the discrete logarithm of h to the base g modulo p.

Exercise 14.2.8 One can simplify Definition 14.2.1 and equation (14.4) by replacing g_j by either g^{u_j} or h^{v_j} (independently for each j). Show that this saves one modular addition in each iteration of the algorithm. Explain why this optimisation should not affect the success of the algorithm as long as the walk uses all values for $S(x_i)$ with roughly equal probability.

Algorithm 15 always terminates, but there are several things that can go wrong:

- The value $(b_1 - b_2)$ may not be invertible modulo r.

  Hence, we can only expect to prove that the algorithm succeeds with a certain probability (extremely close to 1).
- The cycle may be very long (as big as r) in which case the algorithm is slower than brute force search.

  Hence, we can only expect to prove an expected running time for the algorithm. We recall that the expected running time in this case is the average, over all choices for the function walk, of the worst-case running time of the algorithm over all problem instances.

Note that the algorithm always halts, but it may fail to output a solution to the DLP. Hence, this is a Monte Carlo algorithm.

14.2.3 Other cycle finding methods

Floyd cycle finding is not a very efficient way to find cycles. Though any cycle finding method requires computing at least $l_t + l_h$ group operations, Floyd's method needs on average $2.47(l_t + l_h)$ group operations (2.47 is 3 times the expected value of the epact).

Also, the "slower" sequence x_i is visiting group elements which have already been computed during the walk of the "faster" sequence x_{2i}. Brent [93] has given an improved cycle finding method[4] that still only requires storage for two group elements, but which requires fewer group operations. Montgomery has given an improvement to Brent's method in [392].

One can do even better by using more storage, as was shown by Sedgewick, Szymanski and Yao [482], Schnorr and Lenstra [474] (also see Teske [541]) and Nivasch [419]. The rho algorithm using Nivasch cycle finding has the optimal expected running time of $\sqrt{\pi r/2} \approx 1.253\sqrt{r}$ group operations and is expected to require polynomial storage.

Finally, a very efficient way to find cycles is to use distinguished points. More importantly, distinguished points allow us to think about the rho method in a different way and this leads to a version of the algorithm that can be parallelised. We discuss this in the next section. Hence, in practice one always uses distinguished points.

14.2.4 Distinguished points and Pollard rho

The idea of using distinguished points in search problems apparently goes back to Rivest. The first application of this idea to computing discrete logarithms is by van Oorschot and Wiener [423].

Definition 14.2.9 An element $g \in G$ is a **distinguished point** if its binary representation $b(g)$ satisfies some easily checked property. Denote by $\mathcal{D} \subset G$ the set of distinguished points. The probability $\#\mathcal{D}/\#G$ that a uniformly chosen group element is a distinguished point is denoted θ.

A typical example is the following.

Example 14.2.10 Let E be an elliptic curve over $\mathbb{F}_p$. A point $P \in E(\mathbb{F}_p)$ that is not the point at infinity is represented by an x-coordinate $0 \le x_P < p$ and a y-coordinate $0 \le y_P < p$. Let H be a hash function, where the output is interpreted as being in $\mathbb{Z}_{\ge 0}$.

Fix an integer $n_{\mathcal{D}}$. Define $\mathcal{D}$ to be the points $P \in E(\mathbb{F}_p)$ such that the $n_{\mathcal{D}}$ least significant bits of $H(x_P)$ are zero. Note that $\mathcal{O}_E \notin \mathcal{D}$. In other words

$$\mathcal{D} = \{P = (x_P, y_P) \in E(\mathbb{F}_p) : H(x_P) \equiv 0 \pmod{2^{n_{\mathcal{D}}}} \text{ where } 0 \le x_P < p\}.$$

Then $\theta \approx 1/2^{n_{\mathcal{D}}}$.

The rho algorithm with distinguished points is as follows. First, choose integers $0 \le a_1, b_1 < r$ uniformly and independently at random, compute the group element $x_1 = g^{a_1}h^{b_1}$ and run the usual deterministic pseudorandom walk until a distinguished point $x_n = g^{a_n}h^{b_n}$ is found. Store (x_n, a_n, b_n) in some easily searched data structure (searchable on x_n). Then choose a fresh randomly chosen group element $x_1 = g^{a_1}h^{b_1}$ and repeat. Eventually two walks will visit the same group element, in which case their paths will continue to the same

[4] This was originally developed to speed up the Pollard rho factoring algorithm.

distinguished point. Once a distinguished group element is found twice then the DLP can be solved with high probability.

Exercise 14.2.11 Write down pseudocode for this algorithm.

We stress the most significant difference between this method and the method of the previous section: the previous method had one long walk with a tail and a cycle, whereas the new method has many short walks. Note that this algorithm does not require self-collisions in the walk and so there is no ρ shape anymore; the word "rho" in the name of the algorithm is therefore a historical artifact, not an intuition about how the algorithm works.

Note that, since the group is finite, collisions must eventually occur, and so the algorithm halts. But the algorithm may fail to solve the DLP (with low probability). Hence, this is a Monte Carlo algorithm.

In the analysis we assume that we are sampling group elements (we sometimes call them "points") uniformly and independently at random. It is important to determine the expected number of steps before landing on a distinguished point.

Lemma 14.2.12 *Let θ be the probability that a randomly chosen group element is a distinguished point. Then:*

1. *the probability that one chooses α/θ group elements, none of which are distinguished, is approximately $e^{-\alpha}$ when $1/\theta$ is large;*
2. *the expected number of group elements to choose before getting a distinguished point is $1/\theta$;*
3. *if one has already chosen i group elements, none of which are distinguished, then the expected number of group elements to further choose before getting a distinguished point is $1/\theta$.*

Proof The probability that i chosen group elements are not distinguished is $(1-\theta)^i$. So the probability of choosing α/θ points, none of which are distinguished, is

$$(1-\theta)^{\alpha/\theta} \le (e^{-\theta})^{\alpha/\theta} = e^{-\alpha}.$$

The second statement is the standard formula for the expected value of a geometric random variable, see Example A.14.1.

For the final statement,[5] suppose one has already sampled i points without finding a distinguished point. Since the trials are independent, the probability of choosing a further j points which are not distinguished remains $(1-\theta)^j$. Hence, the expected number of extra points to be chosen is still $1/\theta$. □

We now make the following assumption. We believe this is reasonable when r is sufficiently large, $n_S > \log(r)$, distinguished points are sufficiently common and specified

[5] This is the "apparent paradox" mentioned in footnote 7 of [423].

using a good hash function (and hence $\mathcal{D}$ is well-distributed), $\theta > \log(r)/\sqrt{r}$ and when the function walk is chosen at random.

Heuristic 14.2.13

1. Walks reach a distinguished point in significantly fewer than $\sqrt{r}$ steps (in other words, there are no cycles in the walks and walks are not excessively longer than $1/\theta$).[6]
2. The expected number of group elements sampled before a collision is $\sqrt{\pi r/2}$.

Theorem 14.2.14 *Let the notation be as above and assume Heuristic 14.2.13. Then the rho algorithm with distinguished points has expected running time of* $(\sqrt{\pi/2} + o(1))\sqrt{r} \approx (1.253 + o(1))\sqrt{r}$ *group operations. The probability the algorithm fails is negligible.*

Proof Heuristic 14.2.13 states there are no cycles or "wasted" walks (in the sense that their steps do not contribute to potential collisions). Hence, before the first collision, after N steps of the algorithm we have visited N group elements. By Heuristic 14.2.13, the expected number of group elements to be sampled before the first collision is $\sqrt{\pi r/2}$. The collision is not detected until walks hit a distinguished point, which adds a further $2/\theta$ to the number of steps. Hence, the total number of steps (calls to the function walk) in the algorithm is $\sqrt{\pi r/2} + 2/\theta$. Since $2/\theta < 2\sqrt{r}/\log(r) = o(1)\sqrt{r}$, the result follows.

Let $x = g^{a_i}h^{b_i} = g^{a_j}h^{b_j}$ be the collision. Since the starting values $g^{a_0}h^{b_0}$ are chosen uniformly and independently at random, the values b_i and b_j are uniformly and independently random. It follows that $b_i \equiv b_j \pmod{r}$ with probability $1/r$, which is a negligible quantity in the input size of the problem. □

Exercise 14.2.15 Show that if $\theta = \log(r)/\sqrt{r}$ then the expected storage of the rho algorithm, assuming it takes $O(\sqrt{r})$ steps, is $O(\log(r))$ group elements (which is typically $O(\log(r)^2)$ bits).

Exercise 14.2.16 The algorithm requires storing a triple (x_n, a_n, b_n) for each distinguished point. Give some strategies to reduce the number of bits that need to be stored.

Exercise 14.2.17 Let $G = \langle g_1, g_2 \rangle$ be a group of order r^2 and exponent r. Design a rho algorithm which, on input $h \in G$ outputs (a_1, a_2) such that $h = g_1^{a_1} g_2^{a_2}$. Determine the complexity of this algorithm.

Exercise 14.2.18 Show that the Pollard rho algorithm with distinguished points has better average-case running time than the BSGS algorithm (see Exercises 13.3.3 and 13.3.4).

Exercise 14.2.19 Explain why taking $\mathcal{D} = G$ (i.e., all group elements distinguished) leads to an algorithm that is much slower than the BSGS algorithm.

Suppose one is given $g, h_1, \ldots, h_L$ (where $1 < L < r^{1/4}$) and is asked to find all a_i for $1 \leq i \leq L$ such that $h_i = g^{a_i}$. Kuhn and Struik [320] propose and analyse a method to solve

[6] More realistically, one could assume that only a negligibly small proportion of the walks fall into a cycle before hitting a distinguished point.

all L instances of the DLP, using Pollard rho with distinguished points, in roughly $\sqrt{2rL}$ group operations. A crucial trick, attributed to Silverman and Stapleton, is that once the ith DLP is known one can re-write all distinguished points $g^a h_i^b$ in the form $g^{a'}$. As noted by Hitchcock, Montague, Carter and Dawson [260], one must be careful to choose a random walk function that does not depend on the elements h_i (however, the random starting points do depend on the h_i).

Exercise 14.2.20 Write down pseudocode for the Kuhn–Struik algorithm for solving L instances of the DLP, and explain why the algorithm works.

Section 14.2.5 explains why the rho algorithm with distinguished points can be easily parallelised. That section also discusses a number of practical issues relating to the use of distinguished points.

Cheon, Hong and Kim [124] speed up Pollard rho in $\mathbb{F}_p^*$ by using a "look ahead" strategy; essentially they determine in which partition the next value of the walk lies, without performing a full group operation. A similar idea for elliptic curves has been used by Bos, Kaihara and Kleinjung [84].

14.2.5 Towards a rigorous analysis of Pollard rho

Theorem 14.2.14 is not satisfying since Heuristic 14.2.13 is essentially equivalent to the statement "the rho algorithm has expected running time $(1 + o(1))\sqrt{\pi r/2}$ group operations". The reason for stating the heuristic is to clarify exactly what properties of the pseudorandom walk are required. The reason for believing Heuristic 14.2.13 is that experiments with the rho algorithm (see Section 14.4.3) confirm the estimate for the running time.

We now give a brief survey of theoretical results on the rho algorithm. The main results for the original rho walk (with $n_S = 3$) are due to Horwitz and Venkatesan [266], Miller and Venkatesan [382] and Kim, Montenegro, Peres and Tetali [302, 301]. These works all assume that the partition function S is a truly random function. The basic idea is to define the **rho graph**, which is a directed graph with vertex set $\langle g \rangle$ and an edge from x_1 to x_2 if x_2 is the next step of the walk when at x_1. Fix an integer n. Define the distribution $\mathcal{D}_n$ on $\langle g \rangle$ obtained by choosing uniformly at random $x_1 \in \langle g \rangle$, running the walk for n steps, and recording the final point in the walk. The crucial property to study is the **mixing time**, which, informally, is the smallest integer n such that $\mathcal{D}_n$ is "sufficiently close" to the uniform distribution. For these results, the squaring operation in the original walk is crucial. The main result of [301] is that the Pollard rho algorithm solves the DLP with probability $1/2$ in $O(\sqrt{r})$ group operations, where the implied constant in the O is not specified. Note that the idealised model of S being a random function is not implementable with constant (or even polynomial) storage.

Sattler and Schnorr [461] and Teske [542] have considered the additive rho walk. One key feature of their work is to discuss the effect of the number of partitions n_S. Sattler and

Schnorr show (subject to a conjecture) that if $n_S \geq 8$ then the expected running time for the rho algorithm is $c\sqrt{\pi r/2}$ group operations for an explicit constant c. Teske shows, using results of Hildebrand, that the additive walk should approximate the uniform distribution after fewer than $\sqrt{r}$ steps once $n_S \geq 6$. She recommends using the additive walk with $n_S \geq 20$ and, when this is done, conjectures that the expected cycle length is $\leq 1.3\sqrt{r}$ (compared with the theoretical $\approx 1.253\sqrt{r}$).

Further motivation for using large n_S is given by Brent and Pollard [94], Arney and Bender [13] and Blackburn and Murphy [55]. They present heuristic arguments that the expected cycle length when using n_S partitions is $\sqrt{c_{n_S}\pi r/2}$ where $c_{n_S} = n_S/(n_S - 1)$. This heuristic is supported by the experimental results of Teske [542].

Finally, when using equivalence classes (see Section 14.4) there are further advantages in taking n_S to be large.

14.3 Distributed Pollard rho

In this section we explain how the Pollard rho algorithm can be parallelised. Rather than a parallel computing model we consider a **distributed computing** model. In this model there is a **server** and $N_P \geq 1$ **clients** (we also refer to the clients as **processors**). There is no shared storage or direct communication between the clients. Instead, the server can send messages to clients and each client can send messages to the server. In general, we prefer to minimise the amount of communication between server and clients.[7]

To solve an instance of the discrete logarithm problem the server will activate a number of clients, providing each with its own individual initial data. The clients will run the rho pseudorandom walk and occasionally send data back to the server. Eventually, the server will have collected enough information to solve the problem, in which case it sends all clients a termination instruction. The rho algorithm with distinguished points can very naturally be used in this setting.

The best one can expect for any distributed computation is a linear speedup compared with the serial case (since if the overall total work in the distributed case was less than the serial case then this would lead to a faster algorithm in the serial case). In other words, with N_P clients we hope to achieve a running time proportional to $\sqrt{r}/N_P$.

14.3.1 The algorithm and its heuristic analysis

All processors perform the same pseudorandom walk $(x_{i+1}, a_{i+1}, b_{i+1}) = \mathsf{walk}(x_i, a_i, b_i)$ as in Section 14.2.1, but each processor starts from a different random starting point. Whenever a processor hits a distinguished point then it sends the triple (x_i, a_i, b_i) to the server and re-starts its walk at a new random point (x_0, a_0, b_0). If one processor ever visits a

[7] There are numerous examples of such distributed computation over the internet. Two notable examples are the Great Internet Mersenne Primes Search (GIMPS) and the Search for Extraterrestrial Intelligence (SETI). One observes that the former search has been more successful than the latter.

point visited by another processor then the walks from that point agree and both walks end at the same distinguished point. When the server receives two triples (x, a, b) and (x, a', b') for the same group element x but with $b \not\equiv b' \pmod{r}$ then it has $g^a h^b = g^{a'} h^{b'}$ and can solve the DLP as in the serial (i.e., non-parallel) case. The server therefore computes the discrete logarithm problem and sends a terminate signal to all processors. Pseudocode for both server and clients are given by Algorithms 16 and 17. By design, if the algorithm halts then the answer is correct.

Algorithm 16 The distributed rho algorithm: server side

INPUT: $g, h \in G$
OUTPUT: c such that $h = g^c$
1: Randomly choose a walk function walk(x, a, b)
2: Initialise an easily searched structure L (sorted list, binary tree etc) to be empty
3: Start all processors with the function walk
4: **while** DLP not solved **do**
5: Receive triples (x, a, b) from clients and insert into L
6: **if** first coordinate of new triple (x, a, b) matches existing triple (x, a', b') **then**
7: **if** $b' \not\equiv b \pmod{r}$ **then**
8: Send terminate signal to all clients
9: **return** $(a - a')(b' - b)^{-1} \pmod{r}$
10: **end if**
11: **end if**
12: **end while**

Algorithm 17 The distributed rho algorithm: client side

INPUT: $g, h \in G$, function walk
1: **while** terminate signal not received **do**
2: Choose uniformly at random $0 \le a, b < r$
3: Set $x = g^a h^b$
4: **while** $x \notin \mathcal{D}$ **do**
5: $(x, a, b) = $ walk(x, a, b)
6: **end while**
7: Send (x, a, b) to server
8: **end while**

We now analyse the performance of this algorithm. To get a clean result we assume that no client ever crashes, that communications between server and client are perfectly reliable, that all clients have the same computational efficiency and are running continuously (in other words, each processor computes the same number of group operations in any given time period).

It is appropriate to ignore the computation performed by the server and instead to focus on the number of group operations performed by each client running Algorithm 17. Each execution of the function walk(x, a, b) involves a single group operation. We must also count the number of group operations performed in line 3 of Algorithm 17; though this term is negligible if walks are long on average (i.e., if $\mathcal{D}$ is a sufficiently small subset of G).

It is an open problem to give a rigorous analysis of the distributed rho method. Hence, we make the following heuristic assumption. We believe this assumption is reasonable when r is sufficiently large, n_S is sufficiently large, $\log(r)/\sqrt{r} < \theta$, the set $\mathcal{D}$ of distinguished points is determined by a good hash function, the number N_P of clients is sufficiently small (e.g., $N_P < \theta\sqrt{\pi r/2}/\log(r)$, see Exercise 14.3.3) and the function walk is chosen at random.

Heuristic 14.3.1

1. The expected number of group elements to be sampled before the same element is sampled twice is $\sqrt{\pi r/2}$.
2. Walks reach a distinguished point in significantly fewer than $\sqrt{r}/N_P$ steps (in other words, there are no cycles in the walks and walks are not excessively long). More realistically, one could assume that only a negligible proportion of the walks fall into a cycle before hitting a distinguished point.

Theorem 14.3.2 *Let the notation be as above, in particular let N_P be the (fixed, independent of r) number of clients. Let θ the probability that a group element is a distinguished point and suppose $\log(r)/\sqrt{r} < \theta$. Assume Heuristic 14.3.1 and the above assumptions about the reliability and equal power of the processors hold. Then the expected number of group operations performed by each client of the distributed rho method is $(1 + 2\log(r)\theta)\sqrt{\pi r/2}/N_P + 1/\theta$ group operations. This is $(\sqrt{\pi/2}/N_P + o(1))\sqrt{r}$ group operations when $\theta < 1/\log(r)^2$. The storage requirement on the server is $\theta\sqrt{\pi r/2} + N_P$ points.*

Proof Heuristic 14.3.1 states that we expect to sample $\sqrt{\pi r/2}$ group elements in total before a collision arises. Since this work is distributed over N_P clients of equal speed it follows that each client is expected to call the function walk about $\sqrt{\pi r/2}/N_P$ times. The total number of group operations is therefore $\sqrt{\pi r/2}/N_P$ plus $2\log(r)\theta\sqrt{\pi r/2}/N_P$ for the work of line 3 of Algorithm 17. The server will not detect the collision until the second client hits a distinguished point, which is expected to take $1/\theta$ further steps by the heuristic (part 3 of Lemma 14.2.12). Hence, each client needs to run an expected $\sqrt{\pi r/2}/N_P + 1/\theta$ steps of the walk.

Of course, a collision $g^a h^b = g^{a'} h^{b'}$ can be useless in the sense that $b' \equiv b \pmod{r}$. A collision implies $a' + cb' \equiv a + cb \pmod{r}$ where $h = g^c$; there are r such pairs (a', b') for each pair (a, b). Since each walk starts with uniformly random values (a_0, b_0) it follows that the values (a, b) are uniformly distributed over the r possibilities. Hence, the probability of a collision being useless is $1/r$ and the expected number of collisions required is 1.

Each processor runs for $\sqrt{\pi r/2}/N_P + 1/\theta$ steps and therefore is expected to send $\theta\sqrt{\pi r/2}/N_P + 1$ distinguished points in its lifetime. The total number of points to store is therefore $\theta\sqrt{\pi r/2} + N_P$. □

Exercise 14.2.15 shows that the complexity in the case $N_P = 1$ can be taken to be $(1 + o(1))\sqrt{\pi r/2}$ group operations with polynomial storage.

Exercise 14.3.3 When distributing the algorithm it is important to ensure that, with very high probability, each processor finds at least one distinguished point in less than its total expected running time. Show that this will be the case if $1/\theta \leq \sqrt{\pi r/2}/(N_P \log(r))$.

Schulte-Geers [480] analyses the choice of θ and shows that Heuristics 14.2.13 and 14.3.1 are not valid asymptotically if $\theta = o(1/\sqrt{r})$ as $r \to \infty$ (e.g., walks in this situation are more likely to fall into a cycle than to hit a distinguished point). In any case, since each processor only travels a distance of $\sqrt{\pi r/2}/N_P$ it follows we should take $\theta > N_P/\sqrt{r}$. In practice, one tends to determine the available storage first (say, c group elements where $c > 10^9$) and to set $\theta = c/\sqrt{\pi r/2}$ so that the total number of distinguished points visited is expected to be c. The results of [480] validate this approach. In particular, it is extremely unlikely that there is a self-collision (and hence a cycle) before hitting a distinguished point.

14.4 Speeding up the rho algorithm using equivalence classes

Gallant, Lambert and Vanstone [215] and Wiener and Zuccherato [565] showed that one can speed up the rho method in certain cases by defining the pseudorandom walk not on the group $\langle g \rangle$ but on a set of equivalence classes. This is essentially the same thing as working in an algebraic group quotient instead of the algebraic group.

Suppose there is an equivalence relation on $\langle g \rangle$. Denote by $\overline{x}$ the equivalence class of $x \in \langle g \rangle$. Let N_C be the size of a generic equivalence class. We require the following properties:

1. One can define a unique representative $\hat{x}$ of each equivalence class $\overline{x}$.
2. Given (x_i, a_i, b_i) such that $x_i = g^{a_i}h^{b_i}$ one can efficiently compute $(\hat{x}_i, \hat{a}_i, \hat{b}_i)$ such that $\hat{x}_i = g^{\hat{a}_i}h^{\hat{b}_i}$.

We give some examples in Section 14.4.1 below.

One can implement the rho algorithm on equivalence classes by defining a pseudorandom walk function walk(x_i, a_i, b_i) as in Definition 14.2.1. More precisely, set $x_1 = g, a_1 = 1, b_1 = 0$ and define the sequence x_i by (this is the "original walk")

$$x_{i+1} = f(x_i) = \begin{cases} \hat{x}_i^2 & \text{if } S(\hat{x}_i) = 0 \\ \hat{x}_i g_j & \text{if } S(\hat{x}_i) = j,\ j \in \{1, \ldots, n_S - 1\} \end{cases} \tag{14.6}$$

where the selection function S and the values $g_j = g^{u_j}h^{v_j}$ are as in Definition 14.2.1. When using distinguished points one defines an equivalence class to be distinguished if the unique equivalence class representative has the distinguished property.

There is a very serious problem with cycles that we do not discuss yet; see Section 14.4.2 for the details.

Exercise 14.4.1 Write down the formulae for updating the values a_i and b_i in the function walk.

Exercise 14.4.2 Write pseudocode for the distributed rho method on equivalence classes.

Theorem 14.4.3 *Let G be a group and $g \in G$ of order r. Suppose there is an equivalence relation on $\langle g \rangle$ as above. Let N_C be the generic size of an equivalence class. Let C_1 be the number of bit operations to perform a group operation in $\langle g \rangle$ and C_2 the number of bit operations to compute a unique equivalence class representative $\hat{x}_i$ (and to compute $\hat{a}_i, \hat{b}_i$).*

Consider the rho algorithm as above (ignoring the possibility of useless cycles, see Section 14.4.2 below). Under a heuristic assumption for equivalence classes analogous to Heuristic 14.2.13 the expected time to solve the discrete logarithm problem is

$$\left(\sqrt{\frac{\pi}{2N_C}} + o(1)\right)\sqrt{r}\,(C_1 + C_2)$$

bit operations. As usual, this becomes $(\sqrt{\pi/2N_C} + o(1))\sqrt{r}/N_P(C_1 + C_2)$ bit operations per client when using N_P processors of equal computational power.

Exercise 14.4.4 Prove this theorem.

Theorem 14.4.3 assumes a perfect random walk. For walks defined on n_S partitions of the set of equivalence classes it is shown in Appendix B of [24] (also see Section 2.2 of [86]) that one predicts a slightly improved constant than the usual factor $c_{n_S} = n_S/(n_S - 1)$ mentioned at the end of Section 14.2.5.

We mention a potential "paradox" with this idea. In general, computing a unique equivalence class representative involves listing all elements of the equivalence class, and hence needs $\tilde{O}(N_C)$ bit operations. Hence, naively, the running time is $\tilde{O}(\sqrt{N_C \pi r/2})$ bit operations, which is worse than doing the rho algorithm without equivalence classes. However, in practice one only uses this method when $C_2 < C_1$, in which case the speedup can be significant.

14.4.1 Examples of equivalence classes

We now give some examples of useful equivalence relations on some algebraic groups.

Example 14.4.5 For a group G with efficiently computable inverse (e.g., elliptic curves $E(\mathbb{F}_q)$ or algebraic tori $\mathbb{T}_z$ with $n > 1$ (e.g., see Section 6.3)) one can define the equivalence relation $x \equiv x^{-1}$. We have $N_C = 2$ (though note that some elements, namely the identity and elements of order 2, are equal to their inverse so these classes have size 1). If $x_i = g^{a_i} h^{b_i}$ then clearly $x^{-1} = g^{-a_i} h^{-b_i}$. One defines a unique representative $\hat{x}$ for the equivalence class by, for example, imposing a lexicographical ordering on the binary representation of the elements in the class.

We can generalise this example as follows.

Example 14.4.6 Let G be an algebraic group over $\mathbb{F}_q$ with an automorphism group $\mathrm{Aut}(G)$ of size N_C (see examples in Sections 9.4 and 11.3.3). Suppose that for $g \in G$ of order r one has $\psi(g) \in \langle g \rangle$ for each $\psi \in \mathrm{Aut}(G)$. Furthermore, assume that for each $\psi \in \mathrm{Aut}(G)$ one can efficiently compute the eigenvalue $\lambda_\psi \in \mathbb{Z}$ such that $\psi(g) = g^{\lambda_\psi}$. Then for $x \in G$ one can define $\overline{x} = \{\psi(x) : \psi \in \mathrm{Aut}(G)\}$.

Again, one defines $\hat{x}$ by listing the elements of $\overline{x}$ as bitstrings and choosing the first one under lexicographical ordering.

Another important class of examples comes from orbits under the Frobenius map.

Example 14.4.7 Let G be an algebraic group defined over $\mathbb{F}_q$ but with group considered over $\mathbb{F}_{q^d}$ (for examples see Sections 11.3.2 and 11.3.3). Let π_q be the q-power Frobenius map on $G(\mathbb{F}_{q^d})$. Let $g \in G(\mathbb{F}_{q^d})$ and suppose that $\pi_q(g) = g^\lambda \in \langle g \rangle$ for some known $\lambda \in \mathbb{Z}$.

Define the equivalence relation on $G(\mathbb{F}_{q^d})$ so that the equivalence class of $x \in G(\mathbb{F}_{q^d})$ is the set $\overline{x} = \{\pi_q^i(x) : 0 \le i < d\}$. We assume that, for elements x of interest, $\overline{x} \subseteq \langle g \rangle$. Then $N_C = d$, though there can be elements defined over proper subfields for which the equivalence class is smaller.

If one uses a normal basis for $\mathbb{F}_{q^d}$ over $\mathbb{F}_q$ then one can efficiently compute the elements $\pi_q^i(x)$ and select a unique representative of each equivalence class using a lexicographical ordering of binary strings.

Example 14.4.8 For some groups (e.g., Koblitz elliptic curves $E/\mathbb{F}_2$ considered as a group over $\mathbb{F}_{2^m}$; see Exercise 9.10.10) we can combine both equivalence classes above. Let m be prime, $\#E(\mathbb{F}_{2^m}) = hr$ for some small cofactor h, and $P \in E(\mathbb{F}_{2^m})$ of order r. Then $\pi_2(P) \in \langle P \rangle$ and we define the equivalence class $\overline{P} = \{\pm\pi_2^i(P) : 0 \le i < m\}$ of size $2m$. Since m is odd, this class can be considered as the orbit of P under the map $-\pi_2$. The distributed rho algorithm on equivalence classes for such curves is expected to require approximately $\sqrt{\pi 2^m/(4m)}$ group operations.

14.4.2 Dealing with cycles

One problem that can arise is walks that fall into a cycle before they reach a distinguished point. We call these **useless cycles**.

Exercise 14.4.9 Suppose the equivalence relation is such that $x \equiv x^{-1}$. Fix $x_i = \hat{x}_i$ and let $x_{i+1} = \hat{x}_i g$. Suppose $\hat{x}_{i+1} = x_{i+1}^{-1}$ and that $S(\hat{x}_{i+1}) = S(\hat{x}_i)$. Show that $x_{i+2} \equiv x_i$ and so there is a cycle of order 2. Suppose the equivalence classes generically have size N_C. Show, under the assumptions that the function S is perfectly random and that $\hat{x}$ is a randomly chosen element of the equivalence class, that the probability that a randomly chosen x_i leads to a cycle of order 2 is $1/(N_C n_S)$.

A theoretical discussion of cycles was given in [215] and by Duursma, Gaudry and Morain [172]. An obvious way to reduce the probability of cycles is to take n_S to be very large compared with the average length $1/\theta$ of walks. However, as argued by Bos, Kleinjung and Lenstra [86], large values for n_S can lead to slower algorithms (e.g., due to the fact that the precomputed steps do not all fit in cache memory). Hence, as Exercise 14.4.9 shows, useless cycles will be regularly encountered in the algorithm. There are several possible ways to deal with this issue. One approach is to use a "look-ahead" technique to avoid falling in 2-cycles. Another approach is to detect small cycles (e.g., by storing a fixed number of previous values of the walk or, at regular intervals, using a cycle-finding algorithm for a small number of steps) and to design a well-defined exit strategy for short cycles; Gallant, Lambert and Vanstone call this **collapsing the cycle**; see Section 6 of [215]. To collapse a cycle one must be able to determine a well-defined element in it; from there one can take a step (different to the steps used in the cycle from that point) or use squaring to exit the cycle. All these methods require small amounts of extra computation and storage, though Bernstein, Lange and Schwabe [57] argues that the additional overhead can be made negligible. We refer to [45, 86] for further discussion of these issues.

Gallant, Lambert and Vanstone [215] presented a different walk that does not, in general, lead to short cycles. Let G be an algebraic group with an efficiently computable endomorphism ψ of order m (i.e., $\psi^m = \psi \circ \psi \circ \cdots \circ \psi = 1$). Let $g \in G$ of order r be such that $\psi(g) = g^\lambda$ so that $\psi(x) = x^\lambda$ for all $x \in \langle g \rangle$. Define the equivalence classes $\overline{x} = \{\psi^j(x) : 0 \le j < m\}$. We define a pseudorandom sequence $x_i = g^{a_i} h^{b_i}$ by using $\hat{x}$ to select an endomorphism $(1 + \psi^j)$ and then acting on x_i with this map. More precisely, j is some function of $\hat{x}$ (e.g., the function S in Section 14.2.1) and

$$x_{i+1} = (1 + \psi^j)x_i = x_i \psi^j(x_i) = x_i^{1+\lambda^j}$$

(the above equation looks more plausible when the group operation is written additively: $x_{i+1} = x_i + \psi^j(x_i) = (1 + \lambda^j)x_i$). One can check that the map is well-defined on equivalence classes and that $x_{i+1} = g^{a_{i+1}} h^{b_{i+1}}$ where $a_{i+1} = (1 + \lambda^j)a_i \pmod{r}$ and $b_{i+1} = (1 + \lambda^j)b_i \pmod{r}$.

We stress that this approach still requires finding a unique representative of each equivalence class in order to define the steps of the walk in a well-defined way. Hence, one can still use distinguished points by defining a class to be distinguished if its representative is distinguished. One suggestion, originally due to Harley, is to use the Hamming weight of the x-coordinate to derive the selection function.

One drawback of the Gallant, Lambert, Vanstone idea is that there is less flexibility in the design of the pseudorandom walk.

Exercise 14.4.10 Generalise the Gallant–Lambert–Vanstone walk to use $(c + \psi^j)$ for any $c \in \mathbb{Z}$. Why do we prefer to only use $c = 1$?

Exercise 14.4.11 Show that taking $n_S = \log(r)$ means the total overhead from handling cycles is $o(\sqrt{r})$, while the additional storage (group elements for the random walks) is $O(\log(r))$ group elements.

Exercise 14.4.11 together with Exercise 14.2.15 shows that (as long as computing equivalence class representatives is fast) one can solve the discrete logarithm problem using equivalence classes of generic size N_C in $(1 + o(1))\sqrt{\pi r/(2N_C)}$ group operations and $O(\log(r))$ group elements storage.

14.4.3 Practical experience with the distributed rho algorithm

Real computations are not as simple as the idealised analysis above: one does not know in advance how many clients will volunteer for the computation; not all clients have the same performance or reliability; clients may decide to withdraw from the computation at any time; the communications between client and server may be unreliable etc. Hence, in practice one needs to choose the distinguished points to be sufficiently common that even the weakest client in the computation can hit a distinguished point within a reasonable time (perhaps after just one or two days). This may mean that the stronger clients are finding many distinguished points every hour.

The largest discrete logarithm problems solved using the distributed rho method are mainly the Certicom challenge elliptic curve discrete logarithm problems. The current records are for the groups $E(\mathbb{F}_p)$ where $p \approx 2^{108} + 2^{107}$ (by a team coordinated by Chris Monico in 2002) and where $p = (2^{128} - 3)/76439 \approx 2^{111} + 2^{110}$ (by Bos, Kaihara and Montgomery in 2009) and for $E(\mathbb{F}_{2^{109}})$ (again by Monico's team in 2004). None of these computations used the equivalence class $\{P, -P\}$.

We briefly summarise the parameters used for these large computations. For the 2002 result the curve $E(\mathbb{F}_p)$ has prime order so $r \approx 2^{108} + 2^{107}$. The number of processors was over 10,000 and they used $\theta = 2^{-29}$. The number of distinguished points found was 68 228 567, which is roughly 1.32 times the expected number $\theta\sqrt{\pi r/2}$ of points to be collected. Hence, this computation was unlucky in that it ran about 1.3 times longer than the expected time. The computation ran for about 18 months.

The 2004 result is for a curve over $\mathbb{F}_{2^{109}}$ with group order $2r$ where $r \approx 2^{108}$. The computation used roughly 2000 processors, $\theta = 2^{-30}$ and the number of distinguished points found was 16 531 676. This is about 0.79 times the expected number $\theta\sqrt{\pi 2^{108}/2}$. This computation took about 17 months.

The computation by Bos, Kaihara and Montgomery [85] was innovative in that the work was done using a cluster of 200 computer game consoles. The random walk used $n_S = 16$ and $\theta = 1/2^{24}$. The total number of group operations performed was 8.5×10^{16} (which is 1.02 times the expected value) and 5×10^9 distinguished points were stored.

Exercise 14.4.12 Verify that the parameters above satisfy the requirements that θ is much larger than $1/\sqrt{r}$ and N_P is much smaller than $\theta\sqrt{r}$.

There is a close fit between the actual running time for these examples and the theoretical estimates. This is evidence that the heuristic analysis of the running time is not too far from the performance in practice.

14.5 The kangaroo method

This algorithm is designed for the case where the discrete logarithm is known to lie in a short interval. Suppose $g \in G$ has order r and that $h = g^a$ where a lies in a short interval $b \leq a < b + w$ of width w. We assume that the values of b and w are known. Of course, one can solve this problem using the rho algorithm, but if w is much smaller than the order of g then this will not necessarily be optimal.

The kangaroo method was originally proposed by Pollard [437]. Van Oorschot and Wiener [423] greatly improved it by using distinguished points. We present the improved version in this section.

For simplicity, compute $h' = hg^{-b}$. Then $h' \equiv g^x \pmod{p}$ where $0 \le x < w$. Hence, there is no loss of generality by assuming that $b = 0$. Thus, from now on our problem is: given g, h, w, find a such that $h = g^a$ and $0 \le a < w$.

As with the rho method, the kangaroo method relies on a deterministic pseudorandom walk. The steps in the walk are pictured as the "jumps" of the kangaroo, and the group elements visited are the kangaroo's "footprints". The idea, as explained by Pollard, is to "catch a wild kangaroo using a tame kangaroo". The "tame kangaroo" is a sequence $x_i = g^{a_i}$ where a_i is known. The "wild kangaroo" is a sequence $y_j = hg^{b_j}$ where b_j is known. Eventually, a footprint of the tame kangaroo will be the same as a footprint of the wild kangaroo (this is called the "collision"). After this point, the tame and wild footprints are the same.[8] The tame kangaroo lays "traps" at regular intervals (i.e., at distinguished points) and, eventually, the wild kangaroo falls in to one of the traps.[9] More precisely, at the first distinguished point after the collision, one finds a_i and b_j such that $g^{a_i} = hg^{b_j}$ and the DLP is solved as $h = g^{a_i - b_j}$.

There are two main differences between the kangaroo method and the rho algorithm.

- Jumps are "small". This is natural since we want to stay within (or, at least, not too far outside) the interval.
- When a kangaroo lands on a distinguished point one **continues** the pseudorandom walk (rather than restarting the walk at a new randomly chosen position).

14.5.1 The pseudorandom walk

The pseudorandom walk for the kangaroo method has some significant differences to the rho walk: steps in the walk correspond to known small increments in the exponent (in other words, kangaroos make small jumps of known distance in the exponent). We therefore do not include the squaring operation $x_{i+1} = x_i^2$ (as the jumps would be too big) or multiplication by h (we would not know the length of the jump in the exponent). We now describe the walk precisely.

- As in Section 14.2.1, we use a function $S : G \to \{0, \ldots, n_S - 1\}$ which partitions G into sets $S_i = \{g \in G : S(g) = i\}$ of roughly similar size.

8 A collision between two different walks can be drawn in the shape of the letter λ. Hence, Pollard also suggested this be called the "lambda method". However, other algorithms (such as the distributed rho method) have collisions between different walks, so this naming is ambiguous. The name "kangaroo method" emphasises the fact that the jumps are small. Hence, as encouraged by Pollard, we do not use the name "lambda method" in this book.

9 Actually, the wild kangaroo can be in front of the tame kangaroo, in which case it is better to think of each kangaroo trying to catch the other.

- For $0 \le j < n_S$ choose exponents $1 \le u_j \le \sqrt{w}$. Define $m = (\sum_{j=0}^{n_S-1} u_j)/n_S$ to be the **mean step size**. As explained below, we will take $m \approx \sqrt{w}/2$.

  Pollard [437, 438] suggested taking $u_j = 2^j$ as this minimises the chance that two different short sequences of jumps add to the same value. This seems to give good results in practice. An alternative is to choose most of the values u_i to be random and the last few to ensure that m is very close to $c_1\sqrt{w}$.
- The pseudorandom walk is a sequence $x_0, x_1, \ldots$ of elements of G defined by an initial value x_0 (to be specified later) and the formula

$$x_{i+1} = x_i g_{S(x_i)}.$$

The algorithm is not based on the birthday paradox, but instead on the following observations. Footprints are spaced, on average, distance m apart, so along a region traversed by a kangaroo there is, on average, one footprint in any interval of length m. Now, if a second kangaroo jumps along the same region and if the jumps of the second kangaroo are independent of the jumps from the first kangaroo, then the probability of a collision is roughly $1/m$. Hence, one expects a collision between the two walks after about m steps.

14.5.2 The kangaroo algorithm

We need to specify where to start the tame and wild kangaroos, and what the mean step size should be. The wild kangaroo starts at $y_0 = h = g^a$ with $0 \le a < w$. To minimise the distance between the tame and wild kangaroos at the start of the algorithm we start the tame kangaroo at $x_0 = g^{\lfloor w/2 \rfloor}$, which is the middle of the interval. We take alternate jumps and store the values (x_i, a_i) and (y_i, b_i) as above (i.e., so that $x_i = g^{a_i}$ and $y_i = hg^{b_i}$). Whenever x_i (respectively, y_i) is distinguished we store (x_i, a_i) (respectively, (y_i, b_i)) in an easily searched structure. The storage can be reduced by using the ideas of Exercise 14.2.16.

When the same distinguished point is visited twice then we have two entries (x, a) and (x, b) in the structure and so either $hg^a = g^b$ or $g^a = hg^b$. The ambiguity is resolved by seeing which of $a - b$ and $b - a$ lies in the interval (or just testing if $h = g^{a-b}$ or not).

As we will explain in Section 14.5.3, the optimal choice for the mean step size is $m = \sqrt{w}/2$.

Exercise 14.5.1 Write this algorithm in pseudocode.

We visualise the algorithm not in the group G but on a line representing exponents. The tame kangaroo starts at $\lfloor w/2 \rfloor$. The wild kangaroo starts somewhere in the interval $[0, w)$. Kangaroo jumps are small steps to the right. See Figure 14.1 for the picture.

Example 14.5.2 Let $g = 3 \in \mathbb{F}_{263}^*$ which has prime order 131. Let $h = 181 \in \langle g \rangle$ and suppose we are told that $h = g^a$ with $0 \le a < w = 53$. The kangaroo method can be used in this case.

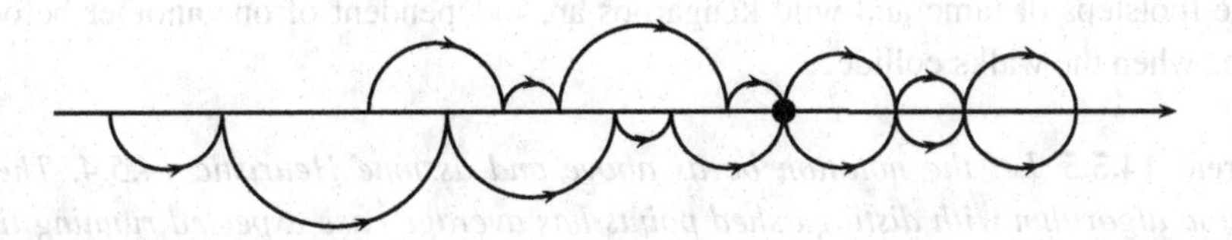

Figure 14.1 Kangaroo walk. Tame kangaroo walk pictured above the axis and wild kangaroo walk pictured below. The dot indicates the first collision.

Since $\sqrt{w}/2 \approx 3.64$ it is appropriate to take $n_S = 4$ and choose steps $\{1, 2, 4, 8\}$. The mean step size is 3.75. The function $S(x)$ is $x \pmod 4$ (where elements of $\mathbb{F}_{263}^*$ are represented by integers in the set $\{1, \ldots, 262\}$).

The tame kangaroo starts at $(x_1, a_1) = (g^{26}, 26) = (26, 26)$. The sequence of points visited in the walk is listed below. A point is distinguished if its representation as an integer is divisible by 3; the distinguished points are written in bold face in the table.

i	0	1	2	3	4
x_i	26	2	**162**	235	**129**
a_i	26	30	34	38	46
$S(x_i)$	2	2	2	3	1
y_i	181	**51**	**75**	2	**162**
b_i	0	2	10	18	22
$S(y_i)$	1	3	3	2	2

The collision is detected when the distinguished point 162 is visited twice. The solution to the discrete logarithm problem is therefore $34 - 22 = 12$.

Exercise 14.5.3 Using the same parameters as Example 14.5.2, solve the DLP for $h = 78$.

14.5.3 Heuristic analysis of the kangaroo method

The analysis of the algorithm does not rely on the birthday paradox; instead, the mean step size is the crucial quantity. We sketch the basic probabilistic argument now. A more precise analysis is given in Section 14.5.6. The following heuristic assumption seems to be reasonable when w is sufficiently large, $n_S > \log(w)$, distinguished points are sufficiently common and specified using a good hash function (and hence are well-distributed), $\theta > \log(w)/\sqrt{w}$ and when the function walk is chosen at random.

Heuristic 14.5.4

1. Walks reach a distinguished point in significantly fewer than $\sqrt{w}$ steps (in other words, there are no cycles in the walks and walks are not excessively longer than $1/\theta$).
2. The footprints of a kangaroo are uniformly distributed in the region over which it has walked with, on average, one footprint in each interval of length m.

3. The footsteps of tame and wild kangaroos are independent of one another before the time when the walks collide.

Theorem 14.5.5 *Let the notation be as above and assume Heuristic 14.5.4. Then the kangaroo algorithm with distinguished points has average case expected running time of* $(2 + o(1))\sqrt{w}$ *group operations. The probability the algorithm fails is negligible.*

Proof We do not know whether the discrete logarithm of h is greater or less than $w/2$. So, rather than speaking of "tame" and "wild" kangaroos we will speak of the "front" and "rear" kangaroos. Since one kangaroo starts in the middle of the interval the distance between the starting point of the rear kangaroo and the starting point of the front kangaroo is between 0 and $w/2$ and is, on average, $w/4$. Hence, on average, $w/(4m)$ jumps are required for the rear kangaro to pass the starting point of the front kangaroo.

After this point, the rear kangaroo is travelling over a region that has already been jumped over by the front kangaroo. By our heuristic assumption, the footprints of the tame kangaroo are uniformly distributed over the region with, on average, one footprint in each interval of length m. Also, the footprints of the wild kangaroo are independent, and with one footprint in each interval of length m. The probability, at each step, that the wild kangaroo does not land on any of the footprints of the tame kangaroo is therefore heuristically $1 - 1/m$. By exactly the same arguments as Lemma 14.2.12 it follows that the expected number of jumps until a collision is m.

Note that there is a miniscule possibility that the walks never meet (this does not require working in an infinite group, it can even happen in a finite group if the "orbits" of the tame and wild walks are disjoint subsets of the group). If this happens then the algorithm never halts. Since the walk function is chosen at random the probability of this eventuality is negligible. On the other hand, if the algorithm halts then its result is correct. Hence, this is a Las Vegas algorithm.

The overall number of jumps made by the rear kangaroo until the first collision is therefore, on average, $w/(4m) + m$. One can easily check that this is minimised by taking $m = \sqrt{w}/2$. The kangaroo is also expected to perform a further $1/\theta$ steps to the next distinguished point. Since there are two kangaroos the expected total number of group operations performed is $2\sqrt{w} + 2/\theta = (2 + o(1))\sqrt{w}$. □

This result is proved by Montenegro and Tetali [390] under the assumption that S is a random function and that the distinguished points are well-distributed. Pollard [438] shows it is valid when the $o(1)$ is replaced by ϵ for some $0 \leq \epsilon < 0.06$.

Note that the expected distance, on average, travelled by a kangaroo is $w/4 + m^2 = w/2$ steps. Hence, since the order of the group is greater than w, we do not expect any self-collisions in the kangaroo walk.

We stress that, as with the rho method, the probability of success is considered over the random choice of pseudorandom walk, not over the space of problem instances. Exercise 14.5.6 considers a different way to optimise the expected running time.

Exercise 14.5.6 Show that, with the above choice of m, the expected number of group operations performed for the worst case of problem instances is $(3 + o(1))\sqrt{w}$. Determine the optimal choice of m to minimise the expected worst-case running time. What is the expected worst-case complexity?

Exercise 14.5.7 A card trick known as **Kruskal's principle** is as follows. Shuffle a deck of 52 playing cards and deal face up in a row. Define the following walk along the row of cards: if the number of the current card is i then step forward i cards (if the card is a King, Queen or Jack then step 5 cards). The magicican runs this walk (in their mind) from the first card and puts a coin on the last card visited by the walk. The magician invites their audience to choose a number j between 1 and 10, then runs the walk from the jth card. The magician wins if the walk also lands on the card with the coin. Determine the probability of success of this trick.

Exercise 14.5.8 Show how to use the kangaroo method to solve Exercises 13.3.8, 13.3.10 and 13.3.11 of Chapter 13.

Pollard's original proposal did not use distinguished points and the algorithm only had a fixed probability of success. In contrast, the method we have described keeps on running until it succeeds (indeed, if the DLP is insoluble then the algorithm would never terminate). Van Oorschot and Wiener (see page 12 of [423]) have shown that repeating Pollard's original method until it succeeds leads to a method with expected running time of approximately $3.28\sqrt{w}$ group operations.

Exercise 14.5.9 Suppose one is given $g \in G$ of order r, an integer w and an instance generator for the discrete logarithm problem that outputs $h = g^a \in G$ such that $0 \leq a < w$ according to some known distribution on $\{0, 1, \ldots, w - 1\}$. Assume that the distribution is symmetric with mean value $w/2$. How should one modify the kangaroo method to take account of this extra information? What is the running time?

14.5.4 Comparison with the rho algorithm

We now consider whether one should use the rho or kangaroo algorithm when solving a general discrete logarithm problem (i.e., where the width w of the interval is equal to, or close to, r). If $w = r$ then the rho method requires roughly $1.25\sqrt{r}$ group operations while the kangaroo method requires roughly $2\sqrt{r}$ group operations. The heuristic assumptions underlying both methods are similar, and in practice they work as well as the theory predicts. Hence, it is clear that the rho method is preferable, unless w is much smaller than r.

Exercise 14.5.10 Determine the interval size below which it is preferable to use the kangaroo algorithm over the rho algorithm.

14.5.5 Using inversion

Galbraith, Ruprai and Pollard [209] showed that one can improve the kangaroo method by exploiting inversion in the group.[10] Suppose one is given g, h, w and told that $h = g^a$ with $0 \leq a < w$. We also require that the order r of g is odd (this will always be the case, due to the Pohlig–Hellman algorithm). Suppose, for simplicity, that w is even. Replacing h by $hg^{-w/2}$ we have $h = g^a$ with $-w/2 \leq a < w/2$. One can perform a version of the kangaroo method with three kangaroos: one tame kangaroo starting from g^u for an appropriate value of u and two wild kangaroos starting from h and h^{-1} respectively.

The algorithm uses the usual kangaroo walk (with mean step size to be determined later) to generate three sequences (x_i, a_i), (y_i, b_i), (z_i, c_i) such that $x_i = g^{a_i}$, $y_i = hg^{b_i}$ and $z_i = h^{-1}g^{c_i}$. The crucial observation is that a collision between any two sequences leads to a solution to the DLP. For example, if $x_i = y_j$ then $h = g^{a_i - b_j}$ and if $y_i = z_j$ then $hg^{b_i} = h^{-1}g^{c_j}$ and so, since g has odd order r, $h = g^{(c_j - b_i)2^{-1} \pmod r}$. The algorithm uses distinguished points to detect a collison. We call this the **three-kangaroo algorithm**.

Exercise 14.5.11 Write down pseudocode for the three-kangaroo algorithm using distinguished points.

We now give a brief heuristic analysis of the three-kangaroo algorithm. Without loss of generality, we assume $0 \leq a \leq w/2$ (taking negative a simply swaps h and h^{-1}, so does not affect the running time). The distance between the starting points of the tame and wild kangaroos is $2a$. The distance between the starting points of the tame and right-most wild kangaroo is $|a - u|$. The extreme cases (in the sense that the closest pair of kangaroos are as far apart as possible) are when $2a = u - a$ or when $a = w/2$. Making all these cases equal leads to the equation $2a = u - a = w/2 - u$. Calling this distance l it follows that $w/2 = 5l/2$ and $u = 3w/10$. The average distance between the closest pair of kangaroos is then $w/10$ and the closest pair of kangaroos can be thought of as performing the standard kangaroo method in an interval of length $2w/5$. Following the analysis of the standard kangaroo method it is natural to take the mean step size to be $m = \frac{1}{2}\sqrt{2w/5} = \sqrt{w/10} \approx 0.316\sqrt{w}$. The average-case expected number of group operations (only considering the closest pair of kangaroos) would be $\frac{3}{2}2\sqrt{2w/5} \approx 1.897\sqrt{w}$. A more careful analysis takes into account the possibility of collisions between any pair of kangaroos. We refer to [209] for the details and merely remark that the correct mean step size is $m \approx 0.375\sqrt{w}$ and the average-case expected number of group operations is approximately $1.818\sqrt{w}$.

Exercise 14.5.12 The distance between $-a$ and a is even, so a natural trick is to use jumps of even length. Since we do not know whether a is even or odd, if this is done we do not know whether to start the tame kangaroo at g^u or g^{u+1}. However, one can consider a variant of the algorithm with two wild kangaroos (one starting from h and one from h^{-1}) and two tame kangaroos (one starting from g^u and one from g^{u+1}) and with jumps of even length.

[10] This research actually grew out of writing this chapter. Sometimes it pays to work slowly.

This is called the **four-kangaroo algorithm**. Explain why the correct choice for the mean step size is $m = 0.375\sqrt{2w}$ and why the heuristic average-case expected number of group operations is approximately $1.714\sqrt{w} = \frac{2\sqrt{2}}{3}1.818\sqrt{w}$.

Galbraith, Pollard and Ruprai [219] have combined the idea of Exercise 14.5.12 and the Gaudry–Schost algorithm (see Section 14.7) to obtain an algorithm for the discrete logarithm problem in an interval of length w that performs $(1.660 + o(1))\sqrt{w}$ group operations.

14.5.6 Towards a rigorous analysis of the kangaroo method

Pollard [438] gave a detailed heuristic analysis of the kangaroo method using jumps that are powers of two. His results suggest the expected running time is asymptotically $(2 + o(1))\sqrt{w}$ group operations. Montenegro and Tetali [390] have analysed the kangaroo method using jumps that are powers of 2, under the assumption that the selection function S is random and that the distinguished points are well-distributed. They prove that the average-case expected number of group operations is $(2 + o(1))\sqrt{w}$ group operations. It is beyond the scope of this book to give the details of these papers.

14.6 Distributed kangaroo algorithm

Let N_P be the number of processors or clients. A naive way to parallelise the kangaroo algorithm is to divide the interval $[0, w)$ into N_P subintervals of size w/N_P and then run the kangaroo algorithm in parallel on each subinterval. This gives an algorithm with running time $O(\sqrt{w/N_P})$ group operations per client, which is not a linear speedup.

Since we are using distinguished points one should be able to do better. But the kangaroo method is not as straightforward to parallelise as the rho method (a good exercise is to stop reading now and think about it for a few minutes). The solution is to use a herd of $N_P/2$ tame kangaroos and a herd of $N_P/2$ wild kangaroos. These are super-kangaroos in the sense that they take much bigger jumps (roughly $N_P/2$ times longer) than in the serial case. The goal is to have a collision between one of the wild kangaroos and one of the tame kangaroos. We imagine that both herds are setting traps, each trying to catch a kangaroo from the other herd (regrettably, they may sometimes catch one of their own kind).

When a kangaroo lands on a distinguished point one continues the pseudorandom walk (rather than restarting the walk at a new randomly chosen position). In other words, the herds march ever onwards with an occasional individual hitting a distinguished point and sending information back to the server. See Figure 14.2 for a picture of the herds in action.

There are two versions of the distributed algorithm, one by van Oorschot and Wiener [423] and another by Pollard [438]. The difference is how they handle the possibility of collisions between kangaroos of the same herd. The former has a mechanism to deal with this, which we will explain later. The latter paper elegantly ensures that there will not be collisions between individuals of the same herd.

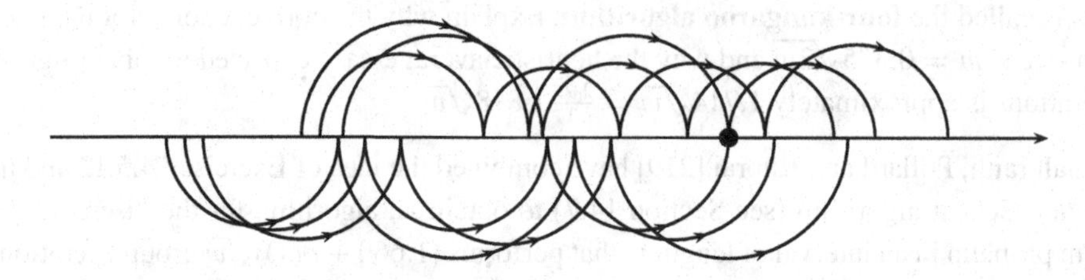

Figure 14.2 Distributed kangaroo walk (van Oorschot and Wiener version). The herd of tame kangaroos is pictured above the axis and the herd of wild kangaroos is pictured below. The dot marks the collision.

14.6.1 Van Oorschot and Wiener version

We first present the algorithm of van Oorschot and Wiener. The herd of tame kangaroos starts around the midpoint of the interval $[0, w)$, and the kangaroos are spaced a (small) distance s apart (as always, we describe kangaroos by their exponent). Similarly, the wild kangaroos start near $a = \log_g(h)$, again spaced a distance s apart. As we will explain later, the mean step size of the jumps should be $m \approx N_P\sqrt{w}/4$.

Here walk(x_i, a_i) is the function which returns $x_{i+1} = x_i g_{S(x_i)}$ and $a_{i+1} = a_i + u_{S(x_i)}$. Each client has a variable type which takes the value 'tame' or 'wild'.

If there is a collision between two kangaroos of the same herd then it will eventually be detected when the second one lands on the same distinguished point as the first. In [423] it is suggested that in this case the server should instruct the second kangaroo to take a jump of random length so that it no longer follows the path of the front kangaroo. Note that Teske [544] has shown that the expected number of collisions within the same herd is 2, so this issue can probably be ignored in practice.

We now give a very brief heuristic analysis of the running time. The following assumption seems to be reasonable when w is sufficiently large, n_S is sufficiently large, $\log(w)/\sqrt{w} < \theta$, the set $\mathcal{D}$ of distinguished points is determined by a good hash function, the number N_P of clients is sufficiently small (e.g., $N_P < \theta\sqrt{\pi r/2}/\log(r)$, see Exercise 14.3.3), the spacing s is independent of the steps in the random walk and is sufficiently large and the function walk is chosen at random.

Heuristic 14.6.1

1. Walks reach a distinguished point in significantly fewer than $\sqrt{w}$ steps (in other words, there are no cycles in the walks and walks are not excessively longer than $1/\theta$).
2. When two kangaroos with mean step size m walk over the same interval, the expected number of group elements sampled before a collision is m.
3. Walks of kangaroos in the same herd are independent.[11]

[11] This assumption is very strong, and indeed is false in general (since there is a chance that walks collide). The assumption is used for only two purposes. First, to "amplify" the second assumption in the heuristic from any pair of kangaroos to the level of herds. Second, to allow us to ignore collisions between kangaroos in the same herd (Teske, in Section 7 of [544], has argued that such collisions are rare). One could replace the assumption of independence by these two consequences.

Algorithm 18 The distributed kangaroo algorithm (van Oorschot and Wiener version): server side

INPUT: $g, h \in G$, interval length w, number of clients N_P
OUTPUT: a such that $h = g^a$

1: Choose n_S, a random function $S : G \to \{0, \ldots, n_S - 1\}$, $m = N_P\sqrt{w}/4$, jumps $\{u_0, \ldots, u_{n_S-1}\}$ with mean m, spacing s
2: **for** $i = 1$ to $N_P/2$ **do** ▷ Start $N_P/2$ tame kangaroo clients
3: Set $a_i = \lfloor w/2 \rfloor + is$
4: Initiate client on $(g^{a_i}, a_i, \text{'tame'})$ with function walk
5: **end for**
6: **for** $j = 1$ to $N_P/2$ **do** ▷ Start $N_P/2$ wild kangaroo clients
7: Set $a_j = js$
8: Initiate client on $(hg^{a_j}, a_j, \text{'wild'})$ with function walk
9: **end for**
10: Initialise an easily sorted structure L (sorted list, binary tree etc) to be empty
11: **while** DLP not solved **do**
12: Receive triples $(x_i, a_i, \text{type}_i)$ from clients and insert into L
13: **if** first coordinate of new triple (x, a_2, type_2) matches existing triple (x, a_1, type_1) **then**
14: **if** $\text{type}_2 = \text{type}_1$ **then**
15: Send message to the sender of (x, a_2, type_2) to take a random jump
16: **else**
17: Send terminate signal to all clients
18: **if** $\text{type}_1 = \text{'tame'}$ **then**
19: **return** $(a_1 - a_2) \pmod r$
20: **else**
21: **return** $(a_2 - a_1) \pmod r$
22: **end if**
23: **end if**
24: **end if**
25: **end while**

Theorem 14.6.2 *Let N_P be the number of clients (fixed, independent of w). Assume Heuristic 14.6.1 and that all clients are reliable and have the same computing power. The average-case expected number of group operations performed by the distributed kangaroo method for each client is* $(2 + o(1))\sqrt{w}/N_P$.

Proof Since we do not know where the wild kangaroo is, we speak of the front herd and the rear herd. The distance (in the exponent) between the front herd and the rear herd is, on average, $w/4$. So it takes $w/(4m)$ steps for the rear herd to reach the starting point of the front herd.

Algorithm 19 The distributed kangaroo algorithm (van Oorschot and Wiener version): client side

INPUT: $(x_1, a_1, \mathsf{type}) \in G \times \mathbb{Z}/r\mathbb{Z}$, function walk

1: **while** terminate signal not received **do**
2: $(x_1, a_1) = \mathsf{walk}(x_1, a_1)$
3: **if** $x_1 \in \mathcal{D}$ **then**
4: Send $(x_1, a_1, \mathsf{type})$ to server
5: **if** Receive jump instruction **then**
6: Choose random $1 < u < 2m$ (where m is the mean step size)
7: Set $a_1 = a_1 + u$, $x_1 = x_1 g^u$
8: **end if**
9: **end if**
10: **end while**

We now consider the footsteps of the rear herd in the region already visited by the front herd of kangaroos. Assuming the $N_P/2$ kangaroos of the front herd are independent, the region already covered by these kangaroos is expected to have $N_P/2$ footprints in each interval of length m. Hence, under our heuristic assumptions, the probability that a random footprint of one of the rear kangaroos lands on a footprint of one of the front kangaroos is $N_P/(2m)$. Since there are $N_P/2$ rear kangaroos, all mutually independent, the probability of one of the rear kangaroos landing on a tame footprint is $N_P^2/(4m)$. By the heuristic assumption the expected number of footprints to be made before a collision occurs is $4m/N_P^2$.

Finally, the collision will not be detected until a distinguished point is visited. Hence, one expects a further $1/\theta$ steps to be made.

The expected number of group operations made by each client in the average case is therefore $w/(4m) + 4m/N_P^2 + 1/\theta$. Ignoring the $1/\theta$ term, this expression is minimised by taking $m = N_P\sqrt{w}/4$. The result follows. □

The remarks made in Section 14.3.1 about parallelisation (for example, Exercise 14.3.3) apply equally for the distributed kangaroo algorithm.

Exercise 14.6.3 The above analysis is optimised for the average-case running time. Determine the mean step size to optimise the worst-case expected running time. Show that the heuristic optimal running time is $(3 + o(1))\sqrt{w}/N_P$ group operations.

Exercise 14.6.4 Give distributed versions of the three-kangaroo and four-kangaroo algorithms of Section 14.5.5.

14.6.2 Pollard version

Pollard's version reduces the computation to essentially a collection of serial versions, but in a clever way so that a linear speed-up is still obtained. One merit of this approach is that the analysis of the serial kangaroo algorithm can be applied; we no longer need the strong heuristic assumption that kangaroos in the same herd are mutually independent.

Let N_P be the number of processors and suppose we can write $N_P = U + V$ where $\gcd(U, V) = 1$ and $U, V \approx N_P/2$. The number of tame kangaroos is U and the number of wild kangaroos is V. The (super) kangaroos perform the usual pseudorandom walk with steps $\{UVu_0, \dots, UVu_{n-1}\}$ having mean $m \approx N_P\sqrt{w}/4$ (this is UV times the mean step size for solving the DLP in an interval of length $w/UV \approx 4w/N_P^2$). As usual, we choose either $u_j \approx 2^j$ or else random values between 0 and $2m/UV$.

The U tame kangaroos start at

$$g^{\lfloor w/2 \rfloor + iV}$$

for $0 \le i < U$. The V wild kangaroos start at hg^{jU} for $0 \le j < V$. Each kangaroo then uses the pseudorandom walk to generate a sequence of values (x_n, a_n) where $x_n = g^{a_n}$ or $x_n = hg^{a_n}$. Whenever a distinguished point is hit the kangaroo sends data to the server and continues the same walk.

Lemma 14.6.5 *Suppose the walks do not cover the whole group, i.e. $0 \le a_n < r$. Then there is no collision between two tame kangaroos or two wild kangaroos. There is a unique pair of tame and wild kangaroos who can collide.*

Proof Each element of the sequence generated by the ith tame kangaroo is of the form

$$g^{\lfloor w/2 \rfloor + iV + lUV}$$

for some $l \in \mathbb{Z}$. To have a collision between two different tame kangaroos one would need

$$\lfloor w/2 \rfloor + i_1 V + l_1 UV = \lfloor w/2 \rfloor + i_2 V + l_2 UV$$

and reducing modulo U implies $i_1 \equiv i_2 \pmod{U}$ which is a contradiction. To summarise, the values a_n for the tame kangaroos all lie in disjoint equivalence classes modulo U. A similar argument shows that wild kangaroos do not collide.

Finally, if $h = g^a$ then $i = (\lfloor w/2 \rfloor - a)V^{-1} \pmod{U}$ and $j = (a - \lfloor w/2 \rfloor)U^{-1} \pmod{V}$ are the unique pair of indices such that the ith tame kangaroo and the jth wild kangaroo can collide. □

The analysis of the algorithm therefore reduces to the serial case, since we have one tame kangaroo and one wild kangaroo who can collide. This makes the heuristic analysis simple and immediate.

Theorem 14.6.6 *Let the notation be as above. Assume Heuristic 14.5.4 and that all clients are reliable and have the same computational power. Then the average-case expected running time for each client is $(1 + o(1))\sqrt{w/UV} = (2 + o(1))\sqrt{w}/N_P$ group operations.*

Proof The action is now constrained to an equivalence class modulo UV, so the clients behave like the serial kangaroo method in an interval of size w/UV (see Exercise 14.5.8 for reducing a DLP in a congruence class to a DLP in a smaller interval). The mean step size is therefore $m \approx UV\sqrt{w/UV}/2 \approx N_P\sqrt{w}/4$. Applying Theorem 14.5.5 gives the result. □

14.6.3 Comparison of the two versions

Both versions of the distributed kangaroo method have the same heuristic running time of $(2 + o(1))\sqrt{w}/N_P$ group operations.[12] So which is to be preferred in practice? The answer depends on the context of the computation. For genuine parallel computation in a closed system (e.g., using special-purpose hardware) either could be used.

In distributed environments both methods have drawbacks. For example, the van Oorschot–Wiener method needs a communication from server to client in response to uploads of distinguished point information (the "take a random jump" instruction); though Teske [544] has remarked that this can probably be ignored.

More significantly, both methods require knowing the number N_P of processors at the start of the computation, since this value is used to specify the mean step size. This causes problems if a large number of new clients join the computation after it has begun.

With the van Oorschot and Wiener method, if further clients want to join the computation after it has begun then they can be easily added (half the new clients tame and half wild) by starting them at further shifts from the original starting points of the herds. With Pollard's method it is less clear how to add new clients. Even worse, since only one pair of "lucky" clients has the potential to solve the problem, if either of them crashes or withdraws from the computation then the problem will not be solved. As mentioned in Section 14.4.3, these are serious issues which do arise in practice.

On the other hand, these issues can be resolved by over-estimating N_P and by issuing clients with fresh problem instances once they have produced sufficiently many distinguished points from their current instance. Note that this also requires communication from server to client.

14.7 The Gaudry–Schost algorithm

Gaudry and Schost [228] give a different approach to solving discrete logarithm problems using pseudorandom walks. As we see in Exercise 14.7.6, this method is slower than the rho method when applied to the whole group. However, the approach leads to low-storage algorithms for the multi-dimensional discrete logarithm problems (see Definition 13.5.1) and the discrete logarithm problem in an interval using equivalence classes. This is interesting since, for both problems, it is not known how to adapt the rho or kangaroo methods to give a low-memory algorithm with the desired running time.

The basic idea of the Gaudry–Schost algorithm is as follows. One has pseudorandom walks in two (or more) subsets of the group such that a collision between walks of different

[12] Though the analysis by van Oorschot and Wiener needs the stronger assumption that the kangaroos in the same herd are mutually independent.

types leads to a solution to the discrete logarithm problem. The sets are smaller than the whole group, but they must overlap (otherwise, there is no chance of a collision). Typically, one of the sets is called a "tame set" and the other a "wild set". The pseudorandom walks are deterministic so that when two walks collide they continue along the same path until they hit a distinguished point and stop. Data from distinguished points is held in an easily searched database held by the server. After reaching a distinguished point, the walks re-start at a freshly chosen point.

14.7.1 Two-dimensional discrete logarithm problem

Suppose we are given $g_1, g_2, h \in G$ and $N \in \mathbb{N}$ (where we assume N is even) and asked to find integers $0 \le a_1, a_2 < N$ such that $h = g_1^{a_1} g_2^{a_2}$. Note that the size of the solution space is N^2, so we seek a low-storage algorithm with number of group operations proportional to N. The basic Gaudry–Schost algorithm for this problem is as follows.

Define the tame set

$$T = \{(x, y) \in \mathbb{Z}^2 : 0 \le x, y < N\}$$

and the wild set

$$\begin{aligned} W &= (a_1 - N/2, a_2 - N/2) + T \\ &= \{(a_1 - N/2 + x, a_2 - N/2 + y) \in \mathbb{Z}^2 : 0 \le x, y < N\}. \end{aligned}$$

In other words, T and W are $N \times N$ boxes centered on $(N/2 - 1, N/2 - 1)$ and (a_1, a_2) respectively. It follows that $\#W = \#T = N^2$ and if $(a_1, a_2) = (N/2 - 1, N/2 - 1)$ then $T = W$, otherwise $T \cap W$ is a proper non-empty subset of T.

Define a pseudorandom walk as follows: first choose $n_S > \log(N)$ random pairs of integers $-M < m_i, n_i < M$ where M is an integer to be chosen later (typically, $M \approx N/(1000\log(N))$) and precompute elements of the form $w_i = g_1^{m_i} g_2^{n_i}$ for $0 \le i < n_S$. Then choose a selection function $S : G \to \{0, 1, \ldots, n_S - 1\}$. The walk is given by the function

$$\mathsf{walk}(g, x, y) = (g w_{S(g)}, x + m_{S(g)}, y + n_{S(g)}).$$

Tame walks are started at $(g_1^x g_2^y, x, y)$ for random elements $(x, y) \in T$ and wild walks are started at $(h g_1^{x-N/2+1} g_2^{y-N/2+1}, x - N/2 + 1, y - N/2 + 1)$ for random elements $(x, y) \in T$. Walks proceed by iterating the function walk until a distinguished element of G is visited, at which time the data (g, x, y), together with the type of walk, is stored in a central database. When a distinguished point is visited, the walk is re-started at a uniformly chosen group element (this is like the rho method, but different from the behaviour of kangaroos). Once two walks of different types visit the same distinguished group element we have a collision of the form

$$g_1^x g_2^y = h g_1^{x'} g_2^{y'}$$

and the two-dimensional DLP is solved.

Exercise 14.7.1 Write pseudocode, for both the client and server, for the distributed Gaudry–Schost algorithm.

Exercise 14.7.2 Explain why the algorithm can be modified to omit storing the type of walk in the database. Show that the methods of Exercise 14.2.16 to reduce storage can also be used in the Gaudry–Schost algorithm.

Exercise 14.7.3 What modifications are required to solve the problem $h = g_1^{a_1} g_2^{a_2}$ such that $0 \le a_1 < N_1$ and $0 \le a_2 < N_2$ for $0 < N_1 < N_2$?

An important practical consideration is that walks will sometimes go outside the tame or wild regions. One might think that this issue can be solved by simply taking the values x and y into account and altering the walk when close to the boundary, but then the crucial property of the walk function (that once two walks collide, they follow the same path) would be lost. By taking distinguished points to be quite common (i.e., increasing the storage) and making M relatively small one can minimise the impact of this problem. Hence, we ignore it in our analysis.

We now briefly explain the heuristic complexity of the algorithm. The key observation is that a collision can only occur in the region where the two sets overlap. Let $A = T \cap W$. If one samples uniformly at random in A, alternately writing elements down on a "tame" and "wild" list, the expected number of samples until the two lists have an element in common is $\sqrt{\pi \# A} + O(1)$ (see, for example, Selivanov [483] or [206]).

The following heuristic assumption seems to be reasonable when N is sufficiently large, $n_S > \log(N)$, distinguished points are sufficiently common and specified using a good hash function (and hence are well-distributed), $\theta > \log(N)/N$, walks are sufficiently "local" that they do not go outside T (respectively, W) but also not too local, and when the function walk is chosen at random.

Heuristic 14.7.4

1. Walks reach a distinguished point in significantly fewer than N steps (in other words, there are no cycles in the walks and walks are not excessively longer than $1/\theta$).
2. Walks are uniformly distributed in T (respectively, W).

Theorem 14.7.5 *Let the notation be as above, and assume Heuristic 14.7.4. Then the average-case expected number of group operations performed by the Gaudry–Schost algorithm is* $(\sqrt{\pi}(2(2-\sqrt{2}))^2 + o(1))N \approx (2.43 + o(1))N$.

Proof We first compute $\#(T \cap W)$. When $(a_1, a_2) = (N/2, N/2)$ then $W = T$ and so $\#(T \cap W) = N^2$. In all other cases, the intersection is less. The extreme case is when $(a_1, a_2) = (0, 0)$ (similar cases are $(a_1, a_2) = (N-1, N-1)$ etc). Then $W = \{(x, y) \in \mathbb{Z}^2 : -N/2 \le x, y < N/2\}$ and $\#(T \cap W) = N^2/4$. By symmetry it suffices to consider the case $0 \le a_1, a_2 < N/2$ in which case we have $\#(T \cap W) \approx (N/2 + a_1)(N/2 + a_2)$ (here we are approximating the number of integer points in a set by its area).

Let $A = T \cap W$. To sample $\sqrt{\pi \# A}$ elements in A it is necessary to sample $\#T/\#A$ elements in T and W. Hence, the number of group elements to be selected overall is

$$\frac{\#T}{\#A}(\sqrt{\pi \# A} + O(1)) = (\#T + o(1))\sqrt{\pi}(\#A)^{-1/2}.$$

The average-case number of group operations is

$$(N^2 + o(1))\sqrt{\pi}\left(\tfrac{2}{N}\right)^2 \int_0^{N/2}\int_0^{N/2} (N-x)^{-1/2}(N-y)^{-1/2}dxdy.$$

Note that

$$\int_0^{N/2} (N-x)^{-1/2}dx = \sqrt{N}(2-\sqrt{2}).$$

The average-case expected number of group operations is therefore

$$(\sqrt{\pi}(2(2-\sqrt{2}))^2 + o(1))N$$

as stated. □

The Gaudry–Schost algorithm has a number of parameters that can be adjusted (such as the type of walks, the sizes of the tame and wild regions etc.). This gives it a lot of flexibility and makes it suitable for a wide range of variants of the DLP. Indeed, Galbraith and Ruprai [210] have improved the running time to $(2.36 + o(1))N$ group operations by using smaller tame and wild sets (also, the wild set is a different shape). One drawback is that it is hard to fine-tune all these parameters to get an implementation which achieves the theoretically optimal running time.

Exercise 14.7.6 Determine the complexity of the Gaudry–Schost algorithm for the standard DLP in G, when one takes $T = W = G$.

Exercise 14.7.7 Generalise the Gaudry–Schost algorithm to the n-dimensional DLP (see Definition 13.5.1). What is the heuristic average-case expected number of group operations?

14.7.2 Discrete logarithm problem in an interval using equivalence classes

Galbraith and Ruprai [211] used the Gaudry–Schost algorithm to solve the DLP in an interval of length $N < r$ faster than is possible using the kangaroo method when the group has an efficiently computable inverse (e.g., elliptic curves or tori). First, shift the discrete logarithm problem so that it is of the form $h = g^a$ with $-N/2 < a \leq N/2$. Define the equivalence relation $u \equiv u^{-1}$ for $u \in G$ as in Section 14.4 and determine a rule that leads to a unique representative of each equivalence class. Design a pseudorandom walk on the set of equivalence classes. The tame set is the set of equivalence classes coming from elements of the form g^x with $-N/2 < x \leq N/2$. Note that the tame set has $1 + N/2$ elements and every equivalence class $\{g^x, g^{-x}\}$ arises in two ways, except the singleton class $\{1\}$ and the class $\{-N/2, N/2\}$.

A natural choice for the wild set is the set of equivalence classes coming from elements of the form hg^x with $-N/2 < x \leq N/2$. Note that the size of the wild set now depends on the discrete logarithm problem: if $h = g^0 = 1$ then the wild set has $1 + N/2$ elements while if $h = g^{N/2}$ then the wild set has N elements. Even more confusingly, sampling from the wild set by uniformly choosing x does not, in general, lead to uniform sampling from the wild set. This is because the equivalence class $\{hg^x, (hg^x)^{-1}\}$ can arise in either one or two ways, depending on h. To analyse the algorithm it is necessary to use a non-uniform version of the birthday paradox (see, for example, Galbraith and Holmes [206]). The main result of [211] is an algorithm that solves the DLP in heuristic average-case expected $(1.36 + o(1))\sqrt{N}$ group operations.

14.8 Parallel collision search in other contexts

Van Oorschot and Wiener [423] propose a general method, motivated by Pollard's rho algorithm, for finding collisions of functions using distinguished points and parallelisation. They give applications to cryptanalysis of hash functions and block ciphers that are beyond the scope of this book. But they also give applications of their method for algebraic meet-in-the-middle attacks, so we briefly give the details here.

First, we sketch the parallel collision search method. Let $f : \mathcal{S} \to \mathcal{S}$ be a function mapping some set $\mathcal{S}$ of size N to itself. Define a set $\mathcal{D}$ of distinguished points in $\mathcal{S}$. Each client chooses a random starting point $x_1 \in \mathcal{S}$, iterates $x_{n+1} = f(x_n)$ until it hits a distinguished point and sends (x_1, x_n, n) to the server. The client then restarts with a new random starting point. Eventually, the server gets two triples (x_1, x, n) and (x_1', x, n') for the same distinguished point. As long as we do not have a "Robin Hood"[13] (i.e., one walk is a subsequence of another), the server can use the values (x_1, n) and (x_1', n') to efficiently find a collision $f(x) = f(y)$ with $x \neq y$. The expected running time for each client is $\sqrt{\pi N/2}/N_P + 1/\theta$, using the notation of this chapter. The storage requirement depends on the choice of θ.

We now consider the application to meet-in-the-middle attacks. A general meet-in-the-middle attack has two sets $\mathcal{S}_1$ and $\mathcal{S}_2$ and functions $f_i : \mathcal{S}_i \to \mathcal{R}$ for $i = 1, 2$. The goal is to find $a_1 \in \mathcal{S}_1$ and $a_2 \in \mathcal{S}_2$ such that $f_1(a_1) = f_2(a_2)$. The standard solution (as in baby-step–giant-step) is to compute and store all $(f_1(a_1), a_1)$ in an easily searched structure and then test, for each $a_2 \in \mathcal{S}_2$, whether $f_2(a_2)$ is in the structure. The running time is $\#\mathcal{S}_1 + \#\mathcal{S}_2$ function evaluations and the storage is proportional to $\#\mathcal{S}_1$.

The idea of [423] is to phrase this as a collision search problem for a single function f. For simplicity we assume that $\#\mathcal{S}_1 = \#\mathcal{S}_2 = N$. We write $I = \{0, 1, \ldots, N-1\}$ and assume one can construct bijective functions $\sigma_i : I \to \mathcal{S}_i$ for $i = 1, 2$. One defines a surjective map

$$\rho : \mathcal{R} \to I \times \{1, 2\}$$

[13] Robin Hood is a character of English folklore who is expert in archery. His prowess allows him to shoot a second arrow on exactly the same trajectory as the first so that the second arrow splits the first. Chinese readers may substitute the name Houyi.

and a set $\mathcal{S} = I \times \{1, 2\}$. Finally, define $f : \mathcal{S} \to \mathcal{S}$ as $f(x, i) = \rho(f_i(\sigma_i(x)))$. Clearly, the desired collision $f_1(a_1) = f_2(a_2)$ can arise from $f(\sigma_1^{-1}(a_1), 1) = f(\sigma_2^{-1}(a_2), 2)$, but collisions can also arise in other ways (for example, due to collisions in ρ). Indeed, since $\#\mathcal{S} = 2N$ one expects there to be roughly $2N$ pairs $(a_1, a_2) \in \mathcal{S}^2$ such that $a_1 \neq a_2$ but $f(a_1) = f(a_2)$. In many applications there is only one collision (van Oorschot and Wiener call it the "golden collision") that actually leads to a solution of the problem. It is therefore necessary to analyse the algorithm carefully to determine the expected time until the problem is solved.

Let N_P be the number of clients and let N_M be the total number of group elements that can be stored on the server. Van Oorschot and Wiener give a heuristic argument that the algorithm finds a useful collision after $2.5\sqrt{(2N)^3/N_M}/N_P$ group operations per client. This is taking $\theta = 2.25\sqrt{N_M/2N}$ for the probability of a distinguished point. We refer to [423] for the details.

14.8.1 The low Hamming weight DLP

Recall the low Hamming weight DLP: given g, h, n, w find x of bit-length n and Hamming weight w such that $h = g^x$. The number of values for x is $M = \binom{n}{w}$ and there is a naive low storage algorithm running in time $\tilde{O}(M)$. We stress that the symbol w here means the Hamming weight, rather than its meaning earlier in this chapter.

Section 13.6 gave baby-step–giant-step algorithms for the low Hamming weight DLP that perform $O(\sqrt{w}\binom{n/2}{w/2})$ group operations. Hence, these methods require time and space roughly proportional to $\sqrt{wM}$.

To solve the low Hamming weight DLP using parallel collision search one sets $\mathcal{R} = \langle g \rangle$ and $\mathcal{S}_1, \mathcal{S}_2$ to be sets of integers of binary length $n/2$ and Hamming weight roughly $w/2$. Define the functions $f_1(a) = g^a$ and $f_2(a) = hg^{-2^{n/2}a}$ so that a collision $f_1(a_1) = f_2(a_2)$ solves the problem. Note that there is a unique choice of (a_1, a_2) such that $f_1(a_1) = f_2(a_2)$, but when one uses the construction of van Oorschot and Wiener to get a single function f then there will be many useless collisions in f. We have $N = \#\mathcal{S}_1 = \#\mathcal{S}_2 \approx \binom{n/2}{w/2} \approx \sqrt{M}$ and so get an algorithm whose number of group operations is proportional to $N^{3/2} = M^{3/4}$ yet requires low storage. This is a significant improvement over the naive low-storage method, but still slower than baby-step–giant-step.

Exercise 14.8.1 Write this algorithm in pseudocode and give a more careful analysis of the running time.

It remains an open problem to give a low memory algorithm for the low Hamming weight DLP with complexity proportional to $\sqrt{wM}$ as with the BSGS methods.

14.9 Pollard rho factoring method

This algorithm was proposed in [436] and was the first algorithm invented by Pollard that exploited pseudorandom walks. As more powerful factoring algorithms exist, we keep the presentation brief. For further details see Section 5.6.2 of Stinson [532] or Section 5.2.1 of Crandall and Pomerance [150].

Let N be a composite integer to be factored and let $p \mid N$ be a prime (usually p is the smallest prime divisor of N). We try to find a relation that holds modulo p, but not modulo other primes dividing N.

The basic idea of the rho factoring algorithm is to consider the pseudorandom walk $x_1 = 2$ and

$$x_{i+1} = f(x_i) \pmod{N}$$

where the usual choice for $f(x)$ is $x^2 + 1$ (or $f(x) = x^2 + a$ for some small integer a). Consider the values $x_i \pmod{p}$ where $p \mid N$. The sequence $x_i \pmod{p}$ is a pseudorandom sequence of residues modulo p, and so after about $\sqrt{\pi p/2}$ steps we expect there to be indicies i and j such that $x_i \equiv x_j \pmod{p}$. We call this a **collision**. If $x_i \not\equiv x_j \pmod{N}$ then we can split N as $\gcd(x_i - x_j, N)$.

Example 14.9.1 Let $p = 11$. Then the rho iteration modulo p is

$$2,\ 5,\ 4,\ 6,\ 4,\ 6,\ 4,\ \dots$$

Let $p = 19$. Then the sequence is

$$2,\ 5,\ 7,\ 12,\ 12,\ 12,\ \dots$$

As with the discrete logarithm algorithms, the walk is deterministic in the sense that a collision leads to a cycle. Let l_t be the length of the tail and l_h be the length of the cycle. Then the first collision is

$$x_{l_t + l_h} \equiv x_{l_t} \pmod{p}.$$

We can use Floyd's cycle finding algorithm to detect the collision. The details are given in Algorithm 20. Note that it is not efficient to compute the gcd in line 5 of the algorithm for each iteration; Pollard [436] gave a solution to reduce the number of gcd computations and Brent [93] gave another.

Algorithm 20 The rho algorithm for factoring

INPUT: N

OUTPUT: A factor of N

1: $x_1 = 2, x_2 = f(x_1) \pmod{N}$
2: **repeat**
3: $\quad x_1 = f(x_1) \pmod{N}$
4: $\quad x_2 = f(f(x_2)) \pmod{N}$
5: $\quad d = \gcd(x_2 - x_1, N)$
6: **until** $1 < d < N$
7: **return** d

We now briefly discuss the complexity of the algorithm. Note that the "algorithm" may not terminate, for example if the length of the cycle and tail are the same for all $p \mid N$ then the gcd will always be either 1 or N. In practice, one would stop the algorithm after a certain number of steps and repeat with a different choice of x_1 and/or $f(x)$. Even if it terminates, the length of the cycle of the rho may be very large. Hence, the usual approach is to make the heuristic assumption that the rho pseudorandom walk behaves like a random walk. To have meaningful heuristics one should analyse the algorithm when the function $f(x)$ is randomly chosen from a large set of possible functions.

Note that the rho method is more general than the $p-1$ method (see Section 12.3), since a random $p \mid N$ is not very likely to be $\sqrt{p}$-smooth.

Theorem 14.9.2 *Let N be composite, not a prime power and not too smooth. Assume that the Pollard rho walk modulo p behaves like a pseudorandom walk for all $p \mid N$. Then the rho algorithm factors N in $O(N^{1/4}\log(N)^2)$ bit operations.*

Proof (Sketch) Let p be a prime dividing N such that $p \le \sqrt{N}$. Define the values l_t and l_h corresponding to the sequence $x_i \pmod{p}$. If the walk behaves sufficiently like a random walk then, by the birthday paradox, we will have $l_h, l_t \approx \sqrt{\pi p/8}$. Similarly, for some other prime $q \mid N$ one expects that the walk modulo q has different values l_h and l_t. Hence, after $O(\sqrt{p})$ iterations of the loop one expects to split N. □

Bach [20] has given a rigorous analysis of the rho factoring algorithm. He proves that if $0 \le x, y < N$ are chosen randomly and the iteration is $x_1 = x$, $x_{i+1} = x_i^2 + y$, then the probability of finding the smallest prime factor p of N after k steps is at least $k(k-1)/2p + O(p^{-3/2})$ as p goes to infinity, where the constant in the O depends on k. Bach's method cannot be used to analyse the rho algorithm for discrete logarithms.

Example 14.9.3 Let $N = 144493$. The values (x_i, x_{2i}) for $i = 1, 2, \ldots, 7$ are

$$(2, 5), (5, 677), (26, 9120), (677, 81496), (24851, 144003), (9120, 117992), (90926, 94594)$$

and one can check that $\gcd(x_{14} - x_7, N) = 131$.

The reason for this can be seen by considering the values x_i modulo $p = 131$. The sequence of values starts

$$2,\ 5,\ 26,\ 22,\ 92,\ 81,\ 12,\ 14,\ 66,\ 34,\ 109,\ 92$$

and we see that $x_{12} = x_5 = 92$. The tail has length $l_t = 5$ and the head has length $l_h = 7$. Clearly, $x_{14} \equiv x_7 \pmod{p}$.

Exercise 14.9.4 Factor the number 576229 using the rho algorithm.

Exercise 14.9.5 The rho algorithm usually uses the function $f(x) = x^2 + 1$. Why do you think this function is used? Why are the functions $f(x) = x^2$ and $f(x) = x^2 - 2$ less suitable?

Exercise 14.9.6 Show that if N is known to have a prime factor $p \equiv 1 \pmod{m}$ for $m > 2$ then it is preferable to use the polynomial $f(x) = x^m + 1$.

Exercise 14.9.7 Floyd's and Brent's cycle finding methods are both useful for the rho factoring algorithm. Explain why one cannot use the other cycle finding methods listed in Section 14.2.2 (Sedgewick–Szymanski–Yao, Schnorr–Lenstra, Nivasch, distinguished points) for the rho factoring method.

15

Factoring and discrete logarithms in subexponential time

One of the most powerful tools in mathematics is linear algebra, and much of mathematics is devoted to solving problems by reducing them to it. It is therefore natural to try to solve the integer factorisation and discrete logarithm problems (DLP) in this way. This chapter briefly describes a class of algorithms that exploit a notion called "smoothness", to reduce factoring or DLP to linear algebra. We present such algorithms for integer factorisation, the DLP in the multiplicative group of a finite field and the DLP in the divisor class group of a curve.

It is beyond the scope of this book to give all the details of these algorithms. Instead, the aim is to sketch the basic ideas. We mainly present algorithms with nice theoretical properties (though often still requiring heuristic assumptions) rather than the algorithms with the best practical performance. We refer to Crandall and Pomerance [150], Shoup [497] and Joux [283] for further reading.

The chapter is arranged as follows. First, we present results on smooth integers, and then sketch Dixon's random squares factoring algorithm. Section 15.2.3 then summarises the important features of all algorithms of this type. We then briefly describe a number of algorithms for the discrete logarithm problem in various groups.

15.1 Smooth integers

Recall from Definition 12.3.1 that an integer is B-smooth if all its prime divisors are at most B. We briefly recall some results on smooth integers; see Granville [243] for a survey of this subject and for further references.

Definition 15.1.1 Let $X, Y \in \mathbb{N}$ be such that $2 \leq Y < X$. Define

$$\Psi(X, Y) = \#\{n \in \mathbb{N} : 1 \leq n \leq X,\ n \text{ is } Y\text{-smooth}\}.$$

It is important for this chapter to have good bounds on $\Psi(X, Y)$. Let $u = \log(X)/\log(Y)$ (as usual log denotes the natural logarithm) so that $u > 1$, $Y = X^{1/u}$ and $X = Y^u$. There is a function $\rho : \mathbb{R}_{>0} \to \mathbb{R}_{>0}$ called the **Dickman–de Bruijn function** (for the exact definition of this function see Section 1.4.5 of [150]) such that, for fixed $u > 1$, $\Psi(X, X^{1/u}) \sim X\rho(u)$, where $f(X) \sim g(X)$ means $\lim_{X \to \infty} f(X)/g(X) = 1$. A crude estimate for $\rho(u)$, as $u \to \infty$ is $\rho(u) \approx 1/u^u$. For further details and references see Section 1.4.5 of [150].

The following result of Canfield, Erdös and Pomerance [110] is the main tool in this subject. This is a consequence of Theorem 3.1 (and the corollary on page 15) of [110].

Theorem 15.1.2 *Let $N \in \mathbb{N}$. Let $\epsilon, u \in \mathbb{R}$ be such that $\epsilon > 0$ and $3 \le u \le (1 - \epsilon)\log(N)/\log(\log(N))$. Then there is a constant c_ϵ (that does not depend on u) such that*

$$\Psi(N, N^{1/u}) = N\exp(-u(\log(u) + \log(\log(u)) - 1 + (\log(\log(u)) - 1)/\log(u) + E(N,u))) \tag{15.1}$$

where $|E(N,u)| \le c_\epsilon(\log(\log(u))/\log(u))^2$.

Corollary 15.1.3 *Let the notation be as in Theorem 15.1.2. Then $\Psi(N, N^{1/u}) = Nu^{-u+o(u)} = Nu^{-u(1+o(1))}$ uniformly as $u \to \infty$ and $u \le (1-\epsilon)\log(N)/\log(\log(N))$ (and hence also $N \to \infty$).*

Exercise 15.1.4 Prove Corollary 15.1.3.
[Hint: Show that the expression inside the exp in equation (15.1) is of the form $-u\log(u) + o(u)\log(u)$.]

We will use the following notation throughout the book.

Definition 15.1.5 Let $0 \le a \le 1$ and $c \in \mathbb{R}_{>0}$. The **subexponential function** for the parameters a and c is

$$L_N(a, c) = \exp(c\log(N)^a \log(\log(N))^{1-a}).$$

Note that taking $a = 0$ gives $L_N(0, c) = \log(N)^c$ (polynomial) while taking $a = 1$ gives $L_N(1, c) = N^c$ (exponential). Hence, $L_N(a, c)$ interpolates exponential and polynomial growth. A complexity $O(L_N(a, c))$ with $0 < a < 1$ is called **subexponential**.

Lemma 15.1.6 *Let $0 < a < 1$ and $0 < c$.*

1. *$L_N(a, c)^m = L_N(a, mc)$ for $m \in \mathbb{R}_{>0}$.*
2. *Let $0 < a_1, a_2 < 1$ and $0 < c_1, c_2$. Then, where the term $o(1)$ is as $N \to \infty$,*

$$L_N(a_1, c_1) \cdot L_N(a_2, c_2) = \begin{cases} L_N(a_1, c_1 + o(1)) & \text{if } a_1 > a_2, \\ L_N(a_1, c_1 + c_2) & \text{if } a_1 = a_2, \\ L_N(a_2, c_2 + o(1)) & \text{if } a_2 > a_1. \end{cases}$$

3.

$$L_N(a_1, c_1) + L_N(a_2, c_2) = \begin{cases} O(L_N(a_1, c_1)) & \text{if } a_1 > a_2, \\ O(L_N(a_1, \max\{c_1, c_2\} + o(1))) & \text{if } a_1 = a_2, \\ O(L_N(a_2, c_2)) & \text{if } a_2 > a_1. \end{cases}$$

4. *Let* $0 < b < 1$ *and* $0 < d$. *If* $M = L_N(a, c)$ *then* $L_M(b, d) = L_N(ab, dc^b a^{1-b} + o(1))$ *as* $N \to \infty$.
5. $\log(N)^m = O(L_N(a, c))$ *for any* $m \in \mathbb{N}$.
6. $L_N(a, c)\log(N)^m = O(L_N(a, c + o(1)))$ *as* $N \to \infty$ *for any* $m \in \mathbb{N}$. *Hence, one can always replace* $\tilde{O}(L_N(a, c))$ *by* $O(L_N(a, c + o(1)))$.
7. $\log(N)^m \le L_N(a, o(1))$ *as* $N \to \infty$ *for any* $m \in \mathbb{N}$.
8. *If* $F(N) = O(L_N(a, c))$ *then* $F(N) = L_N(a, c + o(1))$ *as* $N \to \infty$.
9. $L_N(1/2, c) = N^{c\sqrt{\log(\log(N))/\log(N)}}$.

Exercise 15.1.7 Prove Lemma 15.1.6.

Corollary 15.1.8 *Let* $c > 0$. *As* $N \to \infty$, *the probability that a randomly chosen integer* $1 \le x \le N$ *is* $L_N(1/2, c)$*-smooth is* $L_N(1/2, -1/(2c) + o(1))$.

Exercise 15.1.9 Prove Corollary 15.1.8 (using Corollary 15.1.3).

Exercise 15.1.10 Let $0 < b < a < 1$. Let $1 \le x \le L_N(a, c)$ be a randomly chosen integer. Show that the probability that x is $L_N(b, d)$-smooth is $L_N(a - b, -c(a - b)/d + o(1))$ as $N \to \infty$.

15.2 Factoring using random squares

The goal of this section is to present a simple version of Dixon's random squares factoring algorithm. This algorithm is easy to describe and analyse, and already displays many of the important features of the algorithms in this chapter. Note that the algorithm is not used in practice. We give a complexity analysis and sketch how subexponential running times naturally arise. Further details about this algorithm can be found in Section 16.3 of Shoup [497] and Section 19.5 of von zur Gathen and Gerhard [220].

Let $N \in \mathbb{N}$ be an integer to be factored. We assume in this section that N is odd, composite and not a perfect power. As in Chapter 12, we focus on splitting N into a product of two smaller numbers (neither of which is necessarily prime). The key idea is that if one can find congruent squares

$$x^2 \equiv y^2 \pmod{N}$$

such that $x \not\equiv \pm y \pmod{N}$ then one can split N by computing $\gcd(x - y, N)$.

Exercise 15.2.1 Let N be an odd composite integer and m be the number of distinct primes dividing N. Show that the equation $x^2 \equiv 1 \pmod{N}$ has 2^m solutions modulo N.

A general way to find congruent squares is the following.[1] Select a **factor base** $\mathcal{B} = \{p_1, \ldots, p_s\}$ consisting of the primes $\le B$ for some $B \in \mathbb{N}$. Choose uniformly at random

[1] This idea goes back to Kraitchik in the 1920s; see [439] for some history.

an integer $1 \le x < N$, compute $a = x^2 \pmod{N}$ reduced to the range $1 \le a < N$ and try to factor a as a product in $\mathcal{B}$ (e.g., using trial division).[2] If a is B-smooth then this succeeds, in which case we have a **relation**

$$x^2 \equiv \prod_{i=1}^{s} p_i^{e_i} \pmod{N}. \tag{15.2}$$

The values x for which a relation is found are stored as $x_1, x_2, \ldots, x_t$. The corresponding exponent vectors $\underline{e}_j = (e_{j,1}, \ldots, e_{j,s})$ for $1 \le j \le t$ are also stored. When enough relations have been found we can use linear algebra modulo 2 to obtain congruent squares. More precisely, compute $\lambda_j \in \{0, 1\}$ such that not all $\lambda_j = 0$ and

$$\sum_{j=1}^{t} \lambda_j \underline{e}_j \equiv (0, 0, \ldots, 0) \pmod{2}.$$

Equivalently, this is an integer linear combination

$$\sum_{j=1}^{t} \lambda_j \underline{e}_j = (2f_1, \ldots, 2f_s) \tag{15.3}$$

with not all the f_i equal to zero. Let

$$X \equiv \prod_{j=1}^{t} x_j^{\lambda_j} \pmod{N}, \qquad Y \equiv \prod_{i=1}^{s} p_i^{f_i} \pmod{N}. \tag{15.4}$$

One then has $X^2 \equiv Y^2 \pmod{N}$ and one can hope to split N by computing $\gcd(X - Y, N)$ (note that this gcd could be 1 or N, in which case the algorithm has failed). We present the above method as Algorithm 21.

We emphasise that the random squares algorithm has two distinct stages. The first stage is to generate enough relations. The second stage is to perform linear algebra. The first stage can easily be distributed or parallelised, while the second stage is hard to parallelise.

Example 15.2.2 Let $N = 19 \cdot 29 = 551$ and let $\mathcal{B} = \{2, 3, 5\}$. One finds the following congruences (in general 4 relations would be required, but we are lucky in this case)

$$\begin{aligned} 34^2 &\equiv 2 \cdot 3^3 \pmod{N} \\ 52^2 &\equiv 2^2 \cdot 5^3 \pmod{N} \\ 55^2 &\equiv 2 \cdot 3^3 \cdot 5 \pmod{N}. \end{aligned}$$

These relations are stored as the matrix

$$\begin{pmatrix} 1 & 3 & 0 \\ 2 & 0 & 3 \\ 1 & 3 & 1 \end{pmatrix}.$$

[2] To obtain non-trivial relations one should restrict to integers in the range $\sqrt{N} < x < N - \sqrt{N}$. But it turns out to be simpler to analyse the algorithm for the case $1 \le x < N$. Note that the probability that a randomly chosen integer $1 \le x < N$ satisfies $1 \le x < \sqrt{N}$ is negligible.

Algorithm 21 Random squares factoring algorithm

INPUT: $N \in \mathbb{N}$

OUTPUT: Factor of N

1: Select a suitable $B \in \mathbb{N}$ and construct the **factor base** $\mathcal{B} = \{p_1, \ldots, p_s\}$ consisting of all primes $\leq B$
2: **repeat**
3: Choose an integer $1 \leq x < N$ uniformly at random and compute $a = x^2 \pmod N$ reduced to the range $1 \leq a < N$
4: Try to factor a as a product in $\mathcal{B}$ (e.g., using trial division)
5: **if** a is B-smooth **then**
6: store the value x and the exponent row vector $\underline{e} = (e_1, \ldots, e_s)$ as in equation (15.2) in a matrix
7: **end if**
8: **until** there are $s + 1$ rows in the matrix
9: Perform linear algebra over $\mathbb{F}_2$ to find a non-trivial linear dependence among the vectors $\underline{e}_j$ modulo 2
10: Define X and Y as in equation (15.4)
11: **return** $\gcd(X - Y, N)$

The sum of the three rows is the vector

$$(4,\ 6,\ 4).$$

Let

$$X = 264 \equiv 34 \cdot 52 \cdot 55 \pmod{551} \quad \text{and} \quad Y = 496 \equiv 2^2 \cdot 3^3 \cdot 5^2 \pmod{551}.$$

It follows that

$$X^2 \equiv Y^2 \pmod N$$

and $\gcd(X - Y, N) = 29$ splits N.

Exercise 15.2.3 Factor $N = 3869$ using the above method and factor base $\{2, 3, 5, 7\}$.

15.2.1 Complexity of the random squares algorithm

There are a number of issues to deal with when analysing this algorithm. The main problem is to decide how many primes to include in the factor base. The prime number theorem implies that $s = \#\mathcal{B} \approx B/\log(B)$. If we make B larger then the chances of finding a B-smooth number increase but, on the other hand, we need more relations and the linear algebra takes longer. We will determine an optimal value for B later. First, we must write down an estimate for the running time of the algorithm as a function of s. Already this leads to various issues:

- What is the probability that a random value $x^2 \pmod N$ factors over the factor base $\mathcal{B}$?
- How many relations do we require until we can be sure there is a non-trivial vector $\underline{e}$?
- What are the chances that computing $\gcd(X - Y, N)$ splits N?

We deal with the latter two points first. It is immediate that $s + 1$ relations are sufficient for line 9 of Algorithm 21 to succeed. The question is whether $1 < \gcd(X - Y, N) < N$ for the corresponding integers X and Y. There are several ways the algorithm can fail to split N. For example, it is possible that a relation in equation (15.2) is such that all e_i are even and $x \equiv \pm \prod_i p_i^{e_i/2} \pmod N$. One way that such relations could arise is from $1 \le x < \sqrt{N}$ or $N - \sqrt{N} < x < N$; this situation occurs with negligible probability. If $\sqrt{N} < x < N - \sqrt{N}$ and $a = Y^2$ is a square in $\mathbb{N}$ then $1 \le Y < \sqrt{N}$ and so $x \not\equiv \pm Y \pmod N$ and the relation is useful. The following result shows that all these (and other) bad cases occur with probability at most $1/2$.

Lemma 15.2.4 *The probability to split N using X and Y is at least $\frac{1}{2}$.*

Proof Let X and Y be the integers computed in line 10 of Algorithm 21. We treat Y as fixed, and consider the probability distribution for X. By Exercise 15.2.1, the number of solutions Z to $Z^2 \equiv Y^2 \pmod N$ is 2^m where $m \ge 2$ is the number of distinct primes dividing N. The two solutions $Z = \pm Y$ are useless but the other $2^m - 2$ solutions will all split N.

Since the values for x are chosen uniformly at random it follows that X is a randomly chosen solution to the equation $X^2 \equiv Y^2 \pmod N$. It follows that the probability to split N is $(2^m - 2)/2^m \ge 1/2$. □

Exercise 15.2.5 Show that if one takes $s + l$ relations where $l \ge 2$ then the probability of splitting N is at least $1 - 1/2^l$.

We now consider the probability of smoothness. We first assume the probability that $x^2 \pmod N$ is smooth is the same as the probability that a random integer modulo N is smooth.[3]

Lemma 15.2.6 *Let the notation be as above. Let T_B be the expected number of trials until a randomly chosen integer modulo N is B-smooth. Assuming the above smoothness heuristic, Algorithm 21 has expected running time at most*

$$c_1 \#\mathcal{B}^2 T_B M(\log(N)) + c_2(\#\mathcal{B})^3$$

bit operations for some constants c_1, c_2 (where $M(n)$ is the cost of multiplying two n-bit integers).

[3] Section 16.3 of Shoup [497] gives a modification of the random squares algorithm for which one can avoid this assumption. The trick is to note that at least one of the cosets of $(\mathbb{Z}/N\mathbb{Z})^*/((\mathbb{Z}/N\mathbb{Z})^*)^2$ has at least as great a proportion of smooth numbers as random integers up to N (Shoup credits Rackoff for this trick). The idea is to fix some $1 < \delta < N$ and consider relations coming from smooth values of $\delta x^2 \pmod N$.

Proof Suppose we compute the factorisation of $x^2 \pmod{N}$ over $\mathcal{B}$ by trial division. This requires $O(\#\mathcal{B}M(\log(N)))$ bit operations for each value of x. We need $(\#\mathcal{B}+1)$ relations to have a soluble linear algebra problem. As said above, the expected number of trials of x to get a B-smooth value of $x^2 \pmod{N}$ is T_B. Hence, the cost of finding the relations is $O((\#\mathcal{B}+1)T_B(\#\mathcal{B})M(\log(N)))$, which gives the first term.

The linear algebra problem can be solved using Gaussian elimination (we are ignoring that the matrix is sparse) over $\mathbb{F}_2$, which takes $O((\#\mathcal{B})^3)$ bit operations. This gives the second term. □

It remains to choose B as a function of N to minimise the running time. By the discussion in Section 15.1, it is natural to approximate T_B by u^u where $u = \log(N)/\log(B)$. We now explain how subexponential functions naturally arise in such algorithms. Since increasing B makes the linear algebra slower, but makes relations more likely (i.e., lowers T_B), a natural approach to selecting B is to try to equate both terms of the running time in Lemma 15.2.6. This leads to $u^u = \#\mathcal{B}$. Putting $u = \log(N)/\log(B)$, $\#\mathcal{B} = B/\log(B)$, taking logs and ignoring $\log(\log(B))$ terms gives

$$\log(N)\log(\log(N))/\log(B) \approx \log(B).$$

This implies $\log(B)^2 \approx \log(N)\log(\log(N))$ and so $B \approx L_N(1/2, 1)$. The overall complexity for this choice of B would be $L_N(1/2, 3 + o(1))$ bit operations.

A more careful argument is to set $B = L_N(1/2, c)$ and use Corollary 15.1.8. It follows that $T_B = L_N(1/2, 1/(2c) + o(1))$ as $N \to \infty$. Putting this into the equation of Lemma 15.2.6 gives complexity $L_N(1/2, 2c + 1/(2c) + o(1)) + L_N(1/2, 3c)$ bit operations. The function $x + 1/x$ is minimised at $x = 1$; hence, we should take $c = 1/2$.

Theorem 15.2.7 *Let the notation be as above. Under the same assumptions as Lemma 15.2.6 then Algorithm 21 has complexity*

$$L_N(1/2, 2 + o(1))$$

bit operations as $N \to \infty$.

Proof Put $B = L_N(1/2, 1/2)$ into Lemma 15.2.6. □

We remark that, unlike the Pollard rho or Pollard $p-1$ methods, this factoring algorithm has essentially no dependence on the factors of N. In other words, its running time is essentially the same for all integers of a given size. This makes it particularly suitable for factoring $N = pq$ where p and q are primes of the same size.

15.2.2 The quadratic sieve

To improve the result of the previous section it is necessary to reduce the cost of the linear algebra and to reduce the cost of decomposing smooth elements as products of primes. We sketch the **quadratic sieve** algorithm of Pomerance. We do not have space to present all the details of this algorithm (interested readers should see Section 6.1 of [150] or Section 16.4.2 of [497]).

A crucial idea, which seems to have first appeared in the work of Schroeppel,[4] is **sieving**. The point is to consider a range of values of x and simultaneously determine the decompositions of $x^2 \pmod{N}$ over the factor base. It is possible to do this so that the cost of each individual decomposition is only $O(\log(B))$ bit operations.

Another crucial observation is that the relation matrix is sparse; in other words, rows of the matrix have rather few non-zero entries. In such a case, the cost of linear algebra can be reduced from $O((\#\mathcal{B})^3)$ bit operations to $O((\#\mathcal{B})^{2+o(1)})$ bit operations (as $\#\mathcal{B} \to \infty$). The best methods are due to Lanczos or Wiedemann; see Section 6.1.3 of Crandall and Pomerance [150] or Section 3.4 of Joux [283] for references and discussion.

A further trick is to choose $x = \lfloor \sqrt{N} \rfloor + i$ where $i = 0, 1, -1, 2, -2, \ldots$. The idea is that if $x = \sqrt{N} + \epsilon$ then either $x^2 - N$ or $N - x^2$ is a positive integer of size $2\sqrt{N}|\epsilon|$. Since these integers are much smaller than N they have a much better chance of being smooth than the integers $x^2 \pmod{N}$ in the random squares algorithm. To allow for the case of $\epsilon < 0$ we need to add -1 to our factor base and use the fact that a factorisation $N - x^2 = \prod_{i=1}^{s} p_i^{e_i}$ corresponds to a relation $x^2 \equiv (-1) \prod_{i=1}^{s} p_i^{e_i} \pmod{N}$.

Since we are now only considering values x of the form $\sqrt{N} + \epsilon$ where $|\epsilon|$ is small it is necessary to assume the probability that $x^2 - N$ or $N - x^2$ (as appropriate) is B-smooth is the same as the probability that a randomly chosen integer of that size is B-smooth. This is a rather strong assumption (though it is supported by numerical evidence) and so the running time estimates of the quadratic sieve are only heuristic.

The heuristic complexity of the quadratic sieve is determined in Exercise 15.2.8. Note that, since we will need to test $L_N(1/2, 1 + o(1))$ values (here $o(1)$ is as $N \to \infty$) for smoothness, we have $|\epsilon| = L_N(1/2, 1 + o(1))$. It follows that the integers being tested for smoothness have size $\sqrt{N} L_N(1/2, 1 + o(1)) = N^{1/2+o(1)}$.

Exercise 15.2.8 ★ Let T_B be the expected number of trials until an integer of size $2\sqrt{N} L_N(1/2, 1)$ is B-smooth. Show that the running time of the quadratic sieve is at most

$$c_1 \#\mathcal{B} T_B \log(B) M(\log(N)) + c_2 \#\mathcal{B}^{2+o(1)}$$

bit operations for some constants c_1, c_2 as $N \to \infty$.

Let $B = L_N(1/2, 1/2)$. Show that the natural heuristic assumption (based on Corollary 15.1.8) is that $T_B = L_N(1/2, 1/2 + o(1))$. Hence, show that the heuristic complexity of the quadratic sieve is $L_N(1/2, 1 + o(1))$ bit operations as $N \to \infty$.

Example 15.2.9 Let $N = 2041$ so that $\lfloor \sqrt{N} \rfloor = 45$.

[4] See [333, 439] for some remarks on the history of integer factoring algorithms.

Let $\mathcal{B} = \{-1, 2, 3, 5\}$. Taking $x = 43, 44, 45, 46$ one finds the following factorisations of $x^2 - N$:

x	$x^2 \pmod N$	$\underline{e}$
43	$-2^6 \cdot 3$	$(1, 6, 1, 0)$
44	Not 5-smooth	
45	-2^4	$(1, 4, 0, 0)$
46	$3 \cdot 5^2$	$(0, 0, 1, 2)$

Taking $\underline{e} = \underline{e}_1 + \underline{e}_2 + \underline{e}_3 = (2, 10, 2, 2)$ gives all coefficients even. Putting everything together, we set $X = 43 \cdot 45 \cdot 46 \equiv 1247 \pmod N$ and $Y = -1 \cdot 2^5 \cdot 3 \cdot 5 \equiv 1561 \pmod N$. One can check that $X^2 \equiv Y^2 \pmod N$ and that $\gcd(X - Y, N) = 157$.

Exercise 15.2.10 Show that in the quadratic sieve one can also use values $x = \lfloor \sqrt{kN} \rfloor + i$ where $k \in \mathbb{N}$ is very small and $i = 0, 1, -1, 2, -2, \ldots$.

Exercise 15.2.11 Show that using sieving and fast linear algebra, but not restricting to values $\pm x^2 \pmod N$ of size $N^{1/2+o(1)}$, gives an algorithm with heuristic expected running time of $L_N(1/2, \sqrt{2} + o(1))$ bit operations as $N \to \infty$.

Exercise 15.2.12 A subexponential algorithm is asymptotically much faster than a $\tilde{O}(N^{1/4})$ algorithm. Verify that if $N = 2^{1024}$ then $N^{1/4} = 2^{256}$, while $L_N(1/2, 2) \approx 2^{197}$ and $L_N(1/2, 1) \approx 2^{98.5}$.

The best proven asymptotic complexity for factoring integers N is $L_N(1/2, 1 + o(1))$ bit operations. This result is due to Pomerance and Lenstra [343].

15.2.3 Summary

We briefly highlight the key ideas in the algorithms of this section. The crucial concept of smooth elements of the group $(\mathbb{Z}/N\mathbb{Z})^*$ arises from considering an integer modulo N as an element of $\mathbb{Z}$. The three essential properties of smooth numbers that were used in the algorithm are:

1. One can efficiently decompose an element of the group as a product of smooth elements, or determine that the element is not smooth.
2. The probability that a random element is smooth is sufficiently high.
3. There is a way to apply linear algebra to the relations obtained from smooth elements to solve the computational problem.

We will see analogues of these properties in the algorithms below.

There are other general techniques that can be applied in most algorithms of this type. For example, the linear algebra problems are usually sparse and so the matrices and algorithms should be customised for this. Another general concept is "large prime variation", which, in a nutshell, is to also store "nearly smooth" relations (i.e., elements that are the product of a smooth element with one or two prime elements that are not too large) and perform some elimination of these "large primes" before doing the main linear algebra stage (this is similar to, but more efficient than, taking a larger factor base). Finally, we remark that the first stage of these algorithms (i.e., collecting relations) can always be distributed or parallelised.

15.3 Elliptic curve method revisited

We assume throughout this section that $N \in \mathbb{N}$ is an integer to be factored and that N is odd, composite and not a perfect power. We denote by p the smallest prime factor of N.

The elliptic curve method works well in practice but, as with the Pollard $p - 1$ method, its complexity depends on the size of the smallest prime dividing N. It is not a polynomial-time algorithm because, for any constant $c > 0$ and over all N and $p \mid N$, a randomly chosen elliptic curve over $\mathbb{F}_p$ is not likely to have $O(\log(N)^c)$-smooth order. As we have seen, the theorem of Canfield, Erdös and Pomerance says it is more reasonable to hope that integers have a subexponential probability of being subexponentially smooth. Hence, one might hope that the elliptic curve method has subexponential complexity. Indeed, Lenstra [339] makes the following conjecture (which is essentially that the Canfield–Erdös–Pomerance result holds in small intervals).

Conjecture 15.3.1 *(Lenstra [339], page 670) The probability that an integer, chosen uniformly at random in the range* $(X - \sqrt{X}, X + \sqrt{X})$*, is* $L_X(1/2, c)$*-smooth is* $L_X(1/2, -1/(2c) + o(1))$ *as* X *tends to infinity.*[5]

One can rephrase Conjecture 15.3.1 as saying that if p_s is the probability that a random integer between 1 and X is Y-smooth then $\Psi(X + 2\sqrt{X}, Y) - \Psi(X, Y) \approx 2\sqrt{X} p_s$. More generally, one would like to know that, for sufficiently large[6] X, Y and Z,

$$\Psi(X + Z, Y) - \Psi(X, Y) \sim Z\Psi(X, Y)/X \tag{15.5}$$

or, in other words, that integers in a short interval at X are about as likely to be Y-smooth as integers in a large interval at X.

We now briefly summarise some results in this area; see Granville [243] for details and references. Harman (improved by Lenstra, Pila and Pomerance [342]) showed, for any fixed $\beta > 1/2$ and $X \geq Y \geq \exp(\log(X)^{2/3+o(1)})$, where the $o(1)$ is as $X \to \infty$, that

$$\Psi(X + X^\beta, Y) - \Psi(X, Y) > 0.$$

Obtaining results for the required value $\beta = 1/2$ seems to be hard and the experts refer to the "$\sqrt{X}$ barrier" for smooth integers in short intervals. It is known that this barrier

[5] Lenstra considers the subinterval $(X - \sqrt{X}, X + \sqrt{X})$ of the Hasse interval $[X + 1 - 2\sqrt{X}, X + 1 + 2\sqrt{X}]$ because the distribution of isomorphism classes of randomly chosen elliptic curves is relatively close to uniform when restricted to those whose group order lies in this subinterval. In contrast, elliptic curves whose group orders are near the edge of the Hasse interval arise with lower probability.

[6] The notation $\sim$ means taking a limit as $X \to \infty$, so it is necessary that Y and Z grow in a controlled way as X does.

can be broken most of the time: Hildebrand and Tenenbaum showed that, for any $\epsilon > 0$, equation (15.5) holds when $X \geq Y \geq \exp(\log(X)^{5/6+\epsilon})$ and $Y \exp(\log(X)^{1/6}) \leq Z \leq X$ for all but at most $M/\exp(\log(M)^{1/6-\epsilon})$ integers $1 \leq X \leq M$. As a special case, this result shows that, for almost all primes p, the interval $[p - \sqrt{p}, p + \sqrt{p}]$ contains a Y-smooth integer where $Y = \exp(\log(X)^{5/6+\epsilon})$ (i.e., subexponential smoothness).

Using Conjecture 15.3.1 one obtains the following complexity for the elliptic curve method (we stress that the complexity is in terms of the smallest prime factor p of N, rather than N itself).

Theorem 15.3.2 *(Conjecture 2.10 of [339]) Assume Conjecture 15.3.1. One can find the smallest factor p of an integer N in $L_p(1/2, \sqrt{2} + o(1))M(\log(N))$ bit operations as $p \to \infty$.*

Proof Guess the size of p and choose $B = L_p(1/2, 1/\sqrt{2})$ (since the size of p is not known one actually runs the algorithm repeatedly for slowly increasing values of B). Then each run of Algorithm 11 requires $O(B\log(B)M(\log(N))) = L_p(1/2, 1/\sqrt{2} + o(1))M(\log(N))$ bit operations. By Conjecture 15.3.1 one needs to repeat the process $L_p(1/2, 1/\sqrt{2} + o(1))$ times. The result follows. □

Exercise 15.3.3 Let $N = pq$ where p is prime and $p < \sqrt{N} < 2p$. Show that $L_p(1/2, \sqrt{2} + o(1)) = L_N(1/2, 1 + o(1))$. Hence, in the worst case, the complexity of ECM is the same as the complexity of the quadratic sieve.

For further details on the elliptic curve method we refer to Section 7.4 of [150]. We remark that Lenstra, Pila and Pomerance [342] have considered a variant of the elliptic curve method using divisor class groups of hyperelliptic curves of genus 2. The Hasse–Weil interval for such curves contains an interval of the form $(X, X + X^{3/4})$ and Theorem 1.3 of [342] proves that such intervals contain $L_X(2/3, c_1)$-smooth integers (for some constant c_1) with probability $1/L_X(1/3, 1)$. It follows that there is a rigorous factoring algorithm with complexity $L_p(2/3, c)$ bit operations for some constant c_2. This algorithm is not used in practice, as the elliptic curve method works fine already.

Exercise 15.3.4 Suppose a sequence of values $1 < x < N$ are chosen uniformly at random. Show that one can find such a value that is $L_N(2/3, c)$-smooth, together with its factorisation, in expected $L_N(1/3, c' + o(1))$ bit operations for some constant c'.

Remark 15.3.5 It is tempting to conjecture that the Hasse interval contains a polynomially smooth integer (indeed, this has been done by Maurer and Wolf [367]; see equation (21.9)). This is not relevant for the elliptic curve factoring method, since such integers would be very rare. Suppose the probability that an integer of size X is Y-smooth is exactly $1/u^u$, where $u = \log(X)/\log(Y)$ (by Theorem 15.1.2, this is reasonable as long as $Y^{1-\epsilon} \geq \log(X)$). It is natural to suppose that the interval $[X - 2\sqrt{X}, X + 2\sqrt{X}]$ is likely to contain a Y-smooth integer if $4\sqrt{X} > u^u$. Let $Y = \log(X)^c$. Taking logs of both sides of the inequality gives the condition

$$\log(4) + \tfrac{1}{2}\log(X) > \frac{\log(X)}{c\log(\log(X))}(\log(\log(X)) - \log(c\log(\log(X)))).$$

It is therefore natural to conclude that when $c \geq 2$ there is a good chance that the Hasse interval contains a $\log(p)^c$ smooth integer. Proving such a claim seems to be far beyond the reach of current techniques.

15.4 The number field sieve

The most important integer factorisation algorithm for large integers is the **number field sieve** (NFS). A special case of this method was invented by Pollard.[7] The algorithm requires algebraic number theory and a complete discussion of it is beyond the scope of this book. Instead, we just sketch some of the basic ideas. For full details we refer to Lenstra and Lenstra [334], Section 6.2 of Crandall and Pomerance [150], Section 10.5 of Cohen [127] or Stevenhagen [525].

As we have seen from the quadratic sieve, reducing the size of the values being tested for smoothness yields a better algorithm. Indeed, in the quadratic sieve the numbers were reduced from size $O(N)$ to $O(N^{1/2+o(1)})$ and, as shown by Exercise 15.2.11, this trick alone lowers the complexity from $O(L_N(1/2, \sqrt{2}+o(1)))$ to $O(L_N(1/2, 1+o(1)))$. To break the "$O(L_N(1/2, c))$ barrier" one must make the numbers being tested for smoothness dramatically smaller. A key observation is that if the numbers are of size $O(L_N(2/3, c'))$ then they are $O(L_N(1/3, c''))$ smooth, for some constants c' and c'', with probability approximately $1/u^u = 1/L_N(1/3, c/(3c') + o(1))$. Hence, one can expect an algorithm with running time $O(L_N(1/3, c + o(1)))$ bit operations, for some constant c, by considering smaller values for smoothness.

It seems to be impossible to directly choose values x such that $x^2 \pmod{N}$ is of size $L_N(2/3, c + o(1))$ for some constant c. Hence, the number field sieve relies on two factor bases $\mathcal{B}_1$ and $\mathcal{B}_2$. Using smooth elements over $\mathcal{B}_1$ (respectively, $\mathcal{B}_2$) and linear algebra one finds an integer square u^2 and an algebraic integer square v^2. The construction allows us to associate an integer w modulo N to v such that $u^2 \equiv w^2 \pmod{N}$ and hence one can try to split N.

We briefly outline the ideas behind the algorithm. First, choose a monic irreducible polynomial $P(x) \in \mathbb{Z}[x]$ of degree d (where d grows like $\lfloor (3\log(N)/\log(\log(N)))^{1/3} \rfloor$) with a root $m = \lfloor N^{1/d} \rfloor$ modulo N (i.e., $P(m) \equiv 0 \pmod{N}$). Factor base $\mathcal{B}_1$ is primes up to $B = L_N(1/3, c)$ and factor base $\mathcal{B}_2$ is small prime ideals in the ring $\mathbb{Z}[\theta]$ in the number field $K = \mathbb{Q}(\theta) = \mathbb{Q}[x]/(P(x))$ (i.e., θ is a generic root of $P(x)$). The algorithm exploits, in the final step, the ring homomorphism $\phi : \mathbb{Z}[x]/(P(x)) \to \mathbb{Z}/N\mathbb{Z}$ given by $\phi(\theta) = m \pmod{N}$. Suppose the ideal $(a - b\theta)$ is a product of prime ideals in $\mathcal{B}_2$ (one

[7] The goal of Pollard's method was to factor integers of the form $n^3 + k$ where k is small. The algorithm in the case of numbers of a special form is known as the **special number field sieve**.

factors the ideal $(a - b\theta)$ by factoring its norm in $\mathbb{Z}$), say

$$(a - b\theta) = \prod_{i=1}^{r} \wp_i^{e_i}.$$

Suppose also that $a - bm$ is a smooth integer in $\mathcal{B}_1$, say

$$a - bm = \prod_{j=1}^{s} p_j^{f_j}.$$

If these equations hold then we call $(a - b\theta)$ and $a - bm$ smooth and store a, b and the sequences of e_i and f_j. We do not call this a "relation" as there is no direct relationship between the prime ideals $\wp_i$ and the primes p_j. Indeed, the $\wp_j$ are typically non-principal ideals and do not necessarily contain an element of small norm. Hence, the two products are modelled as being "independent".

It is important to estimate the probability that both the ideal $(a - b\theta)$ and the integer $a - bm$ are smooth. One shows that taking integers $|a|, |b| \le L_N(1/3, c' + o(1))$ for a suitable constant c' gives $(a - b\theta)$ of norm $L_N(2/3, c'' + o(1))$ and $a - bm$ of size $L_N(2/3, c''' + o(1))$ for certain constants c'' and c'''. To obtain a fast algorithm one uses sieving to determine within a range of values for a and b the pairs (a, b) such that both $a - bm$ and $(a - b\theta)$ factor over the appropriate factor base.

Performing linear algebra on both sides gives a set S of pairs (a, b) such that (ignoring issues with units and non-principal ideals)

$$\prod_{(a,b)\in S} (a - bm) = u^2$$

$$\prod_{(a,b)\in S} (a - b\theta) = v^2$$

for some $u \in \mathbb{Z}$ and $v \in \mathbb{Z}[\theta]$. Finally, we can "link" the two factor bases: applying the ring homomorphism $\phi : \mathbb{Z}[\theta] \to \mathbb{Z}$ gives $u^2 \equiv \phi(v)^2 \pmod{N}$ and hence we have a chance to split N. A non-trivial task is computing the actual numbers u and $\phi(v)$ modulo N so that one can compute $\gcd(u - \phi(v), N)$.

Since one is only considering integers $a - bm$ in a certain range (and ideals in a certain range) for smoothness one relies on heuristic assumptions about the smoothness probability. The conjectural complexity of the number field sieve is $O(L_N(1/3, c + o(1)))$ bit operations as $N \to \infty$ where $c = (64/9)^{1/3} \approx 1.923$. Note, comparing with Exercise 15.2.12, that if $N \approx 2^{1024}$ then $L_N(1/3, 1.923) \approx 2^{87}$.

15.5 Index calculus in finite fields

We now explain how similar ideas to the above have been used to find subexponential algorithms for the discrete logarithm problem in finite fields. The original idea is due to Kraitchik [319]. While all subexponential algorithms for the DLP share certain basic concepts, the specific details vary quite widely (in particular, precisely what "linear algebra" is required). We present in this section an algorithm that is very convenient when working in subgroups of prime order r in $\mathbb{F}_q^*$ as it relies only on linear algebra over the field $\mathbb{F}_r$.

Let $g \in \mathbb{F}_q^*$ have prime order r and let $h \in \langle g \rangle$. The starting point is the observation that if one can find integers $0 < Z_1, Z_2 < r$ such that

$$g^{Z_1} h^{Z_2} = 1 \tag{15.6}$$

in $\mathbb{F}_q^*$ then $\log_g(h) = -Z_1 Z_2^{-1} \pmod{r}$. The idea will be to find such a relation using a factor base and linear algebra. Such algorithms go under the general name of **index calculus** algorithms; the reason for this is that **index** is another word for discrete logarithm, and the construction of a solution to equation (15.6) is done by calculations using indices.

15.5.1 Rigorous subexponential discrete logarithms modulo p

We now sketch a subexponential algorithm for the discrete logarithm problem in $\mathbb{F}_p^*$. It is closely related to the random squares algorithm of Section 15.2. Let $g \in \mathbb{F}_p^*$ have order r (we will assume r is prime, but the general case is not significantly different) and let $h \in \mathbb{F}_p^*$ be such that $h \in \langle g \rangle$. We will also assume, for simplicity, that $r^2 \nmid (p-1)$ (we show in Exercise 15.5.8 that this condition can be avoided).

The natural idea is to choose the factor base $\mathcal{B}$ to be the primes in $\mathbb{Z}$ up to B. We let $s = \#\mathcal{B}$. One can take random powers $g^z \pmod{p}$ and try to factor over $\mathcal{B}$. One issue is that the values g^z only lie in a subgroup of $\mathbb{F}_p^*$ and so a strong smoothness heuristic would be required. To get a rigorous algorithm (under the assumption that $r^2 \nmid (p-1)$) write G' for the subgroup of $\mathbb{F}_p^*$ of order $(p-1)/r$, choose a random $\delta \in G'$ at each iteration and try to factor $g^z\delta \pmod{p}$; this is now a uniformly distributed element of $\mathbb{F}_p^*$ and so Corollary 15.1.8 can be applied. We remark that the primes p_i themselves do not lie in the subgroup $\langle g \rangle$.

Exercise 15.5.1 Let $r \mid (p-1)$ be a prime such that $r^2 \nmid (p-1)$. Let $g \in \mathbb{F}_p^*$ have order dividing r and denote by $G' \subseteq \mathbb{F}_p^*$ the subgroup of order $(p-1)/r$. Show that $\langle g \rangle \cap G' = \{1\}$.

Exercise 15.5.2 Give two ways to sample randomly from G'. When would each be used? [Hint: See Section 11.4.]

The algorithm proceeds by choosing random values $1 \le z < r$ and random $\delta \in G'$ and testing $g^z\delta \pmod{p}$ for smoothness. The ith relation is

$$g^{z_i}\delta_i \equiv \prod_{j=1}^{s} p_j^{e_{i,j}} \pmod{p}. \tag{15.7}$$

The values z_i are stored in a vector and the values $\underline{e}_i = (e_{i,1}, \ldots, e_{i,s})$ are stored as a row in a matrix. We need s relations of this form. We also need at least one relation involving h (alternatively we could have used a power of h in every relation in equation (15.7)) so try random values z_{s+1} and $\delta_{s+1} \in G'$ until $g^{z_{s+1}} h \delta_{s+1} \pmod{p}$ is B-smooth. One performs linear algebra modulo r to find integers $0 \le \lambda_1, \ldots, \lambda_{s+1} < r$ such that

$$\sum_{i=1}^{s+1} \lambda_i \underline{e}_i = (rf_1, \ldots, rf_s) \equiv (0, \ldots, 0) \pmod{r}$$

where $f_1, \ldots, f_s \in \mathbb{Z}_{\ge 0}$. In matrix notation, writing $A = (e_{i,j})$, this is $(\lambda_1, \ldots, \lambda_{s+1})A \equiv (0, \ldots, 0) \pmod{r}$. In other words, the linear algebra problem is finding a non-trivial element in the kernel of the matrix A modulo r. Let $Z_1 = \sum_{i=1}^{s+1} \lambda_i z_i \pmod{r}$ and $Z_2 = \lambda_{s+1}$. Then

$$g^{Z_1} h^{Z_2} \left(\prod_i \delta_i^{\lambda_i} \right) \equiv \left(\prod_i p_i^{f_i} \right)^r \pmod{p}. \tag{15.8}$$

Since $g^{Z_1} h^{Z_2} \in \langle g \rangle$ and the other terms are all in G' it follows from Exercise 15.5.1 that $g^{Z_1} h^{Z_2} \equiv 1 \pmod{r}$ as required. We stress that it is not necessary to compute $\prod_i \delta_i^{\lambda_i}$ or the right-hand side of equation (15.8).

The algorithm succeeds as long as $B = \lambda_{s+1} \not\equiv 0 \pmod{r}$ (and if $\lambda_{s+1} = 0$ then there is a linear dependence from the earlier relations, which can be removed by deleting one or more rows of the relation matrix).

Exercise 15.5.3 Show that if one replaces equation (15.7) by $g^{z_{1,i}} h^{z_{2,i}} \delta_i$ for random $z_{1,i}, z_{2,i}$ and δ_i then one obtains an algorithm that succeeds with probability $1 - 1/r$.

Example 15.5.4 Let $p = 223$. Then $g = 15$ has prime order $r = 37$. Suppose $h = 68$ is the instance of the DLP we want to solve. Let $\mathcal{B} = \{2, 3, 5, 7\}$. Choose the element $g_1 = 184$ of order $(p-1)/r = 6$. One can check that we have the following relations.

z	i	Factorisation of $g^z g_1^i \pmod{p}$
1	1	$2^2 \cdot 3 \cdot 7$
33	0	$2^3 \cdot 7$
8	1	$3^2 \cdot 5$
7	0	$2^3 \cdot 3 \cdot 5$

One also finds the relation $hg^7 g_1^2 = 2^3 \cdot 3^2$.

We represent the relations as the vector and matrix

$$\underline{z} = \begin{pmatrix} 1 \\ 33 \\ 8 \\ 7 \\ 7 \end{pmatrix}, \quad \begin{pmatrix} 2 & 1 & 0 & 1 \\ 3 & 0 & 0 & 1 \\ 0 & 2 & 1 & 0 \\ 3 & 1 & 1 & 0 \\ 3 & 2 & 0 & 0 \end{pmatrix}.$$

Now perform linear algebra modulo 37. One finds the non-trivial kernel vector $\underline{v} = (1, 36, 20, 17, 8)$. Computing $Z_1 = \underline{v} \cdot \underline{z} = 7 \pmod{37}$ and $Z_2 = 8$ we find $g^{Z_1} h^{Z_2} \equiv 1 \pmod{223}$ and so the solution is $-Z_1 Z_2^{-1} \equiv 13 \pmod{37}$.

Exercise 15.5.5 Write the above algorithm in pseudocode (using trial division to determine the smooth relations).

Exercise 15.5.6 Let the notation be as above. Let T_B be the expected number of trials of random integers modulo p until one is B-smooth. Show that the expected running time of this algorithm (using naive trial division for the relations and using the Lanczos or Wiedemann methods for the linear algebra) is

$$O((\#\mathcal{B})^2 T_B M(\log(p)) + (\#\mathcal{B})^{2+o(1)} M(\log(r)))$$

bit operations as $p \to \infty$

Exercise 15.5.7 Show that taking $B = L_p(1/2, 1/2)$ is the optimal value to minimise the complexity of the above algorithm, giving a complexity of $O(L_p(1/2, 2 + o(1)))$ bit operations for the discrete logarithm problem in $\mathbb{F}_p^*$ as $p \to \infty$. (Note that, unlike many of the results in this chapter, this result does not rely on any heuristics.)

We remark that, in practice, rather than computing a full exponentiation g^z one might use a pseudorandom walk as done in Pollard rho. For further implementation tricks see Sections 5.1 to 5.5 of Odlyzko [421].

If g does not have prime order (e.g., suppose g is a generator of $\mathbb{F}_p^*$ and has order $p - 1$) then there are several options: one can apply Pohlig–Hellman and reduce to subgroups of prime order and apply index calculus in each subgroup (or at least the ones of large order). Alternatively, one can apply the algorithm as above and perform the linear algebra modulo the order of g. There will usually be difficulties with non-invertible elements in the linear algebra, and there are several solutions, such as computing the Hermite normal form of the relation matrix or using the Chinese remainder theorem, we refer to Section 5.5.2 of Cohen [127] and Section 15.2.1 of Joux [283] for details.

Exercise 15.5.8 Give an algorithm similar to the above that works when $r^2 \mid (p - 1)$.

Exercise 15.5.9 This exercise is about solving many different discrete logarithm instances $h_i = g^{a_i} \pmod{p}$, for $1 \le i \le n$, to the same base g. Once sufficiently many relations are found, determine the cost of solving each individual instance of the DLP. Hence, show that one can solve any constant number of instances of the DLP to a given base $g \in \mathbb{F}_p^*$ in $O(L_p(1/2, 2 + o(1)))$ bit operations as $p \to \infty$.

15.5.2 Heuristic algorithms for discrete logarithms modulo p

To get a faster algorithm it is necessary to improve the time to find smooth relations. It is natural to seek methods to sieve rather than factor each value by trial division, but it is not known how to do this for relations of the form in equation (15.7). It would also be natural to find an analogue to Pomerance's method of considering residues of size about the square-root of random; Exercise 15.5.10 gives an approach to this, but it does not lower the complexity.

Exercise 15.5.10 (Blake, Fuji-Hara, Mullin and Vanstone [57]) Once one has computed $w = g^z \delta \pmod{p}$ one can apply the Euclidean algorithm to find integers w_1, w_2 such that $w_1 w \equiv w_2 \pmod{p}$ and $w_1, w_2 \approx \sqrt{p}$. Since w_1 and w_2 are smaller one would hope that they are much more likely to both be smooth (however, note that both must be smooth). We now make the heuristic assumption that the probability each w_i is B-smooth is independent and the same as the probability that any integer of size $\sqrt{p}$ is B-smooth. Show that the heuristic running time of the algorithm has u^u replaced by $(u/2)^u$ (where $u = \log(p)/\log(B)$) and so the asymptotic running time remains the same.

Coppersmith, Odlyzko and Schroeppel [135] proposed an algorithm for the DLP in $\mathbb{F}_p^*$ that uses sieving. Their idea is to let $H = \lceil \sqrt{p} \rceil$ and define the factor base to be

$$\mathcal{B} = \{q : q \text{ prime}, q < L_p(1/2, 1/2)\} \cup \{H + c : 1 \le c \le L_p(1/2, 1/2 + \epsilon)\}.$$

Since $H^2 \pmod{p}$ is of size $\approx p^{1/2}$ it follows that if $(H + c_1), (H + c_2) \in \mathcal{B}$ then $(H + c_1)(H + c_2) \pmod{p}$ is of size $p^{1/2+o(1)}$. One can therefore generate relations in $\mathcal{B}$. Further, it is shown in Section 4 of [135] how to sieve over the choices for c_1 and c_2. A heuristic analysis of the algorithm gives complexity $L_p(1/2, 1 + o(1))$ bit operations.

The **number field sieve** (NFS) is an algorithm for the DLP in $\mathbb{F}_p^*$ with heuristic complexity $O(L_p(1/3, c + o(1)))$ bit operations. It is closely related to the number field sieve for factoring and requires algebraic number theory. As with the factoring algorithm, there are two factor bases. Introducing the DLP instance requires an extra algorithm (we will see an example of this in Section 15.5.4). We do not have space to give the details and instead refer to Schirokauer, Weber and Denny [467] or Schirokauer [463, 465] for details.

15.5.3 Discrete logarithms in small characteristic

We now consider the discrete logarithm problem in $\mathbb{F}_q^*$ where $q = p^n$, p is relatively small (the case of most interest is $p = 2$) and n is large. We represent such a field with a polynomial basis as $\mathbb{F}_p[x]/(F(x))$ for some irreducible polynomial $F(x)$ of degree n. The natural notion of smoothness of an element $g(x) \in \mathbb{F}_p[x]/(F(x))$ is that it is a product of polynomials of small degree. Since factoring polynomials over finite fields is polynomial-time we expect to more easily get good algorithms in this case. The first work on this topic was due to Hellman and Reyneri but we follow Odlyzko's large paper [421]. First we quote some results on smooth polynomials.

Definition 15.5.11 Let p be prime and $n, b \in \mathbb{N}$. Let $I(n)$ be the number of monic irreducible polynomials in $\mathbb{F}_p[x]$ of degree n. A polynomial $g(x) \in \mathbb{F}_p[x]$ is called ***b*-smooth** if all its irreducible factors have degree $\le b$. Let $N(n, b)$ be the number of b-smooth polynomials of degree exactly equal to n. Let $p(n, b)$ be the probability that a uniformly chosen polynomial of degree at most n is b-smooth.

Theorem 15.5.12 *Let p be prime and $n, b \in \mathbb{N}$.*

1. $I(n) = \frac{1}{n}\sum_{d|n} \mu(d)p^{n/d} = p^n/n + O(p^{n/2}/n)$ *where $\mu(d)$ is the Möbius function.*[8]
2. *If* $n^{1/100} \leq b \leq n^{99/100}$ *then* $N(n, b) = p^n(b/n)^{(1+o(1))n/b}$ *as n tends to infinity.*
3. *If* $n^{1/100} \leq b \leq n^{99/100}$ *then $p(n, b)$ is at least* $u^{-u(1+o(1))}$ *where $u = n/b$ and n tends to infinity.*
4. *If* $n^{1/100} \leq b \leq n^{99/100}$ *then the expected number of trials before a randomly chosen element of $\mathbb{F}_p[x]$ of degree n is b-smooth is* $u^{u(1+o(1))}$ *as* $u \to \infty$.

Proof Statement 1 follows from an elementary counting argument (see, for example, Theorem 3.25 of Lidl and Niederreiter [350]).

Statement 2 in the case $p = 2$ is Corollary A.2 of Odlyzko [421]. The general result was proved by Soundararajan (see Theorems 2.1 and 2.2 of Lovorn Bender and Pomerance [358]). Also see Section 9.15 of [350].

Statement 3 follows immediately from statement 2 and the fact there are p^n monic polynomials of degree at most n (when considering smoothness it is sufficient to study monic polynomials). Statement 4 follows immediately from statement 3. □

The algorithm then follows exactly the ideas of the previous section. Suppose g has prime order $r \mid (p^n - 1)$ and $h \in \langle g \rangle$. The factor base is

$$\mathcal{B} = \{P(x) \in \mathbb{F}_p[x] : P(x) \text{ is monic, irreducible and } \deg(P(x)) \leq b\}$$

for some integer b to be determined later. Note that $\#\mathcal{B} = I(1) + I(2) + \cdots + I(b) \approx p^{b+1}/(b(p-1))$ (see Exercise 15.5.14). We compute random powers of g multiplied by a suitable $\delta \in G'$ (where, if $r^2 \nmid (p^n - 1)$, $G' \subseteq \mathbb{F}_{p^n}^*$ is the subgroup of order $(p^n - 1)/r$; when $r^2 \mid (p^n - 1)$ then use the method of Exercise 15.5.8), reduce to polynomials in $\mathbb{F}_p[x]$ of degree at most n and try to factor them into products of polynomials from $\mathcal{B}$. By Exercise 2.12.11 the cost of factoring the b-smooth part of a polynomial of degree n is $O(bn\log(n)\log(p)M(\log(p))) = O(\log(p^n)^3)$ bit operations (in any case, polynomial-time). As previously, we are generating polynomials of degree n uniformly at random and so, by Theorem 15.5.12, the expected number of trials to get a relation is $u^{u(1+o(1))}$ where $u = n/b$ as $u \to \infty$. We need to obtain $\#\mathcal{B}$ relations in general. Then we obtain a single relation of the form $hg^a\delta = \prod_{P\in\mathcal{B}} P^{e_P}$, perform linear algebra and hence solve the DLP.

Exercise 15.5.13 Write the above algorithm in pseudocode.

Exercise 15.5.14 Show that $\sum_{i=1}^{b} I(b) \leq \frac{1}{b}p^b(1 + 2/(p-1)) + O(bp^{b/2})$. Show that a very rough approximation is $p^{b+1}/(b(p-1))$.

[8] This is the "prime number theorem for polynomials", $I(n) \approx p^n/\log_p(p^n)$.

Exercise 15.5.15 Let the notation be as above. Show that the complexity of this algorithm is at most

$$c_1 \#\mathcal{B} u^{u(1+o(1))} \log(q)^3 + c_2 (\#\mathcal{B})^{2+o(1)} M(\log(r))$$

bit operations (for some constants c_1 and c_2) as $n \to \infty$ in $q = p^n$.

For the complexity analysis it is natural to arrange that $\#\mathcal{B} \approx L_{p^n}(1/2, c)$ for a suitable constant c. Recall that $\#\mathcal{B} \approx p^b/b$. To have $p^b/b = L_{p^n}(1/2, c)$ then, taking logs,

$$b \log(p) - \log(b) = c\sqrt{n \log(p)(\log(n) + \log(\log(p)))}.$$

It follows that $b \approx c\sqrt{n \log(n)/\log(p)}$.

Exercise 15.5.16 Show that one can compute discrete logarithms in $\mathbb{F}_{p^n}^*$ in expected $O(L_{p^n}(1/2, \sqrt{2} + o(1)))$ bit operations for fixed p and as $n \to \infty$. (Note that this result does not rely on any heuristic assumptions.)

Exercise 15.5.17 Adapt the trick of Exercise 15.5.10 to this algorithm. Explain that the complexity of the algorithm remains the same, but is now heuristic.

Lovorn Bender and Pomerance [358] give rigorous complexity $L_{p^n}(1/2, \sqrt{2} + o(1))$ bit operations as $p^n \to \infty$ and $p \le n^{o(n)}$.

15.5.4 Coppersmith's algorithm for the DLP in $\mathbb{F}_{2^n}^*$

This algorithm (inspired by the "systematic equations" of Blake, Fuji-Hara, Mullin and Vanstone [57]) was the first algorithm in computational number theory to have heuristic subexponential complexity of the form $L_q(1/3, c + o(1))$.

The method uses a polynomial basis for $\mathbb{F}_{2^n}$ of the form $\mathbb{F}_2[x]/(F(x))$ for $F(x) = x^n + F_1(x)$ where $F_1(x)$ has very small degree. For example, $\mathbb{F}_{2^{127}} = \mathbb{F}_2[x]/(x^{127} + x + 1)$.

The "systematic equations" of Blake *et al.* are relations among elements of the factor base that come almost for free. For example, in $\mathbb{F}_{2^{127}}$, if $A(x) \in \mathbb{F}_2[x]$ is an irreducible polynomial in the factor base then $A(x)^{128} = A(x^{128}) \equiv A(x^2 + x) \pmod{F(x)}$ and $A(x^2 + x)$ is either irreducible or is a product $P(x)P(x+1)$ of irreducible polynomials of the same degree (Exercise 15.5.18). Hence, for many polynomials $A(x)$ in the factor base one gets a non-trivial relation.

Exercise 15.5.18 Let $A(x) \in \mathbb{F}_2[x]$ be an irreducible polynomial. Show that $A(x^2 + x)$ is either irreducible or a product of two polynomials of the same degree.

Coppersmith [130] extended the idea as follows: let $b \in \mathbb{N}$ be such that $b = cn^{1/3}\log(n)^{2/3}$ for a suitable constant c (later we take $c = (2/(3\log(2)))^{2/3}$), let $k \in \mathbb{N}$ be such that $2^k \approx \sqrt{n/b} \approx \frac{1}{\sqrt{c}}(n/\log(n))^{1/3}$ and let $l = \lceil n/2^k \rceil \approx \sqrt{nb} \approx \sqrt{c}n^{2/3}\log(n)^{1/3}$. Let $\mathcal{B} = \{A(x) \in \mathbb{F}_2[x] : \deg(A(x)) \le b, A(x) \text{ irreducible}\}$. Note that $\#\mathcal{B} \approx 2^b/b$ by Exercise 15.5.14. Suppose $A(x), B(x) \in \mathbb{F}_2[x]$ are such that $\deg(A(x)) = d_A \approx b$ and

$\deg(B(x)) = d_B \approx b$ and define $C(x) = A(x)x^l + B(x)$. In practice, one restricts to pairs $(A(x), B(x))$ such that $\gcd(A(x), B(x)) = 1$. The crucial observation is that

$$C(x)^{2^k} = A(x^{2^k}) \cdot (x^{2^k})^l + B(x^{2^k}) \equiv A(x^{2^k})x^{2^k l - n} F_1(x) + B(x^{2^k}) \pmod{F(x)}. \quad (15.9)$$

Write $D(x)$ for the right-hand side of equation (15.9). We have $\deg(C(x)) \le \max\{d_A + l, d_B\} \approx l \approx n^{2/3}\log(n)^{1/3}$ and $\deg(D(x)) \le \max\{2^k d_A + (2^k l - n) + \deg(F_1(x)), 2^k d_B\} \approx 2^k b \approx n^{2/3}\log(n)^{1/3}$.

Example 15.5.19 (Thomé [546]) Let $n = 607$ and $F_1(x) = x^9 + x^7 + x^6 + x^3 + x + 1$. Let $b = 23$, $d_A = 21$, $d_B = 28$, $2^k = 4$, $l = 152$. The degrees of $C(x)$ and $D(x)$ are 173 and 112 respectively.

We have two polynomials $C(x)$, $D(x)$ of degree $\approx n^{2/3}$ that we wish to be b-smooth where $b \approx n^{1/3}\log(n)^{2/3}$. We will sketch the complexity later under the heuristic assumption that, from the point of view of smoothness, these polynomials are independent. We will also assume that the resulting relations are essentially random (and so with high probability there is a non-trivial linear dependence once $\#\mathcal{B} + 1$ relations have been collected).

Having generated enough relations among elements of the factor base, it is necessary to find some relations involving the elements g and h of the DLP instance. This is not trivial. All DLP algorithms having complexity $L_q(1/3, c + o(1))$ feature a process called **special q-descent** that achieves this. The first step is to express g (respectively, h) as a product $\prod_i G_i(x)$ of polynomials of degree at most $b_1 = n^{2/3}\log(n)^{1/3}$; this can be done by multiplying g (respectively, h) by random combinations of elements of $\mathcal{B}$ and factoring (one can also apply the Blake *et al.* trick as in Exercise 15.5.10). We now have a list of around $2n^{1/3} < n$ polynomials $G_i(x)$ of degree $\approx n^{2/3}$ that need to be "smoothed" further. Section VII of [130] gives a method to do this: essentially one performs the same sieving as earlier except that $A(x)$ and $B(x)$ are chosen so that $G_i(x) \mid C(x) = A(x)x^l + B(x)$ (not necessarily with the same value of l or the same degrees for $A(x)$ and $B(x)$). Defining $D(x) = C(x)^{2^k} \pmod{F(x)}$ (not necessarily the same value of k as before) one hopes that $C(x)/G(x)$ and $D(x)$ are b-smooth. After sufficiently many trials, one has a relation that expresses $G_i(x)$ in terms of elements of $\mathcal{B}$. Repeating for the polynomially many values $G_i(x)$ one eventually has the values g and h expressed in terms of elements of $\mathcal{B}$. One can then do linear algebra modulo the order of g to find integers Z_1, Z_2 such that $g^{Z_1}h^{Z_2} = 1$ and the DLP is solved.

Example 15.5.20 We give an example of Coppersmith's method for $\mathbb{F}_{2^{15}} = \mathbb{F}_2[x]/(F(x))$ where $F(x) = x^{15} + x + 1$. We consider the subgroup of $\mathbb{F}_{2^{15}}^*$ of order $r = 151$ (note that $(2^{15} - 1)/r = 7 \cdot 31 = 217$). Let $g = x^{11} + x^7 + x^5 + x^2 + 1$ and $h = x^{14} + x^{11} + x^{10} + x^9 + 1$ be the DLP instance.

First, note that $n^{1/3} \approx 2.5$ and $n^{2/3} \approx 6.1$. We choose $b = 3$ and so $\mathcal{B} = \{x, x+1, x^2 + x + 1, x^3 + x + 1, x^3 + x^2 + 1\}$. We hope to be testing polynomials of degree around 6 to 8 for smoothness.

First, we find some "systematic equations". We obviously have the relation $x^{15} = x + 1$. We also have $(x+1)^{16} = x^2 + x + 1$ and $(x^3 + x + 1)^{16} = (x^3 + x + 1)(x^3 + x^2 + 1)$.

Now, we do Coppersmith's method. We must choose $2^k \approx \sqrt{n/b} = \sqrt{5} \approx 2.2$ so take $2^k = 2$. Let $l = \lceil n/2^k \rceil = 8$, choose $A(x)$ and $B(x)$ of degree at most 2, set $C(x) = A(x)x^8 + B(x)$ and $D(x) = C(x)^2 \pmod{F(x)}$ and test $C(x)$ and $D(x)$ for smoothness over $\mathcal{B}$. We find the following pairs $(A(x), B(x))$ such that both $C(x)$ and $D(x)$ factor over $\mathcal{B}$.

$A(x)$	$B(x)$	$C(x)$	$D(x)$
1	1	$(x+1)^8$	$x^2 + x + 1$
1	x	$x(x+1)(x^3+x+1)(x^3+x^2+1)$	x
1	x^2	$x^2(x+1)^2(x^2+x+1)^2$	$x(x^3+x+1)$

The first relation in the table is a restatement of $(x+1)^{16} = x^2 + x + 1$. All together we have the relation matrix

$$\begin{pmatrix} 15 & -1 & 0 & 0 & 0 \\ 0 & 16 & -1 & 0 & 0 \\ 0 & 0 & 0 & 15 & -1 \\ 1 & 2 & 0 & 2 & 2 \\ 3 & 4 & 4 & -1 & 0 \end{pmatrix}. \tag{15.10}$$

To solve the DLP one can now try to express g and h over the factor base. One has

$$g^{22} = x(x+1)(x^2+x+1)^2(x^3+x^2+1).$$

For h we find

$$hg^{30} = x^6(x+1)^4 G(x)$$

where $G(x) = x^4 + x + 1$ is a "large prime". To "smooth" $G(x)$ we choose $A(x) = 1$, $B(x) = A(x)x^8 \pmod{G(x)} = x^2 + 1$, $C(x) = A(x)x^8 + B(x)$ and $D(x) = C(x)^2 \pmod{F(x)}$. One finds $C(x) = G(x)^2$ and $D(x) = (x+1)(x^3+x^2+1)$. In other words, $G(x)^4 = (x+1)(x^3+x^2+1)$.

There are now two ways to proceed. Following the algorithm description above we add to the matrix the two rows $(1, 1, 2, 0, 1)$ and $4(6, 4, 0, 0, 0) + (0, 1, 0, 0, 0, 1) = (24, 17, 0, 0, 1)$ corresponding to g^{22} and $h^4 g^{120}$. Finding a non-trivial kernel vector modulo 151, such as $(1, 114, 0, 132, 113, 133, 56)$ gives the relation

$$1 = (g^{22})^{133}(h^4 g^{120})^{56} = g^{133}h^{73}$$

from which we deduce $h = g^{23}$.

An alternative approach to the linear algebra is to diagonalise the system in equation (15.10) using linear algebra over $\mathbb{Z}$ (or at least modulo $2^{15} - 1$) to get $x + 1 = x^{15}$, $x^2 + x + 1 = x^{240}$, $x^3 + x + 1 = x^{1023}$ and $x^3 + x^2 + 1 = x^{15345}$. One then gets

$$g^{22} = x(x+1)(x^2+x+1)^2(x^3+x^2+1) = x^{1+15+2\cdot 240+15345} = x^{15841}$$

and so

$$g = x^{15841 \cdot 22^{-1} \pmod{(2^{15}-1)}} = x^{26040} = (x^{217})^{120}.$$

Similarly, $G(x)^4 = (x+1)(x^3+x^2+1) = x^{15+15345} = x^{15360}$ and so $G(x) = x^{3840}$. Finally,

$$h = g^{-30} x^6 (x+1)^4 G(x) = x^{-30 \cdot 26040 + 6 + 4 \cdot 15 + 3840} = x^{9114} = (x^{217})^{42}$$

and so $h = g^{42 \cdot 120^{-1} \pmod{151}} = g^{23}$.

Conjecture 15.5.21 *Coppersmith's algorithm solves the DLP in $\mathbb{F}_q^*$ where $q = 2^n$ in $L_q(1/3, (32/9)^{1/3} + o(1))$ bit operations as $n \to \infty$.*

Note that, to compare with Exercise 15.2.12, if $q = 2^{1024}$ then $L_q(1/3, (32/9)^{1/3}) \approx 2^{67}$.

This conjecture would hold if the probability that the polynomials $C(x)$ and $D(x)$ are smooth was the same as for independently random polynomials of the same degree. We now give a justification for the constant. Let $b = cn^{1/3}\log(n)^{2/3}$. Note that $2^k \approx \sqrt{n/b} \approx (n/c\log(n))^{1/3}$ and $l \approx \sqrt{nb}$. We need around $2^b/b$ relations, and note that $\log(2^b/b) \approx b\log(2) = c\log(2)n^{1/3}\log(n)^{2/3}$. We have $\deg(C(x)) \approx d_A + l$ and $\deg(D(x)) \approx kd$. The number of trials until $C(x)$ is b-smooth is u^u where $u = (d_A + l)/b \approx l/b \approx \sqrt{n/b} = \frac{1}{\sqrt{c}}(n/\log(n))^{1/3}$. Hence, $\log(u^u) = u\log(u) \approx \frac{1}{3\sqrt{c}} n^{1/3}\log(n)^{2/3}$. Similarly, the number of trials until $D(x)$ is b-smooth is approximately u^u where $u = (2^k d)/b \approx 2^k \approx \sqrt{n/b}$ and the same argument applies. Since both events must occur the expected number of trials to get a relation is $\exp(\frac{2}{3\sqrt{c}}(n/\log(n))^{1/3})$. Hence, total expected time to generate enough relations is

$$\exp\left((c\log(2) + \tfrac{2}{3\sqrt{c}}) n^{1/3} \log(n)^{2/3} \right).$$

This is optimised when $c^{3/2}\log(2) = 2/3$, which leads to the stated complexity for the first stage of the algorithm. The linear algebra is $O((2^b/b)^{2+o(1)} M(\log(r)))$ bit operations, which is the same complexity, and the final stage of solving the DLP has lower complexity (it is roughly the same as the cost of finding polynomially many smooth relations, rather than finding $2^b/b$ of them). For more details about the complexity of Coppersmith's method we refer to Section 2.4 of Thomé [546].

Since one can detect smoothness of polynomials in polynomial-time it is not necessary, from a complexity theory point of view, to sieve. However, in practice sieving can be worthwhile and a method to do this was given by Gordon and McCurley [240].

Coppersmith's idea is a special case of a more general approach to index calculus algorithms known as the **function field sieve**. Note that Coppersmith's algorithm only has one factor base, whereas the function field sieve works using two factor bases.

15.5.5 The Joux–Lercier algorithm

The function field sieve of Adleman is a general algorithm for discrete logarithms in $\mathbb{F}_{p^n}$ where p is relatively small compared with n. Joux and Lercier gave a much simpler and better algorithm. We will sketch this algorithm, but refer to Joux and Lercier [284] and Section 15.2 of [283] for full details. We also refer to [465] for a survey of the function field sieve.

Let p be prime and $n \in \mathbb{N}$. Let $d = \lceil \sqrt{n} \rceil$. Suppose one has monic polynomials $F_1(t), F_2(t) \in \mathbb{F}_p[t]$ such that $\deg(F_1(t)) = \deg(F_2(t)) = d$ and $F_2(F_1(t)) - t$ has an irreducible factor $F(t)$ of degree n. We represent $\mathbb{F}_{p^n}$ with the polynomial basis $\mathbb{F}_p[t]/(F(t))$. Given a prime p and an integer n one can find such polynomials $F_1(t)$ and $F_2(t)$ in very little time.

Exercise 15.5.22 Let $n = 15$. Find polynomials $F_1(t), F_2(t) \in \mathbb{F}_2[t]$ of degree 4 such that $F_2(F_1(t)) - t$ has an irreducible factor of degree 15.

Now consider the polynomial ring $A = \mathbb{F}_p[x, y]$ and two ring homomorphisms $\psi_1 : A \to A_1 = \mathbb{F}_p[x]$ by $\psi_1(y) = F_1(x)$ and $\psi_2 : A \to A_2 = \mathbb{F}_p[y]$ by $\psi_2(x) = F_2(y)$. Define $\phi_1 : A_1 \to \mathbb{F}_{p^n}$ by $\phi_1(x) = t \pmod{F(t)}$ and $\phi_2 : A_2 \to \mathbb{F}_{p^n}$ by $\phi_2(y) = F_1(t) \pmod{F(t)}$.

Exercise 15.5.23 Let the notation be as above and $G(x, y) \in \mathbb{F}_p[x, y]$. Show that $\phi_1(\psi_1(G(x, y))) = \phi_2(\psi_2(G(x, y)))$ in $\mathbb{F}_{p^n}$.

Let $\mathcal{B}_1 \subseteq A_1 = \mathbb{F}_p[x]$ and $\mathcal{B}_2 \subseteq \mathbb{F}_p[y]$ be the sets of linear polynomials. The idea of the algorithm is simply to consider polynomials in $\mathbb{F}_p[x, y]$ of the form $G(x, y) = xy + ax + by + c$. If $\psi_1(G(x, y)) = (x + b)F_1(x) + (ax + c)$ factors over $\mathcal{B}_1$ as $\prod_{i=1}^{d+1}(x - u_i)$ and if $\psi_2(G(x, y)) = (y + a)F_2(y) + (by + c)$ factors over $\mathcal{B}_2$ as $\prod_{j=1}^{d+1}(y - v_j)$ then we have a relation. The point is that such a relation corresponds to

$$\prod_{i=1}^{d+1}(t - u_i) = \prod_{j=1}^{d+1}(F_1(t) - v_j)$$

in $\mathbb{F}_{p^n}$.

One also needs to introduce the DLP instance by using a special q-descent: given an irreducible polynomial $q(x)$ one constructs polynomials $a(x), b(x)$ such that $q(x) \mid (a(x)F_1(x) + b(x))$ and one hopes that $(a(x)F_1(x) + b(x))/q(x)$ has small factors and that $a(F_2(y))y + b(F_2(y))$ has small factors, and hence iterate the process. When enough relations are collected (including at least one "systematic equation" to remove the parasitic solution explained on page 442 of Joux [283]) one can perform linear algebra to solve the DLP. The heuristic complexity of this algorithm is shown in [284] and Section 15.2.1.2 of [283] to be between $L_{p^n}(1/3, 3^{1/3} + o(1))$ and $L_{p^n}(1/3, (32/9)^{1/3} + o(1))$ for $p \leq L_{p^n}(1/3, (4/9)^{1/3} + o(1))$.

15.5.6 Number field sieve for the DLP

Concepts from the number field sieve for factoring have been applied in the setting of the DLP. Again, one uses two factor bases, corresponding to ideals in the ring of integers of some number field (one of the number fields may be $\mathbb{Q}$). As with Coppersmith's method, once sufficiently many relations have been found among elements of the factor bases, special q-descent is needed to solve a general instance of the DLP. We refer to Schirokauer [465] for details of the NFS algorithm for the DLP, and also for the heuristic arguments that one can solve the DLP in $\mathbb{F}_p^*$ in $L_p(1/3, (64/9)^{1/3} + o(1))$ bit operations. When p has a special form (e.g., $p = 2^n \pm 1$) then the **special number field sieve** (SNFS) can be used to solve the DLP in (heuristic) $L_p(1/3, (32/9)^{1/3} + o(1))$ bit operations, see [466].

We should also mention the **special function field sieve** (SFFS) for solving the DLP in $\mathbb{F}_{p^n}^*$, which has heuristic complexity $L_{p^n}(1/3, (32/9)^{1/3} + o(1))$ bit operations as $p^n \to \infty$ as long as $p \le n^{o(\sqrt{n})}$, see Schirokauer [464, 465].

15.5.7 Discrete logarithms for all finite fields

We have sketched algorithms for the DLP in $\mathbb{F}_p^*$ when p is large or $\mathbb{F}_{p^n}^*$ where p is relatively small. We have not considered cases $\mathbb{F}_q^*$ where $q = p^n$ with p large and $n > 1$. The basic concepts can be extended to cover all cases, but ensuring that subexponential complexity is achieved for all combinations of p and n is non-trivial. Adleman and Demarrais [2] were the first to give a heuristic subexponential algorithm for all finite fields. They split the problem space into $p > n$ and $p \le n$; in the latter case they have complexity $L_q(1/2, 3 + o(1))$ bit operations as $q \to \infty$ and in the former case heuristic complexity $L_q(1/2, c + o(1))$ for a non-explicit constant c.

Heuristic algorithms with complexity $L_q(1/3, c + o(1))$ for all finite fields are given by Joux and Lercier [284] and Joux, Lercier, Smart and Vercauteren [285].

15.6 Discrete logarithms on hyperelliptic curves

Some index calculus algorithms for the discrete logarithm problem in finite fields generalise naturally to solving the DLP in the divisor class group of a curve. Indeed, some of these algorithms also apply to the ideal class group of a number field, but we do not explore that situation in this book. An excellent survey of discrete logarithm algorithms for divisor class groups is Chapter VII of [61].

We consider hyperelliptic curves $C : y^2 + H(x)y = F(x)$ over $\mathbb{F}_q$ of genus g, so $\deg(H(x)) \le g + 1$ and $\deg(F(x)) \le 2g + 2$. Recall that elements of the divisor class group have a Mumford representation $(u(x), y - v(x))$ (for curves with a split model there is also an integer $0 \le n \le g - \deg(u(x))$ to take into account the behaviour at infinity). Let D_1 and D_2 be reduced divisors representing divisor classes of order r (where r is a prime such that $r^2 \nmid \#\mathrm{Pic}^0_{\mathbb{F}_q}(C)$). The goal is to compute $a \in \mathbb{Z}/r\mathbb{Z}$ such that $D_2 \equiv [a]D_1$.

Recall from Exercise 10.3.12 that a reduced divisor with Mumford representation $(u(x), v(x))$ is said to be a **prime divisor** if the polynomial $u(x)$ is irreducible over $\mathbb{F}_q$. The **degree** of the effective affine divisor is $\deg(u(x))$. Any effective affine divisor D can be written as a sum of prime effective affine divisors by factoring the $u(x)$ polynomial of its Mumford representation. Hence, it is natural to define D to be b-smooth if it is a sum of prime effective divisors of degree at most b. This suggests selecting the factor base $\mathcal{B}$ to consist of all prime effective affine divisors of degree at most b for some smoothness bound $1 \leq b \leq g$.

One can obtain an algorithm for the DLP of a familiar form, by generating reduced divisors and testing whether they are smooth. One issue is that our smoothness results for polynomials apply when polynomials are sampled uniformly from the set of all polynomials of degree n in $\mathbb{F}_q[x]$, whereas we now need to apply the results to the set of polynomials $u(x) \in \mathbb{F}_q[x]$ of degree g that arise in Mumford's representation. This issue is handled using Theorem 15.6.1.

There are two rather different ways to generate reduced divisors, both of which are useful for the algorithm.

1. One can take random group elements of the form $[n]D_1$ or $[n_1]D_1 + [n_2]D_2$ and compute the Mumford representation of the corresponding reduced effective affine divisor. This is the same approach as used in Section 15.5.1 and, in the context of ideal/divisor class groups, is sometimes called the **Hafner–McCurley algorithm**. If the divisor is $\mathcal{B}$-smooth then we obtain a relation between elements of $\mathcal{B}$ and D_1 and D_2.
2. One can consider the effective affine divisor of the function $a(x) + yb(x)$ for random polynomials $a(x), b(x)$. This idea is due to Adleman, DeMarrais and Huang [4]. Since a principal divisor is equivalent to zero in the ideal class group, if the divisor is $\mathcal{B}$-smooth then we get a relation in $\mathcal{B}$.

To introduce the instance of the DLP into the system it is necessary to have some relations involving D_1 and D_2. This can either be done using the first method, or by choosing $a(x)$ and $b(x)$ so that points in the support of either D_1 or D_2 lie in the support of $\operatorname{div}(a(x) + yb(x))$ (we have seen this kind of idea already, e.g. in Coppersmith's algorithm).

It is convenient to add to $\mathcal{B}$ all points at infinity and all points $P \in C(\overline{\mathbb{F}}_q)$ such that $P = \iota(P)$ (equivalently all $\mathbb{F}_q$-rational prime divisors with this property). Since the latter divisors all have order 2 one automatically obtains relations that can be used to eliminate them during the linear algebra stage of the algorithm. Hence, we say that a reduced divisor $D = \operatorname{div}(u(x), y - v(x))$ in Mumford representation is b-**smooth** if $u(x)$ is b-smooth after any factors corresponding to points of order 2 have been removed.

Let C be a hyperelliptic curve over $\mathbb{F}_q$ of genus g and $1 \leq b < g$. Prime effective affine divisors on C of degree b correspond to irreducible polynomials $u(x)$ of degree b (and for roughly half of all such polynomials $u(x)$ there are two solutions $v(x)$ to $v(x)^2 + v(x)H(x) - F(x) \equiv 0 \pmod{u(x)}$). Hence, it is natural to expect that there are approximately q^b/b such divisors. It follows that $\#\mathcal{B}$ should be around $\sum_{i=1}^{b} q^i/i \approx \frac{1}{b}p^b(1 + 2/(p-1))$ by the same argument as Exercise 15.5.14.

For the analysis one needs to estimate the probability that a randomly chosen reduced divisor is smooth.

Theorem 15.6.1 *(Theorem 6 of Enge and Stein [183]) Let C be a hyperelliptic curve of genus g over $\mathbb{F}_q$. Let $c > 1$ and let $b = \lceil \log_q(L_{q^g}(1/2, c)) \rceil$. Then the number of b-smooth reduced divisors of degree g is at least*

$$\frac{q^g}{L_{q^g}(1/2, 1/(2c) + o(1))}$$

for fixed q and $g \to \infty$.

Note that the smoothness bound in the above result is the ceiling of a real number. Hence one cannot deduce subexponential running time unless the genus is sufficiently large compared with the field size.

15.6.1 Index calculus on hyperelliptic curves

Suppose that $r \mid N = \#\mathrm{Pic}^0_{\mathbb{F}_q}(C)$ and $r^2 \nmid N$. Suppose $\overline{D}_1, \overline{D}_2$ are two divisor classes on C over $\mathbb{F}_q$ of order r represented by reduced divisors D_1 and D_2. The algorithm of Section 15.5.1 immediately applies to solve the DLP: choose the factor base as above; generate random reduced divisors by computing $[n_1]D_1 + [n_2]D_2 + \delta$ (where δ is uniformly chosen[9] from the subgroup $G' \subseteq \mathrm{Pic}^0_{\mathbb{F}_q}(C)$ of order N/r); store the resulting smooth relations; perform linear algebra modulo r to find integers a, b such that $[a]D_1 + [b]D_2 \equiv 0$ (extra care is needed when there are two points at infinity to be sure the relation is correct).

Exercise 15.6.2 Show that the expected running time of this algorithm is (rigorously!) $L_{q^g}(1/2, \sqrt{2} + o(1))$ bit operations as $g \to \infty$.

We refer to Section VII.5 of [61] for practical details of the algorithm. Note that the performance can be improved using the sieving method of Flassenberg and Paulus [189].

15.6.2 The algorithm of Adleman, De Marrais and Huang

This algorithm, from [4], uses the same factor base as the method of the previous section. The main difference is to generate relations by decomposing principal divisors $A(x) + yB(x)$. An advantage of this approach is that group operations are not required.

By Exercise 10.1.21 it is easy to compute $v_P(A(x) + yB(x))$ by computing the norm $A(x)^2 - H(x)A(x)B(x) - F(x)B(x)^2$ and factoring it as a polynomial. If $\deg(A(x)) = d_A < g$ and $\deg(B(x)) = d_B < g$ then the norm has degree at most $\max\{2d_A, (g+1) + d_A + d_B, 2g + 2 + 2d_B\}$, which is much larger in general than the degree g polynomial in a reduced Mumford representation, but still $O(g)$ in practice.

We need to make the heuristic assumption that the probability the norm is b-smooth is the same as the probability that a random polynomial of the same degree is b-smooth. We

[9] We assume that generators for this group are known so that it is easy to sample uniformly from this group.

therefore assume the expected number of trials to get an $L_{q^g}(1/2, c)$-smooth polynomial is $L_{q^g}(1/2, 1/(2c) + o(1))$ as g tends to infinity.

We also need some relations involving D_1 and D_2. Adleman, DeMarrais and Huang do this by first decomposing D_1 and D_2 as a sum of prime divisors. Then they "smooth" each prime divisor $\operatorname{div}(u(x), y - v(x))$ by choosing polynomials $B(x), W(x) \in \mathbb{F}_q[x]$, setting $A'(x) = B(x)(v(x) + H(x)) \pmod{u(x)}$ and then $A(x) = A'(x) + u(x)W(x)$. One computes $N(x) = (A(x)^2 - H(x)A(x)B(x) - F(x)B(x)^2)$. By construction, $u(x) \mid N(x)$ and one continues randomly choosing A and W until $N(x)/u(x)$ is b-smooth.

The details of the algorithm are then the same as the algorithm in Section 15.5.1: one uses linear algebra modulo r to get a relation $[a]D_1 + [b]D_2 \equiv 0$ (again, care is needed when there are two points at infinity). We leave the details as an exercise.

Exercise 15.6.3 Write pseudocode for the Adleman, DeMarrais, Huang algorithm.

The heuristic complexity of the algorithm is of the same form as the earlier algorithm (the cost of smoothing the divisors D_1 and D_2 is heuristically the same as finding less than $2g$ relations, so is negligible. One obtains heuristic asymptotic complexity of $L_{q^g}(1/2, \sqrt{2} + o(1))$ bit operations as g tends to infinity. This is much better than the complexity claimed in [4] since that paper also gives an algorithm to compute the group structure (and so the linear algebra requires computing the Hermite normal form).

These ideas will be used again in Section 15.9.1.

15.6.3 Gaudry's algorithm

Gaudry [223] considered the algorithm of Section 15.6.1 for fixed genus, rather than asympotically as $g \to \infty$. In particular, he chose the smoothness bound $b = 1$ (so the factor base $\mathcal{B}$ only consists of degree 1 prime divisors, i.e. points). Good surveys of Gaudry's algorithm are given in Chapter VII of [61] and Section 21.2 of [16].

Exercise 15.6.4 Let C be a hyperelliptic curve of genus g over a finite field $\mathbb{F}_q$. Show that the number of prime divisors on C of degree 1 is $\#C(\mathbb{F}_q) = q(1 + o(1))$ for fixed g as $q \to \infty$. Hence, show that the probability that a randomly chosen reduced divisor is 1-smooth is $\frac{1}{g!}(1 + o(1))$ as $q \to \infty$.

Exercise 15.6.5 Following Exercise 15.6.4, it is natural to conjecture that one needs to choose $O(g!q(1 + o(1)))$ divisors (again, this is for fixed g as $q \to \infty$, in which case it is more common to write it as $O(q(1 + o(1))))$ to find enough relations to have a non-trivial linear dependence in $\mathcal{B}$. Under this assumption, show that the heuristic expected running time of Gaudry's algorithm is at most

$$c_1 g^2 g! q(1 + o(1)) M(\log(q)) + c_2 g^3 q^2 M(\log(q))) = O(q^2 M(\log(q))(1 + o(1))) \quad (15.11)$$

bit operations (for some constants c_1 and c_2) for fixed g as $q \to \infty$.

The first term in equation (15.11) is the running time for relation generation. If g is fixed then asymptotically this is dominated by the second term, which is the running time for the linear algebra stage. If g is fixed, then the running time is $\tilde{O}(q^2)$ bit operations. Hence Gaudry's algorithm is asymptotically faster than Pollard's rho method for hyperelliptic curves of a fixed genus $g \geq 5$. However, the hidden constant in the expression $\tilde{O}(q^2)$ depends very badly on g. In practice, Gaudry's method seems to be superior to rho for small g (e.g., $g = 5, 6, 7$).

Harley and Thériault (see [545]) suggested reducing the factor base size in Gaudry's algorithm in order to balance the running times of the relation generation and linear algebra stages. Thériault [545] also proposed a "large prime" variant of Gaudry's algorithm. Gaudry, Thériault, Thomé and Diem [229] proposed a "double large prime" variant of Gaudry's algorithm that is based on the double large prime strategy that was successful in accelerating integer factorisation algorithms. The factor base $\mathcal{B}$ is now chosen to be a subset of the degree 1 divisors, and degree 1 divisors that are not in $\mathcal{B}$ are called *large primes*. A divisor is defined to be smooth if it can be written as a sum of prime divisors and at most two large primes. Relations are collected as before, and then combined to eliminate the large primes (we refer to Section 21.3 of [16] for further discussion of large primes and graph methods for eliminating them). It is shown in [229] that, for fixed g, the expected running time of the algorithm is $\tilde{O}(q^{2-\frac{2}{g}})$ bit operations. This is faster than Pollard rho for $g \geq 3$ when q is sufficiently large. Gaudry's approach was generalised to all curves of fixed genus by Diem [164].

15.7 Weil descent

As we have seen, there are subexponential algorithms for the DLP in the divisor class group of a hyperelliptic curve of high genus. A natural approach to solve the DLP on elliptic curves is therefore to transform the problem into a DLP on a high genus curve. However, the naive way to do this embeds a small problem into a big one, and does not help to solve the DLP. Frey proposed[10] to use Weil restriction of scalars to transform the DLP on an elliptic curve $E(\mathbb{F}_{q^n})$ for $n > 1$ to the DLP on a curve of genus $g \geq n$ over $\mathbb{F}_q$. Frey called this idea **Weil descent**.

Geometrically, the principle is to identify the Weil restriction of an open affine subset of $E(\mathbb{F}_{q^n})$ (see Section 5.7) with an open affine subset of an Abelian variety A over $\mathbb{F}_q$ of dimension n. One can then try to find a curve C on A so that there is a map from the Jacobian of C to A. Following Gaudry, Hess and Smart [226] it is more convenient to express the situation in terms of function fields and divisor class groups. We only sketch the details since an excellent survey is provided by Hess in Chapter VIII of [61] and many important details are explained by Diem in [159].

[10] The standard reference is a lecture given by Frey at the ECC 1998 conference. His talk was mostly about a different (constructive) application of Weil restriction of scalars. However, he did mention the possibility of using this idea for an attack. Galbraith and Smart developed the details further in [212] and many works followed.

Let E be an elliptic curve over $\mathbb{K} = \mathbb{F}_{q^n}$ and let $\Bbbk = \mathbb{F}_q$. The function field of E is $\mathbb{K}(E)$. The idea (called in this setting a **covering attack**) is to find a curve C over $\mathbb{K}$ such that $\mathbb{K}(C)$ is a finite extension of $\mathbb{K}(E)$ (so that there is a map $C \to E$ of finite degree) and such that there is an automorphism σ of degree n on $\mathbb{K}(C)$ extending the q-power Frobenius so that the fixed field of $\mathbb{K}(C)$ under $\langle\sigma\rangle$ is $\Bbbk(C^0)$ for some curve C^0. The composition of the conorm map from $E(\mathbb{K})$ to $\mathrm{Pic}^0_C(\mathbb{K})$ and the norm map from $\mathrm{Pic}^0_C(\mathbb{K})$ to $\mathrm{Pic}^0_{C^0}(\Bbbk)$ transfers the DLP from $E(\mathbb{K})$ to $\mathrm{Pic}^0_{C^0}(\Bbbk)$. Hence, as long as the composition of these maps is not trivial, then one has reduced the DLP from $E(\mathbb{K})$ to the divisor class group of a curve C^0 over $\Bbbk$. One can then solve the DLP using an index calculus algorithm, which is feasible if the genus of C^0 is not too large.

A variant of the Weil descent concept that avoids function fields and divisor class groups is to perform index calculus directly on Abelian varieties. This variant is the subject of the following section.

15.8 Discrete logarithms on elliptic curves over extension fields

We now discuss some related algorithms, which can be applied to elliptic curves over extension fields. We start by recalling Semaev's idea of summation polynomials.

15.8.1 Semaev's summation polynomials

Suppose that E is an elliptic curve defined over a prime field $\mathbb{F}_p$, and that elements of $\mathbb{F}_p$ are represented as integers in the interval $[0, p-1]$. Semaev [486] considered a factor base

$$\mathcal{B} = \{(x, y) \in E(\mathbb{F}_p) \,:\, 0 \le x \le p^{1/n}\}$$

for some fixed integer $n \ge 2$. Note that $\#\mathcal{B} \approx p^{1/n}$.

Semaev hoped to perform an index calculus algorithm similar to the one in Section 15.5.1. For random points $R = [a]P + [b]Q$ the task is to write R as a sum of points in $\mathcal{B}$. To accomplish this, Semaev introduced the notion of a summation polynomial.

Definition 15.8.1 Let $E : y^2 = x^3 + a_4x + a_6$ be an elliptic curve defined over $\mathbb{F}_q$, where the characteristic of $\mathbb{F}_q$ is neither 2 nor 3 (this condition can be avoided). The **summation polynomials** $\mathrm{Summ}_n \in \mathbb{F}_q[x_1, x_2, \dots, x_n]$ for $n \ge 2$ are defined as follows:

- $\mathrm{Summ}_2(x_1, x_2) = x_1 - x_2$.
- $\mathrm{Summ}_3(x_1, x_2, x_3) = (x_1 - x_2)^2x_3^2 - 2((x_1 + x_2)(x_1x_2 + a_4) + 2a_6)x_3 + ((x_1x_2 - a_4)^2 - 4a_6(x_1 + x_2))$.
- $\mathrm{Summ}_n(x_1, x_2, \dots, x_n) = R_x(\mathrm{Summ}_{n-1}(x_1, \dots, x_{n-2}, x), \mathrm{Summ}_3(x_{n-1}, x_n, x))$ for $n \ge 4$ where $R_x(F, G)$ is the resultant of the polynomials F and G with respect to the variable x.

For many more details see Section 3 of [163]. The following result is from [486].

Theorem 15.8.2 *Summation polynomials have the following properties:*

- $(x_1, \dots, x_n) \in \overline{\mathbb{F}}_q^n$ *is a root of* Summ_n *if and only if there exists* $(y_1, \dots, y_n) \in \overline{\mathbb{F}}_q^n$ *such that* $P_i = (x_i, y_i) \in E(\overline{\mathbb{F}}_q)$ *and* $\sum_{i=1}^n P_i = \infty$.
- Summ_n *is symmetric.*
- *The degree of* Summ_n *in* x_i *is* 2^{n-2}.

Exercise 15.8.3 Prove Theorem 15.8.2.

One way to decompose $R = (x_R, y_R)$ in $\mathcal{B}$ is to find solutions $(x_1, \dots, x_n) \in \mathbb{Z}^n$ to

$$\mathrm{Summ}_{n+1}(x_1, x_2, \dots, x_n, x_R) \equiv 0 \pmod{p}, \text{ such that } 0 \le x_i \le p^{1/n}. \tag{15.12}$$

If such a solution exists and can be found then one finds the corresponding y-coordinates $\pm y_i$. Suppose that each $y_i \in \mathbb{F}_p$. Then each $P_i = (x_i, y_i)$ is in $\mathcal{B}$ and by Theorem 15.8.2 there exist $s_i \in \{-1, 1\}$ such that $s_1 P_1 + \cdots + s_n P_n = R$. The sign bits s_i can be found by exhaustive search, thereby yielding a relation. Since $\#\{P_1 + P_2 + \cdots + P_n : P_i \in \mathcal{B}\} \approx (p^{1/n})^n/n! = p/n!$ the expected number of points R that have to be selected before a relation is obtained is about $n!$.

Unfortunately, no efficient algorithm is known for solving the polynomial equation (15.12) even for $n = 5$ (in which case the equation has degree 16 in each of its 5 variables). Coppersmith's method (see Section 19.2) seems not to be useful for this task.

In reference to the remarks of Section 15.2.3 we see that all requirements for an index calculus algorithm are met, except that it is not efficient to decompose a smooth element over the factor base.

15.8.2 Gaudry's variant of Semaev's method

Gaudry [225] realised that it might be possible to take roots of summation polynomials if one is working with elliptic curves over extension fields. Gaudry's algorithm may be viewed as doing Weil descent without divisor class groups. Indeed, the paper [225] presents a general approach to index calculus on Abelian varieties and so the results apply in greater generality than just Weil descent of elliptic curves.

Suppose that E is an elliptic curve defined over a finite field $\mathbb{F}_{q^n}$ with $n > 1$. Gaudry [225] defines a factor base

$$\mathcal{B} = \{(x, y) \in E(\mathbb{F}_{q^n}) : x \in \mathbb{F}_q\}$$

so that $\#\mathcal{B} \approx q$. Gaudry considers this as the set of $\mathbb{F}_q$-rational points on the algebraic set F formed by intersecting the Weil restriction of scalars of E with respect to $\mathbb{F}_{q^n}/\mathbb{F}_q$ by $n-1$ hyperplanes $V(x_i)$ for $2 \le i \le n$, where $x = x_1\theta_1 + \cdots + x_n\theta_n$ (with $\theta_1 = 1$) as in Lemma 5.7.1. If the algebraic set F is irreducible then it is a one-dimensional variety F.

In the relation generation stage, one attempts to decompose a randomly selected point $R \in E(\mathbb{F}_{q^n})$ as a sum of points in $\mathcal{B}$. Gaudry observed that this can be accomplished by finding solutions

$$(x_1, x_2, \ldots, x_n) \in \mathbb{F}_q^n \quad \text{such that} \quad \text{Summ}_{n+1}(x_1, x_2, \ldots, x_n, x_R) = 0. \tag{15.13}$$

Note that $\text{Summ}_{n+1}(x_1, \ldots, x_n, x_R) \in \mathbb{F}_{q^n}[x_1, \ldots, x_n]$ since E is defined over $\mathbb{F}_{q^n}$ and $x_R \in \mathbb{F}_{q^n}$. The conditions $x_j \in \mathbb{F}_q$ in equation (15.13) can be expressed algebraically as follows. Select a basis $\{\theta_1, \ldots, \theta_n\}$ for $\mathbb{F}_{q^n}$ over $\mathbb{F}_q$ and write

$$\text{Summ}_{n+1}(x_1, \ldots, x_n, x_R) = \sum_{i=1}^{n} G_i(x_1, \ldots, x_n)\theta_i \tag{15.14}$$

where $G_i(x_1, \ldots, x_n) \in \mathbb{F}_q[x_1, \ldots, x_n]$. Note that the degree of G_i in x_j is at most 2^{n-1}. The polynomials G_i of equation (15.14) define an algebraic set in $X \subseteq \mathbb{A}^n$ and we are interested in the points in $X(\mathbb{F}_q)$ (if there are any). Since $\mathbb{F}_q$ is finite there are only finitely many $\mathbb{F}_q$-rational solutions $(x_1, \ldots, x_n)$ to the system.

Gaudry assumes that X is generically a zero-dimensional algebraic set (Gaudry justifies this assumption by noting that if F is a variety then the variety F^n is n-dimensional, and so the map from F^n to the Weil restriction of E, given by adding together n points in F, is a morphism between varieties of the same dimension, and so generically has finite degree). The $\mathbb{F}_q$-rational solutions can therefore be found by finding a Gröbner basis for the ideal generated by the G_i and then taking roots in $\mathbb{F}_q$ of a sequence of univariate polynomials each of which has degree at most $2^{n(n-1)}$. This is predicted to take $O(2^{cn(n-1)}M(\log(q)))$ bit operations for some constant c. Alternatively, one could add some field equations $x_j^q - x_j$ to the ideal, to ensure it is zero-dimensional, but this could have an adverse effect on the complexity. Gaudry makes a further heuristic assumption, namely that the smoothness probability behaves as expected when using the large prime variant.

The size of the set $\{P_1 + P_2 + \cdots + P_n : P_i \in \mathcal{B}\}$ is approximately $q^n/n!$ and so the expected number of points R that have to be selected before a relation is obtained is about $n!$. One needs approximately $\#\mathcal{B} \approx q$ relations to be able to find a non-trivial element in the kernel of the relation matrix and hence integers a and b such that $[a]D_1 + [b]D_2 \equiv 0$. It follows that the heuristic expected running time of Gaudry's algorithm is

$$\tilde{O}(2^{cn(n-1)}n!qM(\log(q)) + q^{2+o(1)}) \tag{15.15}$$

bit operations as $q \to \infty$. This is exponential in terms of n and $\log(q)$. However, for fixed n, the running time can be expressed as $\tilde{O}(q^2)$ bit operations.

Gaudry's focus was on n fixed and relatively small. For any fixed $n \geq 5$ Gaudry's heuristic algorithm for solving the ECDLP over $\mathbb{F}_{q^n}$ is asymptotically faster than Pollard's rho method. The double large prime variant (mentioned in Section 15.6.3) can also be used in this setting. The complexity therefore becomes (heuristic) $\tilde{O}(q^{2-\frac{2}{n}})$ bit operations. Hence, Gaudry's algorithm is asymptotically faster than Pollard rho even for $n = 3$ and

$n = 4$, namely $\tilde{O}(q^{4/3})$ rather than $\tilde{O}(q^{3/2})$ for $n = 3$ and $\tilde{O}(q^{3/2})$ rather than $\tilde{O}(q^2)$ for $n = 4$.

15.8.3 Diem's algorithm for the ECDLP

Gaudry's focus was on the DLP in $E(\mathbb{F}_{q^n})$ when n is fixed. This yields an exponential-time algorithm. Diem [160, 163] considered the case where n is allowed to grow, and obtained a subexponential-time algorithm.

The crux of Diem's method is remarkably simple: he assumes $n \approx \sqrt{\log(q)}$ and obtains an algorithm for the DLP in $E(\mathbb{F}_{q^n})$ with complexity $O(q^c)$ for some constant c (note that even some exponential-time computations in n are polynomial in q as $e^{n^2} \approx q$). Now, $q^c = \exp(c\log(q))$ and $\log(q^n) = n\log(q) \approx \log(q)^{3/2}$ so $q^c \approx \exp(c\log(q^n)^{2/3}) < L_{q^n}(2/3, c)$.

Diem's algorithm is very similar to Gaudry's. In Gaudry's algorithm, the factor base consists of points whose x-coordinates lie in $\mathbb{F}_q$. Diem defines a function $\varphi = \alpha \circ x$, where α is an automorphism over $\mathbb{F}_{q^n}$ of $\mathbb{P}^1$ that satisfies a certain condition, and defines the factor base to be $\mathcal{B} = \{P \in E(\mathbb{F}_{q^n}) : \varphi(P) \in \mathbb{P}^1(\mathbb{F}_q)\}$. The process of generating relations proceeds in the standard way. Some important contributions of [163] are to prove that the algebraic set defined by the summation polynomials has a good chance of having dimension 0, and that when this is the case the points can be found by taking resultants of multi-homogeneous polynomials in time polynomial in $e^{n^2}\log(q)$ (which is exponential in n but polynomial in q).

The main result of [163] is the following. We stress that this result does not rely on any heuristics.

Theorem 15.8.4 *(Diem) Let $a, b \in \mathbb{R}$ be such that $0 < a < b$. There is an algorithm for the DLP in $\mathbb{F}_{q^n}^*$ such that if q is a prime power and $n \in \mathbb{N}$ is such that*

$$a\sqrt{\log(q)} \le n \le b\sqrt{\log(q)}$$

then the algorithm solves the DLP in $\mathbb{F}_{q^n}^*$ in an expected $e^{O(\log(q^n)^{2/3})}$ bit operations.

15.9 Further results

To end the chapter we briefly mention some methods for non-hyperelliptic curves. It is beyond the scope of the book to present these algorithms in detail. We then briefly summarise the argument that there is no subexponential algorithm for the DLP on elliptic curves in general.

15.9.1 Diem's algorithm for plane curves of low degree

Diem [161] used the Adleman–DeMarrais–Huang idea of generating relations using principal divisors $a(x) - yb(x)$ for the DLP on plane curves $F(x, y) = 0$ of low degree (the degree of such a curve is the total degree of $F(x, y)$ as a polynomial). Such curves are essentially the opposite case to hyperelliptic curves (which have rather high degree in x relative to their genus). The trick is simply to note that if $F(x, y)$ has relatively low degree compared to its genus then so does $b(x)^d F(x, a(x))$ and so the divisor of the function $a - by$ has relatively low weight. The main result is an algorithm with heuristic complexity $\tilde{O}(q^{2-2/(d-2)})$ bit operations for a curve of degree d over $\mathbb{F}_q$.

In the case of non-singular plane quartics (genus 3 curves C over $\mathbb{F}_q$) Diem takes the factor base to be a large set of points $\mathcal{B} \subseteq C(\mathbb{F}_q)$. He generates relations by choosing two distinct points $P_1, P_2 \in \mathcal{B}$ and intersecting the line $y = bx + c$ between them with the curve C. There are two other points of intersection, corresponding to the roots of the quadratic polynomial $F(x, bx + c)/((x - x_{P_1})(x - x_{P_2}))$ and so with probability roughly $1/2$ we expect to get a relation in the divisor class group among points in $C(\mathbb{F}_q)$. Diem shows that the algorithm has complexity $\tilde{O}(q)$ bit operations.

Due to lack of space, and since our focus in this book is hyperelliptic curves (though, it is important to note that Smith [514] has given a reduction of the DLP from hyperelliptic curves of genus 3 to plane quartics), we do not present any further details. Interested readers should see [161, 165].

15.9.2 The algorithm of Enge–Gaudry–Thomé and Diem

The algorithms for the DLP in the divisor class group of a hyperelliptic curve in Sections 15.6.1 and 15.6.2 had complexity $L_{q^g}(1/2, \sqrt{2} + o(1))$ bit operations as $q \to \infty$. A natural problem is to find algorithms with complexity $L_{q^g}(1/3, c + o(1))$, and this is still open in general. However, an algorithm is known for curves of the form $y^n + F(x, y) = 0$ where $\deg_y(F(x, y)) \le n - 1$ and $\deg_x(F(x, y)) = d$ for $n \approx g^{1/3}$ and $d \approx g^{2/3}$. We do not have space to give the details, so simply quote the results and refer to Enge and Gaudry [181], Enge, Gaudry and Thomé [182] and Diem [160]. An algorithm to compute the group structure of $\mathrm{Pic}^0_C(\mathbb{F}_q)$ is given with heuristic complexity of $O(L_{q^g}(1/3, c + o(1)))$ bit operations for some constant c. For the discrete logarithm problem the algorithm has heuristic complexity $O(L_{q^g}(1/3, c' + o(1)))$ bit operations where c' is a constant.

Unlike the $L_N(1/3, c + o(1))$ algorithms for factoring or DLP in finite fields, the algorithm does not use two different factor bases. Instead, the algorithm is basically the same idea as Sections 15.6.2 and 15.9.1 with a complexity analysis tailored for curves of a certain form.

15.9.3 Index calculus for general elliptic curves

In this section we briefly discuss why there does not seem to be a subexponential algorithm for the DLP on general elliptic curves.

An approach to an index calculus algorithm for elliptic curves was already discussed by Miller [384] in the paper that first proposed elliptic curves for cryptography. In particular, he considered "lifting" an elliptic curve E over $\mathbb{F}_p$ to an elliptic curve $\widetilde{E}$ over $\mathbb{Q}$ (i.e., so that reducing the coefficients of $\widetilde{E}$ modulo p yields E). The factor base $\mathcal{B}$ was defined to be the points of small height (see Section VIII.6 of [505] for details of heights) in $\widetilde{E}(\mathbb{Q})$. The theory of descent (see Chapter VIII of Silverman [505]) essentially gives an algorithm to decompose a point as a sum of points of small height (when this is possible). The idea would therefore be to take random points $[a]P + [b]Q \in E(\mathbb{F}_p)$, lift them to $\widetilde{E}(\mathbb{Q})$ and then decompose them over the factor base. There are several obstructions to this method. First, lifting a random point from $E(\mathbb{F}_p)$ to $\widetilde{E}(\mathbb{Q})$ seems to be hard in general. Indeed, Miller argued (see also [507]) that there are very few points of small height in $\widetilde{E}(\mathbb{Q})$ and so (since we are considering random points $[a]P + [b]Q$ from the exponentially large set $E(\mathbb{F}_p)$) it would be necessary to lift to exponentially large points in $\widetilde{E}(\mathbb{Q})$. Second, the lifting itself seems to be a non-trivial computational task (essentially, solving a non-linear diophantine equation over $\mathbb{Z}$).

Silverman propsed the **Xedni calculus** attack,[11] which was designed to solve the lifting problem. This algorithm was analysed in [288], where it is shown that the probability of finding useful relations is too low.

By now, many people have tried and failed to discover an index calculus algorithm for the DLP on general elliptic curves. However, this does not prove that no such algorithm exists, or that a different paradigm could not lead to faster attacks on the elliptic curve DLP.

[11] "Xedni" is "Index" spelled backwards.

PART IV
LATTICES

16
Lattices

The word "lattice" has two different meanings in mathematics. One meaning is related to the theory of partial orderings on sets (for example, the lattice of subsets of a set). The other meaning, which is the one relevant to us, is discrete subgroups of $\mathbb{R}^n$.

There are several reasons for presenting lattices in this book. First, there are hard computational problems on lattices that have been used as a building block for public key cryptosystems (e.g., the Goldreich–Goldwasser–Halevi (GGH) cryptosystem, the NTRU cryptosystem, the Ajtai–Dwork cryptosystem and the LWE cryptosystem); however, we do not present these applications in this book. Second, lattices are used as a fundamental tool for cryptanalysis of public key cryptosystems (e.g., lattice attacks on knapsack cryptosystems, Coppersmith's method for finding small solutions to polynomial equations, attacks on signatures and attacks on variants of RSA). Third, there are applications of lattices to efficient implementation of discrete logarithm systems (such as the GLV method; see Section 11.3.3). Finally, lattices are used as a theoretical tool for security analysis of cryptosystems, for example the bit security of Diffie–Hellman key exchange using the hidden number problem (see Section 21.7) and the security proofs for RSA-OAEP.

Some good references for lattices, applications of lattices and/or lattice reduction algorithms are: Cassels [114], Siegel [504], Cohen [127], von zur Gathen and Gerhard [220], Grötschel, Lovász and Schrijver [245], Nguyen and Stern [414, 415], Micciancio and Goldwasser [378], Hoffstein, Pipher and Silverman [261], Lenstra's chapter in [106], Micciancio and Regev's chapter in [48] and the proceedings of the conference LLL+25.

Notation used in this part

$\mathbb{Z}$, $\mathbb{Q}$, $\mathbb{R}$	Integers, rational, real numbers
$\underline{b}$, $\underline{v}$, $\underline{w}$	Row vectors (usually in $\mathbb{R}^m$)
$\underline{0}$	Zero vector in $\mathbb{R}^m$
$\underline{e}_i$	ith unit vector in $\mathbb{R}^m$
I_n	$n \times n$ identity matrix
$\langle \underline{x}, \underline{x} \rangle$	Inner product
$\Vert\underline{x}\Vert$	Euclidean length (ℓ_2 norm)
$\Vert \cdot \Vert_a$	ℓ_a-norm for $a \in \mathbb{N}$

$\text{span}\{\underline{v}_1, \dots, \underline{v}_n\}$	Span of a set of vectors over $\mathbb{R}$
$\text{rank}(A)$	Rank of a matrix A
$\lfloor x \rceil$	Closest integer to x, $\lfloor 1/2 \rceil = 1$
B	Basis matrix for a lattice
L	Lattice
$\underline{b}_i^*$	Gram–Schmidt vector arising from ordered basis $\{\underline{b}_1, \dots, \underline{b}_n\}$
$\mu_{i,j}$	Gram–Schmidt coefficient $\langle \underline{b}_i, \underline{b}_j^* \rangle / \langle \underline{b}_j^*, \underline{b}_j^* \rangle$
B_i	$\Vert \underline{b}_i^* \Vert^2$
λ_i	Successive minima of a lattice
$\det(L)$	Determinant of a lattice
γ_n	Hermite's constant
X	Bound on the size of the entries in the basis matrix L
$B_{(i)}$	$i \times m$ matrix formed by the first i rows of B
d_i	Determinant of matrix of $\langle \underline{b}_j, \underline{b}_k \rangle$ for $1 \le j, k \le i$
D	Product of d_i
$\mathcal{P}_{1/2}(B)$	Fundamental domain (parallelepiped) for lattice basis B
$F(x), F(x, y)$	Polynomial with "small" root
$G(x), G(x, y)$	Polynomial with "small" root in common with $F(x)$ (resp., $F(x, y)$)
X, Y	Bounds on size of root in Coppersmith's method
b_F	Coefficient vector of polynomial F
$R(F, G), R_x(F(x), G(x))$	Resultant of polynomials
W	Bound in Coppersmith's method
P, R	Constants in noisy Chinese remaindering
$\text{amp}(x)$	The amplitude $\gcd(P, x - R)$ in noisy Chinese remaindering

16.1 Basic notions on lattices

A lattice is a subset of the vector space $\mathbb{R}^m$. We write all vectors as **rows**; be warned that many books and papers write lattice vectors as columns. We denote by $\Vert \underline{v} \Vert$ the Euclidean norm of a vector $\underline{v} \in \mathbb{R}^m$; though some statements also hold for other norms.

Definition 16.1.1 Let $\{\underline{b}_1, \dots, \underline{b}_n\}$ be a linearly independent set of (row) vectors in $\mathbb{R}^m$ ($m \ge n$). The **lattice** generated by $\{\underline{b}_1, \dots, \underline{b}_n\}$ is the set

$$L = \left\{ \sum_{i=1}^{n} l_i \underline{b}_i \;:\; l_i \in \mathbb{Z} \right\}$$

of **integer** linear combinations of the $\underline{b}_i$. The vectors $\underline{b}_1, \dots, \underline{b}_n$ are called a **lattice basis**. The **lattice rank** is n and the **lattice dimension** is m. If $n = m$ then L is said to be a **full rank lattice**.

Let $L \subset \mathbb{R}^m$ be a lattice. A **sublattice** is a subset $L' \subset L$ that is a lattice.

A **basis matrix** B of a lattice L is an $n \times m$ matrix formed by taking the rows to be basis vectors $\underline{b}_i$. Thus $B_{i,j}$ is the jth entry of the row $\underline{b}_i$ and

$$L = \{\underline{x}B : \underline{x} \in \mathbb{Z}^n\}.$$

By assumption, the rows of a basis matrix are always linearly independent.

Example 16.1.2 The lattice in $\mathbb{R}^2$ generated by $\{(1, 0), (0, 1)\}$ is $L = \mathbb{Z}^2$. The corresponding basis matrix is $B = \left(\begin{smallmatrix} 1 & 0 \\ 0 & 1 \end{smallmatrix}\right)$. Any 2×2 integer matrix B of determinant ± 1 is also a basis matrix for L.

We will mainly assume that the basis vectors $\underline{b}_i$ for a lattice have integer entries. In cryptographic applications, this is usually the case. We interchangeably use the words **points** and **vectors** for elements of lattices. The vectors in a lattice form an Abelian group under addition. When $n \geq 2$ there are infinitely many choices for the basis of a lattice.

An alternative approach to lattices is to define $L = \mathbb{Z}^n$ and to have a general length function $q(\underline{v})$. One finds this approach in books on quadratic forms or optimisation problems, e.g. Cassels [113] and Schrijver [478]. In particular, Section 6.2 of [478] presents the LLL algorithm in the context of reducing the lattice $L = \mathbb{Z}^n$ with respect to a length function corresponding to a positive-definite rational matrix.

We now give an equivalent definition of lattice, which is suitable for some applications. A subset $L \subseteq \mathbb{R}^m$ is called **discrete** if, for any real number $r > 0$, the set $\{\underline{v} \in L : \|\underline{v}\| \leq r\}$ is finite. It is clear that a lattice is a subgroup of $\mathbb{R}^m$ that is discrete. The following result shows the converse.

Lemma 16.1.3 *Every discrete subgroup of* $\mathbb{R}^m$ *is a lattice.*

Proof (Sketch) Let $\{\underline{v}_1, \ldots, \underline{v}_n\}$ be a linearly independent subset of L of maximal size. The result is proved by induction. The case $n = 1$ is easy (since L is discrete there is an element of minimal non-zero length). When $n > 1$ consider $V = \text{span}\{\underline{v}_1, \ldots, \underline{v}_{n-1}\}$ and set $L' = L \cap V$. By induction, L' is a lattice and so has a basis $\underline{b}_1, \ldots, \underline{b}_{n-1}$. The set $L \cap \{\sum_{i=1}^{n-1} x_i \underline{b}_i + x_n \underline{v}_n : 0 \leq x_i < 1 \text{ for } 1 \leq i \leq n-1 \text{ and } 0 < x_n \leq 1\}$ is finite and so has an element with smallest x_n, call it $\underline{b}_n$. It can be shown that $\{\underline{b}_1, \ldots, \underline{b}_n\}$ is a basis for L. For full details see Theorem 6.1 of [527]. □

Exercise 16.1.4 Given an $m \times n$ integer matrix A show that $\ker(A) = \{\underline{x} \in \mathbb{Z}^m : \underline{x}A = \underline{0}\}$ is a lattice. Show that the rank of the lattice is $m - \text{rank}(A)$. Given an $m \times n$ integer matrix A and an integer M show that $\{\underline{x} \in \mathbb{Z}^m : \underline{x}A \equiv \underline{0} \pmod{M}\}$ is a lattice of rank m.

In the case $m > n$ it is sometimes convenient to project the lattice L into $\mathbb{R}^n$ using the following construction. The motivation is that a linear map that preserves lengths preserves volumes. Note that if the initial basis for L consists of vectors in $\mathbb{Z}^n$ then the resulting basis does not necessarily have this property.

Lemma 16.1.5 *Let B be an $n \times m$ basis matrix for a lattice L where $m > n$. Then there is a linear map $P : \mathbb{R}^m \to \mathbb{R}^n$ such that $P(L)$ is a rank n lattice and $\|P(\underline{v})\| = \|\underline{v}\|$ for all $\underline{v} \in L$. Furthermore, $\langle \underline{b}_i, \underline{b}_j \rangle = \langle \underline{b}_i P, \underline{b}_j P \rangle$ for all $1 \le i < j \le n$.*

If the linear map is represented by an $m \times n$ matrix P then a basis matrix for the image of L under the projection P is the $n \times n$ matrix BP, which is invertible.

Proof Given the $n \times m$ basis matrix B with rows $\underline{b}_i$, define $V = \mathrm{span}\{\underline{b}_1, \ldots, \underline{b}_n\} \subset \mathbb{R}^m$, which has dimension n by assumption. Choose (perhaps by running the Gram–Schmidt algorithm) a basis $\underline{v}_1, \ldots, \underline{v}_n$ for V that is orthonormal with respect to the inner product in $\mathbb{R}^m$. Define the linear map $P : V \to \mathbb{R}^n$ by $P(\underline{v}_i) = \underline{e}_i$ and $P(V^\perp) = \{0\}$. For $\underline{v} = \sum_{i=1}^n x_i \underline{v}_i \in V$ we have $\|\underline{v}\| = \sqrt{\langle \underline{v}, \underline{v} \rangle} = \sqrt{\sum_{i=1}^n x_i^2} = \|\underline{v} P\|$. Since the vectors $\underline{b}_i$ form a basis for V, the vectors $\underline{b}_i P$ are linearly independent. Hence, BP is an invertible matrix and $P(L)$ is a lattice of rank n. □

We can now prove the following fundamental result.

Lemma 16.1.6 *Two $n \times m$ matrices B and B' generate the same lattice L if and only if B and B' are related by a* **unimodular matrix**, *i.e. $B' = UB$ where U is an $n \times n$ matrix with integer entries and determinant ± 1.*

Proof ($\Rightarrow$) Every row of B' is an integer linear combination

$$\underline{b}'_i = \sum_{j=1}^{n} u_{i,j} \underline{b}_j$$

of the rows in B. This can be represented as $B' = UB$ for an $n \times n$ integer matrix U.

Similarly, $B = U'B' = U'UB$. Now applying the projection P of Lemma 16.1.5 we have $BP = U'UBP$ and, since BP is invertible, $U'U = I_n$ (the identity matrix). Since U and U' have integer entries it follows that $\det(U), \det(U') \in \mathbb{Z}$. From $\det(U)\det(U') = \det(I_n) = 1$ it follows that $\det(U) = \pm 1$.

($\Leftarrow$) Since U is a permutation of $\mathbb{Z}^n$ we have $\{\underline{x}B' : \underline{x} \in \mathbb{Z}^n\} = \{\underline{x}B : \underline{x} \in \mathbb{Z}^n\}$. □

The Hermite normal form is defined in Section A.11. The following result is a direct consequence of Lemma 16.1.6 and the remarks in Section A.11.

Lemma 16.1.7 *If B is the basis matrix of a lattice L then the Hermite normal form of B is also a basis matrix for L.*

The **determinant** of a lattice L is the volume of the fundamental parallelepiped of any basis B for L. When the lattice has full rank then using Definition A.10.7 and Lemma A.10.8 we have $\det(L) = |\det(B)|$. For the case $n < m$ our definition uses Lemma 16.1.5.

Definition 16.1.8 Let the notation be as above. The **determinant** (or **volume**) of a lattice L with basis matrix B is $|\det(BP)|$, where P is a matrix representing the projection of Lemma 16.1.5.

Lemma 16.1.9 *The determinant of a lattice is independent of the choice of basis matrix B and the choice of projection P.*

Proof Let P and P' be two projection matrices corresponding to orthogonal bases $\{\underline{v}_1, \dots, \underline{v}_n\}$ and $\{\underline{v}'_1, \dots, \underline{v}'_n\}$ for $V = \text{span}\{\underline{b}_1, \dots, \underline{b}_n\}$. Then, by Lemma A.10.3, $P' = PW$ for some orthogonal matrix W (hence, $\det(W) = \pm 1$). It follows that $|\det(BP)|$ does not depend on the choice of P.

Let B and B' be two basis matrices for a lattice L. Then $B' = UB$ where U is an $n \times n$ matrix such that $\det(U) = \pm 1$. Then $\det(L) = |\det(BP)| = |\det(UBP)| = |\det(B'P)|$. □

We have seen that there are many different choices of basis for a given lattice L. A fundamental problem is to compute a "nice" lattice basis for L; specifically one where the vectors are relatively short and close to orthogonal. The following exercise shows that these properties are intertwined.

Exercise 16.1.10 Let L be a rank 2 lattice in $\mathbb{R}^2$ and let $\{\underline{b}_1, \underline{b}_2\}$ be a basis for L.

1. Show that
$$\det(L) = \|\underline{b}_1\| \|\underline{b}_2\| |\sin(\theta)| \tag{16.1}$$
where θ is the angle between $\underline{b}_1$ and $\underline{b}_2$.
2. Hence, deduce that the product $\|\underline{b}_1\| \|\underline{b}_2\|$ is minimised over all choices $\{\underline{b}_1, \underline{b}_2\}$ of basis for L when the angle θ is closest to $\pm\pi/2$.

Definition 16.1.11 Let L be a lattice in $\mathbb{R}^m$ of rank n with basis matrix B. The **Gram matrix** of B is BB^T. This is an $n \times n$ matrix whose (i, j)th entry is $\langle \underline{b}_i, \underline{b}_j \rangle$.

Lemma 16.1.12 *Let L be a lattice in $\mathbb{R}^m$ of rank n with basis matrix B. Then* $\det(L) = \sqrt{\det(BB^T)}$.

Proof Consider first the case $m = n$. Then $\det(L)^2 = \det(B)\det(B^T) = \det(BB^T) = \det((\langle \underline{b}_i, \underline{b}_j \rangle)_{i,j})$. Hence, when $m > n$, $\det(L) = \sqrt{\det(B'(B')^T)}$. Now, the (i, j)th entry of $B'(B')^T = (BP)(BP)^T$ is $\langle \underline{b}_i P, \underline{b}_j P \rangle$, which is equal to the (i, j)th entry of BB^T by Lemma 16.1.5. □

Note that an integer lattice of non-full rank may not have integer determinant.

Exercise 16.1.13 Find an example of a lattice of rank 1 in $\mathbb{Z}^2$ whose determinant is not an integer.

Lemma 16.1.14 *Let $\underline{b}_1, \dots, \underline{b}_n$ be an ordered basis for a lattice L in $\mathbb{R}^m$ and let $\underline{b}_1^*, \dots, \underline{b}_n^*$ be the Gram–Schmidt orthogonalisation. Then* $\det(L) = \prod_{i=1}^n \|\underline{b}_i^*\|$.

Proof The case $m = n$ is already proved in Lemma A.10.8. For the general case let $\underline{v}_i = \underline{b}_i^* / \|\underline{b}_i^*\|$ be the orthonormal basis required for the construction of the projection P. Then $P(\underline{b}_i^*) = \|\underline{b}_i^*\| \underline{e}_i$. Write B and B^* for the $n \times m$ matrices formed by the rows $\underline{b}_i$ and

$\underline{b}_i^*$ respectively. It follows that B^*P is an $n \times n$ diagonal matrix with diagonal entries $\|\underline{b}_i^*\|$. Finally, by the Gram–Schmidt construction, $B^* = UB$ for some $n \times n$ matrix U such that $\det(U) = 1$. Combining these facts gives[1]

$$\det(L) = |\det(BP)| = |\det(UBP)| = |\det(B^*P)| = \prod_{i=1}^{n} \|\underline{b}_i^*\|. \qquad \square$$

Exercise 16.1.15 Let $\{\underline{b}_1, \ldots, \underline{b}_n\}$ be an ordered lattice basis in $\mathbb{R}^m$ and let $\{\underline{b}_1^*, \ldots, \underline{b}_n^*\}$ be the Gram–Schmidt orthogonalisation. Show that $\|\underline{b}_i\| \geq \|\underline{b}_i^*\|$ and hence $\det(L) \leq \prod_{i=1}^{n} \|\underline{b}_i\|$.

Definition 16.1.16 Let $\{\underline{b}_1, \ldots, \underline{b}_n\}$ be a basis for a lattice L in $\mathbb{R}^m$. The **orthogonality defect** of the basis is

$$\left(\prod_{i=1}^{n} \|\underline{b}_i\| \right) / \det(L).$$

Exercise 16.1.17 Show that the orthogonality defect of $\{\underline{b}_1, \ldots, \underline{b}_n\}$ is 1 if and only if the basis is orthogonal.

Definition 16.1.18 Let $L \subset \mathbb{R}^m$ be a lattice of rank n. The **successive minima** of L are $\lambda_1, \ldots, \lambda_n \in \mathbb{R}$ such that, for $1 \leq i \leq n$, λ_i is minimal such that there exist i linearly independent vectors $\underline{v}_1, \ldots, \underline{v}_i \in L$ with $\|\underline{v}_j\| \leq \lambda_i$ for $1 \leq j \leq i$.

It follows that $0 < \lambda_1 \leq \lambda_2 \cdots \leq \lambda_n$. In general, there is not a basis consisting of vectors whose lengths are equal to the successive minima, as the following example shows.

Example 16.1.19 Let $L \subset \mathbb{Z}^n$ be the set

$$L = \{(x_1, \ldots, x_n) : x_1 \equiv x_2 \equiv \cdots \equiv x_n \pmod{2}\}.$$

It is easy to check that this is a lattice. The vectors $2\underline{e}_i \in L$ for $1 \leq i \leq n$ are linearly independent and have length 2. Every other vector $\underline{x} \in L$ with even entries has length ≥ 2. Every vector $\underline{x} \in L$ with odd entries has all $x_i \neq 0$ and so $\|\underline{x}\| \geq \sqrt{n}$.

If $n = 2$ the successive minima are $\lambda_1 = \lambda_2 = \sqrt{2}$ and if $n = 3$ the successive minima are $\lambda_1 = \lambda_2 = \lambda_3 = \sqrt{3}$. When $n \geq 4$ then $\lambda_1 = \lambda_2 = \cdots = \lambda_n = 2$. For $n \leq 4$ one can construct a basis for the lattice with vectors of lengths equal to the successive minima. When $n > 4$ there is no basis for L consisting of vectors of length 2, since a basis must contain at least one vector having odd entries.

Exercise 16.1.20 For $n = 2, 3, 4$ in Example 16.1.19 write down a basis for the lattice consisting of vectors of lengths equal to the successive minima.

Exercise 16.1.21 For $n > 4$ in Example 16.1.19 show there is a basis for the lattice such that $\|\underline{b}_i\| = \lambda_i$ for $1 \leq i < n$ and $\|\underline{b}_n\| = \sqrt{n}$.

[1] The formula $BP = U^{-1}(B^*P)$ is the QR decomposition of BP.

Definition 16.1.22 Let $L \subseteq \mathbb{R}^m$ be a lattice and write $V \subseteq \mathbb{R}^m$ for the $\mathbb{R}$-vector space spanned by the vectors in L. The **dual lattice** of L is $L^* = \{\underline{y} \in V : \langle \underline{x}, \underline{y} \rangle \in \mathbb{Z}$ for all $\underline{x} \in L\}$.

Exercise 16.1.23 Show that the dual lattice is a lattice. Let B be a basis matrix of a full rank lattice L. Show that $(B^T)^{-1}$ is a basis matrix for the dual lattice. Hence, show that the determinant of the dual lattice is $\det(L)^{-1}$.

16.2 The Hermite and Minkowski bounds

We state the following results without rigorously defining the term volume and without giving proofs (see Section 1.3 of Micciancio and Goldwasser [378], Chapter 1 of Siegel [504], Chapter 6 of Hoffstein, Pipher and Silverman [261] or Chapter 12 of Cassels [113] for details).

Theorem 16.2.1 *(Blichfeldt) Let L be a lattice in $\mathbb{R}^m$ with basis $\{\underline{b}_1, \dots, \underline{b}_n\}$ and S any measurable set such that $S \subset \mathrm{span}\{\underline{b}_i : 1 \le i \le n\}$. If the volume of S exceeds $\det(L)$ then there exist two distinct points $\underline{v}_1, \underline{v}_2 \in S$ such that $(\underline{v}_1 - \underline{v}_2) \in L$.*

Proof See Theorem 1.3 of [378] or Section III.2.1 of [113]. □

Theorem 16.2.2 *(Minkowski Convex body theorem) Let L be a lattice in $\mathbb{R}^m$ with basis $\{\underline{b}_1, \dots, \underline{b}_n\}$ and let S be any convex set such that $S \subset \mathrm{span}\{\underline{b}_i : 1 \le i \le n\}$, $\underline{0} \in S$ and if $\underline{v} \in S$ then $-\underline{v} \in S$. If the volume of S is $> 2^n \det(L)$ then there exists a non-zero lattice point $\underline{v} \in S \cap L$.*

Proof See Section III.2.2 of Cassels [113], Theorem 6.28 of Hoffstein, Pipher and Silverman [261], Theorem 1.4 of Micciancio and Goldwasser [378], or Theorem 6.1 of Stewart and Tall [527]. □

The convex body theorem is used to prove Theorem 16.2.3. The intuition behind this result is that if the shortest non-zero vector in a lattice is large then the volume of the lattice cannot be small.

Theorem 16.2.3 *Let $n \in \mathbb{N}$. There is a constant $0 < \gamma_n \le n$ such that, for any lattice L of rank n in $\mathbb{R}^n$ having first minimum λ_1 (for the Euclidean norm),*

$$\lambda_1^2 \le \gamma_n \det(L)^{2/n}.$$

Proof See Theorem 1.5 of [378], Theorem 6.25 of [261], or Theorem 12.2.1 of [113]. □

Exercise 16.2.4 Show that the convex body theorem is tight. In other words, find a lattice L in $\mathbb{R}^n$ for some n and a symmetric convex subset $S \subseteq \mathbb{R}^n$ such that the volume of S is $2^n \det(L)$ and yet $S \cap L = \{0\}$.

Exercise 16.2.5 Show that, with respect to the ℓ_∞ norm, $\lambda_1 \le \det(L)^{1/n}$. Show that, with respect to the ℓ_1 norm, $\lambda_1 \le (n! \det(L))^{1/n} \approx n \det(L)^{1/n}/e$.

Exercise 16.2.6★ Let $a, b \in \mathbb{N}$. Show that there is a solution $r, s, t \in \mathbb{Z}$ to $r = as + bt$ such that $s^2 + r^2 \le \sqrt{2}b$.

Definition 16.2.7 Let $n \in \mathbb{N}$. The smallest real number γ_n such that

$$\lambda_1^2 \le \gamma_n \det(L)^{2/n}$$

for all lattices L of rank n is called the **Hermite constant**.

Exercise 16.2.8 This exercise is to show that $\gamma_2 = 2/\sqrt{3}$.

1. Let $\{\underline{b}_1, \underline{b}_2\}$ be a Gauss reduced basis (see Definition 17.1.1 of the next section) for a dimension 2 lattice in $\mathbb{R}^2$. Define the quadratic form $N(x, y) = \|x\underline{b}_1 + y\underline{b}_2\|^2$. Show that, without loss of generality, $N(x, y) = ax^2 + 2bxy + cy^2$ with $a, b, c \ge 0$ and $a \le c$.
2. Using $N(1, -1) \ge N(0, 1)$ (which follows from the property of being Gauss reduced), show that $2b \le a$. Hence, show that $3ac \le 4(ac - b^2)$
3. Show that $\det(L)^2 = |b^2 - ac|$. Hence, deduce that Hermite's constant satisfies $\gamma_2 \le 2/\sqrt{3}$.
4. Show that the lattice $L \subset \mathbb{R}^2$ with basis $\{(1, 0), (-1/2, \sqrt{3}/2)\}$ satisfies $\lambda_1^2 = (2/\sqrt{3})\det(L)$.

   (Optional) Show that L is equal to the ring of algebraic integers of $\mathbb{Q}(\sqrt{-3})$. Show that centering balls of radius $1/2$ at each point of L gives the most dense lattice packing of balls in $\mathbb{R}^2$.

Section 6.5.2 of Nguyen [409] lists the first 8 values of γ_n, gives the bound $\frac{n}{2\pi e} + o(1) \le \gamma_n \le \frac{n}{\pi e}(1 + o(1))$ and gives further references.

Theorem 16.2.9 *(Minkowski) Let L be a lattice of rank n in $\mathbb{R}^n$ with successive minima $\lambda_1, \dots, \lambda_n$ for the Euclidean norm. Then*

$$\left(\prod_{i=1}^{n} \lambda_i\right)^{1/n} < \sqrt{n}\det(L)^{1/n}.$$

Proof See Theorem 12.2.2 of [113]. (The term $\sqrt{n}$ can be replaced by $\sqrt{\gamma_n}$.) □

The **Gaussian heuristic** states that the shortest non-zero vector in a "random" lattice L of dimension n in $\mathbb{R}^n$ is expected to have length approximately

$$\sqrt{\frac{n}{2\pi e}}\det(L)^{1/n}.$$

We refer to Section 6.5.3 of [409] and Section 6.5.3 of [261] for discussion and references.

16.3 Computational problems in lattices

There are several natural computational problems relating to lattices. We start by listing some problems that can be efficiently solved using linear algebra (in particular, the Hermite normal form).

1. **Lattice membership**: Given an $n \times m$ basis matrix B for a lattice $L \subseteq \mathbb{Z}^m$ and a vector $\underline{v} \in \mathbb{Z}^m$ determine whether $\underline{v} \in L$.
2. **Lattice basis**: Given a set of vectors $\underline{b}_1, \ldots, \underline{b}_n$ in $\mathbb{Z}^m$ (possibly linearly dependent) find a basis for the lattice generated by them.
3. **Kernel lattice**: Given an $m \times n$ integer matrix A compute a basis for the lattice $\ker(A) = \{\underline{x} \in \mathbb{Z}^m : \underline{x}A = \underline{0}\}$.
4. **Kernel lattice modulo** M: Given an $m \times n$ integer matrix A and an integer M compute a basis for the lattice $\{\underline{x} \in \mathbb{Z}^m : \underline{x}A \equiv \underline{0} \pmod{M}\}$.

Exercise 16.3.1★ Describe explicit algorithms for the above problems and determine their complexity.

Now we list some computational problems that seem to be hard in general.

Definition 16.3.2 Let L be a lattice in $\mathbb{Z}^m$.

1. The **shortest vector problem (SVP)** is the computational problem: given a basis matrix B for L compute a non-zero vector $\underline{v} \in L$ such that $\|\underline{v}\|$ is minimal (i.e., $\|\underline{v}\| = \lambda_1$).
2. The **closest vector problem (CVP)** is the computational problem: given a basis matrix B for L and a vector $\underline{w} \in \mathbb{Q}^m$ (one can work with high-precision approximations in $\mathbb{R}^m$, but this is essentially still working in $\mathbb{Q}^m$) compute $v \in L$ such that $\|\underline{w} - \underline{v}\|$ is minimal.
3. The **decision closest vector problem (DCVP)** is: given a basis matrix B for a lattice L, a vector $\underline{w} \in \mathbb{Q}^m$ and a real number $r > 0$ decide whether or not there is a vector $\underline{v} \in L$ such that $\|\underline{w} - \underline{v}\| \leq r$.
4. The **decision shortest vector problem** is: given a basis matrix B for a lattice L and a real number $r > 0$ decide whether or not there is a non-zero $\underline{v} \in L$ such that $\|\underline{v}\| \leq r$.
5. Fix $\gamma > 1$. The **approximate SVP** is: given a basis matrix B for L compute a non-zero vector $\underline{v} \in L$ such that $\|\underline{v}\| \leq \gamma\lambda_1$.
6. Fix $\gamma > 1$. The **approximate CVP** is: given a basis matrix B for L and a vector $\underline{w} \in \mathbb{Q}^m$ compute $v \in L$ such that $\|\underline{w} - \underline{v}\| \leq \gamma \|\underline{w} - \underline{x}B\|$ for all $\underline{x} \in \mathbb{Z}^n$.

In general, these computational problems are known to be hard[2] when the rank is sufficiently large. It is known that CVP is NP-hard (this is shown by relating CVP with subset-sum; for details see Chapter 3 of [378]). Also, SVP is NP-hard under randomised reductions and non-uniform reductions (see Chapter 4 of [378] for an explanation of these terms and proofs). Nguyen [409] gives a summary of the complexity results and current best running times of algorithms for these problems.

On the other hand, if a lattice is sufficiently nice then these problems may be easy.

[2] We do not give details of complexity theory in this book; in particular, we do not define the term "NP-hard".

Example 16.3.3 Let $L \subset \mathbb{R}^2$ be the lattice with basis matrix

$$B = \begin{pmatrix} 1001 & 0 \\ 0 & 2008 \end{pmatrix}.$$

Then every lattice vector is of the form $(1001a, 2008b)$ where $a, b \in \mathbb{Z}$. Hence the shortest non-zero vectors are clearly $(1001, 0)$ and $(-1001, 0)$. Similarly, the closest vector to $\underline{w} = (5432, 6000)$ is clearly $(5005, 6024)$.

Why is this example so easy? The reason is that the basis vectors are orthogonal. Even in large dimensions, the SVP and CVP problems are easy if one has an orthogonal basis for a lattice. When given a basis that is not orthogonal it is less obvious whether there exists a non-trivial linear combination of the basis vectors that give a vector strictly shorter than the shortest basis vector. A basis for a lattice that is "as close to orthogonal as it can be" is therefore convenient for solving some computational problems.

17

Lattice basis reduction

The goal of lattice basis reduction is to transform a given lattice basis into a "nice" lattice basis consisting of vectors that are short and close to orthogonal. To achieve this, one needs both a suitable mathematical definition of "nice basis" and an efficient algorithm to compute a basis satisfying this definition.

Reduction of lattice bases of rank 2 in $\mathbb{R}^2$ was given by Lagrange[1] and Gauss. The algorithm is closely related to Euclid's algorithm and we briefly present it in Section 17.1. The main goal of this section is to present the lattice basis reduction algorithm of Lenstra, Lenstra and Lovász, known as the LLL or L^3 algorithm. This is a very important algorithm for practical applications. Some basic references for the LLL algorithm are Section 14.3 of Smart [513], Section 2.6 of Cohen [127] and Chapter 17 of Trappe and Washington [547]. More detailed treatments are given in von zur Gathen and Gerhard [220], Grötschel, Lovász and Schrijver [245], Section 1.2 of Lovász [356], and Nguyen and Vallée [416]. I also highly recommend the original paper [335].

The LLL algorithm generalises the Lagrange–Gauss algorithm and exploits the Gram–Schmidt orthogonalisation. Note that the Gram–Schmidt process is not useful, in general, for lattices since the coefficients $\mu_{i,j}$ do not usually lie in $\mathbb{Z}$ and so the resulting vectors are not usually elements of the lattice. The LLL algorithm uses the Gram–Schmidt vectors to determine the quality of the lattice basis, but ensures that the linear combinations used to update the lattice vectors are all over $\mathbb{Z}$.

17.1 Lattice basis reduction in two dimensions

Let $\underline{b}_1, \underline{b}_2 \in \mathbb{R}^2$ be linear independent vectors and denote by L the lattice for which they are a basis. The goal is to output a basis for the lattice such that the lengths of the basis vectors are as short as possible (in this case, successive minima). Lagrange and Gauss gave the following criteria for a basis to be reduced and then developed Algorithm 22 to compute such a basis.

[1] The algorithm was first written down by Lagrange and later by Gauss, but is usually called the "Gauss algorithm". We refer to [408] for the original references.

Definition 17.1.1 An ordered basis $\underline{b}_1, \underline{b}_2$ for $\mathbb{R}^2$ is **Lagrange–Gauss reduced** if $\|\underline{b}_1\| \le \|\underline{b}_2\| \le \|\underline{b}_2 + q\underline{b}_1\|$ for all $q \in \mathbb{Z}$.

The following theorem shows that the vectors in a Lagrange–Gauss reduced basis are as short as possible. This result holds for any norm, though the algorithm presented below is only for the Euclidean norm.

Theorem 17.1.2 *Let λ_1, λ_2 be the successive minima of L. If L has an ordered basis $\{\underline{b}_1, \underline{b}_2\}$ that is Lagrange–Gauss reduced then $\|\underline{b}_i\| = \lambda_i$ for $i = 1, 2$.*

Proof By definition we have

$$\|\underline{b}_2 + q\underline{b}_1\| \ge \|\underline{b}_2\| \ge \|\underline{b}_1\|$$

for all $q \in \mathbb{Z}$.

Let $\underline{v} = l_1\underline{b}_1 + l_2\underline{b}_2$ be any non-zero point in L. If $l_2 = 0$ then $\|\underline{v}\| \ge \|\underline{b}_1\|$. If $l_2 \ne 0$ then write $l_1 = ql_2 + r$ with $q, r \in \mathbb{Z}$ such that $0 \le r < |l_2|$. Then $\underline{v} = r\underline{b}_1 + l_2(\underline{b}_2 + q\underline{b}_1)$ and, by the triangle inequality

$$\begin{aligned}\|\underline{v}\| &\ge |l_2| \left\|\underline{b}_2 + q\underline{b}_1\right\| - r\|\underline{b}_1\| \\ &= (|l_2| - r)\|\underline{b}_2 + q\underline{b}_1\| + r(\|\underline{b}_2 + q\underline{b}_1\| - \|\underline{b}_1\|) \\ &\ge \|\underline{b}_2 + q\underline{b}_1\| \ge \|\underline{b}_2\| \ge \|\underline{b}_1\|.\end{aligned}$$

This completes the proof. □

Definition 17.1.3 Let $\underline{b}_1, \ldots, \underline{b}_n$ be a list of vectors in $\mathbb{R}^n$. We write[2] $B_i = \|\underline{b}_i\|^2 = \langle \underline{b}_i, \underline{b}_i \rangle$.

A crucial ingredient for the Lagrange–Gauss algorithm is that

$$\|\underline{b}_2 - \mu\underline{b}_1\|^2 = B_2 - 2\mu\langle \underline{b}_1, \underline{b}_2 \rangle + \mu^2 B_1 \tag{17.1}$$

is minimised at $\mu = \langle \underline{b}_1, \underline{b}_2 \rangle / B_1$ (to see this, note that the graph as a function of μ is a parabola and that the minimum can be found by differentiating with respect to μ). Since we are working in a lattice we therefore replace $\underline{b}_2$ by $\underline{b}_2 - \lfloor \mu \rceil \underline{b}_1$ where $\lfloor \mu \rceil$ is the nearest integer to μ. Hence, lines 3 and 9 of Algorithm 22 reduce the size of $\underline{b}_2$ as much as possible using $\underline{b}_1$. In the one-dimensional case, the formula $\underline{b}_2 - \lfloor \mu \rceil \underline{b}_1$ is the familiar operation $r_{i+1} = r_{i-1} - \lfloor r_{i-1}/r_i \rceil r_i$ from Euclid's algorithm.

Lemma 17.1.4 *An ordered basis $\{\underline{b}_1, \underline{b}_2\}$ is Lagrange–Gauss reduced if and only if*

$$\|\underline{b}_1\| \le \|\underline{b}_2\| \le \|\underline{b}_2 \pm \underline{b}_1\|.$$

Proof The forward implication is trivial. For the converse, suppose $\|\underline{b}_2\| \le \|\underline{b}_2 \pm \underline{b}_1\|$. We use the fact that the graph of $F(\mu) = \|\underline{b}_2 + \mu\underline{b}_1\|^2$ is a parabola. It follows that the miniumum of $F(\mu)$ is taken for $-1 < \mu < 1$. Hence, $\|\underline{b}_2\| \le \|\underline{b}_2 + q\underline{b}_1\|$ for $q \in \mathbb{Z}$ such that $|q| > 1$. □

[2] The reader is warned that the notation B_i will have a different meaning when we are discussing the LLL algorithm.

Algorithm 22 gives the Lagrange–Gauss algorithm for lattices in $\mathbb{Z}^2$. Note that the computation of μ is as an exact value in $\mathbb{Q}$. All other arithmetic is exact integer arithmetic.

Lemma 17.1.5 *Algorithm 22 terminates and outputs a Lagrange–Gauss reduced basis for the lattice L.*

Exercise 17.1.6 Prove Lemma 17.1.5.

Example 17.1.7 We run the Lagrange–Gauss algorithm on $\underline{b}_1 = (1, 5)$ and $\underline{b}_2 = (6, 21)$. In the first step, $\mu = 111/26 \approx 4.27$ and so we update $\underline{b}_2 = \underline{b}_2 - 4\underline{b}_1 = (2, 1)$. We then swap $\underline{b}_1$ and $\underline{b}_2$ so that the values in the loop are now $\underline{b}_1 = (2, 1)$ and $\underline{b}_2 = (1, 5)$. This time, $\mu = 7/5 = 1.4$ and so we set $\underline{b}_2 = \underline{b}_2 - \underline{b}_1 = (-1, 4)$. Since $\|\underline{b}_2\| > \|\underline{b}_1\|$ the algorithm halts and outputs $\{(2, 1), (-1, 4)\}$.

Algorithm 22 Lagrange–Gauss lattice basis reduction

INPUT: Basis $\underline{b}_1, \underline{b}_2 \in \mathbb{Z}^2$ for a lattice L
OUTPUT: Basis $(\underline{b}_1, \underline{b}_2)$ for L such that $\|\underline{b}_i\| = \lambda_i$

1: $B_1 = \|\underline{b}_1\|^2$
2: $\mu = \langle \underline{b}_1, \underline{b}_2 \rangle / B_1$
3: $\underline{b}_2 = \underline{b}_2 - \lfloor \mu \rceil \underline{b}_1$
4: $B_2 = \|\underline{b}_2\|^2$
5: **while** $B_2 < B_1$ **do**
6: Swap $\underline{b}_1$ and $\underline{b}_2$
7: $B_1 = B_2$
8: $\mu = \langle \underline{b}_1, \underline{b}_2 \rangle / B_1$
9: $\underline{b}_2 = \underline{b}_2 - \lfloor \mu \rceil \underline{b}_1$
10: $B_2 = \|\underline{b}_2\|^2$
11: **end while**
12: **return** $(\underline{b}_1, \underline{b}_2)$

Exercise 17.1.8 Run the Lagrange–Gauss reduction algorithm on the basis $\{(3, 8), (5, 14)\}$.

Lemma 17.1.9 *Let $\underline{b}_1, \underline{b}_2$ be the initial vectors in an iteration of the Lagrange–Gauss algorithm and suppose $\underline{b}_1' = \underline{b}_2 - m\underline{b}_1$ and $\underline{b}_2' = \underline{b}_1$ are the vectors that will be considered in the next step of the algorithm. Then $\|\underline{b}_1'\|^2 < \|\underline{b}_1\|^2/3$, except perhaps for the last two iterations.*

Proof Note that $m = \lfloor \mu \rceil = \lfloor \langle \underline{b}_1, \underline{b}_2 \rangle / \langle \underline{b}_1, \underline{b}_1 \rangle \rceil = \langle \underline{b}_1, \underline{b}_2 \rangle / \langle \underline{b}_1, \underline{b}_1 \rangle + \epsilon$ where $|\epsilon| \leq 1/2$. Hence

$$\langle \underline{b}_1, \underline{b}_1' \rangle = \langle \underline{b}_1, \underline{b}_2 - \left(\langle \underline{b}_1, \underline{b}_2 \rangle / \langle \underline{b}_1, \underline{b}_1 \rangle + \epsilon \right) \underline{b}_1 \rangle = -\epsilon \langle \underline{b}_1, \underline{b}_1 \rangle = -\epsilon \|\underline{b}_1\|^2.$$

We show that $\|\underline{b}_1'\|^2 < \|\underline{b}_1\|^2/3$ unless we are in the last two iterations of the algorithm. To do this, suppose that $\|\underline{b}_1'\|^2 \geq \|\underline{b}_1\|^2/3$. Then

$$|\langle \underline{b}_1', \underline{b}_2' \rangle| = |\langle \underline{b}_1', \underline{b}_1 \rangle| = |\epsilon| \|\underline{b}_1\|^2 \leq \tfrac{1}{2}\|\underline{b}_1\|^2 \leq \tfrac{3}{2}\|\underline{b}_1'\|^2.$$

It follows that, in the next iteration of the algorithm, we will be taking $m = \lfloor \mu \rceil \in \{-1, 0, 1\}$ and so the next iteration would, at most, replace $\underline{b}_1'$ with $\underline{b}_2' \pm \underline{b}_1' = \underline{b}_1 \pm (\underline{b}_2 - m\underline{b}_1)$. But, if this were smaller than $\underline{b}_1'$ then we would have already computed $\underline{b}_1'$ differently in the current iteration. Hence, the next step is the final iteration. □

Theorem 17.1.10 *Let $X \in \mathbb{Z}_{\geq 2}$ and let $\underline{b}_1, \underline{b}_2$ be vectors in $\mathbb{Z}^2$ such that $\|\underline{b}_i\|^2 \leq X$. Then the Lagrange–Gauss algorithm performs $O(\log(X)^3)$ bit operations.*

Proof Lemma 17.1.9 shows that there are $O(\log(X))$ iterations in the Lagrange–Gauss algorithm. Since the squared Euclidean lengths of all vectors in the algorithm are bounded by X, it follows that entries of vectors are integers bounded by $\sqrt{X}$. Similarly, the numerator and denominator of $\mu \in \mathbb{Q}$ require $O(\log(X))$ bits. The result follows. □

A much more precise analysis of the Lagrange–Gauss reduction algorithm is given by Vallée [551]. Indeed, the algorithm has complexity $O(\log(X)^2)$ bit operations; see Nguyen and Stehlé [408].

The above discussion is for the Euclidean norm, but the Lagrange–Gauss reduction algorithm can be performed for any norm (the only change is how one computes μ). We refer to Kaib and Schnorr [292] for analysis and details.

Finally, we remark that there is a natural analogue of Definition 17.1.1 for any dimension. Hence, it is natural to try to generalise the Lagrange–Gauss algorithm to higher dimensions. Generalisations to dimension three have been given by Vallée [550] and Semaev [485]. There are a number of problems when generalising to higher dimensions. For example, choosing the right linear combination to size reduce $\underline{b}_n$ using $\underline{b}_1, \dots, \underline{b}_{n-1}$ is solving the CVP in a sublattice (which is a hard problem). Furthermore, there is no guarantee that the resulting basis actually has good properties in high dimension. We refer to Nguyen and Stehlé [413] for a full discussion of these issues and an algorithm that works in dimensions 3 and 4.

17.1.1 Connection between Lagrange–Gauss reduction and Euclid's algorithm

The Lagrange–Gauss algorithm is closely related to Euclid's algorithm. We briefly discuss some similarities and differences. Recall that if $a, b \in \mathbb{Z}$ then Euclid's algorithm (using signed remainders) produces a sequence of integers r_i, s_i, t_i such that

$$as_i + bt_i = r_i$$

where $|r_i t_i| < |a|$ and $|r_i s_i| < |b|$. The precise formulae are $r_{i+1} = r_{i-1} - qr_i$ and $s_{i+1} = s_{i-1} - qs_i$ where $q = \lfloor r_{i-1}/r_i \rceil$. The sequence $|r_i|$ is strictly decreasing. The initial values are $r_{-1} = a, r_0 = b, s_{-1} = 1, s_0 = 0, t_{-1} = 0, t_0 = 1$. In other words, the lattice with basis matrix

$$B = \begin{pmatrix} 0 & b \\ 1 & a \end{pmatrix} = \begin{pmatrix} s_0 & r_0 \\ s_{-1} & r_{-1} \end{pmatrix}$$

contains the vectors

$$(s_i, r_i) = (t_i, s_i)B.$$

These vectors are typically shorter than the original vectors of the lattice.

We claim that if s_i is sufficiently small compared with r_i then one step of the Lagrange–Gauss algorithm on B corresponds to one step of Euclid's algorithm (with negative remainders).

To see this, let $\underline{b}_1 = (s_i, r_i)$ and consider the Lagrange–Gauss algorithm with $\underline{b}_2 = (s_{i-1}, r_{i-1})$. First compute the value

$$\mu = \frac{\langle \underline{b}_1, \underline{b}_2 \rangle}{\langle \underline{b}_1, \underline{b}_1 \rangle} = \frac{s_i s_{i-1} + r_i r_{i-1}}{s_i^2 + r_i^2}.$$

If s_i is sufficiently small relative to r_i (e.g., in the first step, when $s_0 = 0$) then

$$\lfloor \mu \rceil = \lfloor r_i r_{i-1}/r_i^2 \rceil = \lfloor r_{i-1}/r_i \rceil = q.$$

Hence, the operation $\underline{v} = \underline{b}_2 - \lfloor \mu \rceil \underline{b}_1$ is $\underline{v} = (s_{i-1} - q s_i, r_{i-1} - q r_i)$, which agrees with Euclid's algorithm. Finally, the Lagrange–Gauss algorithm compares the lengths of the vectors $\underline{v}$ and $\underline{b}_1$ to see if they should be swapped. When s_{i+1} is small compared with r_{i+1} then $\|\underline{v}\|$ is smaller than $\|b_1\|$. Hence, the vectors are swapped and the matrix becomes

$$\begin{pmatrix} s_{i-1} - q s_i & r_{i-1} - q r_i \\ s_i & r_i \end{pmatrix}$$

just as in Euclid's algorithm.

The algorithms start to deviate once s_i become large (this can already happen on the second iteration, as the below example shows). Further, Euclid's algorithm runs until $r_i = 0$ (in which case $s_i \approx b$) whereas Lagrange–Gauss reduction stops when $r_i \approx s_i$.

Example 17.1.11 Let $a = 19$ and $b = 8$. The sequence of remainders in the signed Euclidean algorithm is $3, -1$ while the Lagrange–Gauss lattice basis reduction algorithm computes remainders $3, 2$.

Example 17.1.12 Consider $a = 8239876$ and $b = 1020301$, which have gcd equal to one. Let

$$B = \begin{pmatrix} 0 & b \\ 1 & a \end{pmatrix}.$$

Running the Lagrange–Gauss algorithm on this matrix gives

$$\begin{pmatrix} 540 & 379 \\ 619 & -1455 \end{pmatrix}.$$

One can verify that

$$379 = 540a + t_4 b \text{ where } t_4 = -4361$$

and

$$-1455 = 619a + t_5 b \text{ where } t_5 = -4999.$$

17.2 LLL-reduced lattice bases

This section presents the crucial definition from [335] and some of its consequences. The main result is Theorem 17.2.12, which shows that an LLL-reduced lattice basis does have good properties.

Recall first that if $\underline{b}_1, \dots, \underline{b}_n$ is a set of vectors in $\mathbb{R}^m$ then one can define the Gram–Schmidt orthogonalisation $\underline{b}_1^*, \dots, \underline{b}_n^*$ as in Section A.10.2. We use the notation $\mu_{i,j} = \langle \underline{b}_i, \underline{b}_j^* \rangle / \langle \underline{b}_j^*, \underline{b}_j^* \rangle$ throughout.

As we have noted in Example 16.3.3, computational problems in lattices can be easy if one has a basis that is orthogonal or "sufficiently close to orthogonal". A simple but important observation is that one can determine when a basis is close to orthogonal by considering the lengths of the Gram–Schmidt vectors. More precisely, a lattice basis is "close to orthogonal" if the lengths of the Gram–Schmidt vectors do not decrease too rapidly.

Example 17.2.1 Two bases for $\mathbb{Z}^2$ are $\{(1, 0), (0, 1)\}$ and $\{(23, 24), (24, 25)\}$. In the first case, the Gram–Schmidt vectors both have length 1. In the second case, the Gram–Schmidt vectors are $\underline{b}_1^* = (23, 24)$ and $\underline{b}_2^* = (24/1105, -23/1105)$, which have lengths $\sqrt{1105} \approx 33.24$ and $1/\sqrt{1105} \approx 0.03$ respectively. The fact that the lengths of the Gram–Schmidt vectors dramatically decreases reveals that the original basis is not of good quality.

We now list some easy properties of the Gram–Schmidt orthogonalisation.

Lemma 17.2.2 *Let $\{\underline{b}_1, \dots, \underline{b}_n\}$ be linearly independent in $\mathbb{R}^m$ and let $\{\underline{b}_1^*, \dots, \underline{b}_n^*\}$ be the Gram–Schmidt orthogonalisation.*

1. $\|\underline{b}_i^*\| \le \|\underline{b}_i\|$ *for* $1 \le i \le n$.
2. $\langle \underline{b}_i, \underline{b}_i^* \rangle = \langle \underline{b}_i^*, \underline{b}_i^* \rangle$ *for* $1 \le i \le n$.
3. *Denote the closest integer to* $\mu_{k,j}$ *by* $\lfloor \mu_{k,j} \rceil$. *If* $\underline{b}_k' = \underline{b}_k - \lfloor \mu_{k,j} \rceil \underline{b}_j$ *for* $1 \le k \le n$ *and* $1 \le j < k$ *and if* $\mu_{k,j}' = \langle \underline{b}_k', \underline{b}_j^* \rangle / \langle \underline{b}_j^*, \underline{b}_j^* \rangle$ *then* $|\mu_{k,j}'| \le 1/2$.

Exercise 17.2.3 Prove Lemma 17.2.2.

Definition 17.2.4 Let $\{\underline{b}_1, \dots, \underline{b}_n\}$ be an ordered basis for a lattice. Denote by $\{\underline{b}_1^*, \dots, \underline{b}_n^*\}$ the Gram–Schmidt orthogonalisation and write $B_i = \|\underline{b}_i^*\|^2 = \langle \underline{b}_i^*, \underline{b}_i^* \rangle$. Let

$$\mu_{i,j} = \langle \underline{b}_i, \underline{b}_j^* \rangle / \langle \underline{b}_j^*, \underline{b}_j^* \rangle$$

for $1 \le j < i \le n$ be the coefficients from the Gram–Schmidt process. Fix $1/4 < \delta < 1$. The (ordered) basis is **LLL-reduced** (with factor δ) if the following conditions hold:

- (Size reduced) $|\mu_{i,j}| \le 1/2$ for $1 \le j < i \le n$.
- (Lovász condition)

$$B_i \ge \left(\delta - \mu_{i,i-1}^2\right) B_{i-1}$$

  for $2 \le i \le n$.

It is traditional to choose $\delta = 3/4$ in the Lovász condition.

Exercise 17.2.5 Which of the following basis matrices represents an LLL-reduced basis (with $\delta = 3/4$)?

$$\begin{pmatrix} 1 & 0 \\ 0 & 2 \end{pmatrix}, \begin{pmatrix} 0 & 4 \\ 1 & 0 \end{pmatrix}, \begin{pmatrix} 1 & -2 \\ 3 & 1 \end{pmatrix}, \begin{pmatrix} 5 & 0 \\ 0 & 4 \end{pmatrix}, \begin{pmatrix} 10 & 0 \\ 0 & 9 \end{pmatrix}.$$

Exercise 17.2.6 Prove that an equivalent formulation (more in the flavour of the Lagrange–Gauss method) of the Lovász condition is

$$B_i + \mu_{i,i-1}^2 B_{i-1} = \|\underline{b}_i^* + \mu_{i,i-1}\underline{b}_{i-1}^*\|^2 \ge \delta B_{i-1}.$$

Exercise 17.2.7 Find an ordered basis $\{\underline{b}_1, \underline{b}_2\}$ in $\mathbb{R}^2$ that is LLL-reduced, but has the property that $\|\underline{b}_2\| < \|\underline{b}_1\|$ and that the ordered basis $\{\underline{b}_2, \underline{b}_1\}$ is not LLL-reduced.

For the moment we do not concern ourselves with the question of whether an LLL-reduced basis can exist for every lattice L. In Section 17.4 we will present the LLL algorithm, which constructs such a basis for any lattice (hence, giving a constructive existence proof for an LLL-reduced basis).

The following properties of an LLL-reduced basis hold.

Lemma 17.2.8 *Let $\{\underline{b}_1, \ldots, \underline{b}_n\}$ be an LLL-reduced basis with $\delta = 3/4$ for a lattice $L \subset \mathbb{R}^m$. Let the notation be as above. In particular, $\|\underline{b}\|$ is the Euclidean norm.*

1. $B_j \le 2^{i-j} B_i$ *for* $1 \le j \le i \le n$.
2. $B_i \le \|\underline{b}_i\|^2 \le (\frac{1}{2} + 2^{i-2}) B_i$ *for* $1 \le i \le n$.
3. $\|\underline{b}_j\| \le 2^{(i-1)/2} \|\underline{b}_i^*\|$ *for* $1 \le j \le i \le n$.

Proof

1. The Lovász condition implies $B_i \ge (\frac{3}{4} - \frac{1}{4}) B_{i-1} = \frac{1}{2} B_{i-1}$ and the result follows by induction.

2. From $\underline{b}_i = \underline{b}_i^* + \sum_{j=1}^{i-1} \mu_{i,j}\underline{b}_j^*$ we have

$$\begin{aligned}\|\underline{b}_i\|^2 &= \langle \underline{b}_i, \underline{b}_i \rangle \\ &= \left\langle \underline{b}_i^* + \sum_{j=1}^{i-1} \mu_{i,j}\underline{b}_j^*, \underline{b}_i^* + \sum_{j=1}^{i-1} \mu_{i,j}\underline{b}_j^* \right\rangle \\ &= B_i + \sum_{j=1}^{i-1} \mu_{i,j}^2 B_j,\end{aligned}$$

which is clearly $\geq B_i$. By part 1 this is at most $B_i(1 + \frac{1}{4}\sum_{j=1}^{i-1} 2^{i-j}) = B_i(1 + \frac{1}{4}(2^i - 2)) = B_i(\frac{1}{2} + 2^{i-2})$.

3. Since $j \geq 1$ we have $\frac{1}{2} + 2^{j-2} \leq 2^{j-1}$. Part 2 can therefore be written as $\|b_j\|^2 \leq 2^{j-1}B_j$. By part 1, $B_j \leq 2^{i-j}B_i$ and so $\|b_j\|^2 \leq 2^{j-1}2^{i-j}B_i = 2^{i-1}B_i$. Taking square roots gives the result.

□

We now give the same result for a slightly different value of δ.

Lemma 17.2.9 *Let $\{\underline{b}_1, \ldots, \underline{b}_n\}$ be an LLL-reduced basis with $\delta = 1/4 + 1/\sqrt{2} \approx 0.957$ for a lattice $L \subset \mathbb{R}^m$. Let the notation be as above. In particular, $\|\underline{b}\|$ is the Euclidean norm:*

1. $B_j \leq 2^{(i-j)/2}B_i$ *for* $1 \leq j \leq i \leq n$.
2. $B_i \leq \|\underline{b}_i\|^2 \leq (\frac{1}{6} + 2^{(i-1)/2})B_i$ *for* $1 \leq i \leq n$.
3. $\|\underline{b}_j\| \leq 2^{i/4}\|\underline{b}_i^*\|$ *for* $1 \leq j \leq i \leq n$.

Exercise 17.2.10★ Prove Lemma 17.2.9.

Lemma 17.2.11 *Let $\{\underline{b}_1, \ldots, \underline{b}_n\}$ be an ordered basis for a lattice $L \subset \mathbb{R}^m$ and let $\{\underline{b}_1^*, \ldots, \underline{b}_n^*\}$ be the Gram–Schmidt orthogonalisation. Let λ_1 be the length of the shortest non-zero vector in the lattice. Then*

$$\lambda_1 \geq \min_{1 \leq i \leq n} \|\underline{b}_i^*\|.$$

Furthermore, let $\underline{w}_1, \ldots, \underline{w}_i \in L$ be linearly independent lattice vectors such that $\max\{\|\underline{w}_1\|, \ldots, \|\underline{w}_i\|\} = \lambda_i$, as in the definition of successive minima. Write $\underline{w}_j = \sum_{k=1}^n z_{j,k}\underline{b}_k$. For $1 \leq j \leq i$ denote by $k(j)$ the largest value for k such that $1 \leq k \leq n$ and $z_{j,k} \neq 0$. Then $\|\underline{w}_j\| \geq \|\underline{b}_{k(j)}^*\|$.

Proof Let $\underline{x} = (x_1, \ldots, x_n) \in \mathbb{Z}^n$ be arbitrary such that $\underline{x} \neq \underline{0}$. Let i be the largest index such that $x_i \neq 0$. We will show that $\|\underline{x}B\| \geq \|\underline{b}_i^*\|$, from which the result follows.

We have $\underline{x}B = \sum_{j=1}^i x_j\underline{b}_j$. Since $\underline{b}_i^*$ is orthogonal to the span of $\{\underline{b}_1, \ldots, \underline{b}_{i-1}\}$ we have $\langle \underline{x}B, \underline{b}_i^* \rangle = x_i\langle \underline{b}_i^*, \underline{b}_i^* \rangle = x_i\|\underline{b}_i^*\|^2$. Since $x_i \in \mathbb{Z}$ and $x_i \neq 0$ it follows that $|\langle \underline{x}B, \underline{b}_i^* \rangle| \geq$

$\|\underline{b}_i^*\|^2$. By part 4 of Lemma A.10.3 it follows that

$$\|\underline{x}B\| \geq \|\underline{b}_i^*\|,$$

which completes the proof. □

Theorem 17.2.12 shows that an LLL-reduced lattice basis has good properties. In particular, the first vector of an LLL-reduced lattice basis has length at most $2^{(n-1)/2}$ times the length of a shortest non-zero vector.

Theorem 17.2.12 *Let $\{\underline{b}_1, \ldots, \underline{b}_n\}$ be an LLL-reduced basis with $\delta = 3/4$ for a lattice $L \subset \mathbb{R}^m$. Let the notation be as above. In particular, $\|\underline{b}\|$ is the Euclidean norm.*

1. $\|\underline{b}_1\| \leq 2^{(n-1)/2}\lambda_1$.
2. $\|\underline{b}_j\| \leq 2^{(n-1)/2}\lambda_i$ *for $1 \leq j \leq i \leq n$. (This may look strange, but it tends to be used for fixed i and varying j, rather than the other way around.)*
3. $2^{(1-i)/2}\lambda_i \leq \|\underline{b}_i\| \leq 2^{(n-1)/2}\lambda_i$.
4. $\det(L) \leq \prod_{i=1}^n \|\underline{b}_i\| \leq 2^{n(n-1)/4}\det(L)$.
5. $\|\underline{b}_1\| \leq 2^{(n-1)/4}\det(L)^{1/n}$.

Proof

1. From part 1 of Lemma 17.2.8 we have $\|b_i^*\| \geq 2^{(i-1)/2}\|b_1^*\|$. Hence, part 1 implies

$$\begin{aligned}\lambda_1 &\geq \min_{1\leq i\leq n} \|\underline{b}_i^*\| \\ &\geq \min_{1\leq i\leq n} 2^{(1-i)/2}\|\underline{b}_1^*\| \\ &= 2^{(1-n)/2}\|\underline{b}_1^*\|.\end{aligned}$$

   The result follows since $\underline{b}_1^* = \underline{b}_1$.
2. Let $\underline{w}_1, \ldots, \underline{w}_i \in L$ be linearly independent lattice vectors such that $\max\{\|\underline{w}_1\|, \ldots, \|\underline{w}_i\|\} = \lambda_i$. Let $k(j)$ be defined as in Lemma 17.2.11 so that $\|\underline{w}_j\| \geq \|\underline{b}_{k(j)}^*\|$.

   Renumber the vectors $\underline{w}_j$ so that $k(1) \leq k(2) \leq \cdots \leq k(i)$. We claim that $j \leq k(j)$. If not then $\underline{w}_1, \ldots, \underline{w}_j$ would belong to the span of $\{\underline{b}_1, \ldots, \underline{b}_{j-1}\}$ and would be linearly dependent.

   Finally

$$\|\underline{b}_j\| \leq 2^{(k(j)-1)/2}\|\underline{b}_{k(j)}^*\| \leq 2^{(n-1)/2}\|\underline{w}_j\| \leq 2^{(n-1)/2}\lambda_i,$$

   which proves the result.
3. The upper bound on $\|\underline{b}_i\|$ is given by part 2.

   Since $\{\underline{b}_1, \ldots, \underline{b}_i\}$ are linearly independent we have $\lambda_i \leq \max_{1\leq j\leq i}\|\underline{b}_j\|$ and by part 3 of Lemma 17.2.8 each $\|\underline{b}_j\| \leq 2^{(i-1)/2}\|\underline{b}_i^*\|$. Using $\|\underline{b}_i^*\| \leq \|\underline{b}_i\|$ we obtain the lower bound on $\|\underline{b}_i\|$.
4. By Lemma 16.1.14 we have $\det(L) = \prod_{i=1}^n \|\underline{b}_i^*\|$. The result follows from $\|\underline{b}_i^*\| \leq \|\underline{b}_i\| \leq 2^{(i-1)/2}\|\underline{b}_i^*\|$.

5. By part 3 of Lemma 17.2.8 we have $\|\underline{b}_1\| \le 2^{(i-1)/2}\|\underline{b}_i^*\|$ and so

$$\|\underline{b}_1\|^n \le \prod_{i=1}^{n} 2^{(i-1)/2}\|\underline{b}_i^*\| = 2^{n(n-1)/4}\det(L).$$

□

Corollary 17.2.13 *If $\|\underline{b}_1\| \le \|\underline{b}_i^*\|$ for all $1 \le i \le n$ then $\underline{b}_1$ is a correct solution to SVP.*

Exercise 17.2.14 Prove Corollary 17.2.13.

Exercise 17.2.15 Suppose L is a lattice in $\mathbb{Z}^m$ and let $\{\underline{b}_1, \dots, \underline{b}_n\}$ be an LLL-reduced basis. Rename these vectors as $\underline{v}_1, \dots, \underline{v}_n$ such that $1 \le \|\underline{v}_1\| \le \|\underline{v}_2\| \le \cdots \le \|\underline{v}_n\|$. Show that one does not necessarily have $\|\underline{v}_1\| = \|\underline{b}_1\|$. Show that, for $1 \le i \le n$

$$\|\underline{v}_i\| \le \left(2^{n(n-1)/4}\det(L)\right)^{1/(n+1-i)}.$$

As a final remark, the results in this section have only given upper bounds on the sizes of $\|\underline{b}_i\|$ in an LLL-reduced lattice basis. In many practical instances, one finds that LLL-reduced lattice vectors are much shorter than these bounds might suggest.

17.3 The Gram–Schmidt algorithm

The LLL algorithm requires computing a Gram–Schmidt basis. For the complexity analysis of the LLL algorithm it is necessary to give a more careful description and analysis of the Gram–Schmidt algorithm than was done in Section A.10.2. We present pseudocode in Algorithm 23 (the "downto" in line 4 is not necessary, but we write it that way for future reference in the LLL algorithm).

Algorithm 23 Gram–Schmidt algorithm

INPUT: $\{\underline{b}_1, \dots, \underline{b}_n\}$ in $\mathbb{R}^m$
OUTPUT: $\{\underline{b}_1^*, \dots, \underline{b}_n^*\}$ in $\mathbb{R}^m$

1: $\underline{b}_1^* = \underline{b}_1$
2: **for** $i = 2$ to n **do**
3: $\quad \underline{v} = \underline{b}_i$
4: $\quad$ **for** $j := i - 1$ downto 1 **do**
5: $\quad\quad \mu_{i,j} = \langle \underline{b}_i, \underline{b}_j^* \rangle / \langle \underline{b}_j^*, \underline{b}_j^* \rangle$
6: $\quad\quad \underline{v} = \underline{v} - \mu_{i,j}\underline{b}_j^*$
7: $\quad$ **end for**
8: $\quad \underline{b}_i^* = \underline{v}$
9: **end for**
10: **return** $\{\underline{b}_1^*, \dots, \underline{b}_n^*\}$

When working in $\mathbb{R}$ the standard way to implement this algorithm is using floating-point arithmetic. However, problems can arise (especially if the $\underline{b}_i^*$ decrease quickly in size). Such issues are beyond the scope of this book; we refer to Higham [259] for details.

If the input vectors lie in $\mathbb{Z}^m$ then one can perform Algorithm 23 using exact arithmetic over $\mathbb{Q}$. However, the integers can become very large (this is called **coefficient explosion**). We now analyse the size of the integers and prove the complexity of the exact version of the Gram–Schmidt algorithm. These results are used later when determining the complexity of LLL.

Definition 17.3.1 Let $\underline{b}_1, \ldots, \underline{b}_n$ be an ordered set of vectors in $\mathbb{Z}^m$. Define $B_i = \|\underline{b}_i^*\|^2$ (as before). For $1 \le i \le n-1$ define the $i \times m$ matrix $B_{(i)}$ whose rows are $\underline{b}_1, \ldots, \underline{b}_i$. Define $d_0 = 0$ and, for $1 \le i \le n$

$$d_i = \det(B_{(i)} B_{(i)}^T) = \det((\langle \underline{b}_j, \underline{b}_k \rangle)_{1 \le j,k \le i}) \in \mathbb{Z},$$

which is the square of the volume of the sublattice generated by $B_{(i)}$.

Lemma 17.3.2 *Let the notation be as above.*

1. $d_i = \prod_{j=1}^i B_j$ *for* $1 \le i \le n$.
2. $B_i = d_i / d_{i-1}$ *for* $1 \le i \le n$.
3. $d_{i-1} \underline{b}_i^* \in L \subseteq \mathbb{Z}^n$ *for* $1 \le i \le n$, *where L is the lattice spanned by* $\{\underline{b}_1, \ldots, \underline{b}_n\}$.
4. $d_j \mu_{i,j} \in \mathbb{Z}$ *for* $1 \le j < i \le n$.

Proof

1. Write $L_{(i)}$ for the lattice spanned by the first i vectors (i.e., L is given by the matrix $B_{(i)}$). Then $d_i = \det(L_{(i)})^2 = \prod_{j=1}^i \|\underline{b}_j^*\|^2 = \prod_{j=1}^i B_j$ by Lemma 16.1.14.
2. This property follows immediately from the previous one.
3. Write $\underline{b}_i^* = \underline{b}_i - \sum_{j=1}^{i-1} a_{i,j} \underline{b}_j$ for some $a_{i,j} \in \mathbb{R}$. Note that the sum is over vectors $\underline{b}_j$ not $\underline{b}_j^*$, so the $a_{i,j}$ are not the same as the $\mu_{i,j}$. Since $\langle \underline{b}_l, \underline{b}_i^* \rangle = 0$ for $1 \le l < i$ we have

$$\langle \underline{b}_l, \underline{b}_i \rangle = \sum_{j=1}^{i-1} a_{i,j} \langle \underline{b}_l, \underline{b}_j \rangle,$$

which corresponds to the matrix product

$$(\langle \underline{b}_i, \underline{b}_1 \rangle, \ldots, \langle \underline{b}_i, \underline{b}_{i-1} \rangle) = (a_{i,1}, \ldots, a_{i,i-1}) B_{(i-1)} B_{(i-1)}^T.$$

Inverting $B_{(i-1)} B_{(i-1)}^T$ to solve for the $a_{i,j}$ gives $d_{i-1} a_{i,j} \in \mathbb{Z}$. It follows that $d_{i-1} \underline{b}_i^* \in L \subset \mathbb{Z}^n$ as required.
4. By the previous results we have $d_j \mu_{i,j} = d_{j-1} B_j \langle \underline{b}_i, \underline{b}_j^* \rangle / B_j = \langle \underline{b}_i, d_{j-1} \underline{b}_j^* \rangle \in \mathbb{Z}$.

□

Exercise 17.3.3 Consider the vector $\underline{v} = \underline{b}_i - \sum_{k=j}^{i-1} \mu_{i,k}\underline{b}_k^*$ in line 6 of Algorithm 23 during iteration j. Show that

$$\|\underline{v}\|^2 = \|\underline{b}_i\|^2 - \sum_{k=j}^{i-1} \mu_{i,k}^2 \|\underline{b}_k^*\|^2.$$

Deduce that $\|\underline{v}\| \le \|\underline{b}_i\|$ and that $d_{i-1}\underline{v} \in \mathbb{Z}^m$ throughout the loop in line 4 of the algorithm.

Theorem 17.3.4 *Let $\underline{b}_1, \ldots, \underline{b}_n$ be vectors in $\mathbb{Z}^m$. Let $X \in \mathbb{Z}_{\ge 2}$ be such that $\|\underline{b}_i\|^2 \le X$ for $1 \le i \le n$. Then the Gram–Schmidt algorithm performs $O(n^4 m \log(X)^2)$ bit operations. The output size is $O(n^2 m \log(X))$.*

Proof One runs Algorithm 23 using exact $\mathbb{Q}$ arithmetic for the vectors $\underline{b}_i^*$. Lemma 17.3.2 shows that the denominators in $\underline{b}_i^*$ are all factors of d_{i-1}, which has size $\prod_{j=1}^{i-1} B_j \le \prod_{j=1}^{i-1} \|\underline{b}_j\|^2 \le X^{i-1}$. Also, $\|\underline{b}_i^*\| \le \|\underline{b}_i\| \le X$, so the numerators are bounded by X^i. The size of each vector $\underline{b}_i^*$ and, by Exercise 17.3.3, the intermediate steps $\underline{v}$ in the computation are therefore $O(mi\log(X))$ bits, which gives the output size of the algorithm. The computation $\langle \underline{b}_i, \underline{b}_j^* \rangle$ requires $O(mn\log(X)^2)$ bit operations and the computation $\langle \underline{b}_j^*, \underline{b}_j^* \rangle$ requires $O(mn^2 \log(X)^2)$ bit operations. As there are $O(n^2)$ vector operations to perform, one gets the stated running time. □

Corollary 17.3.5 *Let the notation be as in Theorem 17.3.4 and let L be the lattice in $\mathbb{Z}^m$ with basis $\{\underline{b}_1, \ldots, \underline{b}_n\}$. Then one can compute* $\det(L)^2$ *in $O(n^4 m \log(X)^2)$ bit operations.*[3]

Proof Lemma 16.1.14 implies $\det(L)^2 = \prod_{i=1}^n \|\underline{b}_i^*\|^2$. One computes $\underline{b}_i^*$ using exact (naive) arithmetic over $\mathbb{Q}$ in $O(n^4 m \log(X)^2)$ bit operations. One computes each $\|\underline{b}_i^*\|^2 \in \mathbb{Q}$ in $O(mn^2 \log(X)^2)$ bit operations. Since $\|\underline{b}_i^*\|^2 \le X$ and $d_{i-1}\|\underline{b}_i^*\|^2 \in \mathbb{Z}$ it follows that $\|\underline{b}_i^*\|^2$ is a ratio of integers bounded by X^n. One computes the product of the $\|\underline{b}_i^*\|^2$ in $O(n^3 \log(X)^2)$ bit operations (since the integers in the product are bounded by X^{n^2}). Finally, one can reduce the fraction using Euclid's algorithm and division in $O(n^4 \log(X)^2)$ bit operations. □

17.4 The LLL algorithm

The Lenstra–Lenstra–Lovász (LLL) algorithm is an iterative algorithm that transforms a given lattice basis into an LLL-reduced one. Since the definition of LLL-reduced uses Gram–Schmidt vectors, the algorithm performs the Gram–Schmidt method as a subroutine. The first condition of Definition 17.2.4 is easily met by taking suitable integer linear combinations. If the second condition is not met then $\underline{b}_i$ is not significantly longer than $\underline{b}_{i-1}$. In this case, we swap $\underline{b}_i$ and $\underline{b}_{i-1}$ and backtrack. The swapping of vectors is familiar from the Lagrange–Gauss two-dimensional lattice basis reduction algorithm and also Euclid's algorithm. We give the precise details in Algorithm 24.

[3] Since $\det(L)^2 \in \mathbb{Z}$ while $\det(L)$ may not be rational if $n < m$, we prefer to work with $\det(L)^2$.

Algorithm 24 LLL algorithm with Euclidean norm (typically, choose $\delta = 3/4$)

INPUT: $\underline{b}_1, \dots, \underline{b}_n \in \mathbb{Z}^m$.
OUTPUT: LLL reduced basis $\underline{b}_1, \dots, \underline{b}_n$

1: Compute the Gram–Schmidt basis $\underline{b}_1^*, \dots, \underline{b}_n^*$ and coefficients $\mu_{i,j}$ for $1 \le j < i \le n$
2: Compute $B_i = \langle \underline{b}_i^*, \underline{b}_i^* \rangle = \|\underline{b}_i^*\|^2$ for $1 \le i \le n$
3: $k = 2$
4: **while** $k \le n$ **do**
5: **for** $j = (k-1)$ downto 1 **do** ⊳ Perform size reduction
6: Let $q_j = \lfloor \mu_{k,j} \rceil$ and set $\underline{b}_k = \underline{b}_k - q_j \underline{b}_j$
7: Update the values $\mu_{k,j}$ for $1 \le j < k$
8: **end for**
9: **if** $B_k \ge (\delta - \mu_{k,k-1}^2) B_{k-1}$ **then** ⊳ Check Lovász condition
10: $k = k + 1$
11: **else**
12: Swap $\underline{b}_k$ with $\underline{b}_{k-1}$
13: Update the values $\underline{b}_k^*, \underline{b}_{k-1}^*, B_k, B_{k-1}, \mu_{k-1,j}$ and $\mu_{k,j}$ for $1 \le j < k$, and $\mu_{i,k}, \mu_{i,k-1}$ for $k < i \le n$
14: $k = \max\{2, k-1\}$
15: **end if**
16: **end while**

Lemma 17.4.1 *Throughout the LLL algorithm the values $\underline{b}_i^*$ and B_i for $1 \le i \le n$ and $\mu_{i,j}$ for $1 \le j < i \le n$ are all correct Gram–Schmidt values.*

Exercise 17.4.2 Prove Lemma 17.4.1. In other words, show that line 6 of the LLL algorithm does not change $\underline{b}_i^*$ or B_i for $1 \le i \le n$. Similarly, line 12 of the algorithm does not change any values except those mentioned in line 13.

It is illuminating to compare the LLL algorithm with the Lagrange–Gauss reduction algorithm. The basic concept of size reduction followed by a swap is the same; there are, however, two crucial differences.

1. The size reduction operation in the Lagrange–Gauss algorithm gives the minimal value for $\|\underline{b}_2 + q\underline{b}_1\|$ over $q \in \mathbb{Z}$. In LLL, the coefficient $\mu_{k,j}$ is chosen to depend on $\underline{b}_k$ and $\underline{b}_j^*$ so it does not necessarily minimise $\|\underline{b}_k\|$. Indeed, $\|\underline{b}_k\|$ can grow during the algorithm. Of course, in the two-dimensional case of LLL then $\mu_{2,1}$ is the same as the value used in the Lagrange–Gauss algorithm and so the size reduction step is the same.
2. The size check in LLL (the Lovász condition) is on the lengths of the Gram–Schmidt vectors, unlike the size check in the Lagrange–Gauss algorithm, which is on the length of the basis vectors themselves.

These features of LLL may seem counterintuitive, but they are essential to the proof that the algorithm runs in polynomial-time.

Lemma 17.4.3 *If $\underline{b}_k$ and $\underline{b}_{k-1}$ are swapped then the Gram–Schmidt vectors $\underline{b}_i^*$ for $1 \le i \le n$ are changed as follows:*

1. *For $1 \le i < k-1$ and $k < i < n$, $\underline{b}_i^*$ is unchanged.*
2. *The new value for $\underline{b}_{k-1}^*$ is $\underline{b}_k^* + \mu_{k,k-1}\underline{b}_{k-1}^*$ and the new value for B_{k-1} is $B_k' = B_k + \mu_{k,k-1}^2 B_{k-1}$.*
3. *The new value for $\underline{b}_k^*$ is $(B_k/B_{k-1}')\underline{b}_{k-1}^* - (\mu_{k,k-1}B_{k-1}/B_{k-1}')\underline{b}_k^*$ and the new value for B_k is $B_{k-1}B_k/B_{k-1}'$.*

Proof Denote by $\underline{b}_i'$ the new basis (i.e., $\underline{b}_{k-1}' = \underline{b}_k$ and $\underline{b}_k' = \underline{b}_{k-1}$), $\underline{b}_i'^*$ and $\mu_{i,j}'$ the new Gram–Schmidt values and B_i' the squares of the lengths of the $\underline{b}_i'^*$. Clearly, $\underline{b}_i'^* = \underline{b}_i^*$ for $1 \le i < k-1$ and $\mu_{i,j}' = \mu_{i,j}$ for $1 \le j < i < k-1$. Now

$$\begin{aligned}\underline{b}_{k-1}'^* &= \underline{b}_{k-1}' - \sum_{j=1}^{k-2} \mu_{k-1,j}'\underline{b}_j'^* \\ &= \underline{b}_k - \sum_{j=1}^{k-2} \mu_{k,j}\underline{b}_j^* \\ &= \underline{b}_k^* + \mu_{k,k-1}\underline{b}_{k-1}^*.\end{aligned}$$

Hence, $B_{k-1}' = B_k + \mu_{k,k-1}^2 B_{k-1}$.

Similarly

$$\begin{aligned}\underline{b}_k'^* &= \underline{b}_k' - \sum_{j=1}^{k-1} \mu_{k,j}'\underline{b}_j'^* \\ &= \underline{b}_{k-1} - \sum_{j=1}^{k-2} \mu_{k-1,j}\underline{b}_j^* \;-\; \left(\langle \underline{b}_{k-1}, \underline{b}_{k-1}'^*\rangle / B_{k-1}'\right)\underline{b}_{k-1}'^* \\ &= \underline{b}_{k-1}^* - \left(\langle \underline{b}_{k-1}, \underline{b}_k^* + \mu_{k,k-1}\underline{b}_{k-1}^*\rangle / B_{k-1}'\right)(\underline{b}_k^* + \mu_{k,k-1}\underline{b}_{k-1}^*) \\ &= \underline{b}_{k-1}^* - \left(\mu_{k,k-1}B_{k-1}/B_{k-1}'\right)(\underline{b}_k^* + \mu_{k,k-1}\underline{b}_{k-1}^*) \\ &= \left(1 - \mu_{k,k-1}^2 B_{k-1}/B_{k-1}'\right)\underline{b}_{k-1}^* - \left(\mu_{k,k-1}B_{k-1}/B_{k-1}'\right)\underline{b}_k^*.\end{aligned}$$

The result for $\underline{b}_k'^*$ follows since $1 - \mu_{k,k-1}^2 B_{k-1}/B_{k-1}' = B_k/B_{k-1}'$. Finally

$$B_k' = (B_k^2\langle \underline{b}_{k-1}^*, \underline{b}_{k-1}^*\rangle / B_{k-1}'^2 + \mu_{k,k-1}^2 B_{k-1}^2 \langle \underline{b}_k^*, \underline{b}_k^*\rangle / B_{k-1}'^2 = B_{k-1}B_k/B_{k-1}'. \qquad \square$$

Exercise 17.4.4 Give explicit formulae for updating the other Gram–Schmidt values in lines 7 and 13 of Algorithm 24.

Exercise 17.4.5 Show that the condition in line 9 of Algorithm 24 can be checked immediately after $\mu_{k,k-1}$ has been computed. Hence, show that the cases $1 \le j < k-1$ in the loop in lines 5 to 8 of Algorithm 24 can be postponed to line 10.

Lemma 17.4.6 *If the LLL algorithm terminates then the output basis is LLL-reduced.*

Exercise 17.4.7 Prove Lemma 17.4.6. Indeed, the fact that the Lovász conditions are satisfied is immediate. Prove the bounds on the $\mu_{i,j}$ using the three following steps. Let $1 \le j < k$ and let $b'_k = b_k - \lfloor \mu_{k,j} \rceil b_j$.

1. Prove that $\langle b_j, b_j^* \rangle = \langle b_j^*, b_j^* \rangle$ and $\langle b_j, b_i^* \rangle = 0$ if $j < i$.
2. Hence, writing $\mu'_{k,j} = \langle b'_k, b_j^* \rangle / \langle b_j^*, b_j^* \rangle$, prove that $|\mu'_{k,j}| \le 1/2$ for $1 \le j < k$.
3. For $j < i < k$ denote $\mu'_{k,i} = \langle b'_k, b_i^* \rangle / \langle b_i^*, b_i^* \rangle$. Prove that $\mu'_{k,i} = \mu_{k,i}$.

In the next section we show that the LLL algorithm does terminate. Before then we give an example and some further discussion.

Example 17.4.8 Let L be the lattice with basis matrix

$$B = \begin{pmatrix} 1 & 0 & 0 \\ 4 & 2 & 15 \\ 0 & 0 & 3 \end{pmatrix}.$$

We will perform the LLL algorithm to reduce this lattice basis.

We start with $k = 2$ and compute $\mu_{2,1} = 4/1 = 4$. So $q_1 = 4$ and we define

$$\underline{b}_2 = \underline{b}_2 - 4\underline{b}_1 = (4, 2, 15) - (4, 0, 0) = (0, 2, 15).$$

We now want to check the second LLL condition. To do this we need $\underline{b}_2^*$. We compute $\mu_{2,1} = 0$ and hence $\underline{b}_2^* = \underline{b}_2$. Then $B_1 = 1$ and $B_2 = \langle \underline{b}_2^*, \underline{b}_2^* \rangle = 229$. Clearly, $B_2 > (3/4 - \mu_{2,1}^2)B_1$ and so we set $k = 3$. Now consider $\underline{b}_3$. We compute $\mu_{3,2} = 45/229 \approx 0.19$ and, since $q_2 = 0$ there is no reduction to be performed on $\underline{b}_3$. We compute $\mu_{3,1} = 0$, so again no size reduction is required. We now compute

$$\underline{b}_3^* = \underline{b}_3 - \tfrac{45}{229}\underline{b}_2^* = (0, -90/229, 12/229).$$

We have $B_2 = 229$ and $B_3 = \langle \underline{b}_3^*, \underline{b}_3^* \rangle = 8244/52441 \approx 0.157$. From this, one can check that $B_3 < (3/4 - \mu_{3,2}^2)B_2 \approx 166.1$. Hence, we swap $\underline{b}_2$ and $\underline{b}_3$ and set $k = 2$.

At this point we have the vectors

$$\underline{b}_1 = (1, 0, 0) \text{ and } \underline{b}_2 = (0, 0, 3)$$

and $\underline{b}_1^* = \underline{b}_1$, $\underline{b}_2^* = \underline{b}_2$. First, check that $\mu_{2,1} = 0$ and so no size reduction on $\underline{b}_2$ is required. Second, $B_1 = 1$ and $B_2 = 9$ and one checks that $B_2 > (3/4 - \mu_{2,1}^2)B_1 = 0.75$. Hence, we set $k = 3$. Now

$$\underline{b}_3 = (0, 2, 15)$$

and we compute $\mu_{3,2} = 45/9 = 5$. Hence, we reduce

$$\underline{b}_3 = \underline{b}_3 - 5\underline{b}_2 = (0, 2, 0).$$

Now compute $\mu_{3,1} = 0$, so no reduction is required.

One computes $\mu_{3,2} = 0$, $\underline{b}_3^* = \underline{b}_3$ and $B_3 = 4$. Hence, $B_3 < (3/4 - \mu_{3,2}^2)B_2 = 27/4 = 6.75$ and so we should swap $\underline{b}_2$ and $\underline{b}_3$ and set $k = 2$. One can check that the $k = 2$ phase runs without making any changes. We have $B_1 = 1$ and $B_2 = 4$. Consider now $k = 3$ again. We have $\mu_{3,2} = \mu_{3,1} = 0$ and so $\underline{b}_3$ remains unchanged. Finally, $B_3 = 9 > (3/4 - \mu_{3,2}^2)B_2 = 3$ and so we set $k = 4$ and halt.

Exercise 17.4.9 Perform the LLL algorithm by hand on the basis

$$\{(-1, 5, 0),\ (2, 5, 0),\ (8, 6, 16)\}.$$

Exercise 17.4.10 Perform the LLL algorithm by hand on the basis

$$\{(0, 3, 4), (-1, 3, 3), (5, 4, -7)\}.$$

Remark 17.4.11 Part 1 of Theorem 17.2.12 shows we have $\|\underline{b}_1\| \le 2^{(n-1)/2}\lambda_1$. In other words, the LLL algorithm solves SVP up to an exponential factor, but is not guaranteed to output a shortest vector in the lattice. Hence, LLL does not officially solve SVP.

In practice, at least for relatively small dimensions, the vector $\underline{b}_1$ output by the LLL algorithm is often much closer to the shortest vector than this bound would suggest, and in many cases will be a shortest vector in the lattice. In Example 17.4.8, the theoretical bound gives $\|\underline{b}_1\| \le 2$, so $(0, 2, 0)$ would have been a possible value for $\underline{b}_1$.

17.5 Complexity of LLL

We now show that the LLL algorithm terminates and runs in polynomial-time. The original paper of Lenstra, Lenstra and Lovász [335] proves polynomial termination for any lattice in $\mathbb{R}^m$ but only gives a precise complexity for lattices in $\mathbb{Z}^m$.

Theorem 17.5.1 *Let L be a lattice in $\mathbb{Z}^m$ with basis $\underline{b}_1, \dots, \underline{b}_n$ and let $X \in \mathbb{Z}_{\ge 2}$ be such that $\|\underline{b}_i\|^2 \le X$ for $1 \le i \le n$. Let $1/4 < \delta < 1$. Then the LLL algorithm with factor δ terminates and performs $O(n^2 \log(X))$ iterations.*

Proof We need to bound the number of "backtracks" in Algorithm 24. This number is at most n plus the number of swaps. So it suffices to bound the number of swaps by $O(n^2 \log(X))$.

For $1 \le i \le n - 1$ define the $i \times m$ matrix $B_{(i)}$ formed by the first i basis vectors for the lattice. Define $d_i = \det(B_{(i)} B_{(i)}^T) \in \mathbb{Z}$, which is the square of the volume of the sublattice generated by the rows of $B_{(i)}$. Hence

$$d_i = \prod_{j=1}^{i} B_j = \prod_{j=1}^{i} \|\underline{b}_i^*\|^2 \le \prod_{j=1}^{i} \|\underline{b}_i\|^2 \le X^i.$$

Define

$$D = \prod_{i=1}^{n-1} d_i = \prod_{i=1}^{n-1} B_i^{n-i}.$$

It follows that $D \le X^{(n-1)n/2}$.

Two vectors $\underline{b}_k$ and $\underline{b}_{k-1}$ are swapped when $B_k < (\delta - \mu_{k,k-1}^2)B_{k-1}$. By Lemma 17.4.3, the new values for B_{k-1} and B_k are $B'_{k-1} = B_k + \mu_{k,k-1}^2 B_{k-1}$ and $B'_k = B_{k-1}B_k/B'_{k-1}$. Let d'_i be the new values for the d_i. We have $d'_i = d_i$ when $1 \le i < k-1$. By the Lovász condition $B'_{k-1} \le \delta B_{k-1}$. Hence, $d'_{k-1} \le \delta d_{k-1}$. Finally, since $B'_{k-1}B'_k = B_{k-1}B_k$ we have $d'_i = d_i$ for $k \le i \le n$. Hence, swapping $\underline{b}_k$ and $\underline{b}_{k-1}$ always strictly reduces D.

On the other hand, we always have[4] $d_i \in \mathbb{Z}$ and so $D \ge 1$. It follows that the number of swaps in the LLL algorithm is at most[5] $\log_\delta(X^{(n-1)n/2}) = O(n^2 \log(X))$. Hence, the algorithm requires $O(n^2 \log(X))$ iterations of the main loop. □

Algorithm 24 and Theorem 17.5.1 provide a proof that an LLL-reduced basis exists for every lattice.

Exercise 17.5.2 Let $n \in \mathbb{N}$. Show that Hermite's constant satisfies $\gamma_n \le 2^{(n-1)/4}$ (this bound can be improved to $(4/3)^{(n-1)/2}$; see [335]).

It is clear that if $L \subset \mathbb{Z}^m$ then LLL can be implemented using exact $\mathbb{Q}$ arithmetic, and hence exact integer arithmetic. But we need to show that the size of the integers does not explode. The analysis given already for the Gram–Schmidt algorithm (for example, Lemma 17.3.2) provides most of the tools we need.

Theorem 17.5.3 *Let L be a lattice in $\mathbb{Z}^m$ with basis $\underline{b}_1, \dots, \underline{b}_n$ and let $X \in \mathbb{Z}_{\ge 2}$ be such that $\|\underline{b}_i\|^2 \le X$ for $1 \le i \le n$. Then the LLL algorithm requires arithmetic operations on integers of size $O(n \log(X))$.*

Proof (Sketch) The bounds on the sizes of the $\underline{b}_i^*$ follow the same methods as used in the proof of Theorem 17.3.4. Since $\|\underline{b}_i^*\|$ is never increased during the algorithm (indeed, the vectors are specifically permuted to reduce the $\|\underline{b}_i^*\|$) we have $\|\underline{b}_i^*\| \le X^{1/2}$ at the end of each iteration. Since $d_{i-1}\underline{b}_i^* \in \mathbb{Z}^n$ and $|d_{i-1}| \le X^{i-1}$ it follows that the entries of $\underline{b}_i^*$ can be written as $n_{i,j}^*/d_{i-1}$ where $|n_{i,j}^*| \le X^i$.

Let us now consider the values $\|\underline{b}_i\|^2$ at the end of each iteration. These values all start bounded by X. As the algorithm proceeds, the values are either not yet changed (and hence still bounded by X) or have been modified so that the Gram–Schmidt basis is size reduced (and possibly swapped to an earlier position in the list of vectors). After each size reduction step (and before swapping) we have

$$\underline{b}_i = \underline{b}_i^* + \sum_{j=1}^{i-1} \mu_{i,j}\underline{b}_j^*$$

with $-1/2 \le \mu_{i,j} \le 1/2$ and so

$$\|\underline{b}_i\|^2 = \|\underline{b}_i^*\|^2 + \sum_{j=1}^{i-1} \mu_{i,j}^2 \|\underline{b}_j^*\|^2 \le nX. \tag{17.2}$$

[4] To apply this argument it is necessary to use the square of the determinant. An integer lattice that does not have full rank does not necessarily have integer determinant.

[5] Recall that $1/4 < \delta < 1$ is considered as a fixed constant.

Hence, we have $\|\underline{b}_i\| \le \sqrt{nX}$ at the end of each iteration and so the entries of $\underline{b}_i$ are all integers bounded by $\sqrt{nX}$.

The remaining detail is to bound the sizes of the $\mu_{i,j}$ and the sizes of intermediate values in line 6 of Algorithm 24. We refer to the proof of Proposition 1.26 of [335] for the bounds $|\mu_{i,j}| \le 2^{n-i}(nX^{n-1})^{1/2}$ and for further details. □

Corollary 17.5.4 *Let L be a lattice in $\mathbb{Z}^m$ with basis $\underline{b}_1, \dots, \underline{b}_n$ and let $X \in \mathbb{Z}_{\ge 2}$ be such that $\|\underline{b}_i\|^2 \le X$ for $1 \le i \le n$. Then the LLL algorithm requires $O(n^3 m \log(X))$ arithmetic operations on integers of size $O(n \log(X))$. Using naive arithmetic gives running time $O(n^5 m \log(X)^3)$ bit operations.*

Proof Theorem 17.5.1 implies that the algorithm requires $O(n^2 \log(X))$ iterations of the main loop. Within each iteration there are n operations on vectors of length m. Hence, $O(n^3 m \log(X))$ arithmetic operations. Theorem 17.5.3 implies that all arithmetic operations are on integers of size $O(n \log(X))$. □

Remark 17.5.5

1. Since the input size is $O(nm \log(X))$ and $n \le m$ the running time is cubic in the input size.
2. Note that the bounds on the sizes of integers involved in the LLL algorithm are $O(n \log(X))$ bits for the values $\mu_{i,j}$ and entries of $\underline{b}_i^*$, while only $O(\log(n) + \log(X))$ for the entries in the vectors $\underline{b}_i$. This is not just an artifact of the proof, but is a genuine phenomenon; it can already be seen in Example 17.4.8 where the $\underline{b}_i$ all have very small integer coordinates and yet $\mu_{2,1} = 45/229$.

   This leads to the idea of representing the $\mu_{i,j}$ and $\underline{b}_i^*$ using approximate (floating-point) arithmetic and keeping exact integer arithmetic only for the $\underline{b}_i$. Variants of LLL using floating-point arithmetic for the Gram–Schmidt vectors are much faster than the basic LLL algorithm presented in this chapter. Indeed, the basic algorithm is almost never used in practice.

   A problem with using floating-point approximations is that comparisons now become inexact, and this leads to problems with both termination and correctness of the output. Implementing and analysing floating-point LLL algorithms is beyond the scope of this book. We refer to Stehlé [523] and Schnorr [472] for surveys of this topic.
3. One can show (e.g., using equation (17.2)) that the complexity statement holds also for $X = \max\{\|\underline{b}_i^*\| : 1 \le i \le n\}$, which could be smaller than $\max\{\|\underline{b}_i\| : 1 \le i \le n\}$.
4. Sometimes one is interested in reducing lattice bases that are in $\mathbb{Q}^m$ and not $\mathbb{Z}^m$. Suppose all rational numbers in the basis B have numerator and denominator bounded by X. One can obtain an integer matrix by multiplying B by an integer that clears all denominators, but the resulting integers could be as big as X^{mn}. This gives a worst-case complexity of $O(n^8 m^4 \log(X)^3)$ bit operations for lattice basis reduction.

   Some applications such as simultaneous diophantine approximation (see Section 19.5) and the hidden number problem (see Section 21.7.1) have at most m non-integer entries, giving a complexity of $O(n^5 m^4 \log(X)^3)$ bit operations.

17.6 Variants of the LLL algorithm

There are many refinements of the LLL algorithm that are beyond the scope of the brief summary in this book. We list some of these now.

- As mentioned earlier, it is necessary to use floating-point arithmetic to obtain a fast version of the LLL algorithm. A variant of floating-point LLL whose running time grows quadratically in $\log(X)$ (rather than cubicly, as usual) is the L^2 algorithm of Nguyen and Stehlé [407] (also see Stehlé [523]).
- Schnorr–Euchner "deep insertions". The idea is that, rather than just swapping $\underline{b}_k$ and $\underline{b}_{k-1}$ in the LLL algorithm, one can move $\underline{b}_k$ much earlier in the list of vectors if B_k is sufficiently small. With standard LLL we have shown that swapping $\underline{b}_k$ and $\underline{b}_{k-1}$ changes B_k to $B_k + \mu_{k,k-1}^2 B_{k-1}$. A similar argument shows that inserting $\underline{b}_k$ between $\underline{b}_{i-1}$ and $\underline{b}_i$ for some $1 < i < k$ changes B_k to

$$B = B_k + \sum_{j=i}^{k-1} \mu_{k,j}^2 B_j.$$

Hence, one can let i be the smallest index such that $B < \frac{3}{4} B_i$ and insert $\underline{b}_k$ between $\underline{b}_{i-1}$ and $\underline{b}_i$ (i.e., reorder the vectors $\underline{b}_i, \ldots, \underline{b}_k$ as $\underline{b}_k, \underline{b}_i, \ldots, \underline{b}_{k-1}$). We refer to Schnorr and Euchner [473] and Section 2.6.2 of Cohen [127] for more details.
- Our presentation of the LLL algorithm was for the Euclidean norm. The algorithm has been extended to work with any norm by Lovász and Scarf [357] (also see Kaib and Ritter [291]).

  In practice, if one wants results for a norm other than the Euclidean norm, one usually performs ordinary LLL reduction with respect to the Euclidean norm and then uses the standard relations between norms (Lemma A.10.2) to determine the quality of the resulting vectors.
- Another important approach to lattice basis reduction is the block Korkine–Zolotarev algorithm due to Schnorr [468]. We mention this further in Section 18.5.

18

Algorithms for the closest and shortest vector problems

This chapter presents several algorithms to find lattice vectors close to a given vector. First we consider two methods due to Babai that, although not guaranteed to solve the closest vector problem, are useful in several situations in the book. Then we present an exponential-time algorithm to enumerate all vectors close to a given point. This algorithm can be used to solve the closest and shortest vector problems. We then briefly mention a lattice basis reduction algorithm that is guaranteed to yield better approximate solutions to the shortest vector problem.

The closest vector problem (CVP) was defined in Section 16.3. First, we remark that the shortest distance from a given vector $\underline{w} \in \mathbb{R}^n$ to a lattice vector $\underline{v} \in L$ can be quite large compared with the lengths of short vectors in the lattice.

Example 18.0.1 Consider the lattice in $\mathbb{R}^2$ with basis $(1, 0)$ and $(0, 1000)$. Then $\underline{w} = (0, 500)$ has distance 500 from the closest lattice point, despite the fact that the first successive minimum is 1.

Exercise 18.0.2 Let $L = \mathbb{Z}^n$ and $\underline{w} = (1/2, \ldots, 1/2)$. Show that $\|\underline{w} - \underline{v}\| \geq \sqrt{n}/2$ for all $\underline{v} \in L$. Hence, show that if $n > 4$ then $\|\underline{w} - \underline{v}\| > \lambda_n$ for all $\underline{v} \in L$.

18.1 Babai's nearest plane method

Let L be a full rank lattice given by an (ordered) basis $\{\underline{b}_1, \ldots, \underline{b}_n\}$ and let $\{\underline{b}_1^*, \ldots, \underline{b}_n^*\}$ be the corresponding Gram–Schmidt basis. Let $\underline{w} \in \mathbb{R}^n$. Babai [17] presented a method to inductively find a lattice vector close to $\underline{w}$. The vector $\underline{v} \in L$ output by Babai's method is not guaranteed to be such that $\|\underline{w} - \underline{v}\|$ is minimal. Theorem 18.1.6 shows that if the lattice basis is LLL-reduced then $\|\underline{w} - \underline{v}\|$ is within an exponential factor of the minimal value.

We now describe the method. Define $U = \operatorname{span}\{\underline{b}_1, \ldots, \underline{b}_{n-1}\}$ and let $L' = L \cap U$ be the sublattice spanned by $\{\underline{b}_1, \ldots, \underline{b}_{n-1}\}$. The idea of the nearest plane method is to find a vector $\underline{y} \in L$ such that the distance from $\underline{w}$ to the plane $U + \underline{y}$ is minimal. One then sets $\underline{w}'$ to be the orthogonal projection of $\underline{w}$ onto the plane $U + \underline{y}$ (in other words, $\underline{w}' \in U + \underline{y}$ and $\underline{w} - \underline{w}' \in U^{\perp}$). Let $\underline{w}'' = \underline{w}' - \underline{y} \in U$. Note that if $\underline{w} \notin L$ then $\underline{w}'' \notin L$. One inductively solves the (lower dimensional) closest vector problem of $\underline{w}''$ in L' to get $\underline{y}' \in L'$. The solution to the original instance of the CVP is $\underline{v} = \underline{y} + \underline{y}'$.

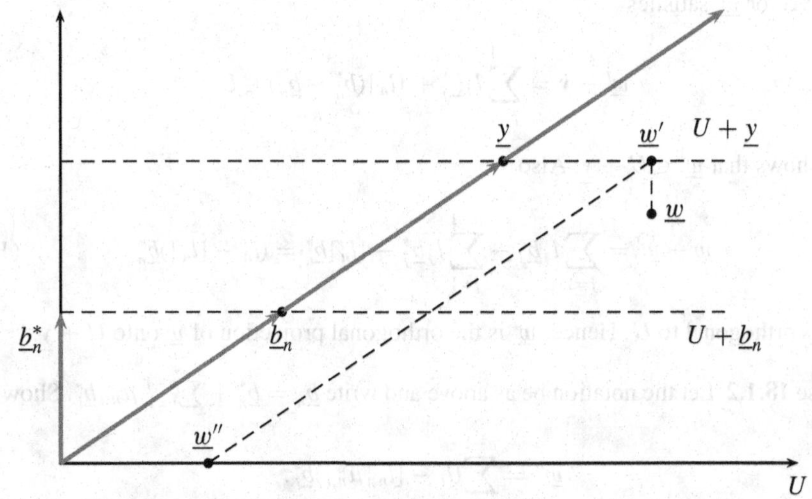

Figure 18.1 Illustration of the Babai nearest plane method. The x-axis represents the subspace U (which has dimension $n-1$) and the y-axis is perpendicular to U.

We now explain how to algebraically find $\underline{y}$ and $\underline{w}'$.

Lemma 18.1.1 *Let*

$$\underline{w} = \sum_{j=1}^{n} l_j \underline{b}_j^* \tag{18.1}$$

with $l_j \in \mathbb{R}$. Define $\underline{y} = \lfloor l_n \rceil \underline{b}_n \in L$ and $\underline{w}' = \sum_{j=1}^{n-1} l_j \underline{b}_j^* + \lfloor l_n \rceil \underline{b}_n^*$. Then $\underline{y}$ is such that the distance between $\underline{w}$ and $U + \underline{y}$ is minimal, and $\underline{w}'$ is the orthogonal projection of $\underline{w}$ onto $U + \underline{y}$.

Proof We use the fact that $U = \operatorname{span}\{\underline{b}_1^*, \dots, \underline{b}_{n-1}^*\}$. The distance from $\underline{w}$ to $U + \underline{y}$ is

$$\inf_{\underline{u} \in U} \|\underline{w} - (\underline{u} + \underline{y})\|.$$

Let $\underline{w}$ be as in equation (18.1) and let $\underline{y} = \sum_{j=1}^{n} l'_j \underline{b}_j$ be any element of L for $l'_j \in \mathbb{Z}$. One can write $\underline{y} = \sum_{j=1}^{n-1} l''_j \underline{b}_j^* + l'_n \underline{b}_n^*$ for some $l''_j \in \mathbb{R}$, $1 \le j \le n-1$.

Lemma A.10.5 shows that, for fixed $\underline{w}$ and $\underline{y}$, $\|\underline{w} - (\underline{u} + \underline{y})\|^2$ is minimised by $\underline{u} = \sum_{j=1}^{n-1} (l_j - l''_j) \underline{b}_j^* \in U$. Indeed

$$\|\underline{w} - (\underline{u} + \underline{y})\|^2 = (l_n - l'_n)^2 \|\underline{b}_n^*\|^2.$$

It follows that one must take $l'_n = \lfloor l_n \rceil$, and so the choice of $\underline{y}$ in the statement of the Lemma is correct (note that one can add any element of L' to $\underline{y}$ and it is still a valid choice).

The vector $\underline{w}'$ satisfies

$$\underline{w}' - \underline{y} = \sum_{j=1}^{n-1} l_j \underline{b}_j^* + \lfloor l_n \rceil (b_n^* - \underline{b}_n) \in U$$

which shows that $\underline{w}' \in U + \underline{y}$. Also

$$\underline{w} - \underline{w}' = \sum_{j=1}^{n} l_j \underline{b}_j^* - \sum_{j=1}^{n-1} l_j \underline{b}_j^* - \lfloor l_n \rceil \underline{b}_n^* = (l_n - \lfloor l_n \rceil) \underline{b}_n^* \qquad (18.2)$$

which is orthogonal to U. Hence, $\underline{w}'$ is the orthogonal projection of $\underline{w}$ onto $U + \underline{y}$. □

Exercise 18.1.2 Let the notation be as above and write $\underline{b}_n = \underline{b}_n^* + \sum_{i=1}^{n-1} \mu_{n,i} \underline{b}_i^*$. Show that

$$\underline{w}'' = \sum_{i=1}^{n-1} (l_i - \lfloor l_n \rceil \mu_{n,i}) \underline{b}_i^*.$$

Exercise 18.1.3 Let $\{\underline{b}_1, \dots, \underline{b}_n\}$ be an ordered basis for a lattice L. Let $\underline{w} \in \mathbb{R}^n$ and suppose that there is an element $\underline{v} \in L$ such that $\|\underline{v} - \underline{w}\| < \frac{1}{2}\|\underline{b}_i^*\|$ for all $1 \leq \underline{i} \leq n$. Prove that the nearest plane algorithm outputs v.

The following Lemma is needed to prove the main result, namely Theorem 18.1.6.

Lemma 18.1.4 *Let $\{\underline{b}_1, \dots, \underline{b}_n\}$ be LLL-reduced (with respect to the Euclidean norm, and with factor $\delta = 3/4$). If $\underline{v}$ is the output of Babai's nearest plane algorithm on input w then*

$$\|\underline{w} - \underline{v}\|^2 \leq \tfrac{2^n - 1}{4} \|\underline{b}_n^*\|^2.$$

Proof We prove the result by induction. Certainly, if $n = 1$ then $\|\underline{w} - \underline{v}\|^2 \leq \frac{1}{4}\|\underline{b}_1^*\|^2$ as required.

Now suppose $n \geq 2$. Recall that the output of the method is $\underline{v} = \underline{y} + \underline{y}'$ where $\underline{y} \in L$ minimises the distance from $\underline{w}$ to $U + \underline{y}$, $\underline{w}'$ is the orthogonal projection of $\underline{w}$ onto $U + \underline{y}$, and $\underline{y}'$ is the output of the algorithm on $\underline{w}'' = \underline{w}' - \underline{y}$ in L'. By the inductive hypothesis we know that $\|\underline{w}'' - \underline{y}'\|^2 \leq \frac{1}{4}(2^{n-1} - 1)\|\underline{b}_{n-1}^*\|^2$. Hence

$$\begin{aligned}
\|\underline{w} - (\underline{y} + \underline{y}')\|^2 &= \|\underline{w} - \underline{w}' + \underline{w}' - (\underline{y} + \underline{y}')\|^2 \\
&= \|\underline{w} - \underline{w}'\|^2 + \|\underline{w}'' - \underline{y}'\|^2 \\
&\leq \tfrac{1}{4}\|\underline{b}_n^*\|^2 + \tfrac{2^{n-1}-1}{4}\|\underline{b}_{n-1}^*\|^2
\end{aligned}$$

using equation (18.2).

Finally, part 1 of Lemma 17.2.8 implies that this is

$$\leq \left(\tfrac{1}{4} + 2\tfrac{2^{n-1}-1}{4} \right) \|\underline{b}_n^*\|^2$$

from which the result follows. □

Exercise 18.1.5 Prove that if $\underline{v}$ is the output of the nearest plane algorithm on input $\underline{w}$ then

$$\|\underline{v} - \underline{w}\|^2 \leq \frac{1}{4} \sum_{i=1}^{n} \|\underline{b}_i^*\|^2.$$

Theorem 18.1.6 *If the basis $\{\underline{b}_1, \ldots, \underline{b}_n\}$ is LLL-reduced (with respect to the Euclidean norm and with factor $\delta = 3/4$) then the output of the Babai nearest plane algorithm on $\underline{w} \in \mathbb{R}^n$ is a vector $\underline{v}$ such that $\|\underline{v} - \underline{w}\| < 2^{n/2}\|\underline{u} - \underline{w}\|$ for all $\underline{u} \in L$.*

Proof We prove the result by induction. For $n = 1$, $\underline{v}$ is a correct solution to the closest vector problem and so the result holds.

Let $n \geq 2$ and let $\underline{u} \in L$ be a closest vector to $\underline{w}$. Let $\underline{y}$ be the vector chosen in the first step of the Babai method. We consider two cases.

1. Case $\underline{u} \in U + \underline{y}$. Then $\|\underline{u} - \underline{w}\|^2 = \|\underline{u} - \underline{w}'\|^2 + \|\underline{w}' - \underline{w}\|^2$ so $\underline{u}$ is also a closest vector to $\underline{w}'$. Hence, $\underline{u} - \underline{y}$ is a closest vector to $\underline{w}'' = \underline{w}' - \underline{y} \in U$. Let $\underline{y}'$ be the output of the Babai nearest plane algorithm on $\underline{w}''$. By the inductive hypothesis

$$\|\underline{y}' - \underline{w}''\| < 2^{(n-1)/2}\|\underline{u} - \underline{y} - \underline{w}''\|.$$

Substituting $\underline{w}' - \underline{y}$ for $\underline{w}''$ gives

$$\|\underline{y} + \underline{y}' - \underline{w}'\| < 2^{(n-1)/2}\|\underline{u} - \underline{w}'\|.$$

Now

$$\|\underline{v} - \underline{w}\|^2 = \|\underline{y} + \underline{y}' - \underline{w}'\|^2 + \|\underline{w}' - \underline{w}\|^2 < 2^{n-1}\|\underline{u} - \underline{w}'\|^2 + \|\underline{w}' - \underline{w}\|^2.$$

Using $\|\underline{u} - \underline{w}'\|, \|\underline{w}' - \underline{w}\| \leq \|\underline{u} - \underline{w}\|$ and $2^{n-1} + 1 \leq 2^n$ gives the result.

2. Case $\underline{u} \notin U + \underline{y}$. Since the distance from $\underline{w}$ to $U + \underline{y}$ is $\leq \frac{1}{2}\|\underline{b}_n^*\|$, we have $\|\underline{w} - \underline{u}\| \geq \frac{1}{2}\|\underline{b}_n^*\|$. By Lemma 18.1.4 we find

$$\tfrac{1}{2}\|\underline{b}_n^*\| \geq \tfrac{1}{2}\sqrt{\frac{4}{2^n - 1}}\|\underline{w} - \underline{v}\|.$$

Hence, $\|\underline{w} - \underline{v}\| < 2^{n/2}\|\underline{w} - \underline{u}\|$.

This completes the proof. □

One can obtain a better result by using the result of Lemma 17.2.9.

Theorem 18.1.7 *If the basis $\{\underline{b}_1, \ldots, \underline{b}_n\}$ is LLL-reduced with respect to the Euclidean norm and with factor $\delta = 1/4 + 1/\sqrt{2}$ then the output of the Babai nearest plane algorithm on $\underline{w} \in \mathbb{R}^n$ is a vector $\underline{v}$ such that*

$$\|\underline{v} - \underline{w}\| < \frac{2^{n/4}}{\sqrt{\sqrt{2} - 1}}\|\underline{u} - \underline{w}\| < (1.6)2^{n/4}\|\underline{u} - \underline{w}\|$$

for all $\underline{u} \in L$.

Exercise 18.1.8 Prove Theorem 18.1.7.
[Hint: First prove that the analogue of Lemma 18.1.4 in this case is $\|\underline{w} - \underline{v}\|^2 \leq (2^{n/2} - 1)/(4(\sqrt{2} - 1))\|\underline{b}_n^*\|^2$. Then follow the proof of Theorem 18.1.6 using the fact that $\left(2^{(n-1)/4}/\sqrt{\sqrt{2}-1}\right)^2 + 1 \leq \left(2^{n/4}/\sqrt{\sqrt{2}-1}\right)^2$.]

Algorithm 25 is the Babai nearest plane algorithm. We use the notation $\underline{y}_n = \underline{y}$, $\underline{w}_n = \underline{w}$, $\underline{w}_{n-1} = \underline{w}''$ etc. Note that Babai's algorithm can be performed using exact arithmetic over $\mathbb{Q}$ or using floating-point arithmetic.

Algorithm 25 Babai nearest plane algorithm

INPUT: $\{\underline{b}_1, \ldots, \underline{b}_n\}$, $\underline{w}$
OUTPUT: $\underline{v}$
  Compute Gram–Schmidt basis $\underline{b}_1^*, \ldots, \underline{b}_n^*$
  Set $\underline{w}_n = \underline{w}$
  for $i = n$ downto 1 **do**
    Compute $l_i = \langle \underline{w}_i, \underline{b}_i^* \rangle / \langle \underline{b}_i^*, \underline{b}_i^* \rangle$
    Set $\underline{y}_i = \lfloor l_i \rceil \underline{b}_i$
    Set $\underline{w}_{i-1} = \underline{w}_i - (l_i - \lfloor l_i \rceil)\underline{b}_i^* - \lfloor l_i \rceil \underline{b}_i$
  end for
  return $\underline{v} = \underline{y}_1 + \cdots + \underline{y}_n$

Exercise 18.1.9 Let $\{\underline{b}_1, \ldots, \underline{b}_n\}$ be a basis for a lattice in $\mathbb{Z}^n$. Let $X \in \mathbb{R}_{>0}$ be such that $\|\underline{b}_i\| \leq X$ for $1 \leq i \leq n$. Show that the complexity of the Babai nearest plane algorithm (not counting LLL) when using exact arithmetic over $\mathbb{Q}$ is $O(n^5 \log(X)^2)$ bit operations.

Exercise 18.1.10 If $\{\underline{b}_1, \ldots, \underline{b}_n\}$ is an ordered LLL-reduced basis then $\underline{b}_1$ is likely to be shorter than $\underline{b}_n$. It would therefore be more natural to start with $\underline{b}_1$ and define U to be the orthogonal complement of $\underline{b}_1$. Why is this not possible?

Example 18.1.11 Consider the LLL-reduced basis

$$B = \begin{pmatrix} 1 & 2 & 3 \\ 3 & 0 & -3 \\ 3 & -7 & 3 \end{pmatrix}$$

and the vector $\underline{w} = (10, 6, 5) \in \mathbb{R}^3$. We perform the nearest plane method to find a lattice vector close to $\underline{w}$.

First compute the Gram–Schmidt basis $\underline{b}_1^* = (1, 2, 3)$, $\underline{b}_2^* = (24/7, 6/7, -12/7)$ and $\underline{b}_3^* = (10/3, -20/3, 10/3)$. Write

$$\underline{w} = \tfrac{37}{14}\underline{b}_1^* + 2\underline{b}_2^* + \tfrac{3}{20}\underline{b}_3^*.$$

Since $\lfloor 3/20 \rceil = 0$ we have $\underline{y}_3 = 0$ and $\underline{w}'' = \underline{w}' = \frac{37}{14}\underline{b}_1^* + 2\underline{b}_2^* = (19/2, 7, 9/2)$. The process is continued inductively, so write $\underline{w} = \underline{w}''$. Then one takes $\underline{y}_2 = 2\underline{b}_2 = (6, 0, -6)$ and

$\underline{w}'' = \underline{w} - \underline{y}_2 = 7/2\underline{b}_1^*$. Since $\lfloor 7/2 \rceil = 3$ we return the solution

$$3\underline{b}_1 + 2\underline{b}_2 = (9, 6, 3).$$

Exercise 18.1.12 Show that the vector $\underline{v}$ output by the Babai nearest plane method lies in the parallelepiped

$$\left\{ \underline{w} + \sum_{j=1}^{n} l_j \underline{b}_j^* : l_j \in \mathbb{R}, |l_j| \le \tfrac{1}{2} \right\}$$

centered on $\underline{w}$. Show that this parallelepiped has volume equal to the volume of the lattice. Hence, show that if $\underline{w}$ does not lie in the lattice then there is exactly one lattice point in this parallelepiped.

Some improvements to the Babai nearest plane algorithm are listed in Section 3.4 of [233]. Similar methods (but using a randomised choice of plane) were used by Klein [306] to solve the CVP when the target vector is particularly close to a lattice point. Another variant of the nearest plane algorithm is given by Lindner and Peikert [352]. The nearest plane algorithm is known by the name "VBLAST" in the communications community (see [396]).

18.2 Babai's rounding technique

An alternative to the nearest plane method is the rounding technique. This is simpler to compute in practice, since it does not require the computation of a Gram–Schmidt basis, but harder to analyse in theory. This method is also not guaranteed to solve CVP. Let $\underline{b}_1, \dots, \underline{b}_n$ be a basis for a full rank lattice in $\mathbb{R}^n$. Given a target $\underline{w} \in \mathbb{R}^n$ we can write

$$\underline{w} = \sum_{i=1}^{n} l_i \underline{b}_i$$

with $l_i \in \mathbb{R}$. One computes the coefficients l_i by solving the system of linear equations (since the lattice is full rank we can also compute the vector $(l_1, \dots, l_n)$ as $\underline{w}B^{-1}$). The rounding technique is simply to set

$$\underline{v} = \sum_{i=1}^{n} \lfloor l_i \rceil \underline{b}_i$$

where $\lfloor l \rceil$ is the closest integer to the real number l. This procedure can be performed using any basis for the lattice. Babai has proved that $\|\underline{v} - \underline{w}\|$ is within an exponential factor of the minimal value if the basis is LLL-reduced. The method trivially generalises to non-full-rank lattices as long as $\underline{w}$ lies in the $\mathbb{R}$-span of the basis.

Theorem 18.2.1 *Let $\underline{b}_1, \dots, \underline{b}_n$ be an LLL-reduced basis (with respect to the Euclidean norm and with factor $\delta = 3/4$) for a lattice $L \subseteq \mathbb{R}^n$. Then the output $\underline{v}$ of the Babai rounding*

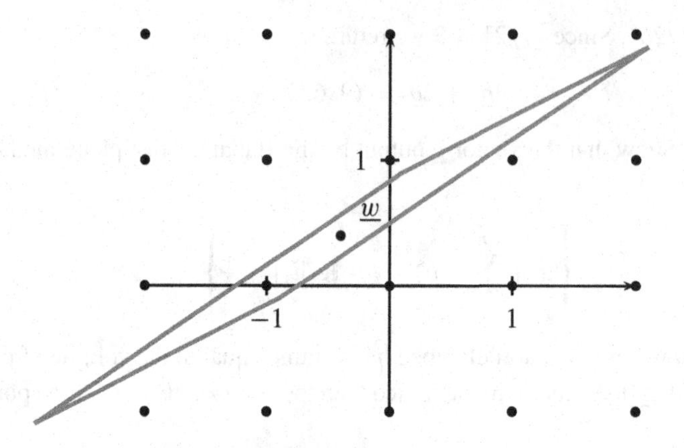

Figure 18.2 Parallelepiped centered at $(-0.4, 0.4)$ corresponding to lattice basis $(3, 2)$ and $(2, 1)$

method on input $\underline{w} \in \mathbb{R}^n$ *satisfies*

$$\|\underline{w} - \underline{v}\| \le (1 + 2n(9/2)^{n/2})\|\underline{w} - \underline{u}\|$$

for all $\underline{u} \in L$.

Proof See Babai [17]. □

Babai rounding gives a lattice point $\underline{v}$ such that $\underline{w} - \underline{v} = \sum_{i=1}^{n} m_i \underline{b}_i$ where $|m_i| \le 1/2$. In other words, $\underline{v}$ lies in the parallelepiped, centered at $\underline{w}$, defined by the basis vectors. Since the volume of the parallelepiped is equal to the volume of the lattice, if $\underline{w}$ is not in the lattice then there is exactly one lattice point in the parallelepiped. The geometry of the parallelepiped determines whether or not an optimal solution to the CVP is found. Hence, though the rounding method can be used with any basis for a lattice, the result depends on the quality of the basis.

Example 18.2.2 Let $\underline{b}_1 = (3, 2)$ and $\underline{b}_2 = (2, 1)$ generate the lattice $\mathbb{Z}^2$. Let $\underline{w} = (-0.4, 0.4)$ so that the solution to CVP is $(0, 0)$. One can verify that $(-0.4, 0.4) = 1.2\underline{b}_1 - 2\underline{b}_2$ and so Babai rounding yields $\underline{b}_1 - 2\underline{b}_2 = (-1, 0)$. Figure 18.2 shows the parallelepiped centered at $\underline{w}$ corresponding to the basis. One can see that $(-1, 0)$ is the only lattice point within that parallelepiped.

Exercise 18.2.3 Consider the vector $\underline{w} = (-0.4, 0.4)$ as in Example 18.2.2 again. Using the basis $\{(1, 0), (0, 1)\}$ for $\mathbb{Z}^2$ use the Babai rounding method to find the closest lattice vector in $\mathbb{Z}^2$ to $\underline{w}$. Draw the parallelepiped centered on $\underline{w}$ in this case.

We stress that the rounding method is not the same as the nearest plane method. The next example shows that the two methods can give different results.

Example 18.2.4 Consider the CVP instance in Example 18.1.11. We have

$$\underline{w} = \frac{141}{40}\underline{b}_1 + \frac{241}{120}\underline{b}_2 + \frac{3}{20}\underline{b}_3.$$

Hence, one sets

$$\underline{v} = 4\underline{b}_1 + 2\underline{b}_2 = (10, 8, 6).$$

Note that this is a different solution to the one found in Example 18.1.11, though both solutions satisfy $\|\underline{w} - \underline{v}\| = \sqrt{5}$.

Exercise 18.2.5 Prove that if $\underline{b}_1, \ldots, \underline{b}_n$ are orthogonal basis vectors for a lattice L then the Babai rounding technique produces a correct solution to the CVP with respect to the Euclidean norm. Show also that the Babai rounding technique gives the same result as the Babai nearest plane method in this case.

Exercise 18.2.6 Show that the nearest plane and rounding methods produce a linear combination of the lattice basis where the vector $\underline{b}_n$ has the same coefficient.

Exercise 18.2.7 Consider the lattice with basis

$$\begin{pmatrix} 7 & 0 & 1 \\ 1 & 17 & 1 \\ -3 & 0 & 10 \end{pmatrix}$$

and let

$$\underline{w} = (100, 205, 305).$$

Find a lattice vector $\underline{v}$ close to $\underline{w}$ using the rounding technique. What is $\|\underline{v} - \underline{w}\|$?

The Babai rounding algorithm is known by the name "zero forcing" in the communications community (see [396]).

18.3 The embedding technique

Another way to solve CVP is the **embedding technique**, due to Kannan (see page 437 onwards of [296]). Let B be a basis matrix for a lattice L and suppose $\underline{w} \in \mathbb{R}^n$ (in practice, we assume $\underline{w} \in \mathbb{Q}^n$). A solution to the CVP corresponds to integers $l_1, \ldots, l_n$ such that

$$\underline{w} \approx \sum_{i=1}^{n} l_i \underline{b}_i .$$

The crucial observation that $\underline{e} = \underline{w} - \sum_{i=1}^{n} l_i \underline{b}_i$ is such that $\|\underline{e}\|$ is small.

The idea of the embedding technique is to define a lattice L' that contains the short vector $\underline{e}$. Let $M \in \mathbb{R}_{>0}$ (for example, $M = 1$). The lattice L' is defined by the vectors (which are a basis for $\mathbb{R}^{n+1}$)

$$(\underline{b}_1, 0), \ldots, (\underline{b}_n, 0), (\underline{w}, M). \tag{18.3}$$

One sees that taking the linear combination of rows with coefficients $(-l_1, \ldots, -l_n, 1)$ gives the vector

$$(\underline{e}, M).$$

Hence, we might be able to find $\underline{e}$ by solving the SVP problem in the lattice L'. One can then solve the CVP by subtracting $\underline{e}$ from $\underline{w}$.

Example 18.3.1 Consider the basis matrix

$$B = \begin{pmatrix} 35 & 72 & -100 \\ -10 & 0 & -25 \\ -20 & -279 & 678 \end{pmatrix}$$

for a lattice in $\mathbb{R}^3$. We solve the CVP instance with $\underline{w} = (100, 100, 100)$.

Apply the LLL algorithm to the basis matrix (taking $M = 1$)

$$\begin{pmatrix} 35 & 72 & -100 & 0 \\ -10 & 0 & -25 & 0 \\ -20 & -279 & 678 & 0 \\ 100 & 100 & 100 & 1 \end{pmatrix}$$

for the lattice L'. This gives the basis matrix

$$\begin{pmatrix} 0 & 1 & 0 & 1 \\ 5 & 0 & 1 & 0 \\ 0 & 5 & 1 & -4 \\ 5 & 5 & -21 & -4 \end{pmatrix}.$$

The first row is $(0, 1, 0, 1)$, so we know that $(0, 1, 0)$ is the difference between $\underline{w}$ and a lattice point $\underline{v}$. One verifies that $\underline{v} = (100, 100, 100) - (0, 1, 0) = (100, 99, 100)$ is a lattice point.

The success of the embedding technique depends on the size of $\underline{e}$ compared with the lengths of short vectors in the original lattice L. As we have seen in Exercise 18.0.2, $\underline{e}$ can be larger than λ_n, in which case the embedding technique is not likely to be a good way to solve the closest vector problem.

Lemma 18.3.2 *Let $\{\underline{b}_1, \ldots, \underline{b}_n\}$ be a basis for a lattice $L \subseteq \mathbb{Z}^n$ and denote by λ_1 the shortest Euclidean length of a non-zero element of L. Let $\underline{w} \in \mathbb{R}^n$ and let $\underline{v} \in L$ be a closest lattice point to $\underline{w}$. Define $\underline{e} = \underline{w} - \underline{v}$. Suppose that $\|\underline{e}\| < \lambda_1/2$ and let $M = \|\underline{e}\|$. Then $(\underline{e}, M)$ is a shortest non-zero vector in the lattice L' of the embedding technique.*

Proof All vectors in the lattice L' are of the form

$$l_{n+1}(\underline{e}, M) + \sum_{i=1}^{n} l_i(\underline{b}_i, 0)$$

for some $l_1, \ldots, l_{n+1} \in \mathbb{Z}$. Every non-zero vector with $l_{n+1} = 0$ is of length at least λ_1. Since

$$\|(\underline{e}, M)\|^2 = \|\underline{e}\|^2 + M^2 = 2M^2 < 2\lambda_1^2/4$$

the vector $(\underline{e}, \pm M)$ has length at most $\lambda_1/\sqrt{2}$. Since $\underline{v}$ is a closest vector to $\underline{w}$ it follows that $\|\underline{e}\| \leq \|\underline{e} + \underline{x}\|$ for all $\underline{x} \in L$, and so every other vector $(\underline{u}, M) \in L'$ has length at least as large. Finally, suppose $|l_{n+1}| \geq 2$. Then

$$\|(\underline{u}, l_{n+1}M)\|^2 \geq \|(0, l_{n+1}M)\|^2 \geq (2M)^2$$

and so $\|(\underline{u}, l_{n+1}M)\| \geq 2\|(\underline{e}, M)\|$. □

Lemma 18.3.2 shows that the CVP can be reduced to SVP as long as the target vector is very close to a lattice vector, and assuming one has a good guess M for the distance. However, when using algorithms such as LLL that solve the approximate SVP it is not possible, in general, to make rigorous statements about the success of the embedding technique. As mentioned earlier, the LLL algorithm often works better than the theoretical analysis predicts. Hence, the embedding technique can potentially be useful even when $\underline{w}$ is not so close to a lattice point. For further discussion see Lemma 6.15 of Kannan [296].

Exercise 18.3.3 Let $\{\underline{b}_1, \ldots, \underline{b}_n\}$ be a basis for a lattice in $\mathbb{R}^n$ and let $\underline{w} \in \mathbb{R}^n$. Let $M = \max_{1 \leq i \leq n} \|\underline{b}_i\|$. Show that the output $(\underline{e}, M)$ of the embedding technique (using LLL) on the basis of equation (18.3) is the same as the output of the Babai nearest plane algorithm when run on the LLL-reduced basis.

Exercise 18.3.4 Solve the following CVP instance using the embedding technique and a computer algebra package.

$$B = \begin{pmatrix} -265 & 287 & 56 \\ -460 & 448 & 72 \\ -50 & 49 & 8 \end{pmatrix}, \qquad \underline{w} = (100, 80, 100).$$

18.4 Enumerating all short vectors

We present a method to enumerate all short vectors in a lattice, given any basis. We will show later that the performance of this enumeration algorithm depends on the quality of the lattice basis. Throughout this section, $\|\underline{v}\|$ denotes the Euclidean norm.

The first enumeration method was given by Pohst in 1981. Further variants were given by Finke and Pohst, Kannan [295, 296], Helfrich [255] and Schnorr and Euchner [473]. These methods are all deterministic and are guaranteed to output a non-zero vector of minimum length. The time complexity is exponential in the lattice dimension, but the storage requirements polynomial. This approach is known by the name "sphere decoding" in the communications community (see [396]).

Exercise 18.4.1 Let $\{\underline{b}_1, \ldots, \underline{b}_n\}$ be an (ordered) basis in $\mathbb{R}^m$ for a lattice and let $\{\underline{b}_1^*, \ldots, \underline{b}_n^*\}$ be the Gram–Schmidt orthogonalisation. Let $\underline{v} \in \mathbb{R}^m$. Show that the projection of $\underline{v}$ onto $\underline{b}_i^*$ is

$$\frac{\langle \underline{v}, \underline{b}_i^* \rangle}{\| \underline{b}_i^* \|^2} \underline{b}_i^*.$$

Show that if $\underline{v} = \sum_{j=1}^n x_j \underline{b}_j$ then this projection is

$$\left(x_i + \sum_{j=i+1}^n x_j \mu_{j,i} \right) \underline{b}_i^*.$$

Lemma 18.4.2 *Let $\{\underline{b}_1, \ldots, \underline{b}_n\}$ be an (ordered) basis for a lattice and let $\{\underline{b}_1^*, \ldots, \underline{b}_n^*\}$ be the Gram–Schmidt orthogonalisation. Fix $A \in \mathbb{R}_{>0}$ and write $B_i = \| \underline{b}_i^* \|^2$. Let $\underline{v} = \sum_{i=1}^n x_i \underline{b}_i$ be such that $\| \underline{v} \|^2 \le A$. For $1 \le i \le n$ define*

$$z_i = x_i + \sum_{j=i+1}^n \mu_{j,i} x_j.$$

Then for $1 \le i < n$

$$\sum_{i=1}^n z_i^2 B_i \le A.$$

Proof Exercise 18.4.1 gives a formula $z_i \underline{b}_i^*$ for the projection of $\underline{v}$ onto each $\underline{b}_i^*$. Since the vectors $\underline{b}_i^*$ are orthogonal we have

$$\| \underline{v} \|^2 = \sum_{i=1}^n \| z_i \underline{b}_i^* \|^2 = \sum_{i=1}^n z_i^2 B_i.$$

The result follows. □

Theorem 18.4.3 *Let the notation be as in Lemma 18.4.2. Then one has $x_n^2 \le A / \| \underline{b}_n^* \|^2$ and, for $1 \le i < n$*

$$\left(x_i + \sum_{j=i+1}^n \mu_{j,i} x_j \right)^2 B_i \le A - \sum_{j=i+1}^n z_j^2 B_j.$$

Proof Note that $z_n = x_n$ and Lemma 18.4.2 implies $z_n^2 B_n \le A$, which proves the first statement. The second statement is also just a re-writing of Lemma 18.4.2. □

We now sketch the enumeration algorithm for finding all short lattice vectors $\underline{v} = \sum_{i=1}^n x_i \underline{b}_i$, which follows from the above results. First, without loss of generality we may assume that $x_n \ge 0$. By Theorem 18.4.3 we know $0 \le x_n \le \sqrt{A/B_i}$. For each candidate x_n one knows that

$$(x_{n-1} + \mu_{n,n-1} x_n)^2 B_{n-1} \le A - x_n^2 B_n$$

and so

$$|x_{n-1} + \mu_{n,n-1} x_n| \le \sqrt{(A - x_n^2 B_n)/B_{n-1}}.$$

To phrase this as a bound on x_{n-1} one uses the fact that, for any $a \in \mathbb{R}, b \in \mathbb{R}_{\ge 0}$, the solutions $x \in \mathbb{R}$ to $|x + a| \le b$ satisfy $-(b + a) \le x \le b - a$. Hence, writing $M_1 = \sqrt{\left(A - x_n^2 B_n\right) / B_{n-1}}$ one has

$$-(M_1 + \mu_{n,n-1} x_n) \le x_{n-1} \le M_1 - \mu_{n,n-1} x_n.$$

Exercise 18.4.4 Generalise the above discussion to show that for $1 \le i < n$ one has

$$-(M_1 + M_2) \le x_i \le M_1 - M_2$$

where

$$M_1 = \sqrt{\left(A - \sum_{j=i+1}^{n} x_j^2 B_j\right) / B_i}$$

and $M_2 = \sum_{j=i+1}^{n} \mu_{j,i} x_j$.

Exercise 18.4.5 Write pseudocode for the algorithm to enumerate all short vectors of a lattice.

The algorithm to find a non-zero vector of minimal length is then straightforward. Set A to be $\|\underline{b}_1\|^2$, enumerate all vectors of length at most A and, for each vector, compute the length. One is guaranteed to find a shortest vector in the lattice. Schnorr and Euchner [473] organised the search in a manner to minimise the running time.

The running time of this algorithm depends on the quality of the basis in several ways. First, it is evidently important to have a good bound A for the length of the shortest vector. Taking $A = \|\underline{b}_1\|^2$ is only sensible if $\underline{b}_1$ is already rather short; alternatively one may choose, say, $A = \sqrt{\frac{n}{2\pi e}} \det(L)^{1/n}$ using the Gaussian heuristic (one can choose a small bound for A and then, if the search fails, increase A accordingly). Second, one sees that if $\underline{b}_n^*$ is very short then the algorithm searches a huge range of values for x_n, and similarly if $\underline{b}_{n-1}^*$ is very short etc. Hence, the algorithm performs best if the values $\|\underline{b}_i^*\|$ descrease rather gently.

To solve SVP in practice using enumeration one first performs LLL and other pre-computation to get a sufficiently nice basis. We refer to Kannan [295, 296], Schnorr and Euchner [473] and Agrell, Eriksson, Vardy and Zeger [7] for details. The best complexity statement in the literature is due to Hanrot and Stehlé.

Theorem 18.4.6 *(Hanrot and Stehlé [249]) There exists a polynomial $p(x, y) \in \mathbb{R}[x, y]$ such that, for any n-dimensional lattice L in $\mathbb{Z}^m$ with basis consisting of vectors with coefficients bounded by B, one can compute all the shortest non-zero vectors in L in at most $p(\log(B), m) n^{n/2e + o(n)}$ bit operations.*

Due to lack of space we refer to the original papers for further details about enumeration algorithms. Pujol and Stehlé [455] give an analysis of issues related to floating-point implementation.

In practice, the most efficient methods to compute the SVP are heuristic "pruning" methods. These methods are still exponential in the lattice dimension, and are not guaranteed to output the shortest vector. The extreme pruning algorithm of Gama, Nguyen and Regev [227] is currently the most practical method.

A quite different approach, leading to non-deterministic algorithms (in other words, the output is a non-zero vector in the lattice that, with high probability, has minimal length) is due to Ajtai, Kumar and Sivakumar (see [322] for a survey). The running time and storage requirements of the algorithm are both exponential in the lattice dimension. For some experimental results we refer to Nguyen and Vidick [417]. Micciancio and Voulgaris [394] have given an improved algorithm, still requiring exponential time and storage.

18.4.1 Enumeration of closest vectors

The above ideas can be adapted to list lattice points close to some $\underline{w} \in \mathbb{R}^n$. Let $A \in \mathbb{R}_{>0}$ and suppose we seek all $\underline{v} \in L$ such that $\|\underline{v} - \underline{w}\|^2 \leq A$. Write $\underline{v} = \sum_{i=1}^n x_i \underline{b}_i = \sum_{i=1}^n z_i \underline{b}_i^*$ as before and write

$$\underline{w} = \sum_{i=1}^{n} y_i \underline{b}_i^*.$$

Then $\|\underline{v} - \underline{w}\|^2 \leq A$ is equivalent to

$$\sum_{i=1}^{n} (z_i - y_i)^2 \|\underline{b}_i^*\|^2 \leq A.$$

It follows that

$$y_n - \sqrt{A/B_n} \leq x_n \leq y_n + \sqrt{A/B_n}$$

and so on.

Lemma 18.4.7 *Let the notation be as above and define*

$$M_i = \sqrt{\left(A - \sum_{j=i+1}^{n} z_j^2 B_j\right) / B_i}$$

and $N_i = \sum_{j=i+1}^n \mu_{j,i} x_j$ *for* $1 \leq i \leq n$. *If* $\underline{v} = \sum_{i=1}^n x_i \underline{b}_i$ *satisfies* $\|\underline{v} - \underline{w}\|^2 \leq A$ *then, for* $1 \leq i \leq n$

$$y_i - M_i - N_i \leq x_i \leq y_i + M_i - N_i.$$

Exercise 18.4.8 Prove Lemma 18.4.7.

The paper by Agrell, Eriksson, Vardy and Zeger [7] gives an excellent survey and comparison of the various enumeration techniques. They conclude that the Schnorr-Euchner variant is much more efficient than the Pohst or Kannan versions.

18.5 Korkine–Zolotarev bases

We present a notion of reduced lattice basis that has better properties than an LLL-reduced basis.

Definition 18.5.1 Let L be a lattice of rank n in $\mathbb{R}^m$. An ordered basis $\{\underline{b}_1, \ldots, \underline{b}_n\}$ for L is **Korkine–Zolotarev reduced**[1] if

1. $\underline{b}_1$ is a non-zero vector of minimal length in L;
2. $|\mu_{i,1}| < 1/2$ for $2 \le i \le n$;
3. the basis $\{\underline{b}_2 - \mu_{2,1}\underline{b}_1, \ldots, \underline{b}_n - \mu_{n,1}\underline{b}_1\}$ is Korkine–Zolotarev reduced (this is the orthogonal projection of the basis of L onto the orthogonal complement of $\underline{b}_1$)

where $\underline{b}_i^*$ is the Gram–Schmidt orthogonalisation and $\mu_{i,j} = \langle \underline{b}_i, \underline{b}_j^* \rangle / \langle \underline{b}_j^*, \underline{b}_j^* \rangle$.

One problem is that there is no known polynomial-time algorithm to compute a Korkine–Zolotarev basis.

Theorem 18.5.2 *Let $\{\underline{b}_1, \ldots, \underline{b}_n\}$ be a Korkine–Zolotarev reduced basis of a lattice L. Then*

1. for $1 \le i \le n$

$$\frac{4}{i+3}\lambda_i^2 \le \|\underline{b}_i\|^2 \le \frac{i+3}{4}\lambda_i^2;$$

2.

$$\prod_{i=1}^{n} \|\underline{b}_i\|^2 \le \left(\gamma_n^n \prod_{i=1}^{n} \frac{i+3}{4} \right) \det(L)^2.$$

Proof See Theorem 2.1 and 2.3 of Lagarias, Lenstra and Schnorr [324]. □

As we have seen, for lattices of relatively small dimension it is practical to enumerate all short vectors. Hence, one can compute a Korkine–Zolotarev basis for lattices of small dimension. Schnorr has developed the **block Korkine–Zolotarev** lattice basis reduction algorithm, which computes a Korkine–Zolotarev basis for small-dimensional projections of the original lattice and combines this with the LLL algorithm. The output basis can be proved to be of a better quality than an LLL-reduced basis. This is the most powerful algorithm for finding short vectors in lattices of large dimension. Due to lack of space we are unable to present this algorithm; we refer to Schnorr [468] for details.

[1] Some authors also call it Hermite–Korkine–Zolotarev (HKV) reduced.

19
Coppersmith's method and related applications

An important application of lattice basis reduction is finding small solutions to polynomial equations $F(x) \equiv 0 \pmod{M}$ of degree $d > 1$. The main purpose of this chapter is to present some results of Coppersmith [131] on this problem. We also discuss finding small roots of bivariate integer polynomials and some other applications of these ideas.

In general, finding solutions to modular equations is easy if we know the factorisation of the modulus, see Section 2.12. However, if the factorisation of the modulus M is not known then finding solutions can be hard. For example, if we can find a solution to $x^2 \equiv 1 \pmod{M}$ that is not $x = \pm 1$ then we can split M. Hence, we do not expect efficient algorithms for finding all solutions to modular equations in general.

Suppose then that the polynomial equation has a "small" solution. It is not so clear that finding the roots is necessarily a hard problem. The example $x^2 \equiv 1 \pmod{M}$ no longer gives any intuition since the two non-trivial roots both have absolute value at least $\sqrt{M}$. As we will explain in this chapter, if $F(x) \equiv 0 \pmod{M}$ of degree d has a solution x_0 such that $|x_0| < M^{1/d-\epsilon}$ for small $\epsilon > 0$ then it can be found in polynomial-time. This result has a number of important consequences.

General references for the contents of this chapter are Coppersmith [131, 132], May [370, 371], Nguyen and Stern [415] and Nguyen [409].

19.1 Coppersmith's method for modular univariate polynomials

19.1.1 First steps to Coppersmith's method

We sketch the basic idea of the method, which goes back to Håstad. Let $F(x) = x^d + a_{d-1}x^{d-1} + \cdots + a_1 x + a_0$ be a monic polynomial of degree d with integer coefficients. Suppose we know that there exist one or more integers x_0 such that $F(x_0) \equiv 0 \pmod{M}$ and that $|x_0| < M^{1/d}$. The problem is to find all such roots.

Since $|x_0^i| < M$ for all $0 \le i \le d$ then, if the coefficients of $F(x)$ are small enough, one might have $F(x_0) = 0$ over $\mathbb{Z}$. The problem of finding integer roots of integer polynomials is easy: we can find roots over $\mathbb{R}$ using numerical analysis (e.g., Newton's method) and then round the approximations of the roots to the nearest integer and check whether they are solutions of $F(x)$.

The problem is therefore to deal with polynomials $F(x)$ having a small solution but whose coefficients are not small. Coppersmith's idea (in the formulation of Howgrave-Graham [268]) is to build from $F(x)$ a polynomial $G(x)$ that still has the same solution x_0, but which has coefficients small enough that the above logic does apply.

Example 19.1.1 Let $M = 17 \cdot 19 = 323$ and let

$$F(x) = x^2 + 33x + 215.$$

We want to find the small solution to $F(x) \equiv 0 \pmod{M}$ (in this case, $x_0 = 3$ is a solution, but note that $F(3) \neq 0$ over $\mathbb{Z}$).

We seek a related polynomial with small coefficients. For this example

$$G(x) = 9F(x) - M(x + 6) = 9x^2 - 26x - 3$$

satisfies $G(3) = 0$. This root can be found using Newton's method over $\mathbb{R}$ (or even the quadratic formula).

We introduce some notation for the rest of this section. Let $M, X \in \mathbb{N}$ and let $F(x) = \sum_{i=0}^{d} a_i x^i \in \mathbb{Z}[x]$. Suppose $x_0 \in \mathbb{Z}$ is a solution to $F(x) \equiv 0 \pmod{M}$ such that $|x_0| < X$. We associate with the polynomial $F(x)$ the row vector

$$b_F = (a_0, a_1 X, a_2 X^2, \cdots, a_d X^d). \tag{19.1}$$

Vice versa, any such row vector corresponds to a polynomial. Throughout this section we will interpret polynomials as row vectors, and row vectors as polynomials, in this way.

Theorem 19.1.2 *(Howgrave-Graham [268]) Let $F(x), X, M, b_F$ be as above (i.e., there is some x_0 such that $|x_0| \leq X$ and $F(x_0) \equiv 0 \pmod{M}$). If $\|b_F\| < M/\sqrt{d+1}$ then $F(x_0) = 0$.*

Proof Recall the Cauchy–Schwarz inequality $(\sum_{i=1}^{n} x_i y_i)^2 \leq (\sum_{i=1}^{n} x_i^2)(\sum_{i=1}^{n} y_i^2)$ for $x_i, y_i \in \mathbb{R}$. Taking $x_i \geq 0$ and $y_i = 1$ for $1 \leq i \leq n$ one has

$$\sum_{i=1}^{n} x_i \leq \sqrt{n \sum_{i=1}^{n} x_i^2}.$$

Now

$$|F(x_0)| = \left| \sum_{i=0}^{d} a_i x_0^i \right| \leq \sum_{i=0}^{d} |a_i||x_0|^i \leq \sum_{i=0}^{d} |a_i| X^i$$
$$\leq \sqrt{d+1}\|b_F\| < \sqrt{d+1} M/\sqrt{d+1} = M$$

where the third inequality is Cauchy–Schwarz, so $-M < F(x_0) < M$. But $F(x_0) \equiv 0 \pmod{M}$ and so $F(x_0) = 0$. □

Let $F(x) = \sum_{i=0}^{d} a_i x^i$ be a monic polynomial. We assume that $F(x)$ has at least one solution x_0 modulo M such that $|x_0| < X$ for some specified integer X. If $F(x)$ is not monic but $\gcd(a_d, M) = 1$ then one can multiply $F(x)$ by $a_d^{-1} \pmod{M}$ to make it monic. If $\gcd(a_d, M) > 1$ then one can split M and reduce the problem to two (typically easier) problems. As explained above, to find x_0 it will be sufficient to find a polynomial $G(x)$ with the same root x_0 modulo M but with sufficiently small coefficients.

To do this, consider the $d+1$ polynomials $G_i(x) = Mx^i$ for $0 \le i < d$ and $F(x)$. They all have the solution $x = x_0$ modulo M. Define the lattice L with basis corresponding to these polynomials (by associating with a polynomial the row vector in equation (19.1)). Therefore, the basis matrix for the lattice L is

$$B = \begin{pmatrix} M & 0 & \cdots & 0 & 0 \\ 0 & MX & \cdots & 0 & 0 \\ \vdots & & & \vdots & \vdots \\ 0 & 0 & \cdots & MX^{d-1} & 0 \\ a_0 & a_1 X & \cdots & a_{d-1}X^{d-1} & X^d \end{pmatrix}. \tag{19.2}$$

Every element of this lattice is a row vector that can be interpreted as a polynomial $F(x)$ (via equation (19.1) such that $F(x_0) \equiv 0 \pmod{M}$.

Lemma 19.1.3 *The dimension of the lattice L defined in equation (19.2) above is $d+1$ and the determinant is*

$$\det(L) = M^d X^{d(d+1)/2}.$$

Exercise 19.1.4 Prove Lemma 19.1.3.

One now runs the LLL algorithm on this (row) lattice basis. Let $G(x)$ be the polynomial corresponding to the first vector $\underline{b}_1$ of the LLL-reduced basis (since every row of B has the form of equation (19.1) then so does $\underline{b}_1$).

Theorem 19.1.5 *Let the notation be as above and let $G(x)$ be the polynomial corresponding to the first vector in the LLL-reduced basis for L. Set $c_1(d) = 2^{-1/2}(d+1)^{-1/d}$. If $X < c_1(d)M^{2/d(d+1)}$ then any root x_0 of $F(x)$ modulo M such that $|x_0| \le X$ satisfies $G(x_0) = 0$ in $\mathbb{Z}$.*

Proof Recall that $\underline{b}_1$ satisfies

$$\|\underline{b}_1\| \le 2^{(n-1)/4} \det(L)^{1/n} = 2^{d/4} M^{d/(d+1)} X^{d/2}.$$

For $\underline{b}_1$ to satisfy the conditions of Howgrave-Graham's theorem (i.e., $\|\underline{b}_1\| < M/\sqrt{d+1}$) it is sufficient that

$$2^{d/4} M^{d/(d+1)} X^{d/2} < M/\sqrt{d+1}.$$

This can be written as

$$\sqrt{d+1} 2^{d/4} X^{d/2} < M^{1/(d+1)},$$

which is equivalent to the condition in the statement of the theorem. □

In other words, if $d = 2$ then it is sufficient that $X \approx M^{1/3}$ to find the small solution using the above method. If $d = 3$ then it is sufficient that $X \approx M^{1/6}$. This is the result of Håstad. Of course, LLL often works better than the worst-case bound, so small solutions x_0 may be found even when x_0 does not satisfy the condition of the Theorem.

Example 19.1.6 Let $M = 10001$ and consider the polynomial

$$F(x) = x^3 + 10x^2 + 5000x - 222.$$

One can check that $F(x)$ is irreducible, and that $F(x)$ has the small solution $x_0 = 4$ modulo M. Note that $|x_0| < M^{1/6}$ so one expects to be able to find x_0 using the above method. Suppose $X = 10$ is the given bound on the size of x_0. Consider the basis matrix

$$B = \begin{pmatrix} M & 0 & 0 & 0 \\ 0 & MX & 0 & 0 \\ 0 & 0 & MX^2 & 0 \\ -222 & 5000X & 10X^2 & X^3 \end{pmatrix}.$$

Running LLL on this matrix gives a reduced basis, the first row of which is

$$(444, 10, -2000, -2000).$$

The polynomial corresponding to this vector is

$$G(x) = 444 + x - 20x^2 - 2x^3.$$

Running Newton's root-finding method on $G(x)$ gives the solution $x_0 = 4$.

19.1.2 The full Coppersmith method

The method in the previous section allows one to find small roots of modular polynomials, but it can be improved further. Looking at the proof of Theorem 19.1.5 one sees that the requirement for success is essentially $\det(L) < M^n$ (more precisely, it is $2^{d/4}M^{d/(d+1)}X^{d/2} < M/\sqrt{d+1}$). There are two strategies to extend the utility of the method (i.e., to allow bigger values for X). The first is to increase the dimension n by adding rows to L that contribute less than M to the determinant. The second is to increase the power of M on the right hand side. One can increase the dimension without increasing the power of M by using the so-called "x-shift" polynomials $xF(x), x^2F(x), \ldots, x^kF(x)$; Example 19.1.7 gives an example of this. One can increase the power of M on the right hand side by using powers of $F(x)$ (since if $F(x_0) \equiv 0 \pmod{M}$ then $F(x_0)^k \equiv 0 \pmod{M^k}$).

Example 19.1.7 Consider the problem of Example 19.1.6. The lattice has dimension 4 and determinant M^3X^3. The condition for LLL to output a sufficiently small vector is

$$2^{3/4}\left(M^3X^6\right)^{1/4} \leq \frac{M}{\sqrt{4}},$$

which, taking $M = 10001$, leads to $X \approx 2.07$. (Note that the method worked for a larger value of x_0; this is because the bound used on LLL only applies in the worst case.)

Consider instead the basis matrix that also includes rows corresponding to the polynomials $xF(x)$ and $x^2F(x)$

$$B = \begin{pmatrix} M & 0 & 0 & 0 & 0 & 0 \\ 0 & MX & 0 & 0 & 0 & 0 \\ 0 & 0 & MX^2 & 0 & 0 & 0 \\ -222 & 5000X & 10X^2 & X^3 & 0 & 0 \\ 0 & -222X & 5000X^2 & 10X^3 & X^4 & 0 \\ 0 & 0 & -222X^2 & 5000X^3 & 10X^4 & X^5 \end{pmatrix}.$$

The dimension is 6 and the determinant is M^3X^{15}. The condition for LLL to output a sufficiently small vector is

$$2^{5/4}\left(M^3X^{15}\right)^{1/6} \leq \frac{M}{\sqrt{6}},$$

which leads to $X \approx 3.11$. This indicates that some benefit can be obtained by using x-shifts.

Exercise 19.1.8 Let $G(x)$ be a polynomial of degree d. Show that taking d x-shifts $G(x), xG(x), \ldots, x^{d-1}G(x)$ gives a method that works for $X \approx M^{1/(2d-1)}$.

Exercise 19.1.8 shows that when $d = 3$ we have improved the result from $X \approx M^{1/6}$ to $X \approx M^{1/5}$. Coppersmith [131] exploits both x-shifts and powers of $F(x)$. We now present the method in full generality.

Theorem 19.1.9 *(Coppersmith) Let $0 < \epsilon < \min\{0.18, 1/d\}$. Let $F(x)$ be a monic polynomial of degree d with one or more small roots x_0 modulo M such that $|x_0| < \frac{1}{2}M^{1/d-\epsilon}$. Then x_0 can be found in time, bounded by a polynomial in d, $1/\epsilon$ and $\log(M)$.*

Proof Let $h > 1$ be an integer that depends on d and ϵ and will be determined in equation (19.3) below. Consider the lattice L corresponding (via the construction of the previous section) to the polynomials $G_{i,j}(x) = M^{h-1-j}F(x)^jx^i$ for $0 \leq i < d$, $0 \leq j < h$. Note that $G_{i,j}(x_0) \equiv 0 \pmod{M^{h-1}}$. The dimension of L is dh. One can represent L by a lower triangular basis matrix with diagonal entries $M^{h-1-j}X^{jd+i}$. Hence, the determinant of L is

$$\det(L) = M^{(h-1)hd/2}X^{(dh-1)dh/2}.$$

Running LLL on this basis outputs an LLL-reduced basis with first vector $\underline{b}_1$ satisfying

$$\|\underline{b}_1\| < 2^{(dh-1)/4}\det(L)^{1/dh} = 2^{(dh-1)/4}M^{(h-1)/2}X^{(dh-1)/2}.$$

This vector corresponds to a polynomial $G(x)$ of degree $dh - 1$ such that $G(x_0) \equiv 0 \pmod{M^{h-1}}$. If $\|\underline{b}_1\| < M^{h-1}/\sqrt{dh}$ then Howgrave-Graham's result applies and we have $G(x_0) = 0$ over $\mathbb{Z}$.

Hence, it is sufficient that

$$\sqrt{dh}2^{(dh-1)/4}M^{(h-1)/2}X^{(dh-1)/2} < M^{h-1}.$$

Rearranging gives

$$\sqrt{dh}2^{(dh-1)/4}X^{(dh-1)/2} < M^{(h-1)/2},$$

which is equivalent to

$$c(d,h)X < M^{(h-1)/(dh-1)}$$

where $c(d,h) = (\sqrt{dh}2^{(dh-1)/4})^{2/(dh-1)} = \sqrt{2}(dh)^{1/(dh-1)}$.

Now

$$\frac{h-1}{dh-1} = \frac{1}{d} - \frac{d-1}{d(dh-1)}.$$

Equating $(d-1)/(d(dh-1)) = \epsilon$ gives

$$h = ((d-1)/(d\epsilon)+1)/d \approx 1/(d\epsilon). \tag{19.3}$$

Note that $dh = 1 + (d-1)/(d\epsilon)$ and so $c(d,h) = \sqrt{2}(1+(d-1)/(d\epsilon))^{d\epsilon/(d-1)}$, which converges to $\sqrt{2}$ as $\epsilon \to 0$. Since $X < \frac{1}{2}M^{1/d-\epsilon}$ we require $\frac{1}{2} \leq \frac{1}{c(d,h)}$. Writing $x = d\epsilon/(d-1)$ this is equivalent to $(1+1/x)^x \leq \sqrt{2}$, which holds for $0 \leq x \leq 0.18$.

Rounding h up to the next integer gives a lattice such that if

$$|x_0| < \tfrac{1}{2}M^{1/d-\epsilon}$$

then the LLL algorithm and polynomial root finding leads to x_0.

Since the dimension of the lattice is $dh \approx 1/\epsilon$ and the coefficients of the polynomials $G_{i,j}$ are bounded by M^h it follows that the running time of LLL depends on d, $1/\epsilon$ and $\log(M)$. □

Exercise 19.1.10 Show that the precise complexity of Coppersmith's method is $O((1/\epsilon)^9 \log(M)^3)$ bit operations (recall that $1/\epsilon > d$). Note that if one fixes d and ϵ and considers the problem as M tends to infinity then one has a polynomial-time algorithm in $\log(M)$.

We refer to Section 3 of [132] for some implementation tricks that improve the algorithm. For example, one can add basis vectors to the lattice corresponding to polynomials of the form $M^{h-1}x(x-1)\cdots(x-i+1)/i!$.

Example 19.1.11 Let $p = 2^{30}+3$, $q = 2^{32}+15$ and $M = pq$. Consider the polynomial

$$\begin{aligned} F(x) &= a_0 + a_1x + a_2x^2 + a_3x^3 \\ &= 1942528644709637042 + 1234567890123456789x \\ &\quad + 987654321987654321x^2 + x^3, \end{aligned}$$

which has a root x_0 modulo M such that $|x_0| \le 2^{14}$. Set $X = 2^{14}$. Note that $X \approx M^{1/4.4}$. One can verify that the basic method in Section 19.1.1 does not find the small root.

Consider the basis matrix (this is of smaller dimension than the lattice in the proof of Theorem 19.1.9 in the case $d = 3$ and $h = 3$)

$$\begin{pmatrix} M^2 & 0 & 0 & 0 & 0 & 0 & 0 \\ 0 & M^2X & 0 & 0 & 0 & 0 & 0 \\ 0 & 0 & M^2X^2 & 0 & 0 & 0 & 0 \\ Ma_0 & Ma_1X & Ma_2X^2 & MX^3 & 0 & 0 & 0 \\ 0 & Ma_0X & Ma_1X^2 & Ma_2X^3 & MX^4 & 0 & 0 \\ 0 & 0 & Ma_0X^2 & Ma_1X^3 & Ma_2X^4 & MX^5 & 0 \\ a_0^2 & 2a_0a_1X & (a_1^2+2a_0a_2)X^2 & (2a_0+2a_1a_2)X^3 & (a_2^2+2a_1)X^4 & 2a_2X^5 & X^6 \end{pmatrix}.$$

The dimension is 7 and the determinant is M^9X^{21}. The first vector of the LLL-reduced basis is

$$(-369928294330603367352173305173409792,$$
$$1451057442025994832259962670402797568, \dots).$$

This corresponds to the polynomial

$$\begin{aligned} &-369928294330603367352173305173409792 \\ &\quad + 88565517701781911148679362207202x - 3439987357258441728608570659x^2 \\ &\quad + 446358057645551896819258x^3 + 4564259979987386926x^4 \\ &\quad - 1728007960413053x^5 - 21177681998x^6, \end{aligned}$$

which has $x_0 = 16384 = 2^{14}$ as a real root.

Exercise 19.1.12 Let $M = (2^{20}+7)(2^{21}+17)$ and $F(x) = x^3 + (2^{25} - 2883584)x^2 + 46976195x + 227$. Use Coppersmith's algorithm to find an integer x_0 such that $|x_0| < 2^9$ and $F(x_0) \equiv 0 \pmod{M}$.

Remark 19.1.13 It is natural to wonder whether one can find roots right up to the limit $X = M^{1/d}$. Indeed, the $-\epsilon$ term can be eliminated by performing an exhaustive search over the top few bits of the root x_0. An alternative way to proceed is to set $\epsilon = 1/\log_2(M)$, break the range $|x_0| < M^{1/d}$ of size $2M^{1/d}$ into $M^{2\epsilon} = 4$ intervals of size $2M^{1/d-2\epsilon} = M^{1/d-\epsilon}$ and perform Coppersmith's algorithm for each subproblem in turn.

Another question is whether one can go beyond the boundary $X = M^{1/d}$. A first observation is that for $X > M^{1/d}$ one does not necessarily expect a constant number of solutions; see Exercise 19.1.14. Coppersmith [132] gives further arguments why $M^{1/d}$ is the best one can hope for.

Exercise 19.1.14 Let $M = p^2$ and consider $F(x) = x^2 + px$. Show that if $X = M^{1/2+\epsilon}$ where $0 < \epsilon < 1/2$ then the number of solutions $|x| < X$ to $F(x) \equiv 0 \pmod{M}$ is $2M^\epsilon$,

Exercise 19.1.15 Let $N = pq$ be a product of two primes of similar size and let $e \in \mathbb{N}$ be a small integer such that $\gcd(e, \varphi(N)) = 1$. Let $1 < a, y < N$ be such that there is an integer $0 \le x < N^{1/e}$ satisfying $(a + x)^e \equiv y \pmod{N}$. Show that, given N, e, a, y, one can compute x in polynomial-time.

19.2 Multivariate modular polynomial equations

Suppose one is given $F(x, y) \in \mathbb{Z}[x, y]$ and integers X, Y and M and is asked to find one or more roots (x_0, y_0) to $F(x, y) \equiv 0 \pmod{M}$ such that $|x_0| < X$ and $|y_0| < Y$. One can proceed using similar ideas to the above, hoping to find two polynomials $F_1(x, y), F_2(x, y) \in \mathbb{Z}[x, y]$ such that $F_1(x_0, y_0) = F_2(x_0, y_0) = 0$ over $\mathbb{Z}$, and such that the resultant $R_x(F_1(x, y), F_2(x, y)) \neq 0$ (i.e., that $F_1(x, y)$ and $F_2(x, y)$ are **algebraically independent**). This yields a heuristic method in general, since it is hard to guarantee the independence of $F_1(x, y)$ and $F_2(x, y)$.

Theorem 19.2.1 *Let $F(x, y) \in \mathbb{Z}[x, y]$ be a polynomial of total degree d (i.e., every monomial $x^i y^j$ satisfies $i + j \le d$). Let $X, Y, M \in \mathbb{N}$ be such that $XY < M^{1/d-\epsilon}$ for some $0 < \epsilon < 1/d$. Then one can compute (in time polynomial in $\log(M)$ and $1/\epsilon > d$) polynomials $F_1(x, y), F_2(x, y) \in \mathbb{Z}[x, y]$ such that, for all $(x_0, y_0) \in \mathbb{Z}^2$ with $|x_0| < X$, $|y_0| < Y$ and $F(x_0, y_0) \equiv 0 \pmod{M}$, one has $F_1(x_0, y_0) = F_2(x_0, y_0) = 0$ over $\mathbb{Z}$.*

Proof We refer to Jutla [290] and Section 6.2 of Nguyen and Stern [415] for a sketch of the details. □

19.3 Bivariate integer polynomials

We now consider $F(x, y) \in \mathbb{Z}[x, y]$ and seek a root $(x_0, y_0) \in \mathbb{Z}^2$ such that both $|x_0|$ and $|y_0|$ are small. Coppersmith has proved the following important result.

Theorem 19.3.1 *Let $F(x, y) \in \mathbb{Z}[x, y]$ and let $d \in \mathbb{N}$ be such that $\deg_x(F(x, y))$, $\deg_y(F(x, y)) \le d$. Write*

$$F(x, y) = \sum_{0 \le i, j \le d} F_{i,j} x^i y^j.$$

For $X, Y \in \mathbb{N}$ define

$$W = \max_{0 \le i, j, \le d} |F_{i,j}| X^i Y^j.$$

If $XY < W^{2/(3d)}$ then there is an algorithm that takes as input $F(x, y)$, X, Y, runs in time (bit operations) bounded by a polynomial in $\log(W)$ and 2^d and outputs all pairs $(x_0, y_0) \in \mathbb{Z}^2$ such that $F(x_0, y_0) = 0$, $|x_0| \leq X$ and $|y_0| \leq Y$.

The condition in Theorem 19.3.1 is somewhat self-referential. If one starts with a polynomial $F(x, y)$ and bounds X and Y on the size of roots, then one can compute W and determine whether or not the algorithm will succeed in solving the problem.

Proof (Sketch) There are two proofs of this theorem, both of which are rather technical. The original by Coppersmith can be found in [131]. We sketch a simpler proof by Coron [139].

As usual, we consider shifts of the polynomial $F(x, y)$. Choose $k \in \mathbb{N}$ (sufficiently large) and consider the k^2 polynomials

$$s_{a,b}(x, y) = x^a y^b F(x, y) \quad \text{for} \quad 0 \leq a, b < k$$

in the $(d+k)^2$ monomials $x^i y^j$ with $0 \leq i, j < d+k$. Coron chooses a certain set of k^2 monomials (specifically of the form $x^{i_0+i} y^{j_0+j}$ for $0 \leq i, j < k$ and fixed $0 \leq i_0, j_0 \leq d$) and obtains a $k^2 \times k^2$ matrix S with non-zero determinant M. (The most technical part of [139] is proving that this can always be done and bounding the size of M.)

One can now consider the $(d+k)^2$ polynomials $Mx^i y^j$ for $0 \leq i, j < d+k$. Writing each polynomial as a row vector of coefficients, we now have a $k^2 + (d+k)^2$ by $(d+k)^2$ matrix. One can order the rows such that the matrix is of the form

$$\begin{pmatrix} S & * \\ MI_{k^2} & 0 \\ 0 & MI_w \end{pmatrix}$$

where $w = (d+k)^2 - k^2$, $*$ represents a $k^2 \times w$ matrix, and I_w denotes the $w \times w$ identity matrix.

Now, since $M = \det(S)$ there exists an integer matrix S' such that $S'S = MI_{k^2}$. Perform the row operations

$$\begin{pmatrix} I_{k^2} & 0 & 0 \\ -S' & I_{k^2} & 0 \\ 0 & 0 & I_w \end{pmatrix} \begin{pmatrix} S & * \\ MI_{k^2} & 0 \\ 0 & MI_w \end{pmatrix} = \begin{pmatrix} S & * \\ 0 & T \\ 0 & MI_w \end{pmatrix}$$

for some $k^2 \times w$ matrix T. Further row operations yield a matrix of the form

$$\begin{pmatrix} S & * \\ 0 & T' \\ 0 & 0 \end{pmatrix}$$

for some $w \times w$ integer matrix T'.

Coron considers a lattice L corresponding to T' (where the entries in a column corresponding to monomial $x^i y^j$ are multiplied by $X^i Y^j$ as in equation (19.2)) and computes the determinant of this lattice. Lattice basis reduction yields a short vector that corresponds to a polynomial $G(x, y)$ with small coefficients such that every root of $F(x, y)$ is a root of $G(x, y)$ modulo M. If (x_0, y_0) is a sufficiently small solution to $F(x, y)$ then, using an analogue of Theorem 19.1.2, one infers that $G(x_0, y_0) = 0$ over $\mathbb{Z}$.

A crucial detail is that $G(x, y)$ has no common factor with $F(x, y)$. To show this suppose $G(x, y) = F(x, y)A(x, y)$ for some polynomial (we assume that $F(x, y)$ is irreducible, if not then apply the method to its factors in turn). Then $G(x, y) = \sum_{0 \le i,j<k} A_{i,j} x^i y^j F(x, y)$ and so the vector of coefficients of $G(x, y)$ is a linear combination of the coefficient vectors of the k^2 polynomials $s_{a,b}(x, y)$ for $0 \le a, b < k$. But this vector is also a linear combination of the rows of the matrix $(0\ T')$ in the original lattice. Considering the first k^2 columns (namely the columns of S) one has a linear dependence of the rows in S. Since $\det(S) \neq 0$ this is a contradiction.

It follows that the resultant $R_x(F, G)$ is a non-zero polynomial, and so one can find all solutions by finding the integer roots of $R_x(F, G)(y)$ and then solving for x.

To determine the complexity it is necessary to compute the determinant of T' and to bound M. Coron shows that the method works if $XY < W^{2/(3d)-1/k} 2^{-9d}$. To get the stated running time for $XY < W^{2/(3d)}$ Coron proposes setting $k = \lfloor \log(W) \rfloor$ and performing exhaustive search on the $O(d)$ highest-order bits of x_0 (i.e., running the algorithm a polynomial in 2^d times). □

Example 19.3.2 Consider $F(x, y) = axy + bx + cy + d = 127xy - 1207x - 1461y + 21$ with $X = 30$, $Y = 20$. Let $M = 127^4$ (see below).

Consider the 13×9 matrix (this is taking $k = 2$ in the above proof and introducing the powers $X^i Y^j$ from the start)

$$B = \begin{pmatrix} aX^2Y^2 & bX^2Y & cXY^2 & dXY & 0 & 0 & 0 & 0 & 0 \\ 0 & aX^2Y & 0 & cXY & bX^2 & 0 & dX & 0 & 0 \\ 0 & 0 & aXY^2 & bXY & 0 & cY^2 & 0 & dY & 0 \\ 0 & 0 & 0 & aXY & 0 & 0 & bX & cY & d \\ MX^2Y^2 & 0 & 0 & 0 & 0 & 0 & 0 & 0 & 0 \\ 0 & MX^2Y & 0 & 0 & 0 & 0 & 0 & 0 & 0 \\ \vdots & & & & & & & & \vdots \\ 0 & 0 & 0 & 0 & 0 & 0 & 0 & 0 & M \end{pmatrix}.$$

We take S to be the matrix

$$\begin{pmatrix} a & b & c & d \\ 0 & a & 0 & c \\ 0 & 0 & a & b \\ 0 & 0 & 0 & a \end{pmatrix}$$

corresponding to the monomials $x^{i_0+i} y^{j_0+j}$ for $0 \le i, j < 2$ and fixed $i_0 = j_0 = 1$. Note that $M = \det(S) = a^4 = 127^4$.

Rather than diagonalising using the method of the proof of Theorem 19.3.1 we compute the Hermite normal form of B. This gives the matrix

$$B' = \left(\begin{array}{cccc|ccccc}
aX^2Y^2 & * & * & * & * & * & * & * & * \\
 & aX^2Y & * & * & * & * & * & * & * \\
 & & aXY^2 & * & * & * & * & * & * \\
 & & & aXY & * & * & * & * & * \\
\hline
 & & & & 16129X^2 & 16129Y^2 & 100125X & 1064641Y & 202558777 \\
 & & & & & 2048383Y^2 & * & * & * \\
 & & & & & & 2048383X & * & * \\
 & & & & & & & 260144641Y & * \\
 & & & & & & & & 260144641 \\
\hline
 & & & & & & & & \\
 & & & & & & & &
\end{array}\right)$$

where blanks are zeroes and $*$ denotes an entry whose value we do not bother to write down. Let L be the 5×5 diagonal matrix formed of columns 5 to 9 of rows 5 to 9 of B'. Performing LLL-reduction on L gives a matrix whose first row is

$$(-16129X^2, -16129Y^2, 1048258X, 983742Y, -28446222)$$

corresponding to the polynomial

$$G(x, y) = -16129x^2 - 16129y^2 + 1048258x + 983742y - 28446222.$$

Clearly, $G(x, y)$ is not a multiple of $F(x, y)$ since it has no xy term. Computing resultants and factoring gives the solutions $(x, y) = (21, 21)$ and $(23, 19)$.

Exercise 19.3.3 The polynomial

$$F(x, y) = 131xy - 1400x + 20y - 1286$$

has an integer solution with $|x| < 30$ and $|y| < 20$. Use Coron's method as in Example 19.3.2 to find (x, y).

The results of this section can be improved by taking into account the specific shape of the polynomial $F(x, y)$. We refer to Blömer and May [67] for details.

Finally, we remark that results are also known for integer polynomials having three or more variables, but these are heuristic in the sense that the method produces a list of polynomials having small roots in common, but there is no guarantee that the polynomials are algebraically independent.

19.4 Some applications of Coppersmith's method

19.4.1 Fixed padding schemes in RSA

As discussed in Chapter 1, it is necessary to use padding schemes for RSA encryption (for example, to increase the length of short messages and to prevent algebraic relationships between the messages and ciphertexts). One simple proposal for κ-bit RSA moduli is to take a κ' bit message and pad it by putting $(\kappa - \kappa' - 1)$ ones to the left-hand side of it. This brings a short message to full length. This padding scheme is sometimes called **fixed pattern padding**; we discuss it further in Section 24.4.5.

Suppose short messages (for example, 128-bit AES keys K) are being encrypted using this padding scheme with $\kappa = 1024$. Then

$$\mathsf{m} = 2^{1024} - 2^{128} + K.$$

Suppose also that the encryption exponent is $e = 3$. Then the ciphertext is

$$\mathsf{c} = \mathsf{m}^3 \pmod{N}.$$

If such a ciphertext is intercepted then the cryptanalyst only needs to find the value for K. In this case, we know that K is a solution to the polynomial

$$F(x) = (2^{1024} - 2^{128} + x)^3 - \mathsf{c} \equiv 0 \pmod{N}.$$

This is a polynomial of degree 3 with a root modulo N of size at most $N^{128/1024} = N^{1/8}$. So Coppersmith's method finds the solution K in polynomial-time.

Example 19.4.1 Let $N = 8873554201598479508804632335361$ (which is a 103 bit integer) and suppose Bob is sending 10-bit keys K to Alice using the padding scheme $\mathsf{m} = 2^{100} - 2^{10} + K$.

Suppose we have intercepted the ciphertext $\mathsf{c} = 8090574557775662005354455491076$ and wish to find K. Let $X = 2^{10}$. We write $F(x) = (x + 2^{100} - 2^{10})^3 - \mathsf{c} = x^3 + a_2x^2 + a_1x + a_0$ and define

$$B = \begin{pmatrix} N & 0 & 0 & 0 \\ 0 & NX & 0 & 0 \\ 0 & 0 & NX^2 & 0 \\ a_0 & a_1X & a_2X^2 & X^3 \end{pmatrix}.$$

Performing lattice reduction and taking the first row vector gives the polynomial with factorisation

$$(x - 987)(-920735567540915376297 + 726745175435904508x + 277605904865853x^2).$$

One can verify that the message is $K = 987$.

19.4.2 Factoring $N = pq$ with partial knowledge of p

Let $N = pq$ and suppose we are given an approximation $\tilde{p}$ to p such that $p = \tilde{p} + x_0$ where $|x_0| < X$. For example, suppose p is a 2κ-bit prime and $\tilde{p}$ is an integer that has the same κ most significant bits as p (so that $|p - \tilde{p}| < 2^\kappa$). Coppersmith used his ideas to get an algorithm for finding p given N and $\tilde{p}$. Note that Coppersmith originally used a bivariate polynomial method, but we present a simpler version following work of Howgrave-Graham, Boneh, Durfee and others.

The polynomial $F(x) = (x + \tilde{p})$ has a small solution modulo p. The problem is that we do not know p, but we do know a multiple of p (namely, N). The idea is to form a lattice corresponding to polynomials that have a small root modulo p and to apply Coppersmith's method.

Theorem 19.4.2 *Let $N = pq$ with $p < q < 2p$. Let $0 < \epsilon < 1/4$, and suppose $\tilde{p} \in \mathbb{N}$ is such that $|p - \tilde{p}| \le \frac{1}{2\sqrt{2}} N^{1/4-\epsilon}$. Then given N and $\tilde{p}$ one can factor N in time polynomial in $\log(N)$ and $1/\epsilon$.*

Proof Write $F(x) = (x + \tilde{p})$ and note that $\sqrt{N/2} \le p \le \sqrt{N}$. Let $X = \lfloor \frac{1}{2\sqrt{2}} N^{1/4-\epsilon} \rfloor$.

We describe the lattice to be used. Let $h \ge 4$ be an integer to be determined later and let $k = 2h$. Consider the $k + 1$ polynomials

$$N^h, N^{h-1}F(x), N^{h-2}F(x)^2, \dots, NF(x)^{h-1}, F(x)^h, xF(x)^h, \dots, x^{k-h}F(x)^h.$$

Note that if $p = \tilde{p} + x_0$ and if $G(x)$ is one of these polynomials then $G(x_0) \equiv 0 \pmod{p^h}$.

Consider the lattice corresponding to the above polynomials. More precisely, a basis for the lattice is obtained by taking each polynomial $G(x)$ above and writing the vector of coefficients of the polynomial $G(x)$ as in equation (19.1). The lattice has dimension $k + 1$ and determinant $N^{h(h+1)/2}X^{k(k+1)/2}$.

Applying LLL gives a short vector and, to apply Howgrave-Graham's result, we need $2^{k/4}\det(L)^{1/(k+1)} < p^h/\sqrt{k+1}$. Hence, since $p > (N/2)^{1/2}$, it is sufficient that $\sqrt{k+1}\, 2^{k/4} N^{h(h+1)/(2(k+1))} X^{k/2} < (N/2)^{h/2}$. Rearranging gives

$$X < N^{h/k - h(h+1)/(k(k+1))} 2^{-h/k} 2^{-1/2} / (k+1)^{1/k}.$$

Since $k \ge 7$ we have $(k+1)^{1/k} = 2^{\log_2(k+1)/k} \le 2^{1/2}$ and so $1/(k+1)^{1/k} \ge 1/\sqrt{2}$.

Now, since $k = 2h$ we find that the result holds if

$$X < N^{1/2(1-(h+1)/(2h+1))} \frac{1}{2\sqrt{2}}.$$

Since $1/2(1 - (h+1)/(2h+1)) = 1/4 - 1/(4(2h+1))$ the result will follow if $1/(4(2h+1)) < \epsilon$. Taking $h \ge \max\{4, 1/(4\epsilon)\}$ is sufficient. □

One can obtain a more general version of Theorem 19.4.2. If $p = N^\alpha$ and $|x| \le N^\beta$ where $0 < \alpha, \beta < 1$ then, ignoring constants, the required condition in the proof is

$$\frac{h(h+1)}{2} + \frac{\beta k(k+1)}{2} < \alpha h(k+1).$$

Taking $h = \sqrt{\beta} k$ and simplifying gives $\beta < \alpha^2$. The case we have shown is $\alpha = 1/2$ and $\beta < 1/4$. For details see Exercise 19.4.5 or Theorem 6 and 7 of [370].

Example 19.4.3 Let $N = 16803551$, $\tilde{p} = 2830$ and $X = 10$.

Let $F(x) = (x + \tilde{p})$ and consider the polynomials N, $F(x)$, $xF(x) = (x^2 + \tilde{p}x)$ and $x^2F(x)$, which all have the same small solution x_0 modulo p.

We build the lattice corresponding to these polynomials (with the usual method of converting a polynomial into a row vector). This lattice has basis matrix

$$\begin{pmatrix} N & 0 & 0 & 0 \\ \tilde{p} & X & 0 & 0 \\ 0 & \tilde{p}X & X^2 & 0 \\ 0 & 0 & \tilde{p}X^2 & X^3 \end{pmatrix}.$$

The first row of the output of the LLL algorithm on this matrix is $(105, -1200, 800, 1000)$, which corresponds to the polynomial

$$G(x) = x^3 + 8x^2 - 120x + 105.$$

The polynomial has the root $x = 7$ over $\mathbb{Z}$. We can check that $p = \tilde{p} + 7 = 2837$ is a factor of N.

Exercise 19.4.4 Let $N = 22461580086470571723189523$ and suppose you are given the approximation $\tilde{p} = 2736273600000$ to p, which is correct up to a factor $0 \le x < X = 50000$. Find the prime factorisation of N using Coppersmith's method.

Exercise 19.4.5 Let $\epsilon > 0$. Let $F(x)$ be a polynomial of degree d such that $F(x_0) \equiv 0 \pmod{M}$ for some $M \mid N$, $M = N^\alpha$ and $|x_0| \le \frac{1}{2}N^{\alpha^2/d - \epsilon}$. Generalise the proof of Theorem 19.4.2 to show that given $F(x)$ and N one can compute x_0 in time polynomial in $\log(N)$, d and $1/\epsilon$.

Exercise 19.4.6 Coppersmith showed that one can factor N in time polynomial in $\log(N)$ given $\tilde{p}$ such that $|p - \tilde{p}| < N^{1/4}$. Prove this result.

Exercise 19.4.7 Use Coppersmith's method to give an integer factorisation algorithm requiring $\tilde{O}(N^{1/4})$ bit operations. (A factoring algorithm with this complexity was also given in Section 12.5.)

Exercise 19.4.8 Show that the method of this section also works if given $\tilde{p}$ such that $|\tilde{p} - kp| < N^{1/4}$ for some integer k such that $\gcd(k, N) = 1$.

Exercise 19.4.9 Coppersmith also showed that one can factor N in time polynomial in $\log(N)$ given $\tilde{p}$ such that $p \equiv \tilde{p} \pmod{M}$ where $M > N^{1/4}$. Prove this result.

Exercise 19.4.10 Let $N = pq$ with $p \approx q$. Show that if one knows half the high order bits of p then one also knows approximately half the high order bits of q as well.

19.4.3 Factoring $p^r q$

As mentioned in Section 24.1.2, moduli of the form $p^r q$, where p and q are distinct primes and $r \in \mathbb{N}$, can be useful for some applications. When r is large then p is relatively small compared with N and so a natural attack is to try to factor N using the elliptic curve method.

Boneh, Durfee and Howgrave-Graham [74] considered using Coppersmith's method to factor integers of the form $N = p^r q$ when r is large. They observed that if one knows r and an approximation $\tilde{p}$ to p then there is a small root of the polynomial equation

$$F(x) = (\tilde{p} + x)^r \equiv 0 \pmod{p^r}$$

and that p^r is a large factor of N. One can therefore apply the technique of Section 19.4.2.

The algorithm is to repeat the above for all $\tilde{p}$ in a suitably chosen set. An analysis of the complexity of the method is given in [74]. It is shown that if $r \geq \log(p)$ then the algorithm runs in polynomial-time and that if $r = \sqrt{\log_2(p)}$ then the algorithm is asymptotically faster than using the elliptic curve method. One specific example mentioned in [74] is that if $p, q \approx 2^{512}$ and $r = 23$ then $N = p^r q$ should be factored more quickly by their method than with the elliptic curve method.

When r is small it is believed that moduli of the form $N = p^r q$ are still hard to factor. For 3076 bit moduli, taking $r = 3$ and $p, q \approx 2^{768}$ should be such that the best-known attack requires at least 2^{128} bit operations.

Exercise 19.4.11 The integer 876701170324027 is of the form $p^3 q$ where $|p - 5000| < 10$. Use the method of this section to factor N.

19.4.4 Chinese remaindering with errors

Boneh [70], building on work of Goldreich, Ron and Sudan, used ideas very similar to Coppersmith's method to give an algorithm for the following problem in certain cases.

Definition 19.4.12 Let $X, p_1, \ldots, p_n, r_1, \ldots, r_n \in \mathbb{Z}_{\geq 0}$ be such that $p_1 < p_2 < \cdots < p_n$ and $0 \leq r_i < p_i$ for all $1 \leq i \leq n$. Let $1 \leq e \leq n$ be an integer. The **Chinese remaindering with errors problem** (or **CRT list decoding problem**) is to compute an integer $0 \leq x < X$ (if it exists) such that

$$x \equiv r_i \pmod{p_i}$$

for all but e of the indices $1 \leq i \leq n$.

Note that it is not assumed that the integers p_i are coprime, though in many applications they will be distinct primes or prime powers. Also note that there is not necessarily a solution to the problem (for example, if X and/or e are too small).

Exercise 19.4.13 A naive approach to this problem is to run the Chinese remainder algorithm for all subsets $S \subseteq \{p_1, \ldots, p_n\}$ such that $\#S = (n - e)$. Determine the complexity of this algorithm. What is the input size of a Chinese remainder with errors instance when $0 \leq r_i < p_i$? Show that this algorithm is not polynomial in the input size if $e > \log(n)$.

The basic idea of Boneh's method is to construct a polynomial $F(x) \in \mathbb{Z}[x]$ such that all solutions x to the Chinese remaindering with errors problem instance are roots of $F(x)$ over $\mathbb{Z}$. This is done as follows. Define $P = \prod_{i=1}^{n} p_i$ and let $0 \leq R < P$ be the solution to the Chinese remainder instance (i.e., $R \equiv r_i \pmod{p_i}$ for all $1 \leq i \leq n$). For an integer x

define the **amplitude** $\mathrm{amp}(x) = \gcd(P, x - R)$ so that, if the p_i are coprime and S is the set of indices $1 \le i \le n$ such that $x \equiv r_i \pmod{p_i}$, $\mathrm{amp}(x) = \prod_{i \in S} p_i$. Write $F(x) = x - R$. The problem is precisely to find an integer x such that $|x| < X$ and $F(x) \equiv 0 \pmod{M}$ for some large integer $M \mid P$. This is the problem solved by Coppersmith's algorithm in the variant of Exercise 19.4.5. Note that $p_1^n \le P \le p_n^n$ and so $n \log(p_1) \le \log(P) \le n \log(p_n)$.

Theorem 19.4.14 *Let $X, e, p_1, \dots, p_n, r_1, \dots, r_n$ be an instance of the Chinese remainder with errors problem, where $p_1 < p_2 < \cdots < p_n$. Let $P = p_1 \cdots p_n$. There is an algorithm to compute all $x \in \mathbb{Z}$ such that $|x| < X$ and $x \equiv r_i \pmod{p_i}$ for all but e values $1 \le i \le n$ as long as*

$$e \le n - n\frac{\log(p_n)}{\log(p_1)}\sqrt{\log(X)/\log(P)}.$$

The algorithm is polynomial-time in the input size.

Proof Boneh [70] gives a direct proof, but we follow Section 4.7 of May [371] and derive the result using Exercise 19.4.5.

Let $0 \le x < X$ be an integer with $M = \mathrm{amp}(x)$ being divisible by at least $n - e$ of the values p_i. We have $n \log(p_1) \le \log(P) \le n \log(p_n)$ and $(n - e)\log(p_1) \le M \le n \log(p_n)$. Write $M = P^{\beta}$. Then Coppersmith's algorithm finds x if $X < P^{\beta^2}$ in polynomial-time in n and $\log(p_n)$ (note that Exercise 19.4.5 states the result for $X < P^{\beta^2 - \epsilon}$, but we can remove the ϵ using the same ideas as Remark 19.1.13). Hence, it is sufficient to give a bound on e so that $\log(X)/\log(P) < \beta^2$ (i.e., $\beta > \sqrt{\log(X)/\log(P)}$). Now, $\beta = \log(M)/\log(P) \ge (n - e)\log(p_1)/(n \log(p_n))$. Hence, it is sufficient that

$$(n - e)\frac{\log(p_1)}{\log(p_n)} \ge n\sqrt{\log(X)/\log(P)},$$

which is equivalent to the equation in the theorem. □

Exercise 19.4.15 Suppose $p_1, \dots, p_n$ are the first n primes. Show that the above algorithm works when $e \approx n - \sqrt{n \log(X) \log(n)}$. Hence, verify that Boneh's algorithm is polynomial-time in situations where the naive algorithm of Exercise 19.4.13 would be superpolynomial-time.

Bleichenbacher and Nguyen [65] discuss a variant of the Chinese remaindering with errors problem (namely, solving $x \equiv r_i \pmod{p_i}$ for small x, where each r_i lies in a set of m possible values) and a related problem in polynomial interpolation. Section 5 of [65] gives some algorithms for this "noisy CRT" problem.

Smooth integers in short intervals

The above methods can be used to find smooth integers in intervals. Let $I = [U, V] = \{x \in \mathbb{Z} : U \le x \le V\}$ and suppose we want to find a B-smooth integer $x \in I$ if one exists (i.e., all primes dividing x are at most B). We assume that $V < 2U$.

Exercise 19.4.16 Show that if $V \ge 2U$ then one can compute a power of 2 in $[U, V]$.

A serious problem is that only rather weak results have been proven about smooth integers in short intervals (see Section 4 of Granville [243], Sections 6.2 and 7.2 of Naccache and Shparlinski [404] or Section 15.3). Hence, we cannot expect to be able to prove anything rigorous in this section. On the other hand, it is natural to conjecture that, at least most of the time, the probability that a randomly chosen integer in an short interval $[U, V]$ is B-smooth is roughly equal to the probability that a randomly chosen integer of size V is B-smooth. Multiplying this probability by the length of the interval gives a rough guide to whether it is reasonable to expect a solution (see Remark 15.3.5). Hence, for the remainder of this section, we assume that such an integer x exists. We now sketch how the previous results might be used to find x.

Let $W = (U + V)/2$ and $X = (V - U)/2$ so that $I = [W - X, W + X]$. We seek all $x \in \mathbb{Z}$ such that $|x| \leq X$ and $x \equiv -W \pmod{p_i^{e_i}}$ for certain prime powers where $p_i \leq B$. Then $W + x$ is a potentially smooth integer in the desired interval (we know that $W + x$ has a large smooth factor, but this may not imply that all prime factors of $W + x$ are small if W is very large). One therefore chooses $P = \prod_{i=1}^{l} p_i^{e_i}$ where $p_1, \ldots, p_l$ are the primes up to B and the e_i are suitably chosen exponents (e.g. $e_i = \lceil \log(W)/(\log(B)\log(p_i)) \rceil$). One then applies Boneh's algorithm. The output is an integer with a large common divisor with P (indeed, this is a special case of the approximate GCD problem considered in Section 19.6). Note that this yields rather "dense" numbers, in the sense that they are divisible by most of the first l primes.

Example 19.4.17 Let $P = 2^4 \cdot 3^2 \cdot 5 \cdot 7 \cdot 11 \cdot 13 \cdot 17 \cdot 19 = 232792560$. Let $W = 100000007 = 10^8 + 7$ and $X = 1000000 = 10^6$. We want to find an integer x between $W - X$ and $W + X$ such that x is divisible by most of the prime powers dividing P.

Taking $R = -W$, $a = 4$ and $a' = 3$ in the notation of Theorem 19.4.14 gives the lattice given by the basis matrix

$$\begin{pmatrix} P^4 & 0 & 0 & 0 & 0 & 0 & 0 \\ -RP^3 & P^3X & 0 & 0 & 0 & 0 & 0 \\ R^2P^2 & -2RP^2X & P^2X^2 & 0 & 0 & 0 & 0 \\ -R^3P & 3R^2PX & -3RPX^2 & PX^3 & 0 & 0 & 0 \\ R^4 & -4R^3X & 6R^2X^2 & -4RX^3 & X^4 & 0 & 0 \\ 0 & R^4X & -4R^3X^2 & 6R^2X^3 & -4RX^4 & X^5 & 0 \\ 0 & 0 & R^4X^2 & -4R^3X^3 & 6R^2X^4 & -4RX^5 & X^6 \end{pmatrix}.$$

The polynomial corresponding to the first row of the LLL-reduced basis is

$$F(x) = -7^4(x + 231767)^4$$

giving the solution $x = -231767$. Indeed

$$W - 231767 = 2^4 \cdot 3^3 \cdot 5 \cdot 11 \cdot 13 \cdot 17 \cdot 19.$$

Note that the algorithm does not output $10^8 = 2^8 \cdot 5^8$ since that number does not have a very large gcd with P.

Exercise 19.4.18 Repeat the above example for $W = 150000001 = 1.5 \cdot 10^8 + 1$ and $W = 46558000$.

If this process fails then one can make adjustments to the value of P (for example, by changing the exponents e_i). Analysing the probability of success of this approach is an open problem.

19.5 Simultaneous Diophantine approximation

Let $\alpha \in \mathbb{R}$. It is well-known that the continued fraction algorithm produces a sequence of rational numbers p/q such that $|\alpha - p/q| < 1/q^2$. This is the subject of **Diophantine approximation**; see Section 1.1 of Lovász [356] for background and discussion. We now define a natural and important generalisation of this problem.

Definition 19.5.1 Let $\alpha_1, \ldots, \alpha_n \in \mathbb{R}$ and let $\epsilon > 0$. Let $Q \in \mathbb{N}$ be such that $Q \geq \epsilon^{-n}$. The **simultaneous Diophantine approximation problem** is to find $q, p_1, \ldots, p_n \in \mathbb{Z}$ such that $0 < q \leq Q$ and

$$|\alpha_i - p_i/q| \leq \epsilon/q \tag{19.4}$$

for all $1 \leq i \leq n$.

A theorem of Dirichlet mentioned in Section 1.1 of [356] and Section 17.3 of [220] shows that there is a solution satisfying the constraints in Definition 19.5.1.

Exercise 19.5.2 Let $\epsilon \geq 1/2$. Prove that integers $p_1, \ldots, p_n$ satisfying equation (19.4) exist for any n and q.

A major application of lattice reduction is to give an algorithm to compute the integers $(q, p_1, \ldots, p_n)$ in Definition 19.5.1. In practice, the real numbers $\alpha_1, \ldots, \alpha_n$ are given to some decimal precision (and so are rational numbers with coefficients of some size). The size of an instance of the simultaneous Diophantine approximation is the sum of the bit lengths of the numerator and denominator of the given approximations to the α_i, together with the bit length of the representation of ϵ and Q. Let X be a bound on the absolute value of all numerators and denominators of the α_i. The computational task is to find a solution $(q, p_1, \ldots, p_n)$ in time that is polynomial in n, $\log(X)$, $\log(1/\epsilon)$ and $\log(Q)$.

Theorem 19.5.3 *Let $\alpha_1, \ldots, \alpha_n \in \mathbb{Q}$ be given as rational numbers with numerator and denominator bounded in absolute value by X. Let $0 < \epsilon < 1$. One can compute in polynomial-time integers $(q, p_1, \ldots, p_n)$ such that $0 < q < 2^{n(n+1)/4}\epsilon^{-(n+1)}$ and $|\alpha_i - p_i/q| \leq \epsilon/q$ for all $1 \leq i \leq n$.*

Proof Let $Q = 2^{n(n+1)/4}\epsilon^{-n}$ and consider the lattice $L \subseteq \mathbb{Q}^{n+1}$ with basis matrix

$$\begin{pmatrix} \epsilon/Q & \alpha_1 & \alpha_2 & \cdots & \alpha_n \\ 0 & -1 & 0 & \cdots & 0 \\ 0 & 0 & -1 & & \\ \vdots & \vdots & & \ddots & \vdots \\ 0 & 0 & & \cdots & -1 \end{pmatrix}. \tag{19.5}$$

The dimension is $n+1$ and the determinant is $\epsilon/Q = 2^{-n(n+1)/4}\epsilon^{n+1}$. Every vector in the lattice is of the form $(q\epsilon/Q, q\alpha_1 - p_1, q\alpha_2 - p_2, \ldots, q\alpha_n - p_n)$. The entries of the lattice are ratios of integers with absolute value bounded by $\max\{X, 2^{n(n+1)/4}/\epsilon^{n+1}\}$.

Note that the lattice L does not have a basis with entries in $\mathbb{Z}$, but rather in $\mathbb{Q}$. By Remark 17.5.5 the LLL algorithm applied to L runs in $O(n^6 \max\{n\log(X), n^2 + n\log(1/\epsilon)\}^3)$ bit operations (which is polynomial in the input size) and outputs a non-zero vector $\underline{v} = (q\epsilon/Q, q\alpha_1 - p_1, \ldots, q\alpha_n - p_n)$ such that

$$\|\underline{v}\| \leq 2^{n/4}\det(L)^{1/(n+1)} = 2^{n/4}2^{-n/4}\epsilon = \epsilon < 1.$$

If $q = 0$ then $\underline{v} = (0, -p_1, \ldots, -p_n)$ with some $p_i \neq 0$ and so $\|\underline{v}\| \geq 1$, and so $q \neq 0$. Without loss of generality $q > 0$. Since $\|\underline{v}\|_\infty \leq \|\underline{v}\|$ it follows that $q\epsilon/Q \leq \epsilon < 1$ and so $0 < q < Q/\epsilon = 2^{n(n+1)/4}\epsilon^{-(n+1)}$. Similarly, $|q\alpha_i - p_i| < \epsilon$ for all $1 \leq i \leq n$. □

Exercise 19.5.4 Let $\alpha_1 = 1.555111, \alpha_2 = 0.771111$ and $\alpha_3 = 0.333333$. Let $\epsilon = 0.01$ and $Q = 10^6$. Use the method of this section to find a good simultaneous rational approximation to these numbers.

See Section 17.3 of [220] for more details and references.

19.6 Approximate integer greatest common divisors

The basic problem is the following. Suppose positive integers a and b exist such that $d = \gcd(a, b)$ is "large". Suppose that one is not given a and b, but only approximations $\tilde{a}, \tilde{b}$ to them. The problem is to find d, a and b. One issue is that there can be surprisingly many solutions to the problem (see Example 19.6.4), so it may not be feasible to compute all solutions for certain parameters. On the other hand, in the case $\tilde{b} = b$ (i.e., one of the values is known exactly, which often happens in practice) there are relatively few solutions.

Howgrave-Graham [269] has considered these problems and has given algorithms that apply in various situations. We present one of the basic ideas. Let $a = \tilde{a} + x$ and $b = \tilde{b} + y$. Suppose $\tilde{a} < \tilde{b}$ and define $q_a = a/d$ and $q_b = b/d$. Then, since $q_a/q_b = a/b$, we have

$$\frac{\tilde{a}}{\tilde{b}} - \frac{q_a}{q_b} = \frac{q_a y - q_b x}{\tilde{b} q_b}. \tag{19.6}$$

If the right hand side of equation (19.6) is small then performing Euclid's algorithm on $\tilde{a}/\tilde{b}$ gives a sequence of possible values for q_a/q_b. For each such value, one can compute

$$\lfloor \tilde{b}/q_b \rceil = \lfloor (dq_b - y)/q_b \rceil = d + \lfloor -y/q_b \rceil.$$

If $|y| < \frac{1}{2}q_b$ then one has computed d exactly and can solve $\tilde{a} + x \equiv \tilde{b} + y \equiv 0 \pmod{d}$. Note that one must use the basic extended Euclidean algorithm, rather than the improved method using negative remainders as in Algorithm 1.

Exercise 19.6.1 Show that if $a < b < \tilde{b}$, $b^{2/3} < d < 2b^{2/3}$ and $|x|, |y| < \frac{1}{4}b^{1/3}$ then the above method finds d, a and b.

Exercise 19.6.2 Let the notation be as above. Suppose $|x|, |y| < \tilde{b}^\beta$ and $d = \tilde{b}^\alpha$. Explain why it is natural to assume $\alpha > \beta$. Show that the above method succeeds if (ignoring constant factors) $\beta < -1 + 2\alpha$ and $\beta < 1 - \alpha$

Exercise 19.6.3 Re-formulate this method in terms of finding a short vector in a 2×2 matrix. Derive the same conditions on α and β as in Exercise 19.6.2.

Example 19.6.4 Let $\tilde{a} = 617283157$ and $\tilde{b} = 630864082$. The first few convergents q_a/q_b to $\tilde{a}/\tilde{b}$ are $1, 45/46, 91/93, 409/418, 500/511, 1409/1440$ and $1909/1951$. Computing approximations to $\tilde{a}/q_a$ and $\tilde{b}/q_b$ for these values (except the first) gives the following table.

$\tilde{a}/q_a$	13717403.5	6783331.4	1509249.8	1234566.3	438100.2	323354.2
$\tilde{b}/q_b$	13714436.6	6783484.8	1509244.2	1234567.7	438100.1	323354.2

Any values around these numbers can be used as a guess for d. For example, taking $d = 13717403$ one finds $\tilde{a} - 22 \equiv \tilde{b} + 136456 \equiv 0 \pmod{d}$, which is a not particularly good solution.

The four values $d_1 = 1234566$, $d_2 = 1234567$, $d_3 = 438100$ and $d_4 = 323354$ lead to the solutions $\tilde{a} - 157 \equiv \tilde{b} - 856 \equiv 0 \pmod{d_1}$, $\tilde{a} + 343 \equiv \tilde{b} - 345 \equiv 0 \pmod{d_2}$, $\tilde{a} - 257 \equiv \tilde{b} - 82 \equiv 0 \pmod{d_3}$ and $\tilde{a} - 371 \equiv \tilde{b} - 428 \equiv 0 \pmod{d_4}$.

Howgrave-Graham gives a more general method for solving the problem that does not require such a strict condition on the size of y. The result relies on heuristic assumptions about Coppersmith's method for bivariate integer polynomials. We state this result as Conjecture 19.6.5.

Conjecture 19.6.5 *(Algorithm 14 and Section 4 of [269]) Let $0 < \alpha < 2/3$ and $\beta < 1 - \alpha/2 - \sqrt{1 - \alpha - \alpha^2/2}$. There is a polynomial-time algorithm that takes as input $\tilde{a} < \tilde{b}$ and outputs all integers $d > \tilde{b}^\alpha$ such that there exist integers x, y with $|x|, |y| < \tilde{b}^\beta$ and $d \mid (\tilde{a} + x)$ and $d \mid (\tilde{b} + y)$.*

Exercise 19.6.6 Let $\tilde{a}, \tilde{b}, X, Y \in \mathbb{N}$ be given with $X < \tilde{a} < \tilde{b}$. Give a brute force algorithm to output all $d > Y$ such that there exist $x, y \in \mathbb{Z}$ with $|x|, |y| \leq X$ and $d = \gcd(\tilde{a} + x, \tilde{b} + y)$. Show that the complexity of this algorithm is $O(X^2 \log(\tilde{b})^2)$ bit operations.

We now mention the case when $\tilde{b} = b$ (in other words, b is known exactly). The natural approach is to consider the polynomial $F(x) = \tilde{a} + x$, which has a small solution to the equation $F(x) \equiv 0 \pmod{d}$ for some $d \mid b$. Howgrave-Graham applies the method used in Section 19.4.2 to solve this problem.

Theorem 19.6.7 *(Algorithm 12 and Section 3 of [269]) Let $0 < \alpha < 1$ and $\beta < \alpha^2$. There is a polynomial-time algorithm that takes as input $\tilde{a}$, b and outputs all integers $d > b^\alpha$ such that there exists an integer x with $|x| < b^\beta$ and $d \mid (\tilde{a} + x)$ and $d \mid b$.*

19.7 Learning with errors

The learning with errors problem was proposed by Regev. There is a large literature on this problem; we refer to Micciancio and Regev [379] and Regev [446] for background and references.

Definition 19.7.1 Let $q \in \mathbb{N}$ (typically prime), $\sigma \in \mathbb{R}_{>0}$, and $n, m \in \mathbb{N}$ with $m > n$.[1] Let $\underline{s} \in (\mathbb{Z}/q\mathbb{Z})^n$. The **LWE distribution** is the distribution on $(\mathbb{Z}/q\mathbb{Z})^{m \times n} \times (\mathbb{Z}/q\mathbb{Z})^m$ corresponding to choosing uniformly at random an $m \times n$ matrix A with entries in $\mathbb{Z}/q\mathbb{Z}$ and a length m vector

$$\underline{c} \equiv A\underline{s} + \underline{e} \pmod{q}$$

where the vector $\underline{e}$ has entries chosen independently from a discretised normal distribution[2] on $\mathbb{Z}$ with mean 0 and standard deviation σ. The **learning with errors** problem (**LWE**) is: given $(A, \underline{c})$ drawn from the LWE distribution to compute the vector $\underline{s}$. The **decision learning with errors** problem (**DLWE**) is: given A as above and a vector $\underline{c} \in (\mathbb{Z}/q\mathbb{Z})^m$ to determine whether $(A, \underline{c})$ is drawn from the uniform distribution or the LWE distribution.

It is necessary to argue that LWE is well-defined since, for any choice $\underline{s}'$, the value $\underline{c} - A\underline{s}' \pmod{q}$ is a possible choice for $\underline{e}$. But, when m is sufficiently large, one value for $\underline{s}$ is much more likely to have been used than any of the others. Hence, LWE is a maximum likelihood problem. Similarly, DLWE is well-defined when m is sufficiently large: if $\underline{c}$ is chosen uniformly at random and independent of A then there is not likely to be a choice for $\underline{s}$ such that $\underline{c} - A\underline{s} \pmod{q}$ is significantly smaller than the other values $\underline{c} - A\underline{s}' \pmod{q}$. We do not make these arguments precise. It follows that m must be significantly larger than n for these problems to be meaningful. It is also clear that increasing m (but keeping n fixed) does not make the LWE problem harder.

We refer to [379] and [446] for surveys of cryptographic applications of LWE and reductions, from computational problems in lattices that are believed to be hard, to LWE. Note that the values m, q and σ in an LWE instance are usually determined by constraints coming from the cryptographic application, while n is the main security parameter.

[1] For theoretical applications one should not assume a fixed number m of rows for A. Instead, the attacker is given an oracle that outputs pairs $(\underline{a}, c)$ where $\underline{a}$ is a row of A and $c = \underline{a}\,\underline{s} + e \pmod{q}$.

[2] In other words, the probability that e_i is equal to $x \in \mathbb{Z}$ is proportional to $e^{-x^2/(2\sigma^2)}$.

Example 19.7.2 Table 3 of Micciancio and Regev [379] suggests the parameters

$$(n, m, q, \sigma) = (233, 4536, 32749, 2.8).$$

Lindner and Peikert [352] suggest (using Figure 4 and the condition $m \geq 2n + \ell$ with $\ell = 128$)

$$(n, m, q, \sigma) = (256, 640, 4093, 3.3).$$

Exercise 19.7.3 Show that if one can determine $\underline{e}$ then one can solve LWE efficiently.

Exercise 19.7.4★ Show that, when q is prime, LWE $\leq_R$ DLWE. Show that DLWE $\leq_R$ LWE.

We now briefly sketch two lattice attacks on LWE. These attacks can be avoided by taking appropriate parameters. For other attacks on LWE see [446].

Example 19.7.5 (Lattice attack on DLWE using short vectors in kernel lattice modulo q.) Suppose one can find a short vector $\underline{w}$ in the lattice

$$\left\{\underline{w} \in \mathbb{Z}^m : \underline{w}A \equiv \underline{0} \pmod{q}\right\}.$$

Then $\underline{w}\,\underline{c} = \underline{w}A\underline{s} + \underline{w}\,\underline{e} \equiv \underline{w}\,\underline{e} \pmod{q}$. If $\underline{w}$ is short enough then one might expect that $\underline{w}\,\underline{e}$ is a small integer. On the other hand, if $\underline{c}$ is independent of A then $\underline{w}\,\underline{c} \pmod{q}$ is a random integer modulo q. Hence, one might be able to distinguish the LWE distribution from the uniform distribution using short enough vectors $\underline{w}$.

Note that one is not obliged to use all the rows of A in this attack, and so one can replace m by a much smaller value m'. For analysis of the best value for m', and for parameters that resist this attack, see Section 5.4.1 (especially equation (10)) of [379].

Example 19.7.6 (Reducing LWE to CVP.) We now consider a natural approach to solving LWE using lattices. Since we always use row lattices, it is appropriate to take the transpose of LWE. Hence, suppose $\underline{c}$, $\underline{s}$ and $\underline{e}$ are row vectors (of lengths m, n and m respectively) such that $\underline{c} = \underline{s}A^T + \underline{e} \pmod{q}$.

Consider the lattice

$$L = \left\{\underline{v} \in \mathbb{Z}^m : \underline{v} \equiv \underline{u}A^T \pmod{q} \text{ for some } \underline{u} \in \mathbb{Z}^n\right\}.$$

Then L has rank m and a basis matrix for it is computed by taking the (row) Hermite normal form of the $(n + m) \times m$ matrix

$$\begin{pmatrix} A^T \\ qI_m \end{pmatrix}$$

where I_m is an $m \times m$ identity matrix. One then tries to find an element $\underline{v}$ of L that is close to $\underline{c}$. Hopefully, $\underline{v} = \underline{c} - \underline{e} \equiv \underline{s}A^T \pmod{q}$.

One can perform lattice basis reduction and apply the nearest plane algorithm. For improved methods and experimental results see Lindner and Peikert [352]. As in Example 19.7.5, one can work with a subset of m' rows of A; see Section 5.1 of [352] for details.

19.8 Further applications of lattice reduction

There are a number of other applications of lattices in cryptography. We briefly list some of them.

- The improvement by Boneh and Durfee of Wiener's attack on small private exponent RSA. This is briefly mentioned in Section 24.5.1.
- Solving the hidden number problem in finite fields and its applications to bit security of Diffie–Hellman key exchange. See Section 21.7.1.
- The attack by Howgrave-Graham and Smart on digital signature schemes in finite fields when there is partial information available about the random nonces. See Section 22.3.
- The deterministic reduction by Coron and May from knowing $\varphi(N)$ to factoring N. This is briefly mentioned in Section 24.1.3.

PART V

CRYPTOGRAPHY RELATED TO DISCRETE LOGARITHMS

20

The Diffie–Hellman problem and cryptographic applications

This chapter introduces some basic applications of the discrete logarithm problem in cryptography, such as Diffie–Hellman key exchange and "textbook" Elgamal encryption. A brief security analysis of these systems is given. This motivates the computational and decisional Diffie–Hellman problems (CDH and DDH). A thorough discussion of these computational problems will be given in Chapter 21.

20.1 The discrete logarithm assumption

The discrete logarithm problem (DLP) was defined in Definition 13.0.1. Our main interest is the DLP in an algebraic group or algebraic group quotient over a finite field $\mathbb{F}_q$ (for example, elliptic curves, the multiplicative group of a finite fields, tori etc.). We always use multiplicative notation for groups in this chapter. As discussed in Section 13.2, in practice we usually restrict to groups of prime order r.

Recall that the difficulty of the DLP is defined with respect to an instance generator that runs on input a security parameter κ. An algorithm to solve the DLP with respect to a given instance generator is only required to succeed with a noticeable probability. The **discrete logarithm assumption** is that there exist instance generators that, on input κ, output instances of the DLP such that no algorithm A running in polynomial-time in κ can solve the DLP apart from with negligible (in κ) probability. The cryptosystems in this chapter rely on the discrete logarithm assumption (and other assumptions).

20.2 Key exchange

20.2.1 Diffie–Hellman key exchange

The starting point of discrete logarithms (indeed, of public key cryptography) is the seminal paper of Diffie and Hellman [166] from 1976 (more recently it became known that this idea was also found by Williamson at GCHQ in 1974).

Suppose Alice and Bob want to agree on a random key K. Assume they both know an algebraic group or algebraic group quotient G and some element $g \in G$ of prime order r (everyone in the world could use the same g). They perform the following protocol:

- Alice chooses a random integer $0 < a < r$ and sends $\mathsf{c}_1 = g^a$ to Bob.
- Bob chooses a random integer $0 < b < r$ and sends $\mathsf{c}_2 = g^b$ to Alice.
- On receiving c_2 Alice computes $K = \mathsf{c}_2^a$.
- On receiving c_1 Bob computes $K = \mathsf{c}_1^b$.

Hence, both players share the key $K = g^{ab}$. One can derive (see Definition 20.2.10 below) a bitstring from the group element K for use as the key of a symmetric encryption scheme. Hence, encryption of data or other functionalities can be implemented using traditional symmetric cryptography. The key K is called the **session key** and the values $\mathsf{c}_1, \mathsf{c}_2$ in the protocol are called **messages** or **ephemeral keys**.

We discuss the security of key exchange protocols (in particular, person-in-the-middle attacks and authenticated key exchange) in Section 20.5. For the remainder of this section we consider the simplest possible attacker. A **passive attacker** or **eavesdropper** (i.e., an attacker who learns g, c_1 and c_2, but does not actively interfere with the protocol) cannot determine K unless they can solve the following computational problem.

Definition 20.2.1 The **Computational Diffie–Hellman problem (CDH)**[1] is: given the triple (g, g^a, g^b) of elements of G to compute g^{ab}.

An extensive discussion of the computational Diffie–Hellman problem will be given in Chapter 21.

Exercise 20.2.2 What is the solution to the CDH instance $(2, 4, 7)$ in the group $\mathbb{F}_{11}^*$?

Suppose one is an eavesdropper on a Diffie–Hellman session and tries to guess the session key K shared by Alice and Bob. The following computational problem is precisely the problem of determining whether the guess for K is correct. This problem arises again later in the chapter in the context of Elgamal encryption.

Definition 20.2.3 Let G be a group and $g \in G$. The **Decisional Diffie–Hellman problem (DDH)** is, given a quadruple (g, g^a, g^b, g^c) of elements in $\langle g \rangle$ to determine whether or not $g^c = g^{ab}$.

Saying that a computational problem such as DDH is hard is slightly less straightforward than with problems like DLP or CDH, since if (g, g^a, g^b, g^c) are chosen uniformly at random in G^4 then the solution to the DDH problem is "no" with overwhelming probability. Clearly, an algorithm that says "no" all the time is not solving the DDH problem, so our notion of success must capture this. The correct approach is to define a DDH solver to be an algorithm that can distinguish two distributions on G^4, namely the distribution of **Diffie–Hellman**

[1] This assumption comes in two flavours, depending on whether g is fixed or variable. We discuss this issue in more detail later. But, as is the convention in this book, whenever we write "Given... compute..." one should understand that all of the inputs are considered as variables.

tuples (i.e., the uniform distribution on tuples of the form $(g, g^a, g^b, g^{ab}) \in G^4$) and the uniform distribution on G^4.

Definition 20.2.4 Let (G_n, r_n) be a family of cyclic groups G_n of order r_n, for $n \in \mathbb{N}$. A **DDH algorithm** for the family G_n is an algorithm A that takes as input a quadruple in G_n^4 and outputs "yes" or "no". The **advantage** of the DDH algorithm A is

$$\begin{aligned}\mathrm{Adv}(A) = \big|&\Pr\left(A\left(g, g^a, g^b, g^{ab}\right) = \text{"yes"} : g \leftarrow G_n,\ a, b \leftarrow \mathbb{Z}/r\mathbb{Z}\right) \\ &- \Pr\left(A\left(g, g^a, g^b, g^c\right) = \text{"yes"} : g \leftarrow G_n,\ a, b, c \leftarrow \mathbb{Z}/r\mathbb{Z}\right)\big| .\end{aligned}$$

A DDH algorithm is called **successful** if the advantage is noticeable. The **DDH assumption** for the family of groups is that all polynomial-time (i.e., running time $O(\log(r_n)^c)$ for some constant c) DDH algorithms have negligible advantage.

Lemma 20.2.5 *DDH $\le_R$ CDH $\le_R$ DLP.*

Exercise 20.2.6 Prove Lemma 20.2.5.

Exercise 20.2.7 Definition 20.2.3 states that r is prime. Show that if (g, g^a, g^b, g^c) is a quadruple of elements such that the order of g is n for some integer n where n has some small factors (e.g., factors $l \mid n$ such that $l \le \log_2(n)$) then one can eliminate some quadruples $(g, g^a, g^b, g^c) \in G^4$ that are not valid DDH tuples by reducing to DDH instances in subgroups of prime order. Show that this is enough to obtain a successful DDH algorithm according to Definition 20.2.4.

20.2.2 Burmester–Desmedt key exchange

In the case of $n > 2$ participants there is a generalisation of Diffie–Hellman key exchange due to Burmester and Desmedt [107] that requires two rounds of broadcast. Let the participants in the protocol be numbered as player 0 to player $n-1$. In the first round, player i (for $0 \le i < n$) chooses a random $1 \le a_i < r$ and sends $\mathsf{c}_i = g^{a_i}$ to the other players (or, at least, to player $i-1 \pmod n$ and $i+1 \pmod n$). In the second round, player i computes

$$\mathsf{t}_i = \left(\mathsf{c}_{i+1 \pmod n}\mathsf{c}_{i-1 \pmod n}^{-1}\right)^{a_i}$$

and sends it to all other players. Finally, player i computes

$$K = \mathsf{c}_{i+1 \pmod n}^{na_i}\mathsf{t}_{i+1 \pmod n}^{n-1}\mathsf{t}_{i+2 \pmod n}^{n-2}\cdots\mathsf{t}_{i+n-1 \pmod n}.$$

Lemma 20.2.8 *Each participant in the Burmester–Desmedt protocol computes*

$$K = g^{a_0a_1 + a_1a_2 + \cdots a_{n-2}a_{n-1} + a_{n-1}a_0}.$$

Exercise 20.2.9 Prove Lemma 20.2.8.

20.2.3 Key derivation functions

The result of Diffie–Hellman key exchange is a group element g^{ab}. Typically this should be transformed into an l-bit string for use as a symmetric key.

Definition 20.2.10 Let G be an algebraic group (or algebraic group quotient) and let l be an integer. A **key derivation function** is a function $\mathsf{kdf} : G \to \{0, 1\}^l$. The **output distribution** of a key derivation function is the probability distribution on $\{0, 1\}^l$ induced by $\mathsf{kdf}(g)$ over uniformly distributed $g \in G$. A key derivation function is **preimage resistant** if there is no polynomial-time algorithm known that, on input $x \in \{0, 1\}^l$, computes $g \in G$ such that $\mathsf{kdf}(g) = x$.

In general, a key derivation function should have output distribution statistically very close to the uniform distribution on $\{0, 1\}^l$. For many applications it is also necessary that kdf be preimage resistant.

A typical instantiation for kdf is to take a binary representation of $K \in G$, apply a cryptographic hash function (see Chapter 3) to obtain a bit string and concatenate/truncate as required. See the IEEE P1363 or ANSI X9.42 standards, Section 8 of Cramer and Shoup [149] or Section 6.1 of Raymond and Stiglic [445] for more details; also see Section 3 of [43] for a specific key derivation function for elliptic curves.

20.3 Textbook Elgamal encryption

In this section we present **textbook Elgamal public key encryption.**[2] This is historically the first public key encryption scheme based on the discrete logarithm problem. As we will see, the scheme has a number of security weaknesses and so is not recommended for practical use. In Chapter 23 we will present secure methods for public key encryption based on computational problems in cyclic groups.

We actually present two "textbook" versions of Elgamal. The first we call "classic textbook Elgamal" as it is essentially the version that appears in [177]. It requires G to be a group (i.e., we cannot use algebraic group quotients) and requires the message m to be encoded as an element of G. Encoding messages as group elements is not difficult, but it is unnatural and inconvenient. The second version, which we call "semi-textbook Elgamal", is more practical as it treats messages as bitstrings. As we will see, the security properties of the two versions are slightly different.

For both schemes, κ denotes a security parameter (so that all attacks should require at least 2^κ bit operations). Box 20.1 gives classic textbook Elgamal and Box 20.2 gives semi-textbook Elgamal. We call the sender Bob and the recipient Alice. Messages in the former scheme are group elements and in the latter are l-bit strings, where l depends on the security parameter. Semi-textbook Elgamal also requires a cryptographic hash function $H : G \to \{0, 1\}^l$ where G is the group.

[2] Some authors write "ElGamal" and others write "El Gamal". Reference [177] uses "ElGamal", but we follow the format apparently used nowadays by Elgamal himself.

KeyGen(κ): Run a parameter generation algorithm on security parameter κ that outputs an algebraic group G over a finite field $\mathbb{F}_q$ such that $\#G(\mathbb{F}_q)$ has a prime divisor r and all known algorithms for the discrete logarithm problem in a subgroup of $G(\mathbb{F}_q)$ of order r require at least 2^κ bit operations.

Compute $g \in G$ of prime order r.

Choose a random integer $0 < a < r$ and set $h = g^a$. The public key is (G, g, h) and the private key is a.

The message space is $\mathsf{M}_\kappa = G$.
The ciphertext space is $\mathsf{C}_\kappa = G \times G$.

Encrypt(m): (where $\mathsf{m} \in G$).

- Obtain the public key h of the recipient, Alice.
- Choose a random $0 < k < r$ and set $\mathsf{c}_1 = g^k$.
- Set $\mathsf{c}_2 = \mathsf{m}h^k$.
- Transmit the ciphertext $(\mathsf{c}_1, \mathsf{c}_2)$.

Decrypt($\mathsf{c}_1, \mathsf{c}_2$): Check that $\mathsf{c}_1, \mathsf{c}_2 \in G$. If so, compute and output

$$\mathsf{m} = \mathsf{c}_2 \mathsf{c}_1^{-a}.$$

Box 20.1 **Classic textbook Elgamal encryption**

KeyGen(κ): Generate an algebraic group or algebraic group quotient G as in Box 20.1. Choose a random $g \in G$ of prime order r.

Choose a message size l and a cryptographic hash function $H : G \to \{0, 1\}^l$.

Choose a random integer $0 < a < r$ and set $h = g^a$. The public key is (G, H, g, h) and the private key is a.

The message space is $\mathsf{M}_\kappa = \{0, 1\}^l$.
The ciphertext space is $\mathsf{C}_\kappa = G \times \{0, 1\}^l$.

Encrypt(m): (where $\mathsf{m} \in \{0, 1\}^l$).

- Obtain the public key of the recipient, Alice.
- Choose a random $0 < k < r$ and set $\mathsf{c}_1 = g^k$.
- Set $\mathsf{c}_2 = \mathsf{m} \oplus H(h^k)$.
- Transmit the ciphertext $(\mathsf{c}_1, \mathsf{c}_2)$.

Decrypt($\mathsf{c}_1, \mathsf{c}_2$): Check that $\mathsf{c}_1 \in G$ and $\mathsf{c}_2 \in \{0, 1\}^l$. If so, compute and output

$$\mathsf{m} = \mathsf{c}_2 \oplus H(\mathsf{c}_1^a).$$

Box 20.2 **Semi-textbook Elgamal encryption**

Remarks

1. Both versions of textbook Elgamal encryption are best understood as a **static Diffie–Hellman** key exchange followed by symmetric encryption. By this we mean that the sender (Bob) is essentially doing a Diffie–Hellman key exchange with the recipient (Alice): he sends g^k and Alice's component is her fixed (i.e., static) public key g^a. Hence, the shared key is g^{ak}, which can then be used as a key for any symmetric encryption scheme (this general approach is known as **hybrid encryption**). The two variants of textbook Elgamal vary in the choice of symmetric encryption scheme: the first uses the map $\mathsf{m} \mapsto \mathsf{m}g^{ak}$ from G to itself, while the second uses the map $\mathsf{m} \mapsto \mathsf{m} \oplus H(g^{ak})$ from $\{0, 1\}^l$ to itself.
2. Elgamal encryption requires two exponentiations in G and decryption requires one. Hence, encryption and decryption are polynomial-time and efficient.
3. Elgamal encryption is randomised, so encrypting the same message with the same public key twice will yield two different ciphertexts in general.
4. Unlike RSA, all users in a system can share the same group G. So typically G and g are fixed for all users, and only the value $h = g^a$ changes. Values that are shared by all users are usually called **system parameters**.

20.4 Security of textbook Elgamal encryption

We now briefly review the security properties for the textbook Elgamal cryptosystem. First, note that the encryption algorithm should use a good pseudorandom number generator to compute the values for k. A simple attack when this is not the case is given in Exercise 20.4.1.

Exercise 20.4.1 Suppose the random values k used by a signer are generated using the linear congruential generator $k_{i+1} = Ak_i + B \pmod{r}$ for some $1 \le A, B < r$. Suppose an adversary knows A and B and sees two classic textbook Elgamal ciphertexts $(\mathsf{c}_1, \mathsf{c}_2)$ and $(\mathsf{c}_1', \mathsf{c}_2')$, for the same public key, generated using consecutive outputs k_i and k_{i+1} of the generator. If both ciphertexts are encryptions of the same message then show how the adversary can compute the message. If both ciphertexts are encryptions of different messages then show how to decrypt both ciphertexts using one query to a decryption oracle.

20.4.1 OWE against passive attacks

Theorem 20.4.2 *The computational problem of breaking OWE security of classic textbook Elgamal under passive attack is equivalent to CDH in $\langle g \rangle$.*

Proof We prove the result only for perfect oracles. To prove OWE-CPA $\le_R$ CDH, let A be a perfect oracle that solves CDH in the subgroup of order r in G. Call $A(g, h_A, \mathsf{c}_2)$ to get u and set $\mathsf{m} = \mathsf{c}_2 u^{-1}$.

To prove CDH $\le_R$ OWE-CPA let A be a perfect adversary that takes an Elgamal public key (g, h_A) and an Elgamal ciphertext $(\mathsf{c}_1, \mathsf{c}_2)$ and returns the corresponding message m.

We will use this to solve CDH. Let the CDH instance be (g, g_1, g_2). Then choose a random element $\mathsf{c}_2 \in \langle g \rangle$ and call $A(g, g_1, g_2, \mathsf{c}_2)$ to get m. Return $\mathsf{c}_2\mathsf{m}^{-1}$ as the solution to the CDH instance. □

One can also consider a non-perfect adversary (for example, maybe an adversary can only decrypt some proportion of the possible ciphertexts). It might be possible to develop methods to "self-correct" the adversary using random self-reductions, but this is considered to be the adversary's job. Instead, it is traditional to simply give a formula for the success probability of the algorithm that breaks the computational assumption in terms of the success probability of the adversary. In the context of Theorem 20.4.2, if the adversary can decrypt with noticeable probability ϵ then we obtain a CDH algorithm that is correct with probability ϵ.

Exercise 20.4.3 Prove OWE-CPA $\leq_R$ CDH for semi-textbook Elgamal. Explain why the proof CDH $\leq_R$ OWE-CPA cannot be applied in this case.

20.4.2 OWE security under CCA attacks

We now show that both variants of textbook Elgamal do not have OWE security against an adaptive (CCA) attacker (and hence not IND-CCA security either). Recall that such an attacker has access to a decryption oracle that will decrypt every ciphertext except the challenge.

Lemma 20.4.4 *Let $(\mathsf{c}_1, \mathsf{c}_2)$ be a ciphertext for classic textbook Elgamal with respect to the public key (G, g, h). Suppose A is a decryption oracle. Then under a CCA attack, one can compute the message corresponding to $(\mathsf{c}_1, \mathsf{c}_2)$.*

Proof Assume that A is perfect. Call A on the ciphertext $(\mathsf{c}_1, \mathsf{c}_2 g) \neq (\mathsf{c}_1, \mathsf{c}_2)$ to obtain a message m'. Then the message corresponding to the original ciphertext is $\mathsf{m} = \mathsf{m}' g^{-1}$.

More generally, if A succeeds only with noticeable probability ϵ then we have a CCA2 attack that succeeds with noticeable probability ϵ. □

Another version of this attack follows from Exercises 23.3.3 and 23.3.2.

Exercise 20.4.5 Show that semi-textbook Elgamal encryption does not have the OWE security property under a CCA attack.

We have seen how a CCA attack can lead to an adversary learning the contents of a message. Exercise 20.4.6 gives an example of a general class of attacks called **small subgroup attacks** or **invalid parameter attacks** that can allow a CCA (even a CCA1) adversary to obtain the private key of a user. Such attacks can be performed in many scenarios. One example is when working in a prime order subgroup of $\mathbb{F}_p^*$ where $p-1$ has many small factors. Another example is when using elliptic curves $E : y^2 = x^3 + a_4x + a_6$; since the addition formula does not feature the value a_6 one can pass an honest user a point of small order on some curve $E'(\mathbb{F}_p)$. A related example is when using x-coordinate only arithmetic on elliptic curves one can choose an x-coordinate corresponding to a point that lies on the quadratic twist. Further discussion is given in Section 4.3 of [248] and a summary of the history of these results is given in Section 4.7 of [248]. We stress that such attacks do not only arise in encryption, but also in authenticated key exchange protocols, undeniable signatures, etc. The general way to avoid such attacks is for all parties to test membership of group elements in every step of the protocol (see Section 11.6).

Exercise 20.4.6 Show how a CCA1 attacker on classic textbook Elgamal can compute u^a for a group element u of their choice where a is the private key of a user. Show that if this attack can be repeated for sufficiently many elements u of coprime small orders then the private key a can be computed.

20.4.3 Semantic security under passive attacks

A serious problem with the classic textbook Elgamal cryptosystem is that, even though encryption is randomised, it does not necessarily provide semantic security under passive attacks.

Example 20.4.7 Consider the case $G = \mathbb{F}_p^*$, $\mathsf{M} = G$. Let $g \in G$ have prime order r. Then the Legendre symbol of g is $(\frac{g}{p}) = 1$. Hence, the Legendre symbol of the message m satisfies

$$\left(\frac{\mathsf{m}}{p}\right) = \left(\frac{\mathsf{c}_2}{p}\right)$$

and so can be computed in polynomial-time from the public key and the ciphertext.

To prevent the attack in Example 20.4.7 one can restrict the message space to elements of $\mathbb{F}_p^*$ with Legendre symbol 1. However, this attack is just a special case of a more general phenomenon. The Legendre symbol is a homomorphism $\mathbb{F}_p^* \to G_1$ where $G_1 = \{-1, 1\} \subseteq \mathbb{F}_p^*$ is the subgroup of order 2. The attack can be performed for any homomorphism onto a subgroup of order coprime to r (this is a slightly different application of the ideas of Section 13.2).

Example 20.4.8 (Boneh, Joux and Nguyen [77]) Let p be a 3072-bit prime and let $r \mid (p-1)$ be a 256-bit prime. Let $g \in \mathbb{F}_p^*$ have order r. Suppose, in violation of the description of classic textbook Elgamal in Section 20.3, one chooses the message space to be

$$\mathsf{M} = \{1, 2, \ldots, 2^{32} - 1\}$$

interpreted as a subset of $\mathbb{F}_p^*$. We identify M with $\{0, 1\}^{32} - \{0\}$. Let $(\mathsf{c}_1 = g^k, \mathsf{c}_2 = \mathsf{m}h^k)$ be a challenge ciphertext for classic textbook Elgamal encryption, where $\mathsf{m} \in \mathsf{M}$. Then

$$\mathsf{c}_2^r = \mathsf{m}^r.$$

One expects that, with overwhelming probability, the 2^{32} values m^r are distinct, and hence one can obtain m with at most 2^{32} exponentiations in $\mathbb{F}_p^*$.

Exercise 20.4.9 (Boneh, Joux and Nguyen) Let p and $r \mid (p-1)$ be prime and let $g \in \mathbb{F}_p^*$ have order r. Suppose one uses classic textbook Elgamal with restricted message space $\mathsf{M} = \{0,1\}^m - \{0\}$ as in Example 20.4.8 where $\#\mathsf{M} = 2^m - 1 < p/r$. Extend the attack of Example 20.4.8 using the baby-step–giant-step method so that it requires $O(2^{m/2+\epsilon})$ exponentiations in G to find m with noticeable probability, for $\epsilon > 0$.

One way to avoid these attacks is to restrict the message space to $\langle g \rangle$. It is then intuitively clear that IND security under passive attacks depends on the decisional Diffie–Hellman problem.

Theorem 20.4.10 *Classic textbook Elgamal with* $\mathsf{M} = \langle g \rangle$ *has IND-CPA security if and only if the DDH problem is hard.*

Proof (for perfect oracles): First, we show IND-CPA $\le_R$ DDH: let A be an oracle to solve DDH. Let $(\mathsf{c}_1, \mathsf{c}_2)$ be a ciphertext that is an encryption of either m_0 or m_1. Call $A(g, \mathsf{c}_1, h_A, \mathsf{c}_2\mathsf{m}_0^{-1})$ and if the answer is 'yes' then the message is m_0 and if the answer is 'no' then the message is m_1.

For the converse (i.e., DDH $\le_R$ IND-CPA of Elgamal): let A be an oracle that breaks indistinguishability of Elgamal. Then A takes as input a public key (g, h), a pair of messages $\mathsf{m}_0, \mathsf{m}_1$ and a ciphertext $(\mathsf{c}_1, \mathsf{c}_2)$ and outputs either 0 or 1. (We assume that A outputs either 0 or 1 even if the ciphertext corresponds to neither message.) Given a DDH instance (g, g_1, g_2, g_3) we repeatedly do the following: choose two random messages m_0 and m_1 in $\langle g \rangle$, choose a random $i \in \{0,1\}$ and call A on the input $(g, g_1, \mathsf{m}_0, \mathsf{m}_1, g_2, \mathsf{m}_i g_3)$. If A outputs i every time then we return 'yes' as the answer to the DDH. If A only outputs the correct answer i about half of the time then we return 'no'. To be sure the decryption oracle is not just being lucky one should repeat the experiment $\Omega(\log(r))$ times. □

If the hash function is sufficiently good then one does not have to make as strong an assumption as DDH to show that semi-textbook Elgamal encryption has IND security. Instead, the IND security intuitively only depends on CDH. Theorem 20.4.11 is a basic example of a security proof in the random oracle model (see Section 3.7 for background on this model). We give the proof as it illustrates one of the ways the random oracle model is used in theoretical cryptography.

Theorem 20.4.11 *In the random oracle model, semi-textbook Elgamal encryption has IND-CPA security if CDH is hard.*

Proof (Sketch) Let A be an adversary for the IND-CPA game on semi-textbook Elgamal encryption. Let g, g^a, g^b be a CDH instance. We will describe a simulator S that will solve the CDH problem using A as a subroutine.

First S runs the adversary A with public key (g, g^a).

The simulator must handle the queries made by A to the random oracle. To do this it stores a list of hash values, initially empty. Let g_i be the input for the ith hash query. If

$g_i = g_j$ for some $1 \le j < i$ then we respond with the same value as used earlier. If not then the simulator chooses uniformly at random an element $H_i \in \{0, 1\}^l$, stores (g_i, H_i) in the list, and answers the query $H(g_i)$ with H_i. This is a perfect simulation of a random oracle, at least until the challenge ciphertext is issued below.

At some time A outputs a pair of messages m_0 and m_1. The simulator sets $\mathsf{c}_1 = g^b$, chooses c_2 uniformly at random in $\{0, 1\}^l$ and responds with the challenge ciphertext $(\mathsf{c}_1, \mathsf{c}_2)$. The adversary A may make further hash function queries (which are answered using the algorithm above) and eventually A outputs $b \in \{0, 1\}$ (of course, A may crash, or run for longer than its specified running time, in which case S treats this as the output 0).

The logic of the proof is as follows: if A never queries the random oracle H on g^{ab} then A has no information on $H(g^{ab})$ and so cannot determine whether the answer should be 0 or 1. Hence, for A to succeed then one of the queries on H must have been on g^{ab}. Once this query is made then the simulator is seen to be fake, as the adversary can check that c_2 is not equal to $\mathsf{m}_b \oplus H(g^{ab})$ for $b \in \{0, 1\}$. However, the simulator is not concerned with this issue since it knows that g^{ab} occurs somewhere in the list of hash queries.

The simulator therefore chooses a random index i and responds with g_i as its solution to the CDH instance. □

Exercise 20.4.12 Fill the gaps in the proof of Theorem 20.4.11 and determine the exact probability of success in terms of the success of the adversary and the number of queries to the random oracle.

The power of the random oracle model is clear: we have been able to "look inside" the adversary's computation.

Exercise 20.4.13 Prove the converse to Theorem 20.4.11.

Indeed, the same technique leads to a much stronger result.

Theorem 20.4.14 *In the random oracle model, semi-textbook Elgamal encryption has OWE-CPA security if CDH is hard.*

Exercise 20.4.15 Prove Theorem 20.4.14.

20.5 Security of Diffie–Hellman key exchange

A discussion of security models for key exchange is beyond the scope of this book. We refer to Bellare and Rogway [36], Bellare, Pointcheval and Rogaway [34], Bellare, Canetti and Krawczyk [30], Canetti and Krawczyk [109], Shoup [495], Boyd and Mathuria [89] and Menezes, van Oorschot and Vanstone [376] for details. However, as a rough approximation we can consider three types of adversary:

- Passive adversary (also called "benign" in [36]). This attacker obtains all messages sent during executions of the key exchange protocol, but does not modify or delete any messages. This attacker is also called an eavesdropper.
- Weak[3] active adversary. This attacker obtains all messages sent during executions of the key exchange protocol and can modify or delete messages. This attacker can also initiate protocol executions with any player.
- Active adversary. This is as above, but the attacker is allowed to corrupt any honest player who has completed an execution of the protocol and thus obtain the agreed key.

There are two possible goals of an adversary:

- To obtain the shared session key.
- To distinguish the session key from a random key. To make this notion more precise consider a game between an adversary and a challenger. The challenger performs one or more executions of the key exchange protocol and obtains a key K. The challenger also chooses uniformly at random a key K' from the space of possible session keys. The challenger gives the adversary either K or K' (with probability $1/2$). The adversary has to decide whether the received key is K or not. This is called **real or random security**.

The Diffie–Hellman key exchange protocol is vulnerable to a person-in-the-middle attack. Unlike similar attacks on public key encryption, the attacker in this case does not need to replace any users' public keys.

Imagine that an adversary Eve can intercept all communication between Alice and Bob. When Alice sends $\mathsf{c}_1 = g^a$ to Bob, Eve stores c_1 and sends g^e to Bob, for some random integer e known to Eve. Similarly, when Bob sends $\mathsf{c}_2 = g^b$ to Alice, Eve stores c_2 and sends g^e to Alice. Alice computes the key g^{ae} and Bob computes the key g^{be}. Eve can compute both keys. If Alice later sends an encrypted message to Bob using the key g^{ae} then Eve can decrypt it, read it, re-encrypt using the key g^{be} and forward to Bob. Hence, Alice and Bob might never learn that their security has been compromised.

One way to overcome person-in-the-middle attacks is for Alice to send a digital signature on her value g^a (and similarly for Bob). As long as Alice and Bob each hold authentic copies of the other's public keys then this attack fails. Note that this solution does not prevent all attacks on the Diffie–Hellman key exchange protocol.

Another solution is given by authenticated key exchange protocols such as STS, KEA, MTI, MQV, etc. (see Chapter 11 of Stinson [532] and the references listed earlier).

We illustrate the basic idea behind most protocols of this type using the **MTI/A0 protocol**: Alice and Bob have public keys $h_A = g^a$ and $h_B = g^b$. We assume that Alice and Bob have authentic copies of each others public keys. They perform Diffie–Hellman key exchange in the usual way (Alice sends g^x and Bob sends g^y). Then the value agreed by both players is

$$g^{ay+bx}.$$

[3] This use of the word "weak" is non-standard.

Exercise 20.5.1 Explain why the person-in-the-middle attack fails for this protocol (assuming the public key authentication process is robust).

Exercise 20.5.2 Consider a key exchange protocol where Alice and Bob have public keys $h_A = g^a$ and $h_B = g^b$, where Alice sends g^x and Bob sends g^y and where the shared key is g^{ab+xy}. Show that if corrupt queries are allowed then this key exchange protocol does not provide authentication.

Exercise 20.5.3 Give a person-in-the-middle attack on the Burmester–Desmedt protocol.

20.6 Efficiency considerations for discrete logarithm cryptography

All cryptographic protocols whose security is related to the DLP involve computations of the form g^a at some stage, and this is usually the most demanding computation in terms of time and computing resources. To make the cryptosystem fast it is natural to try to speed up exponentiation. One could try working in a smaller group; however, it is important to ensure that the security of the system is maintained. Indeed, many of the main topics in this book (e.g., tori, elliptic curves and hyperelliptic curves) are attempts to get the "most efficient" group for a given security level.

A number of methods to speed up exponentiation in certain groups have already been presented. Section 11.1 discussed signed expansions, which are suitable for groups (such as elliptic and hyperelliptic curves or tori) where inversion is very efficient. Section 11.3 presented Frobenius expansions and the GLV method, which are suitable for elliptic curves. Those methods all assume that the exponent a takes any value.

One can also consider methods that do not correspond to values a chosen uniformly at random. Such methods can be much faster than the general methods already mentioned, but understanding the security implications can be more complicated. We do not have space to describe any of these methods in detail, but we briefly mention some of them.

1. Choose a to have low Hamming weight. This is mentioned by Agnew, Mullin, Onyszchuk and Vanstone [5] and Schnorr [470].
2. Choose a to be a random Frobenius expansion of low Hamming weight. This is credited to H. W. Lenstra Jr. in Section 6 of Koblitz [312].
3. Choose a to be given by a random addition chain (or addition–subtraction chain). This is proposed in Section 3.3 of Schroeppel, Orman, O'Malley and Spatscheck [479].
4. Choose a to be a product of integers of low Hamming weight. This was proposed and analysed by Hoffstein and Silverman [262].
5. Choosing a to be a random element in GLV representation, possibly with smaller than typical coefficients.
6. Generate random elements using large amounts of precomputation. A solution that can be used in any group is given by Boyko, Peinado and Venkatesan [91]. The method requires precomputing and storing random powers $g_j = g^{a_j}$. One generates a random pair (a, g^a)

by taking the product of a random subset of the g^{a_j} and setting $a = \sum a_j \pmod{r}$. This method is presented as the "simple solution" in [141].

A more sophisticated method for Koblitz curves is given by Coron, M'Raïhi and Tymen [141]. They use repeated application of sparse Frobenius expansions on elements of the precomputed table. They also give a security analysis.

21
The Diffie–Hellman problem

This chapter gives a thorough discussion of the computational Diffie–Hellman problem (CDH) and related computational problems. We give a number of reductions between computational problems, most significantly reductions from DLP to CDH. We explain self-correction of CDH oracles, study the static Diffie–Hellman problem, and study hard bits of the DLP and CDH. We always use multiplicative notation for groups in this chapter (except for in the Maurer reduction where some operations are specific to elliptic curves).

21.1 Variants of the Diffie–Hellman problem

We present some computational problems related to CDH, and prove reductions among them. The main result is to prove that CDH and Fixed-CDH are equivalent. Most of the results in this section apply to both algebraic groups (AG) and algebraic group quotients (AGQ) of prime order r (some exceptions are Lemma 21.1.9, Lemma 21.1.15 and, later, Lemma 21.3.1). For the algebraic group quotients G considered in this book then one can obtain all the results by lifting from the quotient to the covering group G' and applying the results there.

A subtle distinction is whether the base element $g \in G$ is considered fixed or variable in a CDH instance. To a cryptographer it is most natural to assume the generator is fixed, since that corresponds to the usage of cryptosystems in the real world (the group G and element $g \in G$ are fixed for all users). Hence, an adversary against a cryptosystem leads to an oracle for a fixed generator problem. To a computational number theorist it is most natural to assume the generator is variable, since algorithms in computational number theory usually apply to all problem instances. Hence, both problems are studied in the literature and when an author writes CDH it is sometimes not explicit which of the variants is meant. Definition 20.2.1 was for the case when g varies. Definition 21.1.1 below is the case when g is fixed. This issue is discussed in Section 5 of Shoup [495] and in Sadeghi and Steiner [457] (where it is called "granularity").

Definition 21.1.1 Let G be an algebraic group (AG) or algebraic group quotient (AGQ) and let $g \in G$. The **Fixed-base computational Diffie–Hellman problem (Fixed-CDH)** with respect to g is: given (g^a, g^b) to compute g^{ab}.

In this book the acronym CDH will always refer to the case where g is allowed to vary. Hence, an algorithm for CDH will always take three inputs (formally we should also include a description of the underlying group G, but we assume this is implicit in the specification of g), while an algorithm for Fixed-CDH will always take two inputs.

It is trivial that Fixed-CDH $\leq_R$ CDH, but the reverse implication is less obvious; see Corollary 21.1.17 below.

Analogously, given $g \in G$ one can define Fixed-DLP (namely, given h to find a such that $h = g^a$) and Fixed-DDH (given (g^a, g^b, g^c) determine whether $g^c = g^{ab}$). Though Fixed-DLP is equivalent to DLP (see Exercise 21.1.2) it is not expected that DDH is equivalent to Fixed-DDH (see Section 5.3.4 of [495]).

Exercise 21.1.2 Prove that Fixed-DLP is equivalent to DLP.

Exercise 21.1.3 Let G be a cyclic group of prime order r. Let $h_1, h_2, h_3 \in G$ such that $h_j \neq 1$ for $j = 1, 2, 3$. Show there exists some $g \in G$ such that (g, h_1, h_2, h_3) is a Diffie–Hellman tuple.

We now introduce some other variants of CDH. These are interesting in their own right, but are also discussed as they play a role in the proof of equivalence between CDH and Fixed-CDH.

Definition 21.1.4 Let G be a group or algebric group quotient of prime order r. The computational problem **Inverse-DH** is: given a pair $g, g^a \in G - \{1\}$ of elements of prime order r in G to compute $g^{a^{-1} \pmod r}$. (Clearly, we must exclude the case $a = 0$ from the set of instances.)

Lemma 21.1.5 *Inverse-DH* $\leq_R$ *CDH.*

Proof Suppose O is a perfect oracle for solving CDH. Let $(g, g_1 = g^a)$ be the given Inverse-DH instance. Then

$$g = g_1^{a^{-1}}.$$

Calling $O(g_1, g, g) = O(g_1, g_1^{a^{-1}}, g_1^{a^{-1}})$ gives $g_1^{a^{-2}}$. Finally

$$g_1^{a^{-2}} = (g^a)^{a^{-2}} = g^{a^{-1}}$$

as required. □

Definition 21.1.6 Let G be an AG or AGQ. The computational problem **Square-DH** is: given (g, g^a) where $g \in G$ has prime order r to compute g^{a^2}.

Exercise 21.1.7 Show that Square-DH $\leq_R$ CDH.

Lemma 21.1.8 *Square-DH* $\leq_R$ *Inverse-DH.*

Proof Let O be a perfect oracle that solves Inverse-DH and let $(g, g_1 = g^a)$ be given. If $g_1 = 1$ then return 1. Otherwise, we have

$$O(g_1, g) = O(g_1, g_1^{a^{-1}}) = g_1^a = (g^a)^a = g^{a^2}.$$ □

Hence, Square-DH $\le_R$ Inverse-DH $\le_R$ CDH. Finally, we show CDH $\le_R$ Square-DH and so all these problems are equivalent.

Lemma 21.1.9 *Let G be a group of odd order. Then CDH $\le_R$ Square-DH.*

Proof Let (g, g^a, g^b) be a CDH instance. Let O be a perfect oracle for Square-DH. Call $O(g, g^a)$ to get $g_1 = g^{a^2}$, $O(g, g^b)$ to get $g_2 = g^{b^2}$ and $O(g, g^a g^b)$ to get $g_3 = g^{a^2+2ab+b^2}$.
Now compute

$$(g_3/(g_1 g_2))^{2^{-1} \pmod{r}},$$

which is g^{ab} as required. □

Exercise 21.1.10 Let G be a group of prime order r. Show that Inverse-DH and Square-DH are random self-reducible. Hence, give a self-corrector for Square-DH. Finally, show that Lemma 21.1.9 holds for non-perfect oracles. (Note that it seems to be hard to give a self-corrector for Inverse-DH directly, though one can do this via Lemma 21.1.8.)

Note that the proofs of Lemmas 21.1.5 and 21.1.8 require oracle queries where the first group element in the input is not g. Hence, these proofs do not apply to variants of these problems where g is fixed. We now define the analogous problems for fixed g and give reductions between them.

Definition 21.1.11 Let g have prime order r and let $G = \langle g \rangle$. The computational problem **Fixed-Inverse-DH** is: given $g^a \ne 1$ to compute $g^{a^{-1} \pmod{r}}$. Similarly, the computational problem **Fixed-Square-DH** is: given g^a to compute g^{a^2}.

Exercise 21.1.12 Show that Fixed-Inverse-DH and Fixed-Square-DH are random self-reducible.

Lemma 21.1.13 *Let $g \in G$. Let A be a perfect Fixed-CDH oracle. Let $h = g^a$ and let $n \in \mathbb{N}$. Then one can compute $g^{a^n \pmod{r}}$ using $\le 2\log_2(n)$ queries to A.*

Proof Assume A is a perfect Fixed-CDH oracle. Define $h_i = g^{a^i \pmod{r}}$ so that $h_1 = h$. One has $h_{2i} = A(h_i, h_i)$ and $h_{i+1} = A(h_i, h)$. Hence, one can compute h_n by performing the standard square-and-multiply algorithm for efficient exponentiation. □

Note that the number of oracle queries in Lemma 21.1.13 can be reduced by using window methods or addition chains.

Lemma 21.1.14 *Fixed-Inverse-DH $\le_R$ Fixed-CDH.*

Proof Fix $g \in G$. Let O be a perfect Fixed-CDH oracle. Let g^a be the given Fixed-Inverse-DH instance. Our task is to compute $g^{a^{-1}}$. The trick is to note that $a^{-1} = a^{r-2} \pmod{r}$. Hence, one computes $g^{a^{r-2}}$ using Lemma 21.1.13. The case of non-perfect oracles requires some care, although at least one can check the result using O since one should have $O(g^a, g^{a^{-1}}) = g$. □

Lemma 21.1.15 *Fixed-Square-DH $\le_R$ Fixed-Inverse-DH.*

Proof Let $h = g^a$ be the input Fixed-Square-DH instance and let A be a perfect oracle for the Fixed-Inverse-DH problem. Call $A(gh)$ to get $g^{(1+a)^{-1}}$ and call $A(gh^{-1})$ to get $g^{(1-a)^{-1}}$.
Multiplying these outputs gives

$$w = g^{(1+a)^{-1}} g^{(1-a)^{-1}} = g^{2(1-a^2)^{-1}}.$$

Calling $A(w^{2^{-1} \pmod r})$ gives g^{1-a^2} from which we compute g^{a^2} as required. □

We can now solve a non-fixed problem using an oracle for a fixed problem.

Lemma 21.1.16 *Square-DH $\le_R$ Fixed-CDH.*

Proof Let $g \in G$ be fixed of prime order r and let A be a perfect Fixed-CDH oracle. Let g_1, g_1^b be the input Square-DH problem. Write $g_1 = g^a$. We are required to compute $g_1^{b^2} = g^{ab^2}$.

Call $A(g_1^b, g_1^b)$ to compute $g^{a^2b^2}$. Use the perfect Fixed-CDH oracle as in Lemma 21.1.14 to compute $g^{a^{-1}}$. Then compute $A(g^{a^2b^2}, g^{a^{-1}})$ to get g^{ab^2}. □

Since CDH $\le_R$ Square-DH we finally obtain the main result of this section.

Corollary 21.1.17 *Fixed-CDH and CDH are equivalent.*

Proof We already showed Fixed-CDH $\le_R$ CDH. Now, let A be a perfect Fixed-CDH oracle. Lemma 21.1.16 together with Lemma 21.1.9 gives CDH $\le_R$ Square-DH $\le_R$ Fixed-CDH as required.

Now suppose A only succeeds with noticeable probability $\epsilon > 1/\log(r)^c$ for some fixed c. The reductions CDH $\le_R$ Square-DH $\le_R$ Fixed-CDH require $O(\log(r))$ oracle queries. We perform self-correction (see Section 21.3) to obtain an oracle A for Fixed-CDH that is correct with probability $1 - 1/(\log(r)^{c'})$ for some constant c'; by Theorem 21.3.8 this requires $O(\log(r)^c \log\log(r))$ oracle queries. □

Exercise 21.1.18 It was assumed throughout this section that G has prime order r. Suppose instead that G has order r_1r_2 where r_1 and r_2 are odd primes and that g is a generator for G. Which of the results in this section no longer necessarily hold? Is Fixed-CDH in $\langle g \rangle$ equivalent to Fixed-CDH in $\langle g^{r_1} \rangle$?

We end with a variant of the DDH problem.

Exercise 21.1.19 Let g have prime order r and let $\{x_1, \dots, x_n\} \subset \mathbb{Z}/r\mathbb{Z}$. For a subset $A \subset \{1, \dots, n\}$ define

$$g^A = g^{\prod_{i \in A} x_i}.$$

The **group decision Diffie–Hellman problem** (GDDH) is: given g, g^A for all proper subsets $A \subsetneq \{1, \dots, n\}$, and h, to distinguish $h = g^c$ (where $c \in \mathbb{Z}/r\mathbb{Z}$ is chosen uniformly at random) from $g^{x_1x_2\cdots x_n}$. Show that GDDH $\equiv$ DDH.

21.2 Lower bound on the complexity of CDH for generic algorithms

We have seen (Theorem 13.4.5) that a generic algorithm requires $\Omega(\sqrt{r})$ group operations to solve the DLP in a group of order r. Shoup proved an analogue of this result for CDH. As before, fix $t \in \mathbb{R}_{>0}$ and assume that all group elements are represented by bitstrings of length at most $t\log(r)$.

Theorem 21.2.1 *Let G be a cyclic group of prime order r. Let A be a generic algorithm for CDH in G that makes at most m oracle queries. Then the probability that $A(\sigma(g), \sigma(g^a), \sigma(g^b)) = \sigma(g^{ab})$ over $a, b \in \mathbb{Z}/r\mathbb{Z}$ and an encoding function $\sigma : G \to S \subseteq \{0, 1\}^{\lceil t\log(r)\rceil}$ chosen uniformly at random is $O(m^2/r)$.*

Proof The proof is almost identical to the proof of Theorem 13.4.5. Let $S = \{0, 1\}^{\lceil t\log(r)\rceil}$. The simulator begins by uniformly choosing three distinct $\sigma_1, \sigma_2, \sigma_3$ in S and running $A(\sigma_1, \sigma_2, \sigma_3)$. The encoding function is then specifed at the two points $\sigma_1 = \sigma(g)$ and $\sigma_2 = \sigma(h)$. From the point of view of A, g and h are independent distinct elements of G.

It is necessary to ensure that the encodings are consistent with the group operations. This cannot be done perfectly without knowledge of a and b, but using polynomials as previously ensures there are no "trivial" inconsistencies. The simulator maintains a list of pairs (σ_i, F_i) where $\sigma_i \in S$ and $F_i \in \mathbb{F}_r[x, y]$ (indeed, the $F_i(x, y)$ will always be linear). The initial values are $(\sigma_1, 1)$, (σ_2, x) and (σ_3, y). Whenever A makes an oracle query on (σ_i, σ_j) the simulator computes $F = F_i - F_j$. If F appears as F_k in the list of pairs then the simulator replies with σ_k and does not change the list. Otherwise, an element $\sigma \in S$, distinct from the previously used values, is chosen uniformly at random, (σ, F) is added to the simulator's list and σ is returned to A.

After making at most m oracle queries, A outputs $\sigma_4 \in \mathbb{Z}/r\mathbb{Z}$. The simulator now chooses a and b uniformly at random in $\mathbb{Z}/r\mathbb{Z}$. Algorithm A wins if $\sigma_4 = \sigma(g^{ab})$. Note that if σ_4 is not equal to σ_1, σ_2 or one of the strings output by the oracle then the probability of success is at most $1/(2^{\lceil t\log(r)\rceil} - m - 2)$. Hence, we assume that σ_4 is on the simulator's list.

Let the simulator's list contain precisely k polynomials $\{F_1(x, y), \ldots, F_k(x, y)\}$ for some $k \le m + 3$. Let E be the event that $F_i(a, b) = F_j(a, b)$ for some pair $1 \le i < j \le k$ or $F_i(a, b) = ab$. The probability that A wins is

$$\Pr(A \text{ wins} \mid E)\Pr(E) + \Pr(A \text{ wins} \mid \neg E)\Pr(\neg E). \tag{21.1}$$

For each pair $1 \le i < j \le k$ the probability that $(F_i - F_j)(a, b) = 0$ is $1/r$ by Lemma 13.4.4. Similarly, the probability that $F_i(a, b) - ab = 0$ is $2/r$. Hence, the probability of event E is at most $k(k + 1)/2r + 2k/r = O(m^2/r)$. On the other hand, if event E does not occur then all A "knows" about (a, b) is that it lies in the set

$$\mathcal{X} = \{(a, b) \in (\mathbb{Z}/r\mathbb{Z})^2 : F_i(a, b) \ne F_j(a, b) \text{ and } F_i(a, b) \ne ab \text{ for all } 1 \le i < j \le k\}.$$

Let $N = \#\mathcal{X} \approx r^2 - m^2/2$ Then $\Pr(\neg E) = N/r^2$ and $\Pr(A \text{ wins} \mid \neg E) = 1/N$.

Hence, the probability that A wins is $O(m^2/r)$. □

21.3 Random self-reducibility and self-correction of CDH

We defined random self-reducibility in Section 2.1.4. Lemma 2.1.19 showed that the DLP in a group G of prime order r is random self-reducible. Lemma 2.1.20 showed how to obtain an algorithm with arbitrarily high success probability for the DLP from an algorithm with noticeable success probability.

Lemma 21.3.1 *Let g have order r and let $G = \langle g \rangle$. Then CDH in G is random self-reducible.*

Proof Let $\mathcal{X} = (G - \{1\}) \times G^2$ Let $(g, h_1, h_2) = (g, g^a, g^b) \in \mathcal{X}$ be the CDH instance. Choose uniformly at random $1 \le u < r$ and $0 \le v, w < r$ and consider the triple $(g^u, h_1^u g^{uv}, h_2^u g^{uw}) = (g^u, (g^u)^{a+v}, (g^u)^{b+w}) \in \mathcal{X}$. Then every triple in $\mathcal{X}$ arises from exactly one triple (u, v, w). Hence, the new triples are uniformly distributed in $\mathcal{X}$. If $Z = (g^u)^{(a+v)(b+w)}$ is the solution to the new CDH instance then the solution to the original CDH instance is

$$Z^{u^{-1} \pmod r} h_1^{-w} h_2^{-v} g^{-vw}.$$

□

Exercise 21.3.2 Show that Fixed-CDH is random self-reducible in a group of prime order r.

The following problem[1] is another cousin of the computational Diffie–Hellman problem. It arises in some cryptographic protocols.

Definition 21.3.3 Fix g of prime order r and $h = g^a$ for some $1 \le a < r$. The **static Diffie–Hellman problem (Static-DH)** is: given $h_1 \in \langle g \rangle$ to compute h_1^a.

Exercise 21.3.4 Show that the static Diffie–Hellman problem is random self-reducible.

One can also consider the decision version of static Diffie–Hellman.

Definition 21.3.5 Fix g of prime order r and $h = g^a$ for some $1 \le a < r$. The **decision static Diffie–Hellman problem (DStatic-DH)** is: given $h_1, h_2 \in \langle g \rangle$ to determine whether $h_2 = h_1^a$.

We now show that DStatic-DH is random-self-reducible. This is a useful preliminary to showing how to deal with DDH.

Lemma 21.3.6 *Fix g of prime order r and $h = g^a$ for some $1 \le a < r$. Then the decision static Diffie–Hellman problem is random self-reducible.*

Proof Write $G = \langle g \rangle$. Choose $1 \le w < r$ and $0 \le x < r$ uniformly at random. Given (h_1, h_2) compute $(Z_1, Z_2) = (h_1^w g^x, h_2^w h^x)$. We must show that if (h_1, h_2) is (respectively, is not) a valid Static-DH pair then (Z_1, Z_2) is uniformly distributed over the set of all valid (respectively, invalid) Static-DH pairs.

[1] The Static-DH problem seems to have been first studied by Brown and Gallant [104].

First, we deal with the case of valid Static-DH pairs. It is easy to check that if $h_2 = h_1^a$ then $Z_2 = Z_1^a$. Furthermore, for any pair $Z_1, Z_2 \in G$ such that $Z_2 = Z_1^a$ one can find exactly $(r-1)$ pairs (w, x) such that $Z_1 = h_1^w g^x$.

On the other hand, if $h_2 \neq h_1^a$ then write $h_1 = g^b$ and $h_2 = g^c$ with $c \not\equiv ab \pmod{r}$. For any pair $(Z_1, Z_2) = (g^y, g^z) \in G^2$ such that $z \not\equiv ay \pmod{r}$ we must show that (Z_1, Z_2) can arise from precisely one choice (w, x) above. Indeed

$$\begin{pmatrix} y \\ z \end{pmatrix} = \begin{pmatrix} b & 1 \\ c & a \end{pmatrix} \begin{pmatrix} w \\ x \end{pmatrix}$$

and, since the matrix has determinant $ab - c \not\equiv 0 \pmod{r}$ one can show that there is a unique solution for (w, x) and that $w \not\equiv 0 \pmod{r}$. □

We now tackle the general case of decision Diffie–Hellman.

Lemma 21.3.7 *Let g have prime order r and let $G = \langle g \rangle$. Then DDH in G is random self-reducible.*

Proof Choose $1 \le u, w < r$ and $0 \le v, x < r$ uniformly at random. Given $(g, h_1, h_2, h_3) = (g, g^a, g^b, g^c)$ define the new tuple $(g^u, h_1^u g^{uv}, h_2^{uw} g^{ux}, h_3^{uw} h_1^{ux} h_2^{uvw} g^{uvx})$. One can verify that the new tuple is a valid Diffie–Hellman tuple if and only if the original input is a valid Diffie–Hellman tuple (i.e., $c = ab$). If the original tuple is a valid Diffie–Hellman tuple then the new tuple is uniformly distributed among all Diffie–Hellman tuples. Finally, we show that if the original tuple is not a valid Diffie–Hellman tuple then the new tuple is uniformly distributed among the set of all invalid Diffie–Hellman tuples. To see this, think of (h_2, h_3) as a DStatic-DH instance with respect to the pair (g, h_1). Since $(g^u, h_1^u g^{uv})$ is chosen uniformly at random from $(G - \{1\}) \times G$ we have a uniformly random DStatic-DH instance with respect to a uniformly random static pair. The result then follows from Lemma 21.3.6. □

It is easy to turn a DLP oracle that succeeds with noticeable probability ϵ into one that succeeds with probability arbitrarily close to 1 since one can check whether a solution to the DLP is correct. It is less easy to amplify the success probability for a non-perfect CDH oracle.

A natural (but flawed) approach is just to run the CDH oracle on random self-reduced instances of CDH until the same value appears twice. We now explain why this approach will not work in general. Consider a Fixed-CDH oracle that, on input (g^a, g^b), returns $g^{ab+\xi}$ where $\xi \in \mathbb{Z}$ is uniformly chosen between $-1/\log(r)$ and $1/\log(r)$. Calling the oracle on instances arising from the random self-reduction of Exercise 21.3.2 one gets a sequence of values $g^{ab+\xi}$. Eventually, the correct value g^{ab} will occur twice, but it is quite likely that some other value will occur twice before that time.

We present Shoup's self-corrector for CDH or Fixed-CDH from [494].[2] Also see Cash, Kiltz and Shoup [112].

[2] Maurer and Wolf [365] were the first to give a self-corrector for CDH but Shoup's method is more efficient.

Theorem 21.3.8 *Fix $l \in \mathbb{N}$. Let g have prime order r. Let A be a CDH (respectively, Fixed-CDH) oracle with success probability at least $\epsilon > \log(r)^{-l}$. Let (g, g^a, g^b) be a CDH instance. Let $1 > \epsilon' > 1/r$. Then one can obtain an oracle that solves the CDH (respectively, Fixed-CDH) with probability at least $1 - \epsilon' - \log(2r)^2/(r\epsilon^2)$ and that makes at most $2\lceil \log(2/\epsilon')/\epsilon \rceil$ queries to A (where* log *is the natural logarithm).*

Proof Define $c = \log(2/\epsilon') \in \mathbb{R}$ so that $e^{-c} = \epsilon'/2$. First, call the oracle $n = \lceil c/\epsilon \rceil$ times on random-self-reduced instances (if the oracle is a CDH oracle then use Lemma 21.3.1 and if the oracle is a Fixed-CDH oracle then use Exercise 21.3.2) of the input problem (g, g^a, g^b) and store the resulting guesses $Z_1, \dots, Z_n$ for g^{ab} in a list L_1. Note that $n = O(\log(r)^{l+1})$. The probability that L_1 contains at least one copy of g^{ab} is $\geq 1 - (1-\epsilon)^{c/\epsilon} \geq 1 - e^{-c} = 1 - \epsilon'/2$.

Now choose uniformly at random integers $1 \leq s_1, s_2 < r$ and define $X_2 = g^{s_1}/(g^a)^{s_2}$. One can show that X_2 is uniformly distributed in $G = \langle g \rangle$ and is independent of $X_1 = g^a$. Call the oracle another n times on random-self-reduced versions of the CDH instance (g, X_2, g^b) and store the results $Z'_1, \dots, Z'_n$ in a list L_2.

Hence, with probability $\geq (1 - \epsilon'/2)^2 \geq 1 - \epsilon'$ there is some $Z_i \in L_1$ and some $Z'_j \in L_2$ such that $Z_i = g^{ab}$ and $Z'_j = g^{b(s_1 - as_2)}$. For each $1 \leq i, j \leq n$ test whether

$$Z_i^{s_2} = (g^b)^{s_1}/Z'_j. \tag{21.2}$$

If there is a unique solution (Z_i, Z'_j) then output Z_i, otherwise output $\perp$. Finding Z_i can be done efficiently by sorting L_1 and then, for each $Z'_j \in L_2$, checking whether the value of the right-hand side of equation (21.2) lies in L_1.

We now analyse the probability that the algorithm fails. The probability there is no pair (Z_i, Z'_j) satisfying equation (21.2), or that there are such pairs but none of them has $Z_i = g^{ab}$, is at most ϵ'. Hence, we now assume that a good pair (Z_i, Z'_j) exists and we want to bound the probability that there is a bad pair (i.e., a solution to equation (21.2) for which $Z_i \neq g^{ab}$). Write $X_1 = g^a$, $X_2 = g^{a'}$ (where $a' = s_1 - as_2$) and $Y = g^b$. Suppose (Z, Z') is a pair such that

$$Z^{s_2} Z' = Y^{s_1}. \tag{21.3}$$

We claim that $Z = Y^a$ and $Z' = Y^{a'}$ with probability at least $1 - 1/q$. Note that if equation (21.3) holds then

$$(Z/Y^a)^{s_1} = Y^{a'}/Z'. \tag{21.4}$$

If precisely one of $Z = Y^a$ or $Z' = Y^{a'}$ holds then this equation does not hold. Hence, $Z \neq Y^a$ and $Z' \neq Y^{a'}$, in which case there is precisely one value for s_1 for which equation (21.4) holds. Considering all n^2 pairs $(Z, Z') \in L_1 \times L_2$ it follows there are at most n^2 values for s_1, which would lead to an incorrect output for the self-corrector. Since s_1 is chosen uniformly at random the probability of an incorrect output is at most n^2/r. Since $n \leq \log(2r)/\epsilon$ one gets the result. Note that $\log(2r)^2/(r\epsilon^2) = O(\log(r)^{2+2l}/r)$. □

Exercise 21.3.9 Show that if the conjecture of Stolarsky (see Section 2.8) is true then one can compute g^{a^n} in $\log_2(n) + \log_2(\log_2(n))$ Fixed-CDH oracle queries.

21.4 The den Boer and Maurer reductions

The goal of this section is to discuss reductions from DLP to CDH or Fixed-CDH in groups of prime order r. Despite having proved that Fixed-CDH and CDH are equivalent, we prefer to treat them separately in this section. The first such reduction (assuming a perfect Fixed-CDH oracle) was given by den Boer [156] in 1988. Essentially, den Boer's method involves solving a DLP in $\mathbb{F}_r^*$, and so it requires $r - 1$ to be sufficiently smooth. Hence, there is no hope of this approach giving an equivalence between Fixed-CDH and DLP for all groups of prime order.

The idea was generalised by Maurer [362] in 1994, by replacing the multiplicative group $\mathbb{F}_r^*$ by an elliptic curve group $E(\mathbb{F}_r)$. Maurer and Wolf [365, 366, 368] extended the result to non-perfect oracles. If $\#E(\mathbb{F}_r)$ is sufficiently smooth then the reduction is efficient. Unfortunately, there is no known algorithm to efficiently generate such smooth elliptic curves. Hence, Maurer's result also does not prove equivalence between Fixed-CDH and DLP for all groups. A subexponential-time reduction that conjecturally applies to all groups was given by Boneh and Lipton [78]. An exponential-time reduction (but still faster than known algorithms to solve DLP) that applies to all groups was given by Muzereau, Smart and Vercauteren [402] and Bentahar [39, 40].

21.4.1 Implicit representations

Definition 21.4.1 Let G be a group and let $g \in G$ have prime order r. For $a \in \mathbb{Z}/r\mathbb{Z}$ we call $h = g^a$ an **implicit representation** of a.

In this section we call the usual representation of $a \in \mathbb{Z}/r\mathbb{Z}$ the **explicit representation** of a.

Lemma 21.4.2 *There is an efficient (i.e., computable in polynomial-time) mapping from $\mathbb{Z}/r\mathbb{Z}$ to the implicit representations of $\mathbb{Z}/r\mathbb{Z}$. One can test equality of elements in $\mathbb{Z}/r\mathbb{Z}$ given in implicit representation. If h_1 is an implicit representation of a and h_2 is an implicit representation of b then h_1h_2 is an implicit representation of $a + b$ and h_1^{-1} is an implicit representation of $-a$.*

In other words, we can compute in the additive group $\mathbb{Z}/r\mathbb{Z}$ using implicit representations.

Lemma 21.4.3 *If h is an implicit representation of a and $b \in \mathbb{Z}/r\mathbb{Z}$ is known explicitly, then h^b is an implicit representation of ab.*

Let O be a perfect Fixed-CDH oracle with respect to g. Suppose h_1 is an implicit representation of a and h_2 is an implicit representation of b. Then $h = O(h_1, h_2)$ is an implicit representation of ab.

In other words, if one can solve Fixed-CDH then one can compute multiplication modulo r using implicit representatives.

Exercise 21.4.4 Prove Lemmas 21.4.2 and 21.4.3.

Lemma 21.4.5 *Let g have order r. Let h_1 be an implicit representation of a such that $h_1 \neq 1$ (in other words, $a \not\equiv 0 \pmod{r}$).*

1. *Given a perfect CDH oracle, one can compute an implicit representation for $a^{-1} \pmod{r}$ using one oracle query.*
2. *Given a perfect Fixed-CDH oracle with respect to g, one can compute an implicit representation for $a^{-1} \pmod{r}$ using $\leq 2\log_2(r)$ oracle queries.*

Proof Given a perfect CDH oracle A one calls $A(g^a, g, g) = g^{a^{-1} \pmod{r}}$. Given a perfect Fixed-CDH oracle one computes $g^{a^{r-2} \pmod{r}}$ as was done in Lemma 21.1.14. □

To summarise, since $\mathbb{Z}/r\mathbb{Z} \cong \mathbb{F}_r$, given a perfect CDH or Fixed-CDH oracle then one can perform all field operations in $\mathbb{F}_r$ using implicit representations. Boneh and Lipton [78] call the set of implicit representations for $\mathbb{Z}/r\mathbb{Z}$ a **black box field**.

21.4.2 The den Boer reduction

We now present the **den Boer reduction** [156], which applies when $r-1$ is smooth. The crucial idea is that the Pohlig–Hellman and baby-step–giant-step methods only require the ability to add, multiply and compare group elements. Hence, if a perfect CDH oracle is given then these algorithms can be performed using implicit representations.

Theorem 21.4.6 *Let $g \in G$ have prime order r. Suppose l is the largest prime factor of $r-1$. Let A be a perfect oracle for the Fixed-CDH problem with respect to g. Then one can solve the DLP in $\langle g \rangle$ using $O(\log(r)\log(\log(r)))$ oracle queries, $O(\log(r)(\sqrt{l}/\log(l) + \log(r))$ multiplications in $\mathbb{F}_r$ and $O(\sqrt{l}\log(r)^2/\log(l))$ operations in G (where the constant implicit in the $O(\cdot)$ does not depend on l).*

Proof Let the challenge DLP instance be $g, h = g^a$. If $h = 1$ then return $a = 0$. Hence, we now assume $1 \leq a < r$. We can compute a primitive root $\gamma \in \mathbb{F}_r^*$ in $O(\log(r)\log(\log(r)))$ operations in $\mathbb{F}_r$ (see Section 2.15). The (unknown) logarithm of h satisfies

$$a \equiv \gamma^u \pmod{r} \tag{21.5}$$

for some integer u. To compute a it is sufficient to compute u.[3] The idea is to solve the DLP in equation (21.5) using the implicit representation of a. Since $r-1$ is assumed to be smooth then we can use the Pohlig–Hellman (PH) method, followed by the baby-step–giant-step (BSGS) method in each subgroup. We briefly sketch the details.

[3] It may seem crazy to try to work out u without knowing a, but it works!

Write $r - 1 = \prod_{i=1}^{n} l_i^{e_i}$ where the l_i are prime. The PH method involves projecting a and γ into the subgroup of $\mathbb{F}_r^*$ of order $l_i^{e_i}$. In other words, we must compute

$$h_i = g^{a^{(r-1)/l_i^{e_i}}}$$

for $1 \le i \le n$. Using the Fixed-CDH oracle to perform computations in implicit representation, Algorithm 4 computes all the h_i together in $O(\log(r)\log\log(r))$ oracle queries.[4] A further $O(\log(r))$ oracle queries are required to compute all $g^{a^{(r-1)/l_i^f}}$ where $0 \le f < e_i$. Similarly, one computes all $x_i = \gamma^{(r-1)/l_i^{e_i}}$ in $O(\log(r)\log\log(r))$ multiplications in $\mathbb{F}_r$. We then have

$$h_i = g^{x_i^{u \pmod{l_i^{e_i}}}}.$$

Following Section 13.2 one reduces these problems to $\sum_{i=1}^{n} e_i$ instances of the DLP in groups of prime order l_i. This requires $O(\log(r)^2)$ group operations and field operations overall (corresponding to the computations in line 6 of Algorithm 12).

For the baby-step–giant-step algorithm suppose we wish to solve $g^a = g^{\gamma^u}$ (where, for simplicity, we redefine a and γ so that they now have order l modulo r). Set $m = \lceil \sqrt{l} \rceil$ and write $u = u_0 + m u_1$ where $0 \le u_0, u_1 < m$. From

$$g^a = g^{\gamma^u} = g^{\gamma^{u_0 + m u_1}} = g^{\gamma^{u_0}(\gamma^m)^{u_1}} \tag{21.6}$$

one has

$$(g^a)^{(\gamma^{-m})^{u_1}} = g^{\gamma^{u_0}}. \tag{21.7}$$

We compute and store (in a sorted structure) the baby steps g^{γ^i} for $i = 0, 1, 2, \dots, m-1$ (this involves computing one exponentiation in G at each step, as $g^{\gamma^{i+1}} = (g^{\gamma^i})^{\gamma}$, which is at most $2\log_2(r)$ operations in G).

We then compute the giant steps $(g^a)^{\gamma^{-mj}}$. This involves computing $w_0 = \gamma^{-m} \pmod r$ and then the sequence $w_j = \gamma^{-mj} \pmod r$ as $w_{j+1} = w_i w_0 \pmod r$; this requires $O(\log(m) + m)$ multiplications in $\mathbb{F}_r$. We also must compute $(g^a)^{w_j}$, each of which requires $\le 2\log_2(r)$ operations in G.

When we find a match then we have solved the DLP in the subgroup of order l. The BSGS algorithm for each prime l requires $O(\sqrt{l}\log(r))$ group operations and $O(\sqrt{l} + \log(r))$ operations in $\mathbb{F}_r$. There are $O(\log(r))$ primes l for which the BSGS must be run, but a careful analysis of the cost (using the result of Exercise 13.2.7) gives an overall running time of $O(\log(r)^2\sqrt{l}/\log(l))$ group operations and $O(\log(r)^2 + \log(r)\sqrt{l}/\log(l))$ multiplications in $\mathbb{F}_r$. Note that the CDH oracle is not required for the BSGS algorithm.

Once u is determined modulo all prime powers $l^e \mid (r-1)$ one uses the Chinese remainder theorem to compute $u \in \mathbb{Z}/(r-1)\mathbb{Z}$. Finally, one computes $a = \gamma^u \pmod r$. These final steps require $O(\log(r))$ operations in $\mathbb{F}_r$. □

[4] Remark 2.15.9 does not lead to a better bound, since the value n (which is m in the notation of that remark) is not necessarily large.

Corollary 21.4.7 *Let $A(\kappa)$ be an algorithm that outputs triples (g, h, r) such that r is a κ-bit prime, g has order r, $r - 1$ is $O(\log(r)^2)$-smooth and $h \in \langle g \rangle$. Then DLP $\leq_R$ Fixed-CDH for the problem instances output by A.*

Proof Suppose one has a perfect Fixed-CDH oracle. Putting $l = O(\log(r)^2)$ into Theorem 21.4.6 gives a reduction with $O(\log(r) \log\log(r))$ oracle queries and $O(\log(r)^3)$ group and field operations. □

The same results trivially hold if one has a perfect CDH oracle.

Exercise 21.4.8★ Determine the complexity in Theorem 21.4.6 if one has a Fixed-CDH oracle that only succeeds with probability ϵ.

Cherepnev [126] iterates the den Boer reduction to show that if one has an efficient CDH algorithm for arbitrary groups then one can solve DLP in a given group in subexponential time. This result is of a very different flavour to the other reductions in this chapter (which all use an oracle for a group G to solve a computational problem in the same group G) so we do not discuss it further.

21.4.3 The Maurer reduction

The den Boer reduction can be seen as solving the DLP in the algebraic group $G_m(\mathbb{F}_r)$, performing all computations using implicit representation. Maurer's idea was to replace $G_m(\mathbb{F}_r)$ by any algebraic group $G(\mathbb{F}_r)$, in particular the group of points on an elliptic curve $E(\mathbb{F}_r)$. As with Lenstra's elliptic curve factoring method, even when $r - 1$ is not smooth, there might be an elliptic curve E such that $E(\mathbb{F}_r)$ is smooth.

When one uses a general algebraic group G there are two significant issues that did not arise in the den Boer reduction.

- The computation of the group operation in G may require inversions. This is true for elliptic curve arithmetic using affine coordinates.
- Given $h = g^a$ one must be able to compute an element $P \in G(\mathbb{F}_r)$, in implicit representation, such that once P has been determined in explicit representation one can compute a. For an elliptic curve E one could hope that $P = (a, b) \in E(\mathbb{F}_r)$ for some $b \in \mathbb{F}_r$.

Before giving the main result we address the second of these issues. In other words, we show how to embed a DLP instance into an elliptic curve point.

Lemma 21.4.9 *Let g have prime order r and let $h = g^a$. Let $E : y^2 = x^3 + Ax + B$ be an affine elliptic curve over $\mathbb{F}_r$. Given a perfect Fixed-CDH oracle there is an algorithm that outputs an implicit representation (g^X, g^Y) of a point $(X, Y) \in E(\mathbb{F}_r)$ and some extra data, and makes an expected $O(\log(r))$ oracle queries and is expected to perform $O(\log(r))$ group operations in $\langle g \rangle$. Furthermore, given the explicit value of X and the extra data, one can compute a.*

Proof The idea is to choose uniformly at random $0 \le \alpha < r$ and set $X = a + \alpha$. An implicit representation of X can be computed as $h_1 = hg^{\alpha}$ using $O(\log(r))$ group operations. If we store α then, given X, we can compute a. Hence, the extra data is α.

Given the implicit representation for X one determines an implicit representation for $\beta = X^3 + AX + B$ using two oracle queries. Given g^{β}, one can compute (here $(\frac{\beta}{r}) \in \{-1, 1\}$ is the Legendre symbol)

$$h_2 = g^{(\frac{\beta}{r})} = g^{\beta^{(r-1)/2}} \tag{21.8}$$

using $O(\log(r))$ oracle queries. If $h_2 = g$ then β is a square and so X is an x-coordinate of a point of $E(\mathbb{F}_r)$.

Since there are at least $(r - 2\sqrt{r})/2$ possible x-coordinates of points in $E(\mathbb{F}_r)$, it follows that if one chooses X uniformly at random in $\mathbb{F}_r$ then the expected number of trials until X is the x-coordinate of a point in $E(\mathbb{F}_r)$ is approximately two.

Once β is a square modulo r then one can compute an implicit representation for $Y = \sqrt{\beta} \pmod{r}$ using the Tonelli–Shanks algorithm with implicit representations. We use the notation of Algorithm 3. The computation of the non-residue n is expected to require $O(\log(r))$ operations in $\mathbb{F}_r$ and can be done explicitly. The computation of the terms w and b requires $O(\log(r))$ oracle queries, some of which can be avoided by storing intermediate values from the computation in equation (21.8). The computation of i using a Pohlig–Hellman-style algorithm is done as follows. First compute the sequence $b, b^2, \dots, b^{2^{e-1}}$ using $O(\log(r))$ oracle queries and the sequence $y, y^2, \dots, y^{2^{e-1}}$ using $O(\log(r))$ group operations. With a further $O(\log(r))$ group operations, one can determine the bits of i. □

Theorem 21.4.10 *Let $B \in \mathbb{N}$. Let $g \in G$ have order r. Let E be an elliptic curve over $\mathbb{F}_r$ such that $E(\mathbb{F}_r)$ is a cyclic group. Suppose that the order of $E(\mathbb{F}_r)$ is known and is B-smooth. Given a perfect Fixed-CDH oracle with respect to g, one can solve the DLP in $\langle g \rangle$ using expected $O(\log(r)^2 \log(\log(r)))$ oracle queries.*[5]

Indeed, there are two variants of the reduction, one using exhaustive search and one using the baby-step–giant-step algorithm. One can also consider the case of a perfect CDH oracle. The following table gives the full expected complexities (where the constant implicit in the $O(\cdot)$ is independent of B). We use the abbreviation $l(x) = \log(x)$ so that $l(l(r)) = \log(\log(r))$.

Oracle	*Reduction*	*Oracle queries*	*Group operations*	$\mathbb{F}_r$ *operations*
Fixed-CDH	*PH only*	$O(l(r)^2 l(l(r)))$	$O(Bl(r)^2/l(B))$	$O(Bl(r)^2/l(B))$
Fixed-CDH	*PH+BSGS*	$O(\sqrt{B}l(r)^2/l(B) + l(r)^2 l(l(r)))$	$O(\sqrt{B}l(r)^2/l(B))$	$O(\sqrt{B}l(r)^2/l(B))$
CDH	*PH only*	$O(l(r)l(l(r)))$	$O(Bl(r)^2/l(B))$	$O(Bl(r)^2/l(B))$
CDH	*PH+BSGS*	$O(\sqrt{B}l(r)/l(B) + l(r)l(l(r)))$	$O(\sqrt{B}l(r)^2/l(B))$	$O(\sqrt{B}l(r)^2/l(B))$

Proof Let the discrete logarithm instance be $(g, h = g^a)$. Write $N = \#E(\mathbb{F}_r) = \prod_{i=1}^{k} l_i^{e_i}$. We assume that affine coordinates are used for arithmetic in $E(\mathbb{F}_r)$. Let P be a generator of $E(\mathbb{F}_r)$.

[5] This is improved to $O(\log(r) \log\log(r))$ in Remark 21.4.11.

The reduction is conceptually the same as the den Boer reduction. One difference is that elliptic curve arithmetic requires inversions (which are performed using the method of Lemma 21.1.13 and Lemma 21.1.14) and hence the number of Fixed-CDH oracle queries must increase. A sketch of the reduction in the case of exhaustive search is given in Algorithm 26.

The first step is to use Lemma 21.4.9 to associate with h the implicit representations of a point $Q \in E(\mathbb{F}_r)$. This requires an expected $O(\log(r))$ oracle queries and $O(\log(r))$ group operations for all four variants. Then $Q \in \langle P \rangle$ where P is the generator of the cyclic group $E(\mathbb{F}_r)$.

The idea is again to use Pohlig–Hellman (PH) and baby-step–giant-step (BSGS) to solve the discrete logarithm of Q with respect to P in $E(\mathbb{F}_r)$. If we can compute an integer u such that $Q = [u]P$ (with computations done in implicit representation) then computing $[u]P$ and using Lemma 21.4.9 gives the value a explicitly.

First, we consider the PH algorithm. As with the den Boer reduction, one needs to compute explicit representations (i.e., standard affine coordinates) for $[N/l_i^{e_i}]P$ and implicit representations for $[N/l_i^{e_i}]Q$. It is possible that $[N/l_i^{e_i}]Q = \mathcal{O}_E$ so this case must be handled. As in Section 2.15.1, computing these points requires $O(\log(r)\log\log(r))$ elliptic curve operations. Hence, for the multiples of P we need $O(\log(r)\log\log(r))$ operations in $\mathbb{F}_r$, while for the multiples of Q we need $O(\log(r)^2\log\log(r))$ Fixed-CDH oracle queries and $O(\log(r)\log\log(r))$ group operations. (If a CDH oracle is available then this stage only requires $O(\log(r)\log\log(r))$ oracle queries.) Computing the points $[N/l_i^f]P$ for $1 \le f < e_i$ and all i requires at most a further $2\sum_{i=1}^k e_i \log_2(l_i) = 2\log_2(N) = O(\log(r))$ group operations. Similarly, computing the implicit representations of the remaining $[N/l_i^f]Q$ requires $O(\log(r)^2)$ Fixed-CDH oracle queries and $O(\log(r))$ group operations.

The computation of $u_i P_0$ in line 8 of Algorithm 26 requires $O(\log(r))$ operations in $\mathbb{F}_r$ followed by $O(1)$ operations in G and oracle queries.

The exhaustive search algorithm for the solution to the DLP in a subgroup of prime order l_i is given in lines 9 to 16 of Algorithm 26. The point P_0 in line 8 has already been computed, and computing Q_0 requries only one elliptic curve addition (i.e., $O(\log(r))$ Fixed-CDH oracle queries). The while loop in line 12 runs for $\le B$ iterations, each iteration involves a constant number of field operations to compute $T + P_0$ followed by two exponentiations in the group to compute g^{x_T} and g^{y_T} (an obvious improvement is to use g^{x_T} only). The complexity of lines 9 to 16 is therefore $O(B\log(r))$ group operations and $O(B)$ field operations.

If one uses BSGS the results are similar. Suppose Q and P are points of order l, where P is known explicitly, while we only have an implicit representation (g^{x_Q}, g^{y_Q}) for Q. Let $m = \lceil\sqrt{l}\rceil$ and $P_1 = [m]P$ so that $Q = [u_0]P + [u_1]P_1$ for $0 \le u_0, u_1 < m$. One computes a list of baby steps $[u_0]P$ in implicit representation using $O(\sqrt{B})$ field operations and $O(\sqrt{B}\log(r))$ group operations as above. For the giant steps $Q - [u_1]P_1$ one is required to perform elliptic curve arithmetic with the implicit point Q and the explicit point $[u_1]P_1$, which requires an inversion of an implicit element. Hence, the giant steps require $O(\sqrt{B})$ field operations, $O(\sqrt{B}\log(r))$ group operations and $O(\sqrt{B}\log(r))$ Fixed-CDH oracle queries.

Since $\sum_{i=1}^{k} e_i \le \log_2(N)$ the exhaustive search or BSGS subroutine is performed $O(\log(r))$ times. A more careful analysis using Exercise 13.2.7 means the complexity is multiplied by $\log(r)/\log(B)$. The Chinese remainder theorem and later stages are negligible. The result follows. □

Algorithm 26 Maurer reduction

INPUT: $g, h = g^a, E(\mathbb{F}_r)$

OUTPUT: a

1: Associate to h an implicit representation for a point $Q = (X, Y) \in E(\mathbb{F}_r)$ using Lemma 21.4.9
2: Compute a point $P \in E(\mathbb{F}_r)$ that generates $E(\mathbb{F}_r)$. Let $N = \#E(\mathbb{F}_r) = \prod_{i=1}^{k} l_i^{e_i}$
3: Compute explicit representations of $\{[N/l_i^j]P : 1 \le i \le k, 1 \le j \le e_i\}$
4: Compute implicit representations of $\{[N/l_i^j]Q : 1 \le i \le k, 1 \le j \le e_i\}$
5: **for** $i = 1$ to k **do**
6:     $u_i = 0$
7:     **for** $j = 1$ to e_i **do** ▷ Reducing DLP of order $l_i^{e_i}$ to cyclic groups
8:         Let $P_0 = [N/l_i^j]P$ and $Q_0 = [N/l_i^j]Q - u_i P_0$
9:         **if** $Q_0 \ne \mathcal{O}_E$ **then**
10:             Let $(h_{0,x}, h_{0,y})$ be the implicit representation of Q_0
11:             $P = [N/l_i]P, n = 1, T = P = (x_T, y_T)$
12:             **while** $h_{0,x} \ne g^{x_T}$ or $h_{0,y} \ne g^{y_T}$ **do** ▷ Exhaustive search
13:                 $n = n + 1, T = T + P_0$
14:             **end while**
15:             $u_i = u_i + nl^{j-1}$
16:         **end if**
17:     **end for**
18: **end for**
19: Use Chinese remainder theorem to compute $u \equiv u_i \pmod{l_i^{e_i}}$ for $1 \le i \le k$
20: Compute $(X, Y) = [u]P$ and hence compute a
21: **return** a

Remark 21.4.11 We have seen that reductions involving a Fixed-CDH oracle are less efficient (i.e., require more oracle queries) than reductions using a CDH oracle. A solution[6] to this is to work with projective coordinates for elliptic curves. Line 12 of Algorithm 26 tests whether the point Q_0 given in implicit representation is equal to the point (x_T, y_T) given in affine representation. When $Q_0 = (x_0 : y_0 : z_0)$ then the test $h_{0,x} = g^{x_T}$ in line 12 is replaced with the comparison

$$g^{x_0} = \left(g^{z_0}\right)^{x_T}.$$

[6] This idea is briefly mentioned in Section 3 of [362], but was explored in detail by Bentahar [39].

Hence, the number of oracle queries in the first line of the table in Theorem 21.4.10 can be reduced to $O(\log(r)\log\log(r))$. As mentioned in Remark 13.3.2 one cannot use the BSGS algorithm with projective coordinates, as the non-uniqueness of the representation means one cannot efficiently detect a match between two lists.

Exercise 21.4.12★ Generalise the Maurer algorithm to the case where the group of points on the elliptic curve is not necessarily cyclic. Determine the complexity if l_1 is the largest prime for which $E(\mathbb{F}_r)[l_1]$ is not cyclic and l_2 is the largest prime dividing $\#E(\mathbb{F}_r)$ for which $E(\mathbb{F}_r)[l_2]$ is cyclic.

Exercise 21.4.13 If $r+1$ is smooth then one can use the algebraic group $G_{2,r} \cong \mathbb{T}_2(\mathbb{F}_r)$ (see Section 6.3) instead of $G_m(\mathbb{F}_r)$ or $E(\mathbb{F}_r)$. There are two approaches: the first is to use the usual representation $\{a + b\theta \in \mathbb{F}_{r^2} : \mathrm{N}_{\mathbb{F}_{r^2}/\mathbb{F}_r}(a+b\theta) = 1\}$ for $G_{2,r}$ and the second is to use the representation $\mathbb{A}^1(\mathbb{F}_r)$ for $\mathbb{T}_2(\mathbb{F}_r) - \{1\}$ corresponding to the map decomp_2 from Definition 6.3.7. Determine the number of (perfect) oracle queries in the reductions from Fixed-CDH to DLP for these two representations. Which is better? Repeat the exercise when one has a CDH oracle.

Corollary 21.4.14 *Let $c \in \mathbb{R}_{>1}$. Let (G_n, g_n, r_n) be a family of groups for $n \in \mathbb{N}$ where $g_n \in G_n$ has order r_n and r_n is an n-bit prime. Suppose we are given* ***auxiliary elliptic curves*** *(E_n, N_n) for the family, where E_n is an elliptic curve over $\mathbb{F}_{r_n}$ such that $\#E_n(\mathbb{F}_{r_n}) = N_n$ and N_n is $O(\log(r_n)^c)$-smooth. Then the DLP in $\langle g_n\rangle$ is equivalent to the Fixed-CDH problem in $\langle g_n\rangle$.*

Exercise 21.4.15 Prove Corollary 21.4.14.

We now state the conjecture of Maurer and Wolf that all Hasse intervals contain a polynomially smooth integer. Define $\nu(r)$ to be the minimum, over all integers $n \in [r+1-2\sqrt{r}, r+1+2\sqrt{r}]$, of the largest prime divisor of n. Conjecture 1 of [367] states that

$$\nu(r) = \log(r)^{O(1)}. \tag{21.9}$$

See Remark 15.3.5 for discussion of this. Muzereau, Smart and Vercauteren [402] note that if r is a pseudo-Mersenne prime (as is often used in elliptic curve cryptography) then the Hasse interval usually contains a power of 2. Similarly, as noted by Maurer and Wolf in [365], one can first choose a random smooth integer n and then search for a prime r close to n and work with a group G of order r.

Exercise 21.4.16★ Show how to use the algorithm of Section 19.4.4 to construct a smooth integer in the Hasse interval. Construct a 2^{40}-smooth integer (not equal to 2^{255}) close to $p = 2^{255} - 19$ using this method.

Remark 21.4.17 There are two possible interpretations of Corollary 21.4.14. The first interpretation is: if there exists an efficient algorithm for CDH or Fixed-CDH in a group $G = \langle g\rangle$ of prime order r and if there exists an auxiliary elliptic curve over $\mathbb{F}_r$ with sufficiently smooth order then there exists an efficient algorithm to solve the DLP in G.

Maurer and Wolf [368] (also see Section 3.5 of [369]) claim this gives a non-uniform reduction from DLP to CDH, however the validity of this claim depends on the DLP instance generator.[7]

In other words, if one believes that there does not exist a non-uniform polynomial-time algorithm for DLP in G (for certain instance generators) and if one believes the conjecture that the Hasse interval around r contains a polynomially smooth integer then one must believe there is no polynomial-time algorithm for CDH or Fixed-CDH in G. Hence, one can use the results to justify the assumption that CDH is hard. We stress that this is purely a statement of existence of algorithms; it is independent of the issue of whether or not it is feasible to write the algorithms down.

A second interpretation is that CDH might be easy and that this reduction yields the best algorithm for solving the DLP. If this were the case (or if one wants a uniform reduction) then, in order to solve a DLP instance, the issue of how to implement the DLP algorithm becomes important. The problem is that there is no known polynomial-time algorithm to construct auxiliary elliptic curves $E(\mathbb{F}_r)$ of smooth order. An algorithm to construct smooth curves (based on the CM method) is given in Section 4 of [365] but it has exponential complexity. Hence, if one can write down an efficient algorithm for CDH then the above ideas alone do not allow one to write down an efficient algorithm for DLP.

Boneh and Lipton [78] handle the issue of auxiliary elliptic curves by giving a subexponential-time reduction between Fixed-CDH and DLP. They make the natural assumption (essentially Conjecture 15.3.1, as used to show that the elliptic curve factoring method is subexponential-time) that, for sufficiently large primes, the probability that a randomly chosen integer in the Hasse interval $[r + 1 - 2\sqrt{r}, r + 1 + 2\sqrt{r}]$ is $L_r(1/2, c)$-smooth is $1/L_r(1/2, c')$ for some constants $c, c' > 0$ (see Section 15.3 for further discussion of these issues). By randomly choosing $L_r(1/2, c')$ elliptic curves over $\mathbb{F}_r$ one therefore expects to find one that has $L_r(1/2, c)$-smooth order. One can then perform Algorithm 26 to solve an instance of the DLP in subexponential-time and using polynomially many oracle queries. We refer to [78] for the details.

Maurer and Wolf extend the Boneh–Lipton idea to genus 2 curves and use results of Lenstra, Pila and Pomerance (Theorem 1.3 of [342]) to obtain a reduction with proven complexity $L_r(2/3, c)$ for some constant c (see Section 3.6 of [369]). This is the only reduction from DLP to CDH that does not rely on any conjectures or heuristics. Unfortunately, it is currently impractical to construct suitable genus 2 curves in practice (despite being theoretically polynomial-time).

Muzereau, Smart and Vercauteren [402] go even further than Boneh and Lipton. They allow an exponential-time reduction, with the aim of minimising the number of CDH or

[7] An instance generator for the DLP (see Example 2.1.9) outputs a quadruple (G, r, g, h) where G is a description of a group, $g \in G$ has order r, $h \in \langle g \rangle$ and r is prime. The size of the instance depends on the representation of G and g, but is at least $2\log_2(r)$ bits since one must represent r and h. If one considers the DLP with respect to an instance generator for which r is constant over all instances of a given size n, then a single auxiliary curve is needed for all DLP instances of size n and so Corollary 21.4.14 gives a non-uniform reduction. On the other hand, if there are superpolynomially many r among the outputs of size n of the instance generator (this would be conjecturally true for the instance generator of Example 2.1.9) then the amount of auxiliary data is not polynomially bounded and hence the reduction is not non-uniform.

Fixed-CDH oracle queries. The motivation for this approach is to give tight reductions between CDH and DLP (i.e., to give a lower bound on the running time for an algorithm for CDH in terms of conjectured lower bounds for the running time of an algorithm for DLP). Their results were improved by Bentahar [39, 40]. It turns out to be desirable to have an auxiliary elliptic curve such that $\#E(\mathbb{F}_r)$ is a product of three coprime integers of roughly equal size $r^{1/3}$. The reduction then requires $O(\log(r))$ oracle queries but $O(r^{1/3}\log(r))$ field operations. It is natural to conjecture[8] that suitable auxiliary elliptic curves exist for each prime r. One can construct auxiliary curves by choosing random curves, counting points and factoring; one expects only polynomially many trials, but the factoring computation is subexponential. We refer to [402, 39, 40] for further details.

Exercise 21.4.18 Write down the algorithm for the Muzereau-Smart-Vercauteren reduction using projective coordinates. Prove that the algorithm has the claimed complexity.

Exercise 21.4.19 Show how to generate in heuristic expected polynomial-time primes $r, p \equiv 2 \pmod 3$ such that $r \mid (p+1)$, $r+1$ is κ-smooth, and $2^{\kappa-1} \le r < p \le 2^{\kappa+3}$. Hence, by Exercise 9.10.4, taking $E : y^2 = x^3 + 1$ then $E(\mathbb{F}_p)$ is a group of order divisible by r and $E(\mathbb{F}_r)$ has κ-smooth order and is a suitable auxiliary elliptic curve for the Maurer reduction.

Finally, we remark that the den Boer and Maurer reductions cannot be applied to relate CDH and DLP in groups of unknown order. For example, let N be composite and $g \in (\mathbb{Z}/N\mathbb{Z})^*$ of unknown order M. Given a perfect Fixed-CDH oracle with respect to g one can still compute with the algebraic group $G_m(\mathbb{Z}/M\mathbb{Z})$ in implicit representation (or projective equations for $E(\mathbb{Z}/M\mathbb{Z})$), but if M is not known then the order of $G = G_m(\mathbb{Z}/M\mathbb{Z})$ (respectively, $G = E(\mathbb{Z}/M\mathbb{Z})$) is also not known and so one cannot perform the Pohlig–Hellman algorithm in G.

21.5 Algorithms for static Diffie–Hellman

Brown and Gallant [104] studied the relationship between Static-DH and DLP. Their main result is an algorithm to solve an instance of the DLP using a perfect Static-DH oracle. Cheon [122] independently discovered this algorithm in a different context, showing that a variant of the DLP (namely, the problem of computing a given g, g^a and g^{a^d}; we call this **Cheon's variant of the DLP**) can be significantly easier than the DLP. We now present the algorithm of Brown-Gallant and Cheon, and discuss some of its applications.

Theorem 21.5.1 *Let g have prime order r and let $d \mid (r-1)$. Given $h_1 = g^a$ and $h_d = g^{a^d}$ then one can compute a in $O((\sqrt{(r-1)/d} + \sqrt{d})\log(r))$ group operations, $O(\sqrt{(r-1)/d} + \sqrt{d})$ group elements of storage and $O(\sqrt{(r-1)/d} + \sqrt{d})$ multiplications in $\mathbb{F}_r$.*[9]

[8] This conjecture seems to be possible to prove using current techniques, but I am not aware of any reference for it.

[9] As usual, we are being careless with the $O(\cdot)$-notation. What we mean is that there is a constant c independent of r, d, g and a such that the algorithm requires $\le c(\sqrt{(r-1)/d} + \sqrt{d})\log(r)$ group operations.

Proof First, the case $a \equiv 0 \pmod{r}$ is easy, so we assume $a \not\equiv 0 \pmod{r}$. The idea is essentially the same as the den Boer reduction. Let γ be a primitive root modulo r. Then $a = \gamma^u \pmod{r}$ for some $0 \le u < r - 1$ and it suffices to compute u. The den Boer reduction works by projecting the unknown a into prime order subgroups of $\mathbb{F}_r^*$ using a Diffie–Hellman oracle. In our setting, we already have an implicit representation of the projection a^d into the subgroup of $\mathbb{F}_r^*$ of order $(r-1)/d$.

The first step is to solve $h_d = g^{a^d} = g^{\gamma^{du}}$ for some $0 \le u \le (r-1)/d$. Let $m = \lceil \sqrt{(r-1)/d} \rceil$ and write $u = u_0 + m u_1$ with $0 \le u_0, u_1 < m$. This is exactly the setting of equations (21.6) and (21.7) and hence one can compute (u_0, u_1) using a baby-step–giant-step algorithm. This requires $\le m$ multiplications in $\mathbb{F}_r$ and $\le 2m$ exponentiations in the group. Thus the total complexity is $O(\sqrt{(r-1)/d}\log(r))$ group operations and $O(\sqrt{(r-1)/d})$ field operations.

We now have $a^d = \gamma^{du}$ and so $a = \gamma^{u+v(r-1)/d}$ for some $0 \le v < d$. It remains to compute v. Let

$$h = h_1^{\gamma^{-u}} = g^{a\gamma^{-u}} = g^{\gamma^{v(r-1)/d}}.$$

Set $m = \lceil \sqrt{d} \rceil$ and write $v = v_0 + m v_1$ where $0 \le v_0, v_1 < m$. Using the same ideas as above (since γ is known explicitly the powers are computed efficiently) one can compute (v_0, v_1) using a baby-step–giant-step algorithm in $O(\sqrt{d}\log(r))$ group operations. Finally, we compute $a = \gamma^{u+v(r-1)/d} \pmod{r}$. □

Kozaki, Kutsuma and Matsuo [318] show how to reduce the complexity in the above result to $O(\sqrt{(r-1)/d} + \sqrt{d})$ group operations by using precomputation to speed up the exponentiations to constant time. Note that this trick requires exponential storage and is not applicable when low-storage discrete logarithm algorithms are used (as in Exercise 21.5.5).

The first observation is that if $r - 1$ has a suitable factorisation then Cheon's variant of the DLP can be much easier than the DLP.

Corollary 21.5.2 *Let g have prime order r and suppose $r-1$ has a factor d such that $d \approx r^{1/2}$. Given $h_1 = g^a$ and $h_d = g^{a^d}$ then one can compute a in $O(r^{1/4}\log(r))$ group operations.*

Corollary 21.5.3 *Let g have prime order r and suppose $r - 1 = \prod_{i=1}^n d_i$ where the d_i are coprime. Given $h_1 = g^a$ and $h_{d_i} = g^{a^{d_i}}$ for $1 \le i \le n$ then one can compute a in $O((\sum_{i=1}^n \sqrt{d_i})\log(r))$ group operations.*

Exercise 21.5.4 Prove Corollaries 21.5.2 and 21.5.3.

As noted in [104] and [122], one can replace the baby-step–giant-step algorithms by Pollard methods. Brown and Gallant[10] suggest a variant of the Pollard rho method, but with several non-standard features: one needs to find the precise location of the collision (i.e., steps $x_i \ne x_j$ in the walk such that $x_{i+1} = x_{j+1}$) and there is only a (heuristic) 0.5

[10] See Appendix B.2 of the first version of [104]. This does not appear in the June 2005 version.

probability that a collision leads to a solution of the DLP. Cheon [122] suggests using the Kangaroo method, which is a more natural choice for this application.

Exercise 21.5.5 Design a pseudorandom walk for the Pollard kangaroo method to solve the DLP in implicit representation arising in the proof of Theorem 21.5.1.

Brown and Gallant use Theorem 21.5.1 to obtain the following result.

Theorem 21.5.6 *Let g have prime order r and let $d \mid (r-1)$. Let $h = g^a$ and suppose A is a perfect oracle for the static Diffie–Hellman problem with respect to (g, h) (i.e., $A(h_1) = h_1^a$). Then one can compute a using d oracle queries, $O((\sqrt{(r-1)/d} + \sqrt{d})\log(r))$ group operations and $O((\sqrt{(r-1)/d} + \sqrt{d})\log(r))$ multiplications in $\mathbb{F}_r$.*

Proof Write $h_1 = h = g^a$ and compute the sequence $h_{i+1} = O(h_i) = g^{a^i}$ until g^{a^d} is computed. Then apply Theorem 21.5.1. □

Note that the reduction uses a Static-DH oracle with respect to g^a to compute a. The reduction does not solve a general instance of the DLP using a specific Static-DH oracle; hence, it is not a reduction from DLP to Static-DH. Also recall that Exercise 20.4.6 showed how one can potentially compute a efficiently given access to a Static-DH oracle (with respect to a) that does not check that the inputs are group elements of the correct order. Hence, the Brown–Gallant result is primarily interesting in the case where the Static-DH oracle does perform these checks.

Corollary 21.5.7 *Let g have prime order r and suppose $r-1$ has a factor d such that $d \approx r^{1/3}$. Given $h = g^a$ and a perfect Static-DH oracle with respect to (g, h) then one can compute a in $O(r^{1/3})$ oracle queries and $O(r^{1/3}\log(r))$ group operations.*

Exercise 21.5.8 Prove Corollary 21.5.7.

Brown and Gallant use Theorem 21.5.6 to give a lower bound on the difficulty of Static-DH under the assumption that the DLP is hard.

Exercise 21.5.9 Let g have order r. Assume that the best algorithm to compute a, given $h = g^a$, requires $\sqrt{r}$ group operations. Suppose that $r-1$ has a factor $d = c_1\log(r)^2$ for some constant c_1. Prove that the best algorithm to solve Static-DH with respect to (g, h) requires at least $c_2\sqrt{r}/\log(r)^2$ group operations for some constant c_2.

All the above results are predicated on the existence of a suitable factor d of $r-1$. Of course, $r-1$ may not have a factor of the correct size; for example, if $r-1 = 2l$ where l is prime then we have shown that given (g, g^a, g^{a^2}) one can compute a in $O(\sqrt{r/2}\log(r))$ group operations, which is no better than general methods for the DLP. To increase the applicability of these ideas, Cheon also gives a method for when there is a suitable factor d of $r+1$. The method in this case is not as efficient as the $r-1$ case, and requires more auxiliary data.

Theorem 21.5.10 *Let g have prime order r and let $d \mid (r+1)$. Given $h_i = g^{a^i}$ for $1 \le i \le 2d$ then one can compute a in $O((\sqrt{(r+1)/d} + d)\log(r))$ group operations, $O(\sqrt{(r+1)/d} + \sqrt{d})$ group elements storage and $O((\sqrt{(r+1)/d} + \sqrt{d})\log(r))$ multiplications in $\mathbb{F}_r$.*

Proof As in Exercise 21.4.13, the idea is to work in the algebraic group $G_{2,r}$, which has order $r+1$. Write $\mathbb{F}_{r^2} = \mathbb{F}_r(\theta)$ where $\theta^2 = t \in \mathbb{F}_r$. By Lemma 6.3.10 each element $\alpha \in G_{2,r} - \{1\} \subseteq \mathbb{F}_{r^2}^*$ is of the form $\alpha_0 + \alpha_1\theta$ where

$$\alpha_0 = \frac{a^2 - t}{a^2 + t}, \qquad \alpha_1 = \frac{2a}{a^2 + t}$$

for some $a \in \mathbb{F}_r$. For each $d \in \mathbb{N}$ there exist polynomials $f_{d,0}(x), f_{d,1}(x) \in \mathbb{F}_r[x]$ of degree $2d$ such that, for α as above, one has

$$\alpha^d = \frac{f_{d,0}(a) + \theta f_{d,1}(a)}{(a^2 + t)^d}.$$

The idea is to encode the DLP instance g^a into the element $\beta \in G_{2,r}$ as

$$\beta = \frac{a^2 - t}{a^2 + t} + \theta \frac{2a}{a^2 + t}.$$

We do not know β, but we can compute $(a^2 - t)$, $(a^2 + t)$ and $2a$ in implicit representation.

Let γ be a generator for $G_{2,r}$, known explicitly. Then $\beta = \gamma^u$ for some $0 \le u < r + 1$. It suffices to compute u.

The first step is to project into the subgroup of order $(r+1)/d$. We have $\beta^d = \gamma^{du}$ for some $0 \le u < (r+1)/d$. Let $m = \lceil \sqrt{(r+1)/d} \rceil$ so that $u = u_0 + m u_1$ for $0 \le u_0, u_1 < m$. Write $\gamma^i = \gamma_{i,0} + \theta\gamma_{i,1}$. Then $\beta^d \gamma^{-u_0} = \gamma^{d u_1}$ and so $(f_{d,0}(a) + \theta f_{d,1}(a))(\gamma_{-u_0,0} + \theta\gamma_{-u_0,1}) = (a^2 + t)^d(\gamma_{du_1,0} + \theta\gamma_{du_1,1})$. Hence

$$\left(g^{f_{d,0}(a)}\right)^{\gamma_{-u_0,0}} \left(g^{f_{d,1}(a)}\right)^{\gamma_{-u_0,1}} = \left(g^{(a^2+t)^d}\right)^{\gamma_{du_1,0}}$$

and similarly for the implicit representation of the coefficient of θ. It follows that one can perform the baby-step–giant-step algorithm in this setting to compute (u_0, u_1) and hence $u \pmod{(r+1)/d}$. Note that computing $g^{f_{d,0}(a)}$, $g^{f_{d,1}(a)}$ and $g^{(a^2+t)^d}$ requires $6d$ exponentiations. The stated complexity follows.

For the second stage, we have $\beta = \gamma^{u + v(r+1)/d}$ where $0 \le v < d$. Giving a baby-step–giant-step algorithm here is straightforward and we leave the details as an exercise. □

One derives the following result. Note that it is not usually practical to consider a computational problem whose input is a $O(r^{1/3})$-tuple of group elements; hence, this result is mainly of theoretical interest.

Corollary 21.5.11 *Let g have prime order r and suppose $r+1$ has a factor d such that $d \approx r^{1/3}$. Given $h_i = g^{a^i}$ for $1 \le i \le 2d$ then one can compute a in $O(r^{1/3}\log(r))$ group operations.*

Corollary 21.5.12 *Let g have prime order r and suppose $r+1$ has a factor d such that $d \approx r^{1/3}$. Given $h = g^a$ and a perfect Static-DH oracle with respect to (g, h) then one can compute a in $O(r^{1/3})$ oracle queries and $O(r^{1/3}\log(r))$ group operations.*

Exercise 21.5.13 Fill in the missing details in the proof of Theorem 21.5.10 and prove Corollaries 21.5.11 and 21.5.12.

Satoh [459] extends Cheon's algorithm to algebraic groups of order $\varphi_n(r)$ (essentially, to the groups $G_{n,r}$). He also improves Theorem 21.5.10 in the case of $d \mid (r+1)$ to only require $h_i = g^{a^i}$ for $1 \le i \le d$.

A natural problem is to generalise Theorem 21.5.10 to other algebraic groups, such as elliptic curves. The obvious approach does not seem to work (see Remark 1 of [122]), so it seems a new idea is needed to achieve this. Finally, Section 5.2 of [123] shows that, at least asymptotically, most primes r are such that $r-1$ or $r+1$ has a useful divisor.

Both [104] and [122] remark that a decryption oracle for classic textbook Elgamal leads to a Static-DH oracle: given an Elgamal public key (g, g^a) and any $h_1 \in \langle g \rangle$ one can ask for the decryption of the ciphertext $(c_1, c_2) = (h_1, 1)$ (one can also make this less obvious using random self-reducibility of Elgamal ciphertexts) to get $c_2 c_1^{-a} = h_1^{-a}$. From this, one computes h_1^a. By performing this repeatedly one can compute a sequence $h_i = g^{a^i}$ as required. The papers [104, 122] contain further examples of cryptosystems that provide Static-DH oracles, or computational assumptions that contain values of the form $h_i = g^{a^i}$.

21.6 Hard bits of discrete logarithms

Saying that a computational problem is hard is the same as saying that it is hard to write down a binary representation of the answer. Some bits of a representation of the answer may be easy to compute (at least, up to a small probability of error) but if a computational problem is hard then there must be at least one bit of any representation of the answer that is hard to compute. In some cryptographic applications, it is important to be able to locate some of these "hard bits". Hence, the main challenge is to prove that a specific bit is hard. A potentially easier problem is to determine a small set of bits, at least one of which is hard. A harder problem is to prove that some set of bits are all simultaneously hard (for this concept see Definition 21.6.14).

The aim of this section is to give a rigorous definition for the concept of "hard bits" and to give some easy examples (hard bits of the solution to the DLP). In Section 21.7 we will consider related problems for the CDH problem. We first show that certain individual bits of the DLP, for any group, are as hard to compute as the whole solution.

Definition 21.6.1 Let $g \in G$ have prime order r. The computational problem **DL-LSB** is: given (g, g^a) where $0 \le a < r$ to compute the least significant bit of a.

Exercise 21.6.2 Show that DL-LSB $\le_R$ DLP.

Theorem 21.6.3 *Let G be a group of prime order r. Then DLP $\le_R$ DL-LSB.*

Proof Let A be a perfect oracle that on input (g, g^a) outputs the least significant bit of $0 \leq a < r$. In other words, if the binary expansion of a is $\sum_{i=0}^{m} a_i 2^i$ then A outputs a_0. We will use A to compute a.

The first step is to call $A(g, h)$ to get a_0. Once this has been obtained we set $h' = hg^{-a_0}$. Then $h' = g^{2a_1+4a_2+\cdots}$. Let $u = 2^{-1} = (r+1)/2 \pmod{r}$ and define

$$h_1 = (h')^u.$$

Then $h_1 = g^{a_1+2a_2+\cdots}$ so calling $A(g, h_1)$ gives a_1. For $i = 2, 3, \ldots$ compute $h_i = (h_{i-1}g^{-a_{i-1}})^u$ and $a_i = A(g, h_i)$, which computes the binary expansion of a. This reduction runs in polynomial-time and requires polynomially many calls to the oracle A. □

Exercise 21.6.4 Give an alternative proof of Theorem 21.6.3 based on bounding the unknown a in the range

$$(l-1)r/2^j \leq a < lr/2^j.$$

Initially, one sets $l = 1$ and $j = 0$. At step j, if one has $(l-1)r/2^j \leq a < lr/2^j$ and if a is even then $(l-1)r/2^{j+1} \leq a/2 < lr/2^{j+1}$ and if a is odd then $(2^j + l - 1)r/2^{j+1} \leq (a+r)/2 < (2^j + l)r/2^{j+1}$. Show that when $j = \lceil \log_2(r) \rceil$ one can compute $2^{-j}a \pmod{r}$ exactly and hence deduce a.

Exercise 21.6.5 Since one can correctly guess the least significant bit of the DLP with probability 1/2, why does Theorem 21.6.3 not prove that DLP is easy?

One should also consider the case of a DL-LSB oracle that only works with some noticeable probability ϵ. It is then necessary to randomise the calls to the oracle, but the problem is to determine the LSB of a given the LSBs of some algebraically related values. The trick is to guess some $u = O(\log(1/\epsilon)) = O(\log(\log(r)))$ most significant bits of a and set them to zero (i.e., replace h by $h' = g^{a'}$ where the u most significant bits of a' are zero). One can then call the oracle on $h'g^y$ for random $0 \leq y \leq r - r/2^u$ and take a majority vote to get the result. For details of the argument see Blum and Micali [68].

We conclude that computing the LSB of the DLP is as hard as computing the whole DLP; such bits are called **hardcore bits**.

Definition 21.6.6 Let $f : \{0,1\}^* \to \{0,1\}^*$ be a function computable in polynomial-time (i.e., there is some polynomial $p(n)$ such that for $x \in \{0,1\}^n$ one can compute $f(x)$ in at most $p(n)$ bit operations). A function $b : \{0,1\}^* \to \{0,1\}$ is a **hardcore bit** or **hardcore predicate** for f if, for all probabilistic polynomial-time algorithms A, the **advantage**

$$\mathrm{Adv}_{x \in \{0,1\}^n}\big(A(f(x)) = b(x)\big)$$

is negligible as a function of n.

We now give some candidate hardcore predicates for the DLP. We also restate the meaning of hardcore bit for functions defined on $\{0, 1, \ldots, r-1\}$ rather than $\{0,1\}^*$.

Definition 21.6.7 For all $n \in \mathbb{N}$ let (G_n, g_n, r_n) be such that G_n is a group and $g_n \in G_n$ is an element of order r_n where r_n is an n-bit prime. We call this a **family of groups**. For $n \in \mathbb{N}$ define the function $f_n : \{0, 1, \dots, r_n - 1\} \to G_n$ by $f_n(a) = g_n^a$. For $n \in \mathbb{N}$ define $i(n) \in \{0, 1, \dots, n - 1\}$. The predicate $b_{i(n)} : \{0, 1, \dots, r_n - 1\} \to \{0, 1\}$ is defined so that $b_{i(n)}(a)$ is bit $i(n)$ of a, when a is represented as an n-bit string. Then $b_{i(n)}$ is a **hardcore predicate for the DLP** (alternatively, bit $i(n)$ is a **hardcore bit for the DLP**) if, for all probabilistic polynomial-time algorithms A, the advantage

$$\mathrm{Adv}_{a \in \{0,1,\dots,r_n-1\}}\big(A(f_n(a)) = b_{i(n)}(a)\big)$$

is negligible as a function of n.

The least significant bit (LSB) is the case $i(n) = 0$ in the above definition. If the DLP is hard then Theorem 21.6.3 shows that the LSB is a hardcore bit.

Example 21.6.8 Fix $m \in \mathbb{N}$. Let g have prime order $r > 2^m$. Suppose A is a perfect oracle such that, for $x \in \{0, 1, \dots, r - 1\}$, $A(g^x)$ is the predicate $b_m(x)$ (i.e., bit m of x). One can use A to solve the DLP by guessing the $m - 1$ LSBs of x and then using essentially the same argument as Theorem 21.6.3. Hence, if m is fixed and g varies in a family of groups as in Example 21.6.7 then $b_m(x)$ is a hardcore predicate for the DLP. A similar result holds if m is allowed to grow, but is bounded as $m = O(\log(\log(r)))$.

We now give an example of a hardcore predicate that is not just a bit of the DLP.

Exercise 21.6.9 Let g have prime order r. Let $f : \{0, 1, \dots, r - 1\} \to G$ be $f(x) = g^x$. Define the predicate $b : \{0, 1, \dots, r - 1\} \to \{0, 1\}$ by $b(x) = x_1 \oplus x_0$ where x_0 and x_1 are the two least significant bits of x. Show that b is a hardcore predicate for f.

It is not true that any bit of the DLP is necessarily hardcore. For example, one can consider the most significant bit of a, which is $b_{n-1}(x)$ in Definition 21.6.7.

Example 21.6.10 Let $r = 2^l + u$ be a prime where $0 < u < 2^{l-\kappa}$. Let $0 \le a < r$ be chosen uniformly at random and interpreted as an $(l + 1)$-bit string. Then the most significant bit of a is equal to 1 with probability $u/r < u/2^l < 1/2^\kappa$ and is equal to 0 with probability at least $1 - 1/2^\kappa$. Hence, when $\kappa \le 1$ then the most significant bit is not a hardcore bit for the DLP. Note that the function g^a is not used here; the result merely follows from the distribution of integers modulo r.

Exercise 21.6.11 Let $r = 2^l + 2^{l-1} + u$ where $0 < u < 2^{l/2}$. Let $0 \le a < r$ be uniformly chosen and represented as an $(l + 1)$-bit string. Show that neither the most significant bit (i.e., bit l) nor bit $l - 1$ of a are hardcore for the DLP.

The above examples show that for some primes the most significant bit is easy to predict. For other primes the most significant bit can be hard.

Exercise 21.6.12 Suppose $r = 2^l - 1$ is a Mersenne prime and let g have order r. Fix $0 \le i \le l$. Show that if $O(g, h)$ is a perfect oracle that returns the ith bit of the DLP of h with respect to g then one can compute the whole DLP.

To summarise, low order bits of the DLP are always as hard as the DLP, while high order bits may or may not be hard. However, our examples of cases where the high order bits are easy are due not to any weakness of the DLP, but rather to statistical properties of residues modulo r. One way to deal with this issue is to define a bit as being "hard" if it cannot be predicted better than the natural statistical bias (see, for example, Definition 6.1 of Håstad and Näslund [253]). However, this approach is less satisfactory for cryptographic applications if one wants to use the DLP as a source of unpredictable bits. Hence, it is natural to introduce a more statistically balanced predicate to use in place of high order bits. In practice, it is often more efficient to compute the least significant bit than to evaluate this predicate.

Exercise 21.6.13 Let g have order r. Let $f : \{0, 1, \dots, r-1\} \to G$ be $f(x) = g^x$. Define $b(x) = 0$ if $0 \le x < r/2$ and $b(x) = 1$ if $r/2 \le x < r$. Show, using the method of Exercise 21.6.4, that $b(x)$ is a hardcore bit for f.

We do not cover all results on hard bits for the DLP. See Section 9 of Håstad and Näslund [253] for a general result and further references.

So far we have only discussed showing that single bits of the DLP are hard. There are several approaches to defining the notion of a set of k bits being simultaneously hard. One definition states that the bits are hard if, for every non-constant function $B : \{0, 1\}^k \to \{0, 1\}$, given an oracle that takes as input g^x and computes B on the k bits of x in question, one can use the oracle to solve the DLP. Another definition, which seems to be more useful in practice, is in terms of distinguishing the bits from random.

Definition 21.6.14 Let $f : \{0, 1\}^n \to \{0, 1\}^m$ be a one-way function and let $S \subset \{1, \dots, n\}$. We say the bits labelled by S are **simultaneously hard** if there is no polynomial-time algorithm that given $f(x)$ can distinguish the sequence $(x_i : i \in S)$ from a random $\#S$-bit string.

Peralta [430] (using next-bit-predictability instead of hardcore predicates or Definition 21.6.14) proves that $O(\log(\log(r)))$ least significant bits of the DLP are hard. Schnorr [471] (using Definition 21.6.14) proves that essentially any $O(\log(\log(r)))$ bits of the DLP are simultaneously hard (using the "bits" of Exercise 21.6.13 for the most significant bits).

Patel and Sundaram [428] showed, under a stronger assumption, that many more bits are simultaneously hard. Let g be an element of prime order r, let $l \in \mathbb{N}$ and set $k = \lceil \log_2(r) \rceil - l$. The ideas of Patel and Sundaram lead to the following result. If, given g^x, the k least significant bits of x are not simultaneously hard then there is an efficient algorithm to solve the DLP in an interval of length 2^l (see Exercise 13.3.6 for the definition of this problem). Hence, under the assumption that the DLP in an interval of length 2^l is hard, then one can output many bits. Taking $l = \log(\log(p))^{1+\epsilon}$ gives an essentially optimal asymptotic bit security result for the DLP.

21.6.1 Hard bits for DLP in algebraic group quotients

One can consider hard bits for the DLP in algebraic group quotients. In other words, let O_i be a perfect oracle that on input the equivalence class of an element $[g^a]$ outputs bit i of a. The first problem is that there is more than one value a for each class $[g^a]$ and so the bit is not necessarily well-defined.

Section 7 of Li, Näslund and Shparlinski [349] considers this problem for LUC. To make the problem well-defined they consider an element $g \in \mathbb{F}_{p^2}$ of prime order r and an oracle A such that $A(t) = a_i$ where a_i is the ith bit of a for the unique $0 \le a < r/2$ such that $t = \mathrm{Tr}_{\mathbb{F}_{p^2}/\mathbb{F}_p}(g^a)$. The idea of their method is, given t, to compute the two roots $h_1 = g^a$ and $h_2 = g^{r-a}$ of $X^2 - tX + 1$ in $\mathbb{F}_{p^2}$ then use previous methods (e.g., Theorem 21.6.3 or Exercise 21.6.4) on each of them to compute either a or $r - a$ (whichever is smaller).

Exercise 21.6.15 Work out the details of the Li, Näslund and Shparlinski result for the case of the least significant bit of the DLP in LUC.

Exercise 21.6.16 Consider the algebraic group quotient corresponding to elliptic curve arithmetic using x-coordinates only. Fix $P \in E(\mathbb{F}_q)$ of prime order r. Let A be an oracle that on input $u \in \mathbb{F}_q$ outputs a_0 where a_0 is the 0th bit of a such that $0 \le a < r/2$ and $x([a]P) = u$. Show that the method of Li, Näslund and Shparlinski can be applied to show that this bit is a hard bit for the DLP.

Li, Näslund and Shparlinski remark that it seems to be hard to obtain a similar result for XTR. Theorem 3 of Jiang, Xu and Wang [282] claims to be such a result, but it does not seem to be proved in their paper.

21.7 Bit security of Diffie–Hellman

We now consider which bits of the CDH problem are hard. Since the solution to a CDH instance is a group element it is natural to expect, in contrast with our discussion of the DLP, that the hardcore bits and the proof techniques depend on which group is being studied.

We first consider the case $g \in \mathbb{F}_p^*$ where p is a large prime and g is a primitive root. Our presentation follows Boneh and Venkatesan [80]. We assume every element $x \in \mathbb{F}_p^*$ is represented as an element of the set $\{1, 2, \dots, p-1\}$ and we interpret $x \pmod{p}$ as returning a value in this set.

Definition 21.7.1 Let p be odd. Let $x \in \{1, 2, \dots, p-1\}$. Define

$$\mathrm{MSB}_1(x) = \begin{cases} 0 & \text{if } 1 \le x < p/2 \\ 1 & \text{otherwise.} \end{cases}$$

For $k \in \mathbb{N}$ let $0 \le t < 2^k$ be the integer such that

$$tp/2^k \le x < (t+1)p/2^k$$

and define $\mathrm{MSB}_k(x) = t$.

An alternative definition, which is commonly used in the literature and sometimes used in this book, is $\mathrm{MSB}_k(x) = u \in \mathbb{Z}$ such that $|x - u| < p/2^{k+1}$ (e.g., $u = \lfloor tp/2^k + p/2^{k+1} \rfloor$). For this definition it is unnecessary to assume $k \in \mathbb{N}$ and so one can allow $k \in \mathbb{R}_{>0}$.

Note that these are not bits of the binary representation of x. Instead, as in Exercise 21.6.13, they correspond to membership of x in a certain partition of $\{1, 2, \ldots, p-1\}$.

Ideally we would like to show that, say, MSB_1 is a hardcore bit for CDH. This seems to be out of reach for $\mathbb{F}_p^*$. Instead, we will show that, for $k \approx \sqrt{\log_2(r)}$, if one can compute $\mathrm{MSB}_k(g^{ab} \pmod{p})$ then one can compute $g^{ab} \pmod{p}$. A consequence of this result is that there exists some predicate defined on $\mathrm{MSB}_k(g^{ab} \pmod{p})$ whose value is a hardcore bit for CDH.

The central idea of most results on the bit security of CDH is the following. Let p be an odd prime and let $g \in \mathbb{F}_p^*$ be a primitive root. Let $h_1 = g^a, h_2 = g^b$ be a CDH instance where b is coprime to $p-1$. For $k \in \mathbb{N}$ let A_k be a perfect oracle such that

$$A_k(g, g^a, g^b) = \mathrm{MSB}_k(g^{ab}).$$

Choose a random element $1 \le x < p$ and set $u = A_k(g, h_1 g^x, h_2)$. One has

$$u = \mathrm{MSB}_k(g^{(a+x)b}) = \mathrm{MSB}_k(g^{ab}t) \quad \text{where} \quad t = h_2^x.$$

In other words, the oracle A_k gives the most significant bits of multiples of the unknown g^{ab} by uniformly random elements $t \in \mathbb{F}_p^*$. The problem of using this information to compute g^{ab} is (a special case of) the hidden number problem.

21.7.1 The hidden number problem

Definition 21.7.2 Let p be an odd prime and $k \in \mathbb{R}_{>1}$. Let $\alpha \in \mathbb{F}_p^*$ and let $t_1, \ldots, t_n \in \mathbb{F}_p^*$ be chosen uniformly at random. The **hidden number problem** (**HNP**) is given $(t_i, u_i = \mathrm{MSB}_k(\alpha t_i \pmod{p}))$ for $1 \le i \le n$ to compute α.

Throughout this section we will allow any $k \in \mathbb{R}_{>1}$ and define $\mathrm{MSB}_k(x)$ to be any integer u such that $|x - u| < p/2^{k+1}$.

Before giving the main results we discuss two easy variants of Definition 21.7.2 where the values t_i can be chosen adaptively.

Lemma 21.7.3 *Let p be an odd prime and $1 \le \alpha < p$. Suppose one has a perfect oracle A_1 such that $A_1(t) = \mathrm{MSB}_1(\alpha t \pmod{p})$. Then one can compute α using $O(\log(p))$ oracle queries.*

Exercise 21.7.4 Prove Lemma 21.7.3.

Lemma 21.7.5 *Let p be an odd prime and $1 \le \alpha < p$. Suppose one has a perfect oracle A such that $A(t) = \mathrm{LSB}_1(\alpha t \pmod{p})$, where $\mathrm{LSB}_1(x)$ is the least significant bit of the binary representation of $0 \le x < p$. Then one can compute α using $O(\log_2(p))$ oracle queries.*

Exercise 21.7.6 Prove Lemma 21.7.5.

Lemmas 21.7.3 and 21.7.5 show that the hidden number problem would be easy if the values t_i in Definition 21.7.2 are chosen adaptively. However, it intuitively seems harder to solve the hidden number problem when the t_i are randomly chosen. On the other hand, as k grows the HNP becomes easier, the case $k = \lceil \log_2(p) \rceil$ being trivial. Hence, one could hope to be able to solve the HNP as long as k is sufficiently large. We now explain the method of Boneh and Venkatesan [80] to solve the HNP using lattices.

Definition 21.7.7 Let $(t_i, u_i = \mathrm{MSB}_k(\alpha t_i))$ for $1 \le i \le n$. Define a lattice $L \subseteq \mathbb{R}^{n+1}$ by the rows of the basis matrix

$$B = \begin{pmatrix} p & 0 & 0 & \cdots & 0 & 0 \\ 0 & p & 0 & & 0 & 0 \\ \vdots & \vdots & & & \vdots & \vdots \\ 0 & 0 & 0 & \cdots & p & 0 \\ t_1 & t_2 & t_3 & \cdots & t_n & 1/2^{k+1} \end{pmatrix}.$$

Define the vector $\underline{u} = (u_1, u_2, \ldots, u_n, 0) \in \mathbb{R}^{n+1}$ where $|u_i - (\alpha t_i \pmod{p})| < p/2^{k+1}$.

Lemma 21.7.8 *Let L, $\underline{u}$ and n be as in Definition 21.7.7. Then $\det(L) = p^n/2^{k+1}$ and there exists a vector $\underline{v} \in L$ such that $\|\underline{u} - \underline{v}\| < \sqrt{n+1}\,p/2^{k+1}$.*

Proof The first statement is trivial. For the second, note that $u_i = \mathrm{MSB}_k(\alpha t_i \pmod{p})$ is the same as saying $\alpha t_i = u_i + \epsilon_i + l_i p$ for some $\epsilon_i, l_i \in \mathbb{Z}$ such that $|\epsilon_i| \le p/2^{k+1}$ for $1 \le i \le n$. Now define $\underline{v} \in L$ by

$$\begin{aligned} \underline{v} = (-l_1, -l_2, \ldots, -l_n, \alpha)B &= (\alpha t_1 - l_1 p, \ldots, \alpha t_n - l_n p, \alpha/2^{k+1}) \\ &= (u_1 + \epsilon_1, \ldots, u_n + \epsilon_n, \alpha/2^{k+1}). \end{aligned}$$

The result follows since $\alpha/2^{k+1} < p/2^{k+1}$. □

We now show that, for certain parameters, it is reasonable to expect that any vector in the lattice L that is close to $\underline{u}$ gives the solution α.

Theorem 21.7.9 *Let $p > 2^8$ be prime and let $\alpha \in \mathbb{F}_p^*$. Let $n = 2\lceil \sqrt{\log_2(p)} \rceil \in \mathbb{N}$ and let $k \in \mathbb{R}$ be such that $\log_2(p) - 1 > k > \mu = \frac{1}{2}\sqrt{\log_2(p)} + 3$. Suppose $t_1, \ldots, t_n$ are chosen uniformly and independently at random in $\mathbb{F}_p^*$ and set $u_i = \mathrm{MSB}_k(\alpha t_i)$ for $1 \le i \le n$. Construct the lattice L as above. Let $\underline{u} = (u_1, \ldots, u_n, 0)$. Then, with probability at least $1 - 1/2^n \ge 63/64$ over all choices for $t_1, \ldots, t_n$, any vector $\underline{v} \in L$ such that $\|\underline{v} - \underline{u}\| < p/2^{\mu+1}$ is of the form*

$$\underline{v} = (\beta t_1 \pmod{p}, \ldots, \beta t_n \pmod{p}, \beta/2^{k+1})$$

where $\beta \equiv \alpha \pmod{p}$.

Proof In the first half of the proof we consider $t_1, \ldots, t_n$ as fixed values. Later in the proof we compute a probability over all choices for the t_i.

First, note that every vector in the lattice is of the form

$$\underline{v} = (\beta t_1 - l_1 p, \beta t_2 - l_2 p, \ldots, \beta t_n - l_n p, \beta/2^{k+1})$$

for some $\beta, l_1, \ldots, l_n \in \mathbb{Z}$. If $\beta \equiv \alpha \pmod{p}$ then we are done, so suppose now that $\beta \not\equiv \alpha \pmod{p}$. Suppose also that $\|\underline{v} - \underline{u}\| < p/2^{\mu+1}$, which implies $|(\beta t_i \pmod{p}) - u_i| < p/2^{\mu+1}$ for all $1 \le i \le n$. Note that

$$\begin{aligned}|(\beta - \alpha)t_i \pmod{p}| &= |(\beta t_i \pmod{p}) - u_i + u_i - (\alpha t_i \pmod{p})| \\ &\le |(\beta t_i \pmod{p}) - u_i| + |(\alpha t_i \pmod{p}) - u_i| \\ &< p/2^{\mu+1} + p/2^{\mu+1} = p/2^{\mu}.\end{aligned}$$

We now consider $\gamma = (\beta - \alpha)$ as a fixed non-zero element of $\mathbb{F}_p$ and denote by A the probability, over all $t \in \mathbb{F}_p^*$, that $\gamma t \equiv u \pmod{p}$ for some $u \in \mathbb{Z}$ such that $|u| < p/2^{\mu}$ and $u \ne 0$. Since γt is uniformly distributed over $\mathbb{F}_p^*$ it follows that

$$A = \frac{2\lfloor p/2^{\mu} \rfloor}{p-1} \le \frac{1}{p-1}\left(\frac{2(p-1)+2}{2^{\mu}}\right) < \frac{2}{2^{\mu}} + \frac{2}{p-1} < \frac{4}{2^{\mu}}.$$

Since there are n uniformly and independently chosen $t_1, \ldots, t_n \in \mathbb{F}_p^*$ the probability that $|\gamma t_i \pmod{p}| < p/2^{\mu}$ for all $1 \le i \le n$ is A^n. Finally, there are $p-1$ choices for $\beta \in \{0, 1, \ldots, p-1\}$ such that $\beta \not\equiv \alpha \pmod{p}$. Hence, the probability over all such β and all $t_1, \ldots, t_n$ that $\|\underline{v} - \underline{u}\| < p/2^{\mu+1}$ is at most

$$(p-1)A^n < \frac{(p-1)4^n}{2^{\mu n}} < \frac{2^{\log_2(p)+2n}}{2^{\mu n}}.$$

Now, $\mu n = (\sqrt{\log_2(p)}/2 + 3)2\lceil\sqrt{\log_2(p)}\rceil \ge \log_2(p) + 3n$ so $(p-1)A^n < 2^{-n}$. Since $n \ge 6$ the result follows. □

Corollary 21.7.10 *Let $p > 2^{32}$ be prime, let $n = 2\lceil\sqrt{\log_2(p)}\rceil$ and let $k = \lceil\sqrt{\log_2(p)}\rceil + \lceil\log_2(\log_2(p))\rceil$. Given $(t_i, u_i = \mathrm{MSB}_k(\alpha t_i))$ for $1 \le i \le n$ as in Definition 21.7.2 one can compute α in polynomial-time.*

Proof One constructs the basis matrix B for the lattice L in polynomial-time. Note that $n = O(\sqrt{\log(p)})$ so that the matrix requires $O(\log(p)^2)$ bits storage.

Running the LLL algorithm with factor $\delta = 1/4 + 1/\sqrt{2}$ is a polynomial-time computation (the lattice is not a subset of $\mathbb{Z}^{n+1}$ so Remark 17.5.5 should be applied, noting that only one column has non-integer entries) which returns an LLL-reduced basis. Let $\underline{u}$ be as above. The Babai nearest plane algorithm finds $\underline{v}$ such that $\|\underline{v} - \underline{u}\| < (1.6)2^{(n+1)/4}\sqrt{n+1}p/2^{k+1}$ by Theorem 18.1.7 and Lemma 21.7.8. This computation requires $O(\log(p)^{4.5})$ bit operations by Exercise 18.1.9. To apply Theorem 21.7.9 we need the vector $\underline{v}$ output from the

Babai algorithm to be within $p/2^{\mu+1}$ of $\underline{u}$ where $\mu = \sqrt{\log_2(p)}/2 + 3$. Hence, we need

$$\frac{(1.6)2^{(n+1)/4}\sqrt{n+1}}{2^{k+1}} < \frac{1}{2^{\mu+1}},$$

which is $\mu + \log_2(1.6) + (n+1)/4 + \log_2(\sqrt{n+1}) < k = \lceil\sqrt{\log_2(p)}\rceil + \lceil\log_2(\log_2(p))\rceil$. Since

$$\mu + \log_2(1.6) + (n+1)/4 + \log_2(\sqrt{n+1}) \le \lceil\sqrt{\log_2(p)}\rceil + 3.95 + \tfrac{1}{2}\log_2(n+1)$$

the result follows whenever p is sufficiently large (the reader can check that $p > 2^{32}$ is sufficient).

It follows from Theorem 21.7.9 that, with probability at least $63/64$ the vector $\underline{v} = (v_1, \ldots, v_{n+1}) \in \mathbb{R}^{n+1}$, output by the Babai algorithm is such that $v_{n+1}2^{k+1} \equiv \alpha \pmod{p}$. It follows that the hidden number α can be efficiently computed. □

Note that if $p \approx 2^{160}$ then $\mu \approx 9.32$. In practice, the algorithm works well for primes of this size. For example, Howgrave-Graham and Smart [270] present results of practical experiments where 8 of the most significant bits are provided by an oracle. We stress that these results do not show that all of the $k = \lceil\sqrt{\log_2(p)}\rceil + \lceil\log_2(\log_2(p))\rceil$ most significant bits are hard. Instead, one can only deduce that there is a predicate defined on these k bits that is a hardcore predicate for CDH.

Nguyen and Shparlinski [411] also remark that one could use other methods than LLL and the Babai nearest plane algorithm. They show that if one uses the Ajtai, Kumar and Sivakumar algorithm for CVP then one only needs $k = \lfloor\log(\log(p))\rfloor$ bits to obtain an algorithm for the hidden number problem with complexity of $p^{O(1/\log(\log(p)))}$ bit operations. They further show that if one has a perfect oracle for CVP (with respect to the ℓ_∞ norm) then one can solve the hidden number problem in polynomial-time given only $k = 1 + \epsilon$ bits for any $\epsilon > 0$.

One final remark, the methods in this section assume a perfect oracle that outputs $\mathrm{MSB}_1(\alpha t \pmod{p})$. Since there seems to be no way to determine whether the output of the oracle is correct, it is an open problem to get results in the presence of an oracle that sometimes makes mistakes. For further discussion and applications of the hidden number problem see Shparlinski [500].

21.7.2 Hard bits for CDH modulo a prime

We can finally state a result about hard bits for CDH.

Theorem 21.7.11 *Let $p > 2^{32}$ be prime, let g be a primitive root modulo p and let $k = \lceil\sqrt{\log_2(p)}\rceil + \lceil\log_2(\log_2(p))\rceil$. Suppose there is no polynomial-time algorithm to solve*[11]

[11] As we have seen, to make such a statement precise one needs an instance generator that outputs groups from a family.

CDH in $\mathbb{F}_p^*$. Then there is no polynomial-time algorithm to compute the k most significant bits of g^{ab} when given g, g^a and g^b.

Proof Let (g, g^a, g^b) be an instance of the CDH problem in $\langle g \rangle$ and write $\alpha = g^{ab}$ for the solution. We assume that $\gcd(b, p-1) = 1$ (this requirement is removed by González Vasco and Shparlinski [238]; other work mentioned below allows g to have prime order, in which case this restriction disappears).

Given a polynomial-time algorithm A such that $A(g, g^x, g^y) = \mathrm{MSB}_k(g^{xy} \pmod{p})$ then one can call $A(g, g^a g^r, g^b)$ polynomially many times for uniformly random $r \in \{1, 2, \ldots, p-2\}$ to get $\mathrm{MSB}_k(\alpha t)$ where $t = g^{br} \pmod{p}$. Applying Corollary 21.7.10 gives a polynomial-time algorithm to compute α. □

A number of significant open problems remain:

1. Theorem 21.7.11 shows it is hard to compute all of $\mathrm{MSB}_k(g^{ab})$ but that does not imply that, say, $\mathrm{MSB}_1(g^{ab})$ is hard. A stronger result would be to determine specific hardcore bits for CDH, or at least to extend the results to MSB_k for smaller values of k. Boneh and Venkatesan [81] give a method that works for $k = \lceil 2\log(\log(p)) \rceil$ bits (where g is a primitive root in $\mathbb{F}_p^*$) but which needs a hint depending on p and g; they claim this is a non-uniform result but this depends on the instance generator (see footnote 7 of Section 21.4.3). For $k = \lfloor \log(\log(p)) \rfloor$ one can also consider the approach of Nguyen and Shparlinski [411] mentioned above.

   Akavia [8] uses a totally different approach to prove that MSB_1 is hard for CDH, but the method is again at best non-uniform (i.e., needs polynomial-sized auxiliary information depending on p and g^b).
2. We assumed perfect oracles for computing $\mathrm{MSB}_k(\alpha t)$ in the above results. For non-perfect oracles one can use the above methods to generate a list of candidate values for g^{ab} and then apply the CDH self-corrector of Section 21.3. We refer to González Vasco, Näslund and Shparlinski [237] for details.

   The method of Akavia [8] also works when the oracle for MSB_1 is unreliable.
3. The above results assumed that g is a primitive root modulo p, whereas in practice one chooses g to lie in a small subgroup of $\mathbb{F}_p^*$ of prime order. The proof of Theorem 21.7.11 generates values t that lie in $\langle g \rangle$ and so they are not uniformly at random in $\mathbb{F}_p^*$. González Vasco and Shparlinski have given results that apply when the order of g is less than $p-1$ (see Chapter 14 of [499] for details and references). Shparlinski and Winterhof [501, 502], building on work of Bourgain and Konyagin, have obtained results when the order of g is at least $\log(p)/\log(\log(p))^{1-\epsilon}$.

Exercise 21.7.12 This exercise concerns a static Diffie–Hellman key exchange protocol due to Boneh and Venkatesan [80] for which one can prove that the most significant bit is a hardcore bit. Suppose Alice chooses a prime p, an integer $1 \le a < p-1$ such that $\gcd(a, p-1) = 1$ and sets $g = 2^{a^{-1} \pmod{p-1}} \pmod{p}$. Alice makes p and g public and keeps a private. When Bob wants to communicate with Alice he sends g^x for random

$1 \le x < p-1$ so that Alice and Bob share the key 2^x. Prove that $\mathrm{MSB}_1(2^x)$ is a hardcore bit.

[Hint: Suppose one has a perfect oracle A that on input g^y outputs $\mathrm{MSB}_1(2^y)$. Then one can store Bob's transmission g^x and call $A(g^x g^y)$ to get $\alpha 2^y$, where $\alpha = 2^x$ is the desired hidden number. Then apply Lemma 21.7.3.]

Exercise 21.7.13 Let $g \in \mathbb{F}_p^*$ be a primitive root and let $\epsilon > 0$. Show that if one has a perfect oracle for $\mathrm{MSB}_{1+\epsilon}(g^{ab})$ then one can solve DDH in $\mathbb{F}_p^*$.

21.7.3 Hard bits for CDH in other groups

So far we have only considered CDH in (subgroups of) $\mathbb{F}_p^*$ where p is prime. It is natural to consider CDH in subgroups of $\mathbb{F}_{p^m}^*$, in algebraic tori, in trace systems such as LUC and XTR and in elliptic curves. The first issue is what is meant by "bits" of such a value. In practice, elements in such a group are represented as an n-tuple of elements in $\mathbb{F}_p$ and so it is natural to consider one component in $\mathbb{F}_p$ and take bits of it as done previously. When p is small, one can consider a sequence of bits, each from different components. An early reference for bit security of CDH in this setting is Verheul [555].

It is possible to extend the results to traces relatively easily. The idea is that if $\{\theta_1, \dots, \theta_m\}$ is a basis for $\mathbb{F}_{p^m}$ over $\mathbb{F}_p$, if $\alpha = \sum_{j=1}^m \alpha_j \theta_j$ is hidden and if $t_i = \sum_{j=1}^m t_{i,j}\theta_j$ are known then $\mathrm{Tr}(\alpha t_i)$ is a linear equation in the unknown α_i. Li, Näslund and Shparlinski [349] have studied the bit security of CDH in LUC and XTR. We refer to Chapters 6 and 19 of Shparlinski [499] for further details and references.

Exercise 21.7.14 Let $\mathbb{F}_{2^m}$ be represented using a normal basis and let $g \in \mathbb{F}_{2^m}^*$. Suppose one has a perfect oracle A such that $A(g, g^a, g^b)$ returns the first coefficient of the normal basis representation of g^{ab}. Show how to use A to compute g^{ab}. Hence, conclude that the first coefficient is a hardcore bit for CDH in $\mathbb{F}_{2^m}^*$.

Exercise 21.7.15 Let $\mathbb{F}_{2^m} = \mathbb{F}_2[x]/(F(x))$ and let $g \in \mathbb{F}_{2^m}^*$ have prime order $r > m$. Suppose one has a perfect oracle A such that $A(g, g^a, g^b)$ returns the constant coefficient of the polynomial basis representation of g^{ab}. Show how to use A to compute g^{ab}. Hence, conclude that the constant coefficient is a hardcore bit for CDH in $\mathbb{F}_{2^m}^*$.

Hard bits for elliptic curve Diffie–Hellman

We now consider the case of elliptic curves E over $\mathbb{F}_q$. A typical way to extract bits from an elliptic curve point P is to consider the x-coordinate $x(P)$ as an element of $\mathbb{F}_q$ and then extract bits of this. It seems hard to give results for the bit security of CDH using an oracle $A(P, [a]P, [b]P) = \mathrm{MSB}_k(x([ab]P))$; the natural generalisation of the previous approach is to call $A(P, [a]P + [z]P, [b]P) = \mathrm{MSB}_k(x([ab]P + [zb]P))$ but the problem is that it is difficult to infer anything useful about $x([ab]P)$ from $x([ab]P + [zb]P)$ (similarly, for least significant bits); see Jao, Jetchev and Venkatesan [276] for some results. However, Boneh and Shparlinski [79] had the insight to consider a more general oracle.

Definition 21.7.16 Let p be an odd prime and $k \in \mathbb{N}$. Let $A_{x,k}(A, B, P, [a]P, [b]P)$ be an oracle that returns $\mathrm{LSB}_k(x([ab]P))$ where $P \in E(\mathbb{F}_p)$ for the elliptic curve $E : y^2 = x^3 + Ax + B$. Similarly, let $A_{y,k}(A, B, P, [a]P, [b]P)$ be an oracle that returns $\mathrm{LSB}_k(y([ab]P))$.

The crucial idea is that, given a point $P = (x_P, y_P) \in E(\mathbb{F}_p)$ where $E : y^2 = x^3 + Ax + B$, one can consider an isomorphism $\phi(x, y) = (u^2 x, u^3 y)$ and $\phi(P) \in E'(\mathbb{F}_p)$ where $E' : Y^2 = X^3 + u^4 AX + u^6 B$. Hence, instead of randomising instances of CDH in a way analogous to that done earlier, one calls the oracle $A_{x,k}(u^4 A, u^6 B, \phi(P), \phi([a]P), \phi([b]P))$ to get $\mathrm{LSB}_k(x(\phi([ab]P))) = \mathrm{LSB}_k(u^2 x([ab]P) \pmod{p})$ where u is controlled by the attacker. This is very similar to the easy case of the hidden number problem in $\mathbb{F}_p^*$ from Lemma 21.7.5.

Lemma 21.7.17 *Suppose $p \equiv 2 \pmod 3$. Then $\mathrm{LSB}_1(y([ab]P))$ is a hardcore bit for CDH on elliptic curves over $\mathbb{F}_p$.*

Proof We suppose $A_{y,1}$ is a perfect oracle for $\mathrm{LSB}_1(y([ab]P))$ as above. Calling

$$A_{y,1}(u^4 A, u^6 B, \phi(P), \phi([a]P), \phi([b]P))$$

gives $\mathrm{LSB}_1(u^3 y([ab]P))$. Since $\gcd(3, p-1) = 1$ it follows that cubing is a permutation of $\mathbb{F}_p^*$ and one can perform the method of Lemma 21.7.5 to compute $y([ab]P)$. Given $y([ab]P)$ there are at most 3 choices for $x([ab]P)$ and so CDH is solved with noticeable probability. □

In the general case (i.e., when $p \not\equiv 2 \pmod 3$), Boneh and Shparlinski have to work harder. They use the method of Alexi, Chor, Goldreich and Schnorr [9] or the simplified version by Fischlin and Schnorr [186] to extend the idea to non-perfect oracles.[12] Once this is done, the following trick can be applied to determine $\mathrm{LSB}_1(tx([ab]P))$: when t is a square one calls the oracle for $\mathrm{LSB}_1(u^2 x([ab]P))$ on $u = \sqrt{t} \pmod p$, and when t is not a square one flips a coin. The resulting non-perfect oracle for LSB_1 therefore solves the problem. We refer to [79] for the details.

We make some remarks.

1. A nice feature of the elliptic curve results is that they are independent of the order of the point P and so work for subgroups of any size.
2. The literature does not seem to contain bit security results for CDH on elliptic curves over non-prime fields. This would be a good student project.
3. Jetchev and Venkatesan [281] use isogenies to extend the applicability of the Boneh–Shparlinski method. Their motivation is that if one has an $\mathrm{LSB}_1(x([ab]P))$ oracle that

[12] This is why Boneh and Shparlinski consider least significant bits rather than most significant bits for their result. The technique of Alexi et al is to randomise the query $\mathrm{LSB}_1(t\alpha)$ as $\mathrm{LSB}_1(s\alpha) \oplus \mathrm{LSB}_1((t+s)\alpha)$ for suitable values s. A good student project would be to obtain an analogous result for other bits (e.g., most significant bits).

works with only small (but noticeable) probability then it is possible to have a CDH instance on an elliptic curve E for which the oracle does not work for any twist of E. By moving around the isogeny class they claim that the probability of success increases. However, it is still possible to have a CDH instance on an elliptic curve E for which the oracle does not work for any elliptic curve in the isogeny class of E.

22
Digital signatures based on discrete logarithms

Public key signatures and their security notions were defined in Section 1.3.2. They are arguably the most important topic in public key cryptography (for example, to provide authentication of automatic software updates; see Section 1.1). This chapter gives some digital signature schemes based on the discrete logarithm problem. The literature on this topic is enormous and we only give a very brief summary of the area. RSA signatures are discussed in Section 24.6.

22.1 Schnorr signatures

We assume throughout this section that an algebraic group G and an element $g \in G$ of prime order r are known to all users. The values (G, g, r) are known as **system parameters**. Let $h = g^a$ be a user's public key. A digital signature, on a message m with respect to a public key h, can be generated by a user who knows the private key a. It should be hard to compute a signature for a given public key without knowing the private key.

To explain the Schnorr signature scheme it is simpler to first discuss an identification scheme.

22.1.1 The Schnorr identification scheme

Informally, a **public key identification scheme** is a protocol between a Prover and a Verifier, where the Prover has a public key pk and private key sk, and the Verifier has a copy of pk. The protocol has three communication stages: first, the Prover sends a commitment s_0, then the Verifier sends a challenge s_1 and finally the Prover sends a response s_2. The Verifier either accepts or rejects the proof. The protocol is supposed to convince the Verifier that they are communicating with a user who knows the private key corresponding to the Prover's public key. In other words, the Verifier should be convinced that they are communicating with the Prover.

For the Schnorr scheme [469, 470] the Prover has public key $h = g^a$ where g is an element of an algebraic group of prime order r and $1 \le a < r$ is chosen uniformly at random. The Prover chooses a random integer $0 \le k < r$, computes $\mathsf{s}_0 = g^k$ and sends

s_0 to the Verifier. The Verifier sends a "challenge" $1 \leq \mathsf{s}_1 < r$ to the Prover. The Prover returns $\mathsf{s}_2 = k + a\mathsf{s}_1 \pmod{r}$. The Verifier then checks whether

$$g^{\mathsf{s}_2} = \mathsf{s}_0 h^{\mathsf{s}_1} \tag{22.1}$$

and accepts the proof if this is the case. In other words, the Prover has successfully identified themself to the Verifier if the Verifier accepts the proof.

Exercise 22.1.1 Show that the Verifier in an execution of the Schnorr identification scheme does accept the proof when the Prover follows the steps correctly.

Exercise 22.1.2 Let $p = 311$ and $r = 31 \mid (p-1)$. Let $g = 169$, which has order r. Let $a = 11$ and $h = g^a \equiv 47 \pmod{p}$. Which of the following is a transcript $(\mathsf{s}_0, \mathsf{s}_1, \mathsf{s}_2)$ of a correctly performed execution of the Schnorr identification scheme?

$$(15, 10, 12),\ (15, 10, 27),\ (16, 10, 12),\ (15, 16, 0).$$

Security of the private key

Unlike public key encryption (at least, under passive attacks), with identification schemes and digital signature schemes, a user is always outputting the results of computations involving their private key. Hence, it is necessary to ensure that we do not leak information about the private key. Therefore, we now explain why executions of the Schnorr identification protocol do not leak the private key.

The idea is that, since k is chosen uniformly at random, the private key a is completely hidden. More precisely, for any pair $(\mathsf{s}_1, \mathsf{s}_2)$ and every $1 \leq a < r$ there is some integer $0 \leq k < r$ such that $\mathsf{s}_2 \equiv k + a\mathsf{s}_1 \pmod{r}$. Now, if k were known to the verifier then they could solve for a. But, since the discrete logarithm problem is hard, it is computationally infeasible to determine any significant information about the distribution of k from s_0. Hence, s_2 leaks essentially no information about a. Furthermore, there are no choices for s_1 that more readily allow the verifier to determine a. This concept is known as "zero knowledge" and it is beyond the scope of this book to discuss it further. For security, k must be chosen uniformly at random; see Exercise 22.1.3 and Section 22.3 for attacks if some information on k is known. We stress that such attacks are much stronger than the analogous attacks for Elgamal encryption (see Exercise 20.4.1); there the adversary only learns something about a single message, whereas here they learn the private key!

Exercise 22.1.3 Suppose the random values k used by a prover are generated using the linear congruential generator $k_{i+1} = Ak_i + B \pmod{r}$ for some $1 \leq A, B < r$. Suppose an adversary knows A and B and sees two protocol transcripts $(\mathsf{s}_0, \mathsf{s}_1, \mathsf{s}_2)$ and $(\mathsf{s}_0', \mathsf{s}_1', \mathsf{s}_2')$ generated using consecutive outputs k_i and k_{i+1} of the generator. Show how the adversary can determine the private key a.

A generalisation of Exercise 22.1.3, where the modulus for the linear congruential generator is not r, is given by Bellare, Goldwasser and Micciancio [32].

Security against impersonation

Now we explain why the Verifier is convinced that the prover must know the private key a. The main ideas will also be used in the security proof of Schnorr signatures so we go through the argument in some detail. First, we define an adversary against an identification protocol.

Definition 22.1.4 An **adversary against an identification protocol** (with an honest verifier) is a polynomial-time randomised algorithm A that takes as input a public key, plays the role of the Prover in the protocol with an honest Verifier and tries to make the Verifier accept the proof. The adversary repeatedly and adaptively sends a value s_0, receives a challenge s_1 and answers with s_2 (indeed, the sessions of the protocol can be interleaved). The adversary is successful if the Verifier accepts the proof with noticeable probability (i.e., the probability, over all outputs s_0 by A and all choices for s_1, that the adversary can successfully respond with s_2 is at least one over a polynomial function of the security parameter). The protocol is secure if there is no successful adversary.

An adversary is just an algorithm A so it is reasonable to assume that A can be run in very controlled conditions. In particular, we will assume throughout this section that A can be repeatedly run so that it always outputs the same first commitment s_0 (think of A as a computer programme that calls a function `Random` to obtain random bits and then simply arrange that the function always returns the same values to A). This will allow us to respond to the same commitment with various different challenges s_1. Such an attack is sometimes known as a **rewinding attack** (Pointcheval and Stern [433] call it the **oracle replay attack**): if A outputs s_0, receives a challenge s_1 and answers with s_2 then re-running A on challenge s_1' is the same as "rewinding" the clock back to when A had just output s_0 and then giving it a different challenge s_1'.

Theorem 22.1.5 *The Schnorr identification scheme is secure against impersonation (in the sense of Definition 22.1.4) if the discrete logarithm problem is hard.*

We first prove the result for perfect adversaries (namely, those that impersonate the user successfully every time the protocol is run). Later we discuss the result for more general adversaries.

Proof (In the case of a perfect adversary) We build an expected polynomial-time algorithm (called the simulator) that solves a DLP instance (g, h) where g has prime order r and $h = g^a$ where $0 \le a < r$ is chosen uniformly at random.

The simulator will play the role of the Verifier and will try to solve the DLP by interacting with A. First, the simulator starts A by giving it h as the public key and giving some choice for the function `Random`. The adversary outputs a value s_0, receives a response s_1 (chosen uniformly at random) from the simulator and then outputs s_2. Since A is perfect we assume that $(\mathsf{s}_0, \mathsf{s}_1, \mathsf{s}_2)$ satisfy the verification equation.

First note that if values s_0 and s_2 satisfy equation (22.1) then s_0 lies in the group generated by g and so is of the form $s_0 = g^k$ for some $0 \le k < r$. Furthermore, it then follows that $s_2 \equiv k + as_1 \pmod{r}$.

Now the simulator can re-run A on the same h and the same function `Random` (this is the re-winding). It follows that A will output s_0 again. The simulator then gives A a challenge $s_1' \neq s_1$. Since A is perfect, it responds with s_2' satisfying equation (22.1).

We have $s_2 \equiv k + as_1 \pmod{r}$ and $s_2' \equiv k + as_1' \pmod{r}$. Hence, the simulator can compute $a \equiv (s_2 - s_2')(s_1 - s_1')^{-1} \pmod{r}$ and solve the DLP. In other words, if there is no polynomial-time algorithm for the DLP then there can be no polynomial-time adversary A against the protocol. □

The above proof gives the basic idea, but is not sufficient since we must consider adversaries that succeed with rather small probability. There are various issues to deal with. For example, A may not necessarily succeed on the first execution of the identification protocol. Hence, one must consider many executions $(s_{i,0}, s_{i,1}, s_{i,2})$ for $1 \le i \le t$ and guess the value i into which one introduces the challenge $s_{i,1}'$. Also, A may only succeed for a small proportion of the challenges s_1 for a given s_0 (it is necessary for the proof that A can succeed on two different choices of s_1 for the same value s_0). This latter issue is not a problem (since A succeeds with noticeable probability, it must succeed for a noticeable proportion of values s_1 for most values s_0). The former issue is more subtle and is solved using the Forking Lemma.

The **Forking Lemma** was introduced by Pointcheval and Stern [433]. A convenient generalisation has been given by Bellare and Neven [33]. The Forking Lemma determines the probability that a rewinding attack is successful. More precisely, consider an algorithm A (the adversary against the signature scheme) that takes as input a `Random` function and a list of responses s_1 to its outputs s_0. We will repeatedly choose a `Random` function and run A twice, the first time with a set of values $(s_{1,1}, \dots, s_{t,1})$ being the responses in the protocol and the second time with a set $(s_{1,1}, \dots, s_{j-1,1}, s_{j,1}', \dots, s_{t,1}')$ of responses for some $1 \le j \le t$. Note that A will output the same values $s_{i,0}$ in both runs when $1 \le i \le j$. The Forking Lemma gives a lower bound on the probability that A succeeds in the identification protocol in the jth execution as desired. Lemma 1 of [33] states that the success probability is at least $p(p/t - 1/r)$ where p is the success probability of A, t is the number of executions of the protocol in each game and r is the size of the set of possible responses. Hence, if p is noticeable, t is polynomial and $1/r$ is negligible then the simulator solves the DLP with noticeable probability. We refer to [33, 433] and Section 10.4.1 of Vaudenay [553] for further details.

Exercise 22.1.6 Show that if the challenge values s_1 chosen by a Verifier can be predicted (e.g., because the Verifier is using a weak pseudorandom number generator) then a malicious player can impersonate an honest user in the Schnorr identification scheme.

Exercise 22.1.7 In the Schnorr identification scheme as presented above, the challenge is a randomly chosen integer $1 \le \mathsf{s}_1 < r$. Instead, for efficiency[1] reasons, one could choose $1 \le \mathsf{s}_1 < 2^l$ for some l such that $l \ge \kappa$ (where κ is the security parameter, but where 2^l is significantly smaller than r). Show that the proof of Theorem 22.1.5 still holds in this setting.

Exercise 22.1.8 Explain why the Schnorr identification scheme cannot be implemented in an algebraic group quotient.

22.1.2 Schnorr signatures

We now present the **Schnorr signature scheme** [469, 470], which has very attractive security and efficiency. The main idea is to make the identification protocol of the previous section non-interactive by replacing the challenge s_1 by a random integer that depends on the message being signed. This idea is known as the **Fiat–Shamir transform**. By Exercise 22.1.6 it is important that s_1 cannot be predicted and so it is also necessary to make it depend on s_0.

More precisely, one sets $\mathsf{s}_1 = H(\mathsf{m} \| \mathsf{s}_0)$ where H is a cryptographic hash function from $\{0, 1\}^*$ to $\{0, 1\}^l$ for some parameter l and where m and s_0 are interpreted as binary strings (and where $\|$ denotes concatenation of binary strings as usual).

One would therefore obtain the following signature scheme, which we call **naive Schnorr signatures**: To sign a message m choose a random $0 \le k < r$, compute $\mathsf{s}_0 = g^k$, $\mathsf{s}_1 = H(\mathsf{m} \| \mathsf{s}_0)$ and $\mathsf{s}_2 = k + a\mathsf{s}_1 \pmod{r}$ and send the signature $(\mathsf{s}_0, \mathsf{s}_2)$ together with m. A verifier, given m, $(\mathsf{s}_0, \mathsf{s}_2)$ and the public key h, would compute $\mathsf{s}_1 = H(\mathsf{m} \| \mathsf{s}_0)$ and accept the signature if

$$g^{\mathsf{s}_2} = \mathsf{s}_0 h^{\mathsf{s}_1}. \tag{22.2}$$

Schnorr makes the further observation that instead of sending $(\mathsf{s}_0, \mathsf{s}_2)$ one could send $(\mathsf{s}_1, \mathsf{s}_2)$. This has major implications for the size of signatures. For example, g may be an element of order r in $\mathbb{F}_p^*$ (for example, with $r \approx 2^{256}$ and $p \approx 2^{3072}$). In this case, $\mathsf{s}_0 = g^k$ requires 3072 bits, s_2 requires 256 bits and s_1 may require as little as 128 bits. In other words, signatures would have $3072 + 256 = 3328$ bits in the naive scheme, whereas Schnorr signatures only require $128 + 256 = 384$ bits.

We present the precise Schnorr signature scheme in Box 22.1.

Example 22.1.9 Let $p = 311$ and $r = 31 \mid (p - 1)$. Let $g = 169$ which has order r. Let $a = 11$ and $h = g^a \equiv 47 \pmod{p}$.

To sign a message m (a binary string) let $k = 20$ and $\mathsf{s}_0 = g^k \equiv 225 \pmod{p}$. The binary expansion of s_0 is $(11100001)_2$. We must now compute $\mathsf{s}_1 = H(\mathsf{m} \| (11100001))$. Since we do not want to go into the details of H, let us just suppose that the output length of H is 4 and that s_1 is the binary string 1001. Then s_1 corresponds to the integer 9. Finally, we

[1] One could speed up signature verification using similar methods to Exercise 22.1.13.

KeyGen: This is the same as classic textbook Elgamal encryption. It outputs an algebraic group, an element g of prime order r, a public key $h = g^a$ and a private key $1 \le a < r$ where a is uniformly chosen.

Sign(g, a, m): Choose uniformly at random $0 \le k < r$, compute $\mathsf{s}_0 = g^k$, $\mathsf{s}_1 = H(\mathsf{m} \| \mathsf{s}_0)$ and $\mathsf{s}_2 = k + a\mathsf{s}_1 \pmod{r}$, where the binary string s_1 is interpreted as an integer in the usual way. The signature is $(\mathsf{s}_1, \mathsf{s}_2)$.

Verify($g, h, \mathsf{m}, (\mathsf{s}_1, \mathsf{s}_2)$): Ensure that h is a valid public key for the user in question then test whether

$$\mathsf{s}_1 = H(\mathsf{m} \| g^{\mathsf{s}_2} h^{-\mathsf{s}_1}).$$

Box 22.1 **Schnorr signature scheme**

compute $\mathsf{s}_2 = k + a\mathsf{s}_1 \equiv 20 + 11 \cdot 9 \equiv 26 \pmod{r}$. The signature is $(\mathsf{s}_1, \mathsf{s}_2) = (1001, 26)$. To verify the signature one computes

$$g^{\mathsf{s}_2} h^{-\mathsf{s}_1} = 169^{26} 47^{-9} \equiv 225 \pmod{p}$$

and checks that $\mathsf{s}_1 = H(\mathsf{m} \| 11100001)$.

Exercise 22.1.10 Show that the Verify algorithm does succeed when given a pair $(\mathsf{s}_1, \mathsf{s}_2)$ output by the Sign algorithm.

22.1.3 Security of Schnorr signatures

The security of Schnorr signatures essentially follows from the same ideas as used in Theorem 22.1.5. In particular, the security depends on the discrete logarithm problem (rather than CDH or DDH as is the case for Elgamal encryption). However, since the challenge is now a function of the message m and s_0, the exact argument of Theorem 22.1.5 cannot be used directly.

One approach is to replace the hash function by a random oracle H (see Section 3.7). The simulator can then control the values of H, and the proof of Theorem 22.1.5 can be adapted to work in this setting. A careful analysis of Schnorr signatures in the random oracle model, using this approach and the Forking Lemma, was given by Pointcheval and Stern [433]. We refer to Theorem 14 of their paper for a precise result in the case where the output of H is $(\mathbb{Z}/r\mathbb{Z})^*$. A proof is also given in Section 10.4.2 of Vaudenay [553]. An analysis of the case where the hash function H maps to $\{0, 1\}^l$ where $l < \log_2(r)$ is given by Neven, Smart and Warinschi [406].

There is no known proof of the security of Schnorr signatures in the standard model (even under very strong assumptions about the hash function). Paillier and Vergnaud [426] give evidence that one cannot give a reduction, in the standard model, from signature forgery for Schnorr signatures (with H mapping to $\mathbb{Z}/r\mathbb{Z}$) to DLP. More precisely, they show that if there is a reduction of a certain type (which they call an algebraic reduction) in the standard model from signature forgery for Schnorr signatures to DLP then there is an algorithm for the "one-more DLP". We refer to [426] for the details.

We now discuss some specific ways to attack the scheme:

1. Given a signature $(\mathsf{s}_1, \mathsf{s}_2)$ on message m, if one can find a message m' such that $H(\mathsf{m} \| g^{\mathsf{s}_2} h^{-\mathsf{s}_1}) = H(\mathsf{m}' \| g^{\mathsf{s}_2} h^{-\mathsf{s}_1})$ then one has a signature also for the message m'. This fact can be used to obtain an existential forgery under a chosen-message attack.

   While one expects to be able to find hash collisions after roughly $2^{l/2}$ computations of H (see Section 3.2), what is needed here is not a general hash collision. Instead, we need a collision of the form $H(\mathsf{m} \| R) = H(\mathsf{m}' \| R)$ where $R = g^{\mathsf{s}_2} h^{-\mathsf{s}_1}$ is *not known* until a signature $(\mathsf{s}_1, \mathsf{s}_2)$ on m has been obtained. Hence, the adversary must first output a message m, then get the signature $(\mathsf{s}_1, \mathsf{s}_2)$ on m and then find m' such that $H(\mathsf{m} \| R) = H(\mathsf{m}' \| R)$. This is called the random-prefix second-preimage problem in Definition 4.1 of [406]. When R is sufficiently large it seems that solving this problem is expected to require around 2^l computations of H.
2. There is a passive existential forgery attack on Schnorr signatures if one can compute preimages of H of a certain form. Precisely, choose any $(\mathsf{s}_1, \mathsf{s}_2)$ (for example, if the output of H is highly non-uniform then choose s_1 to be a "very likely" output of H), compute $R = g^{\mathsf{s}_2} h^{-\mathsf{s}_1}$, then find a bitstring m such that $H(\mathsf{m} \| R) = \mathsf{s}_1$. This attack is prevented if the hash function is hard to invert.

Hence, given a security parameter κ (so that breaking the scheme is required to take more than 2^κ bit operations) one can implement the Schnorr signature scheme with $r \approx 2^{2\kappa}$ and $l = \kappa$. For example, taking $\kappa = 128$, $2^{255} < r < 2^{256}$ and $l = 128$ gives signatures of 384 bits.

Exercise 22.1.11★ Fix $g \in G$ of order r and $\mathsf{m} \in \{0, 1\}^*$. Can a pair $(\mathsf{s}_1, \mathsf{s}_2)$ be a Schnorr signature on the same message m for two different public keys? Are there any security implications of this fact?

22.1.4 Efficiency considerations for Schnorr signatures

The Sign algorithm performs one exponentiation, one hash function evaluation and one computation modulo r. The Verify algorithm performs a multi-exponentiation $g^{\mathsf{s}_2} h^{-\mathsf{s}_1}$ where $0 \le \mathsf{s}_2 < r$ and $1 \le \mathsf{s}_1 < 2^l$ and one hash function evaluation. Hence, signing is faster than verifying.

There are a number of different avenues to speed up signature verification, depending on whether g is fixed for all users, whether one is always verifying signatures with respect to the same public key h or whether h varies, etc. We give a typical optimisation in Example 22.1.13. More dramatic efficiency improvements are provided by online/offline signatures (see Section 22.4), server-aided signatures etc.

Exercise 22.1.12 Show how to modify the Schnorr signature scheme (with no loss of security) so that the verification equation becomes

$$\mathsf{s}_1 = H(\mathsf{m} \| g^{\mathsf{s}_2} h^{\mathsf{s}_1}).$$

Example 22.1.13 Suppose a server must verify many Schnorr signatures (using the variant of Exercise 22.1.12), always for the same value of g but for varying values of h. Suppose that $2^{l-1} < \sqrt{r} < 2^l$ (where l is typically also the output length of the hash function). One strategy to speed up signature verification is for the server to precompute and store the group element $g_1 = g^{2^l}$.

Given a signature $(\mathsf{s}_1, \mathsf{s}_2)$ with $0 \le \mathsf{s}_1 < 2^l$ and $0 \le \mathsf{s}_2 < r$ one can write $\mathsf{s}_2 = \mathsf{s}_{2,0} + 2^l \mathsf{s}_{2,1}$ with $0 \le \mathsf{s}_{2,0}, \mathsf{s}_{2,1} < 2^l$. The computation of $g^{\mathsf{s}_2} h^{\mathsf{s}_1}$ is performed as the three-dimensional multi-exponentiation (see Section 11.2)

$$g^{\mathsf{s}_{2,0}} g_1^{\mathsf{s}_{2,1}} h^{\mathsf{s}_1}.$$

The cost is roughly l squarings and $3l/2$ multiplications (the number of multiplications can be easily reduced using window methods, signed representations etc.).

Schnorr [470] presents methods to produce the group elements g^k without having to perform a full exponentiation for each signature (the paper [470] is particularly concerned with making signatures efficient for smartcards). Schnorr's specific proposals were cryptanalysed by de Rooij [155].

22.2 Other public key signature schemes

The Schnorr signature scheme is probably the best public key signature scheme for practical applications.[2] A number of similar schemes have been discovered, the most well-known of which are Elgamal and DSA signatures. We discuss these schemes very briefly in this section.

22.2.1 Elgamal signatures in prime order subgroups

Elgamal [177] proposed the first efficient digital signature based on the discrete logarithm problem. We present the scheme for historical reasons, and because it gives rise to some nice exercises in cryptanalysis. For further details see Section 11.5.2 of [376] or Section 7.3 of [532].

Assume that g is an element of prime[3] order r in an algebraic group G. In this section we always think of G as being the "full" algebraic group (such as $\mathbb{F}_q^*$ or $E(\mathbb{F}_q)$) and assume

[2] However, Schnorr signatures are not very widely used in practice. The reason for their lack of use may be the fact that they were patented by Schnorr.

[3] The original Elgamal signature scheme specifies that g is a primitive root in $\mathbb{F}_p^*$, but for compatibility with all other cryptographic protocols in this book we have converted it to work with group elements of prime order in any algebraic group.

that testing membership $g \in G$ is easy. The public key of user A is $h = g^a$ and the private key is a, where $1 \le a < r$ is chosen uniformly at random.

The Elgamal scheme requires a function $F : G \to \mathbb{Z}/r\mathbb{Z}$. The only property required of this function is that the output distribution of F restricted to $\langle g \rangle$ should be close to uniform (in particular, F is not required to be hard to invert). In the case where $G = \mathbb{F}_p^*$ it is usual to define $F : \{0, 1, \dots, p-1\} \to \{0, 1, \dots, r-1\}$ by $F(n) = n \pmod{r}$. If G is the set of points on an elliptic curve over a finite field then one could define $F(x, y)$ by interpreting x (or x and y) as binary strings, letting n be the integer whose binary expansion is x (or $x \| y$), and then computing $n \pmod{r}$.

To sign a message m with hash $H(\mathsf{m}) \in \mathbb{Z}/r\mathbb{Z}$ one chooses a random integer $1 \le k < r$, computes $\mathsf{s}_1 = g^k$, computes $\mathsf{s}_2 = k^{-1}(H(\mathsf{m}) - aF(\mathsf{s}_1)) \pmod{r}$ and returns $(\mathsf{s}_1, \mathsf{s}_2)$. To verify the signature $(\mathsf{s}_1, \mathsf{s}_2)$ on message m one checks whether $\mathsf{s}_1 \in \langle g \rangle$, $0 \le \mathsf{s}_2 < r$, and

$$h^{F(\mathsf{s}_1)} \mathsf{s}_1^{\mathsf{s}_2} = g^{H(\mathsf{m})}$$

in G. Elgamal signatures are the same size as naive Schnorr signatures.

A striking feature of the scheme is the way that s_1 appears both as a group element and as an exponent (this is why we need the function F). In retrospect, this is a poor design choice for both efficiency and security. The following exercises explore these issues in further detail. Pointcheval and Stern give a variant of Elgamal signatures (the trick is to replace $H(\mathsf{m})$ by $H(\mathsf{m} \| \mathsf{s}_1))$ and prove the security in Sections 3.3.2 and 3.3.3 of [433].

Exercise 22.2.1 Show that the Verify algorithm succeeds if the Sign algorithm is run correctly.

Exercise 22.2.2 Show that one can verify Elgamal signatures by computing a single three-dimensional multi-exponentiation. Show that the check $\mathsf{s}_1 \in \langle g \rangle$ can therefore be omitted if $\gcd(\mathsf{s}_2, \#G) = 1$. Hence, show that the time to verify an Elgamal signature, when F and H map to $\mathbb{Z}/r\mathbb{Z}$, is around twice the time of the method in Example 22.1.13 to verify a Schnorr signature. Explain why choosing F and H to map to l-bit integers where $l \approx \log_2(r)/2$ does not lead to a verification algorithm as fast as the one in Example 22.1.13.

Exercise 22.2.3 (Elgamal [177]) Suppose the hash function H is deleted in Elgamal signatures (i.e., we are signing messages $\mathsf{m} \in \mathbb{Z}/r\mathbb{Z}$). Give a passive existential forgery in this case (i.e., the attack only requires the public key).

Exercise 22.2.4★ Consider the Elgamal signature scheme in $\mathbb{F}_p^*$ with the function $F(n) = n \pmod{r}$. Suppose the function $F(n)$ computes $n \pmod{r}$ for all $n \in \mathbb{N}$ (not just $0 \le n < p$) and that the check $\mathsf{s}_1 \in \langle g \rangle$ does not include any check on the size of the integer s_1 (for example, it could simply be the check that $\mathsf{s}_1^r \equiv 1 \pmod{p}$ or the implicit check of Exercise 22.2.2). Give a passive selective forgery attack.

Exercise 22.2.5 Consider the following variant of Elgamal signatures in a group $\langle g \rangle$ of order r: the signature on a message m for public key h is a pair $(\mathsf{s}_1, \mathsf{s}_2)$ such that $0 \le \mathsf{s}_1, \mathsf{s}_2 < r$, and

$$h^{\mathsf{s}_1} g^{H(\mathsf{m})} = g^{\mathsf{s}_2}.$$

Show how to determine the private key of a user given a valid signature.

Exercise 22.2.6★ (Bleichenbacher [62]) Consider the Elgamal encryption scheme in $\mathbb{F}_p^*$ with the function $F(n) = n \pmod{r}$. Suppose the checks $\mathsf{s}_1 \in \langle g \rangle$ and $0 \le \mathsf{s}_2 < r$ are not performed by the Verify algorithm. Show how an adversary who has maliciously chosen the system parameter g can produce selective forgeries for any public key under a passive attack.

Exercise 22.2.7 (Vaudenay [552]) Let H be a hash function with l-bit output. Show how to efficiently compute an l-bit prime r and messages $\mathsf{m}_1, \mathsf{m}_2$ such that $H(\mathsf{m}_1) \equiv H(\mathsf{m}_2) \pmod{r}$. Hence, show that if one can arrange for an algebraic group with subgroup of order r to be used as the system parameters for a signature scheme then one can obtain a signature on m_1 for any public key h by obtaining from user A a signature on m_2.

A convenient feature of Elgamal signatures is that one can verify a batch of signatures faster than individually verifying each of them. Some details are given in Exercise 22.2.8. Early work on this problem was done by Naccache, M'Raïhi, Vaudenay and Raphaeli [403] (in the context of DSA) and Yen and Laih [570]. Further discussion of the problem is given by Bellare, Garay and Rabin [31].

Exercise 22.2.8 Let $(\mathsf{s}_{1,i}, \mathsf{s}_{2,i})$ be purported signatures on messages m_i with respect to public keys h_i for $1 \le i \le t$. A verifier can choose random integers $1 \le w_i < r$ and verify all signatures together by testing whether $\mathsf{s}_{1,i} \in \langle g \rangle$ and $0 \le \mathsf{s}_{2,i} < r$ for all i and the single equation

$$\left(\prod_{i=1}^{t} h_i^{w_i F(\mathsf{s}_{1,i})} \right) \left(\prod_{i=1}^{t} \mathsf{s}_{1,i}^{w_i \mathsf{s}_{2,i}} \right) = g^{\sum_{i=1}^{t} w_i H(\mathsf{m}_i)}. \tag{22.3}$$

Show that if all the signatures $(\mathsf{s}_{1,i}, \mathsf{s}_{2,i})$ are valid then the batch is declared valid. Show that if there is at least one invalid signature in the batch then the probability the batch is declared valid is at most $1/(r-1)$. Show how to determine, with high probability, the invalid signatures using a binary search.

If one uses the methods of Exercise 22.2.2 then verifying the t signatures separately requires t three-dimensional multi-exponentiations. One can break equation (22.3) into about $2t/3$ three-dimensional multi-exponentiations. So, for groups where testing $\mathsf{s}_{1,i} \in \langle g \rangle$ is easy (e.g., elliptic curves of prime order), the batch is asymptotically verified in about $2/3$ the time of verifying the signatures individually. Show how to speed up verification of a batch of signatures further if the public keys h_i are all equal. How much faster is this than verifying the signatures individually?

Yen and Laih [570] consider batch verification of naive Schnorr signatures as mentioned in Section 22.1.2. Given t signatures $(\mathsf{s}_{0,i}, \mathsf{s}_{2,i})$ on messages m_i and for keys h_i Yen and

Laih choose $1 \le w_i < 2^l$ (for a suitable small value of l, they suggest $l = 15$) and verify the batch by testing $\mathsf{s}_{0,i} \in \langle g \rangle, 0 \le \mathsf{s}_{2,i} < r$ and

$$g^{\sum_{i=1}^{t} w_i \mathsf{s}_2} = \prod_{i=1}^{t} \mathsf{s}_{0,i}^{w_i} \prod_{i=1}^{t} h_i^{w_i H(\mathsf{m}_i \| \mathsf{s}_{0,i})}.$$

Give the verification algorithm when the public keys are all equal. Show that the cost is roughly $l/(3\log_2(r))$ times the cost of verifying t Elgamal signatures individually.

Explain why it seems impossible to verify batches of Schnorr signatures faster than verifying each individually.

22.2.2 DSA

A slight variant of the Elgamal signature scheme was standardised by NIST[4] as a digital signature standard. This is often called **DSA**.[5] In the case where the group G is an elliptic curve then the scheme is often called **ECDSA**.

In brief, the scheme has the usual public key $h = g^a$ where g is an element of prime order r in an algebraic group G and $1 \le a < r$ is chosen uniformly at random. As with Elgamal signatures, a function $F : G \to \mathbb{Z}/r\mathbb{Z}$ is required. To sign a message with hash value $H(\mathsf{m})$ one chooses a random $1 \le k < r$ and computes $\mathsf{s}_1 = F(g^k)$. If $\mathsf{s}_1 = 0$ then repeat[6] for a different value of k. Then compute $\mathsf{s}_2 = k^{-1}(H(\mathsf{m}) + a\mathsf{s}_1) \pmod{r}$ and, if $\mathsf{s}_2 = 0$ then repeat for a different value of k. The signature on message m is $(\mathsf{s}_1, \mathsf{s}_2)$. To verify the signature one first checks that $1 \le \mathsf{s}_1, \mathsf{s}_2 < r$, then computes $u_1 = H(\mathsf{m})\mathsf{s}_2^{-1} \pmod{r}$, $u_2 = \mathsf{s}_1\mathsf{s}_2^{-1} \pmod{r}$, then determines whether or not

$$\mathsf{s}_1 = F(g^{u_1} h^{u_2}). \tag{22.4}$$

Note that a DSA signature is a pair of integers modulo r so is shorter in general than an Elgamal signature but longer in general than a Schnorr signature. Verification is performed using multi-exponentiation.

Exercise 22.2.9 Show that Verify succeeds on values output by Sign.

Exercise 22.2.10 The case $\mathsf{s}_1 = 0$ is prohibited in DSA signatures. Show that if this check was omitted and if an adversary could find an integer k such that $F(g^k) = 0$ then the adversary could forge DSA signatures for any message. Hence, show that, as in Exercise 22.2.6, an adversary who maliciously chooses the system parameters could forge signatures for any message and any public key.

Exercise 22.2.11 The case $\mathsf{s}_2 = 0$ is prohibited in DSA signatures since the Verify algorithm fails when s_2 is not invertible. Show that if a signer outputs a signature $(\mathsf{s}_1, \mathsf{s}_2)$

[4] NIST stands for "National Institute of Standards and Technology" and is an agency that develops technology standards for the USA.

[5] DSA stands for "digital signature algorithm".

[6] The events $\mathsf{s}_1 = 0$ and $\mathsf{s}_2 = 0$ occur with negligible probability and so do not effect the performance of the signing algorithm.

produced by the Sign algorithm but with $\mathsf{s}_2 = 0$ then an adversary would be able to determine the private key a.

We saw in Exercise 22.2.2 that verifying Elgamal signatures is slow compared with verifying Schnorr signatures using the method in Example 22.1.13. Exercise 22.2.12 shows a variant of DSA (analogous to naive Schnorr signatures) that allows signature verification closer in speed to Schnorr signatures.

Exercise 22.2.12 (Antipa, *et al.* [11]) Consider the following variant of the DSA signature scheme: to sign m choose $1 \le k < r$ randomly, compute $\mathsf{s}_0 = g^k$, $\mathsf{s}_2 = k^{-1}(H(m) + aF(\mathsf{s}_0)) \pmod{r}$ and return $(\mathsf{s}_0, \mathsf{s}_2)$. To verify $(\mathsf{m}, \mathsf{s}_0, \mathsf{s}_2)$ one computes $u_1 = H(\mathsf{m})\mathsf{s}_2^{-1} \pmod{r}$, $u_2 = F(\mathsf{s}_0)\mathsf{s}_2^{-1} \pmod{r}$ as in standard DSA and checks whether

$$\mathsf{s}_0 = g^{u_1} h^{u_2}. \tag{22.5}$$

Show that one can also verify the signature by checking, for any $1 \le v < r$, whether

$$\mathsf{s}_0^v = g^{u_1 v} h^{u_2 v}. \tag{22.6}$$

Show that one can efficiently compute an integer $1 \le v < r$ such that the equation (22.6) can be verified more quickly than equation (22.5).

There is no proof of security for DSA signatures in the standard or random oracle model. A proof of security in the random oracle model of a slightly modified version of DSA (the change is to replace $H(\mathsf{m})$ with $H(\mathsf{m} \| \mathsf{s}_1)$) was given by Pointcheval and Vaudenay [434, 98] (also see Section 10.4.2 of [553]). A proof of security for DSA in the generic group model[7] was given by Brown; see Chapter II of [61].

Exercise 22.2.13 Consider a digital signature scheme where a signature on message m with respect to public key h is an integer $0 \le \mathsf{s} < r$ such that

$$\mathsf{s} = H(\mathsf{m} \| h^{\mathsf{s}}).$$

What is the problem with this signature scheme?

22.2.3 Signatures secure in the standard model

None of the signature schemes considered so far has a proof of security in the standard model. Indeed, as mentioned, Paillier and Vergnaud [426] give evidence that Schnorr signatures cannot be proven secure in the standard model. In this section we briefly mention a signature scheme due to Boneh and Boyen [71, 72] that is secure in the standard model. However, the security relies on a very different computational assumption than DLP and the scheme needs groups with an extra feature (namely, a pairing; see Definition 22.2.14). We

[7] The generic group model assumes that any algorithm to attack the scheme is a generic algorithm for the group G. This seems to be a reasonable assumption when using elliptic curves.

present a simple version of their scheme that is unforgeable under a weak chosen-message attack if the q-strong Diffie–Hellman problem holds (these notions are defined below).

We briefly introduce pairing groups (more details are given in Chapter 26). We use multiplicative notation for pairing groups, despite the fact that G_1 and G_2 are typically subgroups of elliptic curves over finite fields and hence are usually written additively.

Definition 22.2.14 (Pairing groups) Let G_1, G_2, G_T be cyclic groups of prime order r. A pairing is a map $e : G_1 \times G_2 \to G_T$ such that:

1. e is non-degenerate and bilinear, i.e. $g_1 \in G_1 - \{1\}$ and $g_2 \in G_2 - \{1\}$ implies $e(g_1, g_2) \neq 1$ and $e(g_1^a, g_2^b) = e(g_1, g_2)^{ab}$ for $a, b \in \mathbb{Z}$,
2. there is a polynominal-time algorithm to compute $e(g_1, g_2)$.

For the Boneh–Boyen scheme we also need there to be an efficiently computable injective group homomorphism $\psi : G_2 \to G_1$ (for example, a distortion map; see Section 26.6.1).

KeyGen: Choose $g_2 \in G_2 - \{1\}$ uniformly at random and set $g_1 = \psi(g_2)$. Let $z = e(g_1, g_2)$. Choose $1 \leq a < r$ and set $u = g_2^a$. The public key is $\mathsf{pk} = (g_1, g_2, u, z)$ and the private key is a.

Sign(m, a): We assume that $\mathsf{m} \in \mathbb{Z}/r\mathbb{Z}$. If $a + \mathsf{m} \equiv 0 \pmod{r}$ then return $\perp$, else compute $\mathsf{s} = g_1^{(a+\mathsf{m})^{-1} \pmod{r}}$ and return s.

Verify(m, s, pk): Check that $\mathsf{s} \in G_1$ and then check that

$$e(\mathsf{s}, u g_2^{\mathsf{m}}) = z.$$

Box 22.2 **Weakly secure Boneh–Boyen signature scheme**

We will assume that elements in G_1 have a compact representation (i.e., requiring not much more than $\log_2(r)$ bits), whereas elements of G_2 do not necessarily have a compact representation. The signature is an element of G_1 and hence is very short. Box 22.2 gives the (weakly secure) Boneh–Boyen Signature Scheme.

Exercise 22.2.15 Show that if the Verify algorithm for weakly secure Boneh–Boyen signatures accepts (m, s) then $\mathsf{s} = g_1^{(\mathsf{m}+a)^{-1} \pmod{r}}$.

The Boneh–Boyen scheme is unforgeable under a weak chosen-message attack if the q-strong Diffie–Hellman problem holds. We define these terms now.

Definition 22.2.16 A **weak chosen-message attack** (called a **generic chosen-message attack** in [235]) on a signature scheme is an adversary A that outputs a list $\mathsf{m}_1, \ldots, \mathsf{m}_t$ of messages, receives a public key and signatures $\mathsf{s}_1, \ldots, \mathsf{s}_t$ on these messages and then must output a message m and a signature s. The adversary wins if Verify(m, s, pk) = "valid" and if $\mathsf{m} \notin \{\mathsf{m}_1, \ldots, \mathsf{m}_t\}$.

Hence, a weak chosen-message attack is closer to a known message attack than a chosen-message attack.

Definition 22.2.17 Let $q \in \mathbb{N}$ (not necessarily prime). Let G_1, G_2, G_T be pairing groups as in Definition 22.2.14. Let $g_1 \in G_1 - \{1\}$ and $g_2 \in G_2 - \{1\}$. The **q-strong Diffie–Hellman problem** (**q-SDH**) is, given $(g_1, g_2, g_2^a, g_2^{a^2}, \dots, g_2^{a^q})$, where $1 \le a < r$ is chosen uniformly at random, to output a pair

$$\left(\mathsf{m}, g_1^{(\mathsf{m}+a)^{-1} \pmod r}\right)$$

where $0 \le \mathsf{m} < r$.

This problem may look rather strange at first sight since the value q can vary. The problem is mainly of interest when q is polynomial in the security parameter (otherwise, reading the problem description is not polynomial-time). Problems (or assumptions) like this are sometimes called parameterised since there is a parameter (in this case q) that determines the size for a problem instance. Such problems are increasingly used in cryptography, though many researchers would prefer to have systems whose security relies on more familiar assumptions.

There is evidence that the computational problem is hard. Theorem 5.1 of Boneh and Boyen [71] shows that a generic algorithm for q-SDH needs to make $\Omega(\sqrt{r/q})$ group operations to have a good chance of success. The algorithms of Section 21.5 can be used to solve q-SDH. In particular, if $q \mid (r-1)$ (and assuming $q < \sqrt{r}$) then Theorem 21.5.1 gives an algorithm to compute a with complexity $O(\sqrt{r/q})$ group operations, which meets the lower bound for generic algorithms.

Exercise 22.2.18 Show that one can use ψ and e to verify that the input to an instance of the q-SDH is correctly formed. Similarly, show how to use e to verify that a solution to a q-SDH instance is correct.

Theorem 22.2.19 *If the q-SDH problem is hard then the weak Boneh–Boyen signature scheme is secure under a weak chosen-message attack, where the adversary requests at most $q-1$ signatures.*

Proof (Sketch) Let $(g_1, g_2, g_2^a, g_2^{a^2}, \dots, g_2^{a^q})$ be a q-SDH instance and let A be an adversary against the scheme. Suppose A outputs messages $\mathsf{m}_1, \dots, \mathsf{m}_t$ with $t < q$.

Without loss of generality $t = q - 1$ (since one can just add dummy messages). The idea of the proof is to choose a public key so that one knows $g_1^{(\mathsf{m}_i+a)^{-1}}$ for all $1 \le i \le t$. The natural way to do this would be to set

$$g_1' = g_1^{\prod_{i=1}^{t}(\mathsf{m}_i+a)}$$

but the problem is that we do not know a. The trick is to note that $F(a) = \prod_{i=1}^{t}(\mathsf{m}_i + a) = \sum_{i=0}^{t} F_i a^i$ is a polynomial in a with explicitly computable coefficients in $\mathbb{Z}/r\mathbb{Z}$. One can therefore compute $g_2' = g_2^{F(a)}$, $g_1' = \psi(g_2')$ and $h = g_2^{aF(a)}$ using, for example

$$g_2' = \prod_{i=0}^{t}\left(g^{a^i}\right)^{F_i}.$$

Similarly, one can compute signatures for all the messages m_i. Hence, the simulator provides to A the public key $(g_1', g_2', h, z' = e(g_1', g_2'))$ and all t signatures.

Eventually, A outputs a forgery (m, s) such that $\mathsf{m} \neq \mathsf{m}_i$ for $1 \leq i \leq t$. If $t < q - 1$ and q is polynomial in the security parameter then m is one of the dummy messages with negligible probability $(q - 1 - t)/r$. One has

$$\mathsf{s} = g_1'^{(\mathsf{m}+a)^{-1} \pmod r} = g_1^{F(a)(\mathsf{m}+a)^{-1} \pmod r}.$$

The final trick is to note that the polynomial $F(a)$ can be written as $G(a)(a + \mathsf{m}) + c$ for some explicitly computable values $G(a) \in (\mathbb{Z}/r\mathbb{Z})[a]$ and $c \in (\mathbb{Z}/r\mathbb{Z})^*$. Hence, the rational function $F(a)/(a + \mathsf{m})$ can be written as

$$\frac{F(a)}{a + \mathsf{m}} = G(a) + \frac{c}{a + \mathsf{m}}.$$

One can therefore deduce $g_1^{(a+\mathsf{m})^{-1} \pmod r}$ as required. □

A fully secure signature scheme is given in [71] and it requires an extra element in the public key and an extra element (of $\mathbb{Z}/r\mathbb{Z}$) in the signature. The security proof is rather more complicated, but the essential idea is the same.

Jao and Yoshida [280] showed the converse result, namely that if one can solve q-SDH then one can forge signatures for the Boneh–Boyen scheme. It follows that one can use the algorithms of Section 21.5 to forge signatures, though this is not a serious problem since those algorithms require exponential time.

22.3 Lattice attacks on signatures

As mentioned earlier, there is a possibility that signatures could leak information about the private key. Indeed, Nguyen and Regev [410] give such an attack on lattice-based signatures.

The aim of this section is to present an attack due to Howgrave-Graham and Smart [270]. They determine the private key when given some signatures and given some bits of the random values k (for example, due to a side-channel attack or a weak pseudorandom number generator). The analysis of their attack was improved by Nguyen and Shparlinski [411, 412] (also see Chapter 16 of [499]).

The attack works for any signature scheme where one can obtain from a valid signature a linear equation modulo r with two unknowns, namely the private key a and the randomness k. We now clarify that this attack applies to the s_2 value for the Schnorr, Elgamal and DSA signature schemes:

Schnorr: $\mathsf{s}_2 \equiv k + a\mathsf{s}_1 \pmod r$ where $\mathsf{s}_1, \mathsf{s}_2$ are known.
Elgamal: $k\mathsf{s}_2 \equiv H(\mathsf{m}) - aF(\mathsf{s}_1) \pmod r$ where $H(\mathsf{m})$, $F(\mathsf{s}_1)$ and s_2 are known.
DSA: $k\mathsf{s}_2 \equiv H(\mathsf{m}) + a\mathsf{s}_1 \pmod r$ where $H(\mathsf{m})$, s_1 and s_2 are known.

Suppose we are given a message m and a valid signature $(\mathsf{s}_1, \mathsf{s}_2)$ and also the l most or least significant bits of the random value k used by the Sign algorithm to generate s_1.

Writing these known bits as k_0 we have, in the case of least significant bits, $k = k_0 + 2^l z$ where $0 \le z < r/2^l$. Indeed, one gets a better result by writing

$$k = k_0 + 2^l \lfloor r/2^{l+1} \rfloor + 2^l z \qquad (22.7)$$

with $-r/2^{l+1} \le z \le r/2^{l+1}$. Then one can re-write any of the above linear equations in the form

$$z \equiv ta - u \pmod{r}$$

for some $t, u \in \mathbb{Z}$ that are known. In other words, we know

$$\left(t, u = \mathrm{MSB}_l(at \pmod{r})\right), \qquad (22.8)$$

which is an instance of the hidden number problem (see Section 21.7.1).

If the values t in equation (22.8) are uniformly distributed in $\mathbb{Z}/r\mathbb{Z}$ then Corollary 21.7.10 directly implies that if $r > 2^{32}$ and if we can determine $l \ge \lceil \sqrt{\log_2(r)} \rceil + \lceil \log_2(\log_2(r)) \rceil$ consecutive bits of the randomness k then one can determine the private key a in polynomial-time given $n = 2\lceil \sqrt{\log_2(r)} \rceil$ message–signature pairs. As noted in [411], in the practical attack the values t arising are not uniformly distributed. We refer to [411, 412] for the full details. In practice, the attack works well for current values for r when $l = 4$, see Section 4.2 of [411].

Exercise 22.3.1 Show how to obtain an analogue of equation (22.7) when the l most significant bits are known.

Bleichenbacher has described a similar attack on a specific implementation of DSA that used a biased random generator for the values k.

22.4 Other signature functionalities

There are many topics that are beyond the scope of this book. We briefly discuss two of them now.

- Online/offline signatures. The goal here is to design public key signature schemes that possibly perform some (slow) precomputations when they are "offline" but that generate a signature on a given message m extremely quickly. The typical application is smart cards or other tokens that may have extremely constrained computing power.

  The first to suggest a solution to this problem seems to have been Schnorr in his paper [469] on efficient signatures for smart cards. The Schnorr signature scheme already has this functionality: if $\mathsf{s}_0 = g^k$ is precomputed during the idle time of the device, then generating a signature on message m only requires computing $\mathsf{s}_1 = H(\mathsf{m} \| \mathsf{s}_0)$ and $\mathsf{s}_2 = k + a\mathsf{s}_1 \pmod{r}$. The computation of s_1 and s_2 is relatively fast since no group operations are performed.

  A simple idea due to Girault [231] (proposed for groups of unknown order, typically $(\mathbb{Z}/N\mathbb{Z})^*$) is to make Schnorr signatures even faster by omitting the modular reduction in the computation of s_2. In other words, k, a, s_1 are all treated as integers and s_2 is computed as the integer $k + a\mathsf{s}_1$. To maintain security it is necessary to take k to be bigger than $2^l r$ (i.e., bigger than any possible value for the integer $a\mathsf{s}_1$). This idea was fully analysed (and generalised to groups of known order) by Girault, Poupard and Stern [232].

- Identity-based signatures. Identity-based cryptography is a concept introduced by Shamir. The main feature is that a user's public key is defined to be a function of their "identity" (for example, their email address) together with some master public key. Each user obtains their private key from a Key Generation Center that possesses the master secret. One application of identity-based cryptography is to simplify public key infrastructures.

  An identity-based signature is a public key signature scheme for which it is not necessary to verify a public key certificate on the signer's key before verifying the signature (though note that it may still be necessary to verify a certificate for the master key of the system). There are many proposals in the literature, but we do not discuss them in this section.

23

Public key encryption based on discrete logarithms

Historically, encryption has been considered the most important part of cryptography. So it is not surprising that there is a vast literature about public key encryption. It is important to note that, in practice, public key encryption is not usually used to encrypt documents. Instead, one uses public key encryption to securely send keys, and the data is encrypted using symmetric encryption.

It is beyond the scope of this book to discuss all known results on public key encryption, or even to sketch all known approaches to designing public key encryption schemes. The goal of this chapter is very modest. We simply aim to give some definitions and to provide two efficient encryption schemes (one secure in the random oracle model and one secure in the standard model). The encryption schemes in this chapter are all based on Elgamal encryption, the "textbook" version of which has already been discussed in Sections 20.3 and 20.4.

Finally, we emphasise that this chapter only discusses confidentiality and not simultaneous confidentiality and authentication. The reader is warned that naively combining signatures and encryption does not necessarily provide the expected security (see, for example, the discussion in Section 1.2.3 of Joux [283]).

23.1 CCA secure Elgamal encryption

Recall that security notions for public key encryption were given in Section 1.3.1. As we have seen, the textbook Elgamal encryption scheme does not have OWE-CCA security, since one can easily construct a related ciphertext whose decryption yields the original message. A standard way to prevent such attacks is to add a message authentication code (MAC); see Section 3.3.

We have also seen (see Section 20.3) that Elgamal can be viewed as static Diffie–Hellman key exchange followed by a specific symmetric encryption. Hence, it is natural to generalise Elgamal encryption so that it works with any symmetric encryption scheme. The scheme we present in this section is known as **DHIES** and, when implemented with elliptic curves, is called **ECIES**. We refer to Abdalla, Bellare and Rogaway [1] or Chapter III of [61] for background and discussion.

Let κ be a security parameter. The scheme requires a symmetric encryption scheme, a MAC scheme and a key derivation function. The symmetric encryption functions Enc and Dec take an l_1-bit key and encrypt messages of arbitrary length. The MAC function MAC takes an l_2-bit key and a message of arbitrary length and outputs an l_3-bit binary string. The key derivation function is a function $\mathsf{kdf} : G \to \{0, 1\}^{l_1+l_2}$. The values l_1, l_2 and l_3 depend on the security parameter. Note that it is important that the MAC is evaluated on the ciphertext not the message, since a MAC is not required to have any confidentiality properties. The DHIES encryption scheme is given in Box 23.1.

KeyGen(κ): Generate an algebraic group or algebraic group quotient G whose order is divisible by a large prime r (so that the discrete logarithm problem in the subgroup of prime order r requires at least 2^κ bit operations).

Choose a random $g \in G$ of exact order r. Choose a random integer $0 < a < r$ and set $h = g^a$.

The public key is (G, g, h) and the private key is a. Alternatively, (G, g) are system parameters that are fixed for all users and only h is the public key.

The message space is $\mathsf{M}_\kappa = \{0, 1\}^*$.
The ciphertext space is $\mathsf{C}_\kappa = G \times \{0, 1\}^* \times \{0, 1\}^{l_3}$.

Encrypt(m, h): (m $\in \{0, 1\}^*$ and h is the authentic public key of the receiver)

- Choose a random $0 < k < r$ and set $\mathsf{c}_1 = g^k$.
- Set $K = \mathsf{kdf}(h^k)$ and parse K as $K_1 \| K_2$ where K_1 and K_2 are l_1 and l_2 bit binary strings respectively.
- Set $\mathsf{c}_2 = \mathsf{Enc}_{K_1}(m)$ and $\mathsf{c}_3 = \mathsf{MAC}_{K_2}(\mathsf{c}_2)$.
- Transmit the ciphertext $(\mathsf{c}_1, \mathsf{c}_2, \mathsf{c}_3)$.

Decrypt($\mathsf{c}_1, \mathsf{c}_2, \mathsf{c}_3, a$):

- Check that $\mathsf{c}_1 \in G$ and that c_3 is an l_3-bit string (if not then return $\perp$ and halt).
- Compute $K = \mathsf{kdf}(\mathsf{c}_1^a)$ and parse it as $K_1 \| K_2$.
- Check whether $\mathsf{c}_3 = \mathsf{MAC}_{K_2}(\mathsf{c}_2)$ (if not then return $\perp$ and halt).
- Return $\mathsf{m} = \mathsf{Dec}_{K_1}(\mathsf{c}_2)$.

Box 23.1 **DHIES public key encryption**

Exercise 23.1.1 Show that decryption does return the message when given a ciphertext produced by the DHIES encryption algorithm.

A variant of DHIES is to compute the key derivation function on the pair of group (or algebraic group quotient) elements (g^k, h^k) rather than just h^k. This case is presented in Section 10 of Cramer and Shoup [149]. As explained in Section 10.7 of [149], this variant can yield a tighter security reduction.

23.1.1 The KEM/DEM paradigm

Shoup introduced a formalism for public key encryption that has proved to be useful. The idea is to separate the "public key" part of the system (i.e., the value c_1 in Box 23.1) from the "symmetric" part (i.e., (c_2, c_3) in Box 23.1). A **key encapsulation mechanism** (or **KEM**) outputs a public key encryption of a random symmetric key (this functionality is very similar to **key transport**; the difference being that a KEM generates a fresh random key as part of the algorithm). A **data encapsulation mechanism** (or **DEM**) is the symmetric part. The name **hybrid encryption** is used to describe an encryption scheme obtained by combining a KEM with a DEM.

More formally, a KEM is a triple of three algorithms (KeyGen, Encrypt and Decrypt)[1] that depend on a security parameter κ. Instead of a message space, there is space $\mathcal{K}_\kappa$ of possible keys to be encapsulated. The randomised algorithm Encrypt takes as input a public key and outputs a ciphertext c and a symmetric key $K \in \mathcal{K}_\kappa$ (where κ is the security parameter). One says that c **encapsulates** K. The Decrypt algorithm for a KEM takes as input a ciphertext c and the private key and outputs a symmetric key K (or $\perp$ if the decryption fails). The Encrypt algorithm for a DEM takes as input a message and a symmetric key K and outputs a ciphertext. The Decrypt algorithm of a DEM takes a ciphertext and a symmetric key K and outputs either $\perp$ or a message.

The simplest way to obtain a KEM from Elgamal is given in Box 23.2. The DEM corresponding to the hybrid encryption scheme in Section 23.1 takes as input m and K, parses K as $K_1 \| K_2$, computes $c_2 = \mathsf{Enc}_{K_1}(m)$ and $c_3 = \mathsf{MAC}_{K_2}(c_2)$ and outputs (c_2, c_3).

> **KeyGen**(κ): This is the same as standard Elgamal; see Box 23.1.
>
> **Encrypt**(h): Choose random $0 \le k < r$ and set $c = g^k$ and $K = \mathsf{kdf}(h^k)$. Return the ciphertext c and the key K.
>
> **Decrypt**(c, a): Return $\perp$ if $c \notin \langle g \rangle$. Otherwise return $\mathsf{kdf}(c^a)$.

Box 23.2 **Elgamal KEM**

Shoup has defined an analogue of IND-CCA security for a KEM. We refer to Section 7 of Cramer and Shoup [149] for precise definitions for KEMs, DEMs and their security properties, but give an informal statement now.

Definition 23.1.2 An IND-CCA adversary for a KEM is an algorithm A that plays the following game: the input to A is a public key. The algorithm A can also query a decryption oracle that will provide decryptions of any ciphertext of its choosing. At some point the adversary requests a challenge, which is a KEM ciphertext c^* together with a key $K^* \in \mathcal{K}_\kappa$. The challenger chooses K^* to be either the key corresponding to the ciphertext c^* or an independently chosen random element of $\mathcal{K}_\kappa$ (both cases with probability 1/2). The

[1] Sometimes the names Encap and Decap are used instead of Encrypt and Decrypt.

game continues with the adversary able to query the decryption oracle with any ciphertext $\mathsf{c} \neq \mathsf{c}^*$. Finally, the adversary outputs a guess for whether K^* is the key corresponding to c^*, or a random key (this is the same as the "real or random" security notion for key exchange in Section 20.5). Denote by $\Pr(A)$ the success probability of A in this game and define the **advantage** $\mathrm{Adv}(A) = |\Pr(A) - 1/2|$. The KEM is IND-CCA secure if every polynomial-time adversary has negligible advantage.

Theorem 5 of Section 7.3 of [149] shows that, if a KEM satisfies IND-CCA security and if a DEM satisfies an analogous security property, then the corresponding hybrid encryption scheme has IND-CCA security. Due to lack of space we do not present the details.

23.1.2 Proof of security in the random oracle model

We now sketch a proof that the Elgamal KEM of Box 23.2 has IND-CCA security. The proof is in the random oracle model. The result requires a strong assumption (namely, the strong-Diffie–Hellman or gap-Diffie–Hellman assumption). Do not be misled by the use of the word "strong"! This computational problem is not harder than the Diffie–Hellman problem. Instead, the assumption that this problem is hard is a *stronger* (i.e., less likely to be true) assumption than the assumption that the Diffie–Hellman problem is hard.

Definition 23.1.3 Let G be a group of prime order r. The **strong Diffie–Hellman problem (strong–DH)** is: given $g, g^a, g^b \in G$ (where $1 \leq a, b < r$), together with a decision static Diffie–Hellman oracle (**DStatic-DH oracle**) $A_{g,g^a}(h_1, h_2)$ (i.e., $A_{g,g^a}(h_1, h_2) = 1$ if and only if $h_2 = h_1^a$), to compute g^{ab}.

An instance generator for strong-DH takes as input a security parameter κ, outputs a group G and an element g of prime order r (with $r > 2^{2\kappa}$) and elements $g^a, g^b \in G$ where $1 \leq a, b < r$ are chosen uniformly at random. As usual, we say that strong-DH is hard for the instance generator if all polynomial-time algorithms to solve the problem have negligible success probability. The **strong Diffie–Hellman assumption** is that there is an instance generator for which the Strong-DH problem is hard.

It may seem artificial to include access to a decision oracle as part of the assumption. Indeed, it is a significant drawback of the encryption scheme that such an assumption is needed for the security. Nevertheless, the problem is well-defined and seems to be hard in groups for which the DLP is hard. A related problem is the **gap Diffie–Hellman problem**: again the goal is to compute g^{ab} given (g, g^a, g^b), but this time one is given a full DDH oracle. In some situations (for example, when using supersingular elliptic or hyperelliptic curves) one can use pairings to provide a DDH oracle and the artificial nature of the assumption disappears. The proof of Theorem 23.1.4 does not require a full DDH oracle and so it is traditional to only make the strong-DH assumption.

Theorem 23.1.4 *The Elgamal KEM of Box 23.2, with the key derivation function replaced by a random oracle, is IND-CCA secure if the strong Diffie–Hellman problem is hard.*

Proof (Sketch) Let (g, g^a, g^b) be the Strong-DH instance and let A_{g,g^a} be the DStatic-DH oracle. Let B be an IND-CCA adversary against the KEM. We want to use B to solve our strong-DH instance. To do this we will simulate the game that B is designed to play. The simulation starts B by giving it the public key (g, g^a). Note that the simulator does not know the corresponding private key.

Since the key derivation function is now a random oracle, it is necessary for B to query the simulator whenever it wants to compute kdf; this fact is crucial for the proof. Indeed, the whole idea of the proof is that when B requests the challenge ciphertext we reply with $\mathsf{c}^* = g^b$ and with a randomly chosen $K^* \in \mathcal{K}_\kappa$. Since kdf is a random oracle, the adversary can have no information about whether or not c^* encapsulates K^* unless the query $\mathsf{kdf}((\mathsf{c}^*)^a)$ is made. Finally, note that $(\mathsf{c}^*)^a = g^{ab}$ is precisely the value the simulator wants to find.

More precisely, let E be the event (on the probability space of strong-DH instances and random choices made by B) that B queries kdf on $(\mathsf{c}^*)^a = g^{ab}$. The advantage of B is $\mathrm{Adv}(B) = |\Pr(B) - \frac{1}{2}|$ where $\Pr(B)$ is the probability that B wins the IND-CCA security game. Note that $\Pr(B) = \Pr(B|E)\Pr(E) + \Pr(B|\neg E)\Pr(\neg E)$. When kdf is a random oracle we have $\Pr(B|\neg E) = 1/2$. Writing $\Pr(B|E) = 1/2 + u$ for some $-1/2 \le u \le 1/2$ we have $\Pr(B) = 1/2 + u\Pr(E)$ and so $\mathrm{Adv}(B) = |u|\Pr(E)$. Since $\mathrm{Adv}(B)$ is assumed to be non-negligible it follows that $\Pr(E)$ is non-negligible. In other words, a successful adversary makes an oracle query on the value g^{ab} with non-negligible probability.

To complete the proof it is necessary to explain how to simulate kdf and Decrypt queries and to analyse the probabilities. The simulator maintains a list of all queries to kdf. The list is initially empty. Every time that $\mathsf{kdf}(u)$ is queried, the simulator first checks whether $u \in G$ and returns $\perp$ if not. Then the simulator checks whether an entry (u, K) appears in the list of queries and, if so, returns K. If no entry appears in the list then use the oracle A_{g,g^a} to determine whether $u = g^{ab}$ (i.e., if $A_{g,g^a}(g^b, u) = 1$). If this is the case then g^{ab} has been computed and the simulation outputs that value and halts. In all other cases, a value K is chosen uniformly and independently at random from $\mathcal{K}_\kappa$, (u, K) is added to the list of kdf queries, and K is returned to B.

Similarly, when a decryption query on ciphertext c is made then one checks, for each pair (u, K) in the list of kdf values, whether $A_{g,g^a}(\mathsf{c}, u) = 1$. If this is the case then return K. If there is no such triple then return a random $K' \in \mathcal{K}$.

One can check that the simulation is sound (in the sense that Decrypt does perform the reverse of Encrypt) and that the outputs of kdf are indistinguishable from random. Determining the advantage of the simulator in solving the strong-DH problem is then straightforward. We refer to Section 10.4 of Cramer and Shoup [149] for a careful proof using the "game hopping" methodology (actually, that proof applies to the variant in Exercise 23.1.5, but it is easily adapted to the general case). □

Exercise 23.1.5 A variant of the scheme has the key derivation function applied to the pair (g^k, h^k) in Encrypt instead of just h^k (respectively, $(\mathsf{c}_1, \mathsf{c}_1^a)$ in Decrypt). Adapt the security proof to this case. What impact does this have on the running time of the simulator?

The IND-CPA security of the Elgamal KEM can be proved in the standard model (the proof is analogous to the proof of Theorem 20.4.10) under the assumption of Definition 23.1.6. The IND-CCA security can also be proved in the standard model under an interactive assumption called the oracle Diffie–Hellman assumption. We refer to Abdalla, Bellare and Rogaway [1] for the details of both these results.

Definition 23.1.6 Let G be a group and $\mathsf{kdf} : G \to \mathcal{K}$ a key derivation function. The **hash Diffie–Hellman problem (hash-DH)** is to distinguish the distributions $(g, g^a, g^b, \mathsf{kdf}(g^{ab}))$ and (g, g^a, g^b, K) where K is chosen uniformly from $\mathcal{K}$. The **hash Diffie–Hellman assumption** is that there exist instance generators such that all polynomial time algorithms for hash-DH have negligible advantage.

Exercise 23.1.7 Let G be a group of prime order r and let $\mathsf{kdf} : G \to \{0, 1\}^l$ be a key derivation function such that $\log_2(r)/2 < l < \log_2(r)$ and such that the output distribution is statistically close to uniform. Show that DDH $\leq_R$ hash-DH $\leq_R$ CDH.

An elegant variant of Elgamal, with IND-CCA security in the random oracle model depending only on CDH rather than strong Diffie–Hellman, is given by Cash, Kiltz and Shoup [112].

23.2 Cramer–Shoup encryption

In their landmark paper [147], Cramer and Shoup gave an encryption scheme with a proof of CCA security in the standard model. Due to lack of space we will only be able to give a sketch of the security analysis of the scheme.

To motivate how they achieve their result, consider the proof of security for the Elgamal KEM (Theorem 23.1.4). The simulator uses the adversary to solve an instance of the CDH problem. To do this, one puts part of the CDH instance in the public key (and, hence, one does not know the private key) and part in the challenge ciphertext. To prove CCA security we must be able to answer decryption queries without knowing the private key. In the proof of Theorem 23.1.4 this requires a DDH oracle (to determine correct ciphertexts from incorrect ones) and also the use of the random oracle model (to be able to "see" some internal operations of the adversary).

In the random oracle model one generally expects to be able to prove security under an assumption of similar flavour to CDH (see Theorem 20.4.11 and Theorem 23.1.4). On the other hand, in the standard model one only expects[2] to be able to prove security under a decisional assumption like DDH (see Theorem 20.4.10). The insight of Cramer and Shoup is to design a scheme where the security depends on DDH and is such that the entire DDH instance can be incorporated into the challenge ciphertext. The crucial consequence is that the simulator can now generate public and private keys for the scheme, run the adversary, and be able to handle decryption queries.

[2] Unless performing "bit by bit" encryption, which is a design approach not considered in this book.

The proof of security hinges (among other things) on the following result.

Lemma 23.2.1 *Let G be a group of prime order r. Let $g_1, g_2, u_1, u_2, h \in G$ with $(g_1, g_2) \neq (1, 1)$. Consider the set*

$$\mathcal{X}_{g_1,g_2,h} = \{(z_1, z_2) \in (\mathbb{Z}/r\mathbb{Z})^2 \ : \ h = g_1^{z_1} g_2^{z_2}\}.$$

Then $\#\mathcal{X}_{g_1,g_2,h} = r$. Let $0 \leq k < r$ be such that $u_1 = g_1^k$. If $u_2 = g_2^k$ then $u_1^{z_1} u_2^{z_2} = h^k$ for all $(z_1, z_2) \in \mathcal{X}_{g_1,g_2,h}$. If $u_2 \neq g_2^k$ then

$$\{u_1^{z_1} u_2^{z_2} : (z_1, z_2) \in \mathcal{X}_{g_1,g_2,h}\} = G.$$

Exercise 23.2.2 Prove Lemma 23.2.1.

Box 23.3 presents the basic **Cramer–Shoup encryption scheme**. The scheme requires a group G of prime order r and the message m is assumed to be an element of G. Of course, it is not necessary to "encode" data into group elements, in practice one would use the Cramer–Shoup scheme as a KEM; we briefly describe a Cramer–Shoup KEM at the end of this section. The scheme requires a target collision resistant hash function $H : G^3 \to \mathbb{Z}/r\mathbb{Z}$ (see Definition 3.1.2) chosen at random from a hash family.

KeyGen(κ): Generate a group G of prime order r as in Box 23.1. Choose random $g_1, g_2 \in G - \{1\}$. Choose integers $0 \leq x_1, x_2, y_1, y_2, z_1, z_2 < r$ uniformly at random and set

$$c = g_1^{x_1} g_2^{x_2}, \quad d = g_1^{y_1} g_2^{y_2}, \quad h = g_1^{z_1} g_2^{z_2}.$$

Choose a target collision resistant hash function H. The public key is $\mathsf{pk} = (G, H, g_1, g_2, c, d, h)$. The private key is $\mathsf{sk} = (x_1, x_2, y_1, y_2, z_1, z_2)$.

Encrypt(m, pk): Choose $0 \leq k < r$ uniformly at random, compute $u_1 = g_1^k$, $u_2 = g_2^k$, $e = h^k \mathsf{m}$, $\alpha = H(u_1, u_2, e)$ and $v = c^k d^{k\alpha \pmod r}$. The ciphertext is $\mathsf{c} = (u_1, u_2, e, v)$.

Decrypt(u_1, u_2, e, v, sk): First check that $u_1, u_2, e \in G$ and output $\perp$ if this is not the case. Next, compute $\alpha = H(u_1, u_2, e)$ and check whether

$$v = u_1^{x_1 + y_1\alpha \pmod r} u_2^{x_2 + y_2\alpha \pmod r}. \tag{23.1}$$

Return $\perp$ if this condition does not hold. Otherwise output

$$\mathsf{m} = e u_1^{-z_1} u_2^{-z_2}.$$

Box 23.3 **Basic Cramer–Shoup encryption scheme**

Exercise 23.2.3 Show that the value $v = c^k d^{k\alpha}$ computed in the Encrypt algorithm does satisfy equation (23.1).

Exercise 23.2.4 Show that the tests $u_1, u_2 \in G$ and equation (23.1) imply that $v \in G$.

Exercise 23.2.5 Show that the final stage of Decrypt in the Cramer–Shoup scheme can be efficiently performed using multi-exponentiation as

$$\mathsf{m} = e u_1^{r-z_1} u_2^{r-z_2}.$$

Example 23.2.6 Let $p = 311$, $r = 31$ and denote by G the subgroup of order r in $\mathbb{F}_p^*$. Take $g_1 = 169$ and $g_2 = 121$. Suppose $(x_1, x_2, y_1, y_2, z_1, z_2) = (1, 2, 3, 4, 5, 6)$ so that the public key is

$$(g_1, g_2, c, d, h) = (169, 121, 13, 260, 224).$$

To encrypt $\mathsf{m} = 265 \in G$ choose, say, $k = 15$ and set $u_1 = g_1^k = 24$, $u_2 = g_2^k = 113$ and $e = \mathsf{m}h^k = 126$. Finally, we must compute α. Since we do not want to get into the details of H, suppose $\alpha = H(u_1, u_2, e) = 20$ and so set $v = c^k d^{k\alpha \pmod{r}} = c^{15} d^{21} = 89$. The ciphertext is $(u_1, u_2, e, v) = (24, 113, 126, 89)$.

To decrypt, one first checks that $u_1^r = u_2^r = e^r = 1$. Then one recomputes α and checks equation (23.1). Since

$$u_1^{x_1 + y_1\alpha \pmod{r}} u_2^{x_2 + y_2\alpha \pmod{r}} = u_1^{30} u_2^{20} = 89$$

the ciphertext passes this test. One then computes $e u_1^{r-z_1} u_2^{r-z_2} = 126 \cdot 24^{26} \cdot 113^{25} = 265$.

Exercise 23.2.7 Using the same private keys as Example 23.2.6, which of the following ciphertexts are valid, and for those that are, what is the corresponding message? Assume that $H(243, 83, 13) = 2$.

$$(243, 83, 13, 97),\ (243, 83, 13, 89),\ (243, 83, 13, 49).$$

We now turn to the security analysis. Note that the condition of equation (23.1) does not imply that the ciphertext (u_1, u_2, e, v) was actually produced by the Encrypt algorithm. However, we now show that, if u_1 and u_2 are not of the correct form, then the probability that a randomly chosen v satisfies this condition is $1/r$. Indeed, Lemma 23.2.8 shows that an adversary who can solve the discrete logarithm problem cannot even construct an invalid ciphertext that satisfies this equation with probability better than $1/r$.

Lemma 23.2.8 *Let G be a cyclic group of prime order r. Let $g_1, g_2, c, d, v \in G$ and $\alpha \in \mathbb{Z}/r\mathbb{Z}$ be fixed. Suppose $u_1 = g_1^k$ and $u_2 = g_2^{k+k'}$ where $k' \not\equiv 0 \pmod{r}$. Then the probability, over all choices (x_1, x_2, y_1, y_2) such that $c = g_1^{x_1} g_2^{x_2}$ and $d = g_1^{y_1} g_2^{y_2}$, that $v = u_1^{x_1+\alpha y_1} u_2^{x_2+\alpha y_2}$ is $1/r$.*

Proof Write $g_2 = g_1^w$, $c = g_1^{w_1}$ and $d = g_1^{w_2}$ for some $0 \le w, w_1, w_2 < r$ with $w \ne 0$. The values c and d imply that $x_1 + wx_2 = w_1$ and $y_1 + wy_2 = w_2$. Now $u_1^{x_1+\alpha y_1} u_2^{x_2+\alpha y_2}$ equals g_1 to the power

$$\begin{aligned} k(x_1 + \alpha y_1) + (k + k')w(x_2 + \alpha y_2) &= k((x_1 + wx_2) + \alpha(y_1 + wy_2)) + k'w(x_2 + \alpha y_2) \\ &= k(w_1 + \alpha w_2) + k'w(x_2 + \alpha y_2). \end{aligned}$$

The values $w, w_1, w_2, k, k', \alpha$ are all uniquely determined but, by Lemma 23.2.1, x_2 and y_2 can take any values between 0 and $r-1$. Hence, $u_1^{x_1+\alpha y_1} u_2^{x_2+\alpha y_2}$ equals any fixed value v for exactly r of the r^2 choices for (x_2, y_2). □

Theorem 23.2.9 *The basic Cramer–Shoup encryption scheme is IND-CCA secure if DDH is hard and if the function H is target collision resistant.*

Proof (Sketch) Let A be an adversary against the Cramer–Shoup scheme. Given a DDH instance (g_1, g_2, u_1, u_2) the simulator performs the KeyGen algorithm using the given values g_1, g_2. Hence, the simulator knows the private key $(x_1, x_2, y_1, y_2, z_1, z_2)$. The simulator runs A with this public key.

The algorithm A makes decryption queries and these can be answered correctly by the simulator since it knows the private key. Eventually, A outputs two messages $(\mathsf{m}_0, \mathsf{m}_1)$ and asks for a challenge ciphertext. The simulator chooses a random $b \in \{0, 1\}$, computes $e = u_1^{z_1} u_2^{z_2} \mathsf{m}_b$, $\alpha = H(u_1, u_2, e)$ and $v = u_1^{x_1+y_1\alpha} u_2^{x_2+y_2\alpha}$. Here, and throughout this proof, u_1 and u_2 denote the values in the DDH instance that was given to the simulator. The simulator returns

$$\mathsf{c}^* = (u_1, u_2, e, v).$$

to A. The adversary A continues to make decryption queries, which are answered as above. Eventually, A outputs a bit b'. The simulator returns "valid" as the answer to the DDH instance if $b = b'$ and "invalid" otherwise.

The central idea is that if the input is a valid DDH tuple then c^* is a valid encryption of m_b and so A ought to be able to guess b correctly with non-negligible probability. On the other hand, if the input is not a valid DDH tuple then, by Lemma 23.2.1, $u_1^{z_1} u_2^{z_2}$ could be any element in G (with equal probability) and so c^* could be an encryption of any message $\mathsf{m} \in G$. Hence, given c^*, both messages m_0 and m_1 are equally likely and so the adversary can do no better than output a random bit. (Of course, A may actually output a fixed bit in this case, such as 0, but this is not a problem since b was randomly chosen.)

There are several subtleties remaining in the proof. First, by Lemma 23.2.8, before the challenge ciphertext has been received there is a negligible probability that a ciphertext that was not produced by the Encrypt algorithm satisfies equation (23.1). Hence, the simulation is correct with overwhelming probability. However, the challenge ciphertext is potentially an example of a ciphertext that satisfies equation (23.1) and yet is not a valid output of the algorithm Encrypt. It is necessary to analyse the probability that A can somehow produce another ciphertext that satisfies equation (23.1) without just running the Encrypt algorithm. The target collision resistance of the hash function enters at this point (since a ciphertext of the form (u_1, u_2, e', v) such that $H(u_1, u_2, e') = H(u_1, u_2, e)$ would pass the test). Due to lack of space we refer to Section 4 of [147] (for a direct proof) or Section 6.2 of [149] (for a proof using "game hopping"). □

A number of variants of the basic scheme are given by Cramer and Shoup [149] and other authors. In particular, one can design a KEM based on the Cramer–Shoup scheme (see

Section 9 of [149]): just remove the component e of the ciphertext and set the encapsulated key to be $K = \mathsf{kdf}(g_1^k, h^k)$. An alternative KEM (with even shorter ciphertext) was proposed by Kurosawa and Desmedt [323]. Their idea was to set $K = \mathsf{kdf}(v)$ where $v = c^k d^{\alpha k}$ for $\alpha = H(u_1, u_2)$. The KEM ciphertext is therefore just $(u_1, u_2) = (g_1^k, g_2^k)$. The security again follows from Lemma 23.2.8: informally, querying the decryption oracle on badly formed (u_1, u_2) gives no information about the key K.

Exercise 23.2.10 Write down a formal description of the Cramer–Shoup KEM.

Exercise 23.2.11 Show that an adversary against the Cramer–Shoup scheme who knows any pair (z_1, z_2) such that $h = g_1^{z_1} g_2^{z_2}$ can decrypt valid ciphertexts.

Exercise 23.2.12 Suppose an adversary against the Cramer–Shoup scheme knows x_1, x_2, y_1, y_2. Show how the adversary can win the OWE-CCA security game.

Exercise 23.2.13 Suppose the checks that $u_1, u_2 \in G$ are omitted in the Cramer–Shoup cryptosystem. Suppose $G \subset \mathbb{F}_p^*$ where $l \mid (p-1)$ is a small prime (say $l < 2^{10}$). Suppose the Decrypt algorithm uses the method of Exercise 23.2.5. Show how to determine, using a decryption oracle, $z_1 \pmod{l}$ and $z_2 \pmod{l}$. Show that if $p - 1$ has many such small factors l then one could recover the values z_1 and z_2 in the private key of the Cramer–Shoup scheme.

Cramer and Shoup [148] have shown how the above cryptosystem fits into a general framework for constructing secure encryption schemes using "universal hash proof systems". We do not have space to present this topic.

23.3 Other encryption functionalities

There are many variants of public key encryption (such as threshold decryption, server-aided decryption, etc.). In this section we briefly sketch two important variants: homomorphic encryption and identity-based encryption.

23.3.1 Homomorphic encryption

Let $\mathsf{c}_1, \dots, \mathsf{c}_k$ be ciphertexts that are encryptions under some public key of messages $\mathsf{m}_1, \dots, \mathsf{m}_k$. The goal of homomorphic encryption is for any user to be able to efficiently compute a ciphertext that encrypts $F(\mathsf{m}_1, \dots, \mathsf{m}_k)$ for any function F, given only a description of the function F and the ciphertexts $\mathsf{c}_1, \dots, \mathsf{c}_k$. An encryption scheme that has this property is called **fully homomorphic**.

Homomorphic encryption schemes allow third parties to perform computations on encrypted data. A common additional security requirement is that the resulting ciphertexts do not reveal to a user with the private key what computation was performed (except its result). A typical application of homomorphic encryption is voting: if users encrypt either 0 or 1 under a certain public key[3] then a trusted third party can compute a ciphertext that is an encryption of the sum of all the users' votes, and then this ciphertext can be decrypted to give the total number of votes. If the user with the private key never sees the individual votes then they cannot determine how an individual user voted. A general survey on homomorphic encryption that gives some references for applications is Fontaine and Galand [191].

For many applications it is sufficient to consider encryption schemes that only allow a user to compute $F(\mathsf{m}_1, \ldots, \mathsf{m}_k)$ for certain specific functions (for example, addition in the voting application). In this section we focus on the case where $F(\mathsf{m}_1, \mathsf{m}_2)$ is a group operation.

Definition 23.3.1 Let G be a group (written multiplicatively). A public key encryption scheme with message space G and ciphertext space C is said to be **homomorphic** for the group G if there is some efficiently computable binary operation $\star$ on C such that, for all $\mathsf{m}_1, \mathsf{m}_2 \in G$, if c_1 is an encryption of m_1 and c_2 is an encryption of m_2 (both with respect to the same public key) then $\mathsf{c}_1 \star \mathsf{c}_2$ is an encryption of $\mathsf{m}_1\mathsf{m}_2$.

Exercise 23.3.2 shows that one cannot have CCA security when using homomorphic encryption. Hence, the usual security requirement of a homomorphic encryption scheme is that it should have IND-CPA security.

Exercise 23.3.2 Show that a homomorphic encryption scheme does not have IND-CCA security.

Exercise 23.3.3 Let $G = \langle g \rangle$ where g is an element of order r in a group. Let $\mathsf{c}_1 = (\mathsf{c}_{1,1}, \mathsf{c}_{1,2}) = (g^{k_1}, \mathsf{m}_1 h^{k_1})$ and $\mathsf{c}_2 = (\mathsf{c}_{2,1}, \mathsf{c}_{2,2}) = (g^{k_2}, \mathsf{m}_2 h^{k_2})$ be classic textbook Elgamal encryptions of $\mathsf{m}_1, \mathsf{m}_2 \in G$. Define $\mathsf{c}_1 \star \mathsf{c}_2 = (\mathsf{c}_{1,1}\mathsf{c}_{2,1}, \mathsf{c}_{1,2}\mathsf{c}_{2,2})$. Show that $\mathsf{c}_1 \star \mathsf{c}_2$ is an encryption of $\mathsf{m}_1\mathsf{m}_2$ and hence that classic textbook Elgamal encryption is homomorphic for the group G.

Exercise 23.3.4 Let $G = \mathbb{F}_2^l \cong \{0, 1\}^l$. Note that G is a group under addition modulo 2 (equivalently, under exclusive-or $\oplus$). For $1 \le i \le 2$ let $\mathsf{c}_i = (\mathsf{c}_{i,1}, \mathsf{c}_{i,2}) = (g^{k_i}, \mathsf{m}_i \oplus H(h^{k_i}))$ be semi-textbook Elgamal encryptions of messages $\mathsf{m}_i \in G$. Consider the operation $\mathsf{c}_1 \star \mathsf{c}_2 = (\mathsf{c}_{1,1}\mathsf{c}_{2,1}, \mathsf{c}_{1,2} \oplus \mathsf{c}_{2,2})$. Show that semi-textbook Elgamal is not homomorphic with respect to this operation.

Exercise 23.3.5 A variant of Elgamal encryption that is homomorphic with respect to addition is to encrypt m as $(\mathsf{c}_1 = g^k, \mathsf{c}_2 = g^{\mathsf{m}} h^k)$. Prove that if $(\mathsf{c}_{i,1}, \mathsf{c}_{i,2})$ are ciphertexts encrypting messages m_i for $i = 1, 2$ then $(\mathsf{c}_{1,1}\mathsf{c}_{2,1}, \mathsf{c}_{1,2}\mathsf{c}_{2,2})$ encrypts $\mathsf{m}_1 + \mathsf{m}_2$. Give a decryption algorithm for this system and explain why it is only practical when the messages m are small integers. Hence, show that this scheme does not strictly satisfy Definition 23.3.1 when the order of g is large.[4]

[3] It is necessary for users to prove that their vote lies in $\{0, 1\}$.

[4] The order of g must be large for the scheme to have IND-CPA security.

23.3.2 Identity-based encryption

Section 22.4 briefly mentioned identity-based signatures. Recall that in identity-based cryptography a user's public key is defined to be a function of their "identity" (for example, their email address). There is a master public key. Each user obtains their private key from a key generation center (which possesses the master secret).

In this section we sketch the **basic Boneh–Franklin scheme** [75] (the word "basic" refers to the fact that this scheme only has security against a chosen plaintext attack). The scheme uses pairing groups (see Definition 22.2.14 and Chapter 26). Hence, let G_1, G_2 and G_T be groups of prime order r and let $e : G_1 \times G_2 \rightarrow G_T$ be a non-degenerate bilinear pairing.

The first task is to determine the master keys, which are created by the key generation center. Let $g \in G_2$ have order r. The key generation center chooses $1 \leq s < r$ and computes $g' = g^s$. The master public key is (g, g') and the master private key is s. The scheme also requires hash functions $H_1 : \{0, 1\}^* \rightarrow G_1$ and $H_2 : G_T \rightarrow \{0, 1\}^l$ (where l depends on the security parameter). The message space will be $\{0, 1\}^l$ and the ciphertext space will be $G_2 \times \{0, 1\}^l$.

The public key of a user with identity $\mathsf{id} \in \{0, 1\}^*$ is $Q_{\mathsf{id}} = H_1(\mathsf{id}) \in G_1$. With overwhelming probability $Q_{\mathsf{id}} \neq 1$, in which case $e(Q_{\mathsf{id}}, g) \neq 1$. The user obtains the private key

$$Q'_{\mathsf{id}} = H_1(\mathsf{id})^s$$

from the key generation center.

To encrypt a message $\mathsf{m} \in \{0, 1\}^l$ to the user with identity id one obtains the master key (g, g'), computes $Q_{\mathsf{id}} = H_1(\mathsf{id})$, chooses a random $1 \leq k < r$ and computes $\mathsf{c}_1 = g^k$, $\mathsf{c}_2 = \mathsf{m} \oplus H_2(e(Q_{\mathsf{id}}, g')^k)$. The ciphertext is $(\mathsf{c}_1, \mathsf{c}_2)$.

To decrypt the ciphertext $(\mathsf{c}_1, \mathsf{c}_2)$ the user with private key Q'_{id} computes

$$\mathsf{m} = \mathsf{c}_2 \oplus H_2(e(Q'_{\mathsf{id}}, \mathsf{c}_1)).$$

This completes the description of the basic Boneh–Franklin scheme.

Exercise 23.3.6 Show that the Decrypt algorithm does compute the correct message when $(\mathsf{c}_1, \mathsf{c}_2)$ are the outputs of the Encrypt algorithm.

Exercise 23.3.7 Show that the basic Boneh–Franklin scheme does not have IND-CCA security.

The security model for identity-based encryption takes into account that an adversary can ask for private keys on various identities. Hence, the IND security game allows an adversary to output a challenge identity id^* and two challenge messages $\mathsf{m}_0, \mathsf{m}_1$. The adversary is not permitted to know the private key for identity id^* (though it can receive private keys for any other identities of its choice). The adversary then receives an encryption with respect to identity id^* of m_b for randomly chosen $b \in \{0, 1\}$ and must output a guess for b.

Exercise 23.3.8 Suppose there is an efficiently computable group homomorphism $\psi : G_2 \to G_1$. Show that if an adversary knows ψ and can compute preimages of the hash function H_1 then it can determine the private key for any identity by making a private key query on a different identity.

If the output of H_2 is indistinguishable from random l-bit strings then it is natural to believe that obtaining the message from a ciphertext under a passive attack requires computing

$$e(Q'_{\mathsf{id}}, \mathsf{c}_1) = e(Q^s_{\mathsf{id}}, g^k) = e(Q_{\mathsf{id}}, g)^{sk}.$$

Hence, it is natural that the security (at least, in the random oracle model) depends on the following computational problem.

Definition 23.3.9 Let G_1, G_2 and G_T be groups of prime order r and let $e : G_1 \times G_2 \to G_T$ be a non-degenerate bilinear pairing. The **bilinear Diffie–Hellman problem** (**BDH**) is: given $Q \in G_1$, $g \in G_2$, g^a and g^b, where $1 \le a, b < r$, to compute

$$e(Q, g)^{ab}.$$

Exercise 23.3.10 Show that if one can solve CDH in G_2 or in G_T then one can solve BDH.

As seen in Exercise 23.3.7, the basic Boneh–Franklin scheme does not have IND-CCA security. To fix this, one needs to provide some extra components in the ciphertext. Alternatively, one can consider the basic Boneh–Franklin scheme as an identity-based KEM: the ciphertext is $\mathsf{c}_1 = g^k$ and the encapsulated key is $K = \mathsf{kdf}(e(Q_{\mathsf{id}}, g')^k)$. In the random oracle model (treating both H_1 and kdf as random oracles) one can show that the Boneh–Franklin identity-based KEM has IND-CCA security (in the security model for identity-based encryption as briefly mentioned above), assuming that the BDH problem is hard. We refer to Boneh and Franklin [75, 76] for full details and security proofs.

There is a large literature on identity-based encryption and its extensions, including schemes that are secure in the standard model. We do not discuss these topics further in this book.

PART VI

CRYPTOGRAPHY RELATED TO INTEGER FACTORISATION

24

The RSA and Rabin cryptosystems

The aim of this chapter is to briefly present some cryptosystems whose security is based on computational assumptions related to the integer factorisation problem. In particular, we study the RSA and Rabin cryptosystems. We also present some security arguments and techniques for efficient implementation.

Throughout the chapter we take 3072 bits as the benchmark length for an RSA modulus. We make the assumption that the cost of factoring a 3072-bit RSA modulus is 2^{128} bit operations. These figures should be used as a very rough guideline only.

24.1 The textbook RSA cryptosystem

Box 24.1 recalls the "textbook" RSA cryptosystem, which was already presented in Section 1.2. We remind the reader that the main application of RSA encryption is to transport symmetric keys, rather than to encrypt actual documents. For digital signatures we always sign a hash of the message, and it is necessary that the hash function used in signatures is collision resistant.

In Section 1.3 we noted that the security parameter κ is not necessarily the same as the bit-length of the RSA modulus. In this chapter it will be convenient to ignore this, and use the symbol κ to denote the bit-length of an RSA modulus N. We always assume that κ is even.

As we have seen in Section 1.2 certain security properties can only be satisfied if the encryption process is randomised. Since the RSA encryption algorithm is deterministic it follows that the message m used in RSA encryption should be obtained from some **randomised padding scheme**. For example, if N is a 3072-bit modulus then the "message" itself may be a 256-bit AES key and may have 2815 random bits appended to it.

Exercise 24.1.1 Give a KeyGen algorithm that takes as input a security parameter κ (assumed to be even) and an l-bit string u (where $l < \kappa/2$) and outputs a κ-bit product $N = pq$ of two $\kappa/2$-bit primes such that the l most significant bits of N are equal to u. In particular, one can ensure that $2^{\kappa-1} < N < 2^{\kappa}$ and so N is a κ-bit integer.

Exercise 24.1.2 Let $N \in \mathbb{N}$. Prove that the Carmichael function $\lambda(N)$ divides the Euler function $\varphi(N)$. Prove that RSA decryption does return the message.

KeyGen(κ): (Assume κ even.) Generate two distinct primes p and q uniformly at random in the range $2^{\kappa/2-1} < p, q < 2^{\kappa/2}$. Set $N = pq$ so that $2^{\kappa-2} < N < 2^{\kappa}$ is represented by κ bits (see Exercise 24.1.1 to ensure that N has leading bit 1).
Choose a random κ-bit integer e coprime to $(p-1)$ and $(q-1)$ (or choose $e = 2^{16}+1 = 65537$ and insist $p, q \not\equiv 1 \pmod{e}$). Define $d = e^{-1} \pmod{\lambda(N)}$ where $\lambda(N) = \mathrm{lcm}(p-1, q-1)$ is the **Carmichael lambda function**. Output the public key $\mathsf{pk} = (N, e)$ and the private key $\mathsf{sk} = (N, d)$.
Renaming p and q, if necessary, we may assume that $p < q$. Then $p < q < 2p$ and so $\sqrt{N/2} < p < \sqrt{N}$.

In textbooks, the message space and ciphertext space are usually taken to be $\mathsf{C}_\kappa = \mathsf{M}_\kappa = (\mathbb{Z}/N\mathbb{Z})^*$, but it fits Definition 1.3.1 better (and is good training) to define them to be $\mathsf{C}_\kappa = \{0,1\}^\kappa$ and $\mathsf{M}_\kappa = \{0,1\}^{\kappa-2}$ or $\{0,1\}^{\kappa-1}$.

Encrypt($\mathsf{m}, (N, e)$): Assume that $\mathsf{m} \in \mathsf{M}_\kappa$.

- Compute $\mathsf{c} = \mathsf{m}^e \pmod{N}$ (see later for padding schemes).
- Return the ciphertext c.

Decrypt($\mathsf{c}, (N, d)$): Compute $\mathsf{m} = \mathsf{c}^d \pmod{N}$ and output either m, or $\perp$ if $\mathsf{m} \notin \mathsf{M}_\kappa$.

Sign($\mathsf{m}, (N, d)$): Compute $\mathsf{s} = \mathsf{m}^d \pmod{N}$.

Verify($\mathsf{m}, \mathsf{s}, (N, e)$): Check whether $\mathsf{m} \equiv \mathsf{s}^e \pmod{N}$.

Box 24.1 **Textbook RSA public key encryption and signature schemes**

24.1.1 Efficient implementation of RSA

As we have seen in Section 12.2, $\kappa/2$-bit probable primes can be found in expected time of $O(\kappa^5)$ bit operations (or $O(\kappa^{4+o(1)})$ using fast arithmetic). One can make this provable using the AKS method, with asymptotic complexity $O(\kappa^{5+o(1)})$ bit operations using fast arithmetic. In any case, RSA key generation is polynomial-time. A more serious challenge is to ensure that encryption and decryption (equivalently, signing and verification) are as fast as possible.

Encryption and decryption are exponentiation modulo N and thus require $O(\log(N)M(\log(N)))$ bit operations, which is polynomial-time. For current RSA key sizes, Karatsuba multiplication is most appropriate; hence, one should assume that $M(\log(N)) = \log(N)^{1.58}$. Many of the techniques mentioned in earlier chapters to speed up exponentiation can be used in RSA, particularly sliding window methods. Since e and d are fixed, one can also precompute addition chains to minimise the cost of exponentiation.

In practice, the following two improvements are almost always used.

- **Small public exponents** e (also called **low-exponent RSA**). Traditionally, $e = 3$ was proposed, but these days $e = 65537 = 2^{16}+1$ is most common. Encryption requires only 16 modular squarings and a modular multiplication.

- Use the Chinese remainder theorem (CRT) to Decrypt.[1] Let $d_p \equiv e^{-1} \pmod{p-1}$ and $d_q \equiv e^{-1} \pmod{q-1}$. These are called the **CRT private exponents**. Given a ciphertext c one computes $\mathsf{m}_p = \mathsf{c}^{d_p} \pmod{p}$ and $\mathsf{m}_q = \mathsf{c}^{d_q} \pmod{q}$. The message m is then computed using the Chinese remainder theorem (it is convenient to use the method of Exercise 2.6.3 for the CRT).

For this system the private key is then $\mathsf{sk} = (p, q, d_p, d_q)$. If we denote by $T = c\log(N)M(\log(N))$ the cost of a single exponentiation modulo N to a power $d \approx N$ then the cost using the Chinese remainder theorem is approximately $2c(\log(N)/2)M(\log(N)/2)$ (this is assuming the cost of the Chinese remaindering is negligible). When using Karatsuba multiplication this speeds up RSA decryption by a factor of approximately 3 (in other words, the new running time is a third of the old running time).

24.1.2 Variants of RSA

There has been significant effort devoted to finding more efficient variants of the RSA cryptosystem. We briefly mention some of these now.

Example 24.1.3 **(Multiprime-RSA[2])** Let $p_1, \dots, p_k$ be primes of approximately κ/k bits and let $N = p_1 \cdots p_k$. One can use N as a public modulus for the RSA cryptosystem. Using the Chinese remainder theorem for decryption has cost roughly the same as k exponentiations to powers of bit-length κ/k and modulo primes of bit-length κ/k. Hence, the speedup is roughly by a factor $k/k^{2.58} = 1/k^{1.58}$.

To put this in context, going from a single exponentiation to using the Chinese remainder theorem in the case of 2 primes gave a speedup by a factor of 3. Using 3 primes gives an overall speedup by a factor of roughly 5.7, which is a further speedup of a factor 1.9 over the 2-prime case. Using 4 primes gives an overall speedup of about 8.9, which is an additional speedup over 3 primes by a factor 1.6.

However, there is a limit to how large k can be, as the complexity of the elliptic curve factoring method mainly depends on the size of the smallest factor of N.

Exercise 24.1.4 (Tunable balancing of RSA) An alternative approach is to construct the public key $(N = pq, e)$ so that the Chinese remainder decryption exponents are relatively short. The security of this system will be discussed in Section 24.5.2.

Let κ, n_e, n_d be the desired bit-lengths of N, e and $d \pmod{p-1}, d \pmod{q-1}$. Assume that $n_e + n_d > \kappa/2$. Give an algorithm to generate primes p and q of bit-length $\kappa/2$, integers d_p and d_q of bit-length n_d and an integer e of bit-length n_e such that $ed_p \equiv 1 \pmod{p-1}$ and $ed_q \equiv 1 \pmod{q-1}$.

The fastest variant of RSA is due to Takagi and uses moduli of the form $N = p^r q$. For some discussion about factoring such integers see Section 19.4.3.

[1] This idea is often credited to Quisquater and Couvreur [443] but it also appears in Rabin [444].

[2] This idea was proposed (and patented) by Collins, Hopkins, Langford and Sabin.

Example 24.1.5 (Takagi–RSA [538]) Let $N = p^r q$ where p and q are primes and $r > 1$. Suppose the public exponent e in RSA is small. One can compute $\mathsf{c}^d \pmod{N}$ as follows. Let $d_p \equiv d \pmod{p-1}$ and $d_q \equiv d \pmod{q-1}$. One first computes $\mathsf{m}_p = \mathsf{c}^{d_p} \pmod{p}$ and $\mathsf{m}_q = \mathsf{c}^{d_q} \pmod{q}$.

To determine $\mathsf{m} \pmod{p^r}$ one uses Hensel lifting. If we have determined $\mathsf{m}_i = \mathsf{m} \pmod{p^i}$ such that $\mathsf{m}_i^e \equiv \mathsf{c} \pmod{p^i}$ then we lift to a solution modulo p^{i+1} by writing $\mathsf{m}_{i+1} = \mathsf{m}_i + xp^i$, where x is a variable. Then

$$\mathsf{m}_{i+1}^e = (\mathsf{m}_i + xp^i)^e \equiv \mathsf{m}_i^e + x(e\mathsf{m}_i^{e-1})p^i \equiv \mathsf{c} \pmod{p^{i+1}} \qquad (24.1)$$

gives a linear equation in x modulo p. Note that computing $\mathsf{m}_i^e \pmod{p^{i+1}}$ in equation (24.1) is only efficient when e is small. If e is large then the Hensel lifting stage is no faster than just computing $\mathsf{c}^{e^{-1} \pmod{\varphi(p^r)}} \pmod{p^r}$ directly.

The total cost is two "full" exponentiations to compute m_p and m_q, $r-1$ executions of the Hensel lifting stage, plus one execution of the Chinese remainder theorem. Ignoring everything except the two big exponentiations, one has an algorithm whose cost is $2/(r+1)^{2.58}$ times faster than naive textbook RSA decryption. Taking $r = 2$ this is about 9 times faster than standard RSA (i.e., about 1.6 times faster than using 3-prime RSA) and taking $r = 3$ is about 18 times faster than standard RSA (i.e., about 2 times faster than using 4-prime RSA).

Exercise 24.1.6 Let $N = (2^{20}+7)^3(2^{19}+21)$ and let $\mathsf{c} = 474776119073176490663504$ be the RSA encryption of a message m using public exponent $e = 3$. Determine the message using the Takagi decryption algorithm.

Exercise 24.1.7 Describe and analyse the RSA cryptosystem using moduli of the form $N = p^r q^s$. Explain why it is necessary that $r \neq s$.

24.1.3 Security of textbook RSA

We have presented "textbook" RSA above. This is unsuitable for practical applications for many reasons. In practice, RSA should only be used with a secure randomised padding scheme. Nevertheless, it is instructive to consider the security of textbook RSA with respect to the security definitions presented earlier.

Exercise 1.3.4 showed that textbook RSA encryption does not have OWE-CCA security and Exercise 1.3.9 showed that textbook RSA signatures do not have existential forgery security even under a passive attack. We recall one more easy attack.

Exercise 24.1.8 Show that one can use the Jacobi symbol to attack the IND-CPA security of RSA encryption.

Despite the fact that RSA is supposed to be related to factoring, the security actually relies on the following computational problem.

Definition 24.1.9 Let N, e be such that $\gcd(e, \lambda(N)) = 1$. The **RSA problem** (also called the *e***th roots problem**) is: given $y \in (\mathbb{Z}/N\mathbb{Z})^*$ to compute x such that $x^e \equiv y \pmod{N}$.

It is clear that the RSA problem is not harder than factoring.

Lemma 24.1.10 *The OWE-CPA security of textbook RSA is equivalent to the RSA problem.*

Proof (Sketch) We show that an algorithm to break OWE-CPA security of textbook RSA can be used to build an algorithm to solve the RSA problem. Let A be an adversary against the OWE-CPA security of RSA. Let (N, e, c) be a challenge RSA problem instance. If $1 < \gcd(\mathsf{c}, N) < N$ then split N and solve the RSA problem. Otherwise, call the adversary A on the public key (N, e) and offer the challenge ciphertext c. If A returns the message m then we are done. If A returns $\perp$ (e.g., because the decryption of c does not lie in M_κ) then replace c by $\mathsf{c}r^e \pmod{N}$ for a random $1 < r < N$ and repeat. When $\mathsf{M}_\kappa = \{0, 1\}^{\kappa-2}$ then with probability at least $1/4$, the reduction will succeed, and so one expects to perform 4 trials. The converse is also immediate. □

Exercise 24.1.11 Show that textbook RSA has selective signature forgery under passive attacks if and only if the RSA problem is hard.

One of the major unsolved problems in cryptography is to determine the relationship between the RSA problem and factoring. There is no known reduction of factoring to breaking RSA. Indeed, there is some indirect evidence that breaking RSA with small public exponent e is not as hard as factoring: Boneh and Venkatesan [82] show that an efficient "algebraic reduction"[3] from FACTOR to low-exponent RSA can be converted into an efficient algorithm for factoring. Similarly, Coppersmith [131] shows that some variants of the RSA problem, where e is small and only a small part of an eth root x is unknown, are easy (see Exercise 19.1.15).

Definition 24.1.12 describes some computational problems underlying the security of RSA. The reader is warned that some of these names are non-standard.

Definition 24.1.12 Let S be a set of integers, for example $S = \mathbb{N}$ or $S = \{pq : p \text{ and } q \text{ are primes such that } p < q < 2p\}$. We call the latter set the **set of RSA moduli**.

FACTOR: Given $N \in S$ to compute the list of prime factors of N.
COMPUTE-PHI: Given $N \in S$ to compute $\varphi(N)$.
COMPUTE-LAMBDA: Given $N \in S$ to compute $\lambda(N)$.
RSA-PRIVATE-KEY: Given $(N, e) \in S \times \mathbb{N}$ to output $\perp$ if e is not coprime to $\lambda(N)$, or d such that $ed \equiv 1 \pmod{\lambda(N)}$.

Exercise 24.1.13 Show that RSA $\leq_R$ RSA-PRIVATE-KEY $\leq_R$ COMPUTE-LAMBDA $\leq_R$ FACTOR for integers $N \in \mathbb{N}$.

[3] We do not give a formal definition. Essentially this is an algorithm that takes as input N, queries an oracle for the RSA problem, and outputs a finite set of short algebraic formulae, one of which splits the integer N.

Exercise 24.1.13 tells us that FACTOR is at least as hard as RSA. A more useful interpretation is that the RSA problem is no harder than factoring. We are interested in the relative difficulty of such problems, as a function of the input size. Lemma 24.1.14 is the main tool for comparing these problems.[4]

Lemma 24.1.14 *Let A be a perfect oracle that takes as input an integer N and outputs a multiple of $\lambda(N)$. Then one can split N in randomised polynomial-time using an expected polynomially many queries to A.*

Proof Let N be the integer to be factored. We may assume that N is composite, not a prime power, odd and has no very small factors. Let M be the output of the oracle A on N. (Note that the case of non-perfect oracles is not harder: one can easily test whether the output of A is correct by taking a few random integers $1 < a < N$ such that $\gcd(a, N) = 1$ and checking whether $a^M \equiv 1 \pmod{N}$.)

Since N is odd we have that M is even. Write $M = 2^r m$ where m is odd. Now choose uniformly at random an integer $1 < a < N$. Check whether $\gcd(a, N) = 1$. If not then we have split N, otherwise compute $a_0 = a^m \pmod{N}$, $a_1 = a_0^2 \pmod{N}, \dots, a_r = a^M \pmod{N}$ (this is similar to the Miller–Rabin test; see Section 12.1.2). We know that $a_r = 1$, so either $a_0 = 1$ or else somewhere along the way there is a non-trivial square root x of 1. If $x \neq -1$ then $\gcd(x + 1, N)$ yields a non-trivial factor of N. All computations require a polynomially bounded number of bit operations.

Let p and q be two distinct prime factors of N. Since a is chosen uniformly at random it follows that $\gcd(a, N) > 1$ or $(\frac{a}{p}) = -(\frac{a}{q})$ with probability at least $1/2$. In either case, the above process splits N. The expected number of trials to split N is therefore at most 2.

Repeating the above process on each of the factors in turn, one can factor N completely. The expected number of iterations is $O(\log(N))$. For a complete anaysis of this reduction see Section 7.7 of Talbot and Welsh [539] or Section 10.6 of Shoup [497]. □

Two special cases of Lemma 24.1.14 are FACTOR $\le_R$ COMPUTE-LAMBDA and FACTOR $\le_R$ COMPUTE-PHI. Note that these reductions are randomised and the running time is only an expected value. Coron and May [140] showed a deterministic polynomial-time reduction FACTOR $\le_R$ RSA-PRIVATE-KEY (also see Section 4.6 of [371]).

Exercise 24.1.15 Restricting attention to integers of the form $N = pq$ where p and q are distinct primes, show that FACTOR $\le_R$ RSA-PRIVATE KEY.

Exercise 24.1.16 The **STRONG-RSA** problem is: given an RSA modulus N and $y \in \mathbb{N}$ to find any pair (x, e) of integers such that $e > 1$ and

$$x^e \equiv y \pmod{N}.$$

[4] The original RSA paper credits this result to G. Miller.

Give a reduction from STRONG-RSA to RSA. This shows that the STRONG-RSA problem is not harder than the RSA problem.[5]

Exercise 24.1.17 Let $N = pq$ be an RSA modulus. Let A be an oracle that takes as input an integer a and returns $a \pmod{\varphi(N)}$. Show how to use A to factor N.

Exercise 24.1.18 Consider the following variant of RSA encryption. Alice has a public key N and two public exponents e_1, e_2 such that $e_1 \neq e_2$ and $\gcd(e_i, \lambda(N)) = 1$ for $i = 1, 2$. To encrypt to Alice one is supposed to send $\mathsf{c}_1 = \mathsf{m}^{e_1} \pmod{N}$ and $\mathsf{c}_2 = \mathsf{m}^{e_2} \pmod{N}$. Show that if $\gcd(e_1, e_2) = 1$ then an attacker can determine the message given the public key and a ciphertext $(\mathsf{c}_1, \mathsf{c}_2)$.

Exercise 24.1.19 Consider the following signature scheme based on RSA. The public key is an integer $N = pq$, an integer e coprime to $\lambda(N)$ and an integer a such that $\gcd(a, N) = 1$. The private key is the inverse of e modulo $\lambda(N)$ as usual. Let H be a collision-resistant hash function. The signature on a message m is an integer s such that

$$\mathsf{s}^e \equiv a^{H(\mathsf{m})} \pmod{N}$$

where $H(\mathsf{m})$ is interpreted as an integer. Explain how the signer can generate signatures efficiently. Find a known message attack on this system that allows an adversary to make selective forgery of signatures.

24.2 The textbook Rabin cryptosystem

The textbook Rabin cryptosystem [444] is given in Box 24.2. Rabin is essentially RSA with the optimal choice of e, namely $e = 2$.[6] As we will see, the security of Rabin is more closely related to factoring than RSA. We first have to deal with the problem that if $N = pq$ where p and q are distinct primes then squaring is a four-to-one map (in general) so it is necessary to have a rule to choose the correct solution in decryption.

Lemma 24.2.1 *Suppose p and q are primes such that $p \equiv q \equiv 3 \pmod 4$. Let $N = pq$ and $1 < x < N$ be such that $\left(\frac{x}{N}\right) = 1$. Then either x or $N - x$ is a square modulo N.*

Exercise 24.2.2 Prove Lemma 24.2.1.

Definition 24.2.3 Let $p \equiv q \equiv 3 \pmod 4$ be primes. Then $N = pq$ is called a **Blum integer**.

Note that, as with RSA, the value m in encryption is actually a symmetric key (passed through a padding scheme) while in signing it is a hash of the message. The choice

[5] The word "strong" is supposed to indicate that the *assumption* that STRONG-RSA is hard is a *stronger assumption* than the assumption that RSA is hard. Of course, the computational problem is weaker than RSA, in the sense that it might be easier to solve STRONG-RSA than RSA.

[6] The original paper [444] proposed encryption as $E_{N,b}(x) = x(x + b) \pmod{N}$ for some integer b. However, there is a gain in efficiency with no loss of security by taking $b = 0$.

KeyGen(κ): Generate two random $\kappa/2$-bit primes p and q such that $p \equiv q \equiv 3 \pmod{4}$ and set $N = pq$. Output the public key $\mathsf{pk} = N$ and the private key $\mathsf{sk} = (p, q)$.

The message space and ciphertext space depend on the redundancy scheme (suppose for the moment that they are $\mathsf{C}_\kappa = \mathsf{M}_\kappa = (\mathbb{Z}/N\mathbb{Z})^*$).

Encrypt(m, N): Compute $\mathsf{c} = \mathsf{m}^2 \pmod{N}$ (with some redundancy or padding scheme).

Decrypt(c, (p, q)): We want to compute $\sqrt{\mathsf{c}} \pmod{N}$, and this is done by the following method: compute $\mathsf{m}_p = \mathsf{c}^{(p+1)/4} \pmod{p}$ and $\mathsf{m}_q = \mathsf{c}^{(q+1)/4} \pmod{q}$ (see Section 2.9). Test that $\mathsf{m}_p^2 \equiv \mathsf{c} \pmod{p}$ and $\mathsf{m}_q^2 \equiv \mathsf{c} \pmod{q}$, and if not then output $\perp$. Use the Chinese remainder theorem (Exercise 2.6.3) to obtain 4 possibilities for $\mathsf{m} \pmod{N}$ such that $\mathsf{m} \equiv \pm \mathsf{m}_p \pmod{p}$ and $\mathsf{m} \equiv \pm \mathsf{m}_q \pmod{q}$. Use the redundancy (see later) to determine the correct value m and return $\perp$ if there is no such value.

Sign(m, (p, q)): Ensure that $\left(\frac{\mathsf{m}}{N}\right) = 1$ (possibly by adding some randomness). Then either m or $N - \mathsf{m}$ is a square modulo N. Compute $\mathsf{s} = \sqrt{\pm \mathsf{m}} \pmod{N}$ by computing $(\pm \mathsf{m})^{(p+1)/4} \pmod{p}$, $(\pm \mathsf{m})^{(q+1)/4} \pmod{q}$ and applying the Chinese remainder theorem.

Verify(m, s, N): Check whether $\mathsf{m} \equiv \pm \mathsf{s}^2 \pmod{N}$.

Box 24.2 **Textbook Rabin**

of $p, q \equiv 3 \pmod{4}$ is to simplify the taking of square roots (and is also used in the redundancy schemes below); the Rabin scheme can be used with other moduli.

24.2.1 Redundancy schemes for unique decryption

To ensure that decryption returns the correct message it is necessary to have some redundancy in the message, or else to send some extra bits. We now describe three solutions to this problem.

- **Redundancy in the message for Rabin**: For example, insist that the least significant l bits (where $l > 2$ is some known parameter) of the binary string m are all ones. (Note 8.14 of [376] suggests repeating the last l bits of the message.) If l is big enough then it is unlikely that two different choices of square root would have the right pattern in the l bits.

  A message m is encoded as $x = 2^l \mathsf{m} + (2^l - 1)$, and so the message space is $\mathsf{M}_\kappa = \{\mathsf{m} : 1 \le \mathsf{m} < N/2^l, \gcd(N, 2^l \mathsf{m} + (2^l - 1)) = 1\}$ (alternatively, $\mathsf{M}_\kappa = \{0, 1\}^{\kappa - l - 2}$). The ciphertext is $\mathsf{c} = x^2 \pmod{N}$. Decryption involves computing the four square roots of c. If none, or more than one, of the roots corresponds to an element of M_κ then decryption fails (return $\perp$). Otherwise output the message $\mathsf{m} = \lfloor x/2^l \rfloor$.

  This method is a natural choice, since some padding schemes for CCA security (such as OAEP) already have sections of the message with a fixed pattern of bits.

  Note that, since N is odd, the least significant bit of $N - x$ is different to the least significant bit of x. Hence, the $l \ge 1$ least significant bits of x and $N - x$ are never equal.

Treating the other two square roots of $x_2 \pmod N$ as random integers it is natural to conjecture that the probability that either of them has a specific pattern of their l least significant bits is roughly $2/2^l$. This conjecture is confirmed by experimental evidence. Hence, the probability of decryption failure is approximately $1/2^{l-1}$.

- **Extra bits for Rabin**: Send two extra bits of information to specify the square root. For example, one could send the value $b_1 = (\frac{\mathsf{m}}{N})$ of the Jacobi symbol (the set $\{-1, 1\}$ can be encoded as a bit under the map $x \mapsto (x + 1)/2$), together with the least significant bit b_2 of the message. The ciphertext space is now $\mathsf{C}_\kappa = (\mathbb{Z}/N\mathbb{Z})^* \times \{0, 1\}^2$ and, for simplicity of exposition, we take $\mathsf{M}_\kappa = (\mathbb{Z}/N\mathbb{Z})^*$.

  These two bits allow unique decryption, since $(\frac{-1}{N}) = 1$, m and $N - \mathsf{m}$ have the same Jacobi symbol, and if m is odd then $N - \mathsf{m}$ is even.

  Indeed, when using the Chinese remainder theorem to compute square roots then one computes m_p and m_q such that $(\frac{\mathsf{m}_p}{p}) = (\frac{\mathsf{m}_q}{q}) = 1$. Then decryption using the bits b_1, b_2 is: if $b_1 = 1$ then the decryption is $\pm CRT(\mathsf{m}_p, \mathsf{m}_q)$ and if $b_1 = -1$ then solution is $\pm CRT(-\mathsf{m}_p, \mathsf{m}_q)$.

  This scheme is close to optimal in terms of ciphertext expansion (though see Exercise 24.2.6 for an improvement) and decryption never fails. The drawbacks are that the ciphertext contains some information about the message (and so the scheme is not IND-CPA secure), and encryption involves computing the Jacobi symbol, which typically requires far more computational resources than the single squaring modulo N.
- **Williams**: Let $N = pq$ where $p, q \equiv 3 \pmod 4$. If $p \not\equiv \pm q \pmod 8$ then $(\frac{2}{N}) = -1$. Hence, for every $1 \le x < N$ exactly one of $x, N - x, 2x, N - 2x$ is a square modulo N (see Exercise 24.6.3). Without loss of generality, we therefore assume that $p \equiv 3 \pmod 8$ and $q \equiv 7 \pmod 8$. The integer N is called a **Williams integer** in this situation.

  Williams [566] suggests encoding a message $1 \le \mathsf{m} < N/8 - 1$ (alternatively, $\mathsf{m} \in \mathsf{M}_\kappa = \{0, 1\}^{\kappa-5}$) as an integer x such that x is even and $(\frac{x}{N}) = 1$ (and so x or $-x$ is a square modulo N) by

$$x = P(\mathsf{m}) = \begin{cases} 4(2\mathsf{m} + 1) & \text{if } \left(\frac{2\mathsf{m}+1}{N}\right) = 1, \\ 2(2\mathsf{m} + 1) & \text{if } \left(\frac{2\mathsf{m}+1}{N}\right) = -1. \end{cases}$$

  The encryption of m is then $\mathsf{c} = P(\mathsf{m})^2 \pmod N$.

  To decrypt one computes square roots to obtain the unique even integer $1 < x < N$ such that $(\frac{x}{N}) = 1$ and $x^2 \equiv \mathsf{c} \pmod N$. If $8 \mid x$ then decryption fails (return $\perp$). Otherwise, return $\mathsf{m} = (x/2 - 1)/2$ if $x \equiv 2 \pmod 4$ and $\mathsf{m} = (x/4 - 1)/2$ if $x \equiv 0 \pmod 4$.

  Unlike the extra bits scheme, this does not reveal information about the ciphertext. It is almost optimal from the point of view of ciphertext expansion. But it still requires the encrypter to compute a Jacobi symbol (hence losing the performance advantage of Rabin over RSA). The Rabin cryptosystem with the Williams padding is sometimes called the **Rabin–Williams cryptosystem.**

Exercise 24.2.4 Prove all the unproved claims in the above discussion of the Williams redundancy scheme.

Exercise 24.2.5 Let $N = (2^{59} + 21)(2^{20} + 7)$. The three ciphertexts below are Rabin encryptions for each of the three redundancy schemes above (in the first case, $l = 5$). Determine the corresponding message in each case.

$$273067682422, \quad (309135051204, -1, 0), \quad 17521752799.$$

Exercise 24.2.6 (Freeman–Goldreich–Kiltz–Rosen–Segev) This is a variant of the "extra bits" method. Let N be a Williams integer. Let $u_1 = -1$ and $u_2 = 2$. To encrypt message $\mathsf{m} \in (\mathbb{Z}/N\mathbb{Z})^*$ one first determines the bits b_1 and b_2 of the "extra bits" redundancy scheme (i.e., $b_1 = 1$ if and only if $(\frac{\mathsf{m}}{N}) = +1$ and b_2 is the least significant bit of m). Compute the ciphertext

$$\mathsf{c} = \mathsf{m}^2 u_1^{1-b_1} u_2^{b_2} \pmod{N}.$$

Show how a user who knows p and q can decrypt the ciphertext. Show that this scheme still leaks the least significant bit of m (and hence is still not IND-CPA secure), but no longer leaks $(\frac{\mathsf{m}}{N})$.

24.2.2 Variants of Rabin

In terms of computational performance, Rabin encryption is extremely fast (as long as encryption does not require Jacobi symbols) while decryption, using the Chinese remainder theorem, is roughly the same speed as RSA decryption.

Exercise 24.2.7 Describe and analyse the Rabin cryptosystem using moduli of the form $N = pqr$ where p, q and r are distinct primes. What are the advantages and disadvantages of Rabin in this setting?

Exercise 24.2.8 (Takagi–Rabin) Describe and analyse the Rabin cryptosystem using moduli of the form $N = p^r q^s$ $(r \neq s)$. Is there any advantage from using Rabin in this setting?

We now discuss compression of Rabin signatures. For further discussion of these ideas, and an alternative method, see Gentry [230].

Example 24.2.9 (Bleichenbacher) Suppose s is a Rabin signature on a message m, so that $\mathsf{s}^2 \equiv \pm H(\mathsf{m}) \pmod{N}$. To compress s to half the size one uses the Euclidean algorithm on (s, N) to compute a sequence of values $r_i, u_i, v_i \in \mathbb{Z}$ such that $r_i = u_i\mathsf{s} + v_i N$. Let i be the index such that $|r_i| < \sqrt{N} < |r_{i-1}|$. Then $r_i \equiv u_i\mathsf{s} \pmod{N}$ and so

$$r_i^2 \equiv u_i^2\mathsf{s}^2 \equiv \pm u_i^2 H(\mathsf{m}) \pmod{N}.$$

One can therefore send u_i as the signature. Verification is to compute $w = \pm u_i^2 H(\mathsf{m}) \pmod{N}$ and check that w is a perfect square in $\mathbb{Z}$ (e.g., using the method of Exercise 2.4.9 or by computing an approximation to the square root). Part 6 of Lemma 2.3.3 states $|r_{i-1}u_i| \le N$ and so $|u_i| < \sqrt{N}$. Hence, this approach compresses the signature to half the size.

24.2.3 Security of textbook Rabin

Since the Rabin cryptosystem involves squaring it is natural to assume the security is related to computing square roots modulo N, which in turn is equivalent to factoring. Hence, an important feature of Rabin compared with RSA is that the hardness of breaking Rabin can be shown to be equivalent to factoring.

Definition 24.2.10 Let $S = \mathbb{N}$ or $S = \{pq : p, q \equiv 3 \pmod{4}, \text{ primes}\}$. The computational problem **SQRT-MOD-N** is: given $N \in S$ and $y \in \mathbb{Z}/N\mathbb{Z}$ to output $\perp$ if y is not a square modulo N, or a solution x to $x^2 \equiv y \pmod{N}$.

Lemma 24.2.11 *SQRT-MOD-N is equivalent to FACTOR.*

Proof Suppose we have a FACTOR oracle and are given a pair (N, y). Then one can use the oracle to factor N and then solve SQRT-MOD-N using square roots modulo p and Hensel lifting and the Chinese remainder theorem. This reduction is polynomial-time.

Conversely, suppose we have a SQRT-MOD-N oracle A and let N be given. First, if $N = p^e$ then we can factor N in polynomial-time (see Exercise 2.2.7). Hence, we may now assume that N has at least two distinct prime factors.

Choose a random $x \in \mathbb{Z}_N^*$ and set $y = x^2 \pmod{N}$. Call A on y to get x'. We have $x^2 \equiv (x')^2 \pmod{N}$ and there are at least four possible solutions x'. All but two of these solutions will give a non-trivial value of $\gcd(x - x', N)$. Hence, since x was chosen randomly, there is probability at least 1/2 that we can split N. Repeating this process splits N (the expected number of trials is at most 2). As in Lemma 24.1.14, one can repeat the process to factor N in $O(\log(N))$ iterations. The entire reduction is therefore polynomial-time. □

An important remark about the above proof is that the oracle A is not assumed to output a random square root x' of y. Indeed, A could be deterministic. The randomness comes from the choice of x in the reduction.

Exercise 24.2.12 Consider the computational problem FOURTH-ROOT: given $y \in \mathbb{Z}_N^*$ compute a solution to $x^4 \equiv y \pmod{N}$ if such a solution exists. Give reductions that show that FOURTH-ROOT is equivalent to FACTOR in the case $N = pq$ with p, q distinct odd primes.

It is intuitively clear that any algorithm that breaks the one-way encryption property (or selective signature forgery) of Rabin under passive attacks must compute square roots modulo N. We have seen that SQRT-MOD-N is equivalent to FACTOR. Thus we expect breaking Rabin under passive atacks to be as hard as factoring. However, giving a precise security proof involves taking care of the redundancy scheme.

Theorem 24.2.13 *Let $N = pq$, where $p \equiv q \equiv 3 \pmod 4$ are primes, and define $S_{N,l} = \{1 \le x < N : \gcd(x, N) = 1, 2^l | (x+1)\}$. Assume the probability, over $x \in \mathbb{Z}_N^* - S_{N,l}$, that there exists $y \in S_{N,l}$ with $x \ne y$ but $x^2 \equiv y^2 \pmod N$, is $1/2^{l-1}$. Then breaking the one-way encryption security property of the Rabin cryptosystem with the "redundancy in the message" redundancy scheme where $l = O(\log(\log(N)))$ under passive attacks is equivalent to factoring Blum integers.*

Theorem 24.2.14 *Breaking the one-way encryption security property of the Rabin cryptosystem with the "extra bits" redundancy scheme under passive attacks is equivalent to factoring products $N = pq$ of primes $p \equiv q \equiv 3 \pmod 4$.*

Theorem 24.2.15 *Breaking the one-way encryption security property of the Rabin cryptosystem with the Williams redundancy scheme under passive attacks is equivalent to factoring products $N = pq$ of primes $p \equiv q \equiv 3 \pmod 4$, $p \not\equiv \pm q \pmod 8$.*

Note that Theorem 24.2.13 gives a strong security guarantee when l is small, but in that case decryption failures are frequent. Indeed, there is no choice of l for the Rabin scheme with redundancy in the message that provides both a tight reduction to factoring and negligible probability of decryption failure.

We prove the first and third of these theorems and leave Theorem 24.2.14 as an exercise.

Proof (Theorem 24.2.13) Let A be an oracle that takes a Rabin public key N and a ciphertext c (with respect to the "redundancy in the message" padding scheme) and returns either the corresponding message m or an invalid ciphertext symbol $\perp$.

Choose a random $x \in \mathbb{Z}_N^*$ such that neither x nor $N - x$ satisfy the redundancy scheme (i.e., the l least significant bits are not all 1). Set $\mathsf{c} = x^2 \pmod N$ and call the oracle A on c. The oracle A answers with either a message m or $\perp$.

According to the (heuristic) assumption in the theorem, the probability that exactly one of the two (unknown) square roots of c modulo N has the correct l least significant bits is $2^{-(l-1)}$. If this is the case then calling the oracle A on c will output a value m such that, writing $x' = 2^l\mathsf{m} + (2^l - 1)$, we have $(x')^2 \equiv x^2 \pmod N$ and $x' \not\equiv \pm x \pmod N$. Hence, $\gcd(x' - x, N)$ will split N.

We expect to require approximately 2^{l-1} trials before factoring N with this method. Hence, the reduction is polynomial-time if $l = O(\log(\log(N)))$. □

Proof (Proof of Theorem 24.2.15; following Williams [566]) Let A be an oracle that takes a Rabin public key N and a ciphertext c (with respect to the Williams redundancy scheme) and returns either the corresponding message m or an invalid ciphertext symbol $\perp$.

Choose a random integer x such that $(\frac{x}{N}) = -1$, e.g., let $x = \pm 2z^2 \pmod{N}$ for random $z \in (\mathbb{Z}/N\mathbb{Z})^*$. Set $\mathsf{c} = x^2 \pmod{N}$ and call A on (N, c). The oracle computes the unique even integer $1 < x' < N$ such that $(x')^2 \equiv \mathsf{c} \pmod{N}$ and $(\frac{x'}{N}) = 1$. The oracle then attempts to decode x' to obtain the message. If $8 \nmid x'$ (which happens with probability 3/4) then decoding succeeds and the corresponding message m is output by the oracle. Given m we can recover the value x' as $2(2\mathsf{m}+1)$ or $4(2\mathsf{m}+1)$, depending on the value of $(\frac{2\mathsf{m}+1}{N})$, and then factor N as $\gcd(x' - x, N)$.

If $8 \mid x'$ then the oracle outputs $\perp$ so we compute $\mathsf{c}' = \mathsf{c}2^{-4} \pmod{N}$ and call the oracle on c'. The even integer x'' computed by the oracle is equal to $x'/4$ and so the above argument may apply. In extremely rare cases one might have to repeat the process $\frac{1}{2}\log_2(N)$ times, but the expected number of trials is constant. □

Exercise 24.2.16 Prove Theorem 24.2.14.

Exercise 24.2.17 Prove Theorem 24.2.13 when the message space is $\{0, 1\}^{\kappa - l - 2}$.

The above theorems show that the hardness guarantee for the Rabin cryptosystem is often stronger than for the RSA cryptosystem (at least, under passive attacks). Hence, the Rabin cryptosystem is very attractive: it has faster public operations and also has a stronger security guarantee than RSA. On the other hand, the ideas used in the proofs of these theorems can also be used to give adaptive (CCA) attacks on the Rabin scheme that allow the attacker to determine the private key (i.e., the factorisation of the modulus). In comparison, a CCA attack on textbook RSA only decrypts a single message rather than learns the private key.

Example 24.2.18 We describe a CCA attacker giving a total break of Rabin with "redundancy in the message".

As in the proof of Theorem 24.2.13, the adversary chooses a random $x \in \mathbb{Z}_N^*$ such that neither x nor $N - x$ satisfy the redundancy scheme (i.e., the l least significant bits are not all 1). Set $\mathsf{c} = x^2 \pmod{N}$ and call the decryption oracle on c. The oracle answers with either a message m or $\perp$. Given m one computes x' such that $\gcd(x' - x, N)$ splits N.

Exercise 24.2.19 Give CCA attacks giving a total break of Rabin when using the other two redundancy schemes ("extra bits" and Williams).

As we have seen, the method to prove that Rabin encryption has one-way security under a passive attack is also the method to give a CCA attack on Rabin encryption. It was remarked by Williams [566] that such a phenomenon seems to be inevitable. This remark has been formalised and discussed in detail by Paillier and Villar [427].

Exercise 24.2.20 Generalise Rabin encryption to $N = pq$ where $p \equiv q \equiv 1 \pmod{3}$ and encryption is $\mathsf{c} = \mathsf{m}^3 \pmod{N}$. How can one specify redundancy? Is the security related to factoring in this case?

Exercise 24.2.21 Consider the following public key cryptosystem related to Rabin: a user's public key is a product $N = pq$ where p and q are primes congruent to 3 modulo 4. To encrypt a message $1 < \mathsf{m} < N$ to the user compute and send

$$\mathsf{c}_1 = \mathsf{m}^2 \pmod{N} \qquad \text{and} \qquad \mathsf{c}_2 = (\mathsf{m}+1)^2 \pmod{N}.$$

Show that if $x^2 \equiv y^2 \pmod{N}$ and $(x+1)^2 \equiv (y+1)^2 \pmod{N}$ then $x \equiv y \pmod{N}$. Hence, show that decryption is well-defined.

Show that this cryptosystem does not have OWE security under a passive attack.

24.3 Homomorphic encryption

Homomorphic encryption was defined in Section 23.3.1. We first remark that the textbook RSA scheme is homomorphic for multiplication modulo N: if $\mathsf{c}_1 \equiv \mathsf{m}_1^e \pmod{N}$ and $\mathsf{c}_2 \equiv \mathsf{m}_2^e \pmod{N}$ then $\mathsf{c}_1\mathsf{c}_2 \equiv (\mathsf{m}_1\mathsf{m}_2)^e \pmod{N}$. Indeed, this property is behind the CCA attack on textbook RSA encryption. Padding schemes can destroy this homomorphic feature.

Exercise 24.3.1 Show that textbook Rabin encryption is not homomorphic for multiplication when using any of the redundancy schemes of Section 24.2.1.

We now give a scheme that is homomorphic for addition, and that allows a much larger range of values for the message compared with the scheme in Exercise 23.3.5.

Example 24.3.2 (Paillier [424]) Let $N = pq$ be a user's public key. To encrypt a message $\mathsf{m} \in \mathbb{Z}/N\mathbb{Z}$ to the user choose a random integer $1 < u < N$ (note that, with overwhelming probability, $u \in (\mathbb{Z}/N\mathbb{Z})^*$) and compute the ciphertext

$$\mathsf{c} = (1 + N\mathsf{m})u^N \pmod{N^2}.$$

To decrypt compute

$$\mathsf{c}^{\lambda(N)} \equiv 1 + \lambda(N)N\mathsf{m} \pmod{N^2}$$

and hence determine $\mathsf{m} \pmod{N}$. (This requires multiplication by $\lambda(N)^{-1} \pmod{N}$).

The homomorphic property is: if c_1 and c_2 are ciphertexts encrypting m_1 and m_2, respectively, then

$$\mathsf{c}_1\mathsf{c}_2 \equiv (1 + N(\mathsf{m}_1 + \mathsf{m}_2))(u_1u_2)^N \pmod{N^2}$$

encrypts $\mathsf{m}_1 + \mathsf{m}_2 \pmod{N}$.

Exercise 24.3.3 Verify the calculations in Example 24.3.2.

As always, one cannot obtain CCA secure encryption using a homomorphic scheme. Hence, one is only interested in passive attacks. To check whether or not a Paillier ciphertext c corresponds to a specific message m is precisely solving the following computational problem.

Definition 24.3.4 Let $N = pq$. The **composite residuosity problem** is: given $y \in \mathbb{Z}/N^2\mathbb{Z}$ to determine whether or not $y \equiv u^N \pmod{N^2}$ for some $1 < u < N$.

Exercise 24.3.5 Show that the Paillier encryption scheme has IND-CPA security if and only if the composite residuosity problem is hard.

Exercise 24.3.6 Show that composite residuosity is not harder than factoring.

Exercise 24.3.7 Show how to use the Chinese remainder theorem to speed up Paillier decryption.

Encryption using the Paillier scheme is rather slow since one needs an exponentiation to the power N modulo N^2. One can use sliding windows for this exponentiation, though since N is fixed one might prefer to use an addition chain optimised for N. Exercises 24.3.8 and 24.3.9 suggest variants with faster encryption. The disadvantage of the scheme in Exercise 24.3.8 is that it requires a different computational assumption. The disadvantage of the scheme in Exercise 24.3.9 is that it is no longer homomorphic.

Exercise 24.3.8★ Consider the following efficient variant of the Paillier cryptosystem. The public key of a user consists of N and an integer $h = u^N \pmod{N^2}$ where $1 < u < N$ is chosen uniformly at random. To encrypt a message m to the user, choose a random integer $0 \le x < 2^k$ (e.g., with $k = 256$) and set

$$\mathsf{c} \equiv (1 + N\mathsf{m})h^x \pmod{N^2}.$$

State the computational assumption underlying the IND-CPA security of the scheme. Give an algorithm to break the IND-CPA security that requires $O(2^{k/2})$ multiplications modulo N^2. Use multi-exponentiation to give an even more efficient variant of the Paillier cryptosystem, at the cost of even larger public keys.

Exercise 24.3.9 (Catalano, Gennaro, Howgrave-Graham, Nguyen [117]) A version of the Paillier cryptosystem for which encryption is very efficient is the following: the public key is (N, e) such that $\gcd(e, N\lambda(N)) = 1$. One thinks of e as being small. To encrypt one chooses a random integer $1 < u < N$ and computes

$$\mathsf{c} = (1 + N\mathsf{m})u^e \pmod{N^2}.$$

Decryption begins by performing RSA decryption of c modulo N to obtain u.

Write down the decryption algorithm for this system. Explain why this encryption scheme is no longer homomorphic.

24.4 Algebraic attacks on textbook RSA and Rabin

The goal of this section is to briefly describe a number of relatively straightforward attacks on the textbook RSA and Rabin cryptosystems. These attacks can all be prevented if one uses a sufficiently good padding scheme. Indeed, by studying these attacks one develops a better idea of what properties are required of a padding scheme.

24.4.1 The Håstad attack

We now present an attack that can be mounted on the RSA or Rabin schemes in a multi-user situation. Note that such attacks are not covered by the standard security model for encryption as presented in Chapter 1.

Example 24.4.1 Suppose three users have RSA public keys N_1, N_2, N_3 and all use encryption exponent $e = 3$. Let $0 < \mathsf{m} < \min\{N_1, N_2, N_3\}$ be a message. If m is encrypted to all three users then an attacker can determine m from the three ciphertexts $\mathsf{c}_1, \mathsf{c}_2$ and c_3 as follows: the attacker uses the Chinese remainder theorem to compute $1 < \mathsf{c} < N_1 N_2 N_3$ such that $\mathsf{c} \equiv \mathsf{m}^3 \pmod{N_i}$ for $1 \le i \le 3$. It follows that $\mathsf{c} = \mathsf{m}^3$ over $\mathbb{Z}$ and so one can determine m using root-finding algorithms.

This attack is easily prevented by using randomised padding schemes (assuming that the encryptor is not so lazy that they re-use the same randomness each time). Nevertheless, this attack seems to be one of the reasons why modern systems use $e = 65537 = 2^{16} + 1$ instead of $e = 3$.

Exercise 24.4.2 Show that the Håstad attack applies when the same message is sent using textbook Rabin encryption (with any of the three redundancy schemes) to two users.

Exercise 24.4.3 Two users have Rabin public keys $N_1 = 144946313$ and $N_2 = 138951937$. The same message m is encrypted using the "extra bits" padding scheme to the two users, giving ciphertexts

$$C_1 = (48806038, -1, 1) \text{ and } C_2 = (14277753, -1, 1).$$

Use the Håstad attack to find the corresponding message.

24.4.2 Algebraic attacks

We already discussed a number of easy algebraic attacks on textbook RSA, all of which boil down to exploiting the multiplicative property

$$\mathsf{m}_1^e \mathsf{m}_2^e \equiv (\mathsf{m}_1 \mathsf{m}_2)^e \pmod{N}.$$

We also noted that, since textbook RSA is deterministic, it can be attacked by trying all messages. Hence, if one knows that $1 \le \mathsf{m} < 2^k$ (for example, if m is a k-bit symmetric key) then one can attack the system in at most 2^k exponentiations modulo N. We now show that one can improve this to roughly $\sqrt{2^k}$ exponentiations in many cases.

Exercise 24.4.4 (Boneh, Joux, Nguyen [77]) Suppose $\mathsf{c} = \mathsf{m}^e \pmod{N}$ where $1 \le \mathsf{m} < 2^k$. Show that if $\mathsf{m} = \mathsf{m}_1\mathsf{m}_2$ for two integers $1 < \mathsf{m}_1, \mathsf{m}_2 < B$ then one can determine m in $O(B)$ exponentiations modulo N. If $B = 2^{k/2+\epsilon}$ then the probability that m splits in this way is noticeable.

24.4.3 *Desmedt–Odlyzko attack*

This is a "lunchtime attack" proposed by Desmedt and Odlyzko in [157] on textbook RSA signatures. It can produce more forgeries than calls to the signing oracle. The basic idea is to query the signing oracle on the first r prime numbers $p_1, \dots, p_r$ to get signatures $\mathsf{s}_1, \dots, \mathsf{s}_r$. Then, for any message m, if m is a product of powers of the first r primes $\mathsf{m} = \prod_{i=1}^r p_i^{f_i}$ then the corresponding signature is

$$\mathsf{s} = \prod_{i=1}^{r} \mathsf{s}_i^{f_i}.$$

This attack is not feasible if messages are random elements between 1 and N (as the probability of smoothness is usually negligible) but it can be effective if messages in the system are rather small.

Exercise 24.4.5 Let $N = 9178628368309$ and $e = 7$ be an RSA public key. Suppose one learns that the signatures of 2, 3 and 5 are 872240067492, 6442782604386 and 1813566093366 respectively. Determine the signatures for messages $\mathsf{m} = 6, 15, 12$ and 100.

An analogous attack applies to encryption: ask for decryptions of the first r primes (treating them as ciphertexts) and then, given a challenge ciphertext c, if $\mathsf{c} \equiv \prod_{i=1}^r p_i^{e_i}$ then one can work out the decryption of c. Since ciphertexts (even of small messages) are of size up to N this attack is usually not faster than factoring the modulus.

This idea, together with a number of other techniques, has been used by Coron, Naccache, Tibouchi and Weinmann [142] to attack real-world signature proposals.

24.4.4 *Related message attacks*

This attack is due to Franklin and Reiter.[7] Consider textbook RSA with small exponent e or textbook Rabin ($e = 2$). Suppose we obtain ciphertexts c_1 and c_2 (with respect to the same public key (N, e)) for messages m and $\mathsf{m} + a$ for some known integer a. Then m is a common root modulo N of the two polynomials $F_1(x) = x^e - \mathsf{c}_1$ and $F_2(x) = (x + a)^e - \mathsf{c}_2$ (for Rabin we may have polynomials like $F_1(x) = (2^l(x + 1) - 1)^2 - \mathsf{c}_1$ or $F_1(x) = (2(2x + 1))^2 - C_1)$. Hence, one can run Euclid's algorithm on $F_1(x)$ and $F_2(x)$ in $(\mathbb{Z}/N\mathbb{Z})[x]$ and this will either lead to a factor of N (since performing polynomial division

[7] The idea was presented at the "rump session" of CRYPTO 1995.

in $(\mathbb{Z}/N\mathbb{Z})[x]$ involves computing inverses modulo N) or will output, with high probability, a linear polynomial $G(x) = x - \mathsf{m}$.

Euclid's algorithm for polynomials of degree e has complexity $O(e^2 M(\log(N)))$ or $O(M(e)\log(e)M(\log(N)))$ bit operations. Hence, this method is feasible only when e is rather small (e.g., $e < 2^{30}$).

Exercise 24.4.6 Extend the Franklin–Reiter attack to ciphertexts c_1 and c_2 (again, for the same public key) where c_1 is an encrypion of m and c_2 is an encryption of $a\mathsf{m} + b$ for known integers a and b.

Exercise 24.4.7 Let $N = 2157212598407$ and $e = 3$. Suppose we have ciphertexts

$$\mathsf{c}_1 = 1429779991932 \quad \text{and} \quad \mathsf{c}_2 = 655688908482$$

such that c_1 is the encryption of m and c_2 is the encryption of $\mathsf{m} + 2^{10}$. Determine the message m.

These ideas have been extended by Coppersmith, Franklin, Patarin and Reiter [134]. Among other things they study how to break related encryptions for any polynomial relation by using resultants (see Exercise 24.4.8).

Exercise 24.4.8 Let (N, e) be an RSA key. Suppose one is given $\mathsf{c}_1 = \mathsf{m}_1^e \pmod{N}$, $\mathsf{c}_2 = \mathsf{m}_2^e \pmod{N}$ and a polynomial $P(x, y) \in \mathbb{Z}[x, y]$ such that $P(\mathsf{m}_1, \mathsf{m}_2) \equiv 0 \pmod{N}$. Let d be a bound on the total degree of $P(x, y)$. Show how to compute m_1 and m_2 in $O((d + e)^3 d^2 M(\log(N)))$ bit operations.

24.4.5 Fixed pattern RSA signature forgery

The aim of this section is to present a simple padding scheme, often called **fixed pattern padding** for RSA. We then sketch why this approach may not be sufficient to obtain RSA signatures secure against adaptive attackers. These ideas originate in the work of De Jonge and Chaum and later work by Girault and Misarsky. We present the more recent attacks by Brier, Clavier, Coron and Naccache [99]. An attack on RSA encryption with fixed padding is given in Section 19.4.1.

Example 24.4.9 Suppose we are using moduli of length 3072 bits and that messages (or message digests) m are of length at most 1000 bits.

The padding scheme uses a fixed value $P = 2^{3071}$ and the signature on the message digest m (such that $0 \le \mathsf{m} < 2^{1000}$) is

$$\mathsf{s} = (P + \mathsf{m})^d \pmod{N}.$$

The verifier computes $\mathsf{s}^e \pmod{N}$ and checks if it is of the correct form $P + \mathsf{m}$ with $0 \le \mathsf{m} < 2^{1000}$.

The following method (from [99]) forges signatures if messages are roughly $N^{1/3}$ in size. We assume that a signing oracle is available (we assume the signing oracle will only generate signatures if the input is correctly padded) and that a hash function is not applied to the messages. Suppose m is the target message, so we want to compute the dth power of $z = P + \mathsf{m}$. The idea is to find small values u, v, w such that

$$z(z+u) \equiv (z+v)(z+w) \pmod{N}. \tag{24.2}$$

Then given signatures on $\mathsf{m}+u, \mathsf{m}+v$ and $\mathsf{m}+w$ (i.e., dth powers of $z+u, z+v$ and $z+w$) one can compute the signature on m as required.

To find small solutions to equation (24.2) we expand and simplify to

$$z(u-v-w) \equiv vw \pmod{N}.$$

Running the extended Euclidean algorithm (the basic version rather than the fast version of Algorithm 1) on z and N gives a number of integers s, r such that

$$zs \equiv r \pmod{N} \quad \text{and} \quad |rs| \approx N.$$

One can run Euclid until a solution with $|s| \approx N^{1/3}$ and $|r| \approx N^{2/3}$ is found. One then tries to factor r as a product $r = vw$ of numbers of a similar size. If this is feasible (for example, if r has a large number of small prime factors) then set $u = s + v + w$ and we have a solution. This approach is reasonable as long as the messages are at least one third of the bit-length of the modulus.

Example 24.4.10 Let $N = 1043957 \approx 2^{20}$.

Suppose $P = 2^{19}$ is the fixed padding and suppose messages are restricted to be 10-bit binary strings. Thus

$$z = P + \mathsf{m}$$

where $0 \le \mathsf{m} < 2^{10}$.

Suppose have access to a signing oracle and would like to forge a signature on the message $\mathsf{m} = 503$ that corresponds to $z = P + 503 = 524791$.

We apply Euclid's algorithm on N and z to solve the congruence $zs \equiv r \pmod{N}$ where $s \approx N^{1/3} \approx 101$.

i	q	r_i	s_i	t_i
−1	–	1043957	1	0
0	–	524791	0	1
1	1	519166	1	−1
2	1	5625	−1	2
3	92	1666	93	−185

This gives the solution $-185z \equiv 1666 \pmod{N}$ and $|-185| \approx N^{1/3}$. So set $s = -185$ and $r = 1666$. We try to factor r and are lucky that $1666 = 2 \cdot 7^2 \cdot 17$. So choose $v = 34$ and $w = 49$. Finally, choose $u = v + w - 185 = -102$. One can check that

$$z(z+u) \equiv (z+v)(z+w) \pmod{N}$$

and that $z + u$, $z + v$ and $z + w$ are all between P and $P + 2^{10}$. Hence, if one obtains signatures $\mathsf{s}_1, \mathsf{s}_2, \mathsf{s}_3$ on $\mathsf{m} + u = 401$, $\mathsf{m} + v = 537$ and $\mathsf{m} + w = 552$ then one has the signature on z as $\mathsf{s}_2\mathsf{s}_3\mathsf{s}_1^{-1} \pmod{N}$.

The success of this attack depends on the cost of factoring r and the probability that it can be written as a product of integers of similar size. Hence, the attack has subexponential complexity. For fixed m the attack may not succeed (since r might not factor as a product of integers of the required size). On the other hand, if m can vary a little (this is now more like an existential forgery) then the attack should succeed. A method for existential forgery that does not require factoring is given in Example 24.4.13.

Exercise 24.4.11 Give a variant of the above attack for the case where messages can be of size $N^{1/2}$ and for which it is only necessary to obtain signatures on two messages.

Exercise 24.4.12 One could consider affine padding $A\mathsf{m} + B$ instead of $P + \mathsf{m}$, where A and B are fixed integers and m is small. Show that, from the point of view of attacks, the two padding schemes are equivalent.

Example 24.4.13 We sketch the existential forgery from [99]. As before, we seek messages $\mathsf{m}_1, \dots, \mathsf{m}_4$ of size $N^{1/3}$ such that

$$(P + \mathsf{m}_1)(P + \mathsf{m}_2) \equiv (P + \mathsf{m}_3)(P + \mathsf{m}_4) \pmod{N}.$$

Writing $\mathsf{m}_1 = x + t$, $\mathsf{m}_2 = y + t$, $\mathsf{m}_3 = t$ and $\mathsf{m}_4 = x + y + z + t$ the equation is seen to be equivalent to $Pz \equiv xy - tz \pmod{N}$. One again uses Euclid to find $s \approx N^{1/3}$, $r \approx N^{2/3}$ such that $Ps \equiv r \pmod{N}$. One sets $z = s$ and then wants to find x, y, t such that $xy = r + tz$. To do this choose a random integer $N^{1/3} < y < 2N^{1/3}$ such that $\gcd(y, z) = 1$ and set $t \equiv -z^{-1}r \pmod{y}$. One then easily solves for the remaining values and one can check that the m_i are roughly of the right size.

For further details and results we refer to Brier, Clavier, Coron and Naccache [99] and Lenstra and Shparlinski [336].

This idea, together with other techniques, has been used to cryptanalyse the ISO/IEC 9796-1 signature standard with great success. We refer to Coppersmith, Coron, Grieu, Halevi, Jutla, Naccache and Stern [133].

24.5 Attacks on RSA parameters

In this section we briefly recall some attacks on certain choices of RSA public key.

24.5.1 Wiener attack on small private exponent RSA

One proposal to speed-up RSA decryption is to choose d to be a small integer. Key generation is performed by first choosing d and then setting $e = d^{-1} \pmod{\lambda(N)}$. This is called **small private exponent RSA**.[8] We present the famous **Wiener attack**, which is a polynomial-time attack on private exponents $d < N^{1/4}$.

Exercise 24.5.1 Give a brute-force attack on small private exponent RSA that tries each odd integer $d > 1$ in turn. What is the complexity of this attack?

We now sketch Wiener's idea [564]. We assume the key generation of Box 24.1 is used so that $N = pq$ where $p < q < 2p$. Consider the equation defining e and d

$$ed = 1 + k\varphi(N)$$

(a similar attack can be mounted using the equation $ed = 1 + k\lambda(N)$, see Exercise 24.5.5). Since $e < \varphi(N)$ we have $k < d$. Now $\varphi(N) = N + 1 - (p + q)$ and $\sqrt{N} \le (p + q) < 3\sqrt{N}$ so $\varphi(N) = N - u$ where $0 \le u = p + q - 1 \le 3\sqrt{N}$. Rearranging gives

$$-ed + kN = (-1 + ku) < 3k\sqrt{N}. \tag{24.3}$$

If d is smaller than $\sqrt{N}/3$ then the right-hand side is $< N$. Hence, one could try to find d by running the extended Euclidean algorithm on (e, N) and testing the coefficient of e to see if it is a candidate value for $\pm d$ (e.g., by testing whether $(x^e)^d \equiv x \pmod{N}$ for a random $1 < x < N$). Note that one must use the basic extended Euclidean algorithm rather than the faster variant of Algorithm 1. We now explain that this method is guaranteed to find d when it is sufficiently small.

Theorem 24.5.2 *Let $N = pq$ where $p < q < 2p$ are primes. Let $e = d^{-1} \pmod{\varphi(N)}$ where $0 < d < N^{1/4}/\sqrt{3}$. Then given (N, e) one can compute d in polynomial-time.*

Proof Using the notation above, $(d, k, uk - 1)$ is a solution to equation (24.3) with $0 < k < d$ and $0 \le u < 3\sqrt{N}$.

The Euclidean algorithm finds all triples (s, t, r) satisfying $es + Nt = r$ with $|sr| < N$ and $|tr| < e$. Hence, if $|d(uk - 1)| < N$ then the required solution will be found. If $0 < d < N^{1/4}/\sqrt{3}$ then

$$|duk| < d^2u < \frac{N^{1/2}}{3} 3\sqrt{N} = N,$$

which completes the proof. □

Example 24.5.3 Let $N = 86063528783122081$ with $d = 8209$. One computes that $e = 14772019882186053$.

One can check that

$$ed = 1 + 1409\varphi(N).$$

Running Euclid's algorithm with $r_{-1} = N$ and $r_0 = e$ and writing s_i, t_i to be such that $r_i = s_iN + t_ie$ one finds the following table of values.

[8] The reader should remember that, in practice, it is more efficient to use the Chinese remainder theorem to speed up RSA decryption than small private exponents.

i	q	r_i	s_i	t_i
-1	$--$	86063528783122081	1	0
0	$--$	14772019882186053	0	1
1	5	12203429372191816	1	-5
2	1	2568590509994237	-1	6
3	4	1929067332214868	5	-29
4	1	639523177779369	-6	35
5	3	10497798876761	23	-134
6	60	9655245173709	-138	8075
7	1	842553703052	1409	-8209

One sees that d is found in only 7 steps.

Exercise 24.5.4 Consider the RSA public key $(N, e) = (11068562742977, 10543583750987)$. Use the Wiener attack to determine the private key.

Exercise 24.5.5 Show how to perform the Wiener attack when $\varphi(N)$ is replaced by $\lambda(N)$. What is the bound on the size of d for which the attack works?

Exercise 24.5.6 Let $(N, e) = (63875799947551, 4543741325953)$ be an RSA public key where $N = pq$ with $\gcd(p-1, q-1) > 2$ and small private exponent d such that $ed \equiv 1 \pmod{\lambda(N)}$. Use the Wiener attack to find d.

Exercise 24.5.7 Show that one can prevent the Wiener attack by adding a sufficiently large multiple of $\varphi(N)$ to e.

Wiener's result has been extended in several ways. Dujella [170] and Verheul and van Tilborg [557] show how to extend the range of d while still using Euclid's algorithm. Their algorithms are exponential time. Boneh and Durfee [73] used lattices to extend the attack to $d < N^{0.284}$ and, with significant further work, extended the range to $d < N^{0.292}$. Blömer and May [66] give a simpler formulation of the Boneh–Durfee attack for $d < N^{0.284}$.

24.5.2 Small CRT private exponents

As mentioned in Section 24.1.1, a common way to speed up RSA decryption is to use the Chinese remainder theorem. Indeed, one can choose the CRT private exponents d_p and d_q to be small (subject to $d_p \equiv d_q \pmod{\gcd(p-1, q-1)}$) and define e such that $ed_p \equiv 1 \pmod{p-1}$ and $ed_q \equiv 1 \pmod{q-1}$. Of course, one should take $d_p \not\equiv d_q$, or else one can just apply the Wiener attack. We now show that these values cannot be taken to be too small.

Exercise 24.5.8 Give a brute-force attack on small private CRT exponents.

We now present a "birthday attack", which is attributed to Pinch in [442]. Let d_p be such that $ed_p \equiv 1 \pmod{p-1}$ and $ed_p \not\equiv 1 \pmod{q-1}$. Suppose we know that $1 < d_p < K$

and let $L = \lceil \sqrt{K} \rceil$. Then $d_p = d_0 + Ld_1$ where $0 \le d_0, d_1 < L$ and, for a random integer $1 < \mathsf{m} < N$, one expects

$$\gcd(\mathsf{m}^{ed_0 - 1}\mathsf{m}^{Led_1} - 1, N) = p.$$

The problem is to detect this match. The idea is to use the method of Section 2.16 for evaluating polynomials. So, define

$$G(x) = \prod_{j=0}^{L-1} (\mathsf{m}^{ej-1}x - 1) \pmod{N}.$$

This polynomial has degree L and can be constructed using the method in the proof of Theorem 2.16.1 in $O(M(L)\log(L)M(\log(N)))$ bit operations. The polynomial $G(x)$ requires $L\log_2(N)$ bits of storage.

Now, compute $\mathsf{c} = \mathsf{m}^{Le} \pmod{N}$. We wish to evaluate $G(\mathsf{c}^{d_1}) \pmod{N}$ for each of the candidate values $0 \le d_1 < L$ (to obtain a list of L values). This can be performed using Theorem 2.16.1 in $O(L\log(L)^2\log(\log(L))M(\log(N)))$ bit operations. For each value $G(\mathsf{c}^{d_1}) \pmod{N}$ in the list we can compute

$$\gcd(G(\mathsf{c}^{d_1}), N)$$

to see if we have split N. The total running time of the attack is $\tilde{O}(\sqrt{K})$ bit operations.

Exercise 24.5.9 (Galbraith, Heneghan and McKee [203]) Suppose one chooses private CRT exponents of bit-length n and Hamming weight w. Use the ideas of this section together with those of Section 13.6 to give an algorithm to compute a CRT private exponent, given n and w, with complexity $O(\sqrt{w}W\log(W)^2\log(\log(W))M(\log(N)))$ bit operations where $W = \binom{n/2}{w/2}$.

When e is also small (e.g., when using the key generation method of Exercise 24.1.4) then there are lattice attacks on small CRT private exponents. We refer to Bleichenbacher and May [64] for details.

24.6 Digital signatures based on RSA and Rabin

There are numerous signature schemes based on RSA and Rabin. Due to lack of space we just sketch two schemes in the random oracle model. Hohenberger and Waters [264] have given an RSA signature scheme in the standard model whose security relies only on the strong-RSA assumption.

24.6.1 Full domain hash

A simple way to design RSA signatures that are secure in the random oracle model is to assume each user has a hash function $H : \{0, 1\}^* \to (\mathbb{Z}/N\mathbb{Z})^*$ where N is their public key.[9] Such a hash function is called a **full domain hash**, since the hash output is the entire domain of the RSA trapdoor permutation. Constructing such a hash function is not completely trivial; we refer to Section 3.6. The signature on a message m in this case is $\mathsf{s} = H(\mathsf{m})^d \pmod{N}$. These ideas were formalised by Bellare and Rogaway, but we present the slightly improved security result due to Coron.

Theorem 24.6.1 *RSA signatures with full domain hash (**FDH-RSA**) have UF-CMA security in the random oracle model (i.e., where the full domain hash function is replaced by a random oracle) if the RSA problem is hard.*

Proof (Sketch) Let A be an perfect[10] adversary playing the UF-CMA game. We build a simulator that takes an instance (N, e, y) of the RSA problem and, using A as a subroutine, tries to solve the RSA problem.

The simulator in this case starts by running the adversary A on input (N, e). The adversary will make queries to the hash function H, and will make decryption queries. The adversary will eventually output a pair $(\mathsf{m}^*, \mathsf{s}^*)$ such that s^* is a valid signature on m^*. To explain the basic idea of the simulator we remark that if one could arrange that $H(\mathsf{m}^*) = y$ then $(\mathsf{s}^*)^e \equiv y \pmod{N}$ and the RSA instance is solved.

The simulator simulates the random oracle H in the following way. First, the simulator will maintain a list of pairs $(\mathsf{m}, H(\mathsf{m}))$ where m was a query to the random oracle and $H(\mathsf{m}) \in (\mathbb{Z}/N\mathbb{Z})^*$ was the value returned. This list is initially empty. For each query m to the random oracle the simulator first checks if m has already been queried and, if so, responds with the same value $H(\mathsf{m})$. If not, the simulator chooses a random element $1 < r < N$, computes $\gcd(r, N)$ (and if this is not 1 then factors N, solves the RSA instance and halts), computes $z = r^e \pmod{N}$ and with some probability $1 - p$ (we determine p at the end of the proof) returns z as $H(\mathsf{m})$ and with probability p returns $yz \pmod{N}$. The information $(\mathsf{m}, H(\mathsf{m}), r)$ is stored.

When the simulator is asked by the adversary to sign a message m it performs the following: first, it computes $H(\mathsf{m})$ and the corresponding value r. If $H(\mathsf{m}) = r^e \pmod{N}$ then the simulator returns $\mathsf{s} = r$. If $H(\mathsf{m}) = yr^e \pmod{N}$ then the simulator fails.

Eventually, the adversary outputs a pair $(\mathsf{m}^*, \mathsf{s}^*)$. If $H(\mathsf{m}^*) = yr^e \pmod{N}$ where r is known to the simulator, and if $(\mathsf{s}^*)^e \equiv H(\mathsf{m}^*) \pmod{N}$, then $y = (\mathsf{s}^* r^{-1})^e \pmod{N}$ and so the simulator returns $\mathsf{s}^* r^{-1} \pmod{N}$. Otherwise, the simulator fails.

To complete the proof it is necessary to argue that the simulator succeeds with non-negligible probability. If the adversary makes q_S signing queries then the probability that the simulator can answer all of them is $(1 - p)^{q_S}$. The probability that the message m^* corresponds to a random oracle query that allows us to solve the RSA problem is p. Hence, the probability of success is (ignoring some other negligible factors) $(1 - p)^{q_S} p$. Assume that $q_S \geq 1$ and that q_S is known (in practice, one can easily learn a rough estimate of q_S by experimenting with the adversary A). Choose $p = 1/q_S$ so that the probability of success

[9] In practice, one designs $H : \{0, 1\}^* \to \{0, 1, \ldots, N - 1\}$ since the probability that a random element of $\mathbb{Z}/N\mathbb{Z}$ does not lie in $(\mathbb{Z}/N\mathbb{Z})^*$ is negligible.

[10] The proof in the general case, where the adversary succeeds with non-negligible probability ϵ, requires minor modifications.

is $(1 - 1/q_S)^{q_S}\frac{1}{q_S}$, which tends to $1/(eq_S)$ for large q_S (where $e = 2.71828\ldots$). Since a polynomial-time adversary can only make polynomially many signature queries the result follows. We refer to Coron [137] for all the details. □

One problem with the full domain hash RSA scheme is the major loss of security (by a factor of q_S) in Theorem 24.6.1. In other words, the reduction is not tight. This can be avoided by including an extra random input to the hash function. In other words, an RSA signature is $(\mathsf{s}_1, \mathsf{s}_2)$ such that $\mathsf{s}_2^e \equiv H(\mathsf{m} \| \mathsf{s}_1) \pmod{N}$. Then, when the simulator is asked to output a signature on message m, it can choose a "fresh" value s_1^* and define $H(\mathsf{m}\|\mathsf{s}_1^*) = (\mathsf{s}_2^*)^e \pmod{N}$ as above. This approach avoids previous queries to $H(\mathsf{m}\|\mathsf{s}_1)$ with high probability. Hence, the simulator can answer "standard" hash queries with $yr^e \pmod{N}$ and "special" hash queries during signature generation with $r^e \pmod{N}$. This scheme is "folklore", but the details are given in Appendix A of Coron [138]. The drawback is that the extra random value s_1 must be included as part of the signature. The **PSS** signature padding scheme was designed by Bellare and Rogaway [38] precisely to allow extra randomness in this way without increasing the size of the signature. We refer to [138] for a detailed analysis of RSA signatures using the PSS padding.

Exercise 24.6.2 Give a security proof for the RSA full domain hash signature scheme with verification equation $H(\mathsf{m}\|\mathsf{s}_1) = \mathsf{s}_2^e \pmod{N}$.

The above results are all proved in the random oracle model. Paillier [425] has given some evidence that full domain hash RSA and RSA using PSS padding cannot be proved secure in the standard model. Theorem 1 of [425] states that if one has a "black box" reduction from the RSA problem to selective forgery for a signature scheme under a passive attack, then, under an adaptive chosen-message attack, one can, in polynomial-time, forge any signature for any message.

24.6.2 Secure Rabin–Williams signatures in the random oracle model

In this section we give a tight security result, due to Bernstein [44], for Rabin signatures. We assume throughout that $N = pq$ is a **Williams integer**; in other words, a product of primes $p \equiv 3 \pmod{8}$ and $q \equiv 7 \pmod{8}$ (such integers were discussed in Section 24.2.1). We assume that $H : \{0,1\}^* \to \{0,1\}^\kappa$ is a cryptographic hash function (which will be modelled as a random oracle) where $2^\kappa < N < 2^{\kappa+O(\log(\kappa))}$.

Exercise 24.6.3 Suppose $p \equiv 3 \pmod{8}$ and $q \equiv 7 \pmod{8}$ are primes and $N = pq$. Then $(\frac{-1}{p}) = (\frac{-1}{q}) = (\frac{2}{p}) = -1$ while $(\frac{2}{q}) = 1$. Show that, for any integer $h \in (\mathbb{Z}/N\mathbb{Z})^*$, there are unique integers $e \in \{1, -1\}$ and $f \in \{1, 2\}$ such that efh is a square modulo N.

The signature scheme for public key N is as follows. For a message $\mathsf{m} \in \{0,1\}^*$ one computes $H(\mathsf{m})$ and interprets it as an integer modulo N (with overwhelming probability, $H(\mathsf{m}) \in (\mathbb{Z}/N\mathbb{Z})^*$). The signer determines the values e and f as in Exercise 24.6.3 and determines the four square roots s_1, s_2, s_3, s_4 satisfying $s_i^2 \equiv H(\mathsf{m})ef \pmod{N}$. The signer then deterministically chooses one of the values s_i (for example, by ordering the roots as integers $s_1 < s_2 < s_3 < s_4$ and then generating an integer $i \in \{1, 2, 3, 4\}$ using a pseudorandom number generator on input m). The signature is the triple $\mathsf{s} = (e, f, (ef)^{-1} s_i \pmod{N})$. It is crucially important that, if one signs the same message twice, then the same signature is output. To verify a signature $\mathsf{s} = (e, f, s)$ for public key N and message m one computes $H(\mathsf{m})$ and then checks that $efs^2 \equiv H(\mathsf{m}) \pmod{N}$.

Exercise 24.6.4 Show that if a signer outputs two signatures (e_1, f_1, s_1) and (e_2, f_2, s_2) for the same message m such that $s_1 \not\equiv \pm s_2 \pmod{N}$ then one can factor the modulus.

Exercise 24.6.5 Show that it is not necessary to compute the Jacobi symbol $(\frac{H(\mathsf{m})}{N})$ when generating Rabin–Williams signatures as above. Instead, one can compute $H(\mathsf{m})^{(p+1)/4} \pmod{p}$ and $H(\mathsf{m})^{(q+1)/4} \pmod{q}$ as is needed to compute the s_i and determine e and f with only a little additional computation.

Theorem 24.6.6 *The Rabin–Williams signature scheme sketched above has UF-CMA security in the random oracle model (i.e., if H is replaced by a random oracle) if factoring Williams integers is hard and if the pseudorandom generator is indistinguishable from a random function.*

Proof (Sketch) Let A be a perfect adversary against the Rabin–Williams signature scheme and let N be a Williams integer to be factored. The simulator runs the adversary A on N.

The simulator must handle the queries made by A to the random oracle. To do this it maintains a list of hash values, which is initially empty. When A queries H on m, the simulator first checks whether m appears on the list of hash values, and, if it does, responds with the same value as previously. If H has not been previously queried on m, the simulator chooses random $s \in (\mathbb{Z}/N\mathbb{Z})^*$, $e \in \{-1, 1\}$, $f \in \{1, 2\}$ and computes $h = efs^2 \pmod{N}$. If $0 \le h < 2^\kappa$ then return h and store (m, e, f, s, h) in the list. If h is too big then repeat with a different choice for (s, e, f). Since $N < 2^{\kappa + O(\log(\kappa))}$ the expected number of trials is polynomial in $\log(N)$.

When A makes a signature query on m, the simulator first queries $H(\mathsf{m})$ and gets the values (e, f, s) from the hash list such that $H(\mathsf{m}) \equiv efs^2 \pmod{N}$. The simulator can therefore answer with (e, f, s), which is a valid signature. (It is necessary to show that the values (e, f, s) output in this way are indistinguishable from the values output in the real cryptosystem, and this requires that the pseudorandom choice of s from among the four possible roots be computationally indistinguishable from random.)

Finally, A outputs a signature (e^*, f^*, s^*) on a message m^*. Recalling the values (e, f, s) from the construction of $H(\mathsf{m}^*)$ we have $e^* = e$, $f^* = f$ and $(s^*)^2 \equiv s^2 \pmod{N}$. With probability $1/2$ we can factor N as $\gcd(N, s^* - s)$. We refer to Section 6 of Bernstein [44] for the full details. □

Exercise 24.6.7 Show that if one can find a collision for the hash function H in Bernstein's variant of Rabin–Williams signatures and one has access to a signing oracle then one can factor N with probability $1/2$.

Exercise 24.6.8 Suppose the pseudorandom function used to select the square root in Rabin–Williams signatures is a function of $H(\mathsf{m})$ rather than m. Show that, in contrast to Exercise 24.6.7, finding a collision in H no longer leads to an algorithm to split N. On the other hand, show that if one can compute a preimage for H and has access to a signing oracle then one can factor N with probability $1/2$.

Exercise 24.6.9 Adapt the proof of Theorem 24.6.6 to the case where H has full domain output.

Exercise 24.6.10 Adapt the proof of Theorem 24.6.1 to the case where $H : \{0,1\}^* \rightarrow \{0,1\}^\kappa$ where $2^\kappa < N < 2^{\kappa+O(\log(\kappa))}$.

24.7 Public key encryption based on RSA and Rabin

To prevent algebraic attacks such as those mentioned in Sections 1.2 and 24.4 it is necessary to use randomised padding schemes for encryption with RSA. We do not have space to discuss secure schemes, but we make some general remarks.

Three goals of padding schemes for RSA are: to introduce randomness; to ensure that algebraic relationships among messages do not lead to algebraic relationships between the corresponding ciphertexts; to ensure that random elements of $\mathbb{Z}_N^*$ do not correspond to valid ciphertexts (so access to a decryption oracle is not useful).

Bellare and Rogaway [37] designed the **OAEP** (Optimal Asymmetric Encryption Padding) scheme for secure encryption using RSA. We refer to Section 13.2.3 of Katz and Lindell [300], Section 5.9.2 of Stinson [532] or Section 17.3.2 of Smart [513] for details. The word "optimal" refers to the length of the padded message compared with the length of the original message: the idea is that the additional bits in an OAEP padding of a message are as small as can be.

The security of RSA-OAEP has a complicated history: Shoup [496] found a flaw in the original security proof (though not an attack on the scheme). Shoup also gave a variant of OAEP (called OAEP+) together with a security proof in the random oracle model. Fujisaki, Okamoto, Pointcheval and Stern [198] were able to give a proof of the security of RSA using OAEP in the random oracle model by exploiting the random self-reducibility of RSA. The reader is referred to these papers, and the excellent survey by Gentry [230] for the details.

Boneh [69] has considered a padding scheme (which he calls **SAEP**) that is simpler than OAEP and is suitable for encrypting short messages with Rabin. Note that the restriction to short messages is not a serious problem in practice, since public key encryption is mainly used to transport symmetric keys; nevertheless, this restriction means that SAEP is not an "optimal" padding. The proof is again in the random oracle model.

Several methods are known for IND-CCA secure encryption based on factoring in the standard model. Cramer and Shoup [148] have given an encryption scheme, using the universal hash proof systems framework, based on the Paillier cryptosystem. Their scheme (using $N = pq$ where p and q are safe primes) has IND-CCA security in the standard model if the composite residuosity problem is hard (see Definition 24.3.4). Hofheinz and Kiltz [263] have shown that the Elgamal encryption scheme of Section 23.1, when implemented in a certain subgroup of $(\mathbb{Z}/N\mathbb{Z})^*$, has IND-CCA security in the random oracle model if factoring is hard, and in the standard model if a certain higher residuosity assumption holds.

Finally, we mention that it is standard in cryptography to perform encryption using a **hybrid** system. A typical scenario is to use public key cryptography to encrypt a random session key $K_1 \| K_2$. The document is encrypted using a symmetric encryption scheme such as AES with the key K_1. Finally a MAC (message authentication code), with key K_2, of the symmetric ciphertext is appended to ensure the integrity of the transmission.

PART VII

ADVANCED TOPICS IN ELLIPTIC AND HYPERELLIPTIC CURVES

25

Isogenies of elliptic curves

Isogenies are a fundamental object of study in the theory of elliptic curves. The definition and basic properties were given in Sections 9.6 and 9.7. In particular, they are group homomorphisms.

Isogenies are used in algorithms for point counting on elliptic curves and for computing class polynomials for the complex multiplication (CM) method. They have applications to cryptanalysis of elliptic curve cryptosystems. They also have constructive applications: prevention of certain side-channel attacks; computing distortion maps for pairing-based cryptography; designing cryptographic hash functions; relating the discrete logarithm problem on elliptic curves with the same number of points. We do not have space to discuss all these applications.

The purpose of this chapter is to present algorithms to compute isogenies from an elliptic curve. The most important result is Vélu's formulae, which compute an isogeny given an elliptic curve and a kernel subgroup G. We also sketch the various ways to find an isogeny given an elliptic curve and the j-invariant of an elliptic curve ℓ-isogenous to E. Once these algorithms are in place we briefly sketch Kohel's results, the isogeny graph and some applications of isogenies. Due to lack of space we are unable to give proofs of most results.

Algorithms for computing isogenies on Jacobians of curves of genus 2 or more are much more complicated than in the elliptic case. Hence, we do not discuss them in this book.

25.1 Isogenies and kernels

Let $E : y^2 + a_1xy + a_3 = x^3 + a_2x^2 + a_4x + a_6$ be an elliptic curve over $\Bbbk$. Recall from Section 9.6 that a non-zero isogeny $\phi : E \to \widetilde{E}$ over $\Bbbk$ of degree d is a morphism of degree d such that $\phi(\mathcal{O}_E) = \mathcal{O}_{\widetilde{E}}$. Such a map is automatically a group homomorphism and has kernel of size dividing d.

Theorem 9.7.5 states that a separable isogeny $\phi : E \to \widetilde{E}$ over $\Bbbk$ may be written in the form

$$\phi(x, y) = (\phi_1(x), cy\phi_1(x)' + \phi_3(x)) \tag{25.1}$$

where $\phi_1(x), \phi_3(x) \in \Bbbk(x)$, where $\phi_1(x)' = d\phi_1(x)/dx$ is the (formal) derivative of the rational function $\phi_1(x)$, where $c \in \overline{\Bbbk}^*$ is a non-zero constant and where (writing $\widetilde{a}_i$ for the coefficients of $\widetilde{E}$)

$$2\phi_3(x) = -\widetilde{a}_1\phi_1(x) - \widetilde{a}_3 + (a_1x + a_3)\phi_1(x)'.$$

Lemma 9.6.13 showed that if $\phi_1(x) = a(x)/b(x)$ in equation (25.1) then the degree of ϕ is $\max\{\deg_x(a(x)), \deg_x(b(x))\}$. The kernel of an isogeny ϕ with $\phi_1(x) = a(x)/b(x)$ is $\{\mathcal{O}_E\} \cup \{P = (x_P, y_P) \in E(\overline{\Bbbk}) : b(x_P) = 0\}$. The kernel of a separable isogeny of degree d has d elements.

Let E be an elliptic curve over a field $\Bbbk$ and G a finite subgroup of $E(\overline{\Bbbk})$ that is defined over $\Bbbk$. Theorem 9.6.19 states that there is a unique elliptic curve $\widetilde{E}$ (up to isomorphism) and a separable isogeny $\phi : E \to \widetilde{E}$ over $\Bbbk$ such that $\ker(\phi) = G$. We sometimes write $\widetilde{E} = E/G$. Let ℓ be a prime such that $\gcd(\ell, \mathrm{char}(\Bbbk)) = 1$. Since $E[\ell]$ is isomorphic (as a group) to the product of two cyclic groups, there are $\ell + 1$ different subgroups of $E[\ell]$ of order ℓ. It follows that there are $\ell + 1$ isogenies of degree ℓ, not necessarily defined over $\Bbbk$, from E to other curves (some of these isogenies may map to the same image curve).

As implied by Theorem 9.6.18 and discussed in Exercise 9.6.20, an isogeny is essentially determined by its kernel. We say that two separable isogenies $\phi_1, \phi_2 : E \to \widetilde{E}$ are **equivalent isogenies** if $\ker(\phi_1) = \ker(\phi_2)$.

Exercise 25.1.1 Let $\phi : E \to \widetilde{E}$ be a separable isogeny. Show that if $\lambda \in \mathrm{Aut}(\widetilde{E})$ then $\lambda \circ \phi$ is equivalent to ϕ. Explain why $\phi \circ \lambda$ is not necessarily equivalent to ϕ for $\lambda \in \mathrm{Aut}(E)$.

Theorem 25.1.2 shows that isogenies can be written as "chains" of prime-degree isogenies. Hence, in practice, one can restrict to studying isogenies of prime degree. This observation is of crucial importance in the algorithms.

Theorem 25.1.2 *Let E and $\widetilde{E}$ be elliptic curves over $\Bbbk$ and let $\phi : E \to \widetilde{E}$ be a separable isogeny that is defined over $\Bbbk$. Then $\phi = \phi_1 \circ \cdots \circ \phi_k \circ [n]$ where $\phi_1, \ldots, \phi_k$ are isogenies of prime degree that are defined over $\Bbbk$ and $\deg(\phi) = n^2 \prod_{i=1}^{k} \deg(\phi_i)$.*

Proof Theorem 9.6.19 states that ϕ is essentially determined by its kernel subgroup G and that ϕ is defined over $\Bbbk$ if and only if G is. We will also repeatedly use Theorem 9.6.18, which states that an isogeny $\phi : E \to \widetilde{E}$ defined over $\Bbbk$ factors as $\phi = \phi_2 \circ \phi_1$ (where $\phi_1 : E \to E_1$ and $\phi_2 : E_1 \to \widetilde{E}$ are isogenies over $\Bbbk$) whenever $\ker(\phi)$ has a subgroup $G = \ker(\phi_1)$ defined over $\Bbbk$.

First, let n be the largest integer such that $E[n] \subseteq G = \ker(\phi)$ and note that $\phi = \phi' \circ [n]$ where $[n] : E \to E$ is the usual multiplication by n map. Set $i = 1$, define $E_0 = E$ and set $G = G/E[n]$. Now, let $\ell \mid \#G$ be a prime and let $P \in G$ have prime order ℓ. There is an isogeny $\phi_i : E_{i-1} \to E_i$ of degree ℓ with kernel $\langle P \rangle$. Let $\sigma \in \mathrm{Gal}(\overline{\Bbbk}/\Bbbk)$. Since $\sigma(P) \in G$ but $E[\ell] \not\subseteq G$ it follows that $\sigma(P) \in \langle P \rangle$ and so $\langle P \rangle$ is defined over $\Bbbk$. Hence, it follows that ϕ_i is defined over $\Bbbk$. Replace G by $\phi_i(G) \cong G/\langle P \rangle$ and repeat the argument. □

Exercise 25.1.3 How must the statement of Theorem 25.1.2 be modified if the requirement that ϕ be separable is removed?

Exercise 25.1.4 Let E be an ordinary elliptic curve. Let $\phi_1 : E \to E_1$ and $\phi_2 : E_1 \to E_2$ be non-zero separable isogenies over $\Bbbk$ of coprime degrees e and f respectively. Show that there is an elliptic curve $\widetilde{E}_1$ over $\Bbbk$, and a pair of non-zero separable isogenies $\psi_1 : E \to \widetilde{E}_1$ and $\psi_2 : \widetilde{E}_1 \to E_2$ of degrees f and e respectively, such that $\phi_2 \circ \phi_1 = \psi_2 \circ \psi_1$.

25.1.1 Vélu's formulae

We now present explicit formulae, due to Vélu [613], for computing a separable isogeny from an elliptic curve E with given kernel G. These formulae work in any characteristic. As motivation for Vélu's formulae, we now revisit Example 9.6.9.

Example 25.1.5 Let $E : y^2 = x^3 + x$ and consider the subgroup of order 2 generated by the point $(0, 0)$. From Example 9.2.4 we know that the translation by $(0, 0)$ map is given by

$$\tau_{(0,0)}(x, y) = \left(\frac{1}{x}, \frac{-y}{x^2} \right).$$

Hence, it follows that functions invariant under this translation map include

$$X = x + 1/x = (x^2 + 1)/x, \qquad Y = y - y/x^2 = y(x^2 - 1)/x^2.$$

One can compute that $X^3 = (x^6 + 3x^4 + 3x^2 + 1)/x^3$ and so

$$\begin{aligned} Y^2 &= y^2(x^2 - 1)^2/x^4 \\ &= (x^6 - x^4 - x^2 + 1)/x^3 \\ &= X^3 - 4X. \end{aligned}$$

It follows that the map

$$\phi(x, y) = \left(\frac{x^2 + 1}{x}, y\frac{x^2 - 1}{x^2} \right)$$

is an isogeny from E to $\widetilde{E} : Y^2 = X^3 - 4X$.

We remark that ϕ can also be written as

$$\phi(x, y) = \left(\frac{y^2}{x^2}, y\frac{x^2 - 1}{x^2} \right)$$

and can be written projectively as

$$\begin{aligned} \phi(x : y : z) &= (x(x^2 + z^2) : y(x^2 - z^2) : x^2 z) \\ &= (y(x^2 + z^2) : xy^2 - x^2 z - z^3 : xyz) \\ &= (y^2 z : y(x^2 - z^2) : x^2 z) \\ &= (xy^2 : y(y^2 - 2xz) : x^3). \end{aligned}$$

Theorem 25.1.6 *(Vélu) Let E be an elliptic curve over $\Bbbk$ defined by the polynomial*

$$F(x, y) = x^3 + a_2 x^2 + a_4 x + a_6 - \left(y^2 + a_1 xy + a_3 y \right) = 0.$$

Let G be a finite subgroup of $E(\overline{\Bbbk})$. Let G_2 be the set of points in $G - \{\mathcal{O}_E\}$ of order 2 and let G_1 be such that $\#G = 1 + \#G_2 + 2\#G_1$ and

$$G = \{\mathcal{O}_E\} \cup G_2 \cup G_1 \cup \{-Q : Q \in G_1\}.$$

Write

$$F_x = \frac{\partial F}{\partial x} = 3x^2 + 2a_2 x + a_4 - a_1 y \text{ and } F_y = \frac{\partial F}{\partial y} = -2y - a_1 x - a_3.$$

For a point $Q = (x_Q, y_Q) \in G_1 \cup G_2$ define the quantities

$$u(Q) = \left(F_y(Q)\right)^2 = \left(-2y_Q - a_1 x_Q - a_3\right)^2$$

and

$$t(Q) = \begin{cases} F_x(Q) & \text{if } Q \in G_2 \\ 2F_x(Q) - a_1 F_y(Q) & \text{if } Q \in G_1. \end{cases}$$

Note that if $Q \in G_2$ then $F_y(Q) = 0$ and so $u(Q) = 0$.

Define

$$t(G) = \sum_{Q \in G_1 \cup G_2} t(Q) \quad \text{and} \quad w(G) = \sum_{Q \in G_1 \cup G_2} (u(Q) + x_Q t(Q))$$

and set

$$A_1 = a_1,\ A_2 = a_2,\ A_3 = a_3,\ A_4 = a_4 - 5t(G),\ A_6 = a_6 - (a_1^2 + 4a_2)t(G) - 7w(G).$$

Then the map $\phi : (x, y) \mapsto (X, Y)$ where

$$X = x + \sum_{Q \in G_1 \cup G_2} \frac{t(Q)}{x - x_Q} + \frac{u(Q)}{(x - x_Q)^2}$$

and

$$Y = y - \sum_{Q \in G_1 \cup G_2} u(Q)\frac{2y + a_1 x + a_3}{(x - x_Q)^3} + t(Q)\frac{a_1(x - x_Q) + y - y_Q}{(x - x_Q)^2} + \frac{a_1 u(Q) - F_x(Q)F_y(Q)}{(x - x_Q)^2}$$

is a separable isogeny from E to

$$\widetilde{E} : Y^2 + A_1 XY + A_3 Y = X^3 + A_2 X^2 + A_4 X + A_6$$

with kernel G. Further, ϕ satisfies

$$\phi^*\left(\frac{dX}{2Y + A_1 X + A_3}\right) = \frac{dx}{2y + a_1 x + a_3}.$$

Proof (Sketch) The basic idea (as used in Example 25.1.5) is that the function

$$X(P) = \sum_{Q \in G} x(P + Q)$$

on E is invariant under G (in the sense that $X = X \circ \tau_Q$ for all $Q \in G$) and so can be considered as "defined on E/G". To simplify some calculations sketched below it turns out to be more convenient to subtract the constant $\sum_{Q \in G - \{\mathcal{O}_E\}} x(Q)$ from X. (Note that $x(Q) = x_Q$.) Let $t_\infty = -x/y$ be a uniformiser on E at $\mathcal{O}_E$ (one could also take $t_\infty = x/y$, but this makes the signs more messy). The function x can be written as $t_\infty^{-2} - a_1 t_\infty^{-1} - a_2 - a_3 t_\infty - (a_1 a_3 + a_4) t_\infty^2 - \cdots$ (for more details about the expansions of x, y and ω_E in terms of power series see Section IV.1 of Silverman [505]). It follows that $X = t_\infty^{-2} - a_1 t_\infty^{-1} - \cdots$ and so $v_{\mathcal{O}_E}(X) = -2$.

One can also show that $y = -t_\infty^{-3} - a_1 t_\infty^{-2} - a_2 t_\infty^{-1} - \cdots$. The function $Y(P) = \sum_{Q \in G} y(P + Q)$ is invariant under G and has $v_{\mathcal{O}_E}(Y) = -3$. One can therefore show (see Section 12.3 of Washington [560]) that the subfield $\Bbbk(X, Y)$ of $\Bbbk(x, y)$ is the function field of an elliptic curve $\widetilde{E}$ (Washington [560] does this in Lemma 12.17 using the Hurwitz genus formula). The map $\phi : (x, y) \mapsto (X, Y)$ is therefore an isogeny of elliptic curves. By considering the expansions in terms of t_∞ one can show that the equation for the image curve is $Y^2 + A_1 XY + A_3 Y = X^3 + A_2 X^2 + A_4 X + A_6$ where the coefficients A_i are as in the statement of the Theorem.

Now, let $\omega_E = dx/(2y + a_1 x + a_3)$. One has $dx = (-2t_\infty^{-3} + a_1 t_\infty^{-2} + \cdots)dt_\infty$ and $2y + a_1 x + a_3 = -2t_\infty^{-3} - a_1 t_\infty^{-2} + \cdots$ and so $\omega_E = (1 - a_1 t_\infty + \cdots)dt_\infty$. Similarly, $\phi^*(\omega_{\widetilde{E}}) = d(X \circ \phi)/(2Y \circ \phi + A_1 X \circ \phi + A_3) = d(t_\infty^{-2} + a_1 t_\infty^{-1} + \cdots)/(-2t_\infty^{-3} + \cdots) = (1 + \cdots)dt_\infty$. It follows that the isogeny is separable and that $\phi^*(\omega_{\widetilde{E}}) = f\omega_E$ for some function f. Further, $\mathrm{div}(\omega_E) = 0$ and $\mathrm{div}(\phi^*(\omega_{\widetilde{E}})) = \phi^*(\mathrm{div}(\omega_{\widetilde{E}})) = 0$ (by Lemma 8.5.36, since ϕ is unramified[1]) and so $\mathrm{div}(f) = 0$. It follows from the power series expansions that $f = 1$ as required.

Write the isogeny as $\phi(x, y) = (\phi_1(x), y\phi_2(x) + \phi_3(x))$. By Theorem 9.7.5 the isogeny is determined by $\phi_1(x)$ (for the case $\mathrm{char}(\Bbbk) = 2$ see Exercise 9.7.6). Essentially, one only has to prove Vélu's formula for $\phi_1(x)$; we do this now. First, change the definition of X to

$$X(P) = x_P + \sum_{Q \in G - \{\mathcal{O}_E\}} (x_{P+Q} - x_Q)$$

where P is a "generic point" (i.e., $P = (x_P, y_P)$ where x_P and y_P are variables) on the elliptic curve and $Q \in G - \{\mathcal{O}_E\}$. Let $F(x, y)$ be as in the statement of the theorem and let $y = l(x)$ be the equation of the line through P and Q (so that $l(x) = \lambda(x - x_Q) + y_Q$ where $\lambda = (y_P - y_Q)/(x_P - x_Q)$). Define

$$F_1(x) = F(x, l(x)) = (x - x_Q)(x - x_P)(x - x_{P+Q}).$$

Further

$$\frac{\partial F_1}{\partial x}(Q) = (x_Q - x_P)(x_Q - x_{P+Q})$$

and

$$\frac{\partial F_1}{\partial x} = \frac{\partial F}{\partial x} + \frac{\partial F}{\partial y}\frac{\partial l}{\partial x} = F_x + F_y \cdot \lambda.$$

[1] This was already discussed in Section 9.6. One can directly see that separable isogenies are unramified since if $\phi(P_1) = P_2$ then the set of preimages under ϕ of P_2 is $\{P_1 + Q : Q \in \ker(\phi)\}$.

Hence, $x_{P+Q} - x_Q = F_x(Q)/(x_P - x_Q) + (y_P - y_Q)F_y(Q)/(x_P - x_Q)^2$. One now considers two cases: when $[2]Q = \mathcal{O}_E$ then $F_y(Q) = 0$. When $[2]Q \neq \mathcal{O}_E$ then it is convenient to consider

$$x_{P+Q} - x_Q + x_{P-Q} - x_{-Q}.$$

Now, $x_{-Q} = x_Q$, $y_{-Q} = y_Q + F_y(Q)$, $F_x(-Q) = F_x(Q) - a_1 F_y(Q)$ and $F_y(-Q) = -F_y(Q)$. The formula for $\phi_1(x)$ follows.

Now we sketch how to obtain the formula for the Y-coordinate of the isogeny in the case $\mathrm{char}(\Bbbk) \neq 2$. Note that $\phi_1(x) = x + \sum_Q [t(Q)/(x - x_Q) + u(Q)/(x - x_Q)^2]$ and so $\phi_1(x)' = 1 - \sum_Q [t(Q)/(x - x_Q)^2 + 2u_Q/(x - x_Q)^3]$. Using $\phi_3(x) = (-A_1\phi_1(x) - A_3 + (a_1x + a_3)\phi_1(x)')/2$ one computes

$$\begin{aligned}
y\phi_2(x) + \phi_3(x) &= y\phi_1(x)' + \phi_3(x)\\
&= y\left(1 - \sum_Q t(Q)/(x - x_Q)^2 + 2u(Q)/(x - x_Q)^3\right) - (a_1x + a_3)/2\\
&\quad - a_1 \sum_Q [t(Q)/(x - x_Q)^2 + 2u(Q)/(x - x_Q)^3] + (a_1x + a_3)/2\\
&\quad + (a_1x + a_3)\sum_Q [-t(Q)/(x - x_Q)^2 - 2u(Q)/(x - x_Q)^3]\\
&= y - \sum_Q \left[t(Q)\frac{y + a_1(x - x_Q) - y_Q}{(x - x_Q)^2} + u(Q)\frac{2y + a_1x + a_3}{(x - x_Q)^3} \right.\\
&\quad \left. + \frac{t(Q)((a_1x_Q + a_3)/2 + y_Q) + a_1u(Q)/2}{(x - x_Q)^2}\right].
\end{aligned}$$

It suffices to show that the numerator of the final term in the sum is equal to $a_1u(Q) - F_x(Q)F_y(Q)$. However, this follows easily by noting that $(a_1x_Q + a_3)/2 + y_Q = -F_y(Q)/2$, $u(Q) = F_y(Q)^2$ and using the facts that $F_y(Q) = 0$ when $[2]Q = \mathcal{O}_E$ and $t(Q) = 2F_x(Q) - a_1F_y(Q)$ otherwise. □

Corollary 25.1.7 *Let E be an elliptic curve defined over $\Bbbk$ and G a finite subgroup of $E(\overline{\Bbbk})$ that is defined over $\Bbbk$. Then there is an elliptic curve $\widetilde{E} = E/G$ defined over $\Bbbk$ and an isogeny $\phi : E \to \widetilde{E}$ defined over $\Bbbk$ with* $\ker(\phi) = G$.

Proof It suffices to show that the values $t(G)$, $w(G)$ and the rational functions X and Y in Theorem 25.1.6 are fixed by any $\sigma \in \mathrm{Gal}(\overline{\Bbbk}/\Bbbk)$. □

Corollary 25.1.8 *Let $\phi : E \to \widetilde{E}$ be a separable isogeny of odd degree ℓ between elliptic curves over $\Bbbk$. Write $\phi(x, y) = (\phi_1(x), \phi_2(x, y))$, where $\phi_1(x)$ and $\phi_2(x, y)$ are rational functions. Then $\phi_1(x, y) = u(x)/v(x)^2$, where $\deg(u(x)) = \ell$ and $\deg(v(x)) = (\ell - 1)/2$. Also, $\phi_2(x, y) = (yw_1(x) + w_2(x))/v(x)^3$, where $\deg(w_1(x)) \leq 3(\ell - 1)/2$ and $\deg(w_2(x)) \leq (3\ell - 1)/2$.*

Exercise 25.1.9 Prove Corollary 25.1.8.

Definition 25.1.10 An isogeny $\phi : E \to \widetilde{E}$ is **normalised** if $\phi^*(\omega_{\widetilde{E}}) = \omega_E$.

Vélu's formulae give a normalised isogeny. Note that normalised isogenies are incompatible with Theorem 9.7.2 (which, for example, implies $[m]^*(\omega_E) = m\omega_E$). For this reason, in many situations one needs to take an isomorphism from $\widetilde{E}$ to obtain the desired isogeny. Example 25.1.12 shows how this works.

Exercise 25.1.11 Let $\phi : E \to \widetilde{E}$ be an isogeny given by rational functions as in equation (25.1). Show that ϕ is normalised if and only if $c = 1$.

Example 25.1.12 Let $E : y^2 + xy + 3y = x^3 + 2x^2 + 4x + 2$ over $\mathbb{F}_{311}$. Then

$$E[2] = \{\mathcal{O}_E, (-1, -1), (115, 252), (117, 251)\} \subset E(\mathbb{F}_{311}).$$

Let $G = E[2]$. Applying the Vélu formulae one computes $t(G) = 8$, $w(G) = 306$, $A_1 = 1$, $A_2 = 2$, $A_3 = 3$, $A_4 = 275$ and $A_6 = 276$. One can check that E and

$$\widetilde{E} : Y^2 + XY + 3Y^2 = X^3 + 2X^2 + 275X + 276$$

have the same j-invariant, but they are clearly not the same Weierstrass equation. Hence, the Vélu isogeny with kernel $E[2]$ is not the isogeny $[2] : E \to E$.

To recover the map [2] one needs to find a suitable isomorphism from $\widetilde{E}$ to E. The isomorphism will have the form $(X, Y) \mapsto (u^2X + r, u^3Y + su^2X + t)$ where we must have $u = 1/2$ to have the correct normalisation for the action of the isogeny on the invariant differential (see Exercise 25.1.13). One can verify that taking $r = 291$, $s = 233$ and $t = 67$ gives the required isomorphism from $\widetilde{E}$ to E and that the composition of the Vélu isogeny and this isomorphism is the map [2].

Exercise 25.1.13 Show that if $\phi : (x, y) \mapsto (u^2x + r, u^3y + su^2x + t)$ is an isomorphism from E to $\widetilde{E}$ then $\phi^*(\omega_{\widetilde{E}}) = \frac{1}{u}\omega_E$.

Exercise 25.1.14 Determine the complexity of constructing and computing the Vélu isogeny. More precisely, show that if $d = \#G$ and $G \subset E(\mathbb{F}_{q^n})$ then $O(dM(n, q))$ bit operations are sufficient, where $M(n, q) = M(n \log(nq))$ is the number of bit operations to multiply two-degree n polynomials over $\mathbb{F}_q$.

Further, show that if d is an odd prime then $n \leq d - 1$ and so the complexity can be written as $O(d^{2+\epsilon} \log(q)^{1+\epsilon})$ bit operations.

Example 25.1.15 Consider $E : y^2 = x^3 + 2x$ over $\mathbb{F}_{37}$, with $j = 26 \equiv 1728 \pmod{37}$. We have $\#E(\mathbb{F}_{37}) = 2 \cdot 5^2$ so there is a unique point $(0, 0)$ of order 2 over $\mathbb{F}_{37}$ giving a 2-isogeny from E. Using Vélu's formulae, one determines that the image of this isogeny is $E_1 : y^2 = x^3 + 29x$, which also has j-invariant 26 and is isomorphic to E over $\mathbb{F}_{37}$.

Now consider the other points of order 2 on E. Let $\alpha \in \mathbb{F}_{37^2}$ satisfy $\alpha^2 = -2$. The isogeny ϕ_2 with kernel $\{\mathcal{O}_E, (\alpha, 0)\}$ maps to $E_2 : y^2 = x^3 + 28\alpha x$, while the isogeny ϕ_3 with kernel $\{\mathcal{O}_E, (-\alpha, 0)\}$ maps to $E_3 : y^2 = x^3 - 28\alpha x$. Note that there is an isomorphism $\psi : E_2 \to E_3$ over $\mathbb{F}_{37^2}$. We have $\hat{\phi}_2 \circ \phi_2 = \hat{\phi}_3 \circ \phi_3 = [2]$ on E. One can also consider

$\hat{\phi}_3 \circ \psi \circ \phi_2$ on E, which must be an element of $\mathrm{End}(E) = \mathbb{Z}[i]$ of degree 4. One can show that it is $i[2]$ where $i(x, y) = (-x, 31y)$.

Kohel [315] and Dewaghe [158] independently gave formulae for the Vélu isogeny in terms of the coefficients of the polynomial defining the kernel, rather than in terms of the points in the kernel. We give these formulae in Lemma 25.1.16 for the case where G has odd order (they are also given in Section 2.4 of [315]). Since a $\Bbbk$-rational subgroup of an elliptic curve can have points defined over an extension of $\Bbbk$, working with the coefficients of the polynomial can be more efficient than working with the points in G.

Lemma 25.1.16 *Let $E : y^2 + (a_1x + a_3)y = x^3 + a_2x^2 + a_4x + a_6$ be an elliptic curve over $\Bbbk$. Let $G \subseteq E(\overline{\Bbbk})$ be a cyclic group of odd order $2d + 1$. Let $G_1 \subseteq G$ be such that $\#G_1 = d$ and $G = \{\mathcal{O}_E\} \cup G_1 \cup \{-Q : Q \in G_1\}$. Define*

$$\psi(x) = \prod_{Q \in G_1} (x - x_Q) = x^d - s_1x^{d-1} + s_2x^{d-2} + \cdots + (-1)^d s_d \tag{25.2}$$

where the s_i are the ith symmetric polynomials in the roots of $\psi(x)$ (equivalently, in the x-coordinates of elements of G_1). Define $b_2 = a_1^2 + 4a_2$, $b_4 = 2a_4 + a_1a_3$ and $b_6 = a_3^2 + 4a_6$. Then there is an isogeny $\phi : E \to \tilde{E}$, with $\ker(\phi) = G$, of the form $\phi(x, y) = (A(x)/\psi(x)^2, B(x, y)/\psi(x)^3)$ where $A(x)$ and $B(x, y)$ are polynomials. Indeed

$$\begin{aligned}\frac{A(x)}{\psi(x)^2} &= (2d + 1)x - 2s_1 - (4x^3 + b_2x^2 + 2b_4x + b_6)(\psi(x)'/\psi(x))' \\ &\quad - (6x^2 + b_2x + b_4)(\psi(x)'/\psi(x)).\end{aligned}$$

The proof of Lemma 25.1.16 is given as a sequence of exercises.

Exercise 25.1.17 Let the notation be as in Lemma 25.1.16. Let $F_x(Q)$, $F_y(Q)$, $t(Q)$ and $u(Q)$ be as in Theorem 25.1.6. Show that

$$t(Q) = 6x_Q^2 + b_2x_Q + b_4 \quad \text{and} \quad u(Q) = 4x_Q^3 + b_2x_Q^2 + 2b_4x_Q + b_6.$$

Exercise 25.1.18 Let the notation be as in Lemma 25.1.16. Let $F_x(Q)$, $F_y(Q)$, $t(Q)$ and $u(Q)$ be as in Theorem 25.1.6. Show that

$$\begin{aligned}
\frac{x_Q}{x - x_Q} &= \frac{x}{x - x_Q} - 1\,, \\
\frac{x_Q}{(x - x_Q)^2} &= \frac{x}{(x - x_Q)^2} - \frac{1}{(x - x_Q)}\,, \\
\frac{x_Q^2}{(x - x_Q)} &= \frac{x^2}{x - x_Q} - x - x_Q\,, \\
\frac{x_Q^2}{(x - x_Q)^2} &= \frac{x^2}{(x - x_Q)^2} - \frac{2x}{(x - x_Q)} + 1\,, \\
\frac{x_Q^3}{(x - x_Q)^2} &= \frac{x^3}{(x - x_Q)^2} - \frac{3x^2}{(x - x_Q)} + 2x + x_Q.
\end{aligned}$$

Exercise 25.1.19 Let the notation be as in Lemma 25.1.16. For $1 \le i \le 3$ define

$$S_i = \sum_{Q \in G_1} \frac{1}{(x - x_Q)^i}.$$

Show that $S_1 = \psi(x)'/\psi(x)$ and that $S_2 = -(\psi'(x)/\psi(x))' = ((\psi(x)')^2 - \psi(x)\psi(x)'')/\psi(x)^2$.

Exercise 25.1.20 Complete the proof of Lemma 25.1.16.

Exercise 25.1.21 Determine the complexity of using Lemma 25.1.16 to compute isogenies over finite fields. More precisely, show that if $G \subseteq E(\mathbb{F}_{q^n})$ is defined over $\mathbb{F}_q$ and $d = \#G$ then one can compute $\psi(x)$ in $O(d^2)$ operations in $\mathbb{F}_q$. Once $\psi(x) \in \mathbb{F}_q[x]$ is computed show that one can compute the polynomials $A(x)$ and $B(x, y)$ for the isogeny in $O(d)$ operations in $\mathbb{F}_q$.

25.2 Isogenies from j-invariants

Vélu's formulae require that one knows the kernel of the desired isogeny. But in some applications one wants to take a $\Bbbk$-rational isogeny of a given degree d (assuming such an isogeny exists) from E to another curve $\widetilde{E}$ (where $\widetilde{E}$ may or may not be known), and one does not know a specific kernel. By Theorem 25.1.2 one can restrict to the case when $d = \ell$ is prime. We usually assume that ℓ is odd, since the case $\ell = 2$ is handled by points of order 2 and Vélu's formulae.

One solution is to choose a random point $P \in E[\ell]$ that generates a $\Bbbk$-rational subgroup of order ℓ. To find such a point, compute the ℓ-division polynomial (which has degree $(\ell^2 - 1)/2$ when ℓ is odd) and find irreducible factors of it in $\Bbbk[x]$ of degree up to $(\ell - 1)/2$. Roots of such factors are points of order ℓ, and one can determine whether or not they generate a $\Bbbk$-rational subgroup by computing all points in the subgroup. Roots of factors of degree $d > (\ell - 1)/2$ cannot give rise to $\Bbbk$-rational subgroups of order ℓ. This approach is expensive when ℓ is large for a number of reasons. For a start, finding roots of degree at most $(\ell - 1)/2$ of a polynomial of degree $(\ell^2 - 1)/2$ in $\mathbb{F}_q[x]$ takes $\Omega(\ell^3 \log(\ell) \log(q))$ bit operations.

A more elegant approach is to use the ℓth modular polynomial. It is beyond the scope of this book to present the theory of modular functions and modular curves (some basic references are Sections 5.2 and 5.3 of Lang [328] and Section 11.C of Cox [145]). The fundamental fact is that there is a symmetric polynomial, called the **modular polynomial**[2] $\Phi_\ell(x, y) \in \mathbb{Z}[x, y]$ such that if E is an elliptic curve over a field $\Bbbk$ and $\widetilde{E}$ is an elliptic curve over $\Bbbk$ then there is a separable isogeny of degree ℓ (where $\gcd(\ell, \mathrm{char}(\Bbbk)) = 1$) with cyclic kernel from E to $\widetilde{E}$ if and only if $\Phi_\ell(j(E), j(\widetilde{E})) = 0$ (see Theorem 5, Section 5.3 of Lang [328]). The modular polynomial $\Phi_\ell(x, y)$ is a singular model for the modular curve $X_0(\ell)$ over $\mathbb{Q}$. This modular curve is a moduli space in the sense that a (non-cusp) point

[2] The reader should not confuse the modular polynomial $\Phi_\ell(x, y)$ with the cyclotomic polynomial $\Phi_m(x)$.

of $X_0(\ell)(\Bbbk)$ corresponds to a pair (E, G) where E is an elliptic curve over $\Bbbk$ and where G is a cyclic subgroup of E, defined over $\Bbbk$, of order ℓ. Note that it is possible to have an elliptic curve E together with two distinct cyclic subgroups G_1 and G_2 of order ℓ such that the image curves E/G_1 and E/G_2 are isomorphic; in this case (E, G_1) and (E, G_2) are distinct points of $X_0(\ell)$ but correspond to a repeated root of $\Phi_\ell(j(E), y)$ (it follows from the symmetry of $\Phi_\ell(x, y)$ that this is a singular point on the model). In other words, a repeated root of $\Phi_\ell(j(E), y)$ corresponds to non-equivalent ℓ-isogenies from E to some elliptic curve $\widetilde{E}$.

Since there are $\ell + 1$ cyclic subgroups of $E[\ell]$ it follows that $\Phi_\ell(j(E), y)$ has degree $\ell + 1$. Indeed, $\Phi_\ell(x, y) = x^{\ell+1} + y^{\ell+1} - (xy)^\ell + \cdots$ with all other terms of lower degree (see Theorem 11.18 of Cox [145] or Theorem 3 of Section 5.2 of Lang [328]). The coefficients of $\Phi_\ell(x, y)$ are large (as seen in Example 25.2.1, even when $\ell = 2$ the coefficients are large).

Example 25.2.1

$$\begin{aligned}\Phi_2(x, y) = x^3 + y^3 - x^2y^2 + 1488(x^2y + xy^2) - 162000(x^2 + y^2) \\ + 40773375xy + 8748000000(x + y) - 157464000000000.\end{aligned}$$

Let ℓ be prime. Cohen [128] showed that the number of bits in the largest coefficient of $\Phi_\ell(x, y)$ is $O(\ell \log(\ell))$ (see Bröker and Sutherland [103] for a more precise bound). Since there are roughly ℓ^2 coefficients it follows that $\Phi_\ell(x, y)$ can be written down using $O(\ell^3 \log(\ell))$ bits, and it is believed that this space requirement cannot be reduced. Hence, one expects to perform at least $O(\ell^3 \log(\ell)) = O(\ell^{3+\epsilon})$ bit operations[3] to compute $\Phi_\ell(x, y)$. Indeed, using methods based on modular functions, one can conjecturally[4] compute $\Phi_\ell(x, y)$ in $O(\ell^{3+\epsilon})$ bit operations (see Enge [179]). Using modular functions other than the j-function can lead to polynomials with smaller coefficients, but this does not affect the asymptotic complexity.

The fastest method to compute modular polynomials is due to Bröker, Lauter and Sutherland [102]. This method exploits some of the ideas explained later in this chapter (in particular, isogeny volcanoes). The method computes $\Phi_\ell(x, y)$ modulo small primes and then determines $\Phi_\ell(x, y)$ by the Chinese remainder theorem. Under the Generalised Riemann Hypothesis (GRH) the complexity is $O(\ell^3 \log(\ell)^3 \log(\log(\ell)))$ bit operations. For the rest of the chapter we abbreviate the cost as $O(\ell^{3+\epsilon})$ bit operations. The method can also be used to compute $\Phi_\ell(x, y)$ modulo p, in which case the space requirements are $O(\ell^2 \log(\ell)^2 + \ell^2 \log(p))$ bits.

The upshot is that, given an elliptic curve E over $\Bbbk$, the j-invariants of elliptic curves $\widetilde{E}$ that are ℓ-isogenous over $\Bbbk$ (where $\gcd(\ell, \mathrm{char}(\Bbbk)) = 1$) are given by the roots of $\Phi_\ell(j(E), y)$ in $\Bbbk$. When E is ordinary, Theorem 25.4.6 implies that $\Phi_\ell(j(E), y)$ has either $0, 1, 2$ or $\ell + 1$ roots in $\Bbbk$ (counted with multiplicities).

[3] Recall that a function $f(\ell)$ is $O(\ell^{3+\epsilon})$ if, for every $\epsilon > 0$, there is some $C(\epsilon), L(\epsilon) \in \mathbb{R}_{>0}$ such that $f(\ell) < C(\epsilon)\ell^{3+\epsilon}$ for all $\ell > L(\epsilon)$.

[4] Enge needs an assumption that rounding errors do not affect the correctness of the output.

Exercise 25.2.2 Given the polynomial $\Phi_\ell(x, y)$ and a value $j \in \mathbb{F}_q$ show that one can compute $F(y) = \Phi_\ell(j, y) \in \mathbb{F}_q[y]$ in $O(\ell^2(\ell \log(\ell) \log(q) + M(\log(q))))$ bit operations. Show also that one can then compute the roots $\tilde{j} \in \mathbb{F}_q$ of $F(y) = \Phi_\ell(j(E), y)$ (or determine that there are no roots) in expected time bounded by $O(\ell^2 \log(\ell) \log(q))$ field operations (which is $O(\ell^{2+\epsilon} \log(q)^3)$ bit operations).

For the rest of this section we consider algorithms to compute an ℓ-isogeny $\phi : E \to \widetilde{E}$ given an elliptic curve E and the j-invariant of $\widetilde{E}$.

Exercise 25.2.3 Let E be an elliptic curve over $\mathbb{F}_q$ and let E' over $\mathbb{F}_q$ be a twist of E. Show that there is an $\mathbb{F}_q$-rational isogeny of degree ℓ from E (to some elliptic curve) if and only if there is an $\mathbb{F}_q$-rational isogeny of degree ℓ from E'. Show that $\mathrm{End}(E) \cong \mathrm{End}(E')$ (where $\cong$ denotes ring isomorphism).

25.2.1 Elkies' algorithm

Let $\ell > 2$ be a prime and let E be an elliptic curve over $\Bbbk$ where $\mathrm{char}(\Bbbk) = 0$ or $\mathrm{char}(\Bbbk) > \ell + 2$. Assume $j(E) \neq 0, 1728$ (for the case $j(E) \in \{0, 1728\}$ one constructs isogenies using the naive method or the methods of the following sections). Let $\tilde{j} \in \Bbbk$ be such that $\Phi_\ell(j(E), \tilde{j}) = 0$. We also assume that $\tilde{j}$ is a simple root of $\Phi_\ell(j(E), y)$ (more precisely, $(\partial \Phi_\ell(x, y)/\partial x)(j, \tilde{j}) \neq 0$ and $(\partial \Phi_\ell(x, y)/\partial y)(j, \tilde{j}) \neq 0$); see page 248 of Schoof [477] for a discussion of why this condition is not too severe.

Elkies gave a method to determine an explicit equation for an elliptic curve $\widetilde{E}$, such that $j(\widetilde{E}) = \tilde{j}$, and a polynomial giving the kernel of an ℓ-isogeny from E to $\widetilde{E}$. Elkies' original motivation (namely, algorithms for point counting) only required computing the kernel polynomial of the isogeny, but, as we have seen, from this information one can easily compute the rational functions describing the isogeny. The method also works when $\ell > 2$ is composite, but that is not of practical relevance. The condition that $\mathrm{char}(\Bbbk)$ not be small (if it is non-zero) is essential.

We use the same notation as in Lemma 25.1.16: $\psi(x)$ is the polynomial of degree $(\ell - 1)/2$ whose roots are the x-coordinates of affine points in the kernel G of the isogeny and s_i are the i-th symmetric polynomials in these roots. We also define

$$p_i = \sum_{P \in G - \{\mathcal{O}_E\}} x_P^i$$

so that $p_1 = 2s_1$ and $p_2 = 2(s_1^2 - 2s_2)$ (these are Newton's formulae; see Lemma 10.7.6). While the value $\tilde{j}$ specifies the equation for the isogenous curve $\widetilde{E}$ (up to isomorphism) it does not, in general, determine the isogeny (see pages 37 and 44 of Elkies [178] for discussion). It is necessary to have some extra information, and for this the coefficient p_1 suffices and can be computed using partial derivatives of the modular polynomial (this is why the condition on the partial derivatives is needed).

The explanation of Elkies' algorithm requires theory that we do not have space to present. We refer to Schoof [477] for a good summary of the details (also see Elkies [178] for further discussion). The basic idea is to use the fact (Deuring lifting theorem) that the isogeny lifts to an isogeny between elliptic curves over $\mathbb{C}$. One can then interpret the ℓ-isogeny in terms of Tate curves[5] $\mathbb{C}^*/q^{\mathbb{Z}}$ (we have not presented the Tate curve in this book; see Section C.14 of [505] or Section 5.3 of [506]) as a map from $q(z)$ to $q(z)^\ell$. As discussed on page 40 of Elkies [178], this isogeny is not normalised. There are a number of relations between the modular j-function, certain Eisenstein series, the equation of the elliptic curve (in short Weierstrass form) and the kernel polynomial of the isogeny. These relations give rise to formulae that must also hold over $\Bbbk$. Hence, one can work entirely over $\Bbbk$ and obtain the kernel polynomial.

The details of this approach are given in Sections 7 and 8 of Schoof [477]. In particular, Theorem 7.3 shows how to get j' (derivative); Proposition 7.1 allows one to compute the coefficients of the elliptic curve; Proposition 7.2 gives the coefficient p_1 of the kernel polynomial (which is a function of values specified in Proposition 7.1 and Theorem 7.3). The coefficients of the kernel polynomial are related to the coefficients of the series expansion of the Weierstrass ζ-function (see Theorem 8.3 of [477]).

The algorithm is organised as follows (see Algorithm 27). One starts with an ordinary elliptic curve $E : y^2 = x^3 + Ax + B$ over $\Bbbk$ and $j = j(E)$. We assume that $j \notin \{0, 1728\}$ and $\mathrm{char}(\Bbbk) = 0$ or $\mathrm{char}(k) > \ell + 2$. Let $\phi_x = (\frac{\partial \Phi_\ell}{\partial x})(j, \tilde{j})$, $\phi_y = (\frac{\partial \Phi_\ell}{\partial y})(j, \tilde{j})$, $\phi_{xx} = (\frac{\partial^2 \Phi_\ell}{\partial x^2})(j, \tilde{j})$, $\phi_{yy} = (\frac{\partial^2 \Phi_\ell}{\partial y^2})(j, \tilde{j})$ and $\phi_{xy} = (\frac{\partial^2 \Phi_\ell}{\partial x \partial y})(j, \tilde{j})$. One computes the derivative j' and the corresponding values for E_4 and E_6. Given $\tilde{j}$, one computes $\tilde{j}'$ and then the coefficients $\tilde{A}$ and $\tilde{B}$ of the image curve $\tilde{E}$. Finally, one computes p_1, from which it is relatively straightforward to compute all the coefficients of the kernel polynomial $\psi(x)$.

Exercise 25.2.4 Show that Elkies' algorithm requires $O(d^2) = O(\ell^2)$ operations in $\Bbbk$.

Bostan, Morain, Salvy and Schost [88] have given algorithms (exploiting fast arithmetic on power series) based on Elkies' methods. The algorithms apply when the characteristic of the field is zero or is sufficiently large compared with ℓ. There is a slight difference in scope: Elkies' starts with only j-invariants whereas Bostan *et al.* assume that one is given elliptic curves E and $\widetilde{E}$ in short Weierstrass form such that there is a normalised isogeny of degree ℓ over $\Bbbk$ from E to $\widetilde{E}$. In general, one needs to perform Elkies' method before one has such an equation for $\widetilde{E}$ and so the computations with modular curves still dominate the cost. Theorem 2 of [88] states that one can compute the rational functions giving the isogeny in $O(M(\ell))$ operations in $\Bbbk$ when $\mathrm{char}(\Bbbk) > 2\ell - 1$ and when the coefficient p_1 is known. Note that Bostan *et al.* are not restricted to prime degree isogenies. An application of the result of Bostan *et al.* is to determine whether there is a normalised isogeny from E to $\widetilde{E}$ *without* needing to compute modular polynomials. Lercier and Sirvent [346] (again, assuming one is given explicit equations for E and $\widetilde{E}$ such that there is a normalised ℓ-isogeny between them) have shown how to achieve a similarly fast method even when the characteristic of the field is small.

[5] The notation q here refers to $q(z) = \exp(2\pi i z)$ and not a prime power.

Algorithm 27 Elkies' algorithm. (Source code provided by Drew Sutherland.)

INPUT: $A, B \in \Bbbk$, $\ell > 2$, $j, \tilde{j} \in \Bbbk$
OUTPUT: $\tilde{A}, \tilde{B}, \psi(x)$

1: Compute $\phi_x, \phi_y, \phi_{xx}, \phi_{yy}$ and ϕ_{xy} ▷ Compute $\tilde{A}$ and $\tilde{B}$
2: Let $m = 18B/A$, let $j' = mj$, and let $k = j'/(1728 - j)$
3: Let $\tilde{j}' = -j'\phi_x/(\ell\phi_y)$, let $\tilde{m} = \tilde{j}'/\tilde{j}$, and let $\tilde{k} = \tilde{j}'/(1728 - \tilde{j})$
4: Let $\tilde{A} = \ell^4\tilde{m}\tilde{k}/48$ and $\tilde{B} = \ell^6\tilde{m}^2\tilde{k}/864$
5: Let $r = -(j'^2\phi_{xx} + 2\ell j'\tilde{j}'\phi_{xy} + \ell^2\tilde{j}'^2\phi_{yy})/(j'\phi_x)$ ▷ Compute p_1
6: Let $p_1 = \ell(r/2 + (k - \ell\tilde{k})/4 + (\ell\tilde{m} - m)/3)$
7: Let $d = (\ell - 1)/2$ ▷ Compute the power sums t_n of the roots of $\psi(x)$
8: Let $t_0 = d$, $t_1 = p_1/2$, $t_2 = ((1 - 10d)A - \tilde{A})/30$, and $t_3 = ((1 - 28d)B - 42t_1A - \tilde{B})/70$
9: Let $c_0 = 0$, $c_1 = 6t_2 + 2At_0$, $c_2 = 10t_3 + 6At_1 + 4Bt_0$
10: **for** $n = 2$ to $d - 1$ **do**
11: Let $s = \sum_{i=1}^{n-1} c_i c_{n-i}$
12: Let
$$c_{n+1} = \frac{3s - (2n-1)(n-1)Ac_{n-1} - (2n-2)(n-2)Bc_{n-2}}{(n-1)(2n+5)}$$
13: **end for**
14: **for** $n = 3$ to $d - 1$ **do**
15: Let
$$t_{n+1} = \frac{c_n - (4n-2)At_{n-1} - (4n-4)Bt_{n-2}}{4n+2}$$
16: **end for**
17: Let $s_0 = 1$ ▷ Compute the symmetric functions s_n of the roots of $\psi(x)$
18: **for** $n = 1$ to d **do**
19: Let $s_n = \frac{-1}{n}\sum_{i=1}^{n}(-1)^i t_i s_{n-i}$
20: **end for**
21: **return** $\psi(x) = \sum_{i=0}^{d}(-1)^i s_i x^{d-i}$

A number of calculations can fail when char($\Bbbk$) is non-zero but small compared with ℓ, due to divisions by small integers arising in the power series expansions for the modular functions. Algorithms for the case of small characteristic will be mentioned in Section 25.2.3.

25.2.2 Stark's algorithm

Stark [522] gave a method to compute the rational function giving the x-coordinate of an endomorphism $\phi : E \to E$ corresponding to a complex number β (interpreting End(E) as a subset of $\mathbb{C}$). The idea is to use the fact that, for an elliptic curve E over the complex numbers given by short Weierstrass form

$$\wp(\beta z) = \frac{A(\wp(z))}{B(\wp(z))} \tag{25.3}$$

where A and B are polynomials and where $\wp(z) = z^{-2} + 3G_4 z^2 + \cdots$ is the Weierstrass function (see Theorem VI.3.5 of Silverman [505]). This isogeny is not normalised (since $\wp(\beta z) = \beta^{-2} z^{-2} + \cdots$ it follows that the pullback of ω_E under ϕ is $\beta\omega_E$). Stark's idea is to express $\wp$ as a (truncated) power series in z; the coefficients of this power series are determined by the coefficients of the elliptic curve E. One computes A and B by taking the continued fraction expansion of the left-hand side of equation (25.3). One can apply this algorithm to curves over finite fields by applying the Deuring lifting theorem. Due to denominators in the power series coefficients of $\wp(z)$ the method only works when $\mathrm{char}(\Bbbk) = 0$ or $\mathrm{char}(\Bbbk)$ is sufficiently large. Stark's paper [522] gives a worked example in the case $\beta = \sqrt{-2}$.

The idea generalises to normalised isogenies $\phi : E \to \widetilde{E}$ by writing $\wp_{\widetilde{E}}(z) = A(\wp_E(z))/B(\wp_E(z))$ where now the power series for $\wp_E(z)$ and $\wp_{\widetilde{E}}(z)$ are different since the elliptic curve equations are different. Note that it is necessary to have actual curve equations for the normalised isogeny, not just j-invariants. We refer to Section 6.2 of Bostan, Morain, Salvy and Schost [88] for further details and complexity estimates.

25.2.3 The small characteristic case

As we have seen, the Elkies and Stark methods require the characteristic of the ground field to be either zero or relatively large since they use lifting to short Weierstrass forms over $\mathbb{C}$ and since the power series expansions have rational coefficients that are divisible by various small primes. Hence, there is a need for algorithms that handle the case when $\mathrm{char}(\Bbbk)$ is small (especially, $\mathrm{char}(\Bbbk) = 2$). A number of such methods have been developed by Couveignes, Lercier, Morain and others. We briefly sketch Couveignes' "second method" [143].

Let p be the characteristic of the field. We assume that p is "small" (in the sense that an algorithm performing p operations is considered efficient). Let E and $\widetilde{E}$ be ordinary[6] elliptic curves over $\mathbb{F}_{p^m}$.

The basic idea is to use the fact that if $\phi : E \to \widetilde{E}$ is an isogeny of odd prime degree $\ell \neq p$ (isogenies of degree p are easy: they are either Frobenius or Verschiebung) then ϕ maps points of order p^k on E to points of order p^k on $\widetilde{E}$. Hence, one can try to determine the rational functions describing ϕ by interpolation from their values on $E[p^k]$. One could interpolate using any torsion subgroup of E, but using $E[p^k]$ is the best choice since it is a cyclic group and so there are only $\varphi(p^k) = p^k - p^{k-1}$ points to check (compared with $\varphi(n^2)$ if using $E[n]$). The method can be applied to any elliptic curve $\widetilde{E}$ in the isomorphism class, so in general it will not return a normalised isogeny.

[6] The restriction to ordinary curves is not a significant problem. In practice, we are interested in elliptic curves over $\mathbb{F}_{p^m}$ where m is large, whereas supersingular curves are all defined over $\mathbb{F}_{p^2}$. Indeed, for small p there are very few supersingular curves, and isogenies of small degree between them can be computed by factoring division polynomials and using Vélu's formulae.

Couveignes' method is as follows: first, compute points $P \in E[p^k] - E[p^{k-1}]$ and $\widetilde{P} \in \widetilde{E}[p^k] - \widetilde{E}[p^{k-1}]$ over $\overline{\mathbb{F}}_p$ and guess that $\phi(P) = \widetilde{P}$. Then try to determine the rational function $\phi_1(x) = u(x)/v(x)$ by interpolating $\phi_1(x([i]P)) = x([i]\widetilde{P})$; if this does not work then try another guess for $\widetilde{P}$. The interpolation is done as follows (we assume $p^k > 2\ell$). First, compute a polynomial $A(x)$ of degree d where $2\ell < d \le p^k$ such that $A(x([i]P)) = x([i]\widetilde{P})$ for $1 \le i \le d$. Also compute $B(x) = \prod_{i=1}^{d}(x - x([i]P))$. If the guess for $\widetilde{P}$ is correct then $A(x) \equiv u(x)/v(x) \pmod{B(x)}$ where $\deg(u(x)) = \ell$, $\deg(v(x)) = \ell - 1$ and $v(x)$ is a square. Writing this equation as $A(x)v(x) = u(x) + B(x)w(x)$ it follows that $u(x)$ and $v(x)$ can be computed using Euclid's algorithm. The performance of the algorithm depends on the method used to determine points in $E[p^k]$, but is dominated by the fact that these points lie in an extension of the ground field of large degree, and that one expects to try around $\frac{1}{2}p^k \approx \ell$ choices for $\widetilde{P}$ before hitting the right one. The complexity is polynomial in ℓ, p and m (where E is over $\mathbb{F}_{p^m}$). When $p = 2$ the fastest method was given by Lercier [344]. For further details we refer to the surveys by Lercier and Morain [345] and De Feo [185].

25.3 Isogeny graphs of elliptic curves over finite fields

Let E be an elliptic curve over $\mathbb{F}_q$. The $\mathbb{F}_q$**-isogeny class** of E is the set of $\mathbb{F}_q$-isomorphism classes of elliptic curves over $\mathbb{F}_q$ that are isogenous over $\mathbb{F}_q$ to E. Tate's isogeny theorem states that two elliptic curves E and $\widetilde{E}$ over $\mathbb{F}_q$ are $\mathbb{F}_q$-isogenous if and only if $\#E(\mathbb{F}_q) = \#\widetilde{E}(\mathbb{F}_q)$ (see Theorem 9.7.4 for one implication).

We have seen in Sections 9.5 and 9.10 that the number of $\mathbb{F}_q$-isomorphism classes of elliptic curves over $\mathbb{F}_q$ is roughly $2q$ and that there are roughly $4\sqrt{q}$ possible values for $\#E(\mathbb{F}_q)$. Hence, if isomorphism classes were distributed uniformly across all group orders one would expect around $\frac{1}{2}\sqrt{q}$ elliptic curves in each isogeny class. The theory of complex multiplication gives a more precise result (as mentioned in Section 9.10.1). We denote by π_q the q-power Frobenius map; see Section 9.10 for its properties. The number of $\mathbb{F}_q$-isomorphism classes of ordinary elliptic curves over $\mathbb{F}_q$ with $q + 1 - t$ points is the Hurwitz class number of the ring $\mathbb{Z}[\pi_q] = \mathbb{Z}[(t + \sqrt{t^2 - 4q})/2]$. This is the sum of the ideal class numbers $h(\mathcal{O})$ over all orders $\mathbb{Z}[\pi_q] \subseteq \mathcal{O} \subseteq \mathcal{O}_K$. It follows (see Remark 9.10.19) that there are $O(q^{1/2}\log(q)\log(\log(q)))$ elliptic curves in each isogeny class. For supersingular curves see Theorem 9.11.11.

Definition 25.3.1 Let E be an elliptic curve over a field $\Bbbk$ of characteristic p. Let $S \subseteq \mathbb{N}$ be a finite set of primes. Define

$$X_{E,\Bbbk,S}$$

to be the (directed) graph[7] with vertex set being the $\Bbbk$-isogeny class of E. Vertices are typically labelled by $j(\widetilde{E})$, though we also speak of "the vertex $\widetilde{E}$".[8] There is a (directed)

[7] Some authors would use the name "multi-graph", since there can be loops and/or multiple edges between vertices.

[8] In the supersingular case one can label vertices as $j(\widetilde{E})$ without ambiguity only when $\Bbbk$ is algebraically closed: when $\Bbbk$ is finite then, in the supersingular case, one has two distinct vertices in the graph for $\widetilde{E}$ and its quadratic twist. For the ordinary case there is no ambiguity by Lemma 9.11.12 (also see Exercise 25.3.8).

edge $(j(E_1), j(E_2))$ labelled by ℓ for each equivalence class of ℓ-isogenies from E_1 to E_2 defined over $\Bbbk$ for some $\ell \in S$. We usually treat this as an undirected graph, since for every ℓ-isogeny $\phi : E_1 \to E_2$ there is a dual isogeny $\widehat{\phi} : E_2 \to E_1$ of degree ℓ (though see Remark 25.3.2 for an unfortunate, though rare, complication).

Remark 25.3.2 Edges in the isogeny graph correspond to equivalence classes of isogenies. It can happen that two non-equivalent isogenies from $E_1 \to E_2$ have equivalent dual isogenies from $E_2 \to E_1$. It follows that there are two directed edges in the graph from E_1 to E_2 but only one directed edge from E_2 to E_1. (Note that this does not contradict the fact that isogenies satisfy $\widehat{\widehat{\phi}} = \phi$, as we are speaking here about isogenies up to equivalence.) Such an issue was already explained in Exercise 25.1.1; it only arises if $\#\mathrm{Aut}(E_2) > 2$ (i.e., if $j(E_2) = 0, 1728$).

Definition 25.3.3 A (directed) graph is k**-regular** if every vertex has (out-)degree k (a loop is considered as having degree 1). A **path** in a graph is a sequence of (directed) edges between vertices, such that the end vertex of one edge is the start vertex of the next. We will also describe a path as a sequence of vertices. A graph is connected if there is a path from every vertex to every other vertex. The **diameter** of a connected graph is the maximum, over all pairs v_1, v_2 of vertices in the graph, of the length of the shortest path from v_1 to v_2.

There are significant differences (both in the structure of the isogeny graph and the way it is used in applications) between the ordinary and supersingular cases. So we present them separately.

25.3.1 Ordinary isogeny graph

Fix an ordinary elliptic curve E over $\mathbb{F}_q$ such that $\#E(\mathbb{F}_q) = q + 1 - t$. The isogeny graph of elliptic curves isogenous over $\mathbb{F}_q$ to E can be identified, using the theory of complex multiplication, with a graph whose vertices are ideal classes (in certain orders). The goal of this section is to briefly sketch this theory in the special case (the general case is given in Section 25.4) of the subgraph where all elliptic curves have the same endomorphism ring, in which case the edges correspond to multiplication by prime ideals. We do not have space to give all the details; good references for the background are Cox [145] and Lang [328].

The endomorphism ring of E (over $\overline{\mathbb{F}}_q$) is an order $\mathcal{O}$ in the quadratic imaginary field $K = \mathbb{Q}(\sqrt{t^2 - 4q})$. (We refer to Section A.12 for background on orders and conductors.) Let $\mathcal{O}_K$ be the ring of integers of K. Then $\mathbb{Z}[\pi_q] = \mathbb{Z}[(t + \sqrt{t^2 - 4q})/2] \subseteq \mathcal{O} \subseteq \mathcal{O}_K$ and if $\mathcal{O}_K = \mathbb{Z}[\theta]$ then $\mathcal{O} = \mathbb{Z}[c\theta]$ where c is the conductor of $\mathcal{O}$. The ideal class group $\mathrm{Cl}(\mathcal{O})$ is defined to be the classes of invertible $\mathcal{O}$-ideals that are prime to the conductor; see Section 7 of [145] or Section 8.1 of [328]. There is an explicit formula for the order $h(\mathcal{O})$ of the ideal class group $\mathrm{Cl}(\mathcal{O})$ in terms of the class number $h(\mathcal{O}_K)$ of the number field; see Theorem 7.24 of [145] or Theorem 8.7 of [328].

There is a one-to-one correspondence between the set of isomorphism classes of elliptic curves E over $\mathbb{F}_q$ with $\mathrm{End}(E) = \mathcal{O}$ and the set $\mathrm{Cl}(\mathcal{O})$. Precisely, to an invertible $\mathcal{O}$-ideal

$\mathfrak{a}$ one associates the elliptic curve $E = \mathbb{C}/\mathfrak{a}$ over $\mathbb{C}$. An $\mathcal{O}$-ideal $\mathfrak{a}'$ is equivalent to $\mathfrak{a}$ in $\mathrm{Cl}(\mathcal{O})$ if and only if $\mathbb{C}/\mathfrak{a}'$ is isomorphic to E. One can show that $\mathrm{End}(E) = \mathcal{O}$. The theory of complex multiplication shows that E is defined over a number field (called the ring class field) and has good reduction modulo the characteristic p of $\mathbb{F}_q$. This correspondence is not canonical, since the reduction modulo p map is not well-defined (it depends on a choice of prime ideal above p in the ring class field).

Let $\mathfrak{a}$ be an invertible $\mathcal{O}$-ideal and $E = \mathbb{C}/\mathfrak{a}$. Let $\mathfrak{l}$ be an invertible $\mathcal{O}$-ideal and, interpreting $\mathfrak{l} \subseteq \mathrm{End}(E)$, consider the set $E[\mathfrak{l}] = \{P \in E(\mathbb{C}) : \phi(P) = \mathcal{O}_E \text{ for all } \phi \in \mathfrak{l}\}$. Since $\mathcal{O} \subseteq \mathbb{C}$ we can interpret $\mathfrak{l} \subseteq \mathbb{C}$, in which case

$$E[\mathfrak{l}] \cong \{z \in \mathbb{C}/\mathfrak{a} : \alpha z \in \mathfrak{a}, \text{ for all } \alpha \in \mathfrak{l}\}$$
$$\cong \mathfrak{l}^{-1}\mathfrak{a}/\mathfrak{a}.$$

It follows that $\#E[\mathfrak{l}]$ is equal to the norm of the ideal $\mathfrak{l}$. The identity map on $\mathbb{C}$ induces the isogeny

$$\mathbb{C}/\mathfrak{a} \to \mathbb{C}/\mathfrak{l}^{-1}\mathfrak{a}$$

with kernel $\mathfrak{l}^{-1}\mathfrak{a}/\mathfrak{a} \cong E[\mathfrak{l}]$. The above remarks apply to elliptic curves over $\mathbb{C}$, but the theory reduces well to elliptic curves over finite fields, and, indeed, every isogeny from E to an elliptic curve $\widetilde{E}$ with $\mathrm{End}(E) = \mathrm{End}(\widetilde{E})$ arises in this way. This shows that not only do ordinary elliptic curves correspond to ideals in $\mathcal{O}$, but so do their isogenies.

Exercise 25.3.4 Show that if $\mathfrak{l} = (\ell)$ where $\ell \in \mathbb{N}$ then the isogeny $\mathbb{C}/\mathfrak{a} \to \mathbb{C}/\mathfrak{l}^{-1}\mathfrak{a}$ is $[\ell]$.

Exercise 25.3.5 Suppose the prime ℓ splits in $\mathcal{O}$ as $(\ell) = \mathfrak{l}_1\mathfrak{l}_2$ in $\mathcal{O}$. Let $\phi : E \to \widetilde{E}$ correspond to the ideal $\mathfrak{l}_1$. Show that $\widehat{\phi}$ corresponds to $\mathfrak{l}_2$.

Let ℓ be a prime. Then ℓ splits in $\mathcal{O}_K = \mathbb{Z}[\theta]$ if and only if the minimal polynomial of θ factors modulo ℓ with two linear factors. If D is the discriminant of K then ℓ splits if and only if the Kronecker symbol satisfies $(\frac{D}{\ell}) = +1$. Note that the Kronecker symbol is the Legendre symbol when ℓ is odd and

$$(\tfrac{D}{2}) = \begin{cases} 0 & D \equiv 0 \pmod 4, \\ 1 & D \equiv 1 \pmod 8, \\ -1 & D \equiv 5 \pmod 8). \end{cases} \tag{25.4}$$

Let E be an elliptic curve over $\mathbb{F}_q$ with $\mathrm{End}(E) = \mathcal{O}$ and let ℓ be coprime to the conductor of $\mathcal{O}$. There are $1 + (\frac{D}{\ell})$ prime ideals $\mathfrak{l}$ above ℓ, and so there are this many isogenies of degree ℓ from E. It follows there are ℓ-isogenies in the isogeny graph for roughly half the primes ℓ.

Let E be an elliptic curve over $\mathbb{F}_q$ corresponding to an $\mathcal{O}$-ideal $\mathfrak{a}$. Let $S \subseteq \mathbb{N}$ be a finite set of primes that are all co-prime to the conductor. Let G be the component of E in the isogeny graph $X_{E,\mathbb{F}_q,S}$ of Definition 25.3.1. Let $S' = \{\mathfrak{l}_1, \ldots, \mathfrak{l}_k\}$ be the set of classes of invertible $\mathcal{O}$-ideals above primes $\ell \in S$ and let $\langle S' \rangle$ be the subgroup of $\mathrm{Cl}(\mathcal{O})$ generated by S'. From the above discussion it follows that G can be identified with the graph whose vertices are the $\mathcal{O}$-ideal classes in the coset $\mathfrak{a}\langle S' \rangle$ and such that, for each $\mathfrak{b} \in \mathfrak{a}\langle S' \rangle$ and each $\mathfrak{l}_i \in S'$, there is an edge between $\mathfrak{b}$ and $\mathfrak{l}_i^{-1}\mathfrak{b}$. Since ideal class groups are well-understood, this correspondence illuminates the study of the isogeny graph. For example, an immediate corollary is that the graph of elliptic curves E with $\mathrm{End}(E) = \mathcal{O}$ is connected if and only if S' generates $\mathrm{Cl}(\mathcal{O})$. A well-known result of Bach states that (assuming the Riemann hypothesis for the Dedekind zeta function of K and Hecke L-functions for characters of $\mathrm{Cl}(\mathcal{O}_K)$) the group $\mathrm{Cl}(\mathcal{O}_K)$ is generated by prime ideals of norm less than $6\log(|\Delta_K|)^2$ (see page 376 of [19]) where Δ_K is the discriminant of $\mathcal{O}_K$. Another immediate corollary is that the graph is regular (i.e., every vertex has the same degree).

Remark 25.3.6 We stress that there is no canonical choice of $\mathcal{O}$-ideal $\mathfrak{a}$ corresponding to an elliptic curve E with $\mathrm{End}(E) = \mathcal{O}$. However, given a pair $(E, \widetilde{E})$ of isogenous elliptic curves with $\mathrm{End}(E) = \mathrm{End}(\widetilde{E}) = \mathcal{O}$ the ideal class corresponding to the isogeny between them is well-defined. More precisely, if E is identified with $\mathbb{C}/\mathfrak{a}$ for some $\mathcal{O}$-ideal $\mathfrak{a}$ then there is a unique ideal class represented by $\mathfrak{b}$ such that $\widetilde{E}$ is identified with $\mathbb{C}/\mathfrak{b}^{-1}\mathfrak{a}$. The only algorithm known to find such an ideal $\mathfrak{b}$ is to compute an explicit isogeny from E to $\widetilde{E}$ (using algorithms presented later in this chapter) and then determine the corresponding isogeny. If one could determine $\mathfrak{b}$ efficiently from E and $\widetilde{E}$ then navigating the ordinary isogeny graph would be much easier.

Exercise 25.3.7 Let E_1 be an elliptic curve with $\mathrm{End}(E_1) = \mathcal{O}$. Let $\mathfrak{l}$ be a prime ideal of $\mathcal{O}$ above ℓ. Suppose $\mathfrak{l}$ has order d in $\mathrm{Cl}(\mathcal{O})$. Show that there is a cycle $E_1 \to E_2 \to \cdots \to E_d \to E_1$ of ℓ-isogenies.

Exercise 25.3.8 Let E be an ordinary elliptic curve over $\mathbb{F}_q$ and let E' be the quadratic twist of E. Show that the graphs $X_{E,\mathbb{F}_q,S}$ and $X_{E',\mathbb{F}_q,S}$ are identical.

Remark 25.3.9 Let ℓ split in $\mathcal{O} = \mathbb{Z}[\theta] \subseteq \mathrm{End}(E)$ and let $\mathfrak{l}_1 = (\ell, a + \theta)$ and $\mathfrak{l}_2 = (\ell, b + \theta)$ be the corresponding prime ideals. Given an isogeny $\phi : E \to \widetilde{E}$ of degree ℓ one can determine whether ϕ corresponds to $\mathfrak{l}_1$ or $\mathfrak{l}_2$ as follows: compute (using the Elkies method if only $j(E)$ and $j(\widetilde{E})$ are known) the polynomial determining the kernel of ϕ; compute an explicit point $P \in \ker(\phi)$; check whether $[a]P + \theta(P) = \mathcal{O}_E$ or $[b]P + \theta(P) = \mathcal{O}_E$. This trick is essentially due to Couveignes, Dewaghe, and Morain (see Section 3.2 of [144]; also see pages 49-50 of Kohel [315] and Galbraith, Hess and Smart [204]).

Remark 25.3.10 The ideas mentioned above show that all elliptic curves over $\mathbb{F}_q$ with the same endomorphism ring are isogenous over $\mathbb{F}_q$. Combined with the results of Section 25.4 one can prove Tate's isogeny theorem, namely that any two elliptic curves over $\mathbb{F}_q$ with the same number of points are isogenous over $\mathbb{F}_q$.

More details about the structure of the ordinary isogeny graph will be given in Section 25.4. In particular, that section will discuss isogenies between elliptic curves whose endomorphism rings are different orders in the same quadratic field.

25.3.2 Expander graphs and Ramanujan graphs

Let X be a finite (directed) graph on vertices labelled $\{1, \ldots, n\}$. The **adjacency matrix** of X is the $n \times n$ integer matrix A with $A_{i,j}$ being the number of edges from vertex i to vertex j. The **eigenvalues of a finite graph** X are defined to be the eigenvalues of its adjacency matrix A. For the rest of this section we assume that all graphs are un-directed. Since the adjacency matrix of an un-directed graph is real and symmetric, the eigenvalues are real.

Lemma 25.3.11 *Let X be a k-regular graph. Then k is an eigenvalue, and all eigenvalues λ are such that $|\lambda| \leq k$.*

Proof The first statement follows since $(1, 1, \ldots, 1)$ is an eigenvector with eigenvalue k. The second statement is also easy (see Proposition 1.1.2 of Davidoff, Sarnak and Valette [153] or Theorem 1 of Murty [400]). □

Let X be a k-regular graph. We denote by $\lambda(X)$ the maximum of the absolute values of all the eigenvalues that are not equal to $\pm k$. Alon and Boppana showed that the lim inf of $\lambda(X)$ over any family of k-regular graphs (as the number of vertices goes to ∞) is at least $2\sqrt{k-1}$ (see Theorem 1.3.1 of [153], Theorem 3.2 of [431] or Theorem 10 of [400]). The graph X is said to be **Ramanujan** if $\lambda(X) \leq 2\sqrt{k-1}$. Define $\lambda_1(X)$ to be the largest eigenvalue that is strictly less than k.

Let G be a finite group and S a subset of G such that $g \in S$ implies $g^{-1} \in S$ (we also allow S to be a multi-set). The **Cayley graph** of G is the graph X with vertex set G and an edge between g and gs for all $g \in G$ and all $s \in S$. Murty [400] surveys criteria for when a Cayley graph is a Ramanujan graph. If certain character sums are small then X may be a Ramanujan graph; see Section 2 of [400].

Definition 25.3.12 Let X be a graph and A a subset of vertices of X. The **vertex boundary** of A in X is

$$\delta_v(A) = \{v \in X - A : \text{ there is an edge between } v \text{ and a vertex in } A\}.$$

Let E_X be the set of edges (x, y) in X. The **edge boundary** of A in X is

$$\delta_e(A) = \{(x, y) \in E_X : x \in A \text{ and } y \in X - A\}.$$

Let $c > 0$ be real. A k-regular graph X with $\#X$ vertices is a **c-expander** if, for all subsets $A \subseteq X$ such that $\#A \leq \#X/2$

$$\#\delta_v(A) \geq c\#A.$$

Exercise 25.3.13 Show that $\delta_v(A) \leq \delta_e(A) \leq k\delta_v(A)$.

Exercise 25.3.14 Let X be a k-regular graph with n vertices that is a c-expander. Show that if n is even then $0 \leq c \leq 1$ and if n is odd then $0 \leq c \leq (n+1)/(n-1)$.

Expander graphs have a number of theoretical applications; one important property is that random walks on expander graphs reach the uniform distribution quickly.

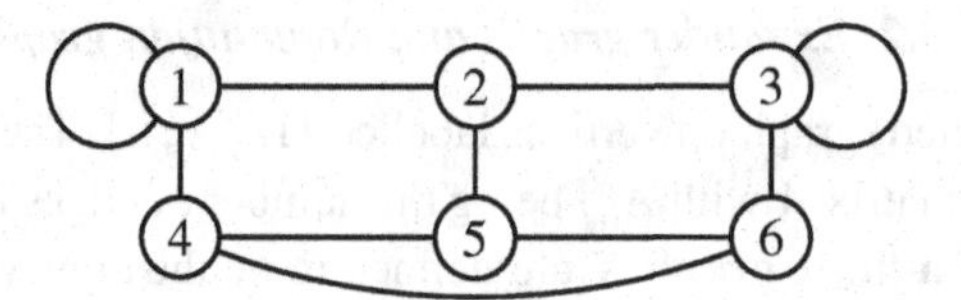

Figure 25.1 A 3-regular graph.

Let X be a k-regular graph. Then

$$\#\delta_e(A) \geq \frac{k - \lambda_1(X)}{2}\#A \tag{25.5}$$

when $\#A \leq \#X/2$ (see Theorem 1.3.1 of Davidoff, Sarnak and Valette [153] or Section 4 of Murty [400][9]). Hence, $\#\delta_v(A) \geq (\frac{1}{2} - \lambda_1(X)/(2k))\#A$ and so Ramanujan graphs are expander graphs. Indeed, small values for $\lambda_1(X)$ give large expansion factors. We refer to [153, 400] for further details and references.

Exercise 25.3.15 Consider the 3-regular graph X in Figure 25.1. Determine the eigenvalues of X. Is this graph Ramanujan? Determine $\delta_v(\{1\})$, $\delta_v(\{1, 2\})$ and $\delta_v(\{1, 3\})$. Verify that $\#\delta_v(A) \geq \#A$ for all subsets A of vertices of X such that $\#A \leq 3$ and so X is an expander.

Exercise 25.3.16 For every $c > 0$ find an integer $n \in \mathbb{N}$, a graph X with n vertices, and a subset A of X such that $\#A \leq n/2$ but $\#\delta_v(A) < c\#A$. (Such a graph is very far from being an expander.)

Consider the ordinary isogeny graph of elliptic curves over $\mathbb{F}_q$ with $\text{End}(E) = \mathcal{O}_K$, the ring of integers in $K = \mathbb{Q}(\sqrt{t^2 - 4q})$. This was shown in the previous section to be a Cayley graph for the ideal class group $\text{Cl}(\mathcal{O}_K)$. Jao, Miller and Venkatesan [278] show, assuming a generalisation of the Riemann hypothesis, that the ordinary isogeny graph is an expander graph (indeed, it is "nearly Ramanujan", i.e. $\lambda_1(X) \leq O(k^\beta)$ for some $0 < \beta < 1$).

25.3.3 Supersingular isogeny graph

For the supersingular isogeny graph we work over $\overline{\mathbb{F}}_p$. The graph is finite. Indeed, Theorem 9.11.11 implies $p/12 - 1 < \#X_{E,\overline{\mathbb{F}}_p,S} < p/12 + 2$. Note that it suffices to consider elliptic curves defined over $\mathbb{F}_{p^2}$ (although the isogenies between them are over $\overline{\mathbb{F}}_p$ in general).

In contrast to the ordinary case, the supersingular graph is always connected using isogenies of any fixed degree. A proof of this result, attributed to Serre, is given in Section 2.4 of Mestre [377].

[9] Note that the proof on page 16 of [400] is for $\delta_e(A)$, not $\delta_v(A)$ as stated.

Theorem 25.3.17 *Let p be a prime and let E and $\widetilde{E}$ be supersingular elliptic curves over $\overline{\mathbb{F}}_p$. Let ℓ be a prime different from p. Then there is an isogeny from E to $\widetilde{E}$ over $\overline{\mathbb{F}}_p$ whose degree is a power of ℓ.*

Proof See Corollary 78 of Kohel [315] or Section 2.4 of Mestre [377]. □

Hence, one can choose any prime ℓ (e.g., $\ell = 2$) and consider the ℓ-isogeny graph $X_{E,\overline{\mathbb{F}}_p,\{\ell\}}$ on supersingular curves over $\overline{\mathbb{F}}_p$. It follows that the graph is $(\ell+1)$-regular and connected.

Exercise 25.3.18 Let $p = 103$. Determine, using Theorem 9.11.11, the number of isomorphism classes of supersingular elliptic curves over $\mathbb{F}_p$ and over $\mathbb{F}_{p^2}$. Determine the 2-isogeny graph whose vertices are supersingular elliptic curves over $\mathbb{F}_{p^2}$.

Exercise 25.3.19 Determine the supersingular 2-isogeny graph over $\mathbb{F}_{11}$. Interpret the results in light of Remark 25.3.2.[10]
[Hint: The isomorphism classes of elliptic curves with $j(E) = 0$ and $j(E) = 1728$ are supersingular modulo 11; this follows from the theory of complex multiplication and the facts that $(\frac{-3}{11}) = (\frac{-4}{11}) = -1$.]

Exercise 25.3.20 Find a prime p such that the set of isomorphism classes of supersingular elliptic curves over $\mathbb{F}_p$ does not form a connected subgraph of $X_{E,\overline{\mathbb{F}}_p,\{2\}}$.

There is a one-to-one correspondence between supersingular elliptic curves E over $\overline{\mathbb{F}}_p$ and projective right modules of rank 1 of a maximal order of the quaternion algebra over $\mathbb{Q}$ ramified at p and ∞ (see Section 5.3 of Kohel [315] or Gross [244]). Pizer has exploited this structure (and connections with Brandt matrices and Hecke operators) to show that the supersingular isogeny graph is a Ramanujan graph. Essentially, the Brandt matrix gives the adjacency matrix of the graph. A good survey is [431], though be warned that the paper does not mention the connection to supersingular elliptic curves.

The supersingular isogeny graph has been used by Charles, Goren and Lauter [120] to construct a cryptographic hash function. It has also been used by Mestre and Oesterlé [377] for an algorithm to compute coefficients of modular forms.

25.4 The structure of the ordinary isogeny graph

This section presents Kohel's results on the structure of the isogeny graph of ordinary elliptic curves over finite fields. Section 25.4.2 gives Kohel's algorithm to compute End(E) for a given ordinary elliptic curve over a finite field.

[10] This example was shown to me by David Kohel.

25.4.1 Isogeny volcanoes

Let E be an ordinary elliptic curve over $\mathbb{F}_q$ and let $\#E(\mathbb{F}_q) = q + 1 - t$. Denote by K the number field $\mathbb{Q}(\sqrt{t^2 - 4q})$ and by $\mathcal{O}_K$ the ring of integers of K. We know that $\mathrm{End}(E) = \mathrm{End}_{\overline{\mathbb{F}}_q}(E)$ is an order in $\mathcal{O}_K$ that contains the order $\mathbb{Z}[\pi_q] = \mathbb{Z}[(t + \sqrt{t^2 - 4q})/2]$ of discriminant $t^2 - 4q$. Let Δ_K be the discriminant of K, namely $\Delta_K = (t^2 - 4q)/c^2$ where c is the largest positive integer such that Δ_K is an integer congruent to 0 or 1 modulo 4. The integer c is the **conductor** of the order $\mathbb{Z}[\pi_q]$.

Suppose E_1 and E_2 are elliptic curves over $\mathbb{F}_q$ such that $\mathrm{End}(E_i) = \mathcal{O}_i$, for $i = 1, 2$, where $\mathcal{O}_1$ and $\mathcal{O}_2$ are orders in K containing $\mathbb{Z}[\pi_q]$. We now present some results about the isogenies between such elliptic curves.

Lemma 25.4.1 *Let $\phi : E \to \widetilde{E}$ be an isogeny of elliptic curves over $\mathbb{F}_q$. If $[\mathrm{End}(E) : \mathrm{End}(\widetilde{E})] = \ell$ (or vice versa) then the degree of ϕ is divisible by ℓ.*

Proof See Propositions 21 and 22 of Kohel [315]. □

Definition 25.4.2 Let ℓ be a prime and E an elliptic curve. Let $\mathrm{End}(E) = \mathcal{O}$. An ℓ-isogeny $\phi : E \to \widetilde{E}$ is called **horizontal** (respectively, **ascending**, **descending**) if $\mathrm{End}(\widetilde{E}) \cong \mathcal{O}$ (respectively, $[\mathrm{End}(\widetilde{E}) : \mathcal{O}] = \ell$, $[\mathcal{O} : \mathrm{End}(\widetilde{E})] = \ell$).

Exercise 25.4.3 Let $\phi : E \to \widetilde{E}$ be an ℓ-isogeny. Show that if ϕ is horizontal (respectively, ascending, descending) then $\widehat{\phi}$ is horizontal (respectively, descending, ascending).

Example 25.4.4 We now give a picture of how the orders relate to one another. Suppose the conductor of $\mathbb{Z}[\pi_q]$ is 6 (e.g., $q = 31$ and $t = \pm 4$) so that $[\mathcal{O}_K : \mathbb{Z}[\pi_q]] = 6$. Write $\mathcal{O}_K = \mathbb{Z}[\theta]$. Then the orders $\mathcal{O}_2 = \mathbb{Z}[2\theta]$ and $\mathcal{O}_3 = \mathbb{Z}[3\theta]$ are contained in $\mathcal{O}_K$ and are such that $[\mathcal{O}_K : \mathcal{O}_i] = i$ for $i = 2, 3$.

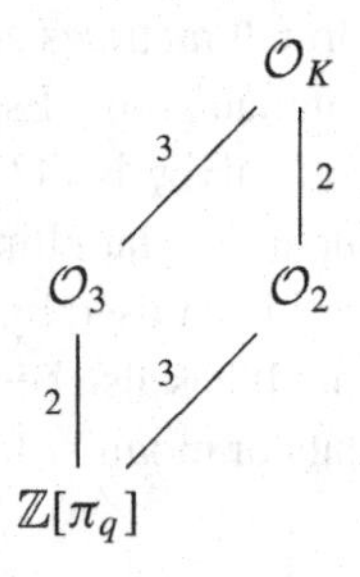

Definition 25.4.5 Let the notation be as above. If $\mathrm{End}(E) = \mathcal{O}_K$ then E is said to be on the **surface** of the isogeny graph.[11] If $\mathrm{End}(E) = \mathbb{Z}[\pi_q]$ then E is said to be on the **floor** of the isogeny graph.

By the theory of complex multiplication, the number of isomorphism classes of elliptic curves over $\mathbb{F}_q$ on the surface is equal to the ideal class number of the ring $\mathcal{O}_K$.

[11] Kohel's metaphor was intended to be aquatic: the floor represents the ocean floor and the surface represents the surface of the ocean.

Theorem 25.4.6 *Let E be an ordinary elliptic curve over $\mathbb{F}_q$ as above and let $\mathcal{O} = \mathrm{End}(E)$ be an order in $\mathcal{O}_K$ containing $\mathbb{Z}[\pi_q]$. Let $c = [\mathcal{O}_K : \mathcal{O}]$ and let ℓ be a prime. Every ℓ-isogeny $\phi : E \to \widetilde{E}$ arises from one of the following cases.*

- *If $\ell \nmid c$ then there are exactly $(1 + (\frac{t^2-4q}{\ell}))$ equivalence classes of horizontal ℓ-isogenies over $\mathbb{F}_q$ from E to other elliptic curves.*[12]
- *If $\ell \mid c$ then there are no horizontal ℓ-isogenies starting at E.*
- *If $\ell \mid c$ there is exactly one ascending ℓ-isogeny starting at E.*
- *If $\ell \mid [\mathcal{O} : \mathbb{Z}[\pi_q]]$ then the number of equivalence classes of ℓ-isogenies starting from E is $\ell + 1$, where the horizontal and ascending isogenies are as described and the remaining isogenies are descending.*
- *If $\ell \nmid [\mathcal{O} : \mathbb{Z}[\pi_q]]$ then there is no desending ℓ-isogeny.*

Proof See Proposition 23 of Kohel [315]. A proof over $\mathbb{C}$ is also given in Appendix A.5 of [200]. □

Corollary 25.4.7 *Let E be an ordinary elliptic curve over $\mathbb{F}_q$ with $\#E(\mathbb{F}_q) = q + 1 - t$. Let c be the conductor of $\mathbb{Z}[\sqrt{t^2 - 4q}]$ and suppose $\ell \mid c$. Then $\ell \nmid [\mathrm{End}(E) : \mathbb{Z}[\pi_q]]$ if and only if there is a single ℓ-isogeny over $\mathbb{F}_q$ starting from E.*

Example 25.4.8 Let $q = 67$ and consider the elliptic curve $E : y^2 = x^3 + 11x + 21$ over $\mathbb{F}_q$. One has $\#E(\mathbb{F}_q) = 64 = q + 1 - t$ where $t = 4$ and $t^2 - 4q = 2^2 \cdot 3^2 \cdot (-7)$. Further, $j(E) = 42 \equiv -3375 \pmod{67}$, so E has complex multiplication by $(1 + \sqrt{-7})/2$. Since the ideal class number of $\mathbb{Q}(\sqrt{-7})$ is 1, it follows that E is the unique elliptic curve up to isomorphism on the surface of the isogeny graph.

Since 2 splits in $\mathbb{Z}[(1 + \sqrt{-7})/2]$ there are two 2-isogenies from E to elliptic curves on the surface (i.e., to E itself) and so there is only one 2-isogeny down from E. Using the modular polynomial we deduce that the 2-isogeny down maps to the isomorphism class of elliptic curves with j-invariant 14. One can verify that the only 2-isogeny over $\mathbb{F}_q$ from $j = 14$ is the ascending isogeny back to $j = 42$.

We have $(\frac{-7}{3}) = -1$ so there are no horizontal 3-isogenies from E. Hence, we expect four 3-isogenies down from E. Using the modular polynomial we compute the corresponding j-invariants to be 33, 35, 51 and 57. One can now consider the 2-isogeny graphs containing these elliptic curves on their surfaces. It turns out that the graph is connected, and that there is a cycle of horizontal 2-isogenies from $j = 33$ to $j = 51$ to $j = 35$ to $j = 57$. For each vertex we therefore only expect one 2-isogeny down to the floor. The corresponding j-invariants are 44, 4, 18 and 32 respectively. Figure 25.2 gives the 2-isogeny graph in this case.

Exercise 25.4.9 Draw the 3-isogeny graph for the elliptic curves in Example 25.4.8. Is $X_{E,\mathbb{F}_q,\{2,3\}}$ connected? If so, what is its diameter?

[12] The symbol $(\frac{t^2-4q}{\ell})$ is the Kronecker symbol as in equation (25.4).

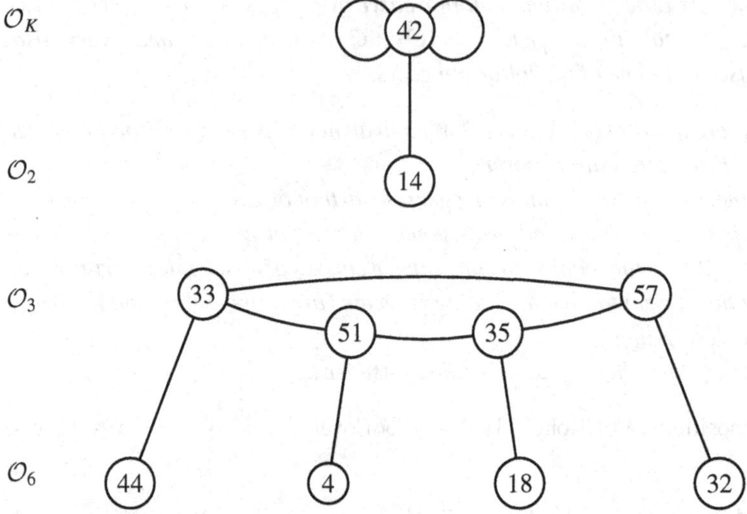

Figure 25.2 A 2-isogeny graph with two volcanoes. The symbols on the left hand side denote the endomorphism ring of curves on that level, using the same notation as Example 25.4.4.

Fix a prime $\ell \mid c$ where c is the conductor of $\mathbb{Z}[\pi_q]$. Consider the subgraph of the isogeny graph corresponding to isogenies whose degree is equal to ℓ. We call this the ℓ-isogeny graph. This graph is often not connected (for example, it is not connected when c is not a power of ℓ or when primes above ℓ do not generate $\mathrm{Cl}(\mathcal{O}_K)$). Even when ℓ splits and c is 1 or a power of ℓ, the graph is often not connected (the graph is connected only when the prime ideals above ℓ generate the ideal class group). Theorem 25.4.6 shows that each component of the ℓ-isogeny graph has a particular shape (that Fouquet and Morain [192] called a **volcano**).

We now give a precise definition for volcanoes. Let $\#E(\mathbb{F}_q) = q + 1 - t$ and let c be the conductor of $\mathbb{Z}[\pi_q]$ and suppose $\ell^m \| c$. Let $K = \mathbb{Q}(\sqrt{t^2 - 4q})$ and denote by $\mathcal{O}_K$ the maximal order in K. A **volcano** is a connected component of the graph $X_{E,\mathbb{F}_q,\{\ell\}}$. A volcano has $m + 1$ "levels" $V_0, \ldots, V_m$, being subgraphs of the ℓ-isogeny graph; where vertices in V_i (i.e., on level i) correspond to isomorphism classes of elliptic curves $\widetilde{E}$ such that $\ell^i \| [\mathcal{O}_K : \mathrm{End}(\widetilde{E})]$. In other words, V_0 is on the surface of this component of the ℓ-isogeny graph (but not necessarily on the surface of the entire isogeny graph $X_{E,\mathbb{F}_q,S}$) and V_m is on the floor of this component of the ℓ-isogeny graph (though, again, not necessarily on the floor of the whole isogeny graph). The surface of a volcano (i.e., V_0) is also called the **crater**. The graph V_0 is a connected regular graph with each vertex of degree at most 2. For all $0 < i \leq m$ and every vertex $v \in V_i$ there is a unique edge from v "up" to a vertex in V_{i-1}. For all $0 \leq i < m$ and every $v \in V_i$ the degree of v is $\ell + 1$. Every vertex in V_m has degree 1.

25.4.2 Kohel's algorithm (ordinary case)

Kohel used the results of Section 25.4.1 to develop deterministic algorithms for computing $\mathrm{End}(E)$ (i.e., determining the level) for an elliptic curve E over $\mathbb{F}_q$. We sketch the algorithm for ordinary curves. Two facts are of crucial importance in Kohel's algorithm. The first (Corollary 25.4.7) is that one can recognise the floor when standing on it. The second fact is that if one starts a chain of ℓ-isogenies with a descending isogeny, and avoids backtracking, then all the isogenies in the chain are descending.

Before going any further, we discuss how to compute a non-backtracking chain of ℓ-isogenies. Given $j(E)$ one can compute the j-invariants of ℓ-isogenous curves over $\mathbb{F}_q$ by computing the roots of $F(y) = \Phi_\ell(j(E), y)$ in $\mathbb{F}_q$. Recall that one computes $\Phi_\ell(x, y)$ in $O(\ell^{3+\epsilon})$ bit operations and finds the roots of $F(y)$ in $\mathbb{F}_q$ in expected time bounded by $O(\ell^2 \log(\ell) \log(q))$ operations in $\mathbb{F}_q$. Let $j_0 = j(E)$ and let j_1 be one of the roots of of $F(y)$. We want to find, if possible, $j_2 \in \mathbb{F}_q$ such that there are ℓ-isogenies from E to E_1 (where $j(E_1) = j_1$) and from E_1 to E_2 (where $j(E_2) = j_2$) and such that $j_2 \neq j_0$ (so the second isogeny is not the dual of the first). The trick is to find roots of $\Phi_\ell(j_1, y)/(y - j_0)$. This process can be repeated to compute a chain $j_0, j_1, j_2, \ldots$ of j-invariants of ℓ-isogenous curves. As mentioned earlier, an alternative approach to walking in the isogeny graph is to find $\mathbb{F}_q$-rational factors of the ℓ-division polynomial and use Vélu's formulae; this is less efficient in general and the method to detect backtracking is to compute the image curve using Vélu's formulae and then compute its j-invariant.

The basic idea of Kohel's algorithm is, for each prime ℓ dividing[13] the conductor of $\mathbb{Z}[\pi_q]$, to find a chain of ℓ-isogenies from E to an elliptic curve on the floor. Suppose ℓ is a prime and $\ell^m \| c$. Kohel (on page 46 of [315]) suggests to take two non-backtracking chains of ℓ-isogenies of length at most m from E. If E is on the floor then this is immediately detected. If E is not on the surface then at least one of the initial ℓ-isogenies is descending, so in at most m steps one finds oneself on the floor. So if after m steps neither chain of isogenies has reached the floor then it follows that we must have started on the surface (and some or all of the ℓ-isogenies in the chain were along the surface). Note that, apart from the algorithm for computing roots of polynomials, the method is deterministic.

Exercise 25.4.10 Let $E : y^2 = x^3 + 3x + 6$ over $\mathbb{F}_{37}$ be an elliptic curve. Note that $\#E(\mathbb{F}_{37}) = 37 + 1 - 4$ and $4^2 - 4 \cdot 37 = -2^4 \cdot 7$. Hence, the conductor is 4. We have $j(E) = 10$. Using the modular polynomial one finds the following j-invariants of elliptic curves 2-isogenous to E: 11, 29, 31. Further, there is a single 2-isogeny from j-invariants 11, 31 (in both cases, back to $j = 10$). But from 29 there is a 2-isogeny to $j = 10$ and two 2-isogenies to $j = 29$. What is End(E)? Give j-invariants for a curve on the floor and a curve on the surface.

The worst case of Kohel's algorithm is when the conductor is divisible by one or more very large primes ℓ (since determining the j-invariant of an ℓ-isogenous curve is polynomial in ℓ and so exponential in the input size). Since c can be as big as $\sqrt{q}$ the above method (i.e., taking isogenies to the floor) would therefore have worst-case complexity of at least $q^{1/2}$ bit operations (indeed, it would be $O(q^{3/2+\epsilon})$ operations in $\mathbb{F}_q$ if one includes the cost of generating modular polynomials). Kohel (pages 53 to 57 of [315]) noted that, when ℓ

[13] It is necessary to find the square factors of $t^2 - 4q$, which can be done in deterministic time $\tilde{O}(q^{1/6})$; see Exercise 12.5.1.

is very large, one can more efficiently resolve the issue of whether or not ℓ divides the conductor by finding elements in the ideal class group that are trivial for the maximal order but non-trivial for an order whose conductor is divisible by ℓ; one can then "test" such a relation using isogenies. Using these ideas Kohel proves in Theorem 24 of [315] that, assuming a certain generalisation of the Riemann hypothesis, his algorithm requires $O(q^{1/3+\epsilon})$ bit operations. Kohel also considers the case of supersingular curves.

Bisson and Sutherland [54] consider a randomised version of Kohel's method using ideas from index calculus algorithms in ideal class groups. Their algorithm has heuristic subexponential expected complexity of $O(L_q(1/2, \sqrt{3}/2))$ bit operations. We do not present the details.

Remark 25.4.11 An important remark is that neither of the two efficient ways to generate elliptic curves over finite fields is likely to lead to elliptic curves E such that the conductor of End(E) is divisible by a large prime.

- When generating elliptic curves by choosing a random curve over a large prime field and counting points, then $t^2 - 4q$ behaves like a random integer and so is extremely unlikely to be divisible by the square of a very large prime
- When using the CM method then it is automatic that the curves have $q + 1 - t$ points where $t^2 - 4q$ has a very large square factor. It is easy to arrange that the square factor is divisible by a large prime. However, the elliptic curve itself output by the CM method has End(E) being the maximal order. To get End(E) to be a non-maximal order, one can either use class polynomials corresponding to non-maximal orders or use descending isogenies. Either way, it is infeasible to compute a curve E such that a very large prime divides the conductor of End(E). Furthermore, Kohel's algorithm is not needed in this case since by construction one already knows End(E).

Hence, in practice, the potential problems with large primes dividing the conductor of End(E) do not arise. It is therefore safe to assume that determining End(E) is easy.

25.5 Constructing isogenies between elliptic curves

The **isogeny problem for elliptic curves** is: given two elliptic curves E and $\widetilde{E}$ over $\mathbb{F}_q$ with the same number of points to compute an isogeny between them. Solving this problem is usually considered in two stages:

1. Performing a pre-computation, that computes a chain of prime-degree isogenies from E to $\widetilde{E}$. The chain is usually computed as a sequence of explicit isogenies, though one could store just the "Elkies information" for each isogeny in the chain.
2. Given a specific point $P \in E(\overline{\mathbb{F}}_q)$ to compute the image of P under the isogeny.

The precomputation is typically slow, while it is desirable that the computation of the isogeny be fast (since it might be performed for a large number of points).

An algorithm to solve the isogeny problem, requiring exponential time and space in terms of the input size, was given by Galbraith [200]. For the case of ordinary elliptic curves, an improved algorithm with low-storage requirements was given by Galbraith, Hess and Smart [204]. We briefly sketch both algorithms in this section.

We now make some preliminary remarks in the ordinary case. Let c_1 be the conductor of $\mathrm{End}(E)$ and c_2 the conductor of $\mathrm{End}(\widetilde{E})$. If there is a large prime ℓ that divides c_1 but not c_2 (or vice versa), then any isogeny between E and $\widetilde{E}$ will have degree divisible by ℓ and hence the isogeny will be slow to compute. Since the conductor is a square factor of $t^2 - 4q$ it can be, in theory, as big as $q^{1/2}$. It follows that one does not expect an efficient algorithm for this problem in general. However, as mentioned in Remark 25.4.11, one can in practice ignore this bad case and assume the primes dividing the conductor are moderate.

For the rest of this section, in the ordinary case, we assume that $\mathrm{End}(E) = \mathrm{End}(\widetilde{E}) = \mathcal{O}$. (If this is not the case then take vertical isogenies from E to reduce to it.) Usually $\mathcal{O}$ is the maximal order. This is desirable, because the class number of the maximal order is typically smaller (and never larger) than the class number of the sub-orders, and so the algorithm to find the isogeny works more quickly in this case. However, for the sake of generality we do not insist that $\mathcal{O}$ is the maximal order. The general case could appear if there is a very large prime dividing the conductor of $\mathcal{O}$.

25.5.1 The Galbraith algorithm

The algorithm of Galbraith [200] finds a path between two vertices in the isogeny graph $X_{E,\Bbbk,S}$ using a breadth-first search (see Section 22.2 of [136]). This algorithm can be used in both the ordinary and supersingular cases. More precisely, one starts with sets $X_0 = \{j(E)\}$ and $Y_0 = \{j(\widetilde{E})\}$ (we are assuming the vertices of the isogeny graph are labelled by j-invariants) and, at step i, computes $X_i = X_{i-1} \cup \delta_v(X_{i-1})$ and $Y_i = Y_{i-1} \cup \delta_v(Y_{i-1})$ where $\delta_v(X)$ is the set of vertices in the graph that are connected to a vertex in X by an edge. Computing $\delta_v(X)$ involves finding the roots in $\Bbbk$ of $\Phi_\ell(j, y)$ for every $j \in X$ and every $\ell \in S$. In the supersingular case, the set S of possible isogeny degrees usually consists of a single prime ℓ. In the ordinary case, S could have as many as $\log(q)$ elements, and one might not compute the whole of $\delta_v(X)$ but just the boundary in a subgraph corresponding to a (random) subset of S. In either case, the cost of computing $\delta_v(X)$ is clearly proportional to $\#X$.[14] The algorithm stops when $X_i \cap Y_i \neq \emptyset$, in which case it is easy to compute the isogeny from E to $\widetilde{E}$.

Exercise 25.5.1 Write pseudocode for the above algorithm.

Under the (probably unreasonable) assumption that new values in $\delta_v(X_i)$ behave like uniformly chosen elements in the isogeny graph, one expects from the birthday paradox that the two sets have non-empty intersection when $\#X_i + \#Y_i \geq \sqrt{\pi \# X_{E,\Bbbk,S}}$. Since the

[14] When all $\ell \in S$ are used at every step, to compute $\delta_v(X_i)$ it suffices to consider only vertices $j \in \delta_v(X_{i-1})$.

graph is an expander, we know that $\#X_i = \#X_{i-1} + \#\delta_v(X_{i-1}) \geq (1+c)\#X_{i-1}$ when X_{i-1} is small, and so $\#X_i \geq (1+c)^i$.

In the supersingular case we have $\#X_{E,\Bbbk,S} = O(q)$ and in the ordinary case we have $\#X_{E,\Bbbk,S} = h(\mathcal{O}) = O(q^{1/2}\log(q))$. In both cases, one expects the algorithm to terminate after $O(\log(q))$ steps. Step i involves, for every vertex $j \in X_i$ (or $j \in \delta_v(X_{i-1})$) and every $\ell \in S$, computing roots of $\Phi_\ell(j, y)$ in $\mathbb{F}_q$. One can check that if all ℓ are polynomially bounded in $\log(q)$ then the expected number of bit operations is bounded by $\sqrt{\#X_{E,\Bbbk,S}}$ times a polynomial in $\log(q)$.

In the supersingular case the algorithm is expected to perform $\tilde{O}(q^{1/2})$ bit operations. In the ordinary case, by Bach's result (and therefore, assuming various generalisations of the Riemann hypothesis) we can restrict to isogenies of degree at most $6\log(q)^2$ and so each step is polynomial-time (the dominant cost of each step is finding roots of the modular polynomial; see Exercise 25.2.2). The total complexity is therefore an expected $\tilde{O}(q^{1/4})$ bit operations. The storage required is expected to be $O(q^{1/4}\log(q)^2)$ bits.

Exercise 25.5.2 Let $m \in \mathbb{N}$. Suppose all $\ell \in S$ are such that $\ell = O(\log(q)^m)$. Let ϕ be the isogeny output by the Galbraith algorithm. Show, under the same heuristic assumptions as above, that the isogeny can be evaluated at $P \in E(\mathbb{F}_q)$ in polynomial-time.

Exercise 25.5.3 Isogenies of small degree are faster to compute than isogenies of large degree. Hence, the average cost to compute an ℓ-isogeny can be used as a weight for the edges in the isogeny graph corresponding to ℓ-isogenies. It follows that there is a well-defined notion of shortest path in the isogeny graph between two vertices. Show how Dijkstra's algorithm (see Section 24.3 of [136]) can be used to find a chain of isogenies between two elliptic curves that can be computed in minimal time. What is the complexity of this algorithm?

25.5.2 The Galbraith–Hess–Smart algorithm

We now restrict to the ordinary isogeny graph and sketch the algorithm of Galbraith, Hess and Smart [204]. The basic idea is to replace the breadth-first search by a random walk, similar to that used in the kangaroo algorithm.

Let H be a hash function from $\mathbb{F}_q$ to a set S of prime ideals of small norm. One starts random walks at $x_0 = j(E)$ and $y_0 = j(\widetilde{E})$ and stores ideals $\mathfrak{a}_0 = (1)$, $\mathfrak{b}_0 = (1)$. One can think of $(x_0, \mathfrak{a}_0)$ as a "tame walk" and $(y_0, \mathfrak{b}_0)$ as a "wild walk". Each step of the algorithm computes new values x_i and y_i from x_{i-1} and y_{i-1}: to compute x_i set $\mathfrak{l} = H(x_{i-1})$ and $\ell = N(\mathfrak{l})$, find the roots of $\Phi_\ell(x_{i-1}, z)$, choose the root corresponding to the ideal $\mathfrak{l}$ (using the trick mentioned in Remark 25.3.9) and call it x_i. The same process is used (with the same function H) for the sequence y_i. The ideals are also updated as $\mathfrak{a}_i = \mathfrak{a}_{i-1}\mathfrak{l}$ (reduced in the ideal class group to some short canonical representation of ideals). If $x_i = y_j$ then the walks follow the same path. We designate certain elements of $\mathbb{F}_q$ as being distinguished points, and if x_i or y_i is distinguished then it is stored together with the corresponding ideal

$\mathfrak{a}_i$ or $\mathfrak{b}_i$. After a walk hits a distinguished point there are two choices: it could be allowed to continue or it could be restarted at a j-invariant obtained by taking a short random isogeny chain (perhaps corresponding to primes not in S) from E or $\widetilde{E}$.

Once a collision is detected one has an isogeny corresponding to ideal $\mathfrak{a}$ from $j(E)$ to some j, and an isogeny corresponding to ideal $\mathfrak{b}$ from $j(\widetilde{E})$ to j. Hence, the ideal $\mathfrak{a}\mathfrak{b}^{-1}$ gives the isogeny from $j(E)$ to $j(\widetilde{E})$.

Stolbunov has noted that, since the ideal class group is Abelian, it is not advisable to choose S such that $\mathfrak{l}, \mathfrak{l}^{-1} \in S$ (since such a choice means that walks remain "close" to the original j-invariant, and cycles in the random walk might arise). It is also faster to use isogenies of small degree more often than those with large degree. We refer to Galbraith and Stolbunov [213] for further details.

The remaining problem is that the ideal $\mathfrak{a}\mathfrak{b}^{-1}$ is typically of large norm. By construction, it is a product of exponentially many small primes. Since the ideal class group is commutative, such a product has a short representation (storing the exponents for each prime), but this leads to an isogeny that requires exponential computation. The proposal from [204] is to represent ideals using the standard representation for ideals in quadratic fields, and to "smooth" the ideal using standard techniques from index calculus algorithms in ideal class groups. It is beyond the scope of this book to discuss these ideas in detail. However, we note that the isogeny then has subexponential length and uses primes ℓ of subexponential degree. Hence, the second stage of the isogeny computation is subexponential-time; this is not as fast as it would be with the basic Galbraith algorithm. The idea of smoothing an isogeny has also been used by Bröker, Charles and Lauter [101] and Jao and Soukharev [279].

Since the ordinary isogeny graph is conjecturally an expander graph, we know that a random walk on it behaves close to the uniform distribution after sufficiently many steps. We make the heuristic assumption that the pseudorandom walk proposed above has this property when the number of different primes used is sufficiently large and the hash function H is good. Then, by the birthday paradox, one expects a collision after $\sqrt{\pi h(\mathcal{O})}$ vertices have been visited. As a result, the heuristic expected running time of the algorithm is $O(q^{1/4})$ isogeny chain steps, and the storage can be made low by making distinguished elements rare. The algorithm can be distributed: using L processors of equal power one solves the isogeny problem in $\tilde{O}(q^{1/4}/L)$ bit operations.

25.6 Relating the discrete logarithm problem on isogenous curves

The main application of the algorithms in Section 25.5 is to relate the discrete logarithm problem on curves with the same number of points. More precisely, let E and $\widetilde{E}$ be elliptic curves over $\mathbb{F}_q$ with $\#E(\mathbb{F}_q) = \#\widetilde{E}(\mathbb{F}_q)$. Let r be a large prime dividing $\#E(\mathbb{F}_q)$. A natural question is whether the discrete logarithm problem (in the subgroup of order r) has the same difficulty in both groups $E(\mathbb{F}_q)$ and $\widetilde{E}(\mathbb{F}_q)$. To study this question one wants to reduce the discrete logarithm problem from $E(\mathbb{F}_q)$ to $\widetilde{E}(\mathbb{F}_q)$. If we have an isogeny $\phi : E \to \widetilde{E}$

of degree not divisible by r and if ϕ can be efficiently computed then we have such a reduction.

As we have seen, if there is a very large prime dividing the conductor of $\mathrm{End}(E)$ but not the conductor of $\mathrm{End}(\widetilde{E})$ (or vice versa) then it is not efficient to compute an isogeny from E to $\widetilde{E}$. In this case, one cannot make any inference about the relative difficulty of the DLP in the two groups. No example is known of elliptic curves E and $\widetilde{E}$ of this form (i.e., with a large conductor gap) but for which the DLP on one is known to be significantly easier than the DLP on another. The nearest we have to an example of this phenomenon is with elliptic curves E with $\#\mathrm{Aut}(E) > 2$ (and so one can accelerate the Pollard rho method using equivalence classes as in Section 14.4) but with an isogeny from E to $\widetilde{E}$ with $\#\mathrm{Aut}(\widetilde{E}) = 2$.

On the other hand, if the conductors of $\mathrm{End}(E)$ and $\mathrm{End}(\widetilde{E})$ have the same very large prime factors (or no large prime factors) then we can (conditional on a generalised Riemann hypothesis) compute an isogeny between them in $\tilde{O}(q^{1/4})$ bit operations. This is not a polynomial-time reduction. But, since the current best algorithms for the DLP on elliptic curves run in $\tilde{O}(q^{1/2})$ bit operations, it shows that from a practical point of view the two DLPs are equivalent.

Jao, Miller and Venkatesan [277] have a different, and perhaps more useful, interpretation of the isogeny algorithms in terms of random self-reducibility of the DLP in an isogeny class of elliptic curves. The idea is that if E is an elliptic curve over $\mathbb{F}_q$ then by taking a relatively short random walk in the isogeny graph one arrives at a "random" (again ignoring the issue of large primes dividing the conductor) elliptic curve $\widetilde{E}$ over $\mathbb{F}_q$ such that $\#\widetilde{E}(\mathbb{F}_q) = \#E(\mathbb{F}_q)$. Hence, one easily turns a specific instance of the DLP (i.e., for a specific elliptic curve) into a random instance. It follows that if there were a "large" set of "weak" instances of the DLP in the isogeny class of E then, after enough trials, one should be able to reduce the DLP from E to one of the elliptic curves in the weak class. One concludes that either the DLP is easy for "most" curves in an isogeny class, or is hard for "most" curves in an isogeny class.

26

Pairings on elliptic curves

This chapter is a very brief summary of the mathematics behind pairings on elliptic curves. Some applications of pairings in elliptic curve cryptography have already been presented in the book (for example, the identity-based encryption scheme of Boneh and Franklin in Section 23.3.2 and the Boneh-Boyen signature scheme in Section 22.2.3). We present several other important applications of pairings, such as the Menezes–Okamoto–Vanstone/Frey–Rück reduction of the discrete logarithm problem from elliptic curves to finite fields.

Due to lack of space we do not give full details of the subject. Good general references for pairings and pairing-based cryptography are Chapters IX and X of [61], Chapters 6, 16 and 24 of [16] and [286].

26.1 Weil reciprocity

The following theorem is an important tool for studying pairings. Recall that a divisor on a curve C over a field $\Bbbk$ is a finite sum $D = \sum_{P \in C(\overline{\Bbbk})} n_P(P)$ (i.e., $n_P = 0$ for all but finitely many $P \in C(\overline{\Bbbk})$). The support of a divisor D is the set of points $\mathrm{Supp}(D) = \{P \in C(\overline{\Bbbk}) : n_P \neq 0\}$. To a function f on a curve one associates the divisor $\mathrm{div}(f)$ as in Definition 7.7.2. If f is a function on a curve and D is a divisor such that the support of D is distinct from the support of $\mathrm{div}(f)$ then $f(D)$ is defined to be $\prod_{P \in C(\overline{\Bbbk}), n_P \neq 0} f(P)^{n_P}$.

Exercise 26.1.1 Let D_1 and D_2 be divisors with disjoint support on a curve C. Suppose D_1 is principal. Show that $f(D_2)$ is well-defined, subject to $\mathrm{div}(f) = D_1$, if and only if D_2 has degree 0.

Theorem 26.1.2 *(Weil reciprocity) Let C be a curve over a field $\Bbbk$. Let $f, g \in \Bbbk(C)$ be functions such that* $\mathrm{Supp}(\mathrm{div}(f)) \cap \mathrm{Supp}(\mathrm{div}(g)) = \varnothing$. *Then*

$$f(\mathrm{div}(g)) = g(\mathrm{div}(f)).$$

Proof (Sketch) One first shows that the result holds for functions on $C = \mathbb{P}^1$. Then take any covering $\phi : C \to \mathbb{P}^1$ and apply the pullback. We refer to the appendix of Chapter IX of [61] for details. A proof over $\mathbb{C}$ is given in the appendix to Section 18.1 of Lang [328]. □

Pages 24–26 of Charlap and Coley [118] present a generalised Weil reciprocity that does not require the divisors to have disjoint support.

26.2 The Weil pairing

The Weil pairing plays an important role in the study of elliptic curves over number fields, but tends to be less important in cryptography. For completeness, we briefly sketch its definition.

Let E be an elliptic curve over $\Bbbk$ and let $n \in \mathbb{N}$ be coprime to char($\Bbbk$). Let $P, Q \in E[n]$. Then there is a function $f \in \overline{\Bbbk}(E)$ such that $\operatorname{div}(f) = n(Q) - n(\mathcal{O}_E)$. Let $Q' \in E(\overline{\Bbbk})$ be any point such that $[n]Q' = Q$, and so $[n^2]Q' = \mathcal{O}_E$. Note that $[n]$ is unramified and the divisor $D = [n]^*((Q) - (\mathcal{O}_E))$ is equal to

$$\sum_{R \in E[n]} (Q' + R) - (R).$$

Since $\sum_{R\in E[n]} R = \mathcal{O}_E$ and $[n^2]Q' = \mathcal{O}_E$ it follows from Theorem 7.9.9 that D is a principal divisor. So there is a function $g \in \overline{\Bbbk}(E)$ such that $\operatorname{div}(g) = D = [n]^*((Q) - (\mathcal{O}_E))$. Now, consider the function $[n]^* f = f \circ [n]$. One has $\operatorname{div}([n]^* f) = [n]^*(\operatorname{div}(f)) = [n]^*(n(Q) - n(\mathcal{O}_E)) = nD$. Hence, the functions $f \circ [n]$ and g^n have the same divisor. Multiplying f by a suitable constant gives $f \circ [n] = g^n$. Now, for any point $U \in E(\overline{\Bbbk})$ such that $[n]U \notin E[n^2]$ we have

$$g(U + P)^n = f([n]U + [n]P) = f([n]U) = g(U)^n.$$

In other words, $g(U + P)/g(U)$ is an nth root of unity in $\overline{\Bbbk}$.

Lemma 26.2.1 *Let the notation be as above. Then $g(U + P)/g(U)$ is independent of the choice of the point $U \in E(\overline{\Bbbk})$.*

Proof See Section 11.2 of Washington [560]. The proof is described as “slightly technical" and uses the Zariski topology. □

Definition 26.2.2 Let E be an elliptic curve over a field $\Bbbk$ and let $n \in \mathbb{N}$ be such that $\gcd(n, \operatorname{char}(\Bbbk)) = 1$. Define

$$\mu_n = \{z \in \overline{\Bbbk}^* : z^n = 1\}.$$

The **Weil pairing** is the function

$$e_n : E[n] \times E[n] \to \mu_n$$

defined (using the notation above) as $e_n(P, Q) = g(U + P)/g(U)$ for any point $U \in E(\overline{\Bbbk})$, $U \notin E[n^2]$ and where $\operatorname{div}(g) = [n]^*((Q) - (\mathcal{O}_E))$.

Theorem 26.2.3 *The Weil pairing satisfies the following properties.*

1. *(Bilinear) For* $P_1, P_2, Q \in E[n]$, $e_n(P_1 + P_2, Q) = e_n(P_1, Q)e_n(P_2, Q)$ *and* $e_n(Q, P_1 + P_2) = e_n(Q, P_1)e_n(Q, P_2)$.
2. *(Alternating) For* $P \in E[n]$, $e_n(P, P) = 1$.
3. *(Non-degenerate) If* $e_n(P, Q) = 1$ *for all* $Q \in E[n]$ *then* $P = \mathcal{O}_E$.
4. *(Galois invariant) If* E *is defined over* $\Bbbk$ *and* $\sigma \in \mathrm{Gal}(\overline{\Bbbk}/\Bbbk)$ *then* $e_n(\sigma(P), \sigma(Q)) = \sigma(e_n(P, Q))$.
5. *(Compatible) If* $P \in E[nm]$ *and* $Q \in E[n]$ *then*

$$e_{nm}(P, Q) = e_n([m]P, Q).$$

Proof See Theorem III.8.1 of Silverman [505] or Theorem 11.7 of Washington [560]. The non-degeneracy proof in [505] is very sketchy (the reference to 4.10(b) only relates the translation map to a Galois map on $\overline{\Bbbk}(E)$), but the treatment in [560] fills in the missing details. The non-degeneracy also needs the fact that the genus of E is not zero, so there is no function with divisor $(P) - (\mathcal{O}_E)$ (see Corollary 8.6.5). □

Exercise 26.2.4 Show that any function $e : E[n] \times E[n] \to \mu_n$ that has the properties of the Weil pairing as in Theorem 26.2.3 also has the following properties.

1. $e(\mathcal{O}_E, P) = e(P, \mathcal{O}_E) = 1$ for all $P \in E[n]$.
2. $e(-P, Q) = e(P, Q)^{-1}$ for all $P, Q \in E[n]$.
3. $e(P, Q) = e(Q, P)^{-1}$ for all $P, Q \in E[n]$.
4. If $\{P, Q\}$ generate $E[n]$ then the values of e on $E[n] \times E[n]$ are uniquely determined by the single value $e(P, Q)$.

Exercise 26.2.5 Let E be an elliptic curve over $\mathbb{F}_q$ and let $n \in \mathbb{N}$. Prove that $E[n] \subseteq E(\mathbb{F}_q)$ implies $n \mid (q - 1)$.

For elliptic curves over $\mathbb{C}$ the Weil pairing has a very simple interpretation. Recall that an elliptic curve over $\mathbb{C}$ is isomorphic (as a manifold) to $\mathbb{C}/L$ where L is a lattice of rank 2 and that this isomorphism also preserves the group structure. Fix a pair $\{z_1, z_2\}$ of generators for L as a $\mathbb{Z}$-module. The points of order n are $\frac{1}{n}L/L$, so are identified with $\{(az_1 + bz_2)/n : 0 \le a, b < n\}$. The function

$$e_n((az_1 + bz_2)/n, (cz_1 + dz_2)/n) = \exp(2\pi i(ad - bc)/n)$$

is easily checked to be bilinear, non-degenerate and alternating. Hence, it is (a power of) the Weil pairing. We refer to the appendix of Section 18.1 of Lang [328] for further details. Connections with the intersection pairing are discussed in Section 12.2 of Husemoller [272] and Edixhoven [174].

There is an alternative definition[1] of the Weil pairing that is more useful for implementation, but for which it is harder to prove non-degeneracy. For $P, Q \in E[n]$ let D_P and D_Q be degree 0 divisors such that $D_P \equiv (P) - (\mathcal{O}_E)$, $D_Q \equiv (Q) - (\mathcal{O}_E)$ and $\mathrm{Supp}(D_P) \cap \mathrm{Supp}(D_Q) = \varnothing$. Let $f_P, f_Q \in \overline{\mathbb{k}}(E)$ be functions such that $\mathrm{div}(f_P) = nD_P$ and $\mathrm{div}(f_Q) = nD_Q$. Then

$$e_n(P, Q) = f_Q(D_P)/f_P(D_Q). \tag{26.1}$$

The equivalence is shown in Theorem 4 of the extended and unpublished version of Hess [256], and in Section 11.6.1 of Washington [560].

The Weil pairing can be generalised from $E[n] \times E[n]$ to $\ker(\phi) \times \ker(\hat{\phi}) \subseteq E[n] \times \widetilde{E}[n]$ where $\phi : E \to \widetilde{E}$ is an isogeny. For details see Exercise 3.15 of Silverman [505] or Garefalakis [219]. For the Weil pairing on Jacobian varieties of curves of genus $g > 1$ we refer to Section 20 of Mumford [398].

26.3 The Tate–Lichtenbaum pairing

Tate defined a pairing for Abelian varieties over local fields and Lichtenbaum showed how to compute it efficiently in the case of Jacobian varieties of curves. Frey and Rück [196] showed how to compute it for elliptic curves over finite fields, and emphasised its cryptographic relevance. This pairing is the basic building block of most pairing-based cryptography.

Exercise 26.3.1 Let E be an elliptic curve over a finite field $\mathbb{F}_q$ and let $n \in \mathbb{N}$ be such that $\gcd(n, q) = 1$ and $n \mid \#E(\mathbb{F}_q)$. Define

$$nE(\mathbb{F}_q) = \{[n]Q : Q \in E(\mathbb{F}_q)\}.$$

Show that $nE(\mathbb{F}_q)$ is a group. Show that $E(\mathbb{F}_q)[n] = \{P \in E(\mathbb{F}_q) : [n]P = \mathcal{O}_E\}$, $E(\mathbb{F}_q)/nE(\mathbb{F}_q) = \{P + nE(\mathbb{F}_q) : P \in E(\mathbb{F}_q)\}$ and $\mathbb{F}_q^*/(\mathbb{F}_q^*)^n$ are finite groups of exponent n.

Let notation be as in Exercise 26.3.1. Let $P \in E(\mathbb{F}_q)[n]$ and $Q \in E(\mathbb{F}_q)$. Then $n(P) - n(\mathcal{O}_E)$ is principal, so there is a function $f \in \mathbb{F}_q(E)$ such that $\mathrm{div}(f) = n(P) - n(\mathcal{O}_E)$. Let D be a divisor on E with support disjoint from $\mathrm{Supp}(\mathrm{div}(f)) = \{\mathcal{O}_E, P\}$ but such that D is equivalent to $(Q) - (\mathcal{O}_E)$ (for example, $D = (Q + R) - (R)$ for some point[2] $R \in E(\overline{\mathbb{F}}_q)$, $R \notin \{\mathcal{O}_E, P, -Q, P - Q\}$). We define the **Tate–Lichtenbaum pairing** to be

$$t_n(P, Q) = f(D). \tag{26.2}$$

We will explain below that

$$t_n : E(\mathbb{F}_q)[n] \times E(\mathbb{F}_q)/nE(\mathbb{F}_q) \to \mathbb{F}_q^*/(\mathbb{F}_q^*)^n.$$

[1] The literature is inconsistent and some of the definitions (for example, Section 18.1 of Lang [328], Exercise 3.16 of Silverman [505] and Section 3 of Miller [383]) are actually for $e_n(Q, P) = e_n(P, Q)^{-1}$. For further discussion of this issue see Remark 11.3 and Section 11.6 of Washington [560]. Also see the "Warning" at the end of Section 4 of Miller [385].

[2] One can usually take $R \in E(\mathbb{F}_q)$, but see page 187 of [61] for an example that shows that this is not always possible.

First, we show that the pairing is well-defined. We sketch the proof, as it is a nice and simple application of Weil reciprocity.

Lemma 26.3.2 *Let the notation be as above. Let $P \in E(\mathbb{F}_q)[n]$ and let $f \in \mathbb{F}_q(E)$ be such that $\operatorname{div}(f) = n(P) - n(\mathcal{O}_E)$. Let D_1, D_2 be divisors on E defined over $\mathbb{F}_q$ with support disjoint from $\{\mathcal{O}_E, P\}$.*

1. *Suppose $D_1 \equiv D_2 \equiv (Q) - (\mathcal{O}_E)$ for some point $Q \in E(\mathbb{F}_q)$. Then $f(D_1)/f(D_2) \in (\mathbb{F}_q^*)^n$.*
2. *Suppose $D_1 \equiv (Q_1) - (\mathcal{O}_E)$ and $D_2 \equiv (Q_2) - (\mathcal{O}_E)$ where $Q_1, Q_2 \in E(\mathbb{F}_q)$ are such that $Q_1 \neq Q_2$ and $Q_1 - Q_2 \in nE(\mathbb{F}_q)$. Then $f(D_1)/f(D_2) \in (\mathbb{F}_q^*)^n$.*

Proof The first statement is a special case of the second, but it is a convenient stepping-stone for the proof. For the first statement write $D_2 = D_1 + \operatorname{div}(h)$ where h is a function on E defined over $\mathbb{F}_q$. Note that $\operatorname{Supp}(\operatorname{div}(h)) \cap \{\mathcal{O}_E, P\} = \varnothing$. We have

$$f(D_2) = f(D_1 + \operatorname{div}(h)) = f(D_1) f(\operatorname{div}(h)).$$

Now, applying Weil reciprocity gives $f(\operatorname{div}(h)) = h(\operatorname{div}(f)) = h(n(P) - n(\mathcal{O}_E)) = (h(P)/h(\mathcal{O}_E))^n \in (\mathbb{F}_q^*)^n$.

For the second statement write $Q_1 - Q_2 = [n]R$ for some $R \in E(\mathbb{F}_q)$. We may assume that $R \neq \mathcal{O}_E$ since the first statement has already been proved. Then $(Q_1) - (Q_2) = n((R + S) - (S)) + \operatorname{div}(h_0)$ for some $h_0 \in \mathbb{F}_q(E)$ and some $S \in E(\mathbb{F}_q)$ with $S \notin \{\mathcal{O}_E, -R, P, P - R\}$.[3] We also have $D_1 = (Q_1) - (\mathcal{O}_E) + \operatorname{div}(h_1)$ and $D_2 = (Q_2) - (\mathcal{O}_E) + \operatorname{div}(h_2)$ for some $h_1, h_2 \in \mathbb{F}_q(E)$. Putting everything together

$$\begin{aligned} f(D_2) &= f(D_1 - n((R + S) - (S)) + \operatorname{div}(h_2) - \operatorname{div}(h_1) - \operatorname{div}(h_0)) \\ &= f(D_1) f((R + S) - (S))^n f(\operatorname{div}(h_2/(h_0 h_1))). \end{aligned}$$

Since $\operatorname{Supp}(\operatorname{div}(h_2/(h_0 h_1))) \subseteq \operatorname{Supp}(D_1) \cup \operatorname{Supp}(D_2) \cup \{R + S, S\}$ is disjoint from $\{\mathcal{O}_E, P\} = \varnothing$ the result follows from Weil reciprocity. □

Theorem 26.3.3 *The Tate–Lichtenbaum pairing satisfies the following properties:*

1. *(Bilinear) For $P_1, P_2 \in E(\mathbb{F}_q)[n]$, and $Q \in E(\mathbb{F}_q)$, $t_n(P_1 + P_2, Q) = t_n(P_1, Q) t_n(P_2, Q)$. For $Q \in E(\mathbb{F}_q)[n]$ and $P_1, P_2 \in E(\mathbb{F}_q)$, $t_n(Q, P_1 + P_2) = t_n(Q, P_1) t_n(Q, P_2)$.*
2. *(Non-degenerate) Assume $\mathbb{F}_q^*$ contains a non-trivial nth root of unity. Let $P \in E(\mathbb{F}_q)[n]$. If $t_n(P, Q) = 1$ for all $Q \in E(\mathbb{F}_q)$ then $P = \mathcal{O}_E$. Let $Q \in E(\mathbb{F}_q)$. If $t_n(P, Q) = 1$ for all $P \in E(\mathbb{F}_q)[n]$ then $Q \in nE(\mathbb{F}_q)$.*
3. *(Galois invariant) If E is defined over $\mathbb{F}_q$ and $\sigma \in \operatorname{Gal}(\overline{\mathbb{F}}_q/\mathbb{F}_q)$ then $t_n(\sigma(P), \sigma(Q)) = \sigma(t_n(P, Q))$.*

[3] Some tedious calculations are required to show that one can choose $S \in E(\mathbb{F}_q)$ rather than $E(\overline{\mathbb{F}}_q)$ in all cases, but the claim is easy when n is large.

Proof Bilinearity can be proved using ideas similar to those used to prove Lemma 26.3.2 (for all the details see Theorem IX.7 of [61]). Non-degeneracy in the case of finite fields was shown by Frey and Rück [196], but simpler proofs can be found in Hess [256] and Section 11.7 of Washington [560]. Galois invariance is straightforward (see Theorem IX.7 of [61]). □

26.3.1 Miller's algorithm

We now briefly explain how to compute the Tate–Lichtenbaum pairing (and hence the Weil pairing via equation (26.1)). The algorithm first appears in Miller [383].

Definition 26.3.4 Let $P \in E(\Bbbk)$ and $i \in \mathbb{N}$. A **Miller function** $f_{i,P} \in \Bbbk(E)$ is a function on E such that $\operatorname{div}(f_{i,P}) = i(P) - ([i]P) - (i-1)(\mathcal{O}_E)$. Furthermore, we assume that Miller functions are "normalised at infinity" in the sense that the power series expansion at infinity with respect to the canonical uniformiser $t_\infty = x/y$ is 1.

Exercise 26.3.5 Show that $f_{1,P} = 1$. Show that if $f_{i,P}$ and $f_{j,P}$ are Miller functions then one can take

$$f_{i+j,P} = f_{i,P} f_{j,P} l(x, y)/v(x, y)$$

where $l(x, y)$ and $v(x, y)$ are the lines arising in the elliptic curve addition of $[i]P$ to $[j]P$ (and so $\operatorname{div}(l(x, y)) = ([i]P) + ([j]P) + (-[i+j]P) - 3(\mathcal{O}_E)$ and $\operatorname{div}(v(x, y)) = ([i+j]P) + (-[i+j]P) - 2(\mathcal{O}_E))$.

We can now give Miller's algorithm to compute $f_{n,P}(D)$ for any divisor D (see Algorithm 28). The basic idea is to compute the Miller function out of smaller Miller functions using a "square-and-multiply" strategy. As usual, we write an integer n in binary as $(1n_{m-1} \dots n_1 n_0)_2$ where $m = \lfloor \log_2(n) \rfloor$. Note that the lines l and v in lines 6 and/or 10 may be simplified if the operation is $[2]T = \mathcal{O}_E$ or $T + P = \mathcal{O}_E$.

The main observation is that Miller's algorithm takes $O(\log_2(n))$ iterations, each of which comprises field operations in $\Bbbk$ if P and all points in the support of D lie in $E(\Bbbk)$. There are a number of important tricks to speed up Miller's algorithm in practice; we mention some of them in the following sections and refer to Chapter IX of [61], Chapter XII of [286] or Section 16.4 of [16] for further details.

Exercise 26.3.6 Give simplified versions of lines 6 and 10 of Algorithm 28 that apply when $[2]T = \mathcal{O}_E$ or $T + P = \mathcal{O}_E$.

26.3.2 The reduced Tate–Lichtenbaum pairing

Definition 26.3.7 Let $n, q \in \mathbb{N}$ be such that $\gcd(n, q) = 1$. Define the **embedding degree** $k(q, n) \in \mathbb{N}$ to be the smallest positive (non-zero) integer such that $n \mid (q^{k(q,n)} - 1)$.

Let E be an elliptic curve over $\mathbb{F}_q$ and suppose $n \mid \#E(\mathbb{F}_q)$ is such that $\gcd(n, q) = 1$. Let $k = k(q, n)$ be the embedding degree. Then $\mu_n \subseteq \mathbb{F}_{q^k}^*$ (in some cases μ_n can lie in a proper subfield of $\mathbb{F}_{q^k}$) and so the Tate–Lichtenbaum pairing maps into $\mathbb{F}_{q^k}^*/(\mathbb{F}_{q^k}^*)^n$. In practice,

Algorithm 28 Miller's algorithm

INPUT: $n = (1n_{m-1}\dots n_1n_0)_2 \in \mathbb{N}$, $P \in E(\Bbbk)$, such that $[n]P = \mathcal{O}_E$, $D \in \mathrm{Div}_{\overline{\Bbbk}}(E)$
OUTPUT: $f_{n,P}(D)$

1: $f = 1$
2: $T = P$
3: $i = m - 1 = \lfloor \log_2(n) \rfloor - 1$
4: **while** $i \geq 0$ **do**
5: Calculate lines l and v for doubling T
6: $f = f^2 \cdot l(D)/v(D)$
7: $T = [2]T$
8: **if** $n_i = 1$ **then**
9: Calculate lines l and v for addition of T and P
10: $f = f \cdot l(D)/v(D)$
11: $T = T + P$
12: **end if**
13: $i = i - 1$
14: **end while**
15: **return** f

it is inconvenient to have a pairing taking values in this quotient group, as cryptographic protocols require well-defined values. To have a canonical representative for each coset in $\mathbb{F}_{q^k}^*/(\mathbb{F}_{q^k}^*)^n$ it would be much more convenient to use μ_n. This is easily achieved using the facts that if $z \in \mathbb{F}_{q^k}^*$ then $z^{(q^k-1)/n} \in \mu_n$, and that the cosets $z_1(\mathbb{F}_{q^k}^*)^n$ and $z_2(\mathbb{F}_{q^k}^*)^n$ are equal if and only if $z_1^{(q^k-1)/n} = z_2^{(q^k-1)/n}$. Also, exponentiation is a group homomorphism from $\mathbb{F}_{q^k}^*/(\mathbb{F}_{q^k}^*)^n$ to μ_n.

For this reason, one usually considers the **reduced Tate–Lichtenbaum pairing**

$$\hat{t}_n(P, Q) = t_n(P, Q)^{(q^k-1)/n},$$

which maps $E(\mathbb{F}_q)[n] \times E(\mathbb{F}_q)/nE(\mathbb{F}_q)$ to μ_n. The exponentiation to the power $(q^k - 1)/n$ is called the **final exponentiation.**

Exercise 26.3.8 Let $n \mid N \mid (q^k - 1)$. Show that

$$t_n(P, Q)^{(q^k-1)/n} = t_N(P, Q)^{(q^k-1)/N}.$$

Exercise 26.3.9 Explain why working in a group whose order has low Hamming weight leads to relatively fast pairings. Suppose $n = E(\mathbb{F}_q)$ has low Hamming weight but $r \mid n$ does not. Explain how to compute the reduced Tate–Lichtenbaum pairing $\hat{t}_r(P, Q)$ efficiently if n/r is small.

In the applications, one usually chooses the elliptic curve E to satisfy the mild conditions in Exercise 26.3.10. In these cases, it follows from the Exercise that we can identify $E(\mathbb{F}_{q^k})/rE(\mathbb{F}_{q^k})$ with $E(\mathbb{F}_{q^k})[r]$. Hence, if the conditions hold, we may interpret the reduced

Tate–Lichtenbaum pairing as a map

$$\hat{t}_r : E[r] \times E[r] \to \mu_r$$

just as the Weil pairing is.

Exercise 26.3.10 Let E be an elliptic curve over $\mathbb{F}_q$ and let r be a prime such that $r \| \#E(\mathbb{F}_q)$, $\gcd(r, q) = 1$, $E[r] \subseteq E(\mathbb{F}_{q^k})$ and $r^2 \| \#E(\mathbb{F}_{q^k})$, where $k = k(q, r)$ is the embedding degree. Show that $E[r]$ is set of representatives for $E(\mathbb{F}_{q^k})/rE(\mathbb{F}_{q^k})$.

In most cryptographic situations one restricts to the case of points of prime order r. Further, one can often insist that $P \in E(\mathbb{F}_q)$ and $Q \in E(\mathbb{F}_{q^k})$. An important observation is that if $k > 1$ and $z \in \mathbb{F}_q^*$ then $z^{(q^k-1)/r} = 1$. This allows us to omit some computations in Miller's algorithm. A further trick, due[4] to Barreto, Kim, Lynn and Scott [27], is given in Lemma 26.3.11 (a similar fact for the Weil pairing is given in Proposition 8 of Miller [385]).

Lemma 26.3.11 *Let E be an elliptic curve over $\mathbb{F}_q$, $P \in E(\mathbb{F}_q)$ a point of prime order r (where $r > 4$ and $\gcd(q, r) = 1$) and $Q \in E(\mathbb{F}_{q^k}) - E(\mathbb{F}_q)$ where $k > 1$ is the embedding degree. Then*

$$\hat{t}_r(P, Q) = f_{r,P}(Q)^{(q^k-1)/r}.$$

Proof A proof for general curves (and without any restriction on r) is given in Lemma 1 of Granger, Hess, Oyono, Thériault and Vercauteren [241]. We give a similar argument.

We have $\hat{t}_r(P, Q) = f_{r,P}((Q + R) - (R))^{(q^k-1)/r}$ for any point $R \in E(\mathbb{F}_{q^k}) - \{\mathcal{O}_E, P, -Q, P - Q\}$. Choose $R \in E(\mathbb{F}_q) - \{\mathcal{O}_E, P\}$. Since $f_{r,P}(R) \in \mathbb{F}_q^*$ and $k > 1$ it follows that

$$\hat{t}_r(P, Q) = f_{r,P}(Q + R)^{(q^k-1)/r}.$$

Now, it is not possible to take $R = \mathcal{O}_E$ in the above argument. Instead we need to prove that $f_{r,P}(Q + R)^{(q^k-1)/r} = f_{r,P}(Q)^{(q^k-1)/r}$ directly. It suffices to prove that

$$f_{r,P}((Q + R) - (Q))^{(q^k-1)/r} = 1.$$

To do this, note that $(Q + R) - (Q) \equiv (R) - (\mathcal{O}_E) \equiv ([2]R) - (R)$. Set, for example, $R = [2]P$ so that $([2]R) - (R)$ has support disjoint from $\{\mathcal{O}_E, P\}$ (this is where the condition $r > 4$ is used). Then there is a function $h \in \mathbb{F}_{q^k}(E)$ such that $(Q + R) - (Q) = ([2]R) - (R) + \operatorname{div}(h)$. We have

$$f_{r,P}((Q + R) - (Q)) = f_{r,p}(([2]R) - (R) + \operatorname{div}(h)) = f_{r,P}(([2]R) - (R))h(\operatorname{div}(f_{r,P})).$$

Finally, note that $f_{r,P}(([2]R) - (R)) \in \mathbb{F}_q^*$ and that $h(\operatorname{div}(f_{r,P})) = (h(P)/h(\mathcal{O}_E))^r \in (\mathbb{F}_{q^k}^*)^r$. The result follows. □

Exercise 26.3.12 Let the embedding degree k be even, $r \nmid (q^{k/2} - 1)$, $P \in E(\mathbb{F}_q)$ and $Q = (x_Q, y_Q) \in E(\mathbb{F}_{q^k})$ points of order r. Suppose $x_Q \in \mathbb{F}_{q^{k/2}}$ (this is usually the case for

[4] Though be warned that the "proof" in [27] is not rigorous.

points of cryptographic interest). Show that all vertical line functions can be omitted when computing the reduced Tate–Lichtenbam pairing.

26.3.3 Ate pairing

Computing pairings on elliptic curves usually requires significantly more effort than exponentiation on an elliptic curve. There has been a concerted research effort to make pairing computation more efficient, and a large number of techniques are known. Due to lack of space we focus on one particular method known as "loop shortening". This idea originates in the work of Duursma and Lee [173] (for hyperelliptic curves) and was further developed by Barreto, Galbraith, Ó hÉigeartaigh and Scott [26]. We present the idea in the ate pairing formulation of Hess, Smart and Vercauteren [258]. Note that the ate pairing is not a "new" pairing. Rather, it is a way to efficiently compute a power, of a restriction to certain subgroups, of the Tate–Lichtenbaum pairing.

Let E be an elliptic curve over $\mathbb{F}_q$ and let r be a large prime such that $r \mid \#E(\mathbb{F}_q) = q + 1 - t$ and $r \mid (q^k - 1)$ for some relatively small integer k, but $r \nmid (q - 1)$. It follows that $\#(E[r] \cap E(\mathbb{F}_q)) = r$. Since the Frobenius map is linear on the $\mathbb{F}_r$-vector space $E[r]$, and its characteristic polynomial satisfies

$$x^2 - tx + q \equiv (x - 1)(x - q) \pmod{r}$$

it follows that π_q has distinct eigenvalues 1 and $q \pmod{r}$ and corresponding eigenspaces (i.e., subgroups)

$$G_1 = E[r] \cap \ker(\pi_q - [1]) , \quad G_2 = E[r] \cap \ker(\pi_q - [q]). \tag{26.3}$$

Since $r \mid (q^k - 1)$ and $q \equiv (t - 1) \pmod{r}$ it follows that $r \mid ((t - 1)^k - 1)$. Let $T = t - 1$ and $N = \gcd(T^k - 1, q^k - 1)$. Note that $r \mid N$. Define the **ate pairing** $a_T : G_2 \times G_1 \to \mu_r$ by

$$a_T(Q, P) = f_{T,Q}(P)^{(q^k-1)/N}.$$

The point is that $|t| \le 2\sqrt{q}$ and, typically, $r \approx q$. Hence, computing the Miller function $f_{T,Q}$ typically requires at most half the number of steps as required to compute $f_{r,P}$. On the downside, the coefficients of the function $f_{T,Q}$ lie in $\mathbb{F}_{q^k}$, rather than $\mathbb{F}_q$ as before. Nevertheless, the ate pairing often leads to faster pairings if carefully implemented (especially when twists are exploited).

Theorem 26.3.13 *Let the notation be as above (in particular, $T = t - 1$ and $N = \gcd(T^k - 1, q^k - 1)$). Let $L = (T^k - 1)/N$ and $c = \sum_{i=0}^{k-1} q^i T^{k-1-i} \pmod{r}$. Then*

$$a_T(Q, P)^c = t_r(Q, P)^{L(q^k-1)/r}.$$

Hence, a_T is bilinear, and a_T is non-degenerate if and only if $r \nmid L$.

Proof (Sketch) Consider $t_r(Q, P)^{(q^k-1)/r} = f_{r,Q}(P)^{(q^k-1)/r}$. Since $r \mid N$, Exercise 26.3.8 implies that this is equal to

$$f_{N,Q}(P)^{(q^k-1)/N}.$$

Indeed

$$t_r(Q, P)^{L(q^k-1)/r} = f_{LN,Q}(P)^{(q^k-1)/N} = f_{T^k-1,Q}(P)^{(q^k-1)/N}.$$

Now, $[T^k - 1]Q = \mathcal{O}_E$ so one can take $f_{T^k,Q} = f_{T^k-1,Q}$. (To prove this, note that $\operatorname{div}(f_{T^k,Q}) = T^k(Q) - ([T^k]Q) - (T^k - 1)(\mathcal{O}_E) = T^k(Q) - (Q) - (T^k - 1)(\mathcal{O}_E) = (T^k - 1)(Q) - (T^k - 1)(\mathcal{O}_E)$.) Hence, the Lth power of the reduced Tate–Lichtenbaum pairing is $f_{T^k,Q}(P)^{(q^k-1)/N}$. Now

$$f_{T^k,Q}(P) = f_{T,Q}(P)^{T^{k-1}} f_{T,[T]Q}(P)^{T^{k-2}} \cdots f_{T,[T^{k-1}]Q}(P), \tag{26.4}$$

which follows by considering the divisors of the left- and right-hand sides. The final step, and the only place we use $\pi_q(Q) = [q]Q = [T]Q$, is to note that

$$f_{T,[T]Q}(P) = f_{T,\pi_q(Q)}(P) = f_{T,Q}^q(P) \tag{26.5}$$

where f^q denotes raising all coefficients of the rational function f to the power q. This follows because E and P are defined over $\mathbb{F}_q$, so $\sigma(f_{T,Q}(P)) = f_{T,\sigma(Q)}(P)$ for all $\sigma \in \operatorname{Gal}(\mathbb{F}_{q^k}/\mathbb{F}_q)$. One therefore computes $f_{T^k,Q}(P) = f_{T,Q}(P)^c$, which completes the proof. □

Exercise 26.3.14 Generalise Theorem 26.3.13 to the case where $T \equiv q^m \pmod{r}$ for some $m \in \mathbb{N}$. What is the corresponding value of c?

26.3.4 Optimal pairings

Lee, Lee and Park [331], Hess [257] and Vercauteren [554] have used combinations of pairings that have the potential for further loop shortening over that provided by the ate pairing.

Ideally, one wants to compute a pairing as $f_{M,Q}(P)$, with some final exponentation, where M is as small as possible. Hess and Vercauteren conjecture that the smallest possible value for $\log_2(M)$, for points of prime order r in an elliptic curve E over $\mathbb{F}_q$ with embedding degree $k(q, r)$, is $\log_2(r)/\varphi(k(q, r))$. For such a pairing, Miller's algorithm would be sped up by a factor of approximately $\varphi(k(q, r))$ compared with the time required when not using loop shortening. The method of Vercauteren actually gives a pairing as a product of $\prod_{i=0}^{l} f_{M_i,Q}(P)^{q^i}$ (together with some other terms) where all the integers M_i are of the desired size; such a pairing is not automatically computed faster than the naive method, but if the integers M_i all have a large common prefix in their binary expansions then such a saving can be obtained. If a pairing can be computed with approximately $\log_2(r)/\varphi(k(q, r))$ iterations in Miller's algorithm then it is called an **optimal pairing**.

The basic principle of Vercauteren's construction is to find a multiple ur, for some $u \in \mathbb{N}$, of the group order that can be written in the form

$$ur = \sum_{i=0}^{l} M_i q^i \tag{26.6}$$

where the $M_i \in \mathbb{Z}$ are "small". One can then show just like with the ate pairing that a certain power of the Tate–Lichtenbaum pairing is

$$\left(\prod_{i=0}^{l} f_{M_i,Q}(P)^{q^i} \prod_{i=1}^{l} g_i(P) \right)^{(q^k-1)/r} \tag{26.7}$$

where the functions g_i take into account additions of certain elliptic curve points. Vercauteren proves that if

$$ukq^{k-1} \not\equiv \frac{(q^k-1)}{r} \left(\sum_{i=0}^{l} i M_i q^{i-1} \right) \pmod{r}$$

then the pairing is non-degenerate. The value of equation (26.7) can be computed efficiently only if all $f_{M_i,Q}(P)$ can, in some sense, be computed simultaneously. This is easiest when all but one of the M_i are small (i.e., in $\{-1, 0, 1\}$) or when the M_i have a large common prefix of most significant bits (possibly in signed binary expansion).

Vercauteren [554] suggests finding solutions to equation (26.6) using lattices. More precisely, given r and q one considers the lattice spanned by the rows of the following matrix

$$B = \begin{pmatrix} r & 0 & 0 & \cdots & 0 \\ -q & 1 & 0 & \cdots & 0 \\ -q^2 & 0 & 1 & \cdots & 0 \\ \vdots & \vdots & \vdots & \ddots & \vdots \\ -q^l & 0 & 0 & \cdots & 1 \end{pmatrix}. \tag{26.8}$$

One sees that $(u, M_1, M_2, \ldots, M_l)B = (M_0, M_1, \ldots, M_l)$ and so candidate values for $u, M_0, \ldots, M_l$ can be found by finding short vectors in the lattice. A demonstration of this method is given in Example 26.6.3.

Note that loop shortening methods should not be confused with the methods, starting with Scott [481], that use an endomorphism on the curve to "recycle" some computations in Miller's algorithm. Such methods do not reduce the number of squarings in Miller's algorithm and, while valuable, do not give the same potential performance improvements as methods that use loop shortening.

26.3.5 Pairing lattices

Hess [257] developed a framework for analysing pairings that is closely related to the framework in the previous section. We briefly sketch the ideas.

Definition 26.3.15 Let notation be as in Section 26.3.3, in particular q is a prime power, r is a prime, q is a primitive kth root of unity modulo r and the groups G_1 and G_2 are as in equation (26.3). Let $s \equiv q^m \pmod{r}$ for some $m \in \mathbb{N}$. For any $h(x) \in \mathbb{Z}[x]$ write $h(x) = \sum_{i=0}^{d} h_i x^i$. Let $P \in G_1$, $Q \in G_2$ and define $f_{s,h(x),Q}$ to be a function normalised at infinity (in the sense of Definition 26.3.4) such that

$$\operatorname{div}(f_{s,h(x),Q}) = \sum_{i=0}^{d} h_i(([s^i]Q) - (\mathcal{O}_E)).$$

Define

$$a_{s,h(x)}(Q, P) = f_{s,h(x),Q}(P)^{(q^k-1)/r}.$$

We stress here that h is a polynomial, not a rational function (as it was in previous sections).

Since $[s]Q = [q^m]Q = \pi_q^m(Q)$, a generalisation of equation (26.5) shows that $f_{h_i,[s^i]Q}(P) = f_{h_i,Q}(P)^{q^{mi}}$. It follows that one can compute $f_{s,h(x),Q}(P)$ efficiently using Miller's algorithm in a similar way to computing the pairings in the previous section. The running time of Miller's algorithm is proportional to $\sum_{i=0}^{d} \log_2(\max\{1, |h_i|\})$ in the worse case (it performs better when the h_i have a large common prefix in their binary expansion).

Hess [257] shows that, for certain choices of $h(x)$, $a_{s,h(x)}$ is a non-degenerate and bilinear pairing. The goal is also to obtain good choices for $h(x)$ so that the pairing can be computed using a short loop. One of the major contributions of Hess [257] is to prove lower bounds on the size of the coefficients of any polynomial $h(x)$ that leads to a non-degenerate, bilinear pairing. This supports the optimality conjecture mentioned in the previous section.

Lemma 26.3.16 *Let notation be as in Definition 26.3.15.*

1. *$a_{s,r}(Q, P)$ is the Tate pairing.*
2. *$a_{s,x-s}(Q, P)$ is a power of the ate pairing.*
3. *$a_{s,h(x)x}(Q, P) = a_{s,h(x)}(Q, P)^s$.*
4. *Let $h(x), g(x) \in \mathbb{Z}[x]$. Then*

$$a_{s,h(x)+g(x)}(Q, P) = a_{s,h(x)}(Q, P)a_{s,g(x)}(Q, P) \quad \textit{and}$$
$$a_{s,h(x)g(x)}(Q, P) = a_{s,h(x)}(Q, P)^{g(s)}.$$

Exercise 26.3.17★ Prove Lemma 26.3.16.

Theorem 26.3.18 *Let notation be as above. Let $s \in \mathbb{N}$ be such that s is a primitive kth root of unity modulo r^2. Let $h(x) \in \mathbb{Z}[x]$ be such that $h(s) \equiv 0 \pmod{r}$ but $r^2 \nmid h(s)$. Then $a_{s,h(x)}$ is a non-degenerate, bilinear pairing on $G_2 \times G_1$.*

Proof Since $s^k \equiv 1 \pmod{r}$ it follows that $s \equiv q^m \pmod{r}$ for some $m \in \mathbb{N}$. Since $h(s) \equiv 0 \pmod{r}$ we can write

$$h(x) = g_1(x)(x - s) + g_2(x)r$$

for some $g_1(x), g_2(x) \in \mathbb{Z}[x]$. It follows from Lemma 26.3.16 that, for some $c \in \mathbb{N}$

$$a_{s,h(x)}(Q, P) = a_T(Q, P)^{cg_1(s)} \hat{t}_r(Q, P)^{g_2(s)}$$

and so $a_{s,h(x)}$ is a bilinear pairing on $G_2 \times G_1$.

Finally, we need to prove non-degeneracy. By assumption, $r^2 \mid (s^k - 1)$ and so, in the version of Theorem 26.3.13 of Exercise 26.3.14, $r \mid L$. It follows that $a_T(Q, P) = 1$. Hence, $a_{s,h(x)}(Q, P) = \hat{t}_r(Q, P)^{g_2(s)}$. To complete the proof, note that $g_2(s) = h(s)/r$, and so $a_{s,h(x)}$ is non-degenerate if and only if $r^2 \nmid h(s)$. □

Hess [257] explains that this construction is "complete" in the sense that every bilinear map coming from functions in a natural class must correspond to some polynomial $h(x)$. Hess also proves that any polynomial $h(x) = \sum_{i=0}^{d} h_i x^i \in \mathbb{Z}[x]$ satisfying the required conditions is such that $\sum_{i=0}^{d} |h_i| \geq r^{1/\varphi(k)}$. Polynomials $h(x)$ that have one coefficient of size $r^{1/\varphi(k)}$ and all other coefficients small satisfy the optimality conjecture. Good choices for the polynomial $h(x)$ are found by considering exactly the same lattice as in equation (26.8) (though in [257] it is written with q replaced by s).

26.4 Reduction of ECDLP to finite fields

An early application of pairings in elliptic curve cryptography was to reduce the discrete logarithm problem in $E(\mathbb{F}_q)[n]$, when $\gcd(n, q) = 1$, to the discrete logarithm problem in the multiplicative group of a finite extension of $\mathbb{F}_q$. Menezes, Okamoto and Vanstone [375] used the Weil pairing to achieve this, while Frey and Rück [196] used the reduced Tate–Lichtenbaum pairing. The case $\gcd(n, q) \neq 1$ will be handled in Section 26.4.1.

The basic idea is as follows: given an instance $P, Q = [a]P$ of the discrete logarithm problem in $E(\mathbb{F}_q)[n]$ and a non-degenerate bilinear pairing e, one finds a point $R \in E(\overline{\mathbb{F}}_q)$ such that $z = e(P, R) \neq 1$. It follows that $e(Q, R) = z^a$ in $\mu_n \subseteq \mathbb{F}_{q^k}^*$ where $k = k(q, n)$ is the embedding degree. When q is a prime power that is not prime then there is the possibility that μ_r lies in a proper subfield of $\mathbb{F}_{q^k}$, in which case re-define k to be the smallest positive rational number such that $\mathbb{F}_{q^k}$ is the smallest field of characteristic char$(\mathbb{F}_q)$ containing μ_n.

The point is that if k is sufficiently small then index calculus algorithms in $\mathbb{F}_{q^k}^*$ could be faster than the baby-step–giant-step or Pollard rho algorithms in $E(\mathbb{F}_q)[n]$. Hence, one has reduced the discrete logarithm problem to a potentially easier problem. The reduction of the DLP from $E(\mathbb{F}_q)$ to a subgroup of $\mathbb{F}_{q^k}^*$ is called the **MOV/FR attack**.

Menezes, Okamoto and Vanstone [375] suggested to use the Weil pairing for the above idea. In this case, the point R can, in principle, be defined over a large extension of $\mathbb{F}_q$. Frey and Rück explained that the Tate–Lichtenbaum pairing is a more natural choice, since it is sufficient to take a suitable point $R \in E(\mathbb{F}_{q^k})$ where $k = k(q, n)$ is the embedding degree. Balasubramanian and Koblitz [25] showed that, in most cases, it is also sufficient to work in $E(\mathbb{F}_{q^k})$ when using the Weil pairing.

Theorem 26.4.1 *Let E be an elliptic curve over $\mathbb{F}_q$ and let r be a prime dividing $\#E(\mathbb{F}_q)$. Suppose that $r \nmid (q-1)$ and that $\gcd(r, q) = 1$. Then $E[r] \subset E(\mathbb{F}_{q^k})$ if and only if r divides $(q^k - 1)$.*

Proof See [25]. □

Balasubramanian and Koblitz also show that a "random" curve is expected to have very large embedding degree. Hence, the MOV/FR attack is not a serious threat to the ECDLP on randomly chosen elliptic curves. However, as noted by Menezes, Okamoto and Vanstone, supersingular elliptic curves are always potentially vulnerable to the attack.

Theorem 26.4.2 *Let E be a supersingular elliptic curve over $\mathbb{F}_q$ and suppose $r \mid \#E(\mathbb{F}_q)$. Then the embedding degree $k(q, r)$ is such that $k(q, r) \le 6$.*

Proof See Corollary 9.11.9. □

26.4.1 Anomalous curves

The discrete logarithm problem on elliptic curves over $\mathbb{F}_p$ with p points (such curves are called **anomalous elliptic curves**) can be efficiently solved. This was first noticed by Semaev [484] and generalised to higher genus curves by Rück [455]. We present their method in this section. An alternative way to view the attack (using p-adic lifting rather than differentials) was given by Satoh and Araki [460] and Smart [511].

The theoretical tool is an observation of Serre [487].

Lemma 26.4.3 *Let $P \in E(\mathbb{F}_p)$ have order p. Let f_P be a function in $\mathbb{F}_p(E)$ with $\mathrm{div}(f_P) = p(P) - p(\mathcal{O}_E)$. Then the map*

$$P \mapsto \frac{df_P}{f_P}$$

is a well-defined group homomorphism from $E(\mathbb{F}_p)[p]$ to $\Omega_{\mathbb{F}_p}(E)$.

Proof First note that f_P is defined up to a constant, and that $d(cf_P)/(cf_P) = df_P/f_P$. Hence, the map is well-defined.

Now let $Q = [a]P$ and let f_P be as in the statement of the lemma. Then there is a function g such that

$$\mathrm{div}(g) = (Q) - a(P) + (a-1)(\mathcal{O}_E).$$

One has

$$\begin{aligned} \mathrm{div}(g^p f_P^a) &= p\mathrm{div}(g) + a\mathrm{div}(f_P) \\ &= p(Q) - ap(P) + p(a-1)(\mathcal{O}_E) + ap(P) - ap(\mathcal{O}_E) \\ &= p(Q) - p(\mathcal{O}_E). \end{aligned}$$

Hence, one can let $f_Q = g^p f_P^a$. Now, using part 4 of Lemma 8.5.17

$$\frac{df_Q}{f_Q} = \frac{d(g^p f_P^a)}{g^p f_P^a} = \frac{g^p df_P^a + f_P^a dg^p}{g^p f_P^a}.$$

Part 6 of Lemma 8.5.17 gives $dg^p = pg^{p-1}dg = 0$ (since we are working in $\mathbb{F}_p$) and $df_P^a = af_P^{a-1}df_P$. Hence

$$\frac{df_Q}{f_Q} = a\,\frac{df_P}{f_P},$$

which proves the result. □

Exercise 26.4.4 Generalise Lemma 26.4.3 to arbitrary curves.

Lemma 26.4.3 therefore maps the DLP in $E(\mathbb{F}_p)[p]$ to a DLP in $\Omega_{\mathbb{F}_p}(E)$. It remains to solve the DLP there.

Lemma 26.4.5 *Let the notation be as in Lemma 26.4.3. Let t be a uniformiser at $\mathcal{O}_E$. Write $f_P = t^{-p} + f_1 t^{-(p-1)} + f_2 t^{-(p-2)} + \cdots$. Then*

$$\frac{df_P}{f_P} = (f_1 + \cdots)dt.$$

Proof Clearly, $f_P^{-1} = t^p - f_1 t^{p+1} + \cdots$. From part 8 of Lemma 8.5.17 we have

$$df_P = \left(\frac{\partial f_P}{\partial t}\right) dt = (-pt^{-p-1} - (p-1)f_1 t^{-p} + \cdots)dt.$$

Since we are working in $\mathbb{F}_p$ we have $df_P = (f_1 t^{-p} + \cdots)dt$. The result follows. □

Putting together Lemma 26.4.3 and Lemma 26.4.5: if $Q = [a]P$ then $df_P/f_P = (f_1 + \cdots)dt$ and $df_Q/f_Q = (af_1 + \cdots)dt$. Hence, as long as one can compute the expansion of df_P/f_P with respect to t, one can solve the DLP. Indeed, this is easy: use Miller's algorithm with power series expansions to compute the power series expansion of f_P and follow the above calculations. Rück [455] gives an elegant formulation (for general curves) that computes only the desired coefficient f_1; he calls it the "additive version of the Tate–Lichtenbaum pairing".

26.5 Computational problems

26.5.1 Pairing inversion

We briefly discuss a computational problem, that is required to be hard for many cryptographic applications of pairings.

Definition 26.5.1 Let G_1, G_2, G_T be groups of prime order r and let $e : G_1 \times G_2 \to G_T$ be a non-degenerate bilinear pairing. The **pairing inversion problem** is: Given $Q \in G_2$, $z \in G_T$ to compute $P \in G_1$ such that $e(P, Q) = z$.

The bilinear Diffie–Hellman problem was introduced in Definition 23.3.9. In additive notation it is: given $P, Q, [a]Q, [b]Q$ to compute $e(P, Q)^{ab}$.

Lemma 26.5.2 *If one has an oracle for pairing inversion then one can solve BDH.*

Proof Given the BDH instance $P, Q, [a]Q, [b]Q$ compute $z_1 = e(P, [a]Q)$ and call the pairing inversion oracle on (Q, z_1) to get P' such that $e(P', Q) = z_1$. It follows that $P' = [a]P$. One then computes $e(P', [b]Q) = e(P, Q)^{ab}$ as required. □

Further discussion of pairing inversion is given by Galbraith, Hess and Vercauteren [205].

Exercise 26.5.3 Show that if one can solve pairing inversion then one can solve the Diffie–Hellman problem in G_1.

Exercise 26.5.4 Show that if one has an oracle for pairing inversion then one can perform passive selective forgery of signatures in the Boneh–Boyen scheme presented in Figure 23.3.9.

Exercise 26.5.5 Show that if one has an oracle for pairing inversion then one can solve the q-SDH problem of Definition 22.2.17.

26.5.2 Solving DDH using Pairings

Pairings can be used to solve the decision Diffie–Hellman (DDH) problem in some cases. First, we consider a variant of DDH that can sometimes be solved using pairings.

Definition 26.5.6 Let $E(\mathbb{F}_q)$ be an elliptic curve and let $P, Q \in E(\mathbb{F}_q)$ have prime order r. The **co-DDH problem** is: given $(P, [a]P, Q, [b]Q)$ to determine whether or not $a \equiv b \pmod{r}$.

Exercise 26.5.7 Show that co-DDH is equivalent to DDH if $Q \in \langle P \rangle$.

Suppose now that $E[r] \subseteq E(\mathbb{F}_q)$, $P \neq \mathcal{O}_E$, and that $Q \notin \langle P \rangle$. Then $\{P, Q\}$ generates $E[r]$ as a group. By non-degeneracy of the Weil pairing, we have $e_r(P, Q) \neq 1$. It follows that

$$e_r([a]P, Q) = e_r(P, Q)^a \quad \text{and} \quad e_r(P, [b]Q) = e_r(P, Q)^b.$$

Hence, the co-DDH problem can be efficiently solved using the Weil pairing.

The above approach cannot be used to solve DDH, since $e_r(P, P) = 1$ by the alternating property of the Weil pairing. In some special cases, the reduced Tate–Lichtenbaum pairing satisfies $\hat{t}_r(P, P) \neq 1$ and so can be used to solve DDH in $\langle P \rangle$. In general, however, DDH cannot be solved by such simple methods.

When E is a supersingular elliptic curve and $P \neq \mathcal{O}_E$ then, even if $\hat{t}_r(P, P) = 1$, there always exists an endomorphism $\psi : E \to E$ such that $\hat{t}_r(P, \psi(P)) \neq 1$. Such an endomorphism is called a **distortion map**; see Section 26.6.1. It follows that DDH is easy on supersingular elliptic curves.

26.6 Pairing-friendly elliptic curves

The cryptographic protocols given in Sections 22.2.3 and 23.3.2 relied on "pairing groups". We now mention the properties needed to have a practical system, and give some popular examples.

For pairing-based cryptography it is desired to have elliptic curves E over $\mathbb{F}_q$ such that:

1. there is a large prime r dividing $\#E(\mathbb{F}_q)$, with $\gcd(r, q) = 1$;
2. the DLP in $E(\mathbb{F}_q)[r]$ is hard;
3. the DLP in $\mathbb{F}_{q^k}^*$ is hard, where $k = k(q, r)$ is the embedding degree;
4. computation in $E(\mathbb{F}_q)$ and $\mathbb{F}_{q^k}^*$ is efficient;
5. elements of $E(\mathbb{F}_q)$ and $\mathbb{F}_{q^k}^*$ can be represented compactly.

Elliptic curves with these properties are called **pairing-friendly curves**. Note that the conditions are incompatible: for the DLP in $\mathbb{F}_{q^k}^*$ to be hard it is necessary that q^k be large (say, at least 3000 bits) to resist index calculus attacks like those in Chapter 15, whereas to represent elements of $\mathbb{F}_{q^k}^*$ compactly we would like q^k to be small. Luckily, we can use techniques such as those in Chapter 6 to represent field elements relatively compactly.

There is a large literature on pairing-friendly elliptic curves, culminating in the "taxonomy" by Freeman, Scott and Teske [194]. We give two examples below.

Example 26.6.1 For $a = 0, 1$ define

$$E_a : y^2 + y = x^3 + x + a$$

over $\mathbb{F}_2$. Then E_a is supersingular and $\#E_a(\mathbb{F}_{2^l}) = 2^l \pm 2^{(l+1)/2} + 1$ when l is odd. Some of these integers have large prime divisors, for example $2^{241} - 2^{121} + 1$ is prime. The embedding degree can be shown to be 4 in general; this follows since

$$(2^l + 2^{(l+1)/2} + 1)(2^l - 2^{(l+1)/2} + 1) = 2^{2l} + 1 \mid (2^{4l} - 1).$$

Example 26.6.2 (Barreto–Naehrig curves [28]) Consider the polynomials

$$p(x) = 36x^4 + 36x^3 + 24x^2 + 6x + 1 \quad \text{and} \quad t(x) = 6x^2 + 1 \tag{26.9}$$

in $\mathbb{Z}[x]$. Note that $t(x)^2 - 4p(x) = -3(6x^2 + 4x + 1)^2$, that $r(x) = p(x) + 1 - t(x)$ is irreducible over $\mathbb{Q}$ and that $r(x) \mid (p(x)^{12} - 1)$. Suppose $x_0 \in \mathbb{Z}$ is such that $p = p(x_0)$ is prime and $r = r(x_0)$ is prime (or is the product of a small integer with a large prime). Then the embedding degree $k(p, r)$ is a divisor of 12 (and is typically equal to 12). Furthermore, one can easily construct an elliptic curve $E/\mathbb{F}_p$ such that $\#E(\mathbb{F}_p) = r$; one of the 6 twists of $y^2 = x^3 + 1$ will suffice. Note that $p \equiv 1 \pmod 3$ and E is an ordinary elliptic curve.

Example 26.6.3 The family of curve parameters in Example 26.6.2 has $t \approx \sqrt{p}$ and so the ate pairing is computed in about half the time of the reduced Tate–Lichtenbaum pairing, as usual. We now demonstrate an optimal pairing with these parameters.

Substituting the polynomials $r(x)$ and $p(x)$ for the values r and q in the matrix of equation (26.8) gives a lattice. Lattice reduction over $\mathbb{Z}[x]$ yields the short vector $(M_0, M_1, M_2, M_3) = (6x+2, 1, -1, 1)$. It is easy to verify that $6x + 2 + p(x) - p(x)^2 + p(x)^3 \equiv 0 \pmod{r(x)}$.

Now $f_{1,Q} = 1$ and $f_{-1,Q} = v_Q$ (and so both can be omitted in pairing computation, by Exercise 26.3.12). The ate pairing can be computed as $f_{6x+2,Q}(P)$ multiplied with three straight line functions, and followed by the final exponentiation; see Section IV of [554]. The point is that Miller's algorithm now runs for approximately one quarter of the iterations as when computing the Tate–Lichtenbaum pairing.

26.6.1 Distortion maps

As noted, when $\hat{t}_r(P, P) = 1$ one can try to find an endomorphism $\psi : E \to E$ such that $\hat{t}_r(P, \psi(P)) \neq 1$.

Definition 26.6.4 Let E be an elliptic curve over $\mathbb{F}_q$, let $r \mid \#E(\mathbb{F}_q)$ be prime, let $e : E[r] \times E[r] \to \mu_r$ be a non-degenerate and bilinear pairing and let $P \in E(\mathbb{F}_q)[r]$. A **distortion map** with respect to E, r, e and P is an endomorphism ψ such that $e(P, \psi(P)) \neq 1$.

Verheul (Theorem 5 of [556]) shows that if E is a supersingular elliptic curve then, for any point $P \in E(\mathbb{F}_{q^k}) - \{\mathcal{O}_E\}$, a distortion map exists. In particular, when $P \in E(\mathbb{F}_q)[r] - \{\mathcal{O}_E\}$ and $k > 1$ then there is an endomorphism ψ (necessarily not defined over $\mathbb{F}_q$) such that $\hat{t}(P, \psi(P)) \neq 1$. Since P is defined over the small field we have a compact representation for all elliptic curve points in the cryptosystem, as well as efficiency gains in Miller's algorithm. For this reason, pairings on supersingular curves are often the fastest choice for certain applications.

Example 26.6.5 Consider again the elliptic curves from Example 26.6.1. An automorphism on E_a is $\psi(x, y) = (x + s^2, y + sx + t)$ where $s \in \mathbb{F}_{2^2}$ and $t \in \mathbb{F}_{2^4}$ satisfy $s^2 = s + 1$ and $t^2 = t + s$. One can represent $\mathbb{F}_{2^{4m}}$ using the basis $\{1, s, t, st\}$. It is clear that if $P \in E_a(\mathbb{F}_{2^l})$ where l is odd then $\psi(P) \in E_a(\mathbb{F}_{2^{4l}})$ and $\psi(P) \notin E_a(\mathbb{F}_{2^{2l}})$, and so ψ is a distortion map for P.

Exercise 26.6.6 Let E be an elliptic curve over $\mathbb{F}_q$ and let $r \mid \#E(\mathbb{F}_q)$ be prime. Let $k = k(q, r) > 1$ be the embedding degree. For any point $P \in E(\mathbb{F}_{q^k})$ define the trace map

$$\mathrm{Tr}(P) = \sum_{\sigma \in \mathrm{Gal}(\mathbb{F}_{q^k}/\mathbb{F}_q)} \sigma(P).$$

Show that $\mathrm{Tr}(P) \in E(\mathbb{F}_q)$. Now, suppose $P \in E[r]$, $P \notin E(\mathbb{F}_q)$ and $\mathrm{Tr}(P) \neq \mathcal{O}_E$. Show that $\{P, \mathrm{Tr}(P)\}$ generates $E[r]$. Deduce that the trace map is a distortion map with respect to E, r, e_r and P.

Exercise 26.6.7 Let notation be as in Exercise 26.6.6. Show that if $Q \in E[r] \cap \ker(\pi_q - [1])$ then $\mathrm{Tr}(Q) = [k]Q$. Show that if $Q \in E[r] \cap \ker(\pi_q - [q])$ then $\mathrm{Tr}(Q) = \mathcal{O}_E$. Hence, deduce that the trace map is not a distortion map for the groups G_1 or G_2 of equation (26.3).

Appendix A
Background mathematics

For convenience, we summarise some notation, conventions, definitions and results that will be used in the book. This chapter is for reference only.

A.1 Basic notation

We write $\mathbb{R}$ for the real numbers and define $\mathbb{R}_{\geq 0} = \{x \in \mathbb{R} : x \geq 0\}$ and similarly for $\mathbb{R}_{>0}$. We write $\mathbb{Z}$ for the integers and $\mathbb{N} = \mathbb{Z}_{>0} = \{n \in \mathbb{Z} : n > 0\} = \{1, 2, 3, \dots\}$ for the natural numbers.

We write $\#S$ for the number of elements of a finite set S. If S, T are sets we write $S - T$ for the set difference $\{s \in S : s \notin T\}$. We denote the empty set by $\varnothing$.

We write $\mathbb{Z}/n\mathbb{Z}$ for the ring of integers modulo n (many authors write $\mathbb{Z}_n$). When n is a prime and we are using the field structure of $\mathbb{Z}/n\mathbb{Z}$ we prefer to write $\mathbb{F}_n$. The statement $a \equiv b \pmod{n}$ means that $n \mid (a - b)$. We follow a common mis-use of this notation by writing $b \pmod{n}$ for the integer $a \in \{0, 1, \dots, n-1\}$ such that $a \equiv b \pmod{n}$. Hence, the statement $a = b \pmod{n}$ is an assignment of a to the value of the operator $b \pmod{n}$ and should not be confused with the predicate $a \equiv b \pmod{n}$.

The word **map** $f : X \to Y$ means a function on some subset of X. In other words, a map is not necessarily defined everywhere. Usually the word **function** implicitly means "defined everywhere on X", though this usage does not apply in algebraic geometry where a rational function is actually a rational map. If $f : X \to Y$ is a map and $U \subset X$ then we write $f|_U$ for the restriction of f to U, which is a map $f|_U : U \to Y$.

If $P = (x_P, y_P)$ is a point and f is a function on points then we write $f(x_P, y_P)$ rather than $f((x_P, y_P))$ for $f(P)$. We write $f \circ g$ for composition of functions (i.e., $(f \circ g)(x) = f(g(x))$); the notation fg will always mean product (i.e., $fg(P) = f(P)g(P)$). The notation f^n usually means exponentiating the value of the function f to the power n, except when f is an endomorphism of an elliptic curve (or Abelian variety), in which context it is standard to write f^n for n-fold composition. Hence, we prefer to write $f(P)^n$ than $f^n(P)$ when denoting powering (and so we write $\log(x)^n$ rather than $\log^n(x)$).

A.2 Groups

Let G be a group and $g \in G$. The **subgroup generated by** g is $\langle g \rangle = \{g^a : a \in \mathbb{Z}\}$. The **order** of the element g is the number of elements in the group $\langle g \rangle$. The **exponent** of a finite group is the smallest positive integer n such that $g^n = 1$ for all $g \in G$.

Let G be a finite Abelian group. The classification of finite Abelian groups (see Theorem II.2.1 of [271] or Section I.8 of [329]) states that G is isomorphic to a direct sum of cyclic groups of orders $m_1, m_2, \dots, m_t$ such that $m_1 \mid m_2 \mid \cdots \mid m_t$.

A.3 Rings

All rings in this book have a multiplicative identity 1. For any ring R, the smallest positive integer n such that $n1 = 0$ is called the **characteristic** of the ring and is denoted $\mathrm{char}(R)$. If there is no such n then we define $\mathrm{char}(R) = 0$.

If R is a ring and $n \in \mathbb{N}$ then we write $M_n(R)$ for the ring of $n \times n$ matrices with entries in R.

If R is a ring then R^* is the multiplicative group of invertible elements of R. The **Euler phi function** $\varphi(n)$ is the order of $(\mathbb{Z}/n\mathbb{Z})^*$. One has

$$\varphi(n) = n \prod_{p \mid n} \left(1 - \frac{1}{p}\right).$$

Theorem A.3.1 *There exists $N \in \mathbb{N}$ such that $\varphi(n) > n/(3\log(\log(n)))$ for all $n \in \mathbb{N}_{>N}$.*

Proof Theorem 328 of [250] states that

$$\liminf_{n \to \infty} \frac{\varphi(n)\log(\log(n))}{n} = e^{-\gamma}$$

where $\gamma \approx 0.57721566$ is the Euler–Mascheroni constant. Since $e^{-\gamma} \approx 0.56 > 1/3$, the result follows from the definition of lim inf. □

An element $a \in R$ is **irreducible** if $a \notin R^*$ and $a = bc$ for $b, c \in R$ implies $b \in R^*$ or $c \in R^*$. We write $a \mid b$ for $a, b \in R$ if there exists $c \in R$ such that $b = ac$. An element $a \in R$ is **prime** if $a \mid bc$ implies $a \mid b$ or $a \mid c$.

An integral domain R is a **unique factorisation domain** (UFD) if each $a \in R$ can be written uniquely (up to ordering and multiplication by units) as a product of irreducibles. In a UFD, an element is prime if and only if it is irreducible.

A.4 Modules

Let R be a ring. An R-module M is an Abelian group, written additively, with an operation rm for $r \in R$ and $m \in M$ such that $(r_1 + r_2)m = r_1 m + r_2 m$ and $r(m_1 + m_2) = rm_1 + rm_2$. An R-module M is **finitely generated** if there is a set $\{m_1, \dots, m_k\} \subset M$ such that $M = \{\sum_{i=1}^{k} r_i m_i : r_i \in R\}$.

A finitely generated R-module M is a **free module** if there is a set $\{m_1, \dots, m_k\}$ that generates M and is such that $0 = \sum_{i=1}^{k} r_i m_i$ if and only if $r_i = 0$ for all $1 \le i \le k$. Such an R-module is said to have **rank** k.

Let R be a commutative ring, M an R-module and $\Bbbk$ a field containing R. Consider the set of all symbols of the form $m \otimes a$ where $m \in M, a \in \Bbbk$ under the equivalence relation $rm \otimes a \equiv m \otimes ra$ for $r \in R$, $(m_1 + m_2) \otimes a = (m_1 \otimes a) + (m_1 \otimes a)$ and $m \otimes (a_1 + a_2) = (m \otimes a_1) + (m \otimes a_2)$. The **tensor product** $M \otimes_R \Bbbk$ is the set of all equivalence classes of such symbols. If M is a finitely generated free R-module with generating set $\{m_1, \dots, m_k\}$ then $M \otimes_R \Bbbk$ is a $\Bbbk$-vector space of dimension k with basis $\{m_1 \otimes 1, \dots, m_k \otimes 1\}$.

A.5 Polynomials

Let $\Bbbk$ be a field. Denote by $\Bbbk[\underline{x}] = \Bbbk[x_1, \dots, x_n]$ the set of polynomials over $\Bbbk$ in n variables. We write $\deg_{x_i}(F(x_1, \dots, x_n))$ to be the degree as a polynomial in x_i with coefficients in $\Bbbk[x_1, \dots, x_{i-1}, x_{i+1}, \dots, x_n]$. For polynomials $F(x) \in \Bbbk[x]$ we write $\deg(F(x))$ for $\deg_x(F(x))$. A degree d polynomial in $\Bbbk[x]$ is **monic** if the coefficient of x^d is 1. The **total degree** of a polynomial $F(\underline{x}) = \sum_{i=1}^{l} F_i x_1^{m_{i,1}} \cdots x_n^{m_{i,n}}$ (with $F_i \neq 0$) is $\deg(F) = \max_{1 \le i \le l} \sum_{j=1}^{n} m_{i,j}$.

A polynomial $F(\underline{x})$ is divisible by $G(\underline{x})$ if there exists a polynomial $H(\underline{x})$ such that $F(\underline{x}) = G(\underline{x})H(\underline{x})$. A polynomial $F(\underline{x}) \in \Bbbk[\underline{x}]$ is **irreducible** over $\Bbbk$ (also called $\Bbbk$-irreducible) if whenever $F(\underline{x}) = G(\underline{x})H(\underline{x})$ with $G(\underline{x}), H(\underline{x}) \in \Bbbk[\underline{x}]$ either G or H is a constant polynomial.

There are various ways to show that a polynomial is irreducible. **Eisenstein's criteria** states that $F(x) = \sum_{i=0}^{n} F_i x^i \in R[x]$, where R is a UFD, is irreducible if there is a prime p in R such that $p \nmid F_n$, $p \mid F_i$ for $0 \le i < n$, and $p^2 \nmid F_0$. We refer to Proposition III.1.14 of [529], Theorem IV.3.1 of [329] or Theorem III.6.15 of [271] for proofs.

If $\Bbbk$ is a field then the polynomial ring $\Bbbk[x_1, \dots, x_n]$ is a UFD (Theorem III.6.14 of [271]). Let $F(x) \in \Bbbk[x]$ be a polynomial in one variable of degree d. Then either $F = 0$ or else $F(x)$ has at most d roots in $\Bbbk$.

Lemma A.5.1 *Let $N_{d,q}$ be the number of monic irreducible polynomials of degree d in $\mathbb{F}_q[x]$. Then $q^d/2d \le N_{d,q} \le q^d/d$.*

Proof See Theorem 20.11 of [497] or Exercise 3.27 of [350]. A more precise result is given in Theorem 15.5.12. □

Let $F(x) \in \Bbbk[x]$. One can define the **derivative** $F'(x)$ by using the rule $(F_n x^n)' = nF_n x^{n-1}$ for $n \ge 0$ for each monomial. This is a formal algebraic operation and does not require an interpretation in terms of calculus.

Lemma A.5.2 *Let $F_1(x), F_2(x) \in \Bbbk[x]$. Then:*

1. $(F_1(x) + F_2(x))' = F_1'(x) + F_2'(x)$.
2. $(F_1(x)F_2(x))' = F_1(x)F_2'(x) + F_2(x)F_1'(x)$.
3. $(F_1(F_2(x))' = F_1'(F_2(x))F_2'(x)$.
4. *If* $\operatorname{char}(\Bbbk) = p$ *then* $F'(x) = 0$ *if and only* $F(x) = G(x)^p$ *for some* $G(x) \in \Bbbk[x]$.

Similarly, the notation $\partial F/\partial x_i$ is used for polynomials $F(\underline{x}) \in \Bbbk[\underline{x}]$ and an analogue of Lemma A.5.2 holds.

A.5.1 Homogeneous polynomials

Definition A.5.3 A non-zero polynomial $F(\underline{x}) \in \Bbbk[\underline{x}]$ is **homogeneous** of degree d if all its monomials have degree d, i.e.

$$F(x_0, \dots, x_n) = \sum_{\substack{i_0, i_1, \dots, i_n \in \mathbb{Z}_{\ge 0} \\ i_0 + i_1 + \cdots i_n = d}} F_{i_0, i_1, \dots, i_n} x_0^{i_0} x_1^{i_1} \cdots x_n^{i_n}.$$

Any polynomial $F(\underline{x}) \in \Bbbk[x_0, \dots, x_n]$ can be written as a **homogeneous decomposition** $\sum_{i=0}^{m} F_i(\underline{x})$ for some $m \in \mathbb{N}$ where $F_i(\underline{x})$ is a homogeneous polynomial of degree i; see Section II.3 of [329].

Lemma A.5.4 *Let R be an integral domain.*

1. *If $F(x) \in R[x_0, \ldots, x_n]$ is homogeneous and $\lambda \in R$ then $F(\lambda x_0, \ldots, \lambda x_n) = \lambda^d F(x_0, \ldots, x_n)$.*
2. *If $F_1, F_2 \in R[x_0, \ldots, x_n]$ are non-zero and homogeneous of degrees r and s respectively then $F_1(\underline{x})F_2(\underline{x})$ is homogeneous of degree $r + s$.*
3. *Let $F_1, F_2 \in R[x_0, \ldots, x_n]$ be non-zero. If $F_1(\underline{x})F_2(\underline{x})$ is homogeneous then $F_1(\underline{x})$ and $F_2(\underline{x})$ are both homogeneous.*

Proof See Exercise 1-1 (page 6) of Fulton [199]. □

A.5.2 Resultants

Let R be a commutative integral domain. Let $F(x) = F_n x^n + F_{n-1}x^{n-1} + \cdots + F_0$ and $G(x) = G_m x^m + G_{n-1}x^{n-1} + \cdots + G_0$ be two polynomials over R with $F_0, F_n, G_0, G_m \neq 0$. The polynomials $F, xF, \ldots, x^{m-1}F, G, xG, \ldots, x^{n-1}G$ can be written as $n+m$ linear combinations of the $n+m$ variables $1, x, \ldots, x^{n+m-1}$ and so the variable x may be eliminated to compute the **resultant** (there should be no confusion between the use of the symbol R for both the ring and the resultant)

$$R(F,G) = R_x(F,G) = \det \begin{pmatrix} F_0 & F_1 & \cdots & F_n & 0 & 0 & \cdots & 0 \\ 0 & F_0 & \cdots & F_{n-1} & F_n & 0 & \cdots & 0 \\ \vdots & & \ddots & & & \ddots & & \vdots \\ 0 & 0 & \cdots & 0 & F_0 & F_1 & \cdots & F_n \\ G_0 & \cdots & G_m & 0 & \cdots & \cdots & \cdots & 0 \\ 0 & G_0 & \cdots & G_m & 0 & \cdots & \cdots & 0 \\ \vdots & & \ddots & & \ddots & & \ddots & \vdots \\ 0 & 0 & \cdots & 0 & 0 & G_0 & \cdots & G_m \end{pmatrix}$$

Theorem A.5.5 *Let $\Bbbk$ be a field and $F(x), G(x) \in \Bbbk[x]$. Write $F(x) = \sum_{i=0}^{n} F_i x^i$ and $G(x) = \sum_{i=0}^{m} G_i x^i$. Suppose $F_0 G_0 \neq 0$. Then $R(F(x), G(x)) = 0$ if and only if $F(x)$ and $G(x)$ have a common root in $\overline{\Bbbk}$.*

Proof See Proposition IV.8.1 and Corollary IV.8.4 of [329]. □

Theorem A.5.5 is generalised to polynomials in $R[x]$ where R is a UFD in Lemma 2.6 on page 41 of Lorenzini [355]. Section IV.2.7 of [355] also describes the relation between $R(F, G)$ and the norm of $G(\alpha)$ in the number ring generated by a root α of $F(x)$.

If $F(x, y), G(x, y) \in \mathbb{Z}[x, y]$ then write $R_x(F, G) \in \mathbb{Z}[y]$ for the resultant, which is a polynomial in y, obtained by treating F and G as polynomials in x over the ring $R = \mathbb{Z}[y]$. If F and G have total degree d in x and y then the degree in y of $R_x(F, G)$ is $O(d^2)$.

A.6 Field extensions

General references for fields and their extensions are Chapter II of Artin [14], Chapter V of Hungerford [271] or Chapter V of Lang [329].

Let $\mathbb{k}$ be a field. An **extension** of $\mathbb{k}$ is any field $\mathbb{k}'$ such that $\mathbb{k} \subseteq \mathbb{k}'$, in which case we write $\mathbb{k}'/\mathbb{k}$. Then $\mathbb{k}'$ is a vector space over $\mathbb{k}$. If this vector space has finite dimension then the **degree** of $\mathbb{k}'/\mathbb{k}$, denoted $[\mathbb{k}' : \mathbb{k}]$, is the vector space dimension of $\mathbb{k}'$ over $\mathbb{k}$.

An element $\theta \in \mathbb{k}'$ is **algebraic** over $\mathbb{k}$ if there is some polynomial $F(x) \in \mathbb{k}[x]$ such that $F(\theta) = 0$. An extension $\mathbb{k}'$ of $\mathbb{k}$ is **algebraic** if every $\theta \in \mathbb{k}'$ is algebraic over $\mathbb{k}$. If $\mathbb{k}'/\mathbb{k}$ is algebraic and $\mathbb{k} \subseteq \mathbb{k}'' \subseteq \mathbb{k}'$ then $\mathbb{k}''/\mathbb{k}$ and $\mathbb{k}'/\mathbb{k}''$ are algebraic. Similarly, if $\mathbb{k}'/\mathbb{k}$ is finite then $[\mathbb{k}' : \mathbb{k}] = [\mathbb{k}' : \mathbb{k}''][\mathbb{k}'' : \mathbb{k}]$.

Lemma A.6.1 *Let $\mathbb{k}$ be a field. Every finite extension of $\mathbb{k}$ is algebraic.*

Proof See Theorem 4 of Section II.3 of [573], Proposition V.1.1 of [329] or Theorem V.1.11 of [271]. □

The **compositum** of two fields $\mathbb{k}$ and $\mathbb{k}'$ is the smallest field that contains both of them. We define $\mathbb{k}(\theta) = \{a(\theta)/b(\theta) : a(x), b(x) \in \mathbb{k}[x], b(\theta) \neq 0\}$ for any element θ. This is the smallest field that contains $\mathbb{k}$ and θ. For example, θ may be algebraic over $\mathbb{k}$ (e.g., $\mathbb{k}(\sqrt{-1})$) or transcendental (e.g., $\mathbb{k}(x)$). More generally, $\mathbb{k}(\theta_1, \dots, \theta_n) = \mathbb{k}(\theta_1)(\theta_2)\cdots(\theta_n)$ is the field generated over $\mathbb{k}$ by $\theta_1, \dots, \theta_n$. A field extension $\mathbb{k}'/\mathbb{k}$ is **finitely generated** if $\mathbb{k}' = \mathbb{k}(\theta_1, \dots, \theta_n)$ for some $\theta_1, \dots, \theta_n \in \mathbb{k}'$.

Theorem A.6.2 *Let $\mathbb{k}$ be a field. Suppose K is field that is finitely generated as a ring over $\mathbb{k}$. Then K is an algebraic extension of $\mathbb{k}$.*

Proof See pages 31–33 of Fulton [199]. □

An **algebraic closure** of a field $\mathbb{k}$ is a field $\overline{\mathbb{k}}$ such that every non-constant polynomial in $\overline{\mathbb{k}}[x]$ has a root in $\overline{\mathbb{k}}$. For details see Section V.2 of [329]. We always assume that there is a fixed algebraic closure of $\mathbb{k}$ and we assume that every algebraic extension $\mathbb{k}'/\mathbb{k}$ is chosen such that $\mathbb{k}' \subset \overline{\mathbb{k}}$ and that $\overline{\mathbb{k}'} = \overline{\mathbb{k}}$. Since the main case of interest is $\mathbb{k} = \mathbb{F}_q$, this assumption is quite natural.

We recall the notions of separable and purely inseparable extensions (see Sections V.4 and V.6 of Lang [329], Section V.6 of Hungerford [271] or Sections A.7 and A.8 of Stichtenoth [529]). An element α, algebraic over a field $\mathbb{k}$, is **separable** (respectively, **purely inseparable**) if the minimal polynomial of α over $\mathbb{k}$ has distinct roots (respectively, one root) in $\overline{\mathbb{k}}$. Hence, α is separable over $\mathbb{k}$ if its minimal polynomial has non-zero derivative. If $\mathrm{char}(\mathbb{k}) = p$ then α is purely inseparable if the minimal polynomial of α is of the form $x^{p^m} - a$ for some $a \in \mathbb{k}$.

Let $\mathbb{k}'/\mathbb{k}$ be a finite extension of fields and let $\alpha \in \mathbb{k}'$. One can define the **norm** and **trace** of α in terms of the matrix representation of multiplication by α as a linear map on the vector space $\mathbb{k}'/\mathbb{k}$ (see Section A.14 of [529] or Section IV.2 of [355]). When $\mathbb{k}'/\mathbb{k}$ is separable, an equivalent definition is to let $\sigma_i : \mathbb{k}' \to \overline{\mathbb{k}}$ be the $n = [\mathbb{k}' : \mathbb{k}]$ distinct embeddings (i.e., injective field homomorphisms), then the norm of $\alpha \in \mathbb{k}'$ is $N_{\mathbb{k}'/\mathbb{k}}(\alpha) = \prod_{i=1}^{n} \sigma_i(\alpha)$ and the **trace** is $\mathrm{Tr}_{\mathbb{k}'/\mathbb{k}}(\alpha) = \sum_{i=1}^{n} \sigma_i(\alpha)$.

An element $x \in K$ is **transcendental** over $\mathbb{k}$ if x is not algebraic over $\mathbb{k}$. Unless there is an implicit algebraic relation between $x_1, \dots, x_n$ we write $\mathbb{k}(x_1, \dots, x_n)$ to mean the field $\mathbb{k}(x_1)(x_2)\cdots(x_n)$ where each x_i is transcendental over $\mathbb{k}(x_1, \dots, x_{i-1})$.

Definition A.6.3 Let K be a finitely generated field extension of $\mathbb{k}$. The **transcendence degree** of $K/\mathbb{k}$, denoted $\mathrm{trdeg}(K/\mathbb{k})$, is the smallest integer n such that there are $x_1, \dots, x_n \in K$ with K algebraic over $\mathbb{k}(x_1, \dots, x_n)$ (by definition x_i is transcendental over $\mathbb{k}(x_1, \dots, x_{i-1})$). Such a set $\{x_1, \dots, x_n\}$ is called a **transcendence basis** for $K/\mathbb{k}$.

Theorem A.6.4 *Let $K/\Bbbk$ be a finitely generated field extension. Then the transcendence degree is well-defined (i.e., all transcendence bases have the same number of elements).*

Proof See Theorem 25 of Section II.12 of [573], Theorem VI.1.8 of [271] or Theorem 1.6.13 of [568]. □

Theorem A.6.5 *Let $K/\Bbbk$ and F/K be finitely generated field extensions. Then* $\mathrm{trdeg}(F/\Bbbk) = \mathrm{trdeg}(F/K) + \mathrm{trdeg}(K/\Bbbk)$.

Proof See Theorem 26 of Section II.12 of [573]. □

Corollary A.6.6 *Let $K/\Bbbk$ be finitely generated with transcendence degree 1 and let $x \in K$ be transcendental over $\Bbbk$. Then K is a finite algebraic extension of $\Bbbk(x)$.*

A **perfect field** is one for which every algebraic extension is separable. A convenient equivalent definition is that a field $\Bbbk$ of characteristic p is perfect if $\{x^p : x \in \Bbbk\} = \Bbbk$ (see Section V.6 of Lang [329]). We restrict to perfect fields for a number of reasons, one of which is that the primitive element theorem does not hold for non-perfect fields, and another is due to issues with fields of definition (see Remark 5.3.7). Finite fields, fields of characteristic zero, and algebraic closures of finite fields are perfect (see Exercise V.7.13 of [271] or Section V.6 of [329]).

Theorem A.6.7 ***(Primitive element theorem)*** *Let $\Bbbk$ be a perfect field. If $\Bbbk'/\Bbbk$ is a finite, separable, extension then there is some $\alpha \in \Bbbk'$ such that $\Bbbk' = \Bbbk(\alpha)$.*

Proof Theorem V.6.15 of [271], Theorem 27 of [14], Theorem V.4.6 of [329]. □

A.7 Galois theory

For an introduction to Galois theory see Chapter V of Hungerford [271], Chapter 6 of Lang [329] or Stewart [526]. An algebraic extension $\Bbbk'/\Bbbk$ is Galois if it is normal (i.e., every irreducible polynomial $F(x) \in \Bbbk[x]$ with a root in $\Bbbk'$ splits completely over $\Bbbk'$) and separable. The Galois group of $\Bbbk'/\Bbbk$ is

$$\mathrm{Gal}(\Bbbk'/\Bbbk) = \{\sigma : \Bbbk' \to \Bbbk' : \sigma \text{ is a field automorphism, and } \sigma(x) = x \text{ for all } x \in \Bbbk\}.$$

Theorem A.7.1 *Let $\Bbbk'/\Bbbk$ be a finite Galois extension. Then there is a one-to-one correspondence between the set of subfields $\{\Bbbk'' : \Bbbk \subseteq \Bbbk'' \subseteq \Bbbk'\}$ and the set of normal subgroups H of $\mathrm{Gal}(\Bbbk'/\Bbbk)$, via*

$$\Bbbk'' = \{x \in \Bbbk' : \sigma(x) = x \text{ for all } \sigma \in H\}.$$

Proof See Theorem V.2.5 of [271]. □

If $\Bbbk$ is a perfect field then $\overline{\Bbbk}$ is a separable extension and hence a Galois extension of $\Bbbk$. If $\Bbbk'$ is any algebraic extension of $\Bbbk$ (not necessarily Galois) then $\overline{\Bbbk}/\Bbbk'$ is Galois. The Galois group $\mathrm{Gal}(\overline{\Bbbk}/\Bbbk)$ can be defined using the notion of an inverse limit (see Chapter 5 of [116]). Topological aspects of $\mathrm{Gal}(\overline{\Bbbk}/\Bbbk)$ are important, but we do not discuss them.

A.7.1 Galois cohomology

One finds brief summaries of **Galois cohomology** in Appendix B of Silverman [505] and Chapter 19 of Cassels [114]. More detailed references are Serre [488] and Cassels and Frölich [116].

Let $K/\Bbbk$ be Galois (we include $K = \overline{\Bbbk}$). Let $G = \mathrm{Gal}(K/\Bbbk)$. Unlike most references we write our Galois groups acting on the left (i.e., as $\sigma(f)$ rather than f^σ). A 1-cocycle in the additive group K is a function[1] $\xi : G \to K$ such that $\xi(\sigma\tau) = \sigma(\xi(\tau)) + \xi(\sigma)$. A 1-coboundary in K is the function $\xi(\sigma) = \sigma(\gamma) - \gamma$ for some $\gamma \in K$. The group of 1-cocycles modulo 1-coboundaries (the group operation is addition $(\xi_1 + \xi_2)(\tau) = \xi_1(\tau) + \xi_2(\tau)$) is denoted $H^1(G, K)$. Similarly, for the multiplicative group K^*, a 1-cocycle satisfies $\xi(\sigma\tau) = \sigma(\xi(\tau))\xi(\sigma)$, a 1-coboundary is $\sigma(\gamma)/\gamma$ and the quotient group is denoted $H^1(G, K^*)$.

Theorem A.7.2 *Let $K/\Bbbk$ be Galois. Then $H^1(\mathrm{Gal}(K/\Bbbk), K) = \{0\}$ and* **(*Hilbert 90*)** *$H^1(\mathrm{Gal}(K/\Bbbk), K^*) = \{1\}$ (i.e., both groups are trivial).*

Proof The case of finite extensions $K/\Bbbk$ is given in Exercise 20.5 of Cassels [114] or Propositions 1 and 2 of Chapter 10 of [488]. For a proof in the infinite case see Propositions 2 and 3 (Sections 2.6 and 2.7) of Chapter 5 of [116]. □

A.8 Finite fields

Let p be a prime. Denote by $\mathbb{F}_p = \mathbb{Z}/p\mathbb{Z}$ the finite field of p elements. The multiplicative group of non-zero elements is $\mathbb{F}_p^*$. Recall that $\mathbb{F}_p^*$ is a cyclic group. A generator for $\mathbb{F}_q^*$ is called a **primitive root**. The number of primitive roots in $\mathbb{F}_q^*$ is $\varphi(q-1)$.

Theorem A.8.1 *Let p be a prime and $m \in \mathbb{N}$. Then there exists a field $\mathbb{F}_{p^m}$ having p^m elements. All such fields are isomorphic. Every finite field can be represented as $\mathbb{F}_p[x]/(F(x))$ where $F(x) \in \mathbb{F}_p[x]$ is a monic irreducible polynomial of degree m; the corresponding vector space basis $\{1, x, \ldots, x^{m-1}\}$ for $\mathbb{F}_{p^m}/\mathbb{F}_p$ is called a* ***polynomial basis****.*

Proof See Corollary V.5.7 of [271] or Section 20.2 of [497]. □

If p is a prime and $q = p^m$ then $\mathbb{F}_{p^{nm}}$ may be viewed as a degree n algebraic extension of $\mathbb{F}_q$.

Theorem A.8.2 *Every finite field $\mathbb{F}_{p^m}$ has a vector space basis over $\mathbb{F}_p$ of the form $\{\theta, \theta^p, \ldots, \theta^{p^{m-1}}\}$; this is called a* ***normal basis****.*

Proof See Theorem 2.35 or Theorem 3.73 of [350, 351] or Exercise 20.14 of [497] (the latter proof works for extensions of $\mathbb{F}_p$, but not for all fields). □

We discuss methods to construct a normal basis in Section 2.14.1.

Theorem A.8.3 *Let q be a prime power and $m \in \mathbb{N}$. Then $\mathbb{F}_{q^m}$ is an algebraic extension of $\mathbb{F}_q$ that is Galois. The Galois group is cyclic of order m and generated by the q-power Frobenius automorphism $\pi : x \mapsto x^q$.*

[1] It is also necessary that ξ satisfy some topological requirements, but we do not explain these here.

Let $\alpha \in \mathbb{F}_{q^m}$. The **trace** and **norm** with respect to $\mathbb{F}_{q^m}/\mathbb{F}_q$ are

$$\mathrm{Tr}_{\mathbb{F}_{q^m}/\mathbb{F}_q}(\alpha) = \sum_{i=0}^{m-1} \alpha^{q^i} \quad \text{and} \quad \mathrm{N}_{\mathbb{F}_{q^m}/\mathbb{F}_q}(\alpha) = \prod_{i=0}^{m-1} \alpha^{q^i}.$$

The **characteristic polynomial** over $\mathbb{F}_q$ of $\alpha \in \mathbb{F}_{q^m}$ is $F(x) = \prod_{i=0}^{m-1}(x - \alpha^{q^i}) \in \mathbb{F}_q[x]$. The trace and norm of $\alpha \in \mathbb{F}_{q^m}$ are (up to sign) the coefficients of x^{m-1} and x^0 in the characteristic polynomial.

An element $\alpha \in \mathbb{F}_q$ is a **square** or **quadratic residue** if the equation $x^2 = \alpha$ has a solution $x \in \mathbb{F}_q$. If g is a primitive root for $\mathbb{F}_q$ then g^a is a square if and only if a is even. Hence, α is a square in $\mathbb{F}_q^*$ if and only if $\alpha^{(q-1)/2} = 1$.

Lemma A.8.4 *Let $g \in \mathbb{F}_{q^m}$, where $m > 1$, be chosen uniformly at random. The probability that g lies in a proper subfield $K \subset \mathbb{F}_{q^m}$ such that $\mathbb{F}_q \subseteq K$ is at most $1/q$.*

Proof If $m = 2$ then the probability is $q/q^2 = 1/q$ so the result is tight in this case. When $m = l^i$ is a power of a prime $l \geq 2$ then all proper subfields of $\mathbb{F}_{q^m}$ that contain $\mathbb{F}_q$ are contained in $\mathbb{F}_{q^{l^{i-1}}}$ so the probability is $q^{l^{i-1}}/q^{l^i} = 1/q^{l^{i-1}(l-1)} \leq 1/q$. Finally, write $m = nl^i$ where $l \geq 2$ is prime, $i \geq 1, n \geq 2$ and $\gcd(n, l) = 1$. Then every proper subfield containing $\mathbb{F}_q$ lies in $\mathbb{F}_{q^{l^i}}$ or $\mathbb{F}_{q^{nl^{i-1}}}$. The probability that a random element lies in either of these fields is

$$\leq q^{l^i}/q^{nl^i} + q^{nl^{i-1}}/q^{nl^i} = 1/q^{l^i(n-1)} + 1/q^{nl^{i-1}(l-1)} \leq 1/q^2 + 1/q^2 \leq 1/q. \qquad \square$$

A.9 Ideals

If R is a commutative ring then an R-**ideal** is a subset $I \subset R$ that is an additive group and is such that, for all $a \in I$ and $r \in R$, $ar \in I$. An R-ideal I is proper if $I \neq R$ and is non-trivial if $I \neq \{0\}$. A **principal ideal** is $(a) = \{ar : r \in R\}$ for some $a \in R$. If $S \subset R$ then (S) is the R-ideal $\{\sum_{i=1}^n s_i r_i : n \in \mathbb{N}, s_i \in S, r_i \in R\}$. An ideal I is **finitely generated** if $I = (S)$ for a finite subset $S \subset R$. The **radical** of an ideal I in a ring R is $\mathrm{rad}_R(I) = \{r \in R : r^n \in I$ for some $n \in \mathbb{N}\}$ (see Definition VIII.2.5 and Theorem VIII.2.6 of Hungerford [271]). If I_1 and I_2 are ideals of R then

$$I_1 I_2 = \left\{ \sum_{i=1}^n a_i b_i : n \in \mathbb{N}, a_i \in I_1, b_i \in I_2 \right\}.$$

Note that $I_1 I_2 \subseteq I_1 \cap I_2$. For $a, b \in R$ one has $(ab) = (a)(b)$.

Let $I_1, \dots, I_n$ be ideals in a ring R such that the ideal $(I_i \cup I_j) = R$ for all $1 \leq i < j \leq n$ (we call such ideals pairwise-coprime). If $a_1, \dots, a_n \in R$ then there exists an element $a \in R$ such that $a \equiv a_i \pmod{I_i}$ (in other words, $a - a_i \in I_i$) for all $1 \leq i \leq n$. This is the **Chinese remainder theorem** for rings; see Theorem III.2.25 of [271] or Theorem II.2.1 of [329].

The following result gives three equivalent conditions for an ideal to be prime.

Lemma A.9.1 *Let I be an ideal of R. The following conditions are equivalent and, if they hold, I is called a* ***prime ideal****:*

1. *If $a, b \in R$ are such that $ab \in I$ then $a \in I$ or $b \in I$.*
2. *R/I is an integral domain (i.e., has no zero divisors).*
3. *If I_1 and I_2 are ideals of R such that $I_1 I_2 \subseteq I$ then $I_1 \subseteq I$ or $I_2 \subseteq I$.*

If $F(\underline{x}) \in \Bbbk[\underline{x}]$ is irreducible then the $\Bbbk[\underline{x}]$-ideal $(F(\underline{x})) = \{F(\underline{x})G(\underline{x}) : G(\underline{x}) \in \Bbbk[\underline{x}]\}$ is a prime ideal.

An R-ideal I is **maximal** if every R-ideal J such that $I \subseteq J \subseteq R$ is such that either $J = I$ or $J = R$.

Lemma A.9.2 *An R-ideal I is maximal if and only if R/I is a field (hence, a maximal R-ideal is prime). If I is a maximal R-ideal and $S \subset R$ is a subring then $I \cap S$ is a prime S-ideal.*

Proof For the first statement see Theorem III.2.20 of [271] or Section II.2 of [329]. The second statement is proved as follows: let I be maximal and consider the injection $S \to R$ inducing $S \to R/I$ with kernel $J = S \cap I$. Then $S/J \to R/I$ is an injective ring homomorphism into a field, so J is a prime S-ideal. □

Let R be a commutative ring. A sequence $I_1 \subset I_2 \subset \cdots$ of R-ideals is called an **ascending chain**. A commutative ring R is **Noetherian** if every ascending chain of R-ideals is finite. Equivalently, a ring is Noetherian if every ideal is finitely generated. For more details see Section VIII.1 of [271] or Section X.1 of [329].

Theorem A.9.3 *(Hilbert basis theorem) If R is a Noetherian ring then $R[x]$ is a Noetherian ring.*

Proof See Theorem 1 page 13 of [199], Theorem VIII.4.9 of [271] Section IV.4 of [329], or Theorem 7.5 of [15]. □

Corollary A.9.4 *$\Bbbk[x_1, \ldots, x_n]$ is Noetherian.*

A **multiplicative subset** of a ring R is a set S such that $1 \in S$, $s_1, s_2 \in S \Rightarrow s_1 s_2 \in S$. The **localisation** of a ring R with respect to a multiplicative subset S is the set

$$S^{-1}R = \{r/s : r \in R, s \in S\}$$

with the equivalence relation $r_1/s_1 \equiv r_2/s_2$ if $r_1 s_2 - r_2 s_1 = 0$. For more details see Chapter 3 of [15], Section 1.3 of [488], Section I.1 of [327], Section II.4 of [329] or Section III.4 of [271]. In the case $S = R^*$, we call $S^{-1}R$ the **field of fractions** of R. If $\mathfrak{p}$ is a prime ideal of R then $S = R - \mathfrak{p}$ is a multiplicative subset and the localisation $S^{-1}R$ is denote $R_{\mathfrak{p}}$.

Lemma A.9.5 *If R is Noetherian and S is a multiplicative subset of R then the localisation $S^{-1}R$ is Noetherian.*

Proof See Proposition 7.3 of [15] or Proposition 1.6 of Section X.1 of [329]. □

A ring R is **local** if it has a unique maximal ideal. If $\mathfrak{m}$ is a maximal idea of a ring R then the localisation $R_{\mathfrak{m}}$ is a local ring. It follows that $R_{\mathfrak{m}}$ is Noetherian.

A.10 Vector spaces and linear algebra

The results of this section are mainly used when we discuss lattices in Chapter 16. A good basic reference is Curtis [151].

Let $\Bbbk$ be a field. We write vectors in $\Bbbk^n$ as row vectors. We interchangeably use the words **points** and **vectors** for elements of $\Bbbk^n$. The zero vector is $\underline{0} = (0, \ldots, 0)$. For $1 \le i \le n$ the ith unit vector is $\underline{e}_i = (e_{i,1}, \ldots, e_{i,n})$ such that $e_{i,i} = 1$ and $e_{i,j} = 0$ for $1 \le j \le n$ and $j \ne i$.

A **linear map** is a function $A : \Bbbk^n \to \Bbbk^m$ such that $A(\lambda \underline{x} + \mu \underline{y}) = \lambda A(\underline{x}) + \mu A(\underline{y})$ for all $\lambda, \mu \in \Bbbk$ and $\underline{x}, \underline{y} \in \Bbbk^n$. Given a basis for $\Bbbk^n$, any linear map can be represented as an $n \times m$ matrix A such that $A(\underline{x}) = \underline{x}A$. We denote the entries of A by $A_{i,j}$ for $1 \le i \le n, 1 \le j \le m$. Denote by I_n the $n \times n$ **identity matrix**. We denote by A^T the **transpose**, which is an $m \times n$ matrix such that $(A^T)_{i,j} = A_{j,i}$. We have $(AB)^T = B^T A^T$.

A fundamental computational problem is to solve the linear system of equations $\underline{x}A = \underline{y}$ and it is well-known that this can be done using Gaussian elimination (see Section 6 of Curtis [151] or Chapter 3 of Schrijver [478]).

The **rank** of an $m \times n$ matrix A (denoted rank(A)) is the maximum number of linearly independent rows of A (equivalently, the maximum number of linearly independent columns). If A is an $n \times n$ matrix then the **inverse** of A, if it exists, is the matrix such that $AA^{-1} = A^{-1}A = I_n$. If A and B are invertible then $(AB)^{-1} = B^{-1}A^{-1}$. One can compute A^{-1} using Gaussian elimination.

A.10.1 Inner products and norms

Definition A.10.1 The **inner product** of two vectors $\underline{v} = (v_1, \dots, v_n)$ and $\underline{w} = (w_1, \dots, w_n) \in \Bbbk^n$ is

$$\langle \underline{v}, \underline{w} \rangle = \sum_{i=1}^{n} v_i w_i.$$

The **Euclidean norm** or **ℓ_2-norm** of a vector $\underline{v} \in \mathbb{R}^n$ is

$$\|\underline{v}\| = \sqrt{\langle \underline{v}, \underline{v} \rangle}.$$

More generally for $\mathbb{R}^n$ one can define the **ℓ_a-norm** of a vector $\underline{v}$ for any $a \in \mathbb{N}$ as $\|\underline{v}\|_a = \left(\sum_{i=1}^n |v_i|^a\right)^{1/a}$. Important special cases are the ℓ_1-norm $\|\underline{v}_i\| = \sum_{i=1}^n |\underline{v}_i|$ and the ℓ_∞-norm $\|\underline{v}\|_\infty = \max\{|v_1|, \dots, |v_n|\}$. (The reader should not confuse the notion of norm in Galois theory with the notion of norm on vector spaces.)

Lemma A.10.2 *Let $\underline{v} \in \mathbb{R}^n$. Then*

$$\|\underline{v}\|_\infty \le \|\underline{v}\|_2 \le \sqrt{n}\|\underline{v}\|_\infty \quad \textit{and} \quad \|\underline{v}\|_\infty \le \|\underline{v}\|_1 \le n\|\underline{v}\|_\infty.$$

Lemma A.10.3 *Let $\underline{v}, \underline{w} \in \mathbb{R}^n$ and let $\|\underline{v}\|$ be the Euclidean norm.*

1. $\|\underline{v} + \underline{w}\| \le \|\underline{v}\| + \|\underline{w}\|$.
2. $\langle \underline{v}, \underline{w} \rangle = \langle \underline{w}, \underline{v} \rangle$.
3. $\|\underline{v}\| = 0$ *implies* $\underline{v} = \underline{0}$.
4. $|\langle \underline{v}, \underline{w} \rangle| \le \|\underline{v}\| \|\underline{w}\|$.
5. *Let A be an $n \times n$ matrix over $\mathbb{R}$. The following are equivalent:*
   (a) $\|\underline{x}A\| = \|\underline{x}\|$ *for all* $\underline{x} \in \mathbb{R}^n$;
   (b) $\langle \underline{x}A, \underline{y}A \rangle = \langle \underline{x}, \underline{y} \rangle$ *for all* $\underline{x}, \underline{y} \in \mathbb{R}^n$;
   (c) $AA^T = I_n$ *(which implies* $\det(A)^2 = 1$*).*
   Such a matrix is called an ***orthogonal matrix****.*

Definition A.10.4 A basis $\{v_1, \dots, v_n\}$ for a vector space is **orthogonal** if

$$\langle v_i, v_j \rangle = 0$$

for all $1 \le i < j \le n$. If we also have the condition $\langle v_i, v_i \rangle = 1$ then the basis is called **orthonormal**.

Lemma A.10.5 *Let $\{\underline{v}_1, \dots, \underline{v}_n\}$ be an orthogonal basis for $\mathbb{R}^n$. If $\underline{v} = \sum_{j=1}^n \lambda_j \underline{v}_j$ then $\|\underline{v}\|^2 = \sum_{j=1}^n \lambda_j^2 \|\underline{v}_j\|^2$.*

If one has an orthogonal basis $\{\underline{v}_1, \dots, \underline{v}_n\}$ then it is extremely easy to decompose an arbitrary vector $\underline{w}$ over the basis. The representation is

$$\underline{w} = \sum_{i=1}^{n} \frac{\langle \underline{w}, \underline{v}_i \rangle}{\langle \underline{v}_i, \underline{v}_i \rangle} \underline{v}_i .$$

This is simpler and faster than solving the linear system using Gaussian elimination.

If $V \subseteq \mathbb{R}^n$ is a subspace then the **orthogonal complement** is $V^\perp = \{\underline{w} \in \mathbb{R}^n : \langle \underline{w}, \underline{v} \rangle = 0 \text{ for all } \underline{v} \in V\}$. The dimension of $V^\perp$ is $n - \dim(V)$. Given a basis $\{\underline{v}_1, \dots, \underline{v}_m\}$ for V (where $m = \dim(V) < n$) one can compute a basis $\{\underline{v}_{m+1}, \dots, \underline{v}_n\}$ for $V^\perp$. The **orthogonal projection** of $\mathbb{R}^n$ to a subspace V is a linear map $P : \mathbb{R}^n \to V$ that is the identity on V and is such that $P(V^\perp) = \{0\}$. In other words, if $\underline{v} \in \mathbb{R}^n$ then $\underline{v} - P(\underline{v}) \in V^\perp$.

A.10.2 Gram–Schmidt orthogonalisation

Given a basis $\underline{v}_1, \dots, \underline{v}_n$ for a vector space, the Gram–Schmidt algorithm iteratively computes an orthogonal basis $\underline{v}_1^*, \dots, \underline{v}_n^*$ (called the **Gram–Schmidt orthogonalisation** or **GSO**). The idea is to set $\underline{v}_1^* = \underline{v}_1$ and then, for $2 \le i \le n$, to compute

$$\underline{v}_i^* = \underline{v}_i - \sum_{j=1}^{i-1} \mu_{i,j} \underline{v}_j^* \text{ where } \mu_{i,j} = \frac{\langle \underline{v}_i, \underline{v}_j^* \rangle}{\langle \underline{v}_j^*, \underline{v}_j^* \rangle}.$$

We discuss this algorithm further in Section 17.3.

A.10.3 Determinants

Let $\underline{b}_1, \dots, \underline{b}_n$ be n vectors in $\Bbbk^n$. One can define the **determinant** of the sequence $\underline{b}_1, \dots, \underline{b}_n$ (or of the matrix B whose rows are $\underline{b}_1, \dots, \underline{b}_n$) in the usual way (see Chapter 5 of Curtis [151] or Section VII.3 of Hungerford [271]).

Lemma A.10.6 *Let $\underline{b}_1, \dots, \underline{b}_n \in \Bbbk^n$.*

1. *Let B be the matrix whose rows are $\underline{b}_1, \dots, \underline{b}_n$. Then B is invertible if and only if* $\det(\underline{b}_1, \dots, \underline{b}_n) \ne 0$.
2. *For $\lambda \in \Bbbk$,* $\det(\underline{b}_1, \dots, \underline{b}_{i-1}, \lambda \underline{b}_i, \underline{b}_{i+1}, \dots, \underline{b}_n) = \lambda \det(\underline{b}_1, \dots, \underline{b}_n)$.
3. $\det(\underline{b}_1, \dots, \underline{b}_{i-1}, \underline{b}_i + \underline{b}_j, \underline{b}_{i+1}, \dots, \underline{b}_n) = \det(\underline{b}_1, \dots, \underline{b}_n)$ *for $i \ne j$.*
4. *If $\{\underline{e}_1, \dots, \underline{e}_n\}$ are the standard unit vectors in $\Bbbk^n$ then* $\det(\underline{e}_1, \dots, \underline{e}_n) = 1$.
5. *If B_1, B_2 are square matrices then* $\det(B_1 B_2) = \det(B_1)\det(B_2)$.
6. $\det(B) = \det(B^T)$.
7. *(Hadamard inequality)* $|\det(\underline{b}_1, \dots, \underline{b}_n)| \le \prod_{i=1}^n \|\underline{b}_i\|$ *(where $\|\underline{b}\|$ is the Euclidean norm).*

Proof See Theorems 16.6, 17.6, 17.15, 18.3 and 19.13 of Curtis [151]. □

Definition A.10.7 Let $\underline{b}_1, \ldots, \underline{b}_n$ be a set of vectors in $\mathbb{R}^n$. The **fundamental parallelepiped** of the set is

$$\left\{ \sum_{i=0}^{n} \lambda_i \underline{b}_i : 0 \leq \lambda_i < 1 \right\}.$$

Lemma A.10.8 *Let the notation be as above.*

1. *The volume of the fundamental parallelepiped of* $\{\underline{b}_1, \ldots, \underline{b}_n\}$ *is* $|\det(\underline{b}_1, \ldots, \underline{b}_n)|$.
2. $|\det(\underline{b}_1, \ldots, \underline{b}_n)| = \prod_{i=1}^{n} \|\underline{b}_i^*\|$ *where* $\underline{b}_i^*$ *are the Gram–Schmidt vectors.*

Proof The first claim is Theorem 19.12 of Curtis [151]. The second claim is Exercise 19.11 (also see Theorem 19.13) of [151]. □

There are two methods to compute the determinant for vectors in $\mathbb{R}^n$. The first is to perform Gaussian elimination to diagonalise and then take the product of the diagonal elements. The second is to apply Gram–Schmidt (using floating-point arithmetic) and then the determinant is just the product of the norms. Over $\mathbb{R}$, both methods only give an approximation to the determinant. To compute the determinant for vectors with entries in $\mathbb{Z}$ or $\mathbb{Q}$ one can use Gaussian elimination or Gram–Schmidt with exact arithmetic in $\mathbb{Q}$ (this gives an exact solution but suffers from coefficient explosion). Alternatively, one can compute the determinant modulo p_i for many small or medium sized primes p_i and use the Chinese remainder theorem.

A.11 Hermite normal form

Definition A.11.1 An $n \times m$ integer matrix $A = (A_{i,j})$ is in (row) **Hermite normal form** (**HNF**) if there is some integer $1 \leq r \leq n$ and a strictly increasing map $f : \{1, \ldots, n - r\} \to \{1, \ldots, m\}$ (i.e., $f(i+1) > f(i)$) such that:

1. the last r rows of A are zero,
2. $0 \leq A_{j,f(i)} < A_{i,f(i)}$ for $1 \leq j < i$ and $A_{j,f(i)} = 0$ for $i < j \leq n$.

In particular, an $n \times n$ matrix that is upper triangular and that satisfies the condition $0 \leq A_{j,i} < A_{i,i}$ for $1 \leq j < i \leq n$ is in Hermite normal form.

The HNF A' of an integer matrix A is unique and there is an $n \times n$ **unimodular matrix** U (i.e., U is a matrix with integer entries and determinant ± 1) such that $A' = UA$. For more details of the Hermite normal form see Section 2.4.2 of Cohen [127] or Section 4.1 of Schrijver [478] (though note that both books use columns rather than rows).

A.12 Orders in quadratic fields

A quadratic field is $\mathbb{Q}(\sqrt{d})$ where $d \neq 0, 1$ is a square-free integer. If $d < 0$ then the field is called an **imaginary quadratic field**. The **discriminant** of $K = \mathbb{Q}(\sqrt{d})$ is $D = d$ if $d \equiv 1 \pmod 4$ or $D = 4d$ otherwise. The **ring of integers** of a quadratic field of discriminant D is $\mathcal{O}_K = \mathbb{Z}[(D + \sqrt{D})/2]$.

An **order** in a field $\Bbbk$ containing $\mathbb{Q}$ is a subring R of $\Bbbk$ that is finitely generated as a $\mathbb{Z}$-module and is such that $R \otimes_{\mathbb{Z}} \mathbb{Q} = \Bbbk$. Every order in a quadratic field is of the form $\mathbb{Z}[c(D+\sqrt{D})/2]$ for some $c \in \mathbb{N}$. The integer c is called the **conductor** and the discriminant of the order is $c^2 D$.

A.13 Binary strings

The binary representation of an integer $a = \sum_{i=0}^{l-1} a_i 2^i$ is written as

$$(a_{l-1} \dots a_1 a_0)_2 \quad \text{or} \quad a_{l-1} \dots a_1 a_0 \tag{A.1}$$

where $a_i \in \{0, 1\}$ and $a_{l-1} = 1$. We say that the **bit-length** of a is l. An integer $a \in \mathbb{N}$ is represented by a binary string of bit-length $\lfloor \log_2(a) \rfloor + 1$. The **least significant bit** of a is $\mathrm{LSB}(a) = a_0 = a \pmod 2$. We call a_i the **ith bit** or **bit** i of a. The "most significant bit" is trivially always one, but in certain contexts one uses different notions of MSB; for example see Definition 21.7.1.

Binary strings of length l are sequences $a_1 a_2 \dots a_l$ with $a_i \in \{0, 1\}$. Such a sequence is also called an **l-bit string**. The ith bit is a_i. There is an ambiguity when one wants to interpret a binary string as an integer; our convention is that a_l is the least significant bit.[2]

We denote by $\{0, 1\}^l$ the set of all length l binary strings and $\{0, 1\}^*$ the set of all binary strings of arbitrary finite length. If a and b are binary strings then the exclusive-or (i.e., **XOR**) $a \oplus b$ is the binary string whose ith bit is $a_i + b_i \pmod 2$ for $1 \le i \le l$.

A.14 Probability and combinatorics

We briefly recall some ideas from probability. Good references are Ross [451], Woodroofe [569] and Chapter 6 of Shoup [497].

The number of ways to choose t items from n without replacement, and where the ordering matters, is $n(n-1)(n-2)\cdots(n-t+1) = n!/(n-t)!$. The number of ways to choose t items from n without replacement, and where the ordering does not matter, is $\binom{n}{t} = n!/(t!(n-t)!)$. The number of ways to choose t items from n with replacement and where the ordering does not matter is $\binom{n+t-1}{t-1}$. We have

$$\left(\frac{n}{m}\right)^m \le \binom{n}{m} \le \left(\frac{ne}{m}\right)^m.$$

Stirling's approximation to the factorial is $n! \approx \sqrt{2\pi n} e^{-n} n^n$ or $\log(n!) \approx n(\log(n) - 1)$ (where log denotes the natural logarithm). For proof see Section 5.4.1 of [569].

Let $[0, 1] = \{x \in \mathbb{R} : 0 \le x \le 1\}$. A **distribution** on a set S is a function Pr mapping "nice"[3] subsets of S to $[0, 1]$, with the properties that $\Pr(\emptyset) = 0$, $\Pr(S) = 1$ and if $A, B \subseteq S$ are disjoint and "nice" then $\Pr(A \cup B) = \Pr(A) + \Pr(B)$. For $s \in S$ we define $\Pr(s) = \Pr(\{s\})$ if $\{s\}$ is "nice". The **uniform distribution** on a finite set S is given by $\Pr(s) = 1/\#S$.

An **event** is a "nice" subset $E \subseteq S$, and $\Pr(E)$ is called the probability of the event. We define $\neg E$ to be $S - E$ so that $\Pr(\neg E) = 1 - \Pr(E)$. We have $\Pr(E_1) \le \Pr(E_1 \cup E_2) \le \Pr(E_1) + \Pr(E_2)$. We define $\Pr(E_1 \text{ and } E_2) = \Pr(E_1 \cap E_2)$.

[2] This means that the ith bit of a binary string is not the ith bit of the corresponding integer. This inconsistency will not cause confusion in the book.

[3] Technically, S must be a set with a measure and the "nice" subsets are the measurable ones. When S is finite or countable then every subset is "nice".

Let S be a finite set with an implicit distribution on it (usually the uniform distribution). In an algorithm we write $s \leftarrow S$ to mean that $s \in S$ is randomly selected from S according to the distribution, i.e. s is chosen with probability $\Pr(s)$.

If $A, E \subseteq S$ and $\Pr(E) > 0$ then the **conditional probability** is

$$\Pr(A \mid E) = \frac{\Pr(A \cap E)}{\Pr(E)}.$$

If $\Pr(A \cap E) = \Pr(A)\Pr(E)$ then A and E are **independent events** (equivalently, if $\Pr(E) > 0$ then $\Pr(A \mid E) = \Pr(A)$). If S is the disjoint union $E_1 \cup E_2 \cup \cdots \cup E_n$ then $\Pr(A) = \sum_{i=1}^{n} \Pr(A \mid E_i)\Pr(E_i)$.

Let S be a set. A **random variable** is a function[4] $X : S \to \mathbb{R}$. Write $\mathcal{X} \subseteq \mathbb{R}$ for the image of X (our applications will always have $\mathcal{X}$ either finite or $\mathbb{N}$). Then X induces a distribution on $\mathcal{X}$, defined for $x \in \mathcal{X}$ by $\Pr(X = x)$ is the measure of $X^{-1}(\{x\})$ (in the case where S is finite or countable, $\Pr(X = x) = \sum_{s \in X^{-1}(x)} \Pr(s)$). Random variables X_1 and X_2 are **independent random variables** if $\Pr(X_1 = x_1 \text{ and } X_2 = x_2) = \Pr(X_1 = x_1)\Pr(X_2 = x_2)$ for all $x_1 \in \mathcal{X}_1$ and $x_2 \in \mathcal{X}_2$.

The **expectation** of a random variable X taking values in a finite or countable set $\mathcal{X} \subseteq \mathbb{R}_{>0}$ is

$$E(X) = \sum_{x \in \mathcal{X}} x \Pr(X = x).$$

If $\mathcal{X} = \mathbb{N}$ then $E(X) = \sum_{n=0}^{\infty} \Pr(X > n)$ (this is shown in the proof of Theorem 14.1.1). Note that if $\mathcal{X}$ is finite then $E(X)$ exists, but for $\mathcal{X}$ countable the expectation only exists if the sum is convergent. If X_1 and X_2 are random variables on S then $E(X_1 + X_2) = E(X_1) + E(X_2)$. If X_1 and X_2 are independent then $E(X_1 X_2) = E(X_1)E(X_2)$.

Example A.14.1 Consider flipping a coin, with probability p of "heads" and probability $1 - p$ of "tails" (where $0 < p < 1$). Assume the coin flips are independent events. What is the expected number of trials until the coin lands "heads"?

Let X be the random variable with values in $\mathbb{N}$ where $\Pr(X = n)$ is the probability that the first head is on the nth throw. Then $\Pr(X > n) = (1 - p)^n$ and $\Pr(X = n) = (1 - p)^{n-1} p$. This gives the **geometric distribution** on $\mathbb{N}$. One can check that $\sum_{n=1}^{\infty} \Pr(X = n) = 1$.

The expectation of X is $E(X) = \sum_{n=1}^{\infty} n \Pr(X = n)$ (the ratio test shows that this sum is absolutely convergent). Write $T = \sum_{n=1}^{\infty} n(1 - p)^{n-1}$. Then

$$\begin{aligned} E(X) = pT = T - (1 - p)T &= \sum_{n=1}^{\infty} n(1 - p)^{n-1} - \sum_{n=1}^{\infty} (n - 1)(1 - p)^{n-1} \\ &= 1 + \sum_{n=2}^{\infty} (1 - p)^{n-1} = \frac{1}{p}. \end{aligned}$$

To define this problem formally one should define the geometric random variable $X : S \to \mathbb{N}$, where S is the (uncountable) set of countable length sequences of bits, such that $X(s_1 s_2 \ldots) > n$ if and only if $s_1 = \cdots s_n =$ "tails". This leads to measure-theoretic technicalities that are beyond the scope of this book, but which are well-understood in probability theory.

[4] Technically, a random variable is defined on a probability space, not an arbitrary set, and is a measurable function; we refer to Woodroofe [569] for the details.

Example A.14.2 Suppose one has a set S of N items and one chooses elements of S (with replacement) uniformly and independently at random. Let X be a random variable taking values in $\mathbb{N}$ such that $\Pr(X = n)$ is the probability that, after sampling n elements from S, the first $n - 1$ elements are distinct and the nth element is equal to one of the previously sampled elements. In other words, X is the number of samples from S until some element is sampled twice. A version of the **birthday paradox** states that the expected value of X is approximately $\sqrt{\pi N/2}$. We discuss this in detail in Section 14.1.

Example A.14.3 A version of the **coupon collector** problem is the following: suppose S is a set of N items and one chooses elements of S (with replacement) uniformly at random.

Let X be a random variable taking values in $\mathbb{N}$ such that $\Pr(X \geq n)$ is the probability that after choosing $n - 1$ elements (sampled uniformly and independently at random from S) one has not yet chosen some element of S. In other words, X is the number of "coupons" to be collected until one has a full set of all N types. The expected value of X is $N(1 + 1/2 + \cdots + 1/(N-1) + 1/N) \approx N\log(N)$ (see Example 7.2j of Ross [451]).

The **statistical distance** (also called the **total variation**) of two distributions $\Pr_1$ and $\Pr_2$ on a finite or countable set S is $\Delta(\Pr_1, \Pr_2) = \frac{1}{2}\sum_{x\in S} |\Pr_1(s) - \Pr_2(s)|$. It is easy to see that $\Delta(\Pr_1, \Pr_1) = 0$ and $0 \leq \Delta(\Pr_1, \Pr_2) \leq 1$ (see Theorem 6.14 of Shoup [497]). Two distributions are **statistically close** if their statistical distance is "negligible" in some appropriate sense (typically, in cryptographic applications, this will mean "negligible in terms of the security parameter").

We end with a result that is often used in the analysis of algorithms.

Theorem A.14.4 *The probability that two uniformly and independently chosen integers* $1 \leq n_1, n_2 < N$ *satisfy* $\gcd(n_1, n_2) = 1$ *tends to* $1/\zeta(2) = 6/\pi^2 \approx 0.608$ *as* N *tends to infinity.*

Proof See Theorem 332 of [250]. □

References

[1] M. Abdalla, M. Bellare and P. Rogaway, *DHIES: An Encryption Scheme based on the Diffie–Hellman Problem*, Preprint, 2001.

[2] L. M. Adleman and J. DeMarrais, A subexponential algorithm for discrete logarithms over all finite fields, *Math. Comp.* **61**(203) (1993), 1–15.

[3] L. M. Adleman, K. L. Manders and G. L. Miller, On taking roots in finite fields, *Foundations of Computer Science (FOCS), IEEE*, 1977, 175–178.

[4] L.M. Adleman, J. De Marrais and M.-D. Huang, A subexponential algorithm for discrete logarithms over the rational subgroup of the Jacobians of large genus hyperelliptic curves over finite fields. In *ANTS I* (L. M. Adleman and M.-D. Huang, eds.), *LNCS*, vol. 877, Springer, 1994, pp. 28–40.

[5] G. B. Agnew, R. C. Mullin, I. M. Onyszchuk and S. A. Vanstone, An implementation for a fast public-key cryptosystem, *J. Crypt.* **3**(2) (1991), 63–79.

[6] M. Agrawal, N. Kayal and N. Saxena, PRIMES is in P, *Ann. of Math.* **160**(2) (2004), 781–793.

[7] E. Agrell, T. Eriksson, A. Vardy and K. Zeger, Closest point search in lattices, *IEEE Trans. Inf. Theory* **48**(8) (2002), 2201–2214.

[8] A. Akavia, Solving hidden number problem with one bit oracle and advice. In *CRYPTO 2009* (S. Halevi, ed.), *LNCS*, vol. 5677, Springer, 2009, pp. 337–354.

[9] W. Alexi, B. Chor, O. Goldreich and C.-P. Schnorr, RSA and Rabin functions: certain parts are as hard as the whole, *SIAM J. Comput.* **17**(2) (1988), 194–209.

[10] W. R. Alford, A. Granville and C. Pomerance, There are infinitely many Carmichael numbers, *Ann. of Math.* **139**(3) (1994), 703–722.

[11] A. Antipa, D. R. L. Brown, R. P. Gallant, R. J. Lambert, R. Struik and S. A. Vanstone, Accelerated verification of ECDSA signatures. In *SAC 2005* (B. Preneel and S. E. Tavares, eds.), *LNCS*, vol. 3897, Springer, 2006, pp. 307–318.

[12] C. Arène, T. Lange, M. Naehrig and C. Ritzenthaler, Faster computation of the Tate pairing, *J. Number Theory* **131**(5) (2011), 842–857.

[13] J. Arney and E. D. Bender, Random mappings with constraints on coalescence and number of origins, *Pacific J. Math.* **103** (1982), 269–294.

[14] E. Artin, *Galois Theory*, 2nd edn, Notre Dame, 1959.

[15] M. F. Atiyah and I. G. Macdonald, *Introduction to Commutative Algebra*, Addison-Wesley, 1969.

[16] R. Avanzi, H. Cohen, C. Doche, G. Frey, T. Lange, K. Nguyen and F. Vercauteren, *Handbook of Elliptic and Hyperelliptic Cryptography*, Chapman & Hall/CRC, 2006.

[17] L. Babai, On Lovász lattice reduction and the nearest lattice point problem, *Combinatorica* **6**(1) (1986), 1–13.

[18] L. Babai and E. Szemerédi, On the complexity of matrix group problems *I*, *Foundations of Computer Science (FOCS)* (1996), 229–240.

[19] E. Bach, Bounds for primality testing and related problems, *Math. Comp.* **55**(191) (1990), 355–380.

[20] ——— Toward a theory of Pollard's rho method, *Inf. Comput.* **90**(2) (1991), 139–155.

[21] E. Bach and J. Shallit, *Algorithmic Number Theory*, MIT Press, 1996.

[22] E. Bach and J. Sorenson, Sieve algorithms for perfect power testing, *Algorithmica* **9** (1993), 313–328.

[23] S. Bai and R. P. Brent, On the efficiency of Pollard's rho method for discrete logarithms, *CATS 2008* (J. Harland and P. Manyem, eds.), Australian Computer Society, 2008, pp. 125–131.

[24] D. V. Bailey, L. Batina, D. J. Bernstein, P. Birkner, J. W. Bos, H.-C. Chen, C.-M. Cheng, G. van Damme, G. de Meulenaer, L. Julian Dominguez Perez, J. Fan, T. Güneysu, F. Gurkaynak, T. Kleinjung, T. Lange, N. Mentens, R. Niederhagen, C. Paar, F. Regazzoni, P. Schwabe, L. Uhsadel, A. Van Herrewege and B.-Y. Yang, *Breaking ECC2K-130*, Cryptology ePrint Archive, Report 2009/541, 2009.

[25] R. Balasubramanian and N. Koblitz, The improbability that an elliptic curve has subexponential discrete log problem under the Menezes–Okamoto–Vanstone algorithm, *J. Crypt.* **11**(2) (1998), 141–145.

[26] P. S. L. M. Barreto, S. D. Galbraith, C. Ó hÉigeartaigh and M. Scott, Efficient pairing computation on supersingular abelian varieties, *Des. Codes Crypt.* **42**(3) (2007), 239–271.

[27] P. S. L. M. Barreto, H. Y. Kim, B. Lynn and M. Scott, Efficient algorithms for pairing-based cryptosystems, *CRYPTO 2002* (M. Yung, ed.), *LNCS*, vol. 2442, Springer, 2002, pp. 354–369.

[28] P. S. L. M. Barreto and M. Naehrig, Pairing-friendly elliptic curves of prime order, *SAC 2005* (B. Preneel and S. E. Tavares, eds.), *LNCS*, vol. 3897, Springer, 2006, pp. 319–331.

[29] A. Bauer, Vers une généralisation rigoureuse des méthodes de Coppersmith pour la recherche de petites racines de polynmes, Ph.D. thesis, Université de Versailles Saint-Quentin-en-Yvelines, 2008.

[30] M. Bellare, R. Canetti and H. Krawczyk, A modular approach to the design and analysis of authentication and key exchange protocols, *Symposium on the Theory of Computing (STOC)*, ACM, 1998, pp. 419–428.

[31] M. Bellare, J. A. Garay and T. Rabin, Fast batch verification for modular exponentiation and digital signatures, *EUROCRYPT 1998* (K. Nyberg, ed.), *LNCS*, vol. 1403, Springer, 1998, pp. 236–250.

[32] M. Bellare, S. Goldwasser and D. Micciancio, "Pseudo-Random" number generation within cryptographic algorithms: the DSS case, *CRYPTO 1997* (B. S. Kaliski Jr., ed.), *LNCS*, vol. 1294, Springer, 1997, pp. 277–291.

[33] M. Bellare and G. Neven, Multi-signatures in the plain public-key model and a general forking lemma, *CCS 2006* (A. Juels, R. N. Wright and S. De Capitani di Vimercati, eds.), ACM, 2006, pp. 390–399.

[34] M. Bellare, D. Pointcheval and P. Rogaway, Authenticated key exchange secure against dictionary attacks, *EUROCRYPT* 2000 (B. Preneel, ed.), *LNCS*, vol. 1807, Springer, 2000, pp. 139–155.

[35] M. Bellare and P. Rogaway, Random oracles are practical: a paradigm for designing efficient protocols, *CCS 1993*, ACM, 1993, pp. 62–73.

[36] ——— Entity authentication and key distribution, *CRYPTO 1993* (D. R. Stinson, ed.), *LNCS*, vol. 773, Springer, 1994, pp. 232–249.

[37] ——— Optimal asymmetric encryption – how to encrypt with RSA, *EUROCRYPT 1994* (A. De Santis, ed.), *LNCS*, vol. 950, Springer, 1995, pp. 92–111.

[38] ——— The exact security of digital signatures – how to sign with RSA and Rabin. In *EUROCRYPT 1996* (U. M. Maurer, ed.), *LNCS*, vol. 1070, Springer, 1996, pp. 399–416.

[39] K. Bentahar, The equivalence between the DHP and DLP for elliptic curves used in practical applications, revisited. In *IMA Cryptography and Coding* (N. P. Smart, ed.), *LNCS*, vol. 3796, Springer, 2005, pp. 376–391.

[40] ——— Theoretical and practical efficiency aspects in cryptography, Ph.D. thesis, University of Bristol, 2008.

[41] D. J. Bernstein, *Pippenger's exponentiation algorithm*, Preprint, 2002.

[42] D. J. Bernstein, T. Lange and P. Schwabe, On the correct use of the negation map in the Pollard rho method. In *PKC 2011* (D. Catalano, N. Fazio, R. Gennaro, and A. Nicolosi, eds.), *LNCS*, vol. 6571, Springer, 2011, pp. 128–146.

[43] ——— Curve 25519: New Diffie–Hellman speed records. In *PKC 2006* (M. Yung, Y. Dodis, A. Kiayias and T. Malkin, eds.), *LNCS*, vol. 3958, Springer, 2006, pp. 207–228.

[44] ——— Proving tight security for Rabin–Williams signatures. In *EUROCRYPT 2008* (N. P. Smart, ed.), *LNCS*, vol. 4965, Springer, 2008, pp. 70–87.

[45] ——— Faster rho for elliptic curves, Rump session talk, *ANTS IX*, 2010.

[46] D. J. Bernstein, P. Birkner, M. Joye, T. Lange and C. Peters, Twisted Edwards curves. In *Africacrypt 2008* (S. Vaudenay, ed.), *LNCS*, vol. 5023, Springer, 2008, pp. 389–405.

[47] D. J. Bernstein, P. Birkner, T. Lange and C. Peters, *ECM using Edwards Curves*, Cryptology ePrint Archive, Report 2008/016, 2008.

[48] D. J. Bernstein, J. Buchmann and E. Dahmen, *Post Quantum Cryptography*, Springer, 2008.

[49] D. J. Bernstein and T. Lange, *Explicit Formulas Database*, 2007 (www.hyperelliptic.org/EFD).

[50] ——— Faster addition and doubling on elliptic curves. In *ASIACRYPT 2007* (K. Kurosawa, ed.), *LNCS*, vol. 4833, Springer, 2007, pp. 29–50.

[51] ——— Analysis and optimization of elliptic-curve single-scalar multiplication, *Contemporary Mathematics* **461** (2008), 1–19.

[52] ——— Type-II optimal polynomial bases. In *WAIFI 2010* (M. A. Hasan and T. Helleseth, eds.), *LNCS*, vol. 6087, Springer, 2010, pp. 41–61.

[53] D. J. Bernstein, T. Lange and R. R. Farashahi, Binary Edwards curves. In *CHES 2008*, (E. Oswald and P. Rohatgi, eds.), *LNCS*, vol. 5154, Springer, 2008, pp. 244–265.

[54] G. Bisson and A. V. Sutherland, Computing the endomorphism ring of an ordinary elliptic curve over a finite field, *J. Number Theory* **131**(5) (2011), 815–831.

[55] S. R. Blackburn and S. Murphy, The number of partitions in Pollard rho, unpublished manuscript, 1998.

[56] S. R. Blackburn and E. Teske, Baby-step–giant-step algorithms for non-uniform distributions. In *ANTS IV* (W. Bosma, ed.), *LNCS*, vol. 1838, Springer, 2000, pp. 153–168.

[57] I. F. Blake, R. Fuji-Hara, R. C. Mullin and S. A. Vanstone, Computing logarithms in finite fields of characteristic two, *SIAM J. Algebraic and Discrete Methods* **5**(2) (1984), 272–285.

[58] I. F. Blake and T. Garefalakis, On the complexity of the discrete logarithm and Diffie–Hellman problems, *J. Complexity* **20**(2–3) (2004), 148–170.

[59] I. F. Blake, T. Garefalakis and I. E. Shparlinski, On the bit security of the Diffie–Hellman key, *Appl. Algebra Eng. Commun. Comput.* **16**(6) (2006), 397–404.

[60] I. F. Blake, G. Seroussi and N. P. Smart, *Elliptic Curves in Cryptography*, Cambridge University Press, 1999.

[61] ——— *Advances in Elliptic Curve Cryptography*, Cambridge University Press, 2005.

[62] D. Bleichenbacher, Generating ElGamal signatures without knowing the secret key, *EUROCRYPT 1996* (U. M. Maurer, ed.), *LNCS*, vol. 1070, Springer, 1996, pp. 10–18.

[63] D. Bleichenbaucher, Compressing Rabin signatures. In *CT-RSA 2004* (T. Okamoto, ed.), *LNCS*, vol. 2964, Springer, 2004, pp. 126–128.

[64] D. Bleichenbacher and A. May, New attacks on RSA with small secret CRT-exponents. In *PKC 2006* (M. Yung, Y. Dodis, A. Kiayias and T. Malkin, eds.), *LNCS*, vol. 3958, Springer, 2006, pp. 1–13.

[65] D. Bleichenbacher and P. Q. Nguyen, Noisy polynomial interpolation and noisy Chinese remaindering. In *EUROCRYPT 2000* (B. Preneel, ed.), *LNCS*, vol. 1807, Springer, 2000, pp. 53–69.

[66] J. Blömer and A. May, Low secret exponent RSA revisited. In *Cryptography and Lattices (CaLC)* (J. H. Silverman, ed.), *LNCS*, vol. 2146, Springer, 2001, pp. 4–19.

[67] J. Blömer and A. May, A tool kit for finding small roots of bivariate polynomials over the integers. In *EUROCRYPT 2005* (R. Cramer, ed.), *LNCS*, vol. 3494, Springer, 2005, pp. 251–267.

[68] M. Blum and S. Micali, How to generate cryptographically strong sequences of pseudo-random bits, *SIAM J. Comput.* **13**(4) (1984), 850–864.

[69] D. Boneh, Simplified OAEP for the RSA and Rabin functions. In *CRYPTO 2001* (J. Kilian, ed.), *LNCS*, vol. 2139, Springer, 2001, pp. 275–291.

[70] ——— Finding smooth integers in short intervals using CRT decoding, *J. Comput. Syst. Sci.* **64**(4) (2002), 768–784.

[71] D. Boneh and X. Boyen, Short signatures without random oracles. In *EUROCRYPT 2004* (C. Cachin and J. Camenisch, eds.), *LNCS*, vol. 3027, Springer, 2004, pp. 56–73.

[72] ——— Short signatures without random oracles and the SDH assumption in bilinear groups, *J. Crypt.* **21**(2) (2008), 149–177.

[73] D. Boneh and G. Durfee, Cryptanalysis of RSA with private key d less than $N^{0.292}$, *IEEE Trans. Inf. Theory* **46**(4) (2000), 1339–1349.

[74] D. Boneh, G. Durfee and N. Howgrave-Graham, Factoring $N = p^r q$ for large r. In *CRYPTO 1999* (M. J. Wiener, ed.), *LNCS*, vol. 1666, Springer, 1999, pp. 326–337.

[75] D. Boneh and M. K. Franklin, Identity based encryption from the Weil pairing. In *CRYPTO 2001* (J. Kilian, ed.), *LNCS*, vol. 2139, Springer, 2001, pp. 213–229.

[76] ——— Identity based encryption from the Weil pairing, *SIAM J. Comput.* **32**(3) (2003), 586–615.

[77] D. Boneh, A. Joux and P. Nguyen, Why textbook ElGamal and RSA encryption are insecure. In *ASIACRYPT 2000* (T. Okamoto, ed.), *LNCS*, vol. 1976, Springer, 2000, pp. 30–43.

[78] D. Boneh and R. J. Lipton, Algorithms for black-box fields and their application to cryptography. In *CRYPTO 1996* (N. Koblitz, ed.), *LNCS*, vol. 1109, Springer, 1996, pp. 283–297.

[79] D. Boneh and I. E. Shparlinski, On the unpredictability of bits of the elliptic curve Diffie–Hellman scheme. In *CRYPTO 2001* (J. Kilian, ed.), *LNCS*, vol. 2139, Springer, 2001, pp. 201–212.

[80] D. Boneh and R. Venkatesan, Hardness of computing the most significant bits of secret keys in Diffie–Hellman and related schemes. In *CRYPTO 1996* (N. Koblitz, ed.), *LNCS*, vol. 1109, Springer, 1996, pp. 129–142.

[81] ——— Rounding in lattices and its cryptographic applications. In *Symposium on Discrete Algorithms (SODA)*, ACM/SIAM, 1997, pp. 675–681.

[82] ——— Breaking RSA may not be equivalent to factoring. In *EUROCRYPT 1998* (K. Nyberg, ed.), *LNCS*, vol. 1403, Springer, 1998, pp. 59–71.

[83] A. Borodin and I. Munro, *The Computational Complexity of Algebraic and Numeric Problems*, Elsevier, 1975.

[84] J. W. Bos, M. E. Kaihara and T. Kleinjung, *Pollard Rho on Elliptic Curves*, Preprint, 2009.

[85] J. W. Bos, M. E. Kaihara and P. L. Montgomery, Pollard rho on the Playstation 3, Handouts of SHARCS 2009, 2009, pp. 35–50.

[86] J. W. Bos, T. Kleinjung and A. K. Lenstra, On the use of the negation map in the Pollard rho method. In *ANTS IX* (G. Hanrot, F. Morain and E. Thomé, eds.), *LNCS*, vol. 6197, Springer, 2010, pp. 66–82.

[87] W. Bosma and H. W. Lenstra Jr., Complete systems of two addition laws for elliptic curves, *J. Number Theory* **53** (1995), 229–240.

[88] A. Bostan, F. Morain, B. Salvy and E. Schost, Fast algorithms for computing isogenies between elliptic curves, *Math. Comp.* **77**(263) (2008), 1755–1778.

[89] C. Boyd and A. Mathuria, Protocols for authentication and key establishment, *Information Security and Cryptography*, Springer, 2003.

[90] X. Boyen, The uber-assumption family. In *Pairing 2008* (S. D. Galbraith and K. G. Paterson, eds.), *LNCS*, vol. 5209, Springer, 2008, pp. 39–56.

[91] V. Boyko, M. Peinado and R. Venkatesan, Speeding up discrete log and factoring based schemes via precomputations. In *EUROCRYPT 1998* (K. Nyberg, ed.), *LNCS*, vol. 1403, Springer, 1998, pp. 221–235.

[92] S. Brands, An efficient off-line electronic cash system based on the representation problem, Tech. report, CWI Amsterdam, 1993, CS-R9323.

[93] R. P. Brent, An improved Monte Carlo factorization algorithm, *BIT* (1980), 176–184.

[94] R. P. Brent and J. M. Pollard, Factorization of the eighth Fermat number, *Math. Comp.* **36**(154) (1981), 627–630.

[95] R. P. Brent and P. Zimmermann, *Modern Computer Arithmetic*, Cambridge University Press, 2010.

[96] ——— An $O(M(n) log n)$ algorithm for the Jacobi symbol. In *ANTS IX* (G. Hanrot, F. Morain and E. Thomé, eds.), *LNCS*, vol. 6197, Springer, 2010, pp. 83–95.

[97] E. Bresson, Y. Lakhnech, L. Mazaré and B. Warinschi, A generalization of DDH with applications to protocol analysis and computational soundness. In *CRYPTO 2007* (A. J. Menezes, ed.), *LNCS*, vol. 4622, Springer, 2007, pp. 482–499.

[98] E. F. Brickell, D. Pointcheval, S. Vaudenay and M. Yung, Design validations for discrete logarithm based signature schemes. In *PKC 2000* (H. Imai and Y. Zheng, eds.), *LNCS*, vol. 1751, Springer, 2000, pp. 276–292.

[99] E. Brier, C. Clavier and D. Naccache, Cryptanalysis of RSA signatures with fixed-pattern padding. In *CRYPTO 2001* (J. Kilian, ed.), *LNCS*, vol. 2139, Springer, 2001, pp. 433– 439.

[100] R. Bröker, Constructing supersingular elliptic curves, *J. Comb. Number Theory* **1**(3) (2009), 269–273.

[101] R. Bröker, D. X. Charles and K. Lauter, Evaluating large degree isogenies and applications to pairing based cryptography. In *Pairing 2008* (S. D. Galbraith and K. G. Paterson, eds.), *LNCS*, vol. 5209, Springer, 2008, pp. 100–112.

[102] R. Bröker, K. Lauter and A. V. Sutherland, Modular polynomials via isogeny volcanoes, http://arxiv.org/abs/1001.0402, 2010, to appear in *Math. Comp.*.

[103] R. Bröker and A. V. Sutherland, An explicit height bound for the classical modular polynomial, *The Ramanujan Journal* **22**(3) (2010), 293–313.

[104] D. R. L. Brown and R. P. Gallant, The static Diffie–Hellman problem, Cryptology ePrint Archive, Report 2004/306, 2004.

[105] B. B. Brumley and K. U. Järvinen, Koblitz curves and integer equivalents of Frobenius expansions. In *SAC 2007* (C. M. Adams, A. Miri and M. J. Wiener, eds.), *LNCS*, vol. 4876, Springer, 2007, pp. 126–137.

[106] J. P. Buhler and P. Stevenhagen, *Algorithmic Number Theory*, MSRI Publications, Cambridge University Press, 2008.

[107] M. Burmester and Y. Desmedt, A secure and efficient conference key distribution system. In *EUROCRYPT 1994* (A. De Santis, ed.), *LNCS*, vol. 950, Springer, 1995, pp. 267–275.

[108] R. Canetti, O. Goldreich and S. Halevi, The random oracle model, revisited. In *Symposium on the Theory of Computing (STOC)*, ACM, 1998, pp. 209–218.

[109] R. Canetti and H. Krawczyk, Analysis of key-exchange protocols and their use for building secure channels. In *EUROCRYPT 2001* (B. Pfitzmann, ed.), *LNCS*, vol. 2045, Springer, 2001, pp. 453–474.

[110] E. R. Canfield, P. Erdös and C. Pomerance, On a problem of Oppenheim concerning "factorisatio numerorum", *J. Number Theory* **17**(1) (1983), 1–28.

[111] D. G. Cantor, Computing in the Jacobian of an hyperelliptic curve, *Math. Comp.* **48**(177) (1987), 95–101.

[112] D. Cash, E. Kiltz and V. Shoup, The twin Diffie–Hellman problem and applications. In *EUROCRYPT 2008* (N. P. Smart, ed.), *LNCS*, vol. 4965, Springer, 2008, pp. 127–145.

[113] J. W. S. Cassels, *An Introduction to the Geometry of Numbers*, Springer, 1959.

[114] ——— *Lectures on Elliptic Curves*, Cambridge University Press, 1991.

[115] J. W. S. Cassels and E. V. Flynn, *Prolegomena to a Middlebrow Arithmetic of Curves of Genus* 2, Cambridge University Press, 1996.

[116] J. W. S. Cassels and A. Frölich, *Algebraic Number Theory*, Academic Press, 1967.

[117] D. Catalano, R. Gennaro, N. Howgrave-Graham and P. Q. Nguyen, Paillier's cryptosystem revisited. In *CCS 2001*, ACM, 2001, pp. 206–214.

[118] L. S. Charlap and R. Coley, An elementary introduction to elliptic curves II, CCR Expository Report 34, Institute for Defense Analysis, 1990.

[119] L. S. Charlap and D. P. Robbins, An elementary introduction to elliptic curves, CRD Expository Report 31, 1988.

[120] D. X. Charles, K. E. Lauter and E. Z. Goren, Cryptographic hash functions from expander graphs, *J. Crypt.* **22**(1) (2009), 93–113.

[121] D. Chaum, E. van Heijst and B. Pfitzmann, Cryptographically strong undeniable signatures, unconditionally secure for the signer. In *CRYPTO 1991* (J. Feigenbaum, ed.), *LNCS*, vol. 576, Springer, 1992, pp. 470–484.

[122] J.-H. Cheon, Security analysis of the strong Diffie–Hellman problem. In *EUROCRYPT 2006* (S. Vaudenay, ed.), *LNCS*, vol. 4004, Springer, 2006, pp. 1–11.

[123] ——— Discrete logarithm problem with auxiliary inputs, *J. Crypt.* **23**(3) (2010), 457–476.

[124] J. H. Cheon, J. Hong and M. Kim, Speeding up the Pollard rho method on prime fields. In *ASIACRYPT 2008* (J. Pieprzyk, ed.), *LNCS*, vol. 5350, Springer, 2008, pp. 471–488.

[125] J. H. Cheon and H.-T. Kim, Analysis of low Hamming weight products, *Discrete Appl. Math.* **156**(12) (2008), 2264–2269.

[126] M. A. Cherepnev, On the connection between the discrete logarithms and the Diffie–Hellman problem, *Discr. Math. Appl.* **6**(4) (1996), 341–349.

[127] H. Cohen, *A Course in Computational Algebraic Number Theory*, *GTM* 138, Springer, 1993.

[128] P. Cohen, On the coefficients of the transformation polynomials for the elliptic modular function, *Math. Proc. Cambridge Philos. Soc.* **95**(3) (1984), 389–402.

[129] S. A. Cook, An overview of computational complexity, *Commun. ACM* **26**(6) (1983), 400–408.

[130] D. Coppersmith, Fast evaluation of logarithms in fields of characteristic 2, *IEEE Trans. Inf. Theory* **30**(4) (1984), 587–594.

[131] ——— Small solutions to polynomial equations, and low exponent RSA vulnerabilities, *J. Crypt.* **10**(4) (1997), 233–260.

[132] ——— Finding small solutions to small degree polynomials. In *Cryptography and Lattices (CaLC)* (J. H. Silverman, ed.) *LNCS*, vol. 2146, Springer, 2001, pp. 20–31.

[133] D. Coppersmith, J.-S. Coron, F. Grieu, S. Halevi, C. Jutla, D. Naccache and J. P. Stern, Cryptanalysis of ISO/IEC 9796-1, *J. Crypt.* **21**(1) (2008), 27–51.

[134] D. Coppersmith, M. K. Franklin, J. Patarin and M. K. Reiter, Low-exponent RSA with related messages. In *EUROCRYPT 1996* (U. M. Maurer, ed.), *LNCS*, vol. 1070, Springer, 1996, pp. 1–9.

[135] D. Coppersmith, A. M. Odlzyko and R. Schroeppel, Discrete logarithms in *GF*(p), *Algorithmica* **1**(1–4) (1986), 1–15.

[136] T. H. Cormen, C. E. Leiserson, R. L. Rivest and C. Stein, *Introduction to Algorithms*, 2nd edn, MIT press, 2001.

[137] J.-S. Coron, On the exact security of full domain hash. In *CRYPTO 2000* (M. Bellare, ed.), *LNCS*, vol. 1880, Springer, 2000, pp. 229–235.

[138] ——— Optimal security proofs for PSS and other signature schemes. In *EUROCRYPT 2002* (L. R. Knudsen, ed.), *LNCS*, vol. 2332, Springer, 2002, pp. 272–287.

[139] ——— Finding small roots of bivariate integer polynomial equations: a direct approach. In *CRYPTO 2007* (A. Menezes, ed.), *LNCS*, vol. 4622, Springer, 2007, pp. 379–394.

[140] J.-S. Coron and A. May, Deterministic polynomial-time equivalence of computing the RSA secret key and factoring, *J. Crypt.* **20**(1) (2007), 39–50.

[141] J.-S. Coron, D. M'Raïhi and C. Tymen, Fast generation of pairs $(k, [k]P)$ for Koblitz elliptic curves. In *SAC 2001* (S. Vaudenay and A. M. Youssef, eds.), vol. 2259, Springer, 2001, pp. 151–164.

[142] J.-S. Coron, D. Naccache, M. Tibouchi and R.-P. Weinmann, Practical cryptanalysis of ISO/IEC 9796-2 and EMV signatures. In *CRYPTO 2009* (S. Halevi, ed.), *LNCS*, vol. 5677, Springer, 2009, pp. 428–444.

[143] J.-M. Couveignes, Computing l-isogenies with the p-torsion. In *ANTS II* (H. Cohen, ed.), *LNCS*, vol. 1122, Springer, 1996, pp. 59–65.

[144] J.-M. Couveignes, L. Dewaghe and F. Morain, Isogeny cycles and the Schoof–Elkies–Atkin algorithm, Research Report LIX/RR/96/03, 1996.

[145] D. A. Cox, *Primes of the Form* $x^2 + ny^2$, Wiley, 1989.

[146] D. A. Cox, J. Little and D. O'Shea, *Ideals, Varieties and Algorithms: an Introduction to Computational Algebraic Geometry and Commutative Algebra*, 2nd edn, Springer, 1997.

[147] R. Cramer and V. Shoup, A practical public key cryptosystem provably secure against adaptive chosen ciphertext attack. In *CRYPTO 1998* (H. Krawczyk, ed.), *LNCS*, vol. 1462, Springer, 1998, pp. 13–25.

[148] ——— Universal hash proofs and a paradigm for adaptive chosen ciphertext secure public-key encryption. In *EUROCRYPT 2002* (L. R. Knudsen, ed.), *LNCS*, vol. 2332, Springer, 2002, pp. 45–64.

[149] ——— Design and analysis of practical public-key encryption schemes secure against adaptive chosen ciphertext attack, *SIAM J. Comput.* **33**(1) (2003), 167–226.

[150] R. Crandall and C. Pomerance, *Prime Numbers: A Computational Perspective*, 2nd edn, Springer, 2005.

[151] C. W. Curtis, *Linear Algebra: An Introductory Approach, Undergraduate Texts in Mathematics*, Springer, 1984.

[152] I. Damgård, On the randomness of Legendre and Jacobi sequences. In *CRYPTO 1988* (S. Goldwasser, ed.), *LNCS*, vol. 403, Springer, 1990, pp. 163–172.

[153] G. Davidoff, P. Sarnak and A. Valette, *Elementary Number Theory, Group Theory, and Ramanujan Graphs*, Cambridge University Press, 2003.

[154] M. Davis and E. J. Weyuker, *Computability, Complexity and Languages*, Academic Press, 1983.

[155] P. de Rooij, On Schnorr's preprocessing for digital signature schemes, *J. Crypt.* **10**(1) (1997), 1–16.

[156] B. den Boer, Diffie–Hellman is as strong as discrete log for certain primes. In *CRYPTO 1988* (S. Goldwasser, ed.), *LNCS*, vol. 403, Springer, 1990, pp. 530–539.
[157] Y. Desmedt and A. M. Odlyzko, A chosen text attack on the RSA cryptosystem and some discrete logarithm schemes. In *CRYPTO 1985* (H. C. Williams, ed.), *LNCS*, vol. 218, Springer, 1986, pp. 516–522.
[158] L. Dewaghe, *Un corollaire aux formules de Vélu*, Preprint, 1995.
[159] C. Diem, The GHS-attack in odd characteristic, *J. Ramanujan Math. Soc.* **18**(1) (2003), 1–32.
[160] ______ On the discrete logarithm problem in elliptic curves over non-prime finite fields, Lecture at ECC 2004, 2004.
[161] ______ An index calculus algorithm for plane curves of small degree. In *ANTS VII* (F. Hess, S. Pauli and M. E. Pohst, eds.), *LNCS*, vol. 4076, Springer, 2006, pp. 543–557.
[162] C. Diem, An index calculus algorithm for non-singular plane curves of high genus, talk at ECC 2006.
[163] ______ On the Discrete Logarithm Problem in Elliptic Curves, *Compositio Math.* **147** (2011), 75–104.
[164] ______ On the discrete logarithm problem in class groups of curves, *Math. Comp.* **80**(273) (2011), 443–475.
[165] C. Diem and E. Thomé, Index calculus in class groups of non-hyperelliptic curves of genus three, *J. Crypt.* **21**(4) (2008), 593–611.
[166] W. Diffie and M. E. Hellman, New directions in cryptography, *IEEE Trans. Inf. Theory* **22** (1976), 644–654.
[167] V. S. Dimitrov, G. A. Jullien and W. C. Miller, Theory and applications of the double-base number system, *IEEE Trans. Computers* **48**(10) (1999), 1098–1106.
[168] V. S. Dimitrov, K. U. Järvinen, M. J. Jacobson, W. F. Chan and Z. Huang, Provably sublinear point multiplication on Koblitz curves and its hardware implementation, *IEEE Trans. Computers* **57**(11) (2008), 1469–1481.
[169] C. Doche, T. Icart and D. R. Kohel, Efficient scalar multiplication by isogeny decompositions. In *PKC 2006* (M. Yung, Y. Dodis, A. Kiayias and T. Malkin, eds.), LNCS, vol. 3958, Springer, 2006, pp. 191–206.
[170] A. Dujella, A variant of Wiener's attack on RSA, *Computing* **85**(1–2) (2009), 77–83.
[171] I. M. Duursma, Class numbers for some hyperelliptic curves. In *Arithmetic, Geometry and Coding Theory* (R. Pellikaan, M. Perret and S. G. Vladut, eds.), Walter de Gruyter, 1996, pp. 45–52.
[172] I. M. Duursma, P. Gaudry and F. Morain, Speeding up the discrete log computation on curves with automorphisms. In *ASIACRYPT 1999* (K. Y. Lam, E. Okamoto and C. Xing, eds.), *LNCS*, vol. 1716, Springer, 1999, pp. 103–121.
[173] I. M. Duursma and H.-S. Lee, Tate pairing implementation for hyperelliptic curves $y^2 = x^p - x + d$. In *ASIACRYPT 2003* (C.-S. Laih, ed.), *LNCS*, vol. 2894, Springer, 2003, pp. 111–123.
[174] S. Edixhoven, Le couplage Weil: de la géométrie à l'arithmétique, Notes from a seminar in Rennes, 2002.
[175] H. M. Edwards, A normal form for elliptic curves, *Bulletin of the AMS* **44** (2007), 393–422.
[176] D. Eisenbud, *Commutative Algebra with a View Toward Algebraic Geometry*, *GTM*, vol. 150, Springer, 1999.
[177] T. ElGamal, A public key cryptosystem and a signature scheme based on discrete logarithms. In *CRYPTO 1984* (G. R. Blakley and D. Chaum, eds.), *LNCS*, vol. 196, Springer, 1985, pp. 10–18.
[178] N. D. Elkies, Elliptic and modular curves over finite fields and related computational issues. In *Computational Perspectives on Number Theory* (D. A. Buell and J. T. Teitelbaum, eds.), Studies in Advanced Mathematics, AMS, 1998, pp. 21–76.

[179] A. Enge, Computing modular polynomials in quasi-linear time, *Math. Comp.* **78**(267) (2009), 1809–1824.

[180] A. Enge and P. Gaudry, A general framework for subexponential discrete logarithm algorithms, *Acta Arith.* **102** (2002), 83–103.

[181] ______ An $L(1/3+\epsilon)$ algorithm for the discrete logarithm problem for low degree curves. In *EUROCRYPT 2007* (M. Naor, ed.), *LNCS*, vol. 4515, Springer, 2007, pp. 379–393.

[182] A. Enge, P. Gaudry and E. Thomé, An $L(1/3)$ discrete logarithm algorithm for low degree curves, *J. Crypt.* **24**(1) (2011), 24–41.

[183] A. Enge and A. Stein, Smooth ideals in hyperelliptic function fields, *Math. Comp.* **71**(239) (2002), 1219–1230.

[184] S. Erickson, M. J. Jacobson Jr., N. Shang, S. Shen and A. Stein, Explicit formulas for real hyperelliptic curves of genus 2 in affine representation, *WAIFI 2007* (C. Carlet and B. Sunar, eds.), *LNCS*, vol. 4547, Springer, 2007, pp. 202–218.

[185] L. De Feo, Fast algorithms for towers of finite fields and isogenies, Ph.D. thesis, L'École Polytechnique, 2010.

[186] R. Fischlin and C.-P. Schnorr, Stronger security proofs for RSA and Rabin bits, *J. Crypt.* **13**(2) (2000), 221–244.

[187] P. Flajolet and A. M. Odlyzko, Random mapping statistics. In *EUROCRYPT 1989* (J.-J. Quisquater and J. Vandewalle, eds.), *LNCS*, vol. 434, Springer, 1990, pp. 329–354.

[188] P. Flajolet and R. Sedgewick, *Analytic Combinatorics*, Cambridge University Press, 2009.

[189] R. Flassenberg and S. Paulus, Sieving in function fields, *Experiment. Math.* **8**(4) (1999), 339–349.

[190] K. Fong, D. Hankerson, J. López and A. J. Menezes, Field inversion and point halving revisited, *IEEE Trans. Computers* **53**(8) (2004), 1047–1059.

[191] C. Fontaine and F. Galand, A survey of homomorphic encryption for nonspecialists, *EURASIP Journal on Information Security* **2007**(15) (2007), 1–10.

[192] M. Fouquet and F. Morain, Isogeny volcanoes and the SEA algorithm. In *ANTS V* (C. Fieker and D. R. Kohel, eds.), *LNCS*, vol. 2369, Springer, 2002, pp. 276–291.

[193] D. M. Freeman, O. Goldreich, E. Kiltz, A. Rosen and G. Segev, More constructions of lossy and correlation-secure trapdoor functions. In *PKC 2010* (P. Q. Nguyen and D. Pointcheval, eds.), *LNCS*, vol. 6065, Springer, 2010, pp. 279–295.

[194] D. Freeman, M. Scott and E. Teske, A taxonomy of pairing-friendly elliptic curves, *J. Crypt.* **23**(2) (2010), 224–280.

[195] G. Grey, How to disguise an elliptic curve, talk at ECC 1998, Waterloo, 1998.

[196] G. Frey and H.-G. Rück, A remark concerning m-divisibility and the discrete logarithm problem in the divisor class group of curves, *Math. Comp.* **62**(206) (1994), 865–874.

[197] M. D. Fried and M. Jarden, *Field Arithmetic*, 3rd edn, Springer, 2008.

[198] E. Fujisaki, T. Okamoto, D. Pointcheval and J. Stern, RSA-OAEP is secure under the RSA assumption, *J. Crypt.* **17**(2) (2004), 81–104.

[199] W. Fulton, *Algebraic Curves*, Addison-Wesley, 1989, Out of print, but freely available here: `www.math.lsa.umich.edu/~wfulton/`.

[200] S. D. Galbraith, Constructing isogenies between elliptic curves over finite fields, *LMS J. Comput. Math.* **2** (1999), 118–138.

[201] ______ Supersingular curves in cryptography. In *ASIACRYPT 2001* (C. Boyd, ed.), *LNCS*, vol. 2248, Springer, 2001, pp. 495–513.

[202] S. D. Galbraith, M. Harrison and D. J. Mireles Morales, Efficient hyperelliptic arithmetic using balanced representation for divisors. In *ANTS VIII* (A. J. van der Poorten and A. Stein, eds.), *LNCS*, vol. 5011, Springer, 2008, pp. 342–356.

[203] S. D. Galbraith, C. Heneghan and J. F. McKee, Tunable balancing of RSA. In *ACISP 2005* (C. Boyd and J. M. González Nieto, eds.), *LNCS*, vol. 3574, Springer, 2005, pp. 280–292.

[204] S. D. Galbraith, F. Hess and N. P. Smart, Extending the GHS Weil descent attack. In *EUROCRYPT 2002* (L. R. Knudsen, ed.), *LNCS*, vol. 2332, Springer, 2002, pp. 29–44.

[205] S. D. Galbraith, F. Hess and F. Vercauteren, Aspects of pairing inversion, *IEEE Trans. Inf. Theory* **54**(12) (2008), 5719–5728.

[206] S. D. Galbraith and M. Holmes, *A Non-Uniform Birthday Problem with Applications to Discrete Logarithms*, Cryptology ePrint Archive, Report 2010/616, 2010.

[207] S. D. Galbraith, X. Lin and M. Scott, Endomorphisms for faster elliptic curve cryptography on a large class of curves. In *EUROCRYPT 2009* (A. Joux, ed.), *LNCS*, vol. 5479, Springer, 2009, pp. 518–535.

[208] S. D. Galbraith and J. F. McKee, The probability that the number of points on an elliptic curve over a finite field is prime, *J. Lond. Math. Soc.* **62**(3) (2000), 671–684.

[209] S. D. Galbraith, J. M. Pollard and R. S. Ruprai, *Computing Discrete Logarithms in an Interval*, Cryptology ePrint Archive, Report 2010/617, 2010.

[210] S. D. Galbraith and R. S. Ruprai, An improvement to the Gaudry–Schost algorithm for multidimensional discrete logarithm problems. In *IMA Cryptography and Coding* (M. G. Parker, ed.), *LNCS*, vol. 5921, Springer, 2009, pp. 368–382.

[211] ——— Using equivalence classes to accelerate solving the discrete logarithm problem in a short interval. In *PKC 2010* (P. Q. Nguyen and D. Pointcheval, eds.), *LNCS*, vol. 6056, Springer, 2010, pp. 368–383.

[212] S. D. Galbraith and N. P. Smart, A cryptographic application of Weil descent. In *IMA Cryptography and Coding* (M. Walker, ed.), *LNCS*, vol. 1746, Springer, 1999, pp. 191–200.

[213] S. D. Galbraith and A. Stolbunov, *Improved Algorithm for the Isogeny Problem for Ordinary Elliptic Curves*, Preprint, 2011.

[214] S. D. Galbraith and E. R. Verheul, An analysis of the vector decomposition problem. In *PKC 2008* (R. Cramer, ed.), *LNCS*, vol. 4939, Springer, 2008, pp. 308–327.

[215] R. P. Gallant, R. J. Lambert and S. A. Vanstone, Improving the parallelized Pollard lambda search on binary anomalous curves, *Math. Comp.* **69** (2000), (232) 1699–1705.

[216] ——— Faster point multiplication on elliptic curves with efficient endomorphisms. In *CRYPTO 2001* (J. Kilian, ed.), *LNCS*, vol. 2139, Springer, 2001, pp. 190–200.

[217] N. Gama, P. Q. Nguyen and O. Regev, Lattice enumeration using extreme pruning. In *EUROCRYPT 2010* (H. Gilbert, ed.), *LNCS*, vol. 6110, Springer, 2010, pp. 257–278.

[218] S. Gao, Normal bases over finite fields, Ph.D. thesis, Waterloo, 1993.

[219] T. Garefalakis, The generalised Weil pairing and the discrete logarithm problem on elliptic curves, *Theor. Comput. Sci.* **321**(1) (2004), 59–72.

[220] J. von zur Gathen and J. Gerhard, *Modern Computer Algebra*, Cambridge University Press, 1999.

[221] J. von zur Gathen and M. Giesbrecht, Constructing normal bases in finite fields, *J. Symb. Comput.* **10**(6) (1990), 547–570.

[222] J. von zur Gathen, I. E. Shparlinski and A. Sinclair, Finding points on curves over finite fields, *SIAM J. Comput.* **32**(6) (2003), 1436–1448.

[223] P. Gaudry, An algorithm for solving the discrete log problem on hyperelliptic curves. In *EUROCRYPT 2000* (B. Preneel, ed.), *LNCS*, vol. 1807, Springer, 2000, pp. 19–34.

[224] ——— Fast genus 2 arithmetic based on theta functions, *J. Math. Crypt.* **1**(3) (2007), 243–265.

[225] ——— Index calculus for abelian varieties of small dimension and the elliptic curve discrete logarithm problem, *J. Symb. Comput.* **44**(12) (2009), 1690–1702.

[226] P. Gaudry, F. Hess and N. P. Smart, Constructive and destructive facets of Weil descent on elliptic curves, *J. Crypt.* **15** (2002), 19–46.

[227] P. Gaudry and D. Lubicz, The arithmetic of characteristic 2 Kummer surfaces and of elliptic Kummer lines, *Finite Fields Appl.* **15** (2009), 246–260.

[228] P. Gaudry and É. Schost, A low-memory parallel version of Matsuo, Chao, and Tsujii's algorithm. In *ANTS VI* (D. A. Buell, ed.), *LNCS*, vol. 3076, Springer, 2004, pp. 208–222.

[229] P. Gaudry, E. Thomé, N. Thériault and C. Diem, A double large prime variation for small genus hyperelliptic index calculus, *Math. Comp.* **76**(257) (2007), 475–492.

[230] C. Gentry, The geometry of provable security: some proofs of security in which lattices make a surprise appearance, *The LLL Algorithm* (P. Q. Nguyen and B. Vallée, eds.), Springer, 2010, pp. 391–426.

[231] M. Girault, An identity-based identification scheme based on discrete logarithms modulo a composite number. In *EUROCRYPT 1990* (I. Damgård, ed.), *LNCS*, vol. 473, Springer, 1991, pp. 481–486.

[232] M. Girault, G. Poupard and J. Stern, On the fly authentication and signature schemes based on groups of unknown order, *J. Crypt.* **19** (2006), 463–487.

[233] O. Goldreich, S. Goldwasser and S. Halevi, Public-key cryptosystems from lattice reduction problems. In *CRYPTO 1997* (B. S. Kaliski Jr., ed.), *LNCS*, vol. 1294, Springer, 1997, pp. 112–131.

[234] O. Goldreich, D. Ron and M. Sudan, Chinese remaindering with errors, *IEEE Trans. Inf. Theory* **46**(4) (2000), 1330–1338.

[235] S. Goldwasser, S. Micali and R. L. Rivest, A digital signature scheme secure against adaptive chosen-message attacks, *SIAM J. Comput.* **17**(2) (1988), 281–308.

[236] G. Gong and L. Harn, Public-key cryptosystems based on cubic finite field extensions, *IEEE Trans. Inf. Theory* **45**(7) (1999), 2601–2605.

[237] M. I. González Vasco, M. Näslund and I. E. Shparlinski, New results on the hardness of Diffie–Hellman bits, *PKC 2004* (F. Bao, R. H. Deng and J. Zhou, eds.), *LNCS*, vol. 2947, Springer, 2004, pp. 159–172.

[238] M. I. González Vasco and I. E. Shparlinski, On the security of Diffie–Hellman bits. In *Cryptography and Computational Number Theory* (H. Wang K. Y. Lam, I. E. Shparlinski and C. Xing, eds.), Progress in Computer Science and Applied Logic, Birkhäuser, 2001, pp. 257–268.

[239] D. M. Gordon, On the number of elliptic pseudoprimes, *Math. Comp.* **52**(185) (1989), 231–245.

[240] D. M. Gordon and K. S. McCurley, Massively parallel computation of discrete logarithms. In *CRYPTO 1992* (E. F. Brickell, ed.), *LNCS*, vol. 740, Springer, 1993, pp. 312–323.

[241] R. Granger, F. Hess, R. Oyono, N. Thériault and F. Vercauteren, Ate pairing on hyperelliptic curves. In *EUROCRYPT 2007* (M. Naor, ed.), *LNCS*, vol. 4515, Springer, 2007, pp. 430–447.

[242] R. Granger and F. Vercauteren, On the discrete logarithm problem on algebraic tori. In *CRYPTO 2005* (V. Shoup, ed.), *LNCS*, vol. 3621, Springer, 2005, pp. 66–85.

[243] A. Granville, Smooth numbers: computational number theory and beyond. In *Algorithmic Number Theory* (J. P. Buhler and P. Stevenhagen, eds.), *MSRI Proceedings*, vol. 44, Cambridge University Press, 2008, pp. 267–323.

[244] B. H. Gross, Heights and the special values of L-series. In *CMS Conf. Proc.*, vol. 7, 1987, pp. 115–187.

[245] M. Grötschel, L. Lovász and A. Schrijver, *Geometric Algorithms and Combinatorial Optimization*, Springer, 1993.

[246] R. K. Guy, *Unsolved Problems in Number Theory*, 2nd edn, Springer, 1994.

[247] J. L. Hafner and K. S. McCurley, Asymptotically fast triangularization of matrices over rings, *SIAM J. Comput.* **20**(6) (1991), 1068–1083.

[248] D. Hankerson, A. Menezes and S. Vanstone, *Guide to Elliptic Curve Cryptography*, Springer, 2004.

[249] G. Hanrot and D. Stehlé, Improved analysis of Kannan's shortest lattice vector algorithm. In *CRYPTO 2007* (A. Menezes, ed.), *LNCS*, vol. 4622, Springer, 2007, pp. 170–186.

[250] G. H. Hardy and E. M. Wright, *An Introduction to the Theory of Numbers*, 5th edn, Oxford University Press, 1980.

[251] R. Harley, *Fast Arithmetic on Genus Two Curves*, Preprint, 2000.

[252] R. Hartshorne, *Algebraic Geometry*, *GTM*, vol. 52, Springer, 1997.

[253] J. Håstad and M. Näslund, The security of all RSA and discrete log bits, *J. ACM* **51**(2) (2004), 187–230.

[254] G. Havas, B. S. Majewski and K. R. Matthews, Extended GCD and Hermite normal form algorithms via lattice basis reduction, *Experimental Math.* **7**(2) (1998), 125–136.

[255] B. Helfrich, Algorithms to construct Minkowski reduced and Hermite reduced lattice bases, *Theor. Comput. Sci.* **41** (1985), 125–139.

[256] F. Hess, A note on the Tate pairing of curves over finite fields, *Arch. Math.* **82** (2004), 28–32.

[257] ______ Pairing lattices. In *Pairing 2008* (S. D. Galbraith and K. G. Paterson, eds.), *LNCS*, vol. 5209, Springer, 2008, pp. 18–38.

[258] F. Hess, N. Smart and F. Vercauteren, The eta pairing revisited, *IEEE Trans. Inf. Theory* **52**(10) (2006), 4595–4602.

[259] N. J. Higham, *Accuracy and Stability of Numerical Algorithms*, 2nd edn, SIAM, 2002.

[260] Y. Hitchcock, P. Montague, G. Carter and E. Dawson, The efficiency of solving multiple discrete logarithm problems and the implications for the security of fixed elliptic curves, *Int. J. Inf. Secur.* **3** (2004), 86–98.

[261] J. Hoffstein, J. Pipher and J. H. Silverman, *An Introduction to Mathematical Cryptography*, Springer, 2008.

[262] J. Hoffstein and J. H. Silverman, Random small Hamming weight products with applications to cryptography, *Discrete Appl. Math.* **130**(1) (2003), 37–49.

[263] D. Hofheinz and E. Kiltz, The group of signed quadratic residues and applications. In *CRYPTO 2009* (S. Halevi, ed.), *LNCS*, vol. 5677, Springer, 2009, pp. 637–653.

[264] S. Hohenberger and B. Waters, Short and stateless signatures from the RSA assumption. In *CRYPTO 2009* (S. Halevi, ed.), *LNCS*, vol. 5677, Springer, 2009, pp. 654–670.

[265] J. E. Hopcroft and J. D. Ullman, *Introduction to Automata Theory, Languages and Computation*, Addison-Wesley, 1979.

[266] J. Horwitz and R. Venkatesan, Random Cayley digraphs and the discrete logarithm. In *ANTS V* (C. Fieker and D. R. Kohel, eds.), *LNCS*, vol. 2369, Springer, 2002, pp. 416–430.

[267] E. W. Howe, On the group orders of elliptic curves over finite fields, *Compositio Mathematica* **85** (1993), 229–247.

[268] N. Howgrave-Graham, Finding small roots of univariate modular equations revisited. In *IMA Cryptography and Coding* (M. Darnell, ed.), *LNCS*, vol. 1355, Springer, 1997, pp. 131– 142.

[269] ______ Approximate integer common divisors, *Cryptography and Lattices (CaLC)* (J. H. Silverman, ed.), *LNCS*, vol. 2146, Springer, 2001, pp. 51–66.

[270] N. Howgrave-Graham and N. P. Smart, Lattice attacks on digital signature schemes, *Des. Codes Crypt.* **23** (2001), 283–290.

[271] T. W. Hungerford, *Algebra*, *GTM* vol. 73, Springer, 1974.

[272] D. Husemöller, *Elliptic Curves*, 2nd edn, GTM, vol. 111, Springer, 2004.

[273] T. Icart, How to hash into elliptic curves. In *CRYPTO 2009* (S. Halevi, ed.), *LNCS*, vol. 5677, Springer, 2009, pp. 303–316.

[274] T. Iijima, K. Matsuo, J. Chao and S. Tsujii, Construction of Frobenius maps of twist elliptic curves and its application to elliptic scalar multiplication. In *Symposium on Cryptography and Information Security (SCIS) 2002*, IEICE Japan, 2002, pp. 699–702.

[275] T. Jager and J. Schwenk, On the equivalence of generic group models. In *ProvSec 2008* (K. Chen J. Baek, F. Bao and X. Lai, eds.), *LNCS*, vol. 5324, Springer, 2008, pp. 200–209.

[276] D. Jao, D. Jetchev and R. Venkatesan, On the bits of elliptic curve Diffie–Hellman keys. In *INDOCRYPT 2007* (K. Srinathan, C. Pandu Rangan and M. Yung, eds.), *LNCS*, vol. 4859, Springer, 2007, pp. 33–47.

[277] D. Jao, S. D. Miller and R. Venkatesan, Do all elliptic curves of the same order have the same difficulty of discrete log?. In *ASIACRYPT 2005* (B. K. Roy, ed.), *LNCS*, vol. 3788, Springer, 2005, pp. 21–40.

[278] ——— Expander graphs based on GRH with an application to elliptic curve cryptography, *J. Number Theory* **129**(6) (2009), 1491–1504.

[279] D. Jao and V. Soukharev, A subexponential algorithm for evaluating large degree isogenies. In *ANTS IX* (G. Hanrot, F. Morain and E. Thomé, eds.), *LNCS*, vol. 6197, Springer, 2010, pp. 219–233.

[280] D. Jao and K. Yoshida, Boneh–Boyen signatures and the strong Diffie–Hellman problem. In *Pairing 2009* (H. Shacham and B. Waters, eds.), *LNCS*, vol. 5671, Springer, 2009, pp. 1–16.

[281] D. Jetchev and R. Venkatesan, Bits security of the elliptic curve Diffie–Hellman secret keys. In *CRYPTO 2008* (D. Wagner, ed.), *LNCS*, vol. 5157, Springer, 2008, pp. 75–92.

[282] Z.-T. Jiang, W.-L. Xu and Y.-M. Wang, Polynomial analysis of DH secrete key and bit security, *Wuhan University Journal of Natural Sciences* **10**(1) (2005), no. 1, 239–242.

[283] A. Joux, *Algorithmic Cryptanalysis*, Chapman & Hall/CRC, 2009.

[284] A. Joux and R. Lercier, The function field sieve in the medium prime case. In *EUROCRYPT 2006* (S. Vaudenay, ed.), *LNCS*, vol. 4004, Springer, 2006.

[285] A. Joux, R. Lercier, N. P. Smart and F. Vercauteren, The number field sieve in the medium prime case. In *CRYPTO 2006* (C. Dwork, ed.), *LNCS*, vol. 4117, Springer, 2006, pp. 326–344.

[286] M. Joye and G. Neven, Identity-based cryptography, *Cryptology and Information Security*, vol. 2, IOS Press, 2008.

[287] M. Joye and S.-M. Yen, Optimal left-to-right binary signed-digit recoding, *IEEE Trans. Computers* **49**(7) (2000), 740–748.

[288] M. J. Jacobson Jr., N. Koblitz, J. H. Silverman, A. Stein and E. Teske, Analysis of the Xedni calculus attack, *Des. Codes Crypt.* **20**(1) (2000), 1–64.

[289] M. J. Jacobson Jr. and A. J. van der Poorten, Computational aspects of NUCOMP. In *ANTS V* (C. Fieker and D. R. Kohel, eds.), *LNCS*, vol. 2369, Springer, 2002, pp. 120–133.

[290] C. S. Jutla, On finding small solutions of modular multivariate polynomial equations. In *EUROCRYPT 1998* (K. Nyberg, ed.), *LNCS*, vol. 1403, Springer, 1998, pp. 158–170.

[291] M. Kaib and H. Ritter, Block reduction for arbitrary norms, Technical Report, Universität Frankfurt am Main, 1994.

[292] M. Kaib and C.-P. Schnorr, The generalized Gauss reduction algorithm, *Journal of Algorithms* **21**(3) (1996), 565–578.

[293] B. S. Kaliski Jr., Elliptic curves and cryptography: a pseudorandom bit generator and other tools, Ph.D. thesis, MIT, 1988.

[294] W. van der Kallen, Complexity of the Havas, Majewski, Matthews LLL Hermite normal form algorithm, *J. Symb. Comput.* **30**(3) (2000), 329–337.

[295] R. Kannan, Improved algorithms for integer programming and related lattice problems. In *Symposium on the Theory of Computing (STOC)*, ACM, 1983, pp. 193–206.

[296] ——— Minkowski's convex body theorem and integer programming, *Mathematics of Operations Research* **12**(3) (1987), 415–440.

[297] R. Kannan and A. Bachem, Polynomial algorithms for computing the Smith and Hermite normal forms of an integer matrix, *SIAM J. Comput.* **8** (1979), 499–507.

[298] M. Katagi, T. Akishita, I. Kitamura and T. Takagi, Some improved algorithms for hyperelliptic curve cryptosystems using degenerate divisors. In *ICISC 2004* (C. Park and S. Chee, eds.), *LNCS*, vol. 3506, Springer, 2004, pp. 296–312.

[299] M. Katagi, I. Kitamura, T. Akishita and T. Takagi, Novel efficient implementations of hyperelliptic curve cryptosystems using degenerate divisors. In *WISA 2004* (C.-H. Lim and M. Yung, eds.), *LNCS*, vol. 3325, Springer, 2004, pp. 345–359.

[300] J. Katz and Y. Lindell, *Introduction to Modern Cryptography*, Chapman & Hall/CRC, 2008.

[301] J. H. Kim, R. Montenegro, Y. Peres and P. Tetali, A birthday paradox for Markov chains, with an optimal bound for collision in the Pollard rho algorithm for discrete logarithm. In *ANTS VIII* (A. J. van der Poorten and A. Stein, eds.), *LNCS*, vol. 5011, Springer, 2008, pp. 402–415.

[302] J. H. Kim, R. Montenegro and P. Tetali, Near optimal bounds for collision in Pollard rho for discrete log. In *Foundations of Computer Science (FOCS)*, IEEE, 2007, pp. 215–223.

[303] B. King, A point compression method for elliptic curves defined over GF(2n). In *PKC 2004* (F. Bao, R. H. Deng, and J. Zhou, eds.), *LNCS*, vol. 2947, Springer, 2004, pp. 333–345.

[304] S. Kim and J.-H. Cheon, A parameterized splitting system and its application to the discrete logarithm problem with low Hamming weight product exponents. In *PKC 2008* (R. Cramer, ed.), *LNCS*, vol. 4939, Springer, 2008, pp. 328–343.

[305] J. F. C. Kingman and S. J. Taylor, *Introduction to Measure Theory and Probability*, Cambridge University Press, 1966.

[306] P. N. Klein, Finding the closest lattice vector when it's unusually close, *Symposium on Discrete Algorithms (SODA)*, ACM/SIAM, 2000, pp. 937–941.

[307] E. W. Knudsen, Elliptic scalar multiplication using point halving. In *ASIACRYPT 1999* (K.-Y. Lam, E. Okamoto and C. Xing, eds.), *LNCS*, vol. 1716, Springer, 1999, pp. 135–149.

[308] D. E. Knuth, *Art of Computer Programming, Volume 2: Semi-Numerical Algorithms*, 3rd edn, Addison-Wesley, 1997.

[309] N. Koblitz, Elliptic curve cryptosystems, *Math. Comp.* **48**(177) (1987), 203–209.

[310] ______ Primality of the number of points on an elliptic curve over a finite field, *Pacific J. Math.* **131**(1) (1988), 157–165.

[311] ______ Hyperelliptic cryptosystems, *J. Crypt.* **1** (1989), 139–150.

[312] ______ CM curves with good cryptographic properties. In *CRYPTO 1991* (J. Feigenbaum, ed.), *LNCS*, vol. 576, Springer, 1992, pp. 279–287.

[313] ______ *A Course in Number Theory and Cryptography*, 2nd edn, *GTM* vol. 114, Springer, 1994.

[314] C. K. Koç and T. Acar, Montgomery multplication in $GF(2^k)$, *Des. Codes Crypt.* **14**(1) (1998), 57–69.

[315] D. R. Kohel, Endomorphism rings of elliptic curves over finite fields, Ph.D. thesis, University of California, Berkeley, 1996.

[316] ______ *Constructive and Destructive Facets of Torus-based Cryptography*, Preprint, 2004.

[317] D. R. Kohel and I. E. Shparlinski, On exponential sums and group generators for elliptic curves over finite fields. In *ANTS IV* (W. Bosma, ed.), *LNCS*, vol. 1838, Springer, 2000, pp. 395–404.

[318] S. Kozaki, T. Kutsuma and K. Matsuo, Remarks on Cheon's algorithms for pairing-related problems, *Pairing 2007* (T. Takagi, T. Okamoto, E. Okamoto and T. Okamoto, eds.), *LNCS*, vol. 4575, Springer, 2007, pp. 302–316.

[319] M. Kraitchik, *Théorie des Nombres, Vol. 1*, Gauthier-Villars, Paris, 1922.

[320] F. Kuhn and R. Struik, Random walks revisited: extensions of Pollard's rho algorithm for computing multiple discrete logarithms. In *SAC 2001* (S. Vaudenay and A. M. Youssef, eds.), *LNCS*, vol. 2259, Springer, 2001, pp. 212–229.

[321] R. M. Kuhn, Curves of genus 2 with split Jacobian, *Trans. Amer. Math. Soc.* **307**(1) (1988), 41–49.

[322] R. Kumar and D. Sivakumar, Complexity of SVP – a reader's digest, *SIGACT News*, Complexity Theory Column **32** (2001), 13.

[323] K. Kurosawa and Y. Desmedt, A new paradigm of hybrid encryption scheme. In *CRYPTO 2004* (M. K. Franklin, ed.), *LNCS*, vol. 3152, Springer, 2004, pp. 426–442.

[324] J. C. Lagarias, H. W. Lenstra Jr. and C.-P. Schnorr, Korkin–Zolotarev bases and successive minima of a lattice and its reciprocal lattice, *Combinatorica* **10**(4) (1990), 333–348.

[325] C. Lanczos, Solution of systems of linear equations by minimized iterations, *J. Res. Nat. Bureau of Standards* **49** (1952), 33–53.

[326] S. Lang, *Introduction to Algebraic Geometry*, Wiley, 1964.

[327] ——— *Algebraic Number Theory*, *GTM*, vol. 110, Springer, 1986.

[328] ——— *Elliptic Functions*, 2nd edn, *GTM*, vol. 112, Springer, 1987.

[329] ——— *Algebra*, 3rd edn, Addison-Wesley, 1993.

[330] T. Lange, Koblitz curve cryptosystems, *Finite Fields Appl.* **11**(2) (2005), 200–229.

[331] E. Lee, H.-S. Lee and C.-M. Park, Efficient and Generalized Pairing Computation on Abelian Varieties, *IEEE Trans. Inf. Theory* **55**(4) (2009), 1793–1803.

[332] A. K. Lenstra, Factorization of polynomials. In *Computational Methods in Number Theory*, (H. W. Lenstra Jr. and R. Tijdeman, eds.) *Mathematical Center Tracts* 154, Mathematisch Centrum Amsterdam, 1984, pp. 169–198.

[333] ——— Integer factoring, *Des. Codes Crypt.* **19**(2/3) (2000), 101–128.

[334] A. K. Lenstra and H. W. Lenstra Jr., *The Development of the Number Field Sieve*, *LNM*, vol. 1554, Springer, 1993.

[335] A. K. Lenstra, H. W. Lenstra Jr. and L. Lovász, Factoring polynomials with rational coefficients, *Math. Ann.* **261** (1982), 515–534.

[336] A. K. Lenstra and I. E. Shparlinski, Selective forgery of RSA signatures with fixed-pattern padding. In *PKC 2002* (D. Naccache and P. Paillier, eds.), *LNCS*, vol. 2274, Springer, 2002, pp. 228–236.

[337] A. K. Lenstra and E. R. Verheul, The XTR public key system. In *CRYPTO 2000* (M. Bellare, ed.), *LNCS*, vol. 1880, Springer, 2000, pp. 1–19.

[338] ——— Fast irreducibility and subgroup membership testing in XTR. In *PKC 2001* (K. Kim, ed.), *LNCS*, vol. 1992, Springer, 2001, pp. 73–86.

[339] H. W. Lenstra Jr., Factoring integers with elliptic curves, *Annals of Mathematics* **126**(3) (1987), 649–673.

[340] ——— Elliptic curves and number theoretic algorithms. In *Proc. International Congr. Math.*, Berkeley 1986, AMS, 1988, pp. 99–120.

[341] ——— Finding isomorphisms between finite fields, *Math. Comp.* **56**(193) (1991), 329–347.

[342] H. W. Lenstra Jr., J. Pila and C. Pomerance, A hyperelliptic smoothness test I, *Phil. Trans. R. Soc. Lond.* A **345** (1993), 397–408.

[343] H. W. Lenstra Jr. and C. Pomerance, A rigorous time bound for factoring integers, *J. Amer. Math. Soc.* **5**(3) (1992), no. 3, 483–516.

[344] R. Lercier, Computing isogenies in F_{2^n}. In *ANTS II* (H. Cohen, ed.), *LNCS*, vol. 1122, Springer, 1996, pp. 197–212.

[345] R. Lercier and F. Morain, Algorithms for computing isogenies between elliptic curves, *Computational Perspectives on Number Theory* (D. A. Buell and J. T. Teitelbaum, eds.), Studies in Advanced Mathematics, vol. 7, AMS, 1998, pp. 77–96.

[346] R. Lercier and T. Sirvent, On Elkies subgroups of ℓ-torsion points in elliptic curves defined over a finite field, *J. Théor. Nombres Bordeaux* **20**(3) (2008), 783–797.

[347] G. Leurent and P. Q. Nguyen, How risky is the random oracle model?. In *CRYPTO 2009* (S. Halevi, ed.), *LNCS*, vol. 5677, Springer, 2009, pp. 445–464.

[348] K.-Z. Li and F. Oort, Moduli of supersingular abelian varieties, *LNM*, vol. 1680, Springer, 1998.

[349] W.-C. Li, M. Näslund and I. E. Shparlinski, Hidden number problem with the trace and bit security of XTR and LUC. In *CRYPTO 2002* (M. Yung, ed.), *LNCS*, vol. 2442, Springer, 2002, pp. 433–448.

[350] R. Lidl and H. Niederreiter, *Introduction to Finite Fields and Their Applications*, Cambridge University Press, 1994.

[351] ——— *Finite Fields*, Cambridge University Press, 1997.

[352] R. Lindner and C. Peikert, Better key sizes (and attacks) for LWE-based encryption, *CT-RSA 2011* (A. Kiayias, ed.), *LNCS*, vol. 6558, 2011, pp. 1–23.

[353] P. Lockhart, On the discriminant of a hyperelliptic curve, *Trans. Amer. Math. Soc.* **342**(2) (1994), 729–752.

[354] D. L. Long and A. Wigderson, The discrete logarithm hides $O(\log n)$ bits, *SIAM J. Comput.* **17**(2) (1988), 363–372.

[355] D. Lorenzini, *An Invitation to Arithmetic Geometry*, Graduate Studies in Mathematics, vol. 106, AMS, 1993.

[356] L. Lovász, *An Algorithmic Theory of Numbers, Graphs and Convexity*, SIAM, 1986.

[357] L. Lovász and H. E. Scarf, The generalized basis reduction algorithm, *Mathematics of Operations Research* **17**(3) (1992), 751–764.

[358] R. Lovorn Bender and C. Pomerance, Rigorous discrete logarithm computations in finite fields via smooth polynomials. In *Computational Perspectives on Number Theory* (D. A. Buell and J. T. Teitelbaum, eds.), Studies in Advanced Mathematics, vol. 7, AMS, 1998, pp. 221–232.

[359] M. Luby, *Pseudorandomness and Cryptographic Applications*, Princeton, 1996.

[360] H. Lüneburg, On a little but useful algorithm. In *AAECC-3, 1985* (J. Calmet, ed.), *LNCS*, vol. 229, Springer, 1986, pp. 296–301.

[361] C. Mauduit and A. Sárközy, On finite pseudorandom binary sequences I: measure of pseudorandomness, the Legendre symbol, *Acta Arith.* **82** (1997), 365–377.

[362] U. M. Maurer, Towards the equivalence of breaking the Diffie–Hellman protocol and computing discrete logarithms. In *CRYPTO 1994* (Y. Desmedt, ed.), *LNCS*, vol. 839, Springer, 1994, pp. 271–281.

[363] ——— Fast generation of prime numbers and secure public-key cryptographic parameters, *J. Crypt.* **8**(3) (1995), 123–155.

[364] ——— Abstract models of computation in cryptography. In *IMA Int. Conf.* (N. P. Smart, ed.), *LNCS*, vol. 3796, Springer, 2005, pp. 1–12.

[365] U. M. Maurer and S. Wolf, Diffie–Hellman oracles. In *CRYPTO 1996* (N. Koblitz, ed.), *LNCS*, vol. 1109, Springer, 1996, pp. 268–282.

[366] ——— On the complexity of breaking the Diffie–Hellman protocol, Technical Report 244, Institute for Theoretical Computer Science, ETH Zurich, 1996.

[367] ——— Lower bounds on generic algorithms in groups. In *EUROCRYPT 1998* (K. Nyberg, ed.), *LNCS*, vol. 1403, Springer, 1998, pp. 72–84.

[368] ——— The relationship between breaking the Diffie–Hellman protocol and computing discrete logarithms, *SIAM J. Comput.* **28**(5) (1999), 1689–1721.

[369] ——— The Diffie–Hellman protocol, *Des. Codes Crypt.* **19**(2/3) (2000), 147–171.

[370] A. May, *New RSA Vulnerabilities Using Lattice Reduction Methods*, Ph.D. thesis, Paderborn, 2003.

[371] ——— Using LLL-reduction for solving RSA and factorization problems: a survey. In *The LLL Algorithm* (P. Q. Nguyen and B. Vallée, eds.), Springer, 2010, pp. 315–348.

[372] J. F. McKee, Subtleties in the distribution of the numbers of points on elliptic curves over a finite prime field, *J. London Math. Soc.* **59**(2) (1999), 448–460.

[373] W. Meier and O. Staffelbach, Efficient multiplication on certain non-supersingular elliptic curves. In *CRYPTO 1992* (E. F. Brickell, ed.), *LNCS*, vol. 740, Springer, 1993, pp. 333– 344.

[374] A. Menezes and S. A. Vanstone, The implementation of elliptic curve cryptosystems. In *AUSCRYPT 1990* (J. Seberry and J. Pieprzyk, eds.), *LNCS*, vol. 453, Springer, 1990, pp. 2–13.

[375] A. J. Menezes, T. Okamoto and S. A. Vanstone, Reducing elliptic curve logarithms to a finite field, *IEEE Trans. Inf. Theory* **39**(5) (1993), 1639–1646.

[376] A.J. Menezes, P.C. van Oorschot and S.A. Vanstone, *Handbook of Applied Cryptography*, CRC Press, 1996.

[377] J.-F. Mestre, La méthode des graphes. exemples et applications, *Proceedings of the International Conference on Class Numbers and Fundamental Units of Algebraic Number fields (Katata, 1986), Nagoya University*, 1986, pp. 217–242.

[378] D. Micciancio and S. Goldwasser, *Complexity of Lattice Problems: A Cryptographic Perspective*, Kluwer, 2002.

[379] D. Micciancio and O. Regev, Lattice-based cryptography. In *Post Quantum Cryptography* (D. J. Bernstein, J. Buchmann and E. Dahmen, eds.), Springer, 2009, pp. 147–191.

[380] D. Micciancio and P. Voulgaris, Faster exponential time algorithms for the shortest vector problem. In *SODA* (M. Charikar, ed.), *SIAM*, 2010, pp. 1468–1480.

[381] D. Micciancio and B. Warinschi, A linear space algorithm for computing the Hermite normal form. In *ISSAC*, 2001, pp. 231–236.

[382] S. D. Miller and R. Venkatesan, Spectral analysis of Pollard rho collisions. In *ANTS VII* (F. Hess, S. Pauli and M. E. Pohst, eds.), *LNCS*, vol. 4076, Springer, 2006, pp. 573–581.

[383] V. S. Miller, Short programs for functions on curves, Unpublished manuscript, 1986.

[384] ______ Use of elliptic curves in cryptography. In *CRYPTO 1985* (H. C. Williams, ed.), *LNCS*, vol. 218, Springer, 1986, pp. 417–426.

[385] ______ The Weil pairing, and its efficient calculation, *J. Crypt.* **17**(4) (2004), 235–261.

[386] A. Miyaji, T. Ono and H. Cohen, Efficient elliptic curve exponentiation. In *ICICS 1997* (Y. Han, T. Okamoto and S. Qing, eds.), *LNCS*, vol. 1334, Springer, 1997, pp. 282–291.

[387] B. Möller, Algorithms for multi-exponentiation. In *SAC 2001* (S. Vaudenay and A. M. Youssef, eds.), *LNCS*, vol. 2259, Springer, 2001, pp. 165–180.

[388] ______ Improved techniques for fast exponentiation. In *ICISC 2002* (P.-J. Lee and C.-H. Lim, eds.), *LNCS*, vol. 2587, Springer, 2003, pp. 298–312.

[389] ______ Fractional windows revisited: improved signed-digit representations for efficient exponentiation. In *ICISC 2004* (C. Park and S. Chee, eds.), *LNCS*, vol. 3506, Springer, 2005, pp. 137–153.

[390] R. Montenegro and P. Tetali, How long does it take to catch a wild kangaroo?. In *Symposium on Theory of Computing (STOC)*, 2009, pp. 553–559.

[391] P. L. Montgomery, Modular multiplication without trial division, *Math. Comp.* **44**(170) (1985), 519–521.

[392] ______ Speeding the Pollard and elliptic curve methods of factorization, *Math. Comp.* **48**(177) (1987), 243–264.

[393] F. Morain and J.-L. Nicolas, *On Cornacchia's Algorithm for Solving the Diophantine Equation* $u^2 + dv^2 = m$, Preprint, 1990.

[394] F. Morain and J. Olivos, Speeding up the computations on an elliptic curve using addition-subtraction chains. In *Theoretical Informatics and Applications*, vol. 24, 1990, pp. 531–543.

[395] C. J. Moreno, *Algebraic Curves over Finite Fields*, Cambridge University Press, 1991.

[396] W. H. Mow, Universal lattice decoding: principle and recent advances, *Wireless Communications and Mobile Computing* **3**(5) (2003), 553–569.

[397] V. Müller, Fast multiplication on elliptic curves over small fields of characteristic two, *J. Crypt.* **11**(4) (1998), 219–234.

[398] D. Mumford, *Abelian Varieties*, Oxford University Press, 1970.

[399] ______ Tata lectures on theta II, *Progess in Mathematics*, vol. 43, Birkhäuser, 1984.

[400] M. R. Murty, Ramanujan graphs, *J. Ramanujan Math. Soc.* **18**(1) (2003), 1–20.

[401] R. Murty and I. E. Shparlinski, Group structure of elliptic curves over finite fields and applications, *Topics in Geometry, Coding Theory and Cryptography* (A. Garcia and H. Stichtenoth, eds.), Springer-Verlag, 2006, pp. 167–194.

[402] A. Muzereau, N. P. Smart and F. Vercauteren, The equivalence between the DHP and DLP for elliptic curves used in practical applications, *LMS J. Comput. Math.* **7** (2004), 50–72.

[403] D. Naccache, D. M'Raïhi, S. Vaudenay and D. Raphaeli, Can D.S.A. be improved? Complexity trade-offs with the digital signature standard. In *EUROCRYPT 1994* (A. De Santis, ed.), *LNCS*, vol. 950, Springer, 1995, pp. 77–85.

[404] D. Naccache and I. E. Shparlinski, Divisibility, smoothness and cryptographic applications, *Algebraic Aspects of Digital Communications* (T. Shaska and E. Hasimaj, eds.) *NATO Series for Peace and Security*, 24 IOS Press, 2009, pp. 115–173.

[405] V. I. Nechaev, Complexity of a determinate algorithm for the discrete logarithm, *Mathematical Notes* **55**(2) (1994), 165–172.

[406] G. Neven, N. P. Smart and B. Warinschi, Hash function requirements for Schnorr signatures, *J. Math. Crypt.* **3**(1) (2009), 69–87.

[407] P. Nguyen and D. Stehlé, Floating-point LLL revisited. In *EUROCRYPT 2005* (R. Cramer, ed.), *LNCS*, vol. 3494, Springer, 2005, pp. 215–233.

[408] ——— Low-dimensional lattice basis reduction revisited, *ACM Transactions on Algorithms* **5**(4:46) (2009), 1–48.

[409] P. Q. Nguyen, Public key cryptanalysis. In *Recent Trends in Cryptography* (I. Luengo, ed.), AMS, 2009, pp. 67–119.

[410] P. Q. Nguyen and O. Regev, Learning a parallelepiped: cryptanalysis of GGH and NTRU signatures. In *EUROCRYPT 2006* (S. Vaudenay, ed.), *LNCS*, vol. 4004, Springer, 2006, pp. 271–288.

[411] P. Q. Nguyen and I. E. Shparlinski, The insecurity of the digital signature algorithm with partially known nonces, *J. Crypt.* **15**(3) (2002), 151–176.

[412] ——— The insecurity of the elliptic curve digital signature algorithm with partially known nonces, *Des. Codes Crypt.* **30**(2) (2003), 201–217.

[413] P. Q. Nguyen and D. Stehlé, Low-dimensional lattice basis reduction revisited. In *ANTS VI* (D. A. Buell, ed.), *LNCS*, vol. 3076, Springer, 2004, pp. 338–357.

[414] P. Q. Nguyen and J. Stern, Lattice reduction in cryptology: an update. In *ANTS IV* (W. Bosma, ed.), *LNCS*, vol. 1838, Springer, 2000, pp. 85–112.

[415] ——— The two faces of lattices in cryptology. In *Cryptography and Lattices (CaLC)* (J. H. Silverman, ed.), *LNCS*, vol. 2146, Springer, 2001, pp. 146–180.

[416] P. Q. Nguyen and B. Vallée, The LLL algorithm: survey and applications, *Information Security and Cryptography*, Springer, 2009.

[417] P. Q. Nguyen and T. Vidick, Sieve algorithms for the shortest vector problem are practical, *J. Math. Crypt.* **2**(2) (2008), 181–207.

[418] H. Niederreiter, A new efficient factorization algorithm for polynomials over small finite fields, *Applicable Algebra in Engineering, Communication and Computing* **4**(2) (1993), 81–87.

[419] G. Nivasch, Cycle detection using a stack, *Inf. Process. Lett.* **90**(3) (2004), 135–140.

[420] I. Niven, H. S. Zuckerman and H. L. Montgomery, *An Introduction to the Theory of Numbers*, 5th edn, Wiley, 1991.

[421] A. M. Odlyzko, Discrete logarithms in finite fields and their cryptographic significance. In *EUROCRYPT 1984* (T. Beth, N. Cot and I. Ingemarsson, eds.), *LNCS*, vol. 209, Springer, 1985, pp. 224–314.

[422] P. C. van Oorschot and M. J. Wiener, On Diffie–Hellman key agreement with short exponents. In *EUROCRYPT 1996* (U. M. Maurer, ed.), *LNCS*, vol. 1070, Springer, 1996, pp. 332–343.

[423] ——— Parallel collision search with cryptanalytic applications, *J. Crypt.* **12**(1) (1999), 1–28.

[424] P. Paillier, Public-key cryptosystems based on composite degree residuosity classes. In *EUROCRYPT 1999* (J. Stern, ed.), *LNCS*, vol. 1592, Springer, 1999, pp. 223–238.

[425] ——— Impossibility proofs for RSA signatures in the standard model. In *CT-RSA 2007* (M. Abe, ed.), *LNCS*, vol. 4377, Springer, 2007, pp. 31–48.

[426] P. Paillier and D. Vergnaud, Discrete-log-based signatures may not be equivalent to discrete log. In *ASIACRYPT 2005* (B. K. Roy, ed.), *LNCS*, vol. 3788, Springer, 2005, pp. 1–20.

[427] P. Paillier and J. L. Villar, Trading one-wayness against chosen-ciphertext security in factoring-based encryption. In *ASIACRYPT 2006* (X. Lai and K. Chen, eds.), *LNCS*, vol. 4284, Springer, 2006, pp. 252–266.

[428] S. Patel and G. S. Sundaram, An efficient discrete log pseudo random generator. In *CRYPTO 1998* (H. Krawczyk, ed.), *LNCS*, vol. 1462, Springer, 1998, pp. 304–317.

[429] S. Paulus and H.-G. Rück, Real and imaginary quadratic representations of hyperelliptic function fields, *Math. Comp.* **68**(227) (1999), 1233–1241.

[430] R. Peralta, Simultaneous security of bits in the discrete log. In *EUROCRYPT 1985* (F. Pichler, ed.), *LNCS*, vol. 219, Springer, 1986, pp. 62–72.

[431] A. K. Pizer, Ramanujan graphs. In *Computational Perspectives on Number Theory* (D. A. Buell and J. T. Teitelbaum, eds.), Studies in Advanced Mathematics, vol. 7, AMS, 1998, pp. 159–178.

[432] S. Pohlig and M. Hellman, An improved algorithm for computing logarithms over GF(p) and its cryptographic significance, *IEEE Trans. Inf. Theory* **24** (1978), 106–110.

[433] D. Pointcheval and J. Stern, Security arguments for digital signatures and blind signatures, *J. Crypt.* **13**(3) (2000), 361–396.

[434] D. Pointcheval and S. Vaudenay, On provable security for digital signature algorithms, Technical report LIENS 96-17, École Normale Supérieure, 1996.

[435] J. M. Pollard, Theorems on factorisation and primality testing, *Proc. Camb. Phil. Soc.* **76** (1974), 521–528.

[436] ——— A Monte Carlo method for factorization, *BIT* **15** (1975), 331–334.

[437] ——— Monte Carlo methods for index computation (mod p), *Math. Comp.* **32**(143) (1978), 918–924.

[438] ——— Kangaroos, monopoly and discrete logarithms, *J. Crypt.* **13**(4) (2000), 437–447.

[439] C. Pomerance, A tale of two sieves, *Notices of the Amer. Math. Soc.* **43** (1996), 1473–1485.

[440] V. R. Pratt, Every prime has a succinct certificate, *SIAM J. Comput.* **4**(3) (1974), 214–220.

[441] X. Pujol and D. Stehlé, Rigorous and efficient short lattice vectors enumeration. In *ASIACRYPT 2008* (J. Pieprzyk, ed.), *LNCS*, vol. 5350, Springer, 2008, pp. 390–405.

[442] G. Qiao and K.-Y. Lam, RSA signature algorithm for microcontroller implementation. In *CARDIS 1998* (J.-J. Quisquater and B. Schneier, eds.), *LNCS*, vol. 1820, Springer, 2000, pp. 353–356.

[443] J. J. Quisquater and C. Couvreur, Fast decipherment algorithm for RSA public-key cryptosystem, *Electronics Letters* (21) (1982), 905–907.

[444] M. O. Rabin, Digitalized signatures and public-key functions as intractable as factorization, Tech. Report MIT/LCS/TR-212, MIT Laboratory for Computer Science, 1979.

[445] J.-F. Raymond and A. Stiglic, *Security Issues in the Diffie–Hellman Key Agreement Protocol*, Preprint, 2000.

[446] O. Regev, The learning with errors problem (invited survey), *25th Annual IEEE Conference on Computational Complexity*, IEEE, 2010, pp. 191–204.

[447] M. Reid, *Undergraduate Algebraic Geometry*, Cambridge University Press, 1988.

[448] ——— Graded rings and varieties in weighted projective space, Chapter of unfinished book, 2002.

[449] G. Reitwiesner, Binary arithmetic, *Advances in Computers* **1** (1960), 231–308.

[450] P. Rogaway, Formalizing human ignorance. In *VIETCRYPT 2006* (P. Q. Nguyen, ed.), *LNCS*, vol. 4341, Springer, 2006, pp. 211–228.

[451] S. Ross, *A First Course in Probability*, 6th edn, Prentice Hall, 2001.

[452] K. Rubin and A. Silverberg, Torus-based cryptography. In *CRYPTO 2003* (D. Boneh, ed.), *LNCS*, vol. 2729, Springer, 2003, pp. 349–365.

[453] ——— Compression in finite fields and torus-based cryptography, *SIAM J. Comput.* **37**(5) (2008), 1401–1428.

[454] H.-G. Rück, A note on elliptic curves over finite fields, *Math. Comp.* **49**(179) (1987), 301–304.

[455] ——— On the discrete logarithm in the divisor class group of curves, *Math. Comp.* **68**(226) (1999), 805–806.

[456] A. Rupp, G. Leander, E. Bangerter, A. W. Dent and A.-R. Sadeghi, Sufficient conditions for intractability over black-box groups: generic lower bounds for generalized DL and DH problems. In *ASIACRYPT 2008* (J. Pieprzyk, ed.), *LNCS*, vol. 5350, Springer, 2008, pp. 489–505.

[457] A.-R. Sadeghi and M. Steiner, Assumptions related to discrete logarithms: why subtleties make a real difference. In *EUROCRYPT 2001* (B. Pfitzmann, ed.), *LNCS*, vol. 2045, Springer, 2001, pp. 244–261.

[458] A. Sárközy and C. L. Stewart, On pseudorandomness in families of sequences derived from the Legendre symbol, *Periodica Math. Hung.* **54**(2) (2007), 163–173.

[459] T. Satoh, *On Generalization of Cheon's Algorithm*, Cryptology ePrint Archive, Report 2009/058, 2009.

[460] T. Satoh and K. Araki, Fermat quotients and the polynomial time discrete log algorithm for anomalous elliptic curves, *Comment. Math. Univ. St. Paul.* **47**(1) (1998), 81–92.

[461] J. Sattler and C.-P. Schnorr, Generating random walks in groups, *Ann. Univ. Sci. Budapest. Sect. Comput.* **6** (1985), 65–79.

[462] A. Schinzel and M. Skałba, On equations $y^2 = x^n + k$ in a finite field, *Bull. Polish Acad. Sci. Math.* **52**(3) (2004), 223–226.

[463] O. Schirokauer, Using number fields to compute logarithms in finite fields, *Math. Comp.* **69**(231) (2000), 1267–1283.

[464] ——— The special function field sieve, *SIAM J. Discrete Math* **16**(1) (2002), 81–98.

[465] ——— The impact of the number field sieve on the discrete logarithm problem in finite fields. In *Algorithmic Number Theory* (J. Buhler and P. Stevenhagen, eds.), MSRI publications, vol. 44, Cambridge University Press, 2008, pp. 397–420.

[466] ——— The number field sieve for integers of low weight, *Math. Comp.* **79**(269) (2010), 583–602.

[467] O. Schirokauer, D. Weber and T. F. Denny, Discrete logarithms: the effectiveness of the index calculus method. In *ANTS II* (H. Cohen, ed.), *LNCS*, vol. 1122, Springer, 1996, pp. 337–361.

[468] C.-P. Schnorr, A hierarchy of polynomial time lattice basis reduction algorithms, *Theor. Comput. Sci.* **53** (1987), 201–224.

[469] ——— Efficient identification and signatures for smart cards. In *CRYPTO 1989* (G. Brassard, ed.), *LNCS*, vol. 435, Springer, 1990, pp. 239–252.

[470] ——— Efficient signature generation by smart cards, *J. Crypt.* **4**(3) (1991), 161–174.

[471] ——— Security of almost all discrete log bits, *Electronic Colloquium on Computational Complexity (ECCC)* **5**(33) (1998), 1–13.

[472] ——— Progress on LLL and lattice reduction. In *The LLL Algorithm* (P. Q. Nguyen and B. Vallée, eds.), Springer, 2010, pp. 145–178.

[473] C.-P. Schnorr and M. Euchner, Lattice basis reduction: improved practical algorithms and solving subset sum problems, *Math. Program.* **66** (1994), 181–199.

[474] C.-P. Schnorr and H. W. Lenstra Jr., A Monte Carlo factoring algorithm with linear storage, *Math. Comp.* **43**(167) (1984), 289–311.

[475] R. Schoof, Elliptic curves over finite fields and the computation of square roots (mod) p, *Math. Comp.* **44**(170) (1985), 483–494.

[476] ______ Nonsingular plane cubic curves over finite fields, *J. Combin. Theory Ser. A* **46** (1987), 183–211.
[477] ______ Counting points on elliptic curves over finite fields, *J. Théor. Nombres Bordeaux* **7** (1995), 219–254.
[478] A. Schrijver, *Theory of Linear and Integer Programming*, Wiley, 1986.
[479] R. Schroeppel, H. K. Orman, S. W. O'Malley and O. Spatscheck, Fast key exchange with elliptic curve systems. In *CRYPTO 1995* (D. Coppersmith, ed.), *LNCS*, vol. 963, Springer, 1995, pp. 43–56.
[480] E. Schulte-Geers, Collision search in a random mapping: some asymptotic results, Presentation at ECC 2000, Essen, Germany, 2000.
[481] M. Scott, Faster pairings using an elliptic curve with an efficient endomorphism. In *INDOCRYPT 2005* (S. Maitra, C. E. V. Madhavan and R. Venkatesan, eds.), *LNCS*, vol. 3797, Springer, 2005, pp. 258–269.
[482] R. Sedgewick, T. G. Szymanski and A. C.-C. Yao, The complexity of finding cycles in periodic functions, *SIAM J. Comput.* **11**(2) (1982), 376–390.
[483] B. I. Selivanov, On waiting time in the scheme of random allocation of coloured particles, *Discrete Math. Appl.* **5**(1) (1995), 73–82.
[484] I. A. Semaev, Evaluation of discrete logarithms in a group of p-torsion points of an elliptic curve in characteristic p, *Math. Comp.* **67**(221) (1998), 353–356.
[485] ______ A 3-dimensional lattice reduction algorithm. In *Cryptography and Lattices (CaLC)* (J. H. Silverman, ed.), *LNCS*, vol. 2146, Springer, 2001, pp. 181–193.
[486] ______ Summation Polynomials and the Discrete Logarithm Problem on Elliptic Curves, Cryptology ePrint Archive, Report 2004/031, 2004.
[487] J.-P. Serre, Sur la topologie des variétés algébriques en charactéristique p. In *Symp. Int. Top. Alg.*, Mexico, 1958, pp. 24–53.
[488] ______ *Local Fields*, *GTM*, vol. 67, Springer, 1979.
[489] I. R. Shafarevich, *Basic Algebraic Geometry*, 2nd edn, Springer, 1995.
[490] J. O. Shallit, *A Primer on Balanced Binary Representations*, Preprint, 1992.
[491] A. Shallue and C. E. van de Woestijne, Construction of rational points on elliptic curves over finite fields. In *ANTS VII* (F. Hess, S. Pauli and M. E. Pohst, eds.), *LNCS*, vol. 4076, Springer, 2006, pp. 510–524.
[492] D. Shanks, Five number-theoretic algorithms, Proceedings of the Second Manitoba Conference on Numerical Mathematics, Congressus Numerantium, No. VII, Utilitas Math., Winnipeg, Man., 1973, pp. 51–70.
[493] M. Shirase, D.-G. Han, Y. Hibino, H.-W. Kim and T. Takagi, Compressed XTR. In *ACNS 2007* (J. Katz and M. Yung, eds.), *LNCS*, vol. 4521, Springer, 2007, pp. 420–431.
[494] V. Shoup, Lower bounds for discrete logarithms and related problems. In *EUROCRYPT 1997* (W. Fumy, ed.), *LNCS*, vol. 1233, Springer, 1997, pp. 256–266.
[495] ______ On formal models for secure key exchange (version 4), 15 November 1999, Tech. report, IBM, 1999, Revision of Report RZ 3120.
[496] ______ OAEP reconsidered. In *CRYPTO 2001* (J. Kilian, ed.), *LNCS*, vol. 2139, Springer, 2001, pp. 239–259.
[497] ______ *A Computational Introduction to Number Theory and Algebra*, Cambridge University Press, 2005.
[498] I. E. Shparlinski, Computing Jacobi symbols modulo sparse integers and polynomials and some applications, *J. Algorithms* **36** (2000), 241–252.
[499] ______ *Cryptographic Applications of Analytic Number Theory*, Birkhauser, 2003.
[500] ______ Playing "hide-and-seek" with numbers: the hidden number problem, lattices and exponential sums. In *Public-Key Cryptography* (P. Garrett and D. Lieman, eds.), *Proceedings of Symposia in Applied Mathematics*, vol. 62, AMS, 2005, pp. 153–177.

[501] I. E. Shparlinski and A. Winterhof, A nonuniform algorithm for the hidden number problem in subgroups. In *PKC 2004* (F. Bao, R. H. Deng and J. Zhou, eds.), *LNCS*, vol. 2947, Springer, 2004, pp. 416–424.

[502] ——— A hidden number problem in small subgroups, *Math. Comp.* **74**(252) (2005), 2073–2080.

[503] A. Sidorenko, Design and analysis of provably secure pseudorandom generators, Ph.D. thesis, Eindhoven, 2007.

[504] C. L. Siegel, *Lectures on the Geometry of Numbers*, Springer, 1989.

[505] J. H. Silverman, *The Arithmetic of Elliptic Curves*, GTM, vol. 106, Springer, 1986.

[506] ——— *Advanced Topics in the Arithmetic of Elliptic Curves*, *GTM*, vol. 151, Springer, 1994.

[507] J. H. Silverman and J. Suzuki, Elliptic curve discrete logarithms and the index calculus. In *ASIACRYPT 1998* (K. Ohta and D. Pei, eds.), *LNCS*, vol. 1514, Springer, 1998, pp. 110–125.

[508] J. H. Silverman and J. Tate, *Rational Points on Elliptic Curves*, Springer, 1994.

[509] M. Sipser, *Introduction to the Theory of Computation*, Course Technology, 2005.

[510] M. Skałba, Points on elliptic curves over finite fields, *Acta Arith.* **117**(3) (2005), 293–301.

[511] N. P. Smart, The discrete logarithm problem on elliptic curves of trace one, *J. Cryptology* **12**(3) (1999), 193–196.

[512] ——— Elliptic curve cryptosystems over small fields of odd characteristic, *J. Crypt.* **12**(2) (1999), 141–151.

[513] ——— *Cryptography: An Introduction*, McGraw-Hill, 2004.

[514] B. A. Smith, Isogenies and the discrete logarithm problem in Jacobians of genus 3 hyperelliptic curves, *J. Crypt.* **22**(4) (2009), 505–529.

[515] P. J. Smith and M. J. J. Lennon, LUC: a new public key system. In *International Conference on Information Security* (E. Graham Dougall, ed.), *IFIP Transactions*, vol. A-37, North-Holland, 1993, pp. 103–117.

[516] P. J. Smith and C. Skinner, A public-key cryptosystem and a digital signature system based on the Lucas function analogue to discrete logarithms. In *ASIACRYPT 1994* (J. Pieprzyk and R. Safavi-Naini, eds.), *LNCS*, vol. 917, Springer, 1994, pp. 357–364.

[517] J. A. Solinas, Efficient arithmetic on Koblitz curves, *Des. Codes Crypt.* **19** (2000), 195–249.

[518] ——— Low-weight binary representations for pairs of integers, Technical Report CORR 2001-41, 2001.

[519] M. Stam, On Montgomery-like representations of elliptic curves over GF(2^k). In *PKC 2003* (Y. G. Desmedt, ed.), *LNCS*, vol. 2567, Springer, 2003, pp. 240–253.

[520] ——— Speeding up subgroup cryptosystems, Ph.D. thesis, Eindhoven, 2003.

[521] M. Stam and A. K. Lenstra, Speeding up XTR. In *ASIACRYPT 2001* (C. Boyd, ed.), *LNCS*, vol. 2248, Springer, 2001, pp. 125–143.

[522] H. M. Stark, Class-numbers of complex quadratic fields. In *Modular Functions of One Variable I* (W. Kuyk, ed.), *LNM*, vol. 320, Springer, 1972, pp. 153–174.

[523] D. Stehlé, Floating point LLL: Theoretical and practical aspects. In *The LLL Algorithm* (P. Q. Nguyen and B. Vallée, eds.), Springer, 2010, pp. 179–213.

[524] D. Stehlé and P. Zimmermann, A binary recursive GCD algorithm. In *ANTS VI* (D. A. Buell, ed.), *LNCS*, vol. 3076, Springer, 2004, pp. 411–425.

[525] P. Stevenhagen, The number field sieve. In *Algorithmic Number Theory* (J. Buhler and P. Stevenhagen, eds.), MSRI publications, Cambridge University Press, 2008, pp. 83–99.

[526] I. Stewart, *Galois Theory*, 3rd edn, Chapman & Hall, 2003.

[527] I. Stewart and D. Tall, *Algebraic Number Theory and Fermat's Last Theorem*, 3rd edn, AK Peters, 2002.

[528] H. Stichtenoth, Die Hasse–Witt Invariante eines Kongruenzfunktionenkörpers, *Arch. Math.* **33** (1979), 357–360.

[529] ——— *Algebraic Function Fields and Codes*, Springer, 1993.

[530] H. Stichtenoth and C. Xing, On the structure of the divisor class group of a class of curves over finite fields, *Arch. Math.* **65** (1995), 141–150.

[531] D. R. Stinson, Some baby-step–giant-step algorithms for the low Hamming weight discrete logarithm problem, *Math. Comp.* **71**(237) (2001), 379–391.

[532] ——— *Cryptography: Theory and Practice*, 3rd edn, Chapman & Hall/CRC, 2005.

[533] A. Storjohann and G. Labahn, Asymptotically fast computation of Hermite normal forms of integer matrices, *ISSAC 1996*, ACM Press, 1996, pp. 259–266.

[534] E. G. Straus, Addition chains of vectors, *American Mathematical Monthly* **71**(7) (1964), 806–808.

[535] A. H. Suk, Cryptanalysis of RSA with lattice attacks, MSc thesis, University of Illinois at Urbana-Champaign, 2003.

[536] A. V. Sutherland, Order computations in generic groups, Ph.D. thesis, MIT, 2007.

[537] ——— Structure computation and discrete logarithms in finite abelian p-groups, *Math. Comp.* **80**(273) (2011), 477–500.

[538] T. Takagi, Fast RSA-type cryptosystem modulo $p^k q$. In *CRYPTO 1998* (H. Krawczyk, ed.), *LNCS*, vol. 1462, Springer, 1998, pp. 318–326.

[539] J. Talbot and D. Welsh, *Complexity and Cryptography: An Introduction*, Cambridge University Press, 2006.

[540] J. Tate, Endomorphisms of abelian varieties over finite fields, *Invent. Math.* **2** (1966), 134–144.

[541] E. Teske, A space efficient algorithm for group structure computation, *Math. Comp.* **67**(224) (1998), 1637–1663.

[542] ——— Speeding up Pollard's rho method for computing discrete logarithms. In *ANTS III* (J. P. Buhler, ed.), *LNCS*, vol. 1423, Springer, 1998, pp. 541–554.

[543] ——— On random walks for Pollard's rho method, *Math. Comp.* **70**(234) (2001), 809–825.

[544] ——— Computing discrete logarithms with the parallelized kangaroo method, *Discrete Appl. Math.* **130** (2003), 61–82.

[545] N. Thériault, Index calculus attack for hyperelliptic curves of small genus. In *ASIACRYPT 2003* (C.-S. Laih, ed.), *LNCS*, vol. 2894, Springer, 2003, pp. 75–92.

[546] E. Thomé, Algorithmes de calcul de logarithmes discrets dans les corps finis, Ph.D. thesis, L'École Polytechnique, 2003.

[547] W. Trappe and L. C. Washington, *Introduction to Cryptography with Coding Theory*, 2nd edn, Pearson, 2005.

[548] M. A. Tsfasman, Group of points of an elliptic curve over a finite field, *Theory of Numbers and Its Applications*, Tbilisi, 1985, pp. 286–287.

[549] J. W. M. Turk, Fast arithmetic operations on numbers and polynomials, *Computational Methods in Number Theory, Part 1, Mathematical Centre Tracts 154*, Amsterdam, 1984.

[550] B. Vallée, Une approche géométrique de la réduction de réseaux en petite dimension, Ph.D. thesis, Université de Caen, 1986.

[551] ——— Gauss' algorithm revisited, *J. Algorithms* **12**(4) (1991), 556–572.

[552] S. Vaudenay, Hidden collisions on DSS. In *CRYPTO 1996* (N. Koblitz, ed.), *LNCS*, vol. 1109, Springer, 1996, pp. 83–88.

[553] ——— *A Classical Introduction to Cryptography*, Springer, 2006.

[554] F. Vercauteren, Optimal pairings, *IEEE Trans. Inf. Theory* **56**(1) (2010), 455–461.

[555] E. R. Verheul, Certificates of recoverability with scale recovery agent security. In *PKC 2000* (H. Imai and Y. Zheng, eds.), *LNCS*, vol. 1751, Springer, 2000, pp. 258–275.

[556] ——— Evidence that XTR is more secure than supersingular elliptic curve cryptosystems, *J. Crypt.* **17**(4) (2004), 277–296.

[557] E. R. Verheul and H. C. A. van Tilborg, Cryptanalysis of 'less short' RSA secret exponents, *Applicable Algebra in Engineering, Communication and Computing* **8**(5) (1997), 425–435.

[558] M.-F. Vignéras, Arithmétique des algèbres de quaternions, *LNM*, vol. 800, Springer, 1980.

[559] J. F. Voloch, A note on elliptic curves over finite fields, *Bulletin de la Société Mathématique de France* **116**(4) (1988), 455–458.

[560] L. C. Washington, *Elliptic Curves: Number Theory and Cryptography*, 2nd edn, CRC Press, 2008.

[561] E. Waterhouse, Abelian varieties over finite fields, *Ann. Sci. École Norm. Sup.* **2** (1969), 521–560.

[562] D. H. Wiedemann, Solving sparse linear equations over finite fields, *IEEE Trans. Inf. Theory* **32** (1986), 54–62.

[563] M. J. Wiener, *Bounds on Birthday Attack Times*, Cryptology ePrint Archive, Report 2005/318.

[564] ______ Cryptanalysis of short RSA secret exponents, *IEEE Trans. Inf. Theory* **36**(3) (1990), 553–558.

[565] M. J. Wiener and R. J. Zuccherato, Faster attacks on elliptic curve cryptosystems. In *SAC 1998* (S. E. Tavares and H. Meijer, eds.), *LNCS*, vol. 1556, Springer, 1998, pp. 190–200.

[566] H. C. Williams, A modification of the RSA public key encryption procedure, *IEEE Trans. Inf. Theory* **26**(6) (1980), 726–729.

[567] ______ A $p + 1$ method of factoring, *Math. Comp.* **39**(159) (1982), 225–234.

[568] D. J. Winter, *The Structure of Fields*, *GTM* 16, Springer, 1974.

[569] M. Woodroofe, *Probability with Applications*, McGraw-Hill, 1975.

[570] S.-M. Yen and C.-S. Laih, Improved digital signature suitable for batch verification, *IEEE Trans. Computers* **44**(7) (1995), 957–959.

[571] S.-M. Yen, C.-S. Laih, and A. K. Lenstra, Multi-exponentiation, *IEEE Proceedings Computers and Digital Techniques* **141**(6) (1994), 325–326.

[572] N. Yui, On the jacobian varieties of hyperelliptic curves over fields of characteristic $p > 2$, *J. Algebra* **52** (1978), 378–410.

[573] O. Zariski and P. Samuel, *Commutative Algebra (Vol. I and II)*, Van Nostrand, Princeton University Press, 1960.

[574] N. Zierler, A conversion algorithm for logarithms on $GF(2^n)$, *J. Pure Appl. Algebra* **4** (1974), 353–356.

[575] P. Zimmermann, Private communication, 10 March 2009.

Author index

Subject index

Printed in the United States
By Bookmasters